D0215617

INTRODUCTION TO
LOGIC DESIGN

INTRODUCTION TO LOGIC DESIGN

Svetlana N. Yanushkevich
University of Calgary, Alberta, Canada

Vlad P. Shmerko
University of Calgary, Alberta, Canada

CRC Press
Taylor & Francis Group
Boca Raton London New York

CRC Press is an imprint of the
Taylor & Francis Group, an **informa** business

CRC Press
Taylor & Francis Group
6000 Broken Sound Parkway NW, Suite 300
Boca Raton, FL 33487-2742

Printed in the United States of America on acid-free paper
10 9 8 7 6 5 4 3 2 1

International Standard Book Number-13: 978-1-4200-6094-2 (Hardcover)

Library of Congress Cataloging-in-Publication Data

Yanushkevich, Svetlana N.
Introduction to logic design / Svetlana N. Yanushkevich and Vlad P. Shmerko. -- 1st ed.
p. cm.
Includes bibliographical references and index.
ISBN 978-1-4200-6094-2 (alk. paper)
1. Logic devices--Design and construction. 2. Logic design. I. Shmerko, Vlad P. II. Title.

TK7872.L64.Y36 2008
621.39'5--dc22 2007044293

Visit the Taylor & Francis Web site at
http://www.taylorandfrancis.com

and the CRC Press Web site at
http://www.crcpress.com

This textbook is dedicated to the memory of Claude Shannon

(1916–2001)

Claude Elwood Shannon's inventive genius, probably more than that of any other single person, has altered man's understanding of communication and digital systems.

Contents

Preface

The Introduction to Logic Design Course in the Electrical Engineering and Computer Science Undergraduate Curriculum

An introductory course on the logic design of discrete devices is fundamental to the study of the many fields that constitute the ever-expanding discipline of computer engineering, computer science, and electrical engineering. Logic design of discrete devices serves as the prerequisite for additional coursework in the study of communications, signal processing, digital system design, and neural networks.

An introductory course on logic design is frequently encountered in the second year of undergraduate programs. Additional courses are then provided that expand on the one-semester course by including a more detailed treatment of digital system design, focusing, in particular, on simulation using hardware description languages.

With respect to the audience, that is, electrical and computer engineering and computer science, the introductory course on logic design can vary significantly. The course on logic design is usually offered at the sophomore or junior level and assumes no background on the part of the reader.

How This Book Satisfies the Essential Needs of This Course

Given the introductory nature of the computing device design course and the diversity of its applications, an appropriate textbook must be easy to read, accurate, and contain an abundance of insightful examples, problems, and computer experiments to expedite learning the fundamentals of logic design of discrete devices and systems. The textbook must reflect the fundamentals of state-of-the-art logic design and must be useful to people from other fields who are interested in the basics of logic network design. This book has been written with all of these objectives in mind.

We have verified in our academic and research work that the material presented in this book is useful for not only computer/electrical engineering and computer science students, but also students from physics and the chemical sciences who are working on applications of physical and molecular phenomena at nanoscale for computing logic functions. This textbook attempts to fill this need with a balanced presentation of classic topics and topics of advanced design, such decision trees and diagrams, focusing on the manipulation of various data structures rather than optimization.

It is the primary purpose of this book to present problems relevance for advanced as well as future technologies that will be educationally informative for students and practicing engineers and researchers. The emphasis will be on the manipulation of various data structures in logic network design. Of course, knowledge of classical minimization techniques is also useful and necessary as background. Such techniques are also included in complete detail.

This book builds on the authors' lectures in various universities, providing a balanced and integrated treatment of the fundamental theory of logic functions and applications in the design of digital devices and systems. This approach has the pedagogical advantage of helping the student see the fundamental similarities and differences between various theoretical approaches to the representation, manipulation, and optimization of Boolean functions, and reflects the integrated nature of the advanced concepts in modern engineering practice.

The text has been written with the aim of offering maximum teaching flexibility in both coverage and order of presentation. A one-semester course sequence would likely cover most, if not all, of the topics in the book.

Structure Designed to Facilitate and Reinforce Learning

We designed this textbook to be up-to-date, comprehensive, and pragmatic in its approach. It aims to provide balanced coverage of concepts, techniques, and practices, and to do so in a way that is, at the same time, both scholarly and supportive of the needs of the typical student reader. We wanted to give students:

- A clear picture of fundamental concepts;
- Effective problem-solving techniques; and
- Appropriate exposure to modern technologies, design techniques, and applications.

Through the mutual support of textual material, worked examples, end-of-chapter problems, and other pedagogical features, we have attempted to show how theoretical ideas, physical devices, and design methodologies can all come together to form a successful design system.

Toward Predictable Technologies

With the advent of new technologies, a major shift has occurred in the field of the design and application of digital systems. As a result, many engineers and scientists have found it necessary to understand the basic operation of digital systems and how these systems can be designed if they carry out particular information-processing tasks associated with their work. This trend has produced a need for an introductory undergraduate course in logic design to provide a unified overview of the interrelationship between digital system design, computer organization, micro- and nano-electronics, and numerical methods.

In this textbook, no specific prerequisites are assumed, nor is any knowledge of electrical circuits or electronics required. In this way, this textbook satisfies the requirements of interdisciplinary interest in logic design.

New Concepts in an Introductory Course

The approach taken in this book is a traditional one. That is, the emphasis is on the presentation of basic principles of logic design and the illustration of each of these principles. However, the following key features distinguish this textbook:

- ► This textbook presents a comprehensive catalog of logic network design techniques, including the most popular decision diagram techniques. Standard textbooks on logic circuit design generally do not discuss decision diagram techniques, either in their entirely or in a tutorial manner.
- ► A central role is reserved for data structures, as a key to applications.
- ► Recent key trends in the theory and practice of logic network design techniques are highlighted.
- ► Novel techniques of advanced logic design are presented with respect to new possibilities in predictable technologies.

In summary, this is a textbook of a new generation of texts for undergraduate students. Although classical in content, this book is different from other introductory books on logic design in emphasizing topics, such as design and data structures, and design and technological requirements. The relationships between various data structures and their manipulation through design represent the most important aspect of contemporary logic design.

This textbook is dedicated to the memory of Claude Shannon. Claude Elwood Shannon, an electrical engineer and mathematician, was a graduate of the University of Michigan, class of 1936. His legendary Master's thesis dealt with the application of Boolean logic to electronic relays, and ushered in the era of the deployment of Boolean logic to ever-evolving electrical circuits. Known for his mathematically rigorous approach, Shannon remarkably contributed to scientific knowledge in the fields of communication and logic design of discrete devices. He is widely regarded as the father of information theory, and his work helped lay the foundation for the electronic age. It is likely that techniques based on Shannon's information theory will become some of the main design tools for nanosystems – computing systems for the age of nanotechnology.

We would like to acknowledge several people for their useful suggestions and discussions:

Acknowledgments

Dr. N. Bartley, University of Calgary, Canada
Dr. D. Bochmann, University of Chemnitz, Germany
Dr. J. T. Butler, Naval Postgraduate School, Monterey, CA, U.S.A.
Dr. G. Dueck, University of New Brunswick, Canada
Dr. D. Green, University of Manchester, UK
Dr. M. Kameyama, Tohoku University, Japan
Dr. T. Luba, Warsaw University of Technology, Poland
Dr. S. E. Lyshevski, Rochester Institute of Technology, Rochester, NY, USA
Dr. D. M. Miller, University of Victoria, Canada
Dr. C. Moraga, Dortmund University, Germany
Dr. J. Muzio, University of Victoria, Canada
Dr. M. Perkowski, Portland State University, OR, U.S.A.
Dr. S. Rudeanu, University of Bucharest, Romania
Dr. T. Sasao, Kyushu Institute of Technology, Japan
Dr. R. Stanković, University of Nis, Serbia
Dr. B. Steinbach, Freiberg University of Mining and Technology, Germany
Dr. A. Stoica, California Institute of Technology, U.S.A.
Dr. H. Watanabe, Soka University, Tokyo, Japan

Also, we would like to acknowledge the efforts of the staff at CRC Press, in particular, Nora Konopka and Prudence Board. We would like to thank our undergraduate and graduate students, and also Ian Pollock, for their valuable suggestions and assistance, which were helpful to us in ensuring the coherence of topics and material delivery "synergy."

Svetlana N. Yanushkevich

Vlad P. Shmerko

Calgary, Canada

1

Design Process and Technology

Design process

- Design levels
- Design hierarchy
- Top-down design methodology
- Bottom-up design methodology

Implementation technologies

- Very large scale integration
- Deep submicron integration

Data structures

- Algebraic
- Graphical

Design methodologies

- Application-specific integrated circuits
- System-on-chip

Predictable technologies

- Energy-efficient design
- Nanoelectronics
- Molecular devices

1.1 Introduction

Digital logic networks are used in all devices that process information in digital form. *Information* – recorded or communicated facts, or *data* – takes a variety of physical forms when being stored, communicated, or manipulated. Information on the nature of a physical phenomenon is conveyed by signals that assume a finite number of discrete values, that is, it is expressed as a finite sequence of symbols. A *signal* is defined as a function of one or more variables, and a *system* is defined as an entity that manipulates one or more signals to accomplish a function, thereby yielding new signals.

In this book, each signal is assumed to have only one of two values, denoted by the symbols 0 and 1. If the signals are constrained to only two values, the system is *binary*.

There are many methodologies for discrete system design. A *discrete system* is a combination of logic networks and discrete devices that is assembled to accomplish a desired result, such as the computing and transferring of data. Discrete system are classified using various criteria; in particular, with respect to applications, power consumption, requirements for reliability, performance, technological criteria, etc.

1.2 Theory of logic design

In 1854, George Boole introduced a systematic treatment of logic in the work *An Investigation of the Laws of Thought* and developed for this purpose an algebraic system, now called *Algebra*. Boole's goal was the development of a formalism to compute the truth or falsehood of complex compound statements from the truth values of their component statements.

Boolean algebra was applied by C. Shannon in 1938 to relay-contact networks, the first switching circuits. This theory is called *switching theory* and has been used ever since in the design of digital *logic networks* or *logic circuits*. Since then the technology has gone from relay-contacts through diode gates, transistor gates, integrated circuits, to future nanotechnologies, and still Boolean algebra is its fundamental and unchanging basis:

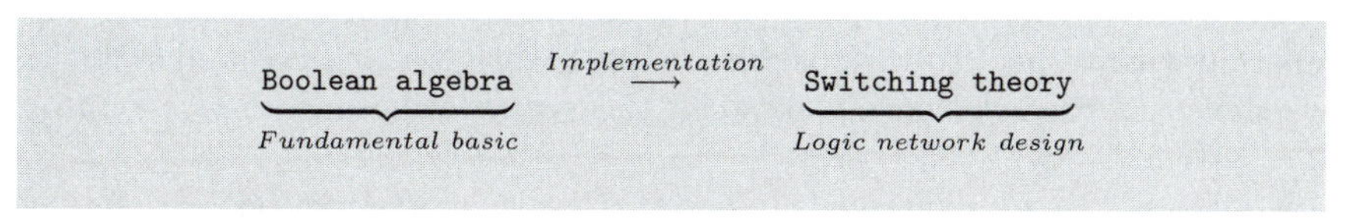

Modern logic design includes methods and techniques from various fields, in particular:

Methods and techniques of modern logic design

- *Digital signal processing* is adopted in logic design for efficient manipulation of data.
- *Communication theory* to solve communication problems between computing components in logic networks.
- *Artificial intelligence methods* and techniques for optimization at logic and physical levels of logic network design.

As far as predictable technologies are concerned, non-traditional computing paradigms that are based on various physical and chemical phenomena are studied in logic design. The assumed stochastic nature of many processes at nanoscale implies that *random* signals must be used instead of *deterministic* signals. Random signals take on random values at any given moment in time and must be modeled probabilistically, while deterministic signals can be modeled as completely specified functions of time. The theoretical base of probabilistic logic signals is called *probabilistic logic*. Schematically, advanced logic design can be viewed as follows:

1.3 Analysis and synthesis

The *specification* of a system is defined as a description of its function. The *implementation* of a system refers to the construction of a system (Figure 1.1). The *analysis* of a system has as its objective to determine its specification from the implementation:

$$\texttt{Logic network} \overset{Analysis}{\longrightarrow} \texttt{Specification}$$

Synthesis, or *design*, consists of obtaining an implementation that satisfies the specification of the system:

$$\texttt{Specification} \overset{Synthesis}{\longrightarrow} \texttt{Logic network}$$

The central task in logic synthesis is to optimize the representation of a logic function with respect to various criteria. In Figure 1.1, analysis and synthesis are shown as the relationship between the *specification* and *implementation* of a system.

A system can be examined at various levels of abstraction. Each such level is called a *design level* (Figure 1.2a). The following design levels are distinguished:

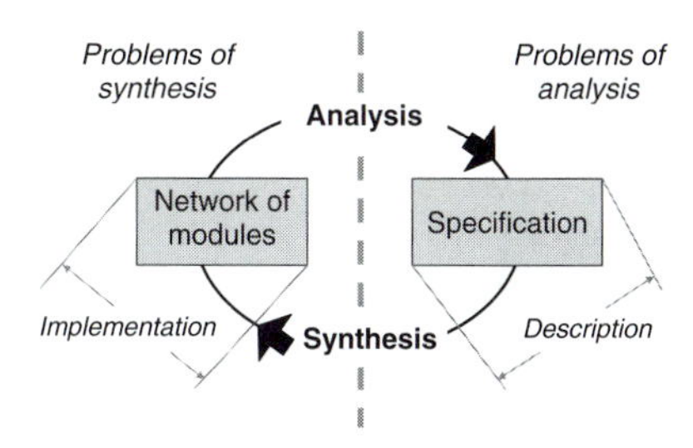

FIGURE 1.1
Design of discrete systems.

- The top design level is called the *architectural* or *system* level.
- The intermediate design level is called the *logic* level; this level is the subject of the present book.
- The bottom design level is called the *physical* level; this level is concerned with the details needed to manufacture or assemble the system.

At the physical level, a system is implemented by a complex interconnection of elements such as transistors, resistors, etc. Because of this complexity, it is impractical to perform design and optimization at this level, motivating a move to the intermediate level of design. At the intermediate level, a modular structure provides a reasonable simplification of design. *Libraries* of standard modules significantly simplify the design of different systems. Assembling modules of increasing complexity into higher hierarchical blocks is achieved at the system level.

1.3.1 Design hierarchy

A *hierarchical* approach to digital system design aims at

- Reducing the cost of the design of a system, and
- Improving the quality of the solutions obtained.

The hierarchical approach to design makes a large system more manageable by reducing complexity and introducing a rational *partitioning* of the design processes. They can be designed, tested, and manufactured separately. This is the basis for *standardization*. The specified components can be mass produced at relatively low cost. In the design process, these components can be composed in standard *libraries* and reused with minor modifications. This facilitates the reduction of overall design time and cost. The *robustness* of the hierarchical approach provides many possibilities for avoiding design errors, design corrections, and repairs after manufacture.

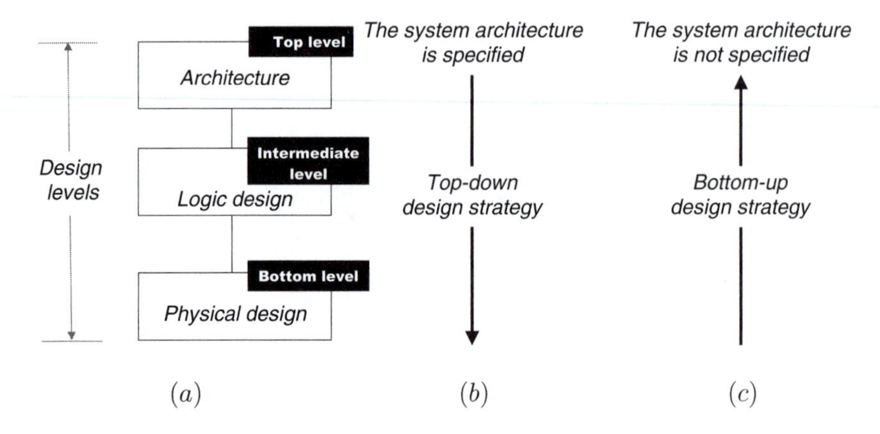

FIGURE 1.2
Design hierarchy (a), top-down (b), and bottom-up (c) design strategies.

Any design process includes a *design loop* that provides the possibility to carry out a *redesign* if errors are detected in simulation. This loop is repeated until the simulation indicates a successful design.

1.3.2 Design methodology

Top-down design methodology

A design that evolves from a generalized or abstract point of view and proceeds in steps to specific components is referred to as a *top-down design methodology* (Figure 1.2b):

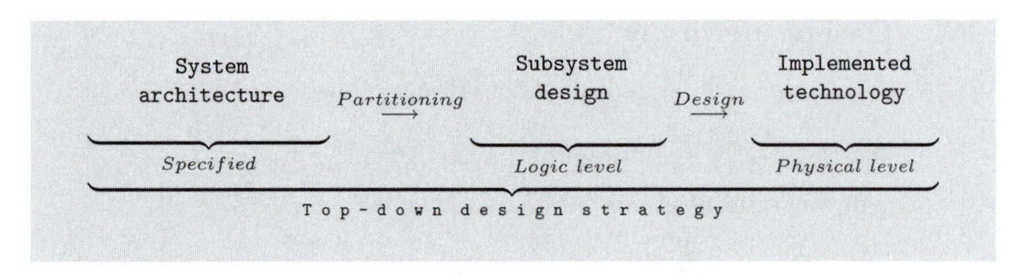

In this approach, the hierarchy tree is traversed from top to bottom. The system architecture is specified at the highest level first. The disadvantage of this approach is that no systematic procedure exists for optimization of the final implementation; that is, optimization at one particular level does not guarantee an optimal final solution. The success of the approach depends mainly on the experience and professional skills of the designer.

An example of an advanced top-down methodology is so-called *platform-based design*. A platform is defined as a family of the designs and not a single design. In this top-down process, the constraints that accompany the specification are mapped into constraints on the components of the platform.

Bottom-up design methodology

An alternative process to the top-down approach is a *bottom-up design methodology* (Figure 1.2c):

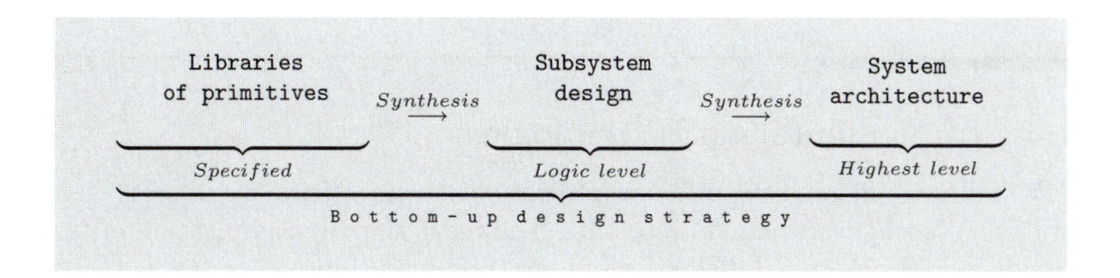

This is the reverse of the top-down design process. One starts with specific components in mind and proceeds by interconnecting these components into a generalized system. In a bottom-up design approach, the components at or near the lowest design level of the hierarchy tree are designed first. The architecture of the entire system is not specified until the top of the tree is reached. Unfortunately, in general, there is no systematic technique that results in correct system specification.

1.3.3 Design styles

In a general sense, the following design styles are distinguished:

Digital device design styles

Full-custom design; this style provides freedom to the designer and is characterized by great flexibility; however, this style is not acceptable for the design of large systems.

Semi-custom design provides more possibilities for automation using, in particular, standardization; such as a *library* of standard cells.

Mixed design styles often provide an acceptable reduction to the flexibility of full-custom style, while opening up possibilities for the automation and optimization of semi-custom style.

Gate-array design meets the requirements of fabrication and simplifies the optimization problem; this style results in regular structures within the chip, that is, connected cells that are placed on a chip in a regular way.

1.3.4 Simulation

The use of software simulation is an important part of any modern design process. The primary uses of a simulator are to check a design for functional correctness and to evaluate its performance. Simulators are key tools in determining whether design goals have been met and whether redesign is necessary. More details on sample simulation tools are given in Chapter 19.

1.4 Implementation technologies

The scaling of microelectronics down to nanoelectronics is the inevitable result of technological evolution (Figure 1.3). The most general classification of the trends in technology is based on grouping computers into *generations*. Using this criterion, five generations of computers are distinguished (Table 1.1). Each computer generation is 8 to 10 years in length.

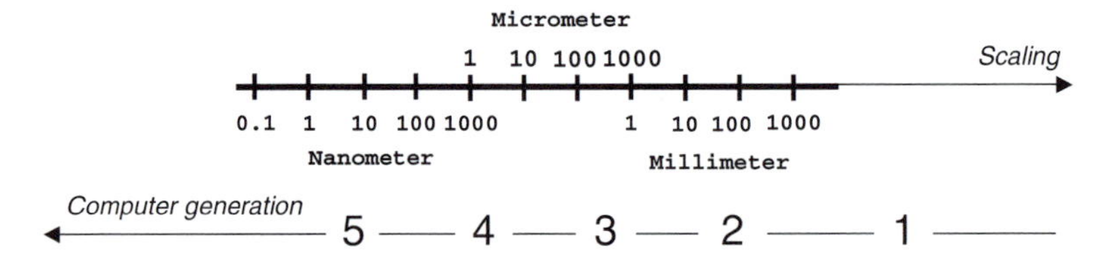

FIGURE 1.3
Progress from micro- to nanosize in computing devices.

TABLE 1.1
Computer generations are determined by the change in the dominant technology.

Generation	Dates	Technology
1	1950–1964	Vacuum tubes (zero-scale integration)
2	1965–1969	Transistors (small-scale integration)
3	1970–1979	Integrated circuits (medium-scale integration)
4	1980–2004	Large, very large, ultra large-scale integration
5	2005–	Nanotechnology (giga-scale integration)

The following can be compared against this scale:

The scaling of microelectronics down to nanoelectronics

- ▶ The size of an atom is approximately 10^{-10} m. Atoms are composed of subatomic particles; e.g., protons, neutrons and electrons. Protons and neutrons form the nucleus, with a diameter of approximately 10^{-15} m.
- ▶ 2-dimensional molecular assembly (1 nm).
- ▶ 3-dimensional functional nanoICs topology with doped carbon molecules ($2 \times 2 \times 2$ nm).
- ▶ 3-dimensional nanobioICs (10 nm).
- ▶ *E.coli* bacteria (2 mm) and ants (5 mm) have complex and high-performance integrated nanobiocircuitry.
- ▶ 1.5×1.5 cm 478-pin Intel® Pentium® processor with millions of transistors, and Intel 4004 Microprocessor (made in 1971 with 2,250 transistors).

Binary states are encountered in many different physical forms in various technologies. The standard approach is to use the digit symbols 0 and 1 to represent the two possible values of binary quantity. These symbols are referred to as *bits*.

Binary arithmetic has the following advantages:

(*a*) It can be implemented using on-off switches, the simplest binary devices.

(*b*) It provides for the simplest decision-making such as YES (1) and NO (0).

(*c*) Binary signals are more reliable than those formed by more than two quantization levels.

Significant evolutionary progress has been achieved in microelectronics. This progress (miniaturization, optimal design and technology enhancement) has been achieved by scaling down microdevices, approaching 45 nm sizing features for structures, while increasing the integration level (Figure 1.4). Complementary metal-oxide semiconductor (CMOS) technology is being enhanced, as nanolithography, advanced etching, enhanced deposition, novel materials, and modified processes are all currently used to fabricate ICs. The channel length of metal-oxide-semiconductor field effect transistors (MOSFETs) has decreased from

- ▶ 50 μm in 1960, to
- ▶ 1 μm in 1990, and to
- ▶ 130 nm in 2004, 65 nm in 2006, and 45 nm in 2007.

This progress in miniaturization and integration can be observed, for example, on Intel processors:

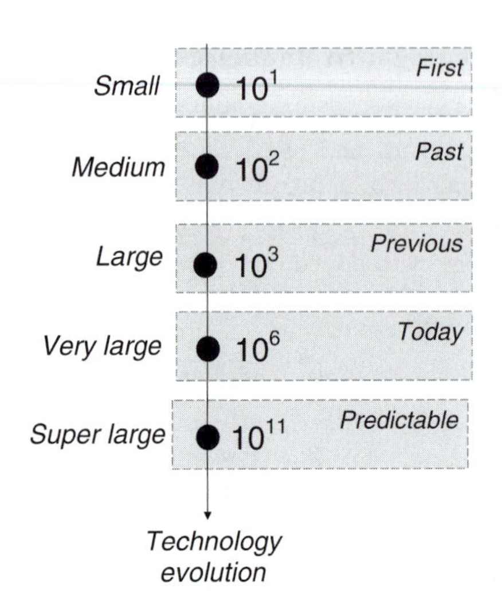

Small-scale integration (SSI):

- *1960s, dozens of gates in a package*

Medium-scale integration (MSI):

- *1970s, hundreds of gates in a package*

Large-scale integration (LSI):

- *1980s, thousands of gates in a package*

Very large-scale integration (VLSI):

- *1980s and 1990s, hundreds of thousands of gates in a package*

Giga-scale integration:

- *Today, millions of gates in a package*

Tera-scale integration:

- *Expected, hundred millions of gates in a package*

FIGURE 1.4
Evolution of technologies: levels of integration of chips.

1971–1982: From Intel 4004 (1971, 2,250 transistors), to Intel 286 (1982, 120,000 transistors),

1993–2007: From Pentium (1993, 3,100,000 transistors), to Pentium 4 (2000, 42,000,000 transistors), to Itanium™ 2 Processor (2002), Pentium® M Processor (2003) with hundreds of millions of transistors, and Pentium Dual-Core in 2007.

Typical signal integrity effects include interconnect delay, crosstalk (in closely coupled lines the phenomenon of crosstalk can be observed), power supply integrity, and noise-on-delay effects. In the early days of very large-scale integration (VLSI) design, these effects were negligible because of relatively slow chip speed and low integration density. However, with the introduction of technological generations working at about 0.25 μm scale and below, there have been many significant changes in wiring and electrical performance.

As the number of computational and functional units on a single chip increases, the need for communication between those units also increases. Interconnection has started to become a dominant factor in chip performance. As chip speed continually increases, the increasing inductance of interconnections affects the signal parameters.

Noise is a deviation of a signal from its intended or ideal value. Most noise in VLSI circuits is created by the system itself. Electromagnetic interference is an external noise source between subsystems. In deep submicron circuits, the noise created by parasitic components with digital switching exhibits the strongest effect on system performance.

Programmable versus ASIC methodology

Application-specific integrated circuits (ASIC) design methodology has been very successful in a wide range of applications. The system-on-chip (SoC) is where the integration of a complete electronic system including all its periphery occurs. The reduction in size and cost that would come with this high level of integration opens the door to truly ubiquitous electronics.

1.5 Predictable technologies

The key to the predictable technologies of the future is changing the *computing paradigm*, that is, the model of computation and the physical effects for the realization of this model:

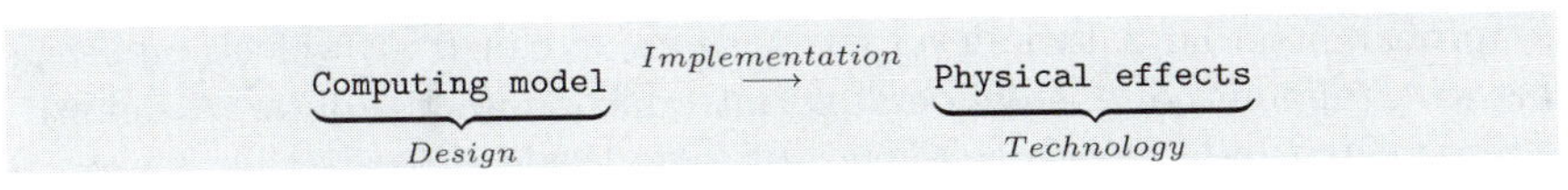

There are fundamental technological differences among

- Nanoelectronic devices vs. microelectronic ones (which can even be nanometers in size),
- Nanoelectronics vs. microelectronics; e.g., nano integrated circuits vs. integrated circuits.

These enormous differences are due to differences in basic physics and other phenomena. The dimensions of nanodevices that have been made and characterized are a hundred times less than even newly designed microelectronic devices (including nanoFETs with 10 nm gate length). Nanoelectronics sizing leads to volume reduction by a factor of 1,000,000 in packaging, not to mention revolutionary enhanced functionality due to multiterminal and spatial features. For example, molecular electronics focuses on the development of electronic devices using small molecules with feature sizes on the order of a few nanometers.

A wide variety of factors, such as voltage scaling and thermal noise, dramatically reduce the reliability of integrated systems. In nanotechnologies, process variations and quantum fluctuations occur in the operations of very-deep submicron transistors. Any computer with nanoscale components will contain a significant number of defects, as well as massive numbers of wires and switches for communication purposes. It therefore makes sense to consider architectural issues and defect tolerance early in the development of a new paradigm.

Two main aspects are critical to the design of nanodevices: *the probabilistic behavior* of nanodevices (electrons, molecules); this means that a valid switching function can be calculated with some probability; and *the high defect rates* of nanodevices; this means that because many of the fabricated devices have defects, their logic correctness is distorted.

There are two types of fault tolerances exhibited by a nanosystem: fault tolerance with respect to (a) data that are noisy, distorted, or incomplete, which results from the manner in which data are organized and represented in the nanosystem, and (b) physical degradation of the nanosystem itself. If certain nanodevices or parts of a network are destroyed, the network will continue to function properly. When the damage becomes extensive, the network will only affect the system's performance, as opposed to causing a complete failure. Self-assembling nanosystems are capable of this type of fault tolerance because they store information in a distributed (redundant) manner, in contrast to traditional storage of data in a specific memory location in which data can be lost in the case of a hardware fault.

The methods of stochastic computing provide another approach to overcoming the problem of the design of reliable computers from unreliable elements, i.e., nanodevices. For example, a signal may be represented by the probability that a logic level has a value of 1 or 0 at a clock pulse. In this way, random noise is deliberately introduced into the data. A quantity is represented by a clocked sequence of logic levels generated by a random process. Operations are performed via completely random data.

1.6 Contemporary CAD of logic networks

The goal of the computer aided design (CAD) of logic network tools is to automatically transform a description of logic networks in the algorithmic or behavioral domains to one in the physical domain, i.e., down to a layout mask for chip production.

A CAD system has to produce logically correct circuits correct results, but because of the complexity of the design process, verifying the correctness of results is a necessary phase of design. Usually, formal verification techniques deal with different data structures and descriptions. Computer aided design tools are intended to support all phases of digital design: description (specification), synthesis (design), including various optimizations to reduce cost and improve performance, and verification of the design with respect to its specification. A CAD system for logic synthesis and analysis employs a suitable representation that allows Boolean functions to be manipulated. These representations are realized internally by means of suitable *data structures*.

Data structures are used for the description or specification of a digital system at various levels. The first approach to describing digital systems consists of a description of their structure through graphical data structures. This description provides a *logic diagram* of the system at different levels, showing the modules and their interconnections. This process is supported by libraries of standard components. The second approach is the use of *hardware description language* (HDL). Both approaches are considered throughout this textbook.

1.7 Summary of design process and technology

This chapter introduces the strategic approaches undertaken in the contemporary design of discrete devices and systems. These approaches and trends are being directed by advances in technology. The key aspects of this chapter are as follows:

(*a*) The ***hierarchical*** approach to design makes a large system more manageable by reducing complexity. In this approach, components can be designed, tested, and manufactured separately. ***Standard*** components can be mass produced at relatively low cost. This facilitates the reduction of overall design time and cost.

(*b*) Both ***top-down*** and ***bottom-up*** design methodologies are employed. A design that evolves from an abstract point of view and proceeds to specific components is a top-down design methodology. In this approach, the system architecture is specified at the highest level first. An alternative process, bottom-up design methodology, requires the components at or near the lowest design level of the hierarchy tree to be designed first.

(*c*) ***Data structures*** are the fundamental components of logic design. No universal data structure exists that can be efficient in all applications. Choosing an appropriate data structure based on its advantages and disadvantages, as well as the relationships between these structures, in order to satisfy design requirements, are crucial points of logic design.

Summary (continuation)

The topics of this chapter are summarized as follows:

- ***Information*** – recorded or communicated facts or ***data*** – takes a variety of physical forms when being stored, communicated, or manipulated. Digital logic networks are characterized by the fact that signals assume a finite number of discrete values. A ***signal*** is defined as a function of one or more variables that conveys information on the nature of a physical phenomenon. A ***system*** is an entity that manipulates one or more signals to accomplish a function, thereby yielding new signals.
- ***Logic design*** emphasizes the development of techniques for the design of logic networks. The basis of logic design is two-valued ***Boolean algebra***, also called ***switching algebra***. Modern logic design includes methods and techniques from various fields.
- The ***specification*** of a system is defined as a description of its function, including various parameters for the evaluation of a system. The ***implementation*** refers to the construction of the system. The ***analysis*** of the system has as its objective the determination of the system's specification from its implementation.
- Complex systems are modeled at various levels of detail, called ***design levels***: (*a*) ***architectural*** or ***system*** level, (*b*) ***logic*** level, and (*c*) ***physical*** level.
- The following design styles are distinguished: (*a*) ***full-custom*** design (provides freedom to the designer), (*b*) ***semi-custom*** design (more possibilities for automation), and (*c*) ***mixed*** design style and ***gate-array*** design (meets the requirements of fabrication).
- Progress in microelectronics striving for miniaturization, optimal design, and technology enhancement is achieved by the scaling down of microdevices. For example, the channel length of metal-oxide-semiconductor field effect transistors (MOSFETs) decreased from 50 μm in 1960 to 1 μm in 1990, and to 45 nm in 2007.
- New horizons in design processes and technology are introduced in the next section, "Further study." For instance, an essential feature of logic design for predictable technologies of the future is that ***random*** signals be used instead of ***deterministic*** signals. Random signals are signals that take on random values at any given instant and must be modeled ***probabilistically***.

1.8 Further study

Historical perspective

1930: Alan Turing is often considered to be the father of modern computer science. Turing provided an influential formalisation of the concept of the algorithm and computation with so-called *Turing machine. Turing test* contributed to the debate regarding artificial intelligence: whether it will ever be possible to say that a machine is conscious and can think. During the Second World War Turing worked at Bletchley Park, Britain's codebreaking centre, the section responsible for German naval cryptanalysis. He devised a number of techniques for breaking German ciphers.

1947: A device called a *transistor*, which has several applications in radio where a vacuum tube ordinarily is employed, was demonstrated for the first time at Bell Telephone Laboratories.

1955: Von Neumann's classic work on probabilistic logic and reliable computation upon nonreliable computing elements: "Probabilistic logics and the synthesis of reliable organisms from unreliable components", in C. E. Shannon and J. McCarthy, Eds., *Automata Studies,* pages 329–378, Princeton University Press, Princeton, NJ, 1955.

1958: Jack S. Kilby, employee at Texas Instruments, put all the circuit elements – transistors, resistors, and capacitors, along with their interconnecting wiring – into a single piece of germanium. His rough prototype was a thin piece of germanium about one-half inch long containing five separate components linked together by tiny wires. Since then, the first commercial ICs began to emerge at the beginning of 1960.

1968: Robert Noyce and Gordon Moore established Intel in Santa Clara, California. Intel produced the first 1K RAM (random access memory). Gordon Moore formulated an empirical law stating that the performance of an integrated circuits, including the number of components on it, doubled every 18–24 months with the same chip price. This became known as *Moore's rule.* It is still holding up.

1971: Ted Hoff at Intel invented Intel's first microprocessor (4004). The 4-bit 4004 ran at 108 kHz and contained 2300 transistors. In 1972, the 8008 microprocessor was developed, twice as powerful as the 4004.

1980: The first programmable integrated circuits were developed. These devices contain circuits whose logical function and connectivity can be programmed by the user, rather than being fixed by the integrated circuit manufacturer. This allows a single chip to be programmed to implement different logic functions. Current devices, called field programmable gate arrays (FPGAs) can now implement tens of thousands of logic networks in parallel and operate at up to 550 MHz.

1981: IBM PC was released.

1985: The Intel 80386 32-bit microprocessor featured 275,000 transistors – more than 100 times as many as the original 4004. It could run multiple programs simultaneously.

1985: David Deutsch described how a computer might run using rules of quantum mechanics and why such a computer differs fundamentally from ordinary computers.

1989: The Intel 486TM processor was the first built-in math coprocessor, which speeded up computing because it offered complex math functions from the central processor, greatly speeding up transcendental functions.

1995: Released in the fall of 1995 the Pentium Pro processor, with 5.5 million transistor and with a second speed-enhancing cache memory chips, was designed to support 32-bit server and workstation-level applications, enabling fast computer-aided design, mechanical engineering and scientific computation.

1997: The 7.5 million-transistor Pentium II processor incorporated Intel MMX technology, which was designed specifically to process video, audio and graphics data efficiently.

2003: The 130 nanometer technology was announced.

2005: The 65 nanometer chip manufacturing process was announced. AMD and NEC have started using a 65 nanometer process.

2007: Intel, IBM, NEC, and AMD started using 45 nanometers for their CPU chips.

Advanced topics of logic design and technology

Topic 1: Specific-area design. The theory and techniques for logic network design, developed to solve problems of computation and information processing, are being very rapidly adopted by those solving problems of communication, control, and artificial intelligence for various specific-area applications such as humanoid robotics, decision-making support systems, bio-medical applications, and biometric-based security systems (in particular, artificial intelligence, encryption, watermarking, and humanoid robotics).

Topic 2: Energy-efficient design of logic networks requires specialized techniques and tools. Mobile devices contain integrated circuits and employ battery-powered systems. The lifetime of the battery decreases as the power consumption of the integrated circuits grows. In energy-efficient design techniques, the most significant power savings are achieved at high levels of abstraction and during early phases of the design. In particular, portable computers, mobile navigation, and robotics. One possible the approach to energy-efficient design is the minimization of the *switching activity* of logic networks. Switching activity is defined as the expected number of logic transitions during one clock cycle.

Topic 3: Trends and potential application for nanoscale devices design. Selected topics in the theory and technique of logic network design are found useful in the design of computing devices based on nanotechnology. As a result, researchers from physics and the chemical sciences have an urgent desire and need to become familiar with the theory and practice of contemporary logic design. For earlier technologies, the relevant problems were primary concerned with component minimization. With the development of VLSI and ULSI logic network technologies and the advent of nanoscale technologies in digital system design, the problems concerned with the minimization of components have become less relevant. These types of problems have been replaced by less well defined and much more difficult problems, such as physical design including partitioning, layout and routing, structural simplicity, and uniformity of modules. The last problem, uniformity, and related data structures are of particular interest in logic design in nanoscale. Many of these problems cannot be solved based on traditional approaches. However, a great amount of theoretical and practical work has been done in these areas.

Further reading

A. Advanced logic design

Digital Design: Principles and Practices by Richard S. Sandige, Prentice Hall, 2002.

Digital Design Essentials by John F. Wakerly, Prentice Hall, 2001.

"Electronic Design Automation at the Turn of Century", special issue of the *IEEE Transactions on Computer-Aided Design of Integrated Circuits and Systems*, volume 19, number 12, 2000.

Fundamentals of Logic Design by Jr. C. H. Roth, 5th Edition, Thomson Brooks/Cole, 2004.

Fundamentals of Digital Logic with VHDL Design by S. Brown and Z. Vranesic, New York, McGraw-Hill, 2000.

Introduction to Digital Logic Design by John P. Hayes, Addison-Wesley, 1993.

Introduction to Digital Systems by M. D. Ercegovac, T. Lang, and J. H. Moreno, John Wiley & Sons, 1999.

"Logic Design of Computational Nanostructures" by S. Yanushkevich, *Journal of Computational and Theoretical Nanoscience*, American Scientific Publishers, volume 4, number 3, pages 384–407, May 2007.

"Logic Design of Nanodevices" by S. Yanushkevich, In: *Handbook of Theoretical and Computational Nanotechnology* by M. Rieth, and W. Schommers, Editors, Scientific American Publishers, Chapter 110, pages 1–52, 2007.

Logic Design Principles: With Emphasis on Testable Semicustom Circuits by Edward J. McCluskey, Prentice Hall, 1986.

Logic Synthesis and Verification edited by S. Hassoun and T. Sasao, Consulting Editor R. K. Brayton, Kluwer, 2002.

Synthesis and Optimization of Digital Circuits by G. De Micheli, McGraw-Hill, 1994.

Switching Theory for Logic Synthesis by T. Sasao, Kluwer, 1999.

"The Future of Logic Synthesis and Verification" by R. K. Brayton, in *Logic Synthesis and Verification* Edited by S. Hassoun and T. Sasao, Consulting Editor R. K. Brayton, Kluwer, Dordrecht, 2002.

B. Predictable technologies and computing devices

"Computing with Molecules" by M. A. Reed and J. M. Tour, *Scientific American*, pages 86–93, June 2000.

Introduction to Nanotechnology by C. P. Jr. Poole and F. J. Owens, John Wiley & Sons, New York, 2003.

Logic Design of NanoICs by S. Yanushkevich, V. Shmerko, and S. E. Lyshevski, CRC Press, Boca Raton, FL, 2005.

"The Topsy Turvy World of Quantum Computing" by J. Mullins, *IEEE Spectrum*, pages 42–49, February 2001.

Molecular Electronics, Circuits and Processing Platforms by S. E. Lyshevski, CRC Press, Taylor & Francis Group, Boca Raton, FL, 2007.

"Ultimate Theoretical Models of Nanocomputers" by M. P. Frank and T. F. Jr. Knight, *Nanotechnology*, volume 9, pages 162–176, 1998.

2

Number Systems

Computer Arithmetic

- Binary
- Octal
- Decimal
- Hexadecimal

Binary arithmetic

- Sign and magnitude
- 1's complement
- 2's complement
- Addition and subtraction

Residue arithmetics

- Modular adder
- Modular subtractor
- Modular multiplier

Binary codes

- Gray code and Hamming distance
- Binary-coded decimal codes
- Weighted codes

Advanced topics

- Number systems and cryptography
- Numbers and information

2.1 Introduction

The binary number system is the most important number system in digital design. This is because it is suited to the binary nature of the phenomena used in dominant microelectronic technology. Even in situations where the binary number system is not used as such, binary codes are employed to represent information at the signal level. For example, multi-valued logic values are often encoded using binary representations. However, humans prefer decimal numbers, – thus, that is, binary numbers must be converted into decimal numbers.

In this chapter, various number systems are examined that are used in digital data structures. These number systems, such as octal and hexadecimal, are used to simplify the manipulation of binary numbers.

2.2 Positional numbers

A number system is defined by

- Its basic symbols, called *digits* or *numbers*, and
- The ways in which the digits can be combined to represent the full range of numbers we need.

2.2.1 The decimal system

The ten digits $0, 1, 2, \ldots, 9$ can be combined in various ways to represent any number. The fundamental method of constructing a number is to form a *sequence* or *string* of digits or *coefficients*:

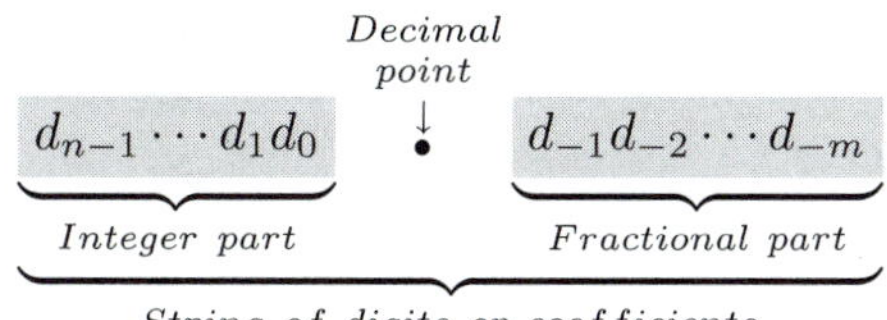

where integer and fractional parts are represented by n and m digits to the left and to the right of the *decimal point*, respectively. The subscript $i = -m, m-1, \ldots, 0, 1, \ldots, n$ gives the position of the digit. Depending on the position of digits in the string, each digit has an associated value of an integer raised to the power of 10 as follows:

The decimal system

$$N = \underbrace{d_{n-1}d_{n-2}\cdots d_1 d_0 \overset{\text{Decimal point}}{\bullet} d_{-1}d_{-2}\cdots d_{-m}}_{\text{String of coefficients}}$$

$$= \underbrace{d_{n-1}\times 10^{n-1} + d_{n-2}\times 10^{n-2} + \cdots + d_1\times 10^1 + d_0\times 10^0}_{\text{Computing the integer part}}$$

$$= \underbrace{d_{-1}\times 10^{-1} + d_{-2}\times 10^{-2} + \cdots + d_{-m}\times 10^{-m}}_{\text{Computing the fractional part}}$$

$$= \sum_{i=-m}^{n-1} d_i 10^i$$

This method of representing numbers is called the *decimal system*. In the positional representation of digits:

Positional representation

- Each digit has a fixed value, or *weight*, determined by its position.
- All the weights used in the decimal number system are powers of 10.
- Each decimal digit d_i ranges between 0 and 9.
- The weighting of the digits is defined relative to the *decimal point*. This symbol means that digits to the left are weighted by positive powers of 10, giving integer values, while digits to the right are weighted by negative powers of 10, giving fractional values.
- Fractions are denoted by sequences of digits whose weights are negative powers of 10.

Example 2.1 **(Decimal numbers.)** *The four digits in the number 2008 represent, from left to right, thousands (digit 2), hundreds (number 0), tens (number 0), and ones (number 8). Hence, this four-digit number can be represented in the following form:*

$$2008 = 2\times \overset{Weight}{10^3} + 0\times \overset{Weight}{10^2} + 0\times \overset{Weight}{10^1} + 8\times \overset{Weight}{10^0}$$

Practice problem **2.1.** **(Decimal numbers.)** Write the decimal number 747 in positional form.
Answer: $747 = 7\times 10^2 + 4\times 10^1 + 7\times 10^0$.

Example 2.2 (Integer and fraction.) *The decimal number 12.3456 consists of an integer part (12) and a fractional part (3456) separated by the decimal point. Thus, this number can be represented in the following form:*

$$12.3456 = \underbrace{1 \times 10^1 + 2 \times 10^0}_{Integer\ part} + \underbrace{3 \times 10^{-1} + 4 \times 10^{-2} + 5 \times 10^{-3} + 5 \times 10^{-4}}_{Fractional\ part}$$

The number 0.34_{10} *is represented as*

$$N = \sum_{i=-2}^{-1} 10^i \times d_i = 3 \times 10^{-1} + 4 \times 10^{-2} = {}^{34}/_{100}$$

Practice problem **2.2. (Fraction.)** What does the number 7.53_8 represent?
Answer is given in "Solutions to practice problems."

2.2.2 Number radix

In general, an n-digit number in radix r consists of n digits, each taking one of r values: $\underbrace{0,\ 1,\ 2,\ \ldots,\ r-1}_{Radix\ r\ system}$. A general number N in a positional number system is represented by the following formula (Figure 2.1):

Number radix

$$N = \overbrace{\underbrace{a_{n-1}a_{n-2}\cdots a_1a_0}_{Integer\ part} \overset{\substack{Radix\\ point\\ \downarrow}}{\bullet} \underbrace{a_{-1}a_{-2}\cdots a_{-m}}_{Fractional\ part}}^{n+m\ digits}$$
$$= \underbrace{a_{n-1} \times r^{n-1} + a_{n-2} \times r^{n-2} + \cdots + a_1 \times r^1 + a_0 \times r^0}_{Integer\ part}$$
$$+ \underbrace{a_{-1} \times r^{-1} + a_{-2} \times r^{-2} + \cdots + a_{-m} \times r^{-m}}_{Fractional\ part} = \sum_{i=-m}^{n-1} a_i r^i \quad (2.1)$$

where a_i denotes a digit in the number system such that

$$0 \le a_i \le (r-1),$$

where r is the base of the number system, n is the number of digits in the integer part of N, and m is the number of digits in the fractional part of N.

The integer part is separated from the fractional part by the *radix point*. The digits a_{n-1} and a_{-m} are referred to as the *most significant digits* (MSD) and the *least significant digits* (LSD) of the number N, respectively.

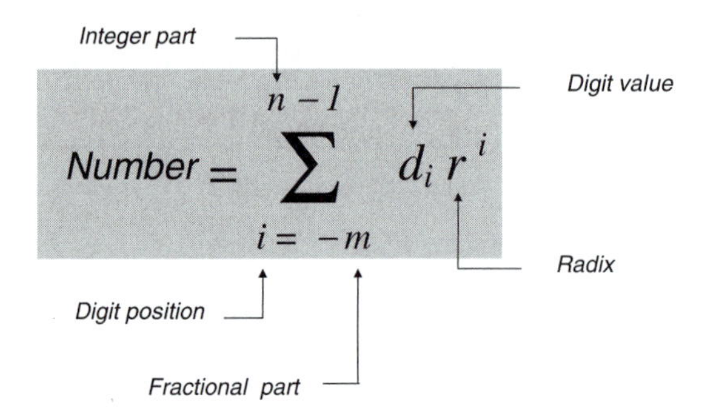

FIGURE 2.1
Number representation in the positional system.

A number system is said to be of *base*, or *radix* r, because the digits are multiplied by powers of r, and this system uses r distinct digits.

> **Example 2.3 (Radix.)** *The decimal number system is said to be of* base, *or* radix, 10, *because the digits are multiplied by powers of 10, and this system uses 10 distinct digits.*

To avoid possible confusion, the radix of the number system is often written as a decimal subscript appended to the number; that is, the subscript is placed after the LSD to indicate the radix of the number. When the context makes the radix obvious, it is not necessary to indicate the radix.

> **Example 2.4 (Radix.)** *The radix of the number system is written as a decimal subscript as follows. The binary* ($r = 2$) *number 10110 can be written in the form* 10110_2. *The octal* ($r = 8$) *number 67344.25 is indicated in the form* 67344.25_8. *The decimal* ($r = 10$) *number 67390.845 is indicated in the form* 67390.845_{10}.

Equation 2.1 is used in number representation as follows:

Algorithm for representing a number in the radix r system
Step 1. `Choose the radix` r `of a number system`
Step 2. `Choose the number of digits` n `in the integer part of` N
Step 3. `Choose the number of digits` m `in the fractional part of` N
Step 4. `Write the number` N `in the radix` r `number system`

Example 2.5 (Positional number systems.) *In Table 2.1, the most useful number systems are listed. Observe that for those number systems of base less than 10, a subset of the digit symbols of the decimal number system is used. For example, maximal two digit numbers in various radix are as follows:*$11_2 = 3_{10}$, $22_3 = 8_{10}$, $33_4 = 15_{10}$, $77_8 = 63_{10}$, 99_{10}, $FF_{16} = 16 \times F + F$.

TABLE 2.1
The most important positional number systems used in data representation and computing (Example 2.5).

Base r	Number system	Digit symbols
2	Binary	0,1
3	Ternary	0,1,2
4	Quaternary	0,1,2,3
8	Octal	0,1,2,3,4,5,6,7
10	Decimal	0,1,2,3,4,5,6,7,8,9
16	Hexadecimal	0,1,2,3,4,5,6,7,8,9,A,B,C,D,E,F

Practice problem 2.3. (Positional number systems.) Using digit symbols for the hexadecimal number system, write digit symbols for the *duodecimal* (base 12) system.
Answer: 0,1,2,3,4,5,6,7,8,9,A, and B.

Example 2.6 (Positional number system.) *Given* $n = 3$*, Equation 2.1 represents a number in hexadecimal, decimal, octal, and binary number systems by the following:*

$$N_{16} = \sum_{i=0}^{2} 16^i \times h_i = h_2 \times 16^2 + h_1 \times 16^1 + h_0 \times 16^0$$

$$N_{10} = \sum_{i=0}^{2} 10^i \times d_i = d_2 \times 10^2 + d_1 \times 10^1 + d_0 \times 10^0$$

$$N_8 = \sum_{i=0}^{2} 8^i \times o_i = o_2 \times 8^2 + o_1 \times 8^1 + o_0 \times 8^0$$

$$N_2 = \sum_{i=0}^{2} 2^i \times b_i = b_2 \times 2^2 + b_1 \times 2^1 + b_0 \times 2^0$$

The largest numbers that can be represented by three digits in these systems are $Max(N_{16}) = FFF_{16} = 4081_{10}$, $Max(N_{10}) = 999_{10}$, $Max(N_8) = 777_8 = 511_{10}$, *and* $Max(N_2) = 111_2 = 7_{10}$.

2.2.3 Fractional binary numbers

In the binary number positional representation, $B = \sum_{i=-n}^{m} 2^i \times b_i$, each binary digit or bit, b_i, is 0 or 1. The symbol "." becomes a *binary point*, separating bits on the left being weighted by positive powers of two, and those on the right being weighted by negative powers of two.

Example 2.7 (Binary numbers.) *The binary number* 101.11_2 *is written as follows:*

$$N = \sum_{i=-2}^{1} 2^i \times b_i$$
$$= \underbrace{1 \times 2^2 + 0 \times 2^1 + 1 \times 2^0}_{Integer\ part} + \underbrace{1 \times 2^{-1} + 1 \times 2^{-2}}_{Fractional\ part} = 5\ ^3/_4$$

Practice problem 2.4. (Binary numbers.) What does the num-ber 10.01_2 represent?
Answer is given in "Solutions to practice problems."

Shifting the binary point one position to the left has the effect of dividing the number by two.

Example 2.8 (Dividing by two.) *The binary number* 101.11_2 *represents the decimal number* $5\ ^3/_4$. *Shifting the point one position to the left results in* 10.111_2, *that is,*

$$10.111_2 = 2 + 0 + ^1/_2 + ^1/_4 + ^1/_8 = 2\ ^7/_8$$

Practice problem 2.5. (Dividing by two.) Divide the number $2\ ^1/_4 = 10.01_2$ by two.
Answer is given in "Solutions to practice problems."

Similarly, shifting the binary point one position to the right has the effect of multiplying the number by two. In computing, replacing arithmetic by shifts can occur when multiplying by constants.

Example 2.9 (Multiplying by two.) *The binary number* 101.11_2 *represents the decimal number* $5\ ^3/_4$:

$$101.11_2 = (1 \times 2^2) + (0 \times 2^1) + (1 \times 2^0) + (1 \times 2^{-1}) + (1 \times 2^{-2})$$
$$= 5\ ^3/_4$$

Shifting the point one position to the right results in $1011.1_2 = 11\ ^1/_2$, *that is,*

$$1011.1_2 = (1 \times 2^3) + (0 \times 2^2) + (1 \times 2^1) + (1 \times 2^0) + (1 \times 2^{-1})$$
$$= 8 + 0 + 2 + 1 + ^1/_2 = 11\ ^1/_2$$

Practice problem 2.6. (Multiplying by two.) Multiply the number $1\,{}^{1}/_{8}$ by two.

Answer is given in "Solutions to practice problems."

Example 2.10 (Fractional part.) *Samples of integer and fractional parts of binary numbers and their corresponding decimal equivalents are given in Table 2.2.*

TABLE 2.2
Integer and fractional parts of binary numbers (Example 2.10).

Binary	Fraction (decimal)	Integer (decimal)
1.		1
10.		2
100.		4
1000.		8
10000.		16
100000.		32
0.1	${}^{1}/_{2}$	0.5
0.01	${}^{1}/_{4}$	0.25
0.001	${}^{1}/_{8}$	0.125
0.0001	${}^{1}/_{16}$	0.0625

2.2.4 Word size

In discrete systems, a *word* refers to a set of bits. *Word size* is defined as the number of bits in the binary numbers. Word size is typically a power of two and ranges from 8 bits, called a *byte*, to 32, 64, 128, or even 256 in some computers.

Example 2.11 (Byte.) *A 16-bit (two-byte) representation of a binary number with 8 bits for the integer part and 8 bits for the fractional part is shown below:*

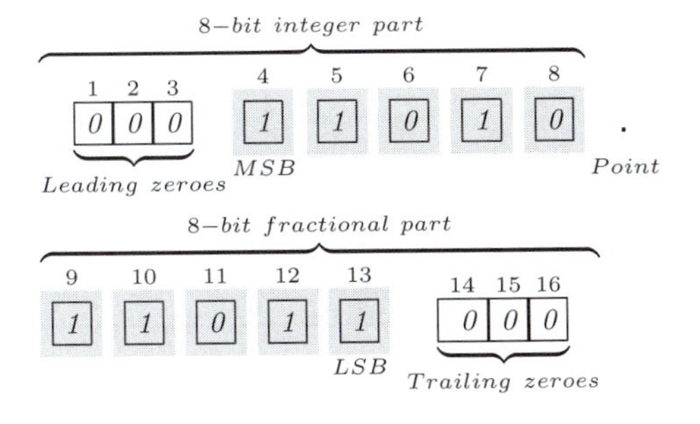

Practice problem **2.7. (Byte.)** Represent the number 7.0625_{10} using an 8-bit (byte) number (four bits for the integer part and four bits for the fractional part).
Answer is given in "Solutions to practice problems."

2.3 Counting in a positional number system

Counting is the fundamental operation in digital systems. A positional number system is well-suited for counting; that is, the counting is well-automated and implemented in software and hardware.

Example 2.12 (Positional number systems.) *In Table 2.3, the counting process is illustrated in various number systems in order to show regularities, which are useful for automation.*

Practice problem **2.8. (Positional number systems.)** Using Table 2.3, write the number 16_{10} in the binary, ternary, octal, and hexadecimal number systems.
Answer: $16_{10} = 10000_2 = 121_3 = 20_8 = 10_{16}$.

2.4 Basic arithmetic operations in various number systems

The four basic arithmetic operations:

- ▶ Addition,
- ▶ Subtraction,
- ▶ Multiplication, and
- ▶ Division

can be performed in various positional number systems.

Example 2.13 (Basic operations.) *Addition and subtraction on the integer numbers in the binary, octal, decimal, and hexadecimal number systems are given in Table 2.4.*

TABLE 2.3
The first 16 integers in the binary, ternary, octal, decimal, and hexadecimal number systems (Example 2.12).

Binary	Ternary	Octal	Decimal	Hexadecimal
0	0	0	0	0
1	1	1	1	1
10	2	2	2	2
11	10	3	3	3
100	11	4	4	4
101	12	5	5	5
110	20	6	6	6
111	21	7	7	7
1000	22	10	8	8
1001	100	11	9	9
1010	101	12	10	A
1011	102	13	11	B
1100	110	14	12	C
1101	111	15	13	D
1110	112	16	14	E
1111	120	17	15	F

TABLE 2.4
Some arithmetic operations with unsigned numbers.

Techniques for computing with unsigned numbers

Radix	Technique	
Binary ($r = 2$)	$\begin{array}{lr} (3_{10}) & 1\ 1\ 1\ 1_2 \\ (2_{10}) & +\ 0\ 1\ 1\ 0_2 \\ \hline (5_{10}) & 1\ 0\ 1\ 0\ 1_2 \end{array}$	$\begin{array}{lr} (5_{10}) & 1\ 1\ 1\ 1_2 \\ (6_{10}) & -\ 0\ 1\ 1\ 0_2 \\ \hline (11_{10}) & 1\ 0\ 0\ 1_2 \end{array}$
Octal ($r = 8$)	$\begin{array}{lr} (15_{10}) & 1\ 7_8 \\ (6_{10}) & +\ \ 6_8 \\ \hline (21_{10}) & 2\ 5_8 \end{array}$	$\begin{array}{lr} (15_{10}) & 1\ 7_8 \\ (6_{10}) & -\ \ 6_8 \\ \hline (9_{10}) & 1\ 1_8 \end{array}$
Decimal ($r = 10$)	$\begin{array}{r} 1\ 5_{10} \\ +\ \ 6_{10} \\ \hline 2\ 1_{10} \end{array}$	$\begin{array}{r} 1\ 5_{10} \\ -\ \ 6_{10} \\ \hline 9_{10} \end{array}$
Hexadecimal ($r = 16$)	$\begin{array}{lr} (15_{10}) & F_{16} \\ (6_{10}) & +\ 6_{16} \\ \hline (21_{10}) & 1\ 5_{16} \end{array}$	$\begin{array}{lr} (15_{10}) & F_{16} \\ (6_{10}) & -\ 6_{16} \\ \hline (9_{10}) & 9_{16} \end{array}$

2.5 Binary arithmetic

In the decimal system, the sign of a number is indicated by a special symbol, "+" or "-". In the binary system, the sign of a number is denoted by the left-most bit. Positive numbers are represented using the positional number representation. The so-called *unsigned* representation (magnitude only) is used to denote positive numbers.

Negative numbers can be represented in different ways. The most commonly used are sign-and-magnitude and complemented, which can be 1's complement or 2's complement notation (Figure 2.2).

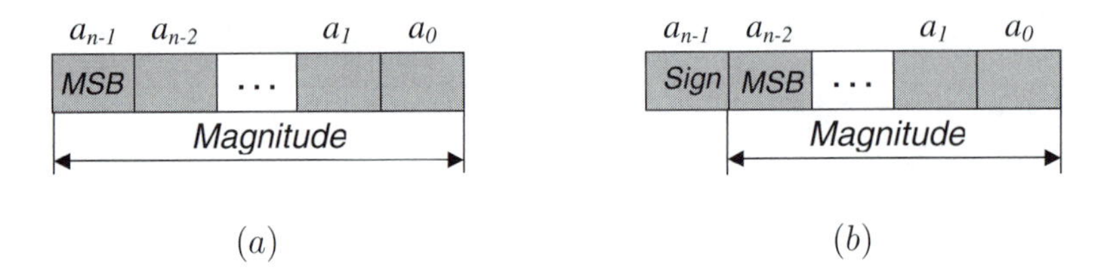

FIGURE 2.2
Unsigned number format (a) and sign-and-magnitude format (b).

Sign-and-magnitude is a two-point binary notation (one bit for sign and the rest for magnitude):

NUMBER = <SIGN><MAGNITUDE>

The *sign bit* is the left-most bit, called the *most significant bit*, and it is equal to 0 for positive numbers and 1 for negative numbers.

While performing addition, the magnitudes are added, and the resulting sum is given the sign of the operands. If the operands have opposite signs, it is necessary to subtract the smaller number from the larger one (logic networks that compare and subtract numbers are needed). The range of signed integers is $-(2^{n-1}-1) \leq x \leq 2^{n-1}-1$. In such a system, zero has two representations: positive zero $00\ldots0$ and negative zero $10\ldots0$.

Example 2.14 (Sign-and-magnitude.) *The 8-bit sign-and-magnitude representation of the number* 18_{10} *is*

$$18_{10} = \boxed{0}\ \overbrace{0010010}^{Magnitude}$$

The 8-bit sign-and-magnitude form of the number -18_{10} *is*

$$-18_{10} = \boxed{1}\ \overbrace{0010010}^{Magnitude}.$$

Practice problem **2.9.** **(Sign-and-magnitude.)** Represent the

number -7_{10} in 8-bit sign-and-magnitude form.
Answer: $-7_{10} = 10000111_2$.

Techniques for manipulation of binary numbers in sign-and-magnitude format are given in Table 2.5.

TABLE 2.5
Sign-and-magnitude techniques.

Techniques for computing in sign-and-magnitude format

Example		Technique
(3) 0 0 1 1 (2) + 0 0 1 0 (5) 0 1 0 1	(7) 0 1 1 1 (7) + 0 1 1 1 (−6) 1 1 1 0 Incorrect	If sign bits are both 0, perform addition of two binary numbers. If sum of the magnitudes is greater then $2^{n-1} - 1$, the result is incorrect.
(+3) 0 0 1 1 (−2) + 1 0 1 0 (1) 0 0 0 1	(−3) 1 0 1 1 (+2) + 0 0 1 0 (−1) 1 0 0 1	If signs are different, compare the magnitudes. Subtract the smallest magnitude from the greater, and assign to the result the sign of the greater magnitude.
(−3) 1 0 1 1 (−2) + 1 0 1 0 (−5) 1 1 0 1	(−7) 1 1 1 1 (−1) + 1 0 0 1 (0) 1 0 0 0 0 Incorrect	If sign bits are both 1, perform addition of both magnitudes; the sign bit of the result is 1. If sum of the magnitudes is greater then $2^{n-1} - 1$, the result is incorrect.

2.6 Radix-complement representations

Consider the number D that consists of n digits d_i, $i = 1, 2, \ldots, n$ in the radix r number system. There are two types of *radix-complements* for representation of the number:

The r radix-complement

$$\overline{D} = r^n - D \tag{2.2}$$

The $r - 1$ radix-complement

$$\overline{D} = (r^n - 1) - D \tag{2.3}$$

Example 2.15 (Radix-complement.) *Given the binary number* $D = 101_2$ *(*$n = 3$*), the 2's and 1's complements are* $\overline{D} = 2^3 - 101 = 1000 - 101 = 011_2$ *and* $\overline{D} = (2^3 - 1) - 101 = 111 - 101 = 010_2$*. Given the the decimal number* $D = 125_{10}$ *(*$n = 3$*), the 10's and 9's complements are* $\overline{D} = 10^3 - 125 = 875_{10}$ *and* $\overline{D} = 10^3 - 1 - 125 = 874_{10}$*.*

While the sign-and-magnitude system makes a number negative by changing its sign, a *complement number system* makes a number negative by taking its complement. Two numbers in a complement number system can be added or subtracted directly without the sign and magnitude checks required by the sign-and-magnitude system.

The *radix-complement* representation for decimal ($r = 10$) and binary ($r = 2$) number systems is shown in Figure 2.3.

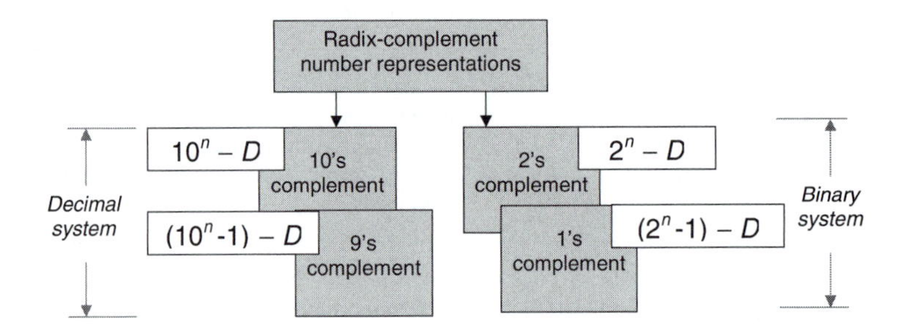

FIGURE 2.3
Radix-complement representations of decimal and binary numbers.

2.6.1 10's and 9's complement systems

In 10's and 9's complement systems, positive numbers are represented using the same binary code as for unsigned numbers. Negative numbers are represented in a complement form.

10's complement system

In the decimal number system, the radix-complement is called *10's complement.*

The rule for forming the 10's complement

Given: A decimal number $D = d_{n-1}d_n \dots d_0$.
Step 1: Subtract each d_i from 9: $(9-d_{n-1}), (9-d_{n-2}), \dots, (9-d_0)$.
Step 2: Add 1 to the resulting number.

Example 2.16 (10's complement.) *Given the decimal number 125, its 10's complement is calculated as follows:*

$$\begin{aligned}\overline{125} &= (9-d_2)(9-d_1)(9-d_0)+1\\ &= (9-1)(9-2)(9-5)+1 = 874+1 = 875\end{aligned}$$

which is the same result as the that obtained by the calculation in Example 2.15.

Practice problem 2.10. **(10's complement.)** Find the 10's complement of the number 17_{10}.
Answer: $\overline{17}_{10} = (9-1)(9-7)+1 = 82+1 = 83_{10}$.

9's complement

The 9's complement of a decimal number is formed as follows:

The rule for forming the 9's complement

Given: `A decimal number` $D = d_{n-1}d_n \ldots d_0$.
Step 1: `Subtract each` d_i `from 9:` $(9-d_{n-1}), (9-d_{n-2}), \ldots, (9-d_0)$.

Example 2.17 (9's complement.) *Given the decimal number 46, 125 and 5329, their 9's complements are calculated as follows:*

$$\begin{aligned}\overline{46} &= (9-d_1)(9-d_0) = (9-4)(9-6) = 53\\ \overline{125} &= (9-d_2)(9-d_1)(9-d_0) = (9-1)(9-2)(9-5) = 1874\\ \overline{5329} &= (9-d_3)(9-d_2)(9-d_1)(9-d_0)\\ &= (9-5)(9-3)(9-2)(9-9) = 4670\end{aligned}$$

Practice problem 2.11. **(9's complement.)** Find the 9's complement of the number 17_{10}.
Answer: $\overline{17}_{10} = (9-1)(9-7) = 82_{10}$.

2.6.2 1's complement system

In 1's complement system, positive numbers are represented in the same way as unsigned numbers. Let a negative number $-P$ be given, where P is the magnitude of the number. An n-bit negative number K is obtained by subtracting the positive number P (magnitude) from $2^n - 1$:

1's complement system

$$K = (2^n - 1) - P \tag{2.4}$$

An advantage of 1's complement representation is that a negative number is generated by complementing all bits of the corresponding positive number (magnitude). The addition of 1's complement numbers may require a correction, and the time needed to add two 1's complement numbers may be twice as long as the time needed to add two unsigned numbers.

Rule for forming the 1's complement

Given: A binary number $B = b_{n-1}b_n \ldots b_0$.
Step 1: Complement each digit of the binary number:

$$(1 - b_{n-1}), (1 - b_{n-2}), \ldots, (1 - b_0)$$

Step 2: Add 1 to the result.

Example 2.18 (1's complement.) *An n-bit binary representation of numbers between +7 and -7 with 1's complement representation of negative numbers is given below.*

Binary number	Decimal number
0111	*+7*
0110	*+6*
0101	*+5*
0100	*+4*
0011	*+3*
0010	*+2*
0001	*+1*
0000	*+0*

Binary number	Decimal number
1111	*-0*
1110	*-1*
1101	*-2*
1100	*-3*
1011	*-4*
1010	*-5*
1001	*-6*
1000	*-7*

For example, the negative number -7_{10} *is represented as a 4-bit number using the equation* $K = (2^4 - 1) - 7 = 8_{10} = \boxed{1}000_2$. *Alternatively, given* $-7_{10} = 0111_2$, *its complement is the desired number* $-7_{10} = \overline{0111}_2 = \boxed{1}000_2$.

Practice problem 2.12. (1's complement.) Represent the numbers −5 and 5 using 4-bit binary code.
Answer: $-5_{10} = (2^4 - 1) - 5 = 10_{10} = 1010_2$, or $-5_{10} = \overline{0101}_2 = 1010_2$. $+5_{10} = 0101_2$.

2.6.3 2's complement

A code for representing an n-bit negative number K is obtained by subtracting its equivalent positive number P from 2^n, i.e., $K = 2^n - P$. An advantage of 2's complement representation is that when the numbers are added, the result is always correct. If there is a carry-out from the sign-bit position, it is simply ignored.

The rule for forming the 2's complement

Given: `The binary code for the magnitude of the negative number` $B = b_{n-1}b_n \dots b_0$

Step 1: `Complement each digit of the code`

$$(1 - b_{n-1}), (1 - b_{n-2}), \dots, (1 - b_0)$$

Example 2.19 (2's complement.) *Binary representation of numbers between +7 and −7 with 2's complement representation of negative numbers is given below. Note that the binary codes of the positive numbers are represented exactly like the unsigned numbers.*

Binary number	Decimal number
0111	+7
0110	+6
0101	+5
0100	+4
0011	+3
0010	+2
0001	+1
0000	0

Binary number	Decimal number
1111	-1
1110	-2
1101	-3
1100	-4
1011	-5
1010	-6
1001	-7

For example, the negative number -7_{10} *is represented in 2's complement form as*

$$-7_{10} = 2^4 - 7 = 9_{10} = 1001_2$$

Alternatively,

$$-7_{10} = \overline{111}1_2 + 1 = 1000_2 + 1 = 1001_2$$

Practice problem 2.13. (2's complement.) Represent the numbers -6_{10} and 6_{10} in 2's complement form.
Answer: $-6_{10} = \overline{0110}_2 + 1 = 1010_2$. $6_{10} = 0110_2$.

Example 2.20 (2's complement.) *Techniques for the addition of 2's complement binary numbers are demonstrated in Table 2.6.*

TABLE 2.6
Techniques for representation and addition of binary numbers: unsigned, sign-and-magnitude, 1's and 2's complement.

Techniques for computing binary numbers

Radix	Technique	
Unsigned	$\begin{array}{rr} (3) & 0\,0\,1\,1 \\ (2) & +\,0\,0\,1\,0 \\ \hline (5) & 0\,1\,0\,1 \end{array}$	$\begin{array}{rr} (5) & 0\,1\,0\,1 \\ (6) & +\,0\,1\,1\,0 \\ \hline (11) & 1\,0\,1\,1 \end{array}$
Sign-and-magnitude	$\begin{array}{rr} (+7) & 0\,1\,1\,1 \\ (-2) & +\,1\,0\,1\,0 \\ \hline (+5) & 0\,1\,0\,1 \end{array}$	$\begin{array}{rr} (-5) & 1\,1\,0\,1 \\ (-2) & +\,1\,0\,1\,0 \\ \hline (-7) & 1\,1\,1\,1 \end{array}$
1's complement $K = (2^n - 1) - P$ $-5 = \underbrace{1111}_{2^4-1} - \underbrace{0101}_{P=5} = 1010$ $-9 = \underbrace{11111}_{2^5-1} - \underbrace{01001}_{P=9} = 10110$	$\begin{array}{rr} (+5) & 0\,1\,0\,1 \\ +(-2) & +\,1\,1\,0\,1 \\ \hline (+3) & \boxed{1}\,0\,0\,1\,0 \\ & + \longrightarrow 1 \\ \hline & 0\,0\,1\,1 \end{array}$	$\begin{array}{rr} (-5) & 1\,0\,1\,0 \\ +(-2) & +\,1\,1\,0\,1 \\ \hline (-7) & \boxed{1}\,0\,1\,1\,1 \\ & + \longrightarrow 1 \\ \hline & 1\,0\,0\,0 \end{array}$
2's complement $K = 2^n - P$ $-5 = \underbrace{1000}_{2^4} - \underbrace{0101}_{P=5} = 1011$ $-9 = \underbrace{10000}_{2^5} - \underbrace{01001}_{P=9} = 10111$	$\begin{array}{rr} (+5) & 0\,1\,0\,1 \\ +(-2) & +\,1\,1\,1\,0 \\ \hline (+3) & \boxed{1}\,0\,0\,1\,1 \end{array}$ $\uparrow$ *Ignore*	$\begin{array}{rr} (-5) & 1\,0\,1\,1 \\ +(-2) & +\,1\,1\,1\,0 \\ \hline (-7) & \boxed{1}\,1\,0\,0\,1 \end{array}$ $\uparrow$ *Ignore*

2.7 Conversion of numbers in various radices

Consider a radix r number that includes a radix point. First, the number must be separated into an integer part and a fractional part, since the parts must be converted differently. There are various algorithms for the conversion of numbers between systems. In Figure 2.4, possible conversions between binary, octal, hexadecimal, and decimal systems are shown.

Conversion of numbers from decimal to other radices

Given a decimal number, the conversion of this number to a radix r number is as follows:

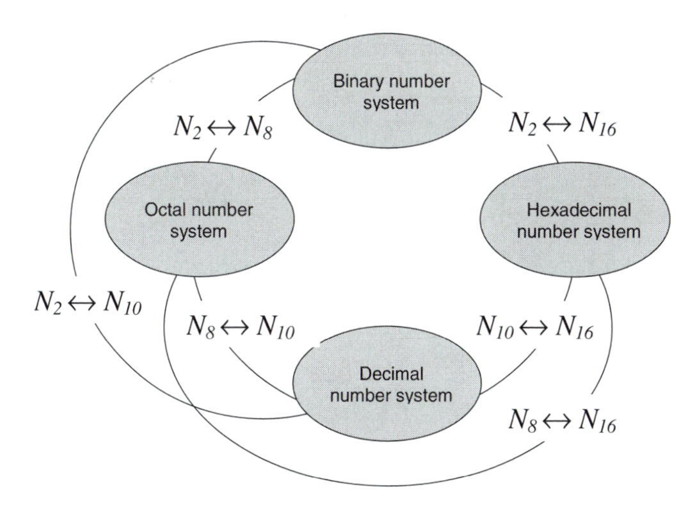

FIGURE 2.4
Conversion of numbers in various radices $N_{r_i} \leftrightarrow N_{r_j}$.

Conversion of numbers from decimal to other radices

- ▶ The conversion of a decimal integer to a radix r number is achieved by dividing the number and all successive quotients by r and accumulating the remainders.
- ▶ The conversion of a fraction to a radix r fraction is accomplished by multiplying by radix r to give the integer and fraction; the new fraction is multiplied by r to give a new integer and a new fraction. This process is continued until the fractional part equals 0 or until there are enough digits to achieve sufficient accuracy.

Conversion of decimal integers to binary integers

For the conversion of decimal integers to binary integers, the following steps are needed:

Conversion of decimal integers to binary integers

Given: `A decimal integer`
Step 1: `Divide the decimal number by two to give a quotient and a remainder`
Step 2: `Divide the quotient by two to give a new quotient and a remainder`
Step 3: `Repeat division until the fractional part is 0 or until required number of digits is achieved`
Result: `The remainders are written bottom-up (from recent to the first obtained) of the desired binary number`

Example 2.21 (Conversion of a decimal into a binary number.) *Convert the decimal integer number 50_{10} into a binary number. Figure 2.5a illustrates the conversion:*

$$\begin{aligned} 50 : 2 &= 25 + \boxed{0} \\ 25 : 2 &= 12 + \boxed{1} \\ 12 : 2 &= 6 + \boxed{0} \\ 6 : 2 &= 3 + \boxed{0} \\ 3 : 2 &= 1 + \boxed{1} \\ 1 : 2 &= 0 + \boxed{1} \end{aligned}$$

The sequence of remainders is 0,1,0,0,1,1. The result of conversion is obtained by reading the remainders in reverse order (from the bottom up in Figure 2.5),

$$50_{10} = 110010_2$$

Practice problem 2.14. (Conversion of a decimal into a binary number.) Convert the decimal number 2159 into a binary number.
Answer is given in "Solutions to practice problems."

Conversion of a decimal fraction to a binary fraction

For the conversion of a decimal fraction to a binary fraction, the following steps are needed:

Conversion of a decimal fraction to a binary fraction

Given: `A decimal fraction`
Step 1: `Multiply the fraction by 2`
Step 2: `Write down the obtained integer part. Multiply the fractional part of the obtained result by 2`
Step 3: `Repeat until 0 is obtained as a fractional part, or until the required accuracy is achieved`

Example 2.22 (Conversion of a decimal into a binary number.) *Convert the decimal fraction 0.4_{10} into a binary fraction. Represent the result using 8 bits. Figure 2.5b illustrates the conversion. The result of conversion is obtained by reading the remainders from the bottom-up, that is, $0.4_{10} = 01100110_2$.*

Practice problem 2.15. (Conversion of a decimal into a binary number.) Convert $+12.0625_{10}$ into a binary number.
Answer is given in "Solutions to practice problems."

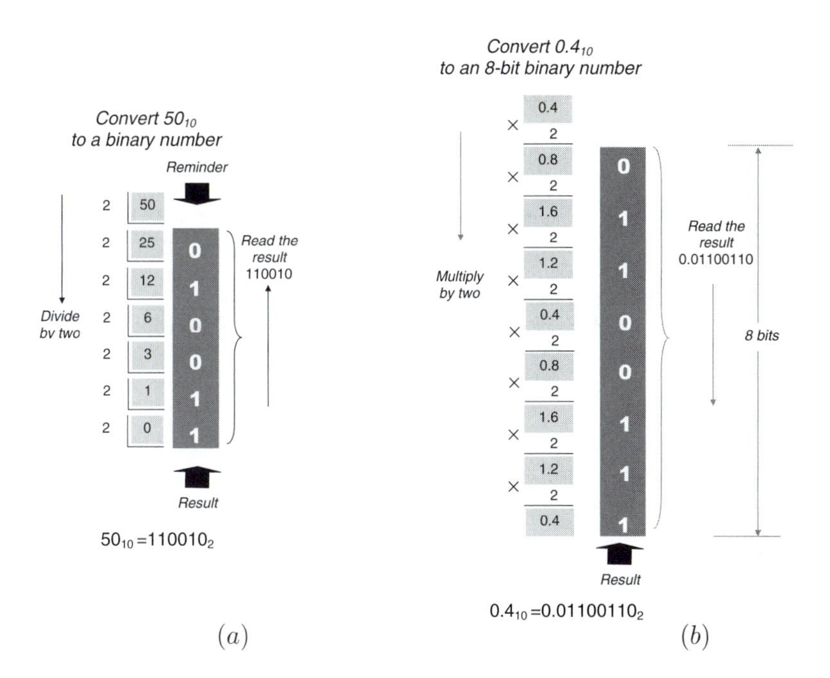

FIGURE 2.5
Conversion of a integer decimal number into a binary one (a) and conversion of a decimal fraction into a binary one (b) (Examples 2.21 and 2.22).

Conversion of decimal integers to octal integers

The following steps are needed to convert a decimal integer to an octal integer:

> **Conversion of decimal integer to octal integer**
>
> **Given:** A decimal integer.
>
> **Step 1:** Divide the decimal number by 8 to give a quotient and a remainder.
>
> **Step 2:** Divide the quotient by 8 to give new quotient and a remainder.
>
> **Step 3:** Repeat division until the fractional part is 0 or until the required number of digits is achieved.
>
> **Result:** The remainders written bottom-up (from recent to the first obtained) of the desired octal number.

Practice problem **2.16. (Conversion of a decimal into an octal number.)** Convert the decimal number 2159 into an octal number using the intermediate binary number system (use the result of Practice problem 2.15).

Answer is given in "Solutions to practice problems."

Conversion of decimal integers to hexadecimal integers

The following steps are needed to convert decimal integer into hexadecimal integer:

Conversion of decimal integer to hexdecimal integer

Given: A decimal integer.
Step 1: Divide the decimal number by 16 to give a quotient and a remainder.
Step 2: Divide the quotient by 16 to give a new quotient and a remainder.
Step 3: Repeat division until the fractional part is 0 or until the required number of digits is achieved.
Result: The remainders written bottom-up (from most recent to first obtained) of the desired hexadecimal number.

Practice problem **2.17. (Conversion of a decimal into a hexadecimal number.)** Convert the decimal number 2159 into a hexadecimal number using the intermediate binary number system (use the result of Practice problem 2.15).
Answer is given in "Solutions to practice problems."

A binary number can be converted to a decimal one via an intermediate system (octal or hexadecimal) (Figure 2.6).

Practice problem **2.18. (Conversion of decimal into octal and hexadecimal numbers.)**

(*a*) Convert the decimal number 746_{10} into an octal number.
(*b*) Convert the decimal number 746_{10} into a hexadecimal number.

Answer is given in "Solutions to practice problems."

Techniques for conversion between binary, octal, and hexadecimal number systems are given in Table 2.7.

2.8 Overflow

The operations under consideration here are executed within the binary system as well other systems of restricted word length or number of digits.

TABLE 2.7
Techniques for conversion between binary, octal, and hexadecimal numbers systems.

Techniques for conversion between numbers

Example	Technique
Binary to octal	
$1101001_2 = \underbrace{001}_{1}\,\underbrace{101}_{5}\,\underbrace{001}_{1} = 151_8$	Separate the bits into groups of three, starting from the right.
$10101.0101_2 = \underbrace{010}_{2}\,\underbrace{101}_{5}\,.\,\underbrace{010}_{2}\,\underbrace{100}_{4} = 25.24_8$	For the fractional part, start from the left.
Octal to binary	
$457.24_8 = \underbrace{100}_{4}\,\underbrace{101}_{5}\,\underbrace{111}_{7}\,.\,\underbrace{010}_{2}\,\underbrace{100}_{4}$	Replace each octal digit with a 3-bit string. For the fractional part, start from the left.
Binary to hexadecimal	
$1011011101_2 = \underbrace{0001}_{1}\,\underbrace{0110}_{6}\,\underbrace{1101}_{\mathsf{D}} = 16\mathsf{D}_{16}$	Separate the bits into groups of four, starting from the right.
$10111.11_2 = \underbrace{0001}_{1}\,\underbrace{0111}_{7}\,.\,\underbrace{0011}_{3} = 17.3_{16}$	For the fractional part, start from the left.
Hexadecimal to binary	
$5\mathsf{E}4_{16} = \underbrace{0101}_{5}\,\underbrace{1110}_{\mathsf{E}}\,\underbrace{0100}_{4}$	Replace each hexadecimal digit with the corresponding 4-bit string.
$\mathsf{C}2.\mathsf{A}1_{16} = \underbrace{1100}_{\mathsf{C}}\,\underbrace{0010}_{2}\,.\,\underbrace{1010}_{\mathsf{A}}\,\underbrace{0001}_{1}$	For the fractional part, start replacing from the left.
Octal to hexadecimal	
$1352_8 = \underbrace{001}_{1}\,\underbrace{011}_{3}\,\underbrace{101}_{5}\,\underbrace{110}_{6} = \underbrace{0010}_{2}\,\underbrace{1110}_{\mathsf{E}}\,\underbrace{1010}_{\mathsf{A}} = 2\mathsf{EA}_{16}$	Separate the bits into groups of three, starting from the right, and then re-group into groups of four.
Hexadecimal to octal	
$2\mathsf{EA}_{16} = \underbrace{0010}_{2}\,\underbrace{1110}_{\mathsf{E}}\,\underbrace{1010}_{\mathsf{A}} = \underbrace{001}_{1}\,\underbrace{011}_{3}\,\underbrace{101}_{5}\,\underbrace{010}_{2} = 1352_8$	Replace each hexadecimal digit with a 4-bit binary string. Separate the bits into groups of four, starting from the right, and then re-group into groups of three.

Techniques for conversion between numbers

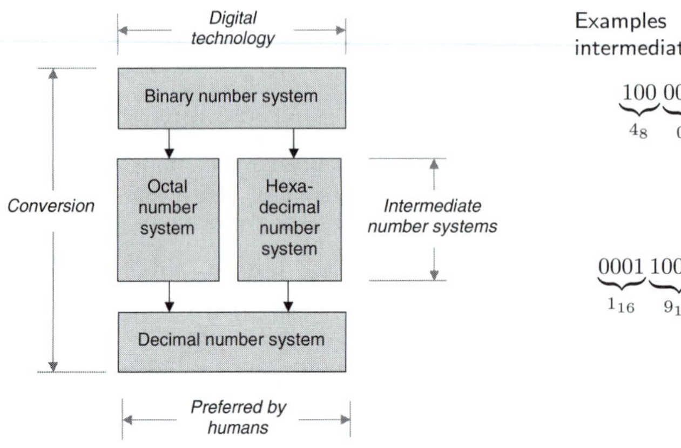

Examples of conversion using intermediate number systems:

$$\underbrace{100}_{4_8}\,\underbrace{000_2}_{0_8} = 40_8$$

$$= 4 \times 8^1 + 0 \times 8^0$$

$$= 32_{10}$$

$$\underbrace{0001}_{1_{16}}\,\underbrace{1001_2}_{9_{16}} = 19_{16}$$

$$= 1 \times 16^1 + 9 \times 16^0$$

$$= 25_{10}$$

FIGURE 2.6
Binary numbers are converted into decimal ones using octal and hexadecimal systems.

Below we will focus on the binary number system. Binary word size determines the range of the allowed number values of both operands and results.

Example 2.23 (The range.) *Given a 4-bit word, the binary numbers that can be represented are varied: (a) from* $-7_{10} = 1111_2$ *to* $+7_{10} = 0111_2$ *for unsigned numbers; (b) from* $-7_{10} = 1111_2$ *to* $+7_{10} = 0111_2$ *in the sign-magnitude system; (c) from* $-8_{10} = 1000_2$ *to* $+7_{10} = 0111_2$ *for the 2's complement system.*

In particular, if an addition operation in the 2's complement system produces a result that does not fit the range -2^{n-1} to $2^{n-1} - 1$, then we say that *arithmetic overflow* has occurred. To ensure the correct operation, it is important to be able to detect the occurrence of overflow.

The following rules are used for detecting overflow:

The rules for detecting overflow

Rule 1: An addition overflows if:

(a) The signs of the addends are the same and
(b) The sign of the sum is different from the addends' sign

Rule 2: An addition overflows (Figure 2.7) if the carry-in, C_{in}, and the carry-out of the most significant bit in 2's complement representation are different, $C_{in} \neq C_{out}$.

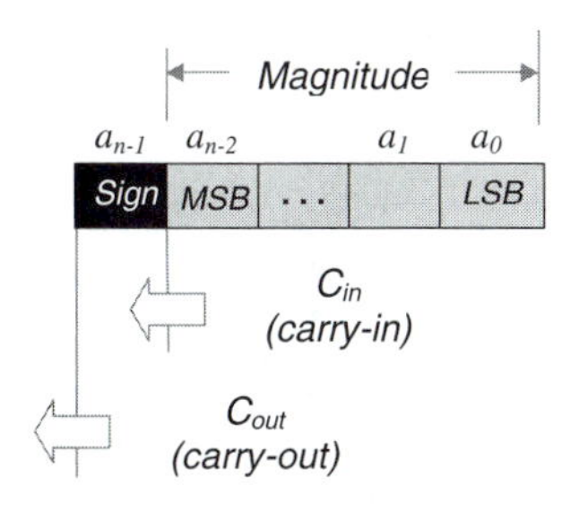

An addition overflows if the carry bit C_{in} into and carry bit C_{out} out of the sign position are different:

$$C_{in} \neq C_{out}$$

There is no overflow if

$$C_{in} = C_{out}$$

FIGURE 2.7
Overflow detection using carries.

Example 2.24 (Overflow.) *Figure 2.8 illustrates four cases of overflow detection using carry bits C_{in} and C_{out} for the two 4-bit binary numbers. Note that in all cases the sign bit is included in computing the sum.*

Practice problem **2.19. (Detection of overflow.)** Find which of the following operations result in overflow using 6-bit 2's complement codes of the given decimal numbers:

(*a*) $14_{10} + 22_{10}$ (*c*) $(-18_{10}) + (-27_{10})$
(*b*) $(-18_{10}) + (-11_{10})$ (*d*) $25_{10} + (-17_{10})$

Answer is given in "Solutions to practice problems."

2.9 Residue arithmetic

Residue arithmetic offers the alternative number format. An arithmetic operation performed on n-bit numbers may produce a result that is too long to be represented completely by n bits; that is, overflow occurs. In residue arithmetic, the results of all arithmetic operations are confined to some fixed set of m values such as $0, 1, \ldots, m-1$. Residue arithmetic ensures a finite word size in computing and is widely used in computing devices.

2.9.1 The basics of residue arithmetic

A *residue* is defined as the remainder after a division. Given the representation of an integer N,

$$N = Im + r$$

Techniques for overflow detection

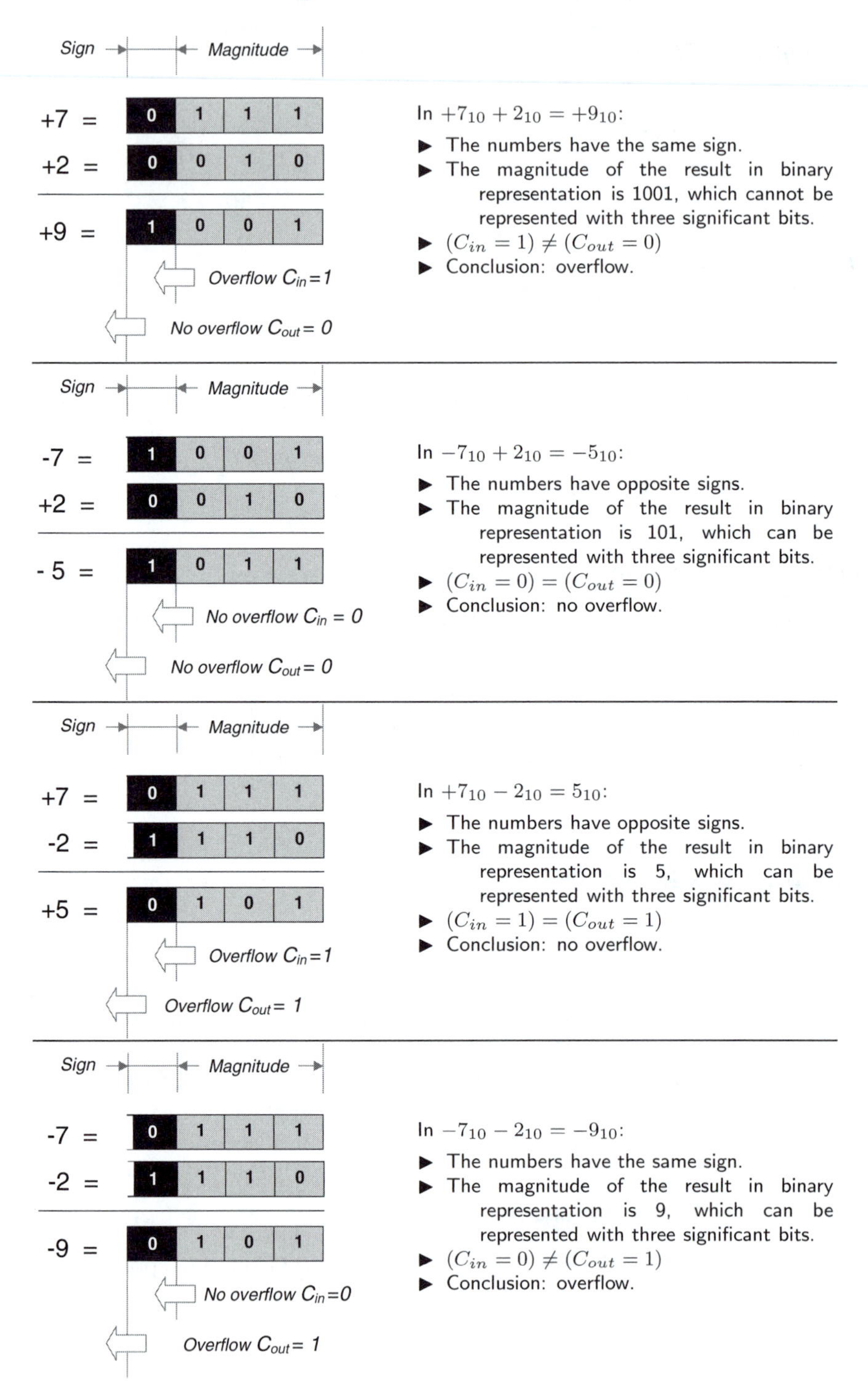

Sign
Magnitude
+7 = 0 1 1 1
+2 = 0 0 1 0
+9 = 1 0 0 1
Overflow C_{in}=1
No overflow C_{out}= 0
In $+7_{10} + 2_{10} = +9_{10}$:
▶ The numbers have the same sign.
▶ The magnitude of the result in binary representation is 1001, which cannot be represented with three significant bits.
▶ $(C_{in} = 1) \neq (C_{out} = 0)$
▶ Conclusion: overflow.
Sign
Magnitude
-7 = 1 0 0 1
+2 = 0 0 1 0
- 5 = 1 0 1 1
No overflow C_{in} = 0
No overflow C_{out}= 0
In $-7_{10} + 2_{10} = -5_{10}$:
▶ The numbers have opposite signs.
▶ The magnitude of the result in binary representation is 101, which can be represented with three significant bits.
▶ $(C_{in} = 0) = (C_{out} = 0)$
▶ Conclusion: no overflow.
Sign
Magnitude
+7 = 0 1 1 1
-2 = 1 1 1 0
+5 = 0 1 0 1
Overflow C_{in}=1
Overflow C_{out}= 1
In $+7_{10} - 2_{10} = 5_{10}$:
▶ The numbers have opposite signs.
▶ The magnitude of the result in binary representation is 5, which can be represented with three significant bits.
▶ $(C_{in} = 1) = (C_{out} = 1)$
▶ Conclusion: no overflow.
Sign
Magnitude
-7 = 0 1 1 1
-2 = 1 1 1 0
-9 = 0 1 0 1
No overflow C_{in}=0
Overflow C_{out}= 1
In $-7_{10} - 2_{10} = -9_{10}$:
▶ The numbers have the same sign.
▶ The magnitude of the result in binary representation is 9, which can be represented with three significant bits.
▶ $(C_{in} = 0) \neq (C_{out} = 1)$
▶ Conclusion: overflow.

FIGURE 2.8
Overflow detection using carries C_{in} and C_{out}.

where m is a check base and I is an integer, so that $0 \leq r \leq m$, and N is said to be equivalent to modulo m to r:

$$r \equiv N \ (mod\ m)$$

The numbers a and b are said to be *equivalent to modulo m* if the remainder obtained when a is divided by m is the same as the remainder that is obtained when b is divided by m:

$$a \equiv b \ \ (mod\ m)$$

(read as "a is congruent b modulo m").

Example 2.25 (Residue arithmetic.) *If* $a = 10$ *and* $b = 18$, *then* $10 \equiv 18 \ \ (mod\ 8)$ *since*

$$10 = 1 \times 8 + \overbrace{\boxed{2}}^{Remainder} \quad and \quad 18 = 2 \times 8 - \overbrace{\boxed{2}}^{Remainder}$$

Example 2.26 (Residue arithmetic.)

(*a*) $3 \equiv 17 \ (mod\ 7)$ *because* $3 - 17 = -14$ *is divisible by 7.*

(*b*) $-2 \equiv 13 \ (mod\ 3)$ *because* $-2 - 13 = -15$ *is divisible by 3.*

(*c*) $60 \equiv 10 \ (mod\ 25)$ *because* $60 - 10 = -50$ *is divisible by 25.*

(*d*) $-4 \equiv -49 \ (mod\ 9)$ *because* $-4 - (-49) = 45$ *is divisible by 9.*

It follows from these examples that an integer a is congruent to 0, $a \equiv 0 \ (mod\ m)$, if and only if it is divisible by m. Additional properties of congruence are:

- $a \equiv a \ (mod\ m)$ (reflexive property),
- If $a \equiv b \ (mod\ m)$, then $b \equiv a \ (mod\ m)$ (symmetry property), and
- If $a \equiv b \ (mod\ m)$ and $b \equiv c \ (mod\ m)$, then $a \equiv c \ (mod\ m)$ (transitive property).

2.9.2 Addition in residue arithmetic

If $a \equiv b$ (mod m) and $c \equiv d$ (mod m), then

Addition in residue arithmetic

$$a + c \equiv b + d \ (mod\ m)$$
$$a - c \equiv b - d \ (mod\ m)$$

Example 2.27 (Addition.) *Find the sum* $1017 + 2876 \ (mod\ 7)$. *There are two ways to calculate the sum. First approach:* $1017+2876 = 3893 \equiv 1 \ (mod\ 7)$. *Second approach:* $1017 \equiv 2 \ (mod\ 7)$, $2876 \equiv 6 \ (mod\ 7)$, *and* $1017 + 2876 \equiv 2+6 = 8 \equiv 1 \ (mod\ 7)$. *The second approach is preferable because it keeps the numbers involved small.*

Practice problem 2.20. **(Addition.)** Find the sum of two congruences: $13 \equiv 4 \pmod 9$ and $16 \equiv -2 \pmod 9$.
Answer: $13 + 16 \equiv 4 - 2 = 29 \equiv 2 \pmod 9$.

2.9.3 Multiplication in residue arithmetic

If $a \equiv b \pmod m$ and $c \equiv d \pmod m$, then

Multiplication in residue arithmetic

$$ac \equiv bd \ (mod\ m)$$

Example 2.28 (Multiplication.) *Find the product* $1017 \times 2876 \ (mod\ 7)$. *Solution:* $1017 \equiv 2 \ (mod\ 7)$, $2876 \equiv 6 \ (mod\ 7)$, *and* $1017 \times 2876 \equiv 2 \times 6 = 12 \equiv 5 \ (mod\ 7)4$.

Practice problem 2.21. **(Multiplication.)** Find the multiplication of two congruences: $13 \equiv 4 \pmod 9$ and $16 \equiv -2 \pmod 9$.
Answer: $13 \times 16 \equiv 4 \times (-2) = 208 \equiv -8 \pmod 9$.

Example 2.29 (Addition and multiplication modulo m=5.) *Addition and multiplication modulo 5 of two residues are given in the following tables:*

ADDITION

$a + c \equiv b + d \ (mod\ 5)$

+	0	1	2	3	4
0	0	1	2	3	4
1	1	2	3	4	0
2	2	3	4	0	1
3	3	4	0	1	2
4	4	0	1	2	3

SUBTRACTION

$a - c \equiv b - d \ (mod\ 5)$

−	0	1	2	3	4
0	0	4	3	2	1
1	1	0	4	3	2
2	2	1	0	4	3
3	3	2	1	0	4
4	4	3	2	1	0

MULTIPLICATION

$ac \equiv bd \ (mod\ m)$

×	0	1	2	3	4
0	0	0	0	0	0
1	0	1	2	3	4
2	0	2	4	1	3
3	0	3	1	4	2
4	0	4	3	2	1

For example, the residue 3 subtracted from the residue 1 modulo 5 is calculated as $1-3$ *equals* -2, *and* $-2 \equiv 3 \ (mod\ 5)$, *so the difference is the residue 3. The product of the two residues 3 and 4 modulo5 is the residue 2, because* $3 \times 4 \equiv 2 \ (mod\ 5)$.

2.9.4 Computing powers in residue arithmetic

If $a \equiv b \pmod m$, then

> **Computing powers in residue arithmetic**
>
> $a^n \equiv b^n \ (mod\ m)$ for every positive integer n

Example 2.30 **(Computing powers.)** *Find the products* $1017^2\ (mod\ 7)$, $1017^3\ (mod\ 7)$, $1017^4\ (mod\ 7)$, $1017^5\ (mod\ 7)$, $1017^{12}\ (mod\ 7)$. *Since* $1017 \equiv 2\ (mod\ 7)$,

$$\begin{aligned}
1017^2 &\equiv 2^2\ (mod\ 7) \\
1017^3 &= 1017^2 \times 1017 \equiv 4 \times 2 = 8 \equiv 1\ (mod\ 7) \\
1017^4 &= 1017^3 \times 1017 \equiv 1 \times 2 = 2\ (mod\ 7) \\
1017^5 &= 1017^4 \times 1017 \equiv 2 \times 2 = 4\ (mod\ 7) \\
1017^{12} &= ((1017^4))^3 \equiv 2^3 = 8 \equiv 1\ (mod\ 7)
\end{aligned}$$

Practice problem **2.22. (Computing powers.)** Compute 3^{13}.
Answer is given in "Solutions to practice problems."

The following property of a residue arithmetic is useful, for example, in electronic cash systems. The quadratic residues of a set modulo n are the elements that have a square root in the set.

Example 2.31 **(Computing powers.)** *For* $n = 11$, *the number 4 is a quadratic residue because* $2^2\ (mod\ 11) = 4$. *The number 5 is also a quadratic residue because* $7^2\ (mod\ 11) = 5$. *If* n *is prime, then there are* $(n-1)/2$ *quadratic residues. In the case of* $n = 11$, *the residues are:*

$$\begin{aligned}
1^2 &= 1\ (mod\ 11) & 6^2 &= 3\ (mod\ 11) \\
2^2 &= 4\ (mod\ 11) & 7^2 &= 5\ (mod\ 11) \\
3^2 &= 9\ (mod\ 11) & 8^2 &= 9\ (mod\ 11) \\
4^2 &= 5\ (mod\ 11) & 9^2 &= 4\ (mod\ 11) \\
5^2 &= 3\ (mod\ 11) & 10^2 &= 1\ (mod\ 11)
\end{aligned}$$

Each quadratic residue has two square roots if n is prime. One of them is smaller than $n/2$. The other is larger.

2.9.5 Solving modular equations

To solve a congruence or a system of congruences involving one or more unknowns means to find all possible values of the unknowns which make the congruence true.

Example 2.32 (Modular equation.) *Solution of the congruence $3x \equiv 1$ (mod 5) can be found by trying all possible values of x modulo 5:*

$$\begin{aligned}
&\text{IF } x = 0, \text{ THEN } 3x = 0 \equiv 0 \ (mod\ 5)\\
&\text{IF } x = 1, \text{ THEN } 3x = 3 \equiv 3 \ (mod\ 5)\\
&\text{IF } x = 2, \text{ THEN } 3x = 6 \equiv 1 \ (mod\ 5)\\
&\text{IF } x = 3, \text{ THEN } 3x = 9 \equiv 4 \ (mod\ 5)\\
&\text{IF } x = 4, \text{ THEN } 3x = 12 \equiv 2 \ (mod\ 5)
\end{aligned}$$

Since the modulus is 5, the integer x is in the range $0 \leq x \leq 5$. Thus, the only solution to the congruence is $x = 2$.

Practice problem 2.23. (Modular equation.) Solve the following congruences if possible: (a) $3x \equiv 1$ (*mod* 6) and (b) $3x \equiv 3$ (*mod* 6).
Answer is given in "Solutions to practice problems."

2.9.6 Complete residue systems

In residue arithmetic, an integer is represented as a set of residues with respect to a set of relatively prime integers called *moduli*. Residue arithmetic is defined in terms of a set of relatively prime moduli $\{r_1, r_2, \ldots, r_s\}$, where the greatest common divisor is equal to 1 for each pair of moduli. The set of integers $\{r_1, r_2, \ldots, r_s\}$ is called a *complete residue system modulo m* if $r_i \neq r_j$ (mod m) whenever $i \neq j$, and for each integer n exists a corresponding r_i such that $n = r_i$ (mod m).

Example 2.33 (Complete residue system.) *The sets $\{1, 2, 3\}$, $\{-1, 0, 1\}$, and $\{1, 7, 9\}$ are all complete residue systems modulo 3. The set $\{0, 1, 2, 3, 4, 5\}$ is a complete residue system modulo 6. It should be noted that this set can be reduced to $\{1, 5\}$.*

While in ordinary arithmetic there is an infinite number of integers $0, 1, 2, \ldots,$ in modular arithmetic there is essentially only a finite number of integers.

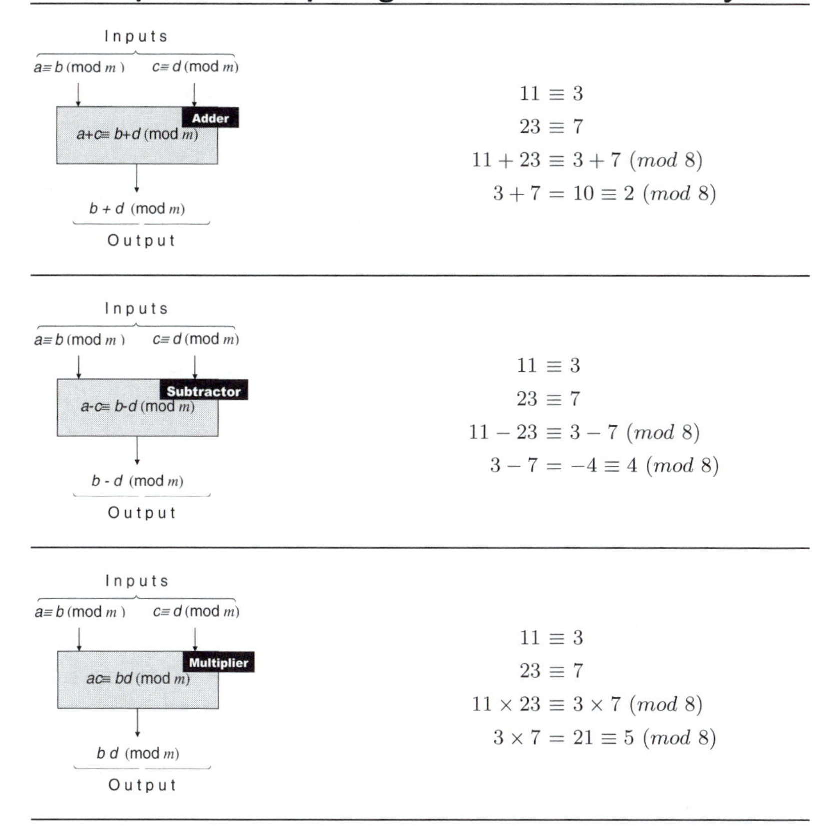

FIGURE 2.9
Modular adder, subtractor, and multiplier.

Example 2.34 (Complete residue system.) *Given the arithmetic 8, there are only eight distinct integers* $0, 1, 2, \ldots, 7$. *Any other integer is congruent to one of these eight integers. For example, 8 is congruent to 0 (mod 8),* $9 \equiv 1$ *(mod 8), etc. Thus addition, subtraction, multiplication, and division may be used to yield*

ADDITION:	$3 + 7 \equiv 2$	$(mod\ 8)$
SUBTRACTION:	$3 - 4 \equiv -1 \equiv 7$	$(mod\ 8)$
MULTIPLICATION:	$3 \cdot 7 \equiv 5$	$(mod\ 8)$
DIVISION:	$3 : 7 \equiv 5$	$(mod\ 8)$

The choice of moduli sets and the conversion of the residue to binary numbers are important issues in residue arithmetic. Residue arithmetic based on the set moduli $\{2^n - 1,\ 2^n,\ 2^n + 1\}$ is popular in digital signal processing. These converters use $2n$-bit or n-bit adders.

An example of the application of a residue number system is public key cryptography. Two keys are used in the encryption and decryption of messages: one that must be kept secret and one that may be made public. These two keys are related mathematically by a so-called "one-way" function. A one-way function is easy to compute in one direction but very hard (computationally infeasible) to compute in the other direction. An example of a one-way function is multiplication versus factorization. It is simple to multiply two large prime numbers, but very hard to factor the result (see details in "Further study" section).

2.10 Other binary codes

Specific binary codes are used in various tasks of logic design of discrete devices. Examples of such codes are the Gray code and binary-to-decimal code.

2.10.1 Gray code

The *Gray code* is used for encoding the indexes of the nodes. There are several reasons to encode the indexes. The most important of them is to simplify analysis, synthesis, and embedding of topological structures. The Gray code is referred to as a *unit-distance* code. Let $b_n \dots b_1 b_0$ be a binary representation of a positive integer number B and $g_n \dots g_1 g_0$ be its Gray code. There is a relationship between the two codes: `Binary code` $b_n \dots b_1 b_0 \Longleftrightarrow$ `Gray code` $g_n \dots g_1 g_0$ (Table 2.8).

TABLE 2.8
The relationship between binary and Gray code, $n = 3$.

Binary code	Gray code		Gray code	Binary code
000	000		000	000
001	001		001	001
010	011		011	010
011	010		010	011
100	110		110	100
101	111		111	101
110	101		101	110
111	100		100	111

Suppose that $B = b_n \dots b_1 b_0$ is given; then the i-th bit of the corresponding binary Gray code is generated as follows:

$$g_i = b_i \oplus b_{i+1} \tag{2.5}$$

where $b_{n+1} = 0$.

Given the Gray code $G = g_n \cdots g_1 g_0$, the corresponding binary representation is derived as follows:

$$b_i = g_0 \oplus g_1 \oplus \dots g_{n-i} = \bigoplus_{i=0}^{n-i} g_i \tag{2.6}$$

Table 2.8 illustrates the above transformation for $n = 3$.

Example 2.35 (Gray code.) *Binary to Gray and Gray to binary transformations are illustrated in Figure 2.10 for $n = 3$.*

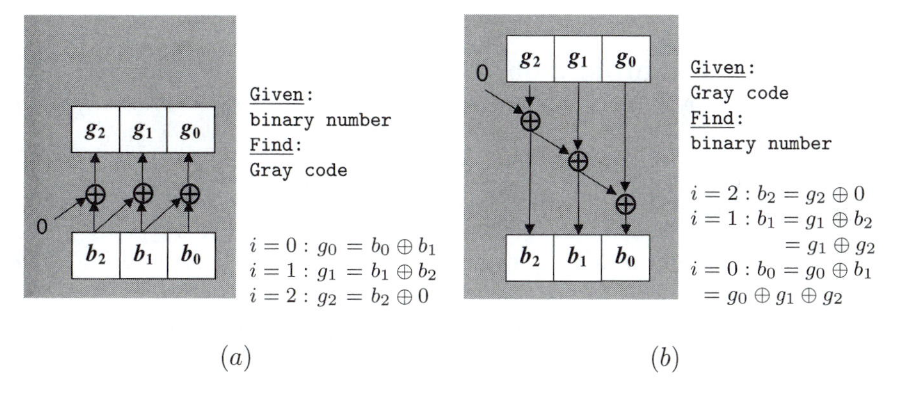

FIGURE 2.10
Data flowgraph and formal equation for binary to Gray code (a) and inverse transformation (b) (Example 2.35).

Hamming distance

The *Hamming distance* is a useful measure in hypercube topology. The Hamming sum is defined as the bitwise operation of two codes, $g_{d-1} \cdots g_0$ and $g'_{d-1} \cdots g'_0$:

$$(g_{d-1} \cdots g_0) \oplus (g'_{d-1} \cdots g'_0) = (g_{d-1} \oplus g'_{d-1}), \dots, (g_1 \oplus g'_1), (g_0 \oplus g'_0) \tag{2.7}$$

where $\oplus$ is an exclusive OR operation. If the sum is equal to 1, then the Hamming distance is said to be 1, that is, the codes G and G'.

In the hypercube, two nodes are connected by a link if and only if they have labels that differ by exactly one bit. The number of bits by which labels g_i and g_j differ is denoted by $h(g_i, g_j)$; this is the Hamming distance between the nodes.

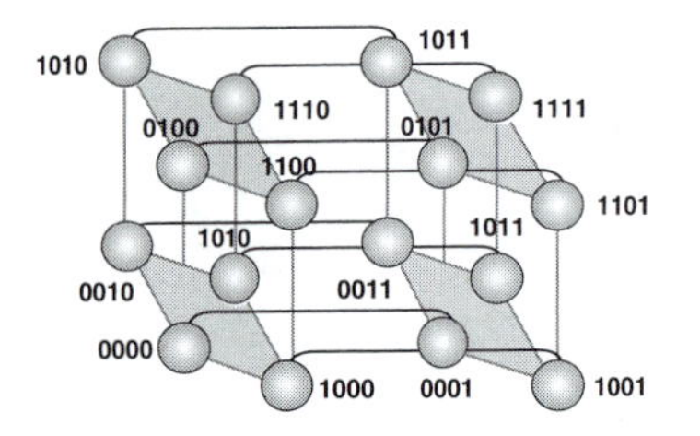

$$
\begin{aligned}
0000 \oplus 0001 &= 0001\\
0010 \oplus 0011 &= 0001\\
0100 \oplus 0101 &= 0001\\
0110 \oplus 0111 &= 0001\\
1000 \oplus 1001 &= 0001\\
1010 \oplus 1011 &= 0001\\
1100 \oplus 1101 &= 0001\\
1110 \oplus 1111 &= 0001
\end{aligned}
$$

FIGURE 2.11
Hamming sum operations and corresponding products of variables of a 4-dimensional hypercube (Example 6.20).

Example 2.36 (Hamming sum.) *The Hamming sum operation on two hypercubes for three-variable Boolean functions results in a 4-dimensional hypercube (Figure 6.17).*

2.10.2 Binary-coded decimal

In Table 2.9, several possible binary codes for the decimal digits $0, 1, 2, \ldots 9$ are given. The 8-4-2-1 (BCD) and 6-3-1-1 codes are examples of weighted codes. A 4-bit weighted code represents a decimal number as follows:

$$\text{WEIGHTED CODE} = w_3a_3 + w_2a_2 + w_1a_1 + w_0a_0$$

where w_0, w_1, w_2 and w_3 are the weights of the code a_0, a_1, a_2 and a_3.

Example 2.37 (Weighted code.) *Consider the codes* $a_3a_2a_1a_0 = 1000_{(8-4-2-1)}$ *and* $a_3a_2a_1a_0 = 1011_{(6-3-1-1)}$ *with weights* $w_3 = 8, w_2 = 4, w_1 = 2, w_0 = 1$ *and* $w_3 = 6, w_2 = 3, w_1 = 1, w_0 = 1$*, respectively. The corresponding decimal numbers are computed as follows:*

$$
\begin{aligned}
N_{(8-4-2-1)} &= 8 \times a_3 + 4 \times a_2 + 2 \times a_1 + 1 \times a_0\\
&= 8 \times 1 + 8 \times 0 + 8 \times 0 + 8 \times 0 = 8\\
N_{(6-3-1-1)} &= 6 \times a_3 + 3 \times a_2 + 1 \times a_1 + 1 \times a_0\\
&= 6 \times 1 + 3 \times 0 + 1 \times 1 + 1 \times 1 = 8
\end{aligned}
$$

TABLE 2.9
Binary codes for decimal numbers.

Decimal digit	8-4-2-1 code	6-3-1-1 code	Excess-3 code	Gray code
0	0000	0000	0011	0000
1	0001	0001	0100	0001
2	0010	0011	0101	0011
3	0011	0100	0110	0010
4	0100	0101	0111	0110
5	0101	0111	1000	1110
6	0110	1000	1001	1010
7	0111	1001	1010	1011
8	1000	1011	1011	1001
9	1001	1100	1100	1000

The Excess-3 code is obtained from 8-4-2-1 code by adding 3 (0011_2) to each group of four binary numbers of the corresponding 8-4-2-1 code.

Example 2.38 (Excess-3 code.) *Given the 8-4-2-1 code 1001 1000, its equivalent Excess-3 code is constructed as follows:*

$$\begin{array}{rl} 1\,0\,0\,1\;\;1\,0\,0\,0 \longleftarrow & 8-4-2-1 \textit{ code} \\ +\,0\,0\,1\,1\;\;0\,0\,1\,1 \longleftarrow & \textit{adding } 3 \\ \hline 1\,1\,0\,0\;\;1\,0\,1\,1 \longleftarrow & \textit{Excess}-3 \textit{ code} \end{array}$$

Example 2.39 (Coded decimal numbers.) *Convert decimal numbers 28 into 8-4-2-1, 6-3-1-1, Excess-3, and Gray code:*

$$28_{(8-4-2-1)} = \underbrace{\boxed{0010}}_{2}\;\underbrace{\boxed{1000}}_{8} \qquad 28_{(6-3-1-1)} = \underbrace{\boxed{0011}}_{2}\;\underbrace{\boxed{1011}}_{8}$$

$$28_{(Excess-3)} = \underbrace{\boxed{0101}}_{2}\;\underbrace{\boxed{1011}}_{8} \qquad 28_{(Gray)} = \underbrace{\boxed{0011}}_{2}\;\underbrace{\boxed{1001}}_{8}$$

2.11 Summary of number systems

This chapter introduces various number systems that are used in data structure descriptions. In digital technology, various number systems are used to simplify the manipulations of binary numbers. The key aspects of this chapter are as follows:

(*a*) The ***binary*** number system is the dominant number system in digital technology, as it is well suited to the binary nature of the phenomena used in today's technology.
(*b*) Four basic ***arithmetic*** operations are used in binary systems: addition, subtraction, multiplication, and division.
(*c*) ***Residue*** arithmetic is an alternative number system.
(*c*) Special binary ***codes*** are used for particular tasks of logic design.

The key statements of this chapter are as follows:

- ► A number system is defined by its basic symbols, called ***digits***. The ***radix***, or ***base*** of a number system is the number of various digits used in the system.

  A digital system processes numbers with a fixed number of digits, called ***word size***. The range of the representable numbers can be expanded beyond the range of ***fixed-point*** numbers by the employment of ***floating-point*** numbers.
- ► The binary system employs the two digits 0 and 1, which are referred to as ***bits***.
- ► Using ***1's complement*** representation, negative numbers are defined according to a subtraction operation involving positive numbers. An advantage of 1's complement is that a negative number is generated by complementing all bits of the corresponding positive number. An n-bit negative number K is obtained by subtracting its equivalent positive number P from $2^n - 1$; i.e., $K = (2^n - 1) - P$. The addition of 1's complement numbers may require a correction to be performed, and so the time needed to add two 1's complement numbers may be twice as long as the time needed to add two unsigned numbers.
- ► Using ***2's complement*** representation, negative numbers are defined according to a subtraction operation involving positive numbers. An n-bit negative number K is obtained by subtracting its equivalent positive number P from 2^n; i.e., $K = 2^n - P$. An advantage of 2's complement representation is that when the numbers are added, the result is always correct.

Summary (continuation)

- The ***octal*** and ***hexadecimal*** number systems are used due to the simplicity of the conversion between the numbers in some assembly languages.
- An arithmetic operation performed on n-bit numbers may produce a result that is too long to be represented completely by n bits; this is the overflow condition. In ***residue*** arithmetic, the results of all arithmetic operations are confined to some fixed set of m values such as $0, 1, \ldots, m - 1$.
- ***Coding*** is the representation of information (signals, numbers, messages, etc.) by code symbols or sequences of code symbols. Information is said to be placed into code form by ***encoding*** and extracted from code form by ***decoding***. ***Binary codes*** have been designed for a variety of applications.

  (*a*) Decimal numbers can be converted to binary format by mapping each decimal digit into a suitable binary code. In the ***binary-coded decimal (BCD)*** scheme, each decimal digit is replaced by the equivalent 4-bit binary number.

  (*b*) In ***error-detecting*** and ***error-correcting*** codes, one or more ***parity check*** bits are used for the data being guarded.

  (*c*) ***Gray code*** is used, in particular, for encoding of states (in sequential systems systems) so that the neighbor states differ by one digit only.
- For advances in number systems, we refer the reader to the "Further study" section.

2.12 Further study

Historical perspective

17th century: Blaise Pascal (1623–1662) studied the special tabulated form of integer functions known as *Pascal triangle*. In logic design, the interest in Pascal's triangle concerns its representation of Boolean functions.

Father Marin Mersenne was interested in integers of the form $2^n - 1$. He showed that these could only be prime if n itself was prime. In Mersenne's honor, primes of the form $2^n - 1$ are called *Mersenne primes*.

Pierre de Fermat (1601–1665) studied the class of prime numbers in the form $2^{2^n}+1$. He stated that $2^{2^n}+1$ is prime for all $n \geq 0$. Leonhard Euler showed that $2^{2^5}+1$ is not a prime, but is divisible by 641. To this day, no further Fermat primes have been discovered.

The binary system (base 2), was propagated in the 17th century by Gottfried Leibniz. Binary numbers came into common use in the 20th century because of computer applications.

Karl Friedrich Gauss (1777–1855) introduced a kind of equation, known as a *congruence*, $a \equiv b \ (mod\ m)$, which means that $a - b$ is divisible by m or, equivalently, a divided by m leaves the remainder b. The trick of many modern cryptographic systems is to work with remainders: a secret information is locked up in remainders.

1949: C. E. Shannon ushered in the era of *scientific secret-key cryptography* ("Communication Theory of Secrecy Systems," *Bell System Technical Journal*, volume 28, number 4, pages 656–715, 1949). The Shannon communication theory has been deployed for many applications in logic network design.

1965: J. W. Cooley and J. W. Tukey developed the Fast Fourier Transform (FFT) for evaluating the Fourier transform ("An algorithm for the Machine Calculation of Complex Fourier Series," *Math. Comput.*, vol. 19, pp. 297–301, 1965). This paper greatly accelerated the applications of number theory in the development of digital devices for signal processing.

1976: Whitfield Diffie and Martin E. Helmann of Stanford University showed for the first time that secret communication based on residue arithmetic ("New Directions in Cryptography," *IEEE Transactions on Information Theory*, volume 22, pages 644–454, 1976) is possible without any transfer of secret key between sender and receiver. This work establishes the direction known as *public-key cryptography*.

Advanced topics of number systems

Topic 1: Numbers and information. *Information* can be defined as recorded or communicated facts or data. Information takes various of physical forms when being stored, communicated, or manipulated. Information in digital form is represented using a finite sequence of symbols. Processing digital information consists of forming a digit sequence or replacing the digits of a sequence with other digits. An *alphabet* is a finite collection of digits. Many alphabets can be used for representation of information. For example, an alphabet of ten digits is commonly employed to express numeric information. Because finite alphabets are used, a single digit can convey only a finite amount of information. Digits of a binary alphabet (0,1) convey a minimum of information.

Information theory provides a unit for measuring information and the mathematical means of computing the information content of a message. In this theory, the unit of information is defined as one *bit*. Each digit from a binary alphabet conveys exactly one bit of information. Each digit of an alphabet of k digits can convey as much as $\log_2\ k$ bits of information. Sequences of digits are used to convey a greater amount of information that is provided by a single digit. The fundamental theorem of information theory is known as *Shannon's theorem.* This theorem predicts error-free data transmission in the presence of noise. Thus, the techniques of information theory are applied to the problems of the extraction of information from systems containing an element of randomness.

Topic 2: Floating-point number system is a numerical-representation system, defined by the IEEE 754 Standard. The main difficulty of fixed-point arithmetic (the word "fixed" refers to the fact that the radix point is placed at a fixed place in each number) is that the range of numbers that could be represented is limited. An alternative representation of numbers, known as the *floating-point format,* may be employed to eliminate the scaling factor problem. In this system, a string of digits (or bits) represents a real number.

The name "floating-point" refers to the fact that the radix point (decimal point, or binary point) can be placed anywhere relative to the digits within the string. This position is indicated separately in the internal representation. Numbers in float-point format consist of two parts: a *fraction* and an *exponent.* The advantage of floating-point representation is that it supports a much wider range of values than integers represented in computers using the *fixed point notation.* For example, while a fixed-point representation that allocates 8 decimal digits and 2 decimal places can represent the numbers 123456.78, 8765.43, 123.00, and so on, a floating-point representation with 8 decimal digits could also represent 1.2345678, 1234567.8, 0.000012345678, 12345678000000000, and so on. Unlike integers, which can represent exactly every number between the smallest and largest number, floating-point numbers are normally approximations for a numbers they cannot really represent.

The speed of performing floating point operations is used as performance measurement for computers. It is measured in "Megaflops" (MFLOPs) (million floating-point operations per second)

$$\text{MFLOPs} = \frac{\text{Number of floating-point operations}}{\text{Execution time} \times 10^6},$$

"Gigaflops" (GFLOPs), "Teraflops" (TFLOPs) etc. A floating-point operations are an addition, subtraction, multiplication, or division operations applied to a number in a single or double precision floating-point representation. Such data are used in scientific calculations. For example, the `i860` processor (announced by Intel in 1989) was able to execute up to two floating-point operations and claimed to offer 100 MFLOPs.

Topic 3: The logarithmic number system is an arithmetic system used for representing real numbers. It was introduced as an alternative to the floating point number system. In logarithmic number system, a number, N, is represented by the logarithm of its absolute value, n, as follows:

$$N \rightarrow \{s, n = \log_b(|N|)\},$$

where n is a bit denoting the sign of N ($s = 0$ if $N > 0$ and $s = 1$ if $N < 0$). The number n is represented by a binary word which usually is in the two's complement format. On the contrary, the operations of addition and subtraction are more complicated.

Topic 4: Cryptography is the study of ways in which messages can be coded so that a third party, intercepting the code, will have great difficulty recovering the original text. *Cryptology* consists of *cryptography* and *cryptanalysis.* Cryptanalysis deals with breaking secret messages.

Advanced*encryption standards* such as RSA (Rivest-Shamir-Adleman) and AES (Advanced Encryption Standard) are based on number theory, encoding, and logic design techniques. Special hardware devices to implement these encryption algorithms have been manufactured.

The RSA encryption system relies upon the mathematics of modulo arithmetic. Both encryption and decryption are completed by raising numbers to a power modulo a number which is the product of two large primes. The two primes are kept secret and the system can be broken if the two primes are recovered by factoring. The factoring is a process that has proven to be extremely difficult.

To encode a message using RSA, a user needs to create a *public* and *secret key* (Figure 2.12). They are chosen through several steps:

Step 1: Two large primes, p and q, are chosen (200–1000 bits). These numbers are chosen at random.
Step 2: The primes are multiplied together to yield $n = p \times q$. This is often at least 512 bit in practice.
Step 3: The secret key e is chosen. The greatest common denominator of e and $(p-1)(q-1)$ should be 1.
Step 4: The public key, d, is the inverse of $d\ mod\ (p-1)(q-1)$.
Step 5: The secret key is the pair of values n and e.

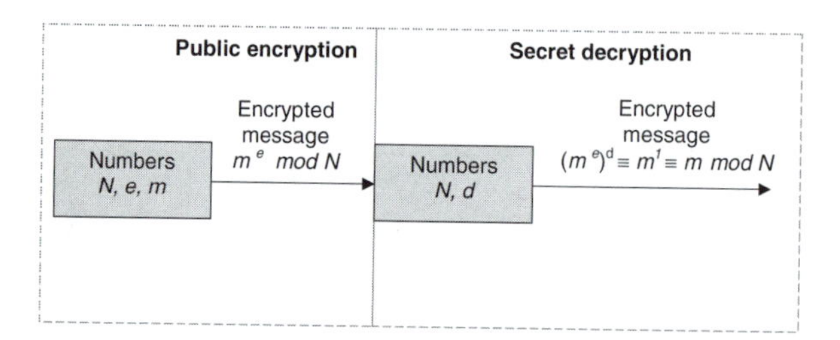

FIGURE 2.12
Using residue arithmetic in public key encryption.

Encryption is as follow. A message is converted into a number m less that n and decrypted by computing $(m^d \times mod\ n)^e \times mod\ n$. Specifically, this decryption works as follows:

$$(m^d \times mod\ n)^e \times mod\ n = m^{d^e} \times mod\ n$$
$$= m^{de} \times mod\ n = m \times mod\ n$$

Typical applications of cryptography is a *digital signature.* Digital signature is a message-dependent quantity that can be computed only by the sender of the message on the basis of some private information. It allows authentification of messages by guaranteeing that no one can forge the sender's signature and the sender cannot deny a message he sent. Applications requiring such signatures include business transactions, for example, between a bank and its customers. Digital signatures can also be used in environments where the interests of the parties involved are not necessary conflicting.

Further reading

Elementary Number Theory with Applications by T. Koshy, Harcourt/Academic Press, 2002.

"Fast Conversion Technique for Binary-Residue Number Systems" by B. Vinnakota and V. V. B. Rao, *IEEE Transactions on Circuits and Systems - I*, volume 41, number 12, pages 927–929, December 1994.

Handbook of Applied Cryptography by A. J. Menezes, P. C. van Oorschot, and S. A. Vanstone, CRC Press, 1996.

"Novel High-Radix Residue Number System Processors" by V. Paliouras, and T. Stouraitis, *IEEE Transactions on Circuits and Systems-Part II*, volume 47, number 10, pages 1059–1073, October 2000.

"Residue-to-Binary Converters Based on New Chinese Remainder Theorems" by Y. Wang, *IEEE Transactions on Circuits and Systems - II*, pages 197–206, March 2000.

"Semi-Logarithmic Number Systems" by J.-M. Muller, A. Scherbyna, and A. Tisserand, *IEEE Transactions on Computers*, volume 47, number 2, pages 145–151, February 1998.

"A Mathematical Theory of Communication," by C. Shannon, *Bell Systems Technical Journal* volume 27, pages 379–423, 623–656, 1948.

"The Sign/Logarithm Number System" by E. E. Swartzlander and A. G. Alexopoulos, *IEEE Transactions on Computers,"* volume 24, pages 1238--1242, December 1975.

2.13 Solutions to practice problems

Practice problem	Solution
2.2.	This is an octal number, therefore, $D = \sum_{i=-1}^{0} 8^i \times d_i$ $= 7 \times 8^0 + 3 \times 8^{-1} + 3 \times 8^{-2} = 7 + {}^5/_8 + {}^3/_{64} = 7\ {}^{43}/_{64}$.
2.4.	Since $B = \sum_{i=-1}^{1} 2^i \times b_i$, $10.01_2 = 1\times2^1+0\times2^0+0\times2^{-1}+1\times2^{-2} = 2\ {}^1/_4$, i.e., it represents the decimal number $2\ {}^1/_4$.
2.5.	Shifting the point in 10.01_2 left yields $1.001_2 = 1 + {}^1/_8 = 1\ {}^1/_8$. Indeed, $1\ {}^1/_8$ is twice as small as $2\ {}^1/_4$.
2.6.	Shifting the point in $1\ {}^1/_8 = 1.001_2$ right yields $1.001_2 = 2\ {}^1/_4$, which is twice as big as $1\ {}^1/_8$.
2.7.	$7.0625_{10} = 1 \times 2^2 + 1 \times 2^1 + 1 \times 2^0 + 0 \times 2^{-1} + 0 \times 2^{-2} + 0 \times 2^{-3} + 1 \times 2^{-4}$ That is, $7.0625_{10} = 0111.0001_2$
2.14.	$2159 : 2 = 1079 + \boxed{1}$ $33 : 2 = 16 + \boxed{0}$ $1079 : 2 = 539 + \boxed{1}$ $16 : 2 = 8 + \boxed{0}$ $539 : 2 = 269 + \boxed{1}$ $8 : 2 = 4 + \boxed{0}$ $269 : 2 = 134 + \boxed{1}$ $4 : 2 = 2 + \boxed{0}$ $134 : 2 = 67 + \boxed{0}$ $2 : 2 = 1 + \boxed{0}$ $167 : 2 = 33 + \boxed{1}$ $1 : 2 = 0 + \boxed{1}$ The sequence of remainders is 1,1,1,1,0,1,1,0,0,0,0,1. The result of conversion is obtained by reading remainders in reverse order, $2159_{10} = 100001101111_2$.
2.15.	The integer and fractional parts must be converted separately: $12_{10} = 1100_2$; 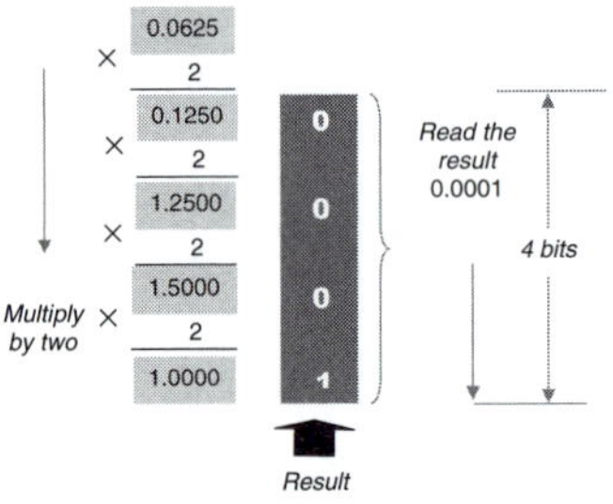 0.0625 × 2 = 0.1250 → 0 0.1250 × 2 = 1.2500 → 0 1.2500 × 2 = 1.5000 → 0 1.5000 × 2 = 1.0000 → 1 Multiply by two; Read the result 0.0001; 4 bits; Result $0.0625_{10} = 0.0001_2$ Thus, $12.0625_{10} = 1100.0001_2$.

Solutions to practice problems (continuation)

Practice problem	Solution
2.16.	Given the binary code of the number, divide it into groups of three bits, starting at the right: $2159_{10} = \underbrace{100}_{4_8}\ \underbrace{001}_{1_8}\ \underbrace{101}_{5_8}\ \underbrace{111}_{7_8}{}_2$. Thus $2159_{10} = 4 \times 8^3 + 1 \times 8^2 + 5 \times 8^1 + 7 \times 8^0 = 4157_8$.
2.17.	Given the binary code of the number, divide it into groups of four bits, starting at the right: $2159_{10} = \underbrace{1000}_{8_{16}}\ \underbrace{0110}_{6_{16}}\ \underbrace{111}_{F_{16}}$. Thus $2159_{10} = 8 \times 16^3 + 6 \times 16^2 + F \times 16^1 = 86F_{16}$.
2.18.	(a) $746_{10} = 1352_8$ — Divide by 8: 8	746, 8	93, 8	11, 8	1; Reminder: 2, 5, 3, 1; Result (read upward). (b) $746_{10} = 2EA_{16}$ — Divide by 16: 16	746, 16	46, 16	2, 16	0; Reminder: A, E, 2; Result (read upward).
2.19.	$14_{10} + 22_{10} = 36_{10}$ (14) [0] 0 1 1 1 0 (22) + [0] 1 0 1 1 0 (36) [1] 0 0 1 0 0 The sum (36) is not in the range $-2^{6-1} = 32$ to $2^{6-1} - 1 = 31$. The resulting number has 1 in the sign bit, which means the result is negative. Overflow occurs, since $C_{in} \neq C_{out}$
	$(-18_{10}) + (-11_{10}) = -29_{10}$ (−18) [1] 0 1 1 1 0 (−11) + [1] 1 0 1 0 1 (−29) [1] 0 0 0 1 1 The sum −29 is within the range −32 to 31. The result is **correct**. There is no overflow, since $C_{in} = C_{out} = 1$
	$(-18_{10}) + (-27_{10}) = -45_{10}$ (−18) [1] 0 1 1 1 0 (−27) + [1] 0 0 1 0 1 (−45) [0] 1 0 0 1 1 The sum −45 is out of the range −32 to 31. The result points to the positive number. Overflow occurs, since $C_{in} \neq C_{out}$

Solutions to practice problems (continuation)

Practice problem	Solution
2.19.	(Continuation) $25_{10} + (-17_{10}) = 8_{10}$ (25) 0 1 1 0 0 1 (−17) + 1 0 1 1 1 1 (8) 0 0 1 0 0 0 The sum 8 is within the range −32 to 31. The result is **correct**. Overflow does not occur, since $C_{in} = C_{out} = 1$
2.22.	$3^{13} = 3^{8+4+1} = 3^8 \times 3^4 \times 3^1 \equiv 16 \times 13 \times 3\ (mod\ 17)$ $\equiv 16 \times 39 \equiv 16 \times 5 \equiv 80 \equiv 12\ (mod\ 17)$
2.23.	(a) There is no solution to the congruence $3x \equiv 1\ (mod\ 6)$ because the values of $3x$ modulo 6 are just 0 and 3: If $x = 0$, then $3 \times 0 = 0 \equiv 0\ (mod\ 6)$, if $x = 1$, then $3 \times 1 = 3 \equiv 3\ (mod\ 6)$ if $x = 2$, then $3 \times 2 = 6 \equiv 0\ (mod\ 6)$, if $x = 3$, then $3 \times 3 = 9 \equiv 3\ (mod\ 6)$ if $x = 4$, then $3 \times 4 = 12 \equiv 0\ (mod\ 6)$, If $x = 5$, then $3 \times 5 = 15 \equiv 3\ (mod\ 6)$ (b) The equation $3x \equiv 3\ (mod\ 6)$ has solutions $x = 1$, $x = 3$, and $x = 5$.

2.14 Problems

Problem 2.1 What are the largest positive and the smallest negative binary numbers that can be expressed with the following number of bits: (a) 8 bits, (b) 32 bits, (c) 64 bits, and (d) 128 bits.

Problem 2.2 A register contains the 12-bit word W = 010101011100. What is N if it represents: (a) An unsigned number, (b) A signed-and-magnitude number, (c) A 2's complement number, and (d) A 1's complement number.

Problem 2.3 Perform the addition in the binary number system: (a) $1110 + 0101$, (b) $0.1010 + 0.0101$, and (c) $101.1010 + 001.0101$.

Problem 2.4 Perform the subtraction in the binary number system: (a) $11101 - 101$, (b) $1110.10 - 101.01$, and (c) $101.1010 - 001.0101$.

Problem 2.5 Perform the multiplication in the binary number system:(a) 110×101, (b) 1110.10×101.01, and (c) 101.101×1.011.

Problem 2.6 Perform the operations in the ternary number system:(a) 102+1011, (b) 102 × 12, and (c) 21121.01 – 102.21.

Problem 2.7 Determine the radix b of the number system for the following operations to be correct: (a) $24_b + 17_b = 40_{10}$, (b) $225_b = 89_{(10)}$, and (c) $({}^{54}/_4)_b = 13_{10}$.

Problem 2.8 Convert the binary number into its equivalent in the octal and hexadecimal systems:(a) 1101110, (b) 1011100.10101, and (c) 0.10101.

Problem 2.9 Convert the octal number into its equivalent in the binary system:(a) 375, (b) 256.055, and (c) 0.675.

Problem 2.10 Convert the hexadecimal number into its equivalent in the binary system: (a) 5FA9, (b) 256.FF1, (c) 0.ABC, and (d) A5CE.9FE.

Problem 2.11 Form the 1's complement and 2's complement for the following binary numbers (assume 8-bit representation): (a) 101010, (b) 0.101010, (c) 10.0101, and (d) 101011.0.

Problem 2.12 Form the 9's complement and 10's complement for the following decimal numbers: (a) 12345, (b) 0.9876, (c) 1357.9876, and (d) 5674.0.

Problem 2.13 Perform subtraction on the following unsigned binary numbers using the 2's complement of the subtrahend. If the result is negative, 2's complement it and affix a minus sign: (a) 11011−11001, (b) 110100−10101, and (c) 1010−11000.

Problem 2.14 Convert the sign-magnitude decimal numbers into the signed 10's complement form and perform the following operations: (a) (+9826) + (+801), (b) (+9826) + (−801), and (c) (−9826) + (+801).

Problem 2.15 Convert the decimal numbers to binary in two ways: (a)convert directly to binary, and (b) convert first to hexadecimal, then from hexadecimal to binary. Compare these approaches.(a) 125, (b) 0.765, (c) 567.53, and (d) 456.0.

Problem 2.16 Perform the following calculations on the decimal numbers using the indicated modular arithmetic: (a) 89 + 15 (mod 13), (b) 89 – 95 (mod 8), and (c) 12 × 24 (mod 15)

Problem 2.17 Convert the following BCD numbers to binary and decimal form: (a) 010100011001, (b) 011100011001, and (c) 100100011001, (d) 100000000010.

Problem 2.18 Detect overflow in the sum of the following numbers X and Y:

Sign Magnitude
X 0 0 1 0 1 1
Y 1 1 1 0 1 1
(a)

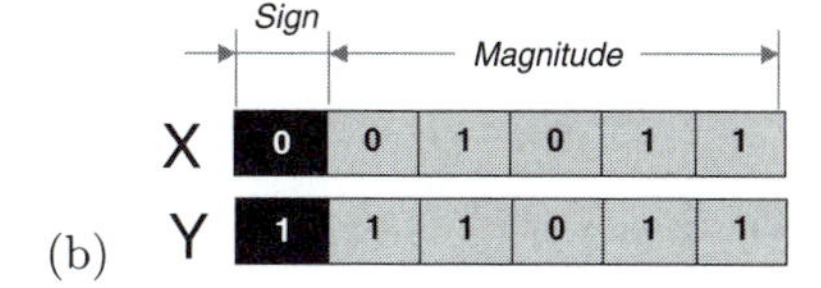

(b)

Problem 2.19 Prove or disprove that (a) $7 \equiv 0\ (mod\ 7)$, (b) $7 \equiv -3\ (mod\ 10)$, and (c) $18 \equiv 36\ (mod\ 9)$.

Problem 2.20 Find (a) $846 + 17\ (mod\ 7)$, (b) $100 - 12\ (mod\ 6)$, and (c) $127 \times 35\ (mod\ 5)$.

Problem 2.21 Derive the Gray code given the binary one: (a) 10010110 and (b) 10000000.

3

Graphical Data Structures

Graphs

- Directed and undirected
- Isomorphic
- Planar
- Cubes and hypercubes

Binary trees

- Nodes, root, links, paths
- The adjacency matrix
- H-trees
- Codes and graphs

Decision trees and functions

- Spanning trees
- Functions and decision trees
- Lattice topology
- Embedding
- Evaluation

Advanced topics

- Topology and placement
- Topology and computing

3.1 Introduction

Graphical data structures are suitable for representing and computing logic functions. Graphs show the relationship between the variables of a logic function. The manipulation of the variables and terms of a function is replaced by the manipulation of graphical components (nodes, links, and paths):

$$< \texttt{Algebraic data structure} > \Leftrightarrow < \texttt{Graphical data structure} >$$

3.2 Graphs in discrete device and system design

A logic network is modeled using an abstraction, which shows the relevant features without their associated details. In the design of digital devices, there are three levels of abstraction:

- The *architectural* level, which represents a set of devices performing a set of operations. Graphs such as block diagrams are used at this level for viewing such features as parallelism of processing, data transfer, and position of devices.
- The *logical* level shows how a network evaluates a set of logic functions. Graphical data structures such as directed graphs, or networks of nodes representing gates, are used at this level for representation, manipulation, and computing.
- The *physical*, or *geometric*, level, which represents a network as a set of geometric entities. Graphical data structures such as placement and routing diagrams are used for a representation of components that is suitable for implementation of the available technology.

3.2.1 Graphs at the logical level

The use of graphs in logic design implies compliance with graph theory. Graph theory includes the classification of graphs, their properties, and techniques for their manipulation. The type of graph used in logic design depends on the specific problem; in particular:

- If the direction of dataflow is important, *directed* graphs are used,
- If data are transmitted using operations, bipartite graphs are appropriate.
- If a problem is represented by a decision table, trees and diagrams are used.

The latter graphical data structure – decision trees and diagrams, – can be used at almost any level of logic design. It is well suited to represent logic functions, since any function that takes several values depending on its arguments (variables) can be represented by a graph called a *decision tree*.

Various algebraic forms of Boolean functions correspond to specific types of decision trees and diagrams. While the manipulation of Boolean functions in algebraic form is based on mathematical relations and theorems, their graphical equivalents are represented in the form of topological objects such as sets of nodes, levels, subtrees, and shapes, as well as in the form of functional characteristics such as distribution of nodes, weights, balance, and homogeneity. Various problems of logic analysis and synthesis have found an efficient solutions through the manipulation of these graphical objects.

Consider the general graph-based representation of a function. Information about the function is carried by the following graph components (Figure 3.1):

Nodes: *intermediate* nodes operate on data, and *terminal* nodes indicate the values of computing functions;

Edges, or *links*, between nodes that are associated with data flow; and

Topology (local and global), which specifies the geometric properties and spatial relations of these nodes and edges, and, therefore, is unaffected by changes in the shape or size of the graph.

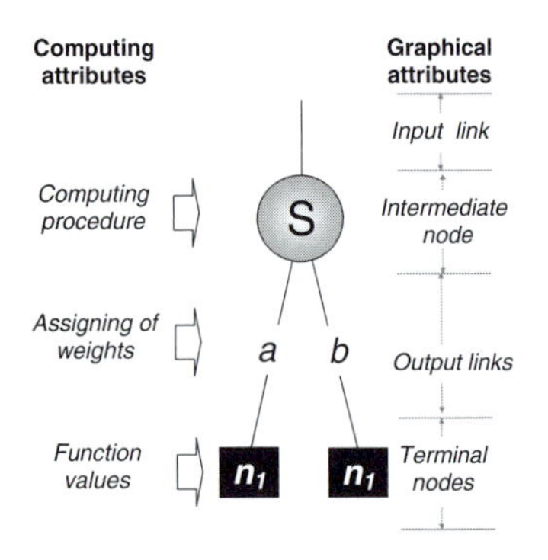

Elements of decision tree encoding:

- *Intermediate nodes indicate the computing procedures*
- *Links show the relationships between variables and computing procedures in intermediate nodes; links can be also labeled, for example, indices of variables or their weights*
- *Terminal nodes are assigned the resulting computed values*

FIGURE 3.1
The elements of a graph representing a function.

3.2.2 Graphs at the physical design level

Physical design, or geometric-level synthesis, combines the specification of all geometric patterns of gates and their positions. It defines the physical layout of the chip. The layers of the layout correspond to the *masks* used for chip fabrication. The physical layout is the final target of a logic network.

The result of logic synthesis is a logic network, that is, a collection of interconnected parts. Each part is represented at the physical level by a shape and the area it occupies on the board, together with the location of one or more pins, where the electrical connections are made. Electrical connections are implemented by a tree or a routing graph.

The major steps in physical design are *placement* and *wiring*, also called *routing*. Placement means the positioning of components on the board in a manner that is feasible with respect to practical given technological limitations. Next, the connections between the pins of the components have to be routed onto the board. The main objective of placement is to minimize the total wire length of all connections.

3.3 Basic definitions

A *graph* consists of two sets V and E, where V is nonempty and each element of E is an unordered pair of distinct elements of V. A graph is defined as $G = (V, E)$. The following terminology is used in the representation of graphs:

- The elements of V are called *vertices* or *nodes*; the elements of E are called *edges* or *links*.
- The vertices v and w are said to be *incident* with the edge (vw).
- Two vertices are *adjacent* if they are the end vertices of an edge; two edges are *adjacent* if they have a vertex in common.

The terms *link*, *edge*, *connection*, and *interconnection* are used interchangeably. Note that the functional elements of a designed logic network correspond to the nodes, and the communication links correspond to the edges in the graph.

In constructing graphs for the representation of computing structures, the following characteristics of graphs are used:

Characteristics of graphs

- The number of intermediate and terminal nodes.
- The *degree* of a node v, written $deg(v)$, is the number of edges incident with v.
- A link in E between the nodes v and u is specified as an unordered pair (v, u), and v and u are said to be *adjacent* to each other or are called *neighbors*.
- The *distance* between two nodes i and j in a graph G is the number of edges in G on the shortest path connecting i and j.
- The *diameter* of a graph G is the maximum distance between two nodes in G.
- A graph G is *connected* if a path exists between any pair of nodes i and j in G.

Example 3.1 (Graphs.) *A graph and its components are represented in Figure 3.2.*

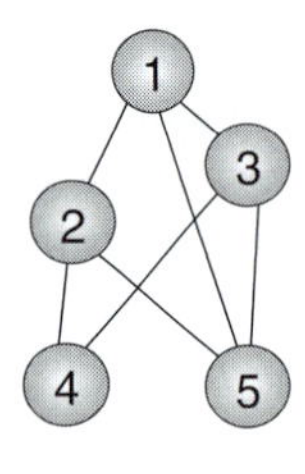

- *The number of nodes: 5*
- *The degrees of the nodes v, $deg(v_i)$, are*

  $$deg(1) = 3,\ deg(2) = 3,\ deg(3) = 3,$$
  $$deg(4) = 2,\ deg(5) = 3$$

- *The diameter is 2*
- *The graph G is connected (a path exists between any pair of nodes)*
- *All distances are 1, except for the distance 2 between nodes 1 and 4, 2 and 3, and 4 and 5*

FIGURE 3.2
A graph's components and characteristics (Example 3.1).

A graph may contain loops at vertices and/or multiple edges. A *loop* is an edge that is incident with only one vertex. *Multiple edges* are several edges incident with the same two vertices.

3.3.1 Directed graphs

A *directed* graph, or *digraph*, G consists of a finite set V of nodes and a set E of directed edges between the vertices, and is characterized by the following:

Characteristics of directed graphs

- The *indegree* of a node i is the number of edges in G leading to i. The *outdegree* of a node i is the number of edges outgoing from i.
- A node is called a *terminal node* if it has an outdegree of 0. If the outdegree of v is greater than 0, v is called an *internal* node. A node is called a *root* if it has an indegree of 0.
- The *adjacency* matrix $A = (a_{ij})$ is defined as a square matrix of size <NUMBER OF NODES> such that element $a_{ij} = 1$ if $(i, j) \in E$, and $a_{ij} = 0$ otherwise.

The adjacency matrix contains only two elements, 0 and 1. The graph and its adjacency matrix contain the same information; they are simply two alternative representations or data structures. A permutation of any two rows or columns in an adjacency matrix corresponds to relabeling the vertices and edges of the same graph.

Example 3.2 (Directed graph.) *The properties of the directed graph with four nodes and its adjacency matrix A are illustrated in Figure 3.3, where $in(v_i)$ and $out(v_i)$ are the indegrees and outdegrees of a node v_i.*

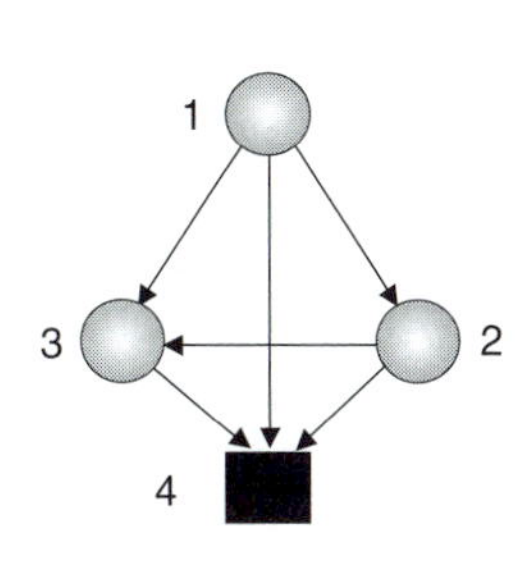

(a)

- 4 nodes with indegree and outdegree:
  in(v_1)=0, out(v_1)=3
  in(v_2)=1, out(v_2)=2
  in(v_3)=2, out(v_3)=1
  in(v_4)=3, out(v_4)=0
- v_1 is a root
- v_4 is a terminal node
- v_2 and v_3 are internal nodes
- Diameter = 3

(b)

$$\mathsf{A} = \begin{array}{c} \\ 1 \\ 2 \\ 3 \\ 4 \end{array} \begin{bmatrix} 1 & 2 & 3 & 4 \\ & 1 & 1 & 1 \\ & & 1 & 1 \\ & & & 1 \\ & & & \end{bmatrix}$$

(c)

FIGURE 3.3
The directed graph (a), its properties (b), and an adjacency matrix (c) (Example 3.2).

Practice problem 3.1. (Directed graph.) Given the directed graph (Figures 3.4a and b), find its nodes' indegrees and outdegrees, root and terminal nodes, diameter, and adjacency matrix.
Answer is given in "Solutions to practice problems."

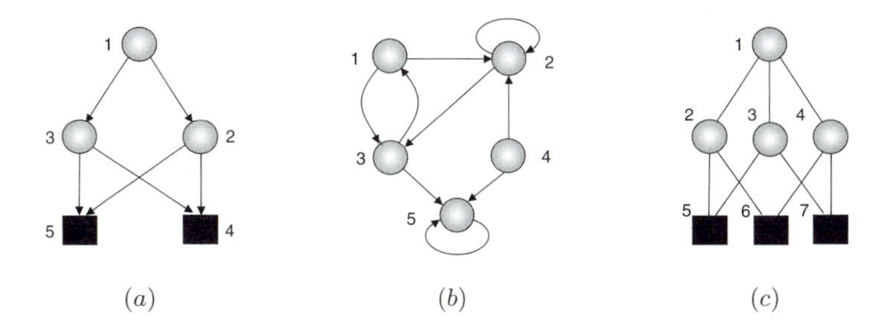

FIGURE 3.4
The directed graphs for Practice problems 3.1 and 3.3.

3.3.2 Flow graphs

A *bipartite* graph is one whose vertices can be partitioned into two (disjoint) nonempty sets S_1 and S_2 in such a way that every edge joints with a vertex in S_1 and a vertex in S_2. A *complete* bipartite graph is a bipartite graph in which every vertex in S_1 is joined to every vertex in S_2.

Typically, the edges in graphs are labeled by numbers, which are used in computing. A *weighted* graph is a graph in which each edge is assigned a non-negative real number, called a *weight.* A *flow* graph is a directed *bipartite* graph. A flow graphs can be derived from matrix equations. In this case, the structure of a flow graph corresponds to the rule of matrix multiplication and the structure of the matrix (distribution of non-zero and zero elements).

Example 3.3 (Deriving flow graph.) *Let a matrix transformation be represented by the equation* $\begin{bmatrix} 1 & 1 \\ 1 & 1 \end{bmatrix} \begin{bmatrix} a_1 \\ a_2 \end{bmatrix}$ *and* $\begin{bmatrix} 1 & 0 \\ 1 & 1 \end{bmatrix} \begin{bmatrix} a_1 \\ a_2 \end{bmatrix}$. *In Figures 3.5a and b, these equations (multiplications of a vector by matrix) are represented by a flow graphs.*

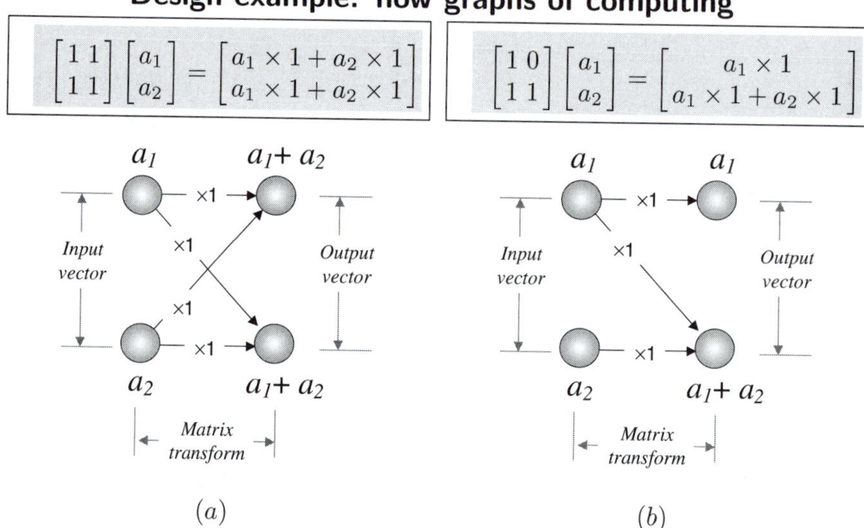

FIGURE 3.5
The structure of the flow graph corresponds to matrix multiplication: the matrix structure specifies the flow graph (Example 3.3).

Practice problem **3.2.** **(Deriving flow graph.)** Draw the flow graph for the matrix multiplication $\left[\begin{array}{cc|cc} 0 & 1 & 0 & 0 \\ 1 & 1 & 0 & 0 \\ \hline 0 & 1 & 0 & 1 \\ 1 & 1 & 1 & 1 \end{array}\right] \begin{bmatrix} a_1 \\ a_2 \\ a_3 \\ a_4 \end{bmatrix}$.

Answer is given in "Solutions to practice problems."

3.3.3 Undirected graphs

In the case of undirected graphs, edges are considered unordered pairs, and therefore have no distinguishable direction. The *degree* of a node i in a graph G is the number of edges in G that are incident with i, i.e., where the outdegree and the indegree coincide.

Example 3.4 (Undirected graphs.) *Figure 3.6 illustrates the properties of the undirected graph. Its adjacency matrix A is equal to the transposed matrix A, $A = A^T$.*

Practice problem 3.3. (Undirected graphs.) Given the undirected graph (Figure 3.4b), find its nodes' indegrees and outdegrees, root and terminal nodes, diameter, and adjacency matrix.
Answer is given in "Solutions to practice problems."

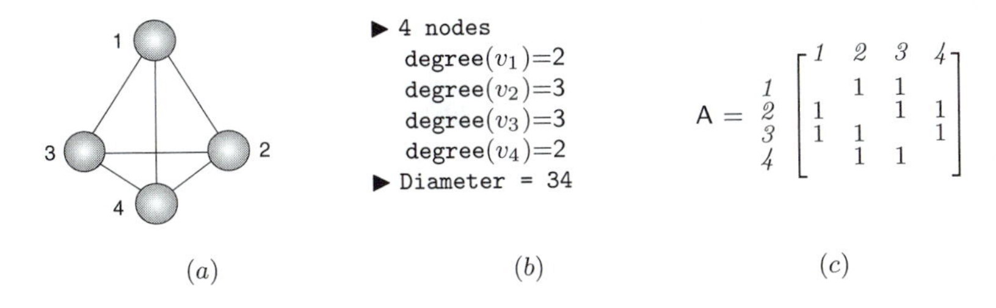

FIGURE 3.6
The undirected graph (a), its properties (b), and an adjacency matrix (c) (Example 3.4).

3.3.4 A path in a graph

A *path* is a chain of edges between two vertices. A path can be measured by a *length* t, which is the number of vertices in the path.

In the modeling of logic networks, the number associated with an edge is some unit used in the problem to be solved, for example, a unit of time, distance, or cost. The problem is formulated as finding the shortest path, or minimizing the lengths of the paths. A *path* in a graph is a sequence of edges and nodes leading from the root node to a terminal node. The *path length* in a graph is the number of non-terminal nodes in the path.

Example 3.5 (The path.) *A decision tree with 6 paths is given in Figure 3.7. For example, the paths $1 \to 2 \to 7$ and $1 \to 3 \to 5 \to 10$ are characterized by different features, such as the number of nodes and links.*

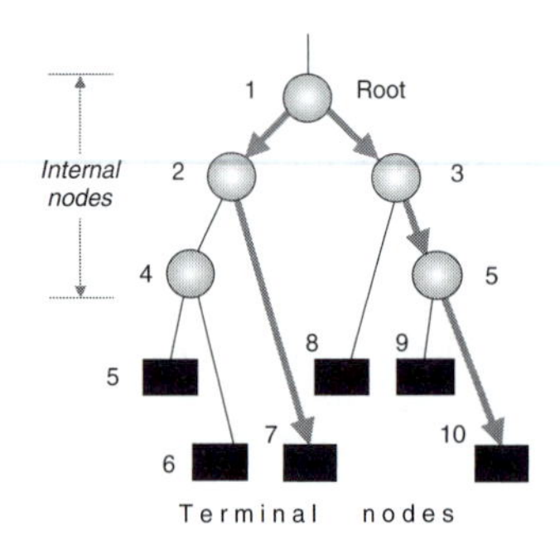

There are 6 paths from the root node to the terminal nodes. For example, the two highlighted paths are derived as follows:

- *Path of length 2 from the root node through the node 2 to the terminal node 7*
- *Path of length 3 from the root node through the internal nodes 3 and 5 to the terminal node 10*

FIGURE 3.7
A directed graph as a decision tree, with two paths highlighted (Example 3.5).

3.3.5 Isomorphism

Two graphs $G_1(V_1, E_2)$ and $G_2(V_1, E_2)$ are isomorphic if there exists a bijection $f : V_1 \to V_2$ such that $(u, v) \in E_1$ if and only if $(f(u), f(v)) \in E_2$.

Given the adjacency matrices A_1 and A_2 of the graphs G_1 and G_2, respectively, these graphs are isomorphic if and only if $A_2 = PA_1P^T$ for some permutation matrix P. That is, two graphs are isomorphic if their vertices can be labeled in such way that the corresponding adjacency matrices are equal. Two isomorphic graphs have

- The same number of nodes,
- The same number of nodes with a given degree, and
- The same number of links.

Example 3.6 (Isomorphic graphs.) *Figure 3.8 shows two isomorphic graphs, where $f(1) = D, f(2) = B, f(3) = C, f(4) = A, f(5) = E$. Graph G_2 is obtained by relabeling the vertices of G_1, maintaining the corresponding edges in G_1 and G_2.*

Practice problem 3.4. (Isomorphic graphs.) Prove, using adjacency matrices, that the graphs depicted in Figure 3.9 are isomorphic.
Answer is given in "Solutions to practice problems."

Example 3.7 (Isomorphic graphs.) *Two graphs representing 4-dimensional hypercubes are isomorphic (Figure 3.10), but they have different topologies.*

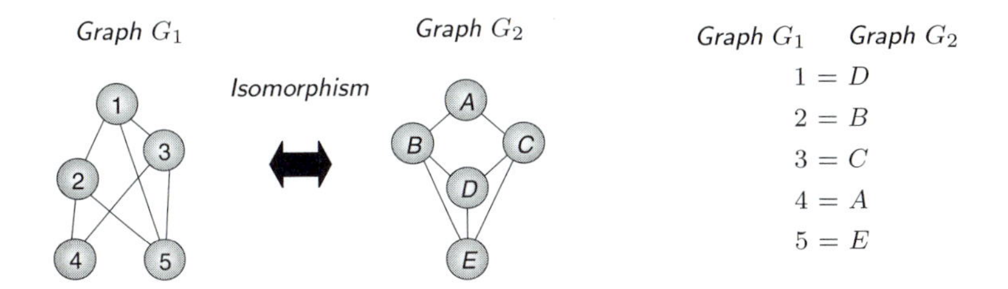

FIGURE 3.8
Examples of isomorphic graphs (Example 3.6).

FIGURE 3.9
Graphs for practice problem 3.4.

3.3.6 A subgraph and spanning tree

A graph $G_2(V_2, E_2)$ is a *subgraph* of $G_1(V_1, E_2)$ if $V_2 \subseteq V_1$ and $E_2 \subseteq E_1$.

> **Example 3.8 (Subgraphs.)** *Figure 3.11 shows a graph (a) and a subgraph (b).*

A *spanning tree* of a connected graph G is a subgraph that is a tree and that includes every vertex of G. These trees are used in the routing problem at the physical level of discrete device design.

> **Example 3.9 (Spanning trees.)** *Figure 3.11c shows a graph and all its corresponding spanning trees.*

3.3.7 Cartesian product

The *Cartesian product* of graphs provides a framework in which it is convenient to analyze as well as to construct new graphs:

$$\underbrace{\texttt{Graph } G_1 \overset{\textit{Cartesian product}}{\times} \texttt{Graph } G_1}_{\textit{New graph}}$$

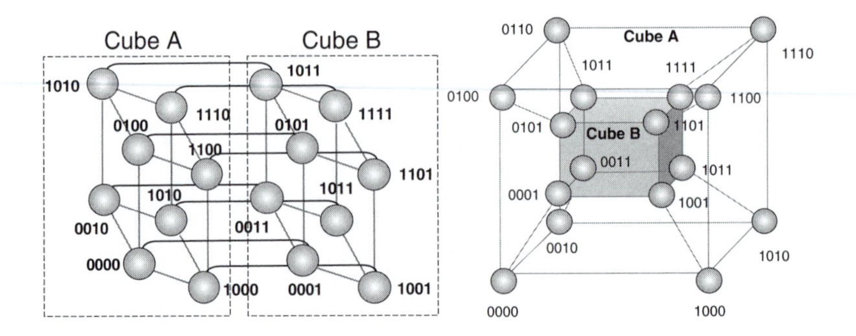

FIGURE 3.10
Isomorphic graphs (Example 3.7).

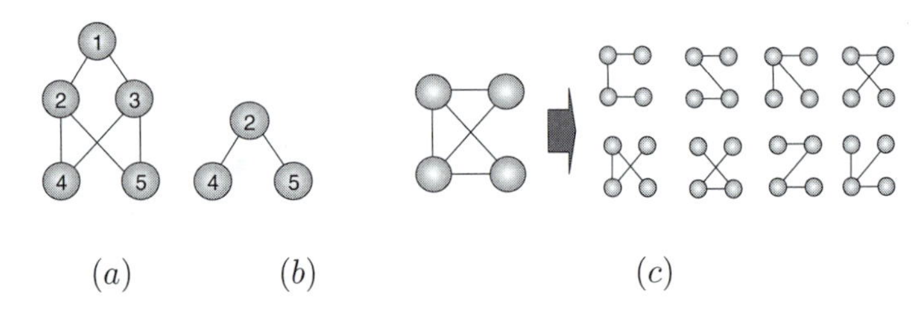

FIGURE 3.11
Examples of a subgraph (b) of a graph (a) (Example 3.8), and a graph and its eight spanning trees (c) (Example 3.9).

Let $G_1 = (V_1, E_1)$ and $G_2 = (V_2, E_2)$ be two graphs. The product of G_1 and G_2, denoted $G_1 \times G_2 = (V_1 \times V_2, E)$, is a graph where the set of nodes is the product set

$$V_1 \times V_2 = \{x_1x_2 | x_1 \in V_1,\ x_2 \in V_2\}$$
$$E = \{\langle x_1x_2, y_1, y_2\rangle | (x_1 = y_1,\ \langle x_2, y_2\rangle \in E_2) \textit{ or } (x_2 = y_2,\ \langle x_1, y_1\rangle \in E_1)\}$$

It can be shown that a hypercube can be defined as the product of n copies of the complete graph with two vertices, K_2. That is,

$$H_n = H_{n-1} \times K_2$$

3.3.8 Planarity

Logic networks are modeled using graphs. In practical implementation, crossings are often expensive; they occupy space, require additional channels, and cause various other unwanted effects. Planar graph-based models can alleviate these difficulties.

A graph G is *planar* if it is isomorphic to a graph G' such that:

(a) The vertices and edges of G' are contained in the same plane, and
(b) At most one vertex occupies or at most one edge passes through any point on the plane.

In other words, a graph is planar if it can be drawn on a plane with no two edges intersecting.

Example 3.10 (Planar graphs.) *In Figure 3.12a, a planar graph is shown. There are no crossings between edges in this graph. The nonplanar graph G_2 in Figure 3.12b can be planarized, that is, converted to an isomorphic planar graph G_3 as depicted in Figure 3.12c.*

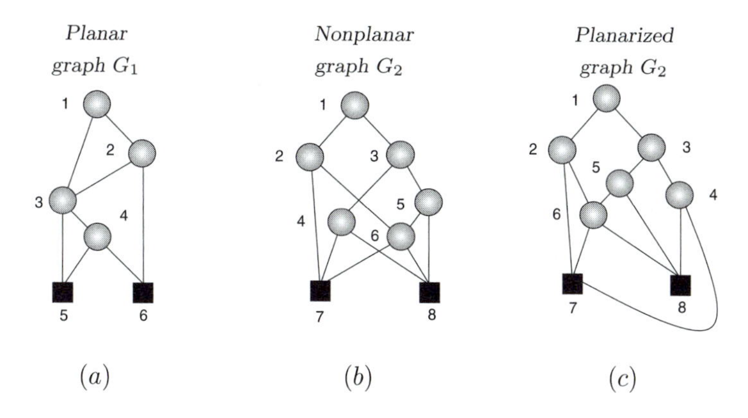

FIGURE 3.12
A planar graph (a); graph G_2 (b) can be planarized (c) (Example 3.10).

Practice problem 3.5. (Planar graphs.) Determine which of the graphs depicted in Figure 3.13 are planar.
Answer: Graph (b).

3.3.9 Operations on graphs

The *union* of two graphs $G_1 = (V_1, E_1)$ and $G_2 = (V_2, E_2)$ is another graph $G_3 = G_1 \cup G_2$, whose vertex set $V_3 = V_1 \cup V_2$ and edge set $E_3 = E_1 \cap E_2$:

$$\underbrace{\texttt{Graph } G_1 \overset{Union}{\cup} \texttt{Graph } G_1}_{New\ graph\ G_1 \cup G_2}$$

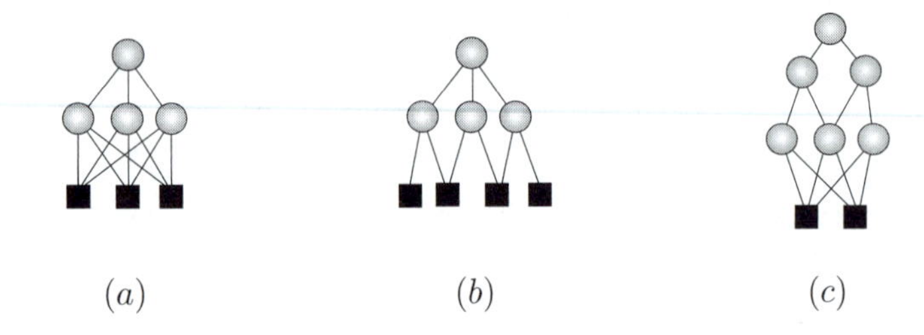

FIGURE 3.13
Graphs for practice problem 3.5.

The *intersection* of two graphs $G_1 = (V_1, E_1)$ and $G_2 = (V_2, E_2)$ is another graph $G_4 = G_1 \cap G_2$, whose vertex set $V_4 = V_1 \cap V_2$ and edge set $E_4 = E_1 \cap E_2$, i.e., consisting only of those vertices and edges that are in both G_1 and G_2:

$$\underbrace{\texttt{Graph } G_1 \quad \overset{\textit{Intersection}}{\cap} \quad \texttt{Graph } G_1}_{\textit{New graph } G_1 \cap G_2}$$

The *ring sum* of the graphs G_1 and G_2, $G_1 \oplus G_2$, is a graph consisting of the vertex set $V_1 \cup V_2$ and of edges that are in either G_1 or G_2, but not in both:

$$\underbrace{\texttt{Graph } G_1 \quad \overset{\textit{Ring sum}}{\oplus} \quad \texttt{Graph } G_1}_{\textit{New graph } G_1 \oplus G_2}$$

These operations can be extended to include any finite number of graphs. A pair of vertices x, y in a graph G is said to be *merged* if the two vertices can be replaced by a single new vertex z, such that every edge that was incident on either x or y or on both is incident on the vertex z. This operation does not alter the number of edges, but reduces the number of vertices by one.

If the graphs G_1 and G_2 are edge disjoint, then

- $G_1 \cap G_2$ is a null graph, and
- $G_1 \oplus G_2 = G_1 \cup G_2$.

If G_1 and G_2 are vertex disjoint, then $G_1 \cap G_2$ is empty. For any graph G,

- $G \cup G = G \cap G = G$, and
- $G \oplus G$ is a null graph.

If g is a subgraph of G, then $G \oplus g$ is that subgraph of G which remains after all the edges in g have been removed from G; i.e., $G \oplus g = G - g$.

Example 3.11 (Operations on graphs.) *Various operations on graphs are given in Table 3.1.*

TABLE 3.1
Operations on graphs (Example 3.11).

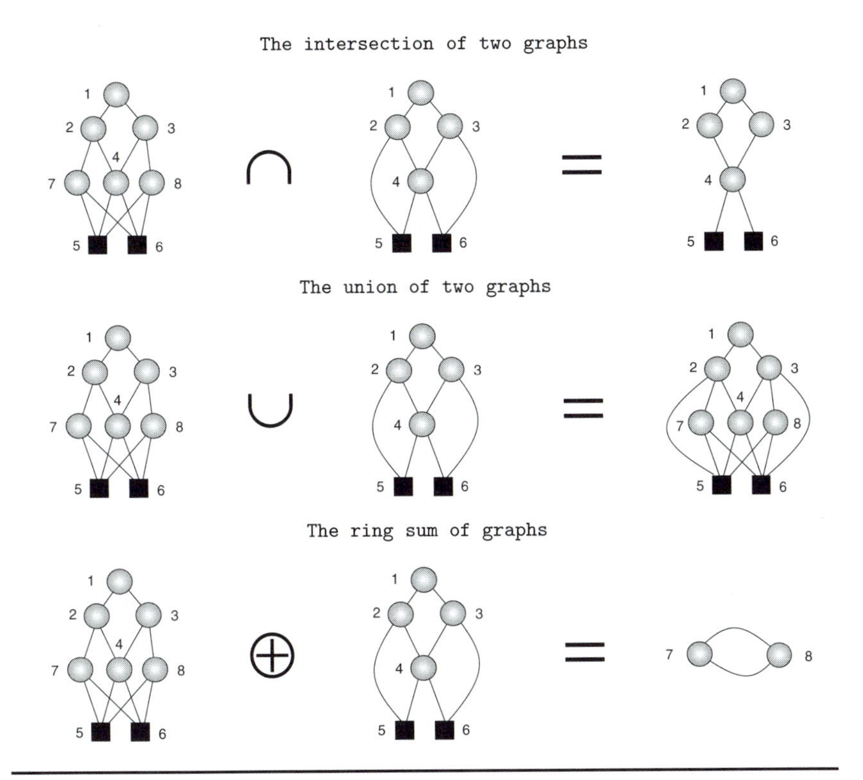

3.3.10 Embedding

An embedding of a *guest* graph G into a *host* graph H is a one-to-one mapping φ: $V(G) \rightarrow V(H)$, along with a mapping α that maps an edge $(u, v) \in E(G)$ to a path between $\varphi(u)$ and $\varphi(v)$ in H:

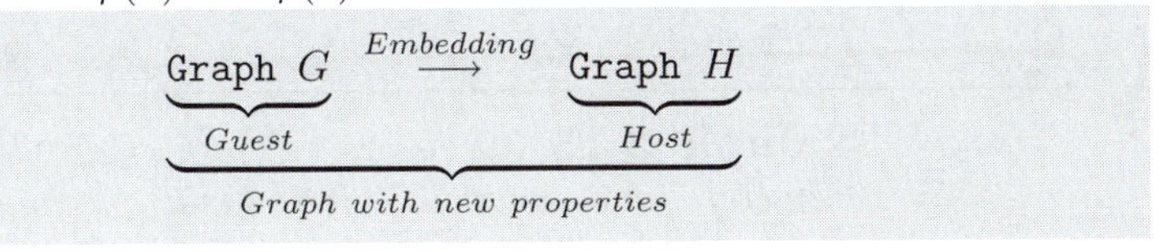

3.4 Tree-like graphs and decision trees

One class of graphical data structure, decision trees is a convenient way to represent binary or multi-valued logic functions. The number of possible values determines the tree's topology.

Example 3.12 (Trees.) *Figure 3.14 demonstrates a binary tree for the representation of Boolean functions (a), a ternary tree for the representation of ternary logic functions (b), and a mixed topology tree (c).*

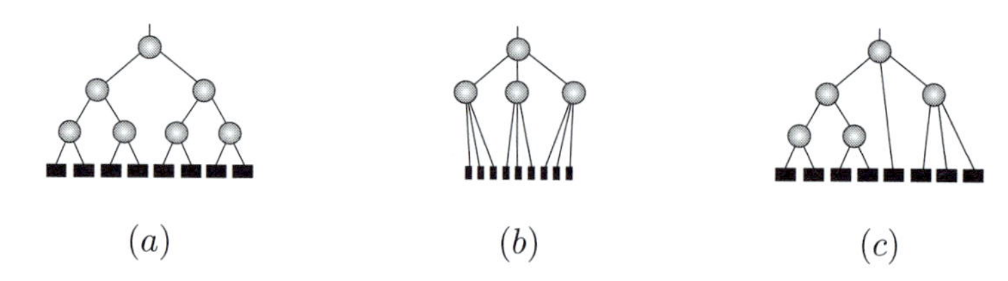

FIGURE 3.14
Binary (a), ternary (b), and mixed (c) trees (Example 3.12).

3.4.1 Basic definitions

A *tree* is a rooted acyclic graph in which every node but the root has an indegree of 1. A special class of rooted trees is called *binary* trees. The *path length* of a tree can be defined as the sum of the path lengths from the root to all terminal vertices. Trees have several useful characteristics and properties, in particular.

Characteristics of trees

- For every node v there exists a unique path from the root to v. The length of this path is called the *depth* or *level* of v.
- The *height* of a tree is equal to the greatest depth of any node in the tree.
- A node with no children is a *terminal* (*external*) node or *leaf*. A non-leaf node is called an *internal* node.

A *complete* n-level p-tree is a tree with p^k nodes per level k for $k = 0, \ldots, n-1$. A p^n-leaf complete tree has a level hierarchy (levels $0, 1, \ldots, n$); the root is associated with the level zero, and its p children are on level one. This link type describes the direction of data transmission between the child and the parent, so that data are sent in only one direction at a time, up or down.

Example 3.13 (Trees.) *Figure 3.15 shows how the parameters of a complete 3-level binary decision tree can be measured.*

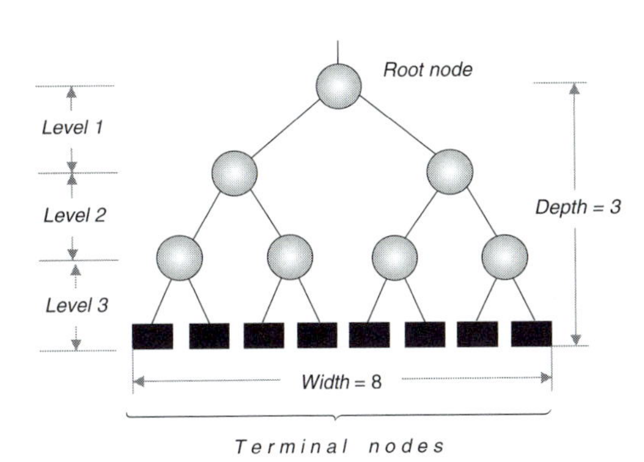

FIGURE 3.15
Measurements of the parameters of a complete 3-level binary decision tree (Example 3.13).

3.4.2 Lattice topology of graphs

Certain classes of logic functions can be represented by trees with a lattice topology. This allows for embedding trees into regular graphical structure called a *lattices.*

Example 3.14 (Lattice structures.) *Figure 3.16 demonstrates lattice-like trees and the topology they can be embedded into.*

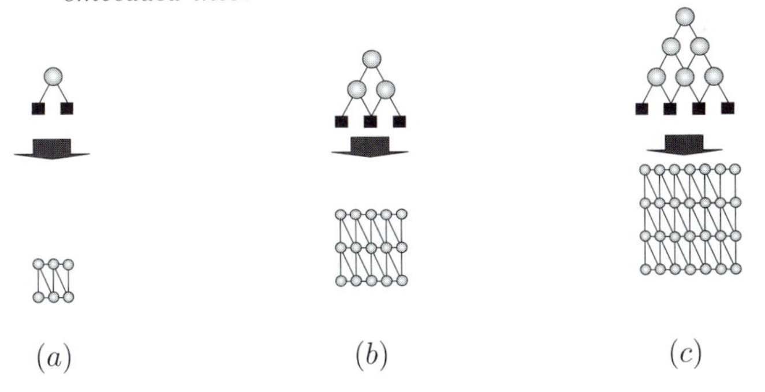

FIGURE 3.16
Embedding complete lattice-like trees into lattices (Example 3.14).

Example 3.15 (Embedding.) *A decision diagram can be extended to a lattice-like tree and then embedded into a lattice (Figure 9.17).*

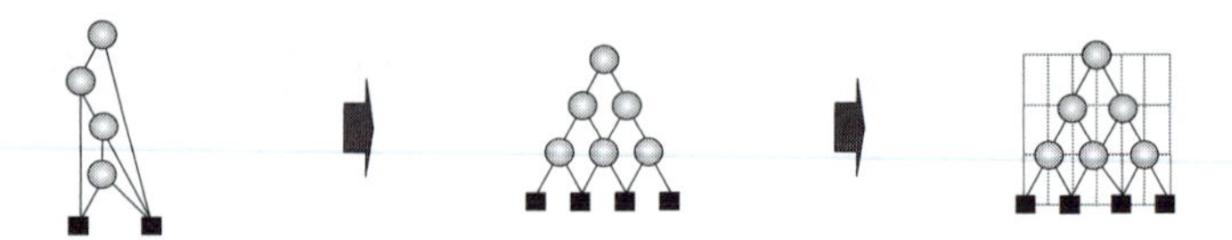

FIGURE 3.17
Embedding complete lattice-like trees into lattices (Example 9.15).

3.4.3 H-trees

An H-tree is a recursive topological construction of H_1-trees, where H_1 is defined as in Figure 3.18a. An H_{k+1}-tree can be constructed by replacing the leaves of H_k with H_1-trees. The number of terminal nodes in an H_k-tree is equal to 4^k. An H-tree makes optimal use of area and wire length. H-trees are the most common style for physical design of logic networks, in particular, clock routing.

Example 3.16 **(H-trees.)** *In Figure 3.18, the $H_{k+1} = H_2$ tree is shown for $k = 1$.*

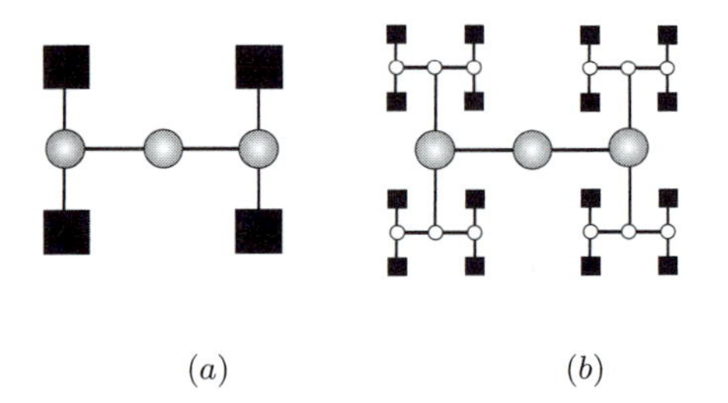

- *4 terminal nodes,*
- *3 non-terminal nodes*

The H_2-tree is constructed by replacing terminal nodes with H_1-trees; the H_1-tree includes:

- $4 \times 4 = 16$ *terminal nodes,*
- $4 \times 3 + 3 = 15$ *non-terminal nodes*

FIGURE 3.18
H_1- (a) and H_2- (b) trees (Example 3.16).

3.4.4 Binary decision trees and functions

A *discrete* function f of n variables is defined as a function in which each variable takes exactly k values. A discrete function is *constant* if it assumes the same value wherever it is defined; it is *completely specified* if it is defined everywhere. A *decision tree* is a graphical model of the evaluation of a discrete function, wherein the value of a variable is determined and the next action is chosen accordingly:

$$\underbrace{\texttt{Discrete function } f \overset{Mapping}{\longrightarrow} \texttt{Graph } G}_{Graphical\ representation\ of\ f}$$

In logic design, the focus is on the construction of decision trees from a function description, evaluating the efficiency of this representation, and the manipulation of the function represented by this tree.

There are other parameters of decision trees. For example, the distance $d(v_i, v_j)$ between two nodes v_i and v_j is the length of the shortest path between them, i.e., the number of edges in the shortest path. The *diameter* of a tree is defined as the length of the longest path in this tree. The notation of the *topology* of decision trees is used to distinguish their different shapes.

3.4.5 The relationship between decision trees and cube-like graphs

The terminal nodes of a complete decision tree correspond to the 2^n values of a Boolean function of n variables. On the other hand, these values can be represented by a hypercube. This implies one-to-one correspondence between the complete binary tree and the hypercube of a function.

Example 3.17 (Hypercubes and trees.) *Table 3.2 shows the complete decision trees and the corresponding hypercubes for* $n = 1, 2, 3, 4$.

The data structure, called *hypercube-like* topology (see details in "Further study" section), corresponds to the embedding of a complete binary tree into a multidimensional hypercube.

3.4.6 The simplification of graphs

The simplification of a graph means deleting some of its vertices or edges. When a vertex is removed from a graph, all edges incident with that vertex must also be removed. The result of deleting a vertex is a new graph or several new graphs. In logic design, the main task is the simplification of decision trees.

Example 3.18 (Tree simplification.) *In Figure 3.19, the effect of deleting nodes is illustrated.*

Practice problem 3.6. (Tree simplification.) Simplify the following complete decision tree by deleting the nodes 4 and 6:

TABLE 3.2
The relationship between the complete binary decision trees and hypercubes

Design example: The relationship between the decision trees and hypercubes

Decision Tree	Cube Structure

A complete 1-level decision tree and 1-dimensional cube

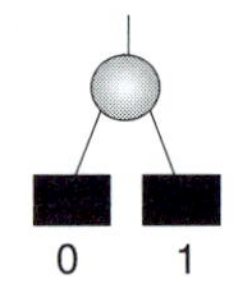

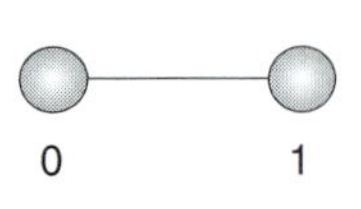

A complete 2-level decision tree and 2-dimensional cube

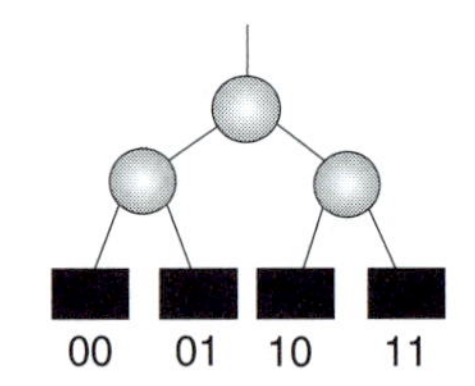

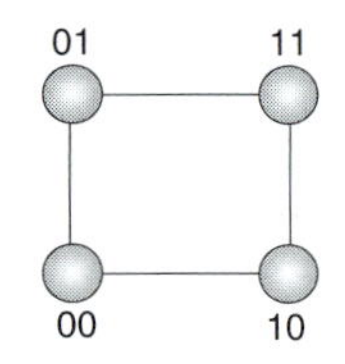

A complete 3-level decision tree and 3-dimensional cube

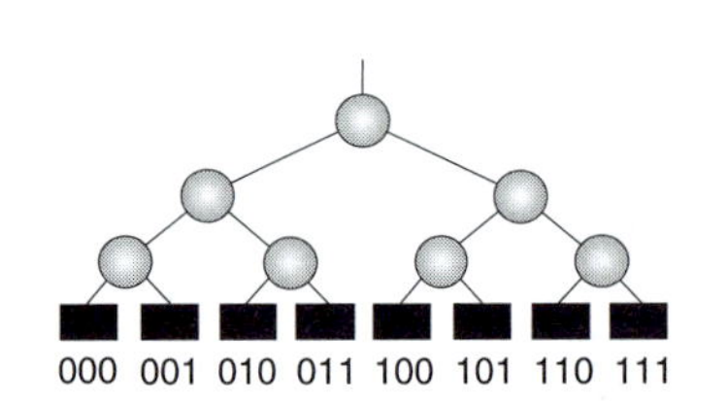

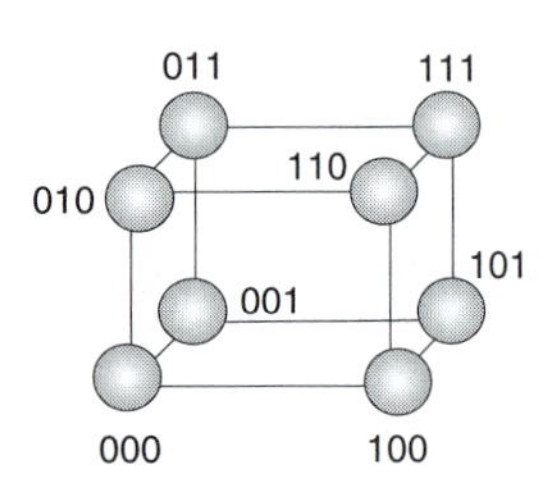

A complete 4-level decision tree and a 4-dimensional cube

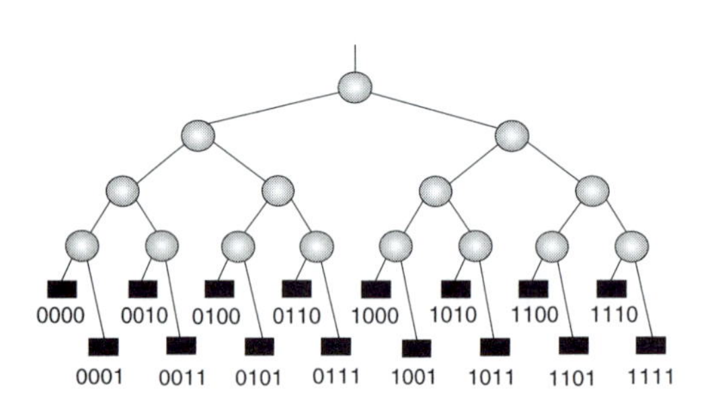

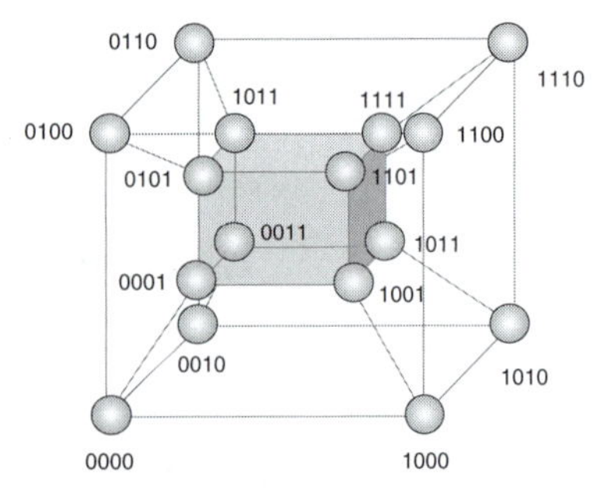

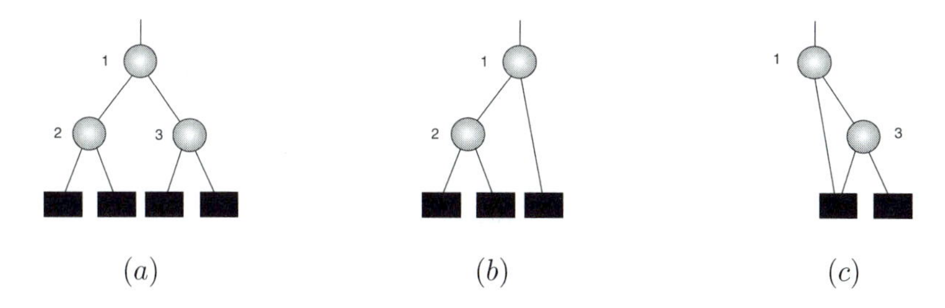

FIGURE 3.19
A complete decision tree (a) and its simplification by deleting the nodes labeled by 3 (b) and by 2 (c) (Example 3.18).

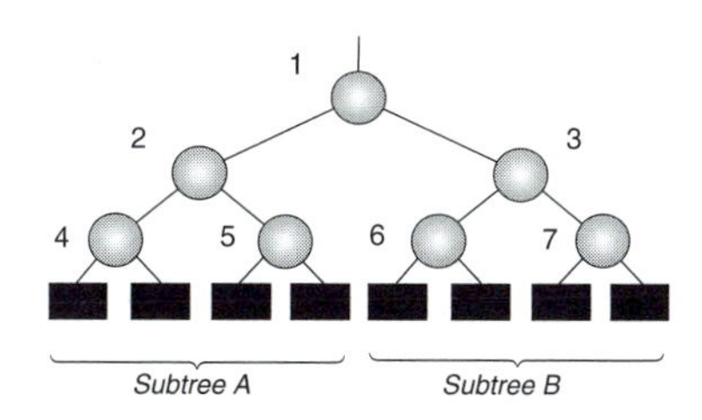

Answer is given in "Solutions to practice problems."

3.5 Summary of graphical data structures

This chapter introduces the basics of graphical data structures for the logic design of discrete devices. Many of the problems of logic analysis and synthesis can be efficiently solved by manipulating graphs. Graphs are employed for ***interpretation*** and better ***understanding*** of various design process steps. The key aspects of this chapter are given below:

(*a*) ***Decision trees*** and ***decision diagrams*** are used for the ***representation***, ***manipulation***, and ***optimization*** of data; and ***implementation*** in the form of multiplexer-based logic networks.

(*b*) ***Logic diagrams*** informally describe a logic network. A logic diagram is a ***visual syntax*** for describing a logic network as an interconnection of logic gates, each of which performs one particular input-to-output transformation of a logic signal.

3.6 Summary (continuation)

The main topics of this chapter are summarized as follows:

- Graphs are ***directed*** and ***undirected.*** Directed graphs are useful when the direction of transmitted data is important. In other cases, undirected graphs are used.
- There is a correspondence between ***algebraic*** and ***graphical*** representations of functions in terms of graphical components such as sets of ***nodes***, ***levels***, ***subgraphs***, and ***shapes***, and also functional characteristics, such as logical relationships expressed using links in the graph.
- Information about functions represented as graphical data structures is carried by nodes (***intermediate*** and ***terminal***), ***edges*** or ***links*** between nodes, and ***topology***, which specifies the geometric properties and spatial relations of nodes and edges.
- The algebraic rules for the manipulation of a function have ***graphical equivalents*** in the form of topological objects such as sets of nodes, levels, subtrees, shapes, etc., and also functional characteristics such as the distribution of nodes, weights, homogeneity, etc.
- A graph is called ***rooted*** if there exists exactly one node with an ***indegree*** of 0, – the root. A ***tree*** is a rooted acyclic graph in which every node but the root has an indegree of 1. Using trees, an arbitrary data processing procedure can be represented. A special class of rooted trees exists, called ***binary*** trees. The ***path length*** of a tree can be defined as the sum of the path lengths from the root to all terminal vertices.
- Graphs show the relationship between the variables of a function. The manipulation of the variables and terms of a function is replaced in graphs by the manipulation of graphical components (nodes, links, and paths).
- There are several useful characteristics and properties of a tree, in particular: for every node v there exists a unique path from the root to v; a node with no children is a ***terminal*** (***external***) node, or ***leaf***; a nonleaf node is called an ***internal*** node; a ***complete*** n-level p-tree, is a tree with p^k nodes per level k for $k = 0, \ldots, n-1$.
- A ***decision tree*** is a graphical model of the evaluation of a discrete function. In logic design, the focus is the construction of decision trees from function descriptions, evaluating the efficiency of this representation, and the manipulation of the function represented by this tree.
- For advances in the application of graphical data structures, we refer the reader to the "Further study" section.

3.7 Further study

Historical perspective

1736: Graph theory originated in 1736, when Leonhard Euler, the Swiss mathematician, proved that it was impossible to cross each of the seven bridges in the city of Königsberg exactly once and return to the starting point.

1860–1930: Common development of graph theory and topology, using the techniques of modern algebra. The example of the application of these results is the work of the physicist Gustav Kirchhoff, who published in 1845 his Kirchhoff's circuit laws for calculating the voltage and current in electric circuits.

1960: Graphs with weights, or weighted graphs, were used to represent structures in which pairwise connections have some numerical values. A digraph with weighted edges in the context of graph theory is called a network.

Advanced topics of graphical data structures

Topic 1: Graphical data structures in logic design. The following problems have received major consideration:

The Steiner tree problem is formulated as follows: Given an edge-weighted graph and a non-empty subset of nodes called *terminal nodes* T, find a minimal weight tree in G that spans T. The Steiner tree problem corresponds to placement in physical design; that is, positioning of the components on the board in a way that is feasible with respect to practical given technological limitations.

The bigraph crossing problem. Another possible objective during placement is to minimize the number of wire crossings. This can be modeled as a *bigraph crossing problem* in the following way. Let $G = (V, E)$ be a bipartite graph with partitions V_1 and V_2. The solution is obtained by embedding G in the plane so that the nodes in V_i occupy distinct positions on the line $y = i$ and the edges are straight lines.

Recursive minimum cut placement. The problem of recursive minimum cut placement is stated by representing the logic network as a hypergraph. A *hypergraph* is like an undirected graph, but each *hyperedge*, rather than connecting two vertices, connects an arbitrary subset of vertices. The components correspond to the nodes and the nets correspond to the hyperedges connecting these components (nodes). A decomposition of the logic network into two sub-networks is formulated as the problem of finding a balanced bipartition of the hypergraph with minimal cuts. This problem is also known as the *balanced hypergraph bipartition problem.*

Graph topology. The primary topic of interest regarding graphs that is used for physical or geometrical design is the manipulation of graph topology. The most often referred topologies include: hypercube topology, cube-connected cycles known as CCC-topology, pyramid topology, X-hypercube topology, hybrid topologies, and specific-purpose topologies (hyper-Peterson, hyper-star, Fibonacci cube, etc.).

(*a*) *Hypercube topology* has received considerable attention in classical logic design due mainly to its ability to interpret logic formulas and logic computations (Figure 3.20a). Hypercube-based structures are at the forefront of massively parallel computation because of the unique characteristics of hypercubes (fault tolerance, ability to efficiently permit the embedding of various topologies, such as lattices and trees).

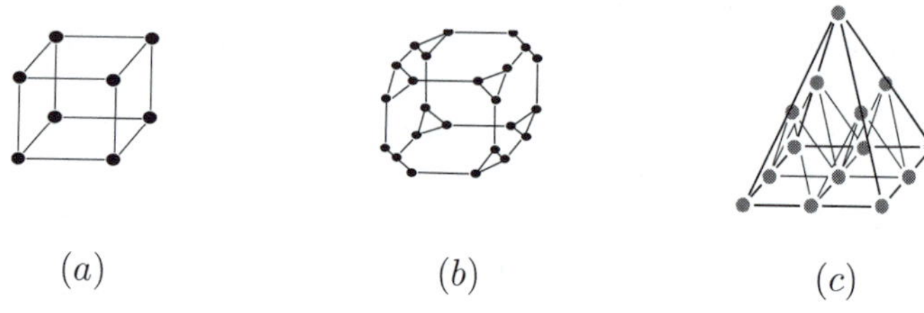

FIGURE 3.20
Spatial configurations: hypercube (a), CCC-hypercube (b), and pyramid (c).

(*b*) *Pyramid topology* is suitable for many computations based on the principle of hierarchical control, for example, decision trees and decision diagrams (Figure 3.20c). Some topologies are effective in particular cases, for example, for symmetric functions, partially specified functions, threshold functions, etc.

Topic 2: Computational paradigm based on embedding. Mapping abstract data structures into computing structures can be viewed as embedding some *guest* data structure into a given *host* data structure. Examples are matrix equations, flow graphs of algorithms, decision trees, and decision diagrams, which can be embedded into spatial dimensions using various techniques.

There are two formulations of the embedding problem:

(*a*) Given a guest data structure, find an appropriate topology for its representation; this is a *direct* problem.

(*b*) Given a topological structure, find the corresponding data structure that is suitably represented by this topology; this is an *inverse* problem (using the properties of the host representation, specify possible data structures that satisfy it with respect to certain criteria).

The direct and inverse problems of embedding address different problems of computing. In direct problems, the topology is not specified, and the designer can map a representation of a logic function into spatial dimensions without strong topological limitations. The inverse formulation of the problem assumes that the topology is constrained by technology. That is, the question is how to use a given topology in the efficiently representation of logic functions.

Nanostructures are associated with a molecular/atomic physical platform. They have a truly three-dimensional structure, instead of the three-dimensional layout of silicon integrated circuits, composed of two-dimensional layers with interconnections forming the third dimension. Figure 3.21 shows how a computing (guest) structure can be embedded into another (host) computing structure.

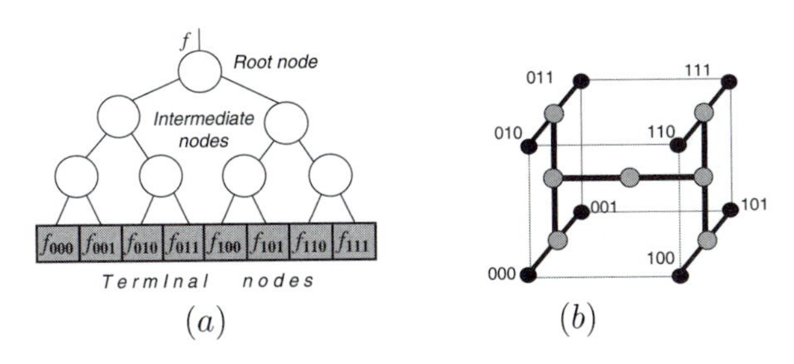

FIGURE 3.21
Embedding a decision tree into a hypercube-like structure.

Further reading

Algorithmic Graph Theory by A. Gibbons, Cambridge University Press, Cambridge, U.K., 1987.

Decision Diagram Techniques for Micro- and Nanoelectronic Design by S. N. Yanushkevich, D. M. Miller, V. P. Shmerko, and R. S. Stanković, CRC/Taylor & Francis Group, Boca Raton, FL, 2006.

"Generalized Hypercube and Hyperbus Structures for a Computer Network" by N. Bhuyan and D. P. Agrawal, *IEEE Transactions on Computers*, volume 33, number 1, pages 323–333, 1984.

Discrete Mathematics with Graph Theory by E. G. Goodaire and M. M. Parmenter, Prentice-Hall, New York, 1998.

"Incomplete Hypercubes: Embeddings of Tree-Related Networks" by S. Öhring and S. K. Das, *Journal of Parallel and Distributed Computing*, volume 26, pages 36–47, 1995.

Introduction to Algorithms by T. H. Cormen, C. E. Leiserson, R. L. Riverst, and C. Stein, MIT Press, Cambridge, MA, 2001.

Introduction to Parallel Algorithms and Architectures: Arrays, Trees, and Hypercubes by F. T. Leighton, Morgan Kaufmann, San Mateo, CA, 1991.

"Topological Properties of Hypercubes" by Y. Saad, and M. H. Schultz, *IEEE Transactions on Computers*, volume 37, number 7, pages 867–872, 1988.

Solutions to practice problems

Practice problem	Solution
3.1a.	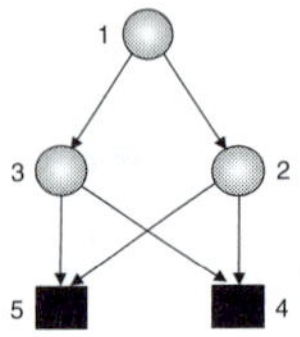 ▶ 5 nodes in(v_1)=0, out(v_1)=2 in(v_2)=1, out(v_2)=1 in(v_3)=1, out(v_3)=2 in(v_4)=2, out(v_4)=0 in(v_5)=2, out(v_5)=0 ▶ v_1 is a root, v_2, v_3 are the internal nodes, and v_4, v_5 are the terminal nodes ▶ Diameter = 2 $A = \begin{array}{c} \\ 1 \\ 2 \\ 3 \\ 4 \\ 5 \end{array} \begin{bmatrix} 1 & 2 & 3 & 4 & 5 \\ & 1 & 1 & & \\ & & & 1 & 1 \\ & & & 1 & 1 \\ & & & & \\ & & & & \end{bmatrix}$
3.1b.	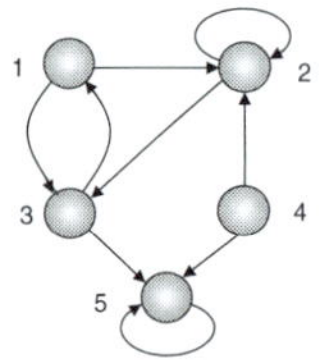 ▶ 5 nodes in(v_1)=1, out(v_1)=2 in(v_2)=3, out(v_2)=2 in(v_3)=1, out(v_3)=2 in(v_4)=2, out(v_4)=0 in(v_5)=3, out(v_5)=1 ▶ v_4 is a root, v_2, v_3, v_4, v_5 are the internal nodes, v_2, v_5 have loops, there are no terminal nodes ▶ Diameter = 3 $A = \begin{array}{c} \\ 1 \\ 2 \\ 3 \\ 4 \end{array} \begin{bmatrix} 1 & 2 & 3 & 4 \\ & 1 & 1 & 1 \\ & & 1 & 1 \\ & & & 1 \\ & & & \end{bmatrix}$
3.2.	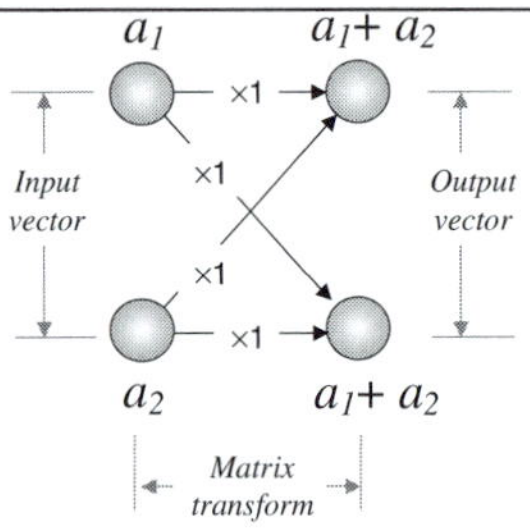 a_1, $a_1 + a_2$, a_2, $a_1 + a_2$ ×1 ×1 ×1 ×1 Input vector Output vector Matrix transform
3.3.	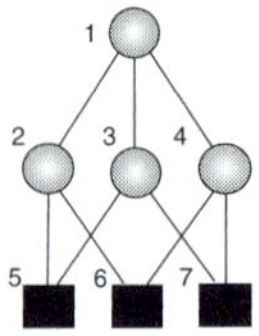 ▶ 7 nodes deg(v_1)=3, deg(v_2)=3 deg(v_3)=3, deg(v_4)=3 deg(v_5)=2, deg(v_6)=2 deg(v_7)=2 ▶ v_1 is a root, v_2, v_3, v_4 are the internal nodes, v_5, v_6, v_7 are the terminal nodes ▶ Diameter = 2 $A = \begin{array}{c} \\ 1 \\ 2 \\ 3 \\ 4 \\ 5 \\ 6 \\ 7 \end{array} \begin{bmatrix} 1 & 2 & 3 & 4 & 5 & 6 & 7 \\ & 1 & 1 & 1 & & & \\ 1 & & & & 1 & 1 & \\ 1 & & & & 1 & & 1 \\ 1 & & & & & 1 & 1 \\ & 1 & 1 & & & & \\ & 1 & & 1 & & & \\ & & 1 & 1 & & & \end{bmatrix}$

Solutions to practice problems (continuation)

Practice problem	Solution
3.4.	The adjacency matrices for both graphs are the same: $A = \begin{array}{c} \\ 1 \\ 2 \\ 3 \\ 4 \\ 5 \end{array} \begin{array}{c} \begin{array}{ccccc} 1 & 2 & 3 & 4 & 5 \end{array} \\ \left[\begin{array}{ccccc} & 1 & 1 & & \\ 1 & & & 1 & 1 \\ 1 & & & 1 & 1 \\ & 1 & 1 & & \\ & 1 & 1 & & \end{array}\right] \end{array}$
3.6.	1 2 3 5 7 Subtree A Subtree B

3.8 Problems

Problem 3.1 Derive the adjacency matrices for the following directed graphs:

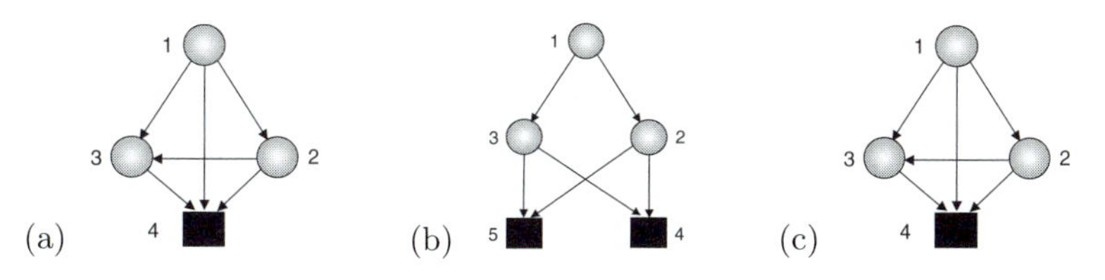

Problem 3.2 Derive the adjacency matrices for the following undirected graphs:

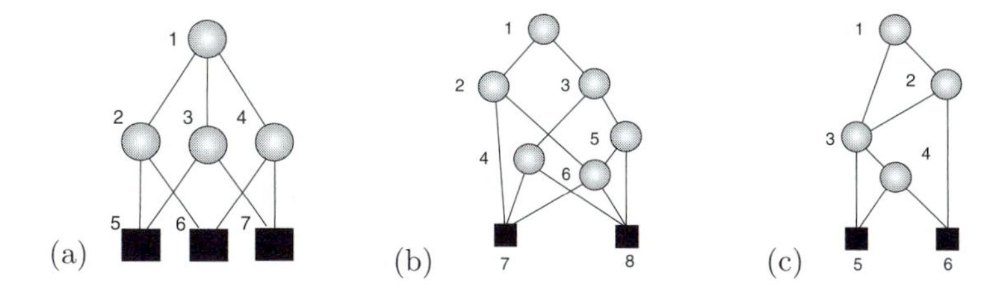

Problem 3.3 Derive the direct graphs using the following adjacency matrices:

$$A = \begin{array}{c} \\ 1 \\ 2 \\ 3 \\ 4 \end{array} \begin{array}{c} \begin{array}{cccc} 1 & 2 & 3 & 4 \end{array} \\ \left[\begin{array}{cccc} & 1 & 1 & 1 \\ & & 1 & 1 \\ & & & 1 \\ & & & \end{array}\right] \end{array} \quad A = \begin{array}{c} \\ 1 \\ 2 \\ 3 \\ 4 \end{array} \begin{array}{c} \begin{array}{cccc} 1 & 2 & 3 & 4 \end{array} \\ \left[\begin{array}{cccc} 1 & 1 & 1 & 1 \\ & 1 & 1 & 1 \\ & & 1 & 1 \\ & & & 1 \end{array}\right] \end{array} \quad A = \begin{array}{c} \\ 1 \\ 2 \\ 3 \\ 4 \end{array} \begin{array}{c} \begin{array}{cccc} 1 & 2 & 3 & 4 \end{array} \\ \left[\begin{array}{cccc} 1 & 1 & & \\ & 1 & & 1 \\ & & 1 & 1 \\ & & & 1 \end{array}\right] \end{array}$$

Problem 3.4 Derive the undirected graphs using the following adjacency matrices:

$$A = \begin{array}{c} \\ 1 \\ 2 \\ 3 \\ 4 \end{array} \begin{array}{c} \begin{array}{cccc} 1 & 2 & 3 & 4 \end{array} \\ \left[\begin{array}{cccc} & 1 & 1 & 1 \\ 1 & & 1 & 1 \\ 1 & 1 & & 1 \\ 1 & 1 & 1 & \end{array}\right] \end{array} \quad A = \begin{array}{c} \\ 1 \\ 2 \\ 3 \\ 4 \end{array} \begin{array}{c} \begin{array}{cccc} 1 & 2 & 3 & 4 \end{array} \\ \left[\begin{array}{cccc} & 1 & 1 & 1 \\ 1 & & & \\ 1 & & & 1 \\ 1 & & 1 & \end{array}\right] \end{array} \quad A = \begin{array}{c} \\ 1 \\ 2 \\ 3 \\ 4 \end{array} \begin{array}{c} \begin{array}{cccc} 1 & 2 & 3 & 4 \end{array} \\ \left[\begin{array}{cccc} 1 & 1 & 1 & \\ 1 & & & 1 \\ 1 & & & 1 \\ & 1 & 1 & \end{array}\right] \end{array}$$

Problem 3.5 Derive the isomorphic graphs for the following graphs:

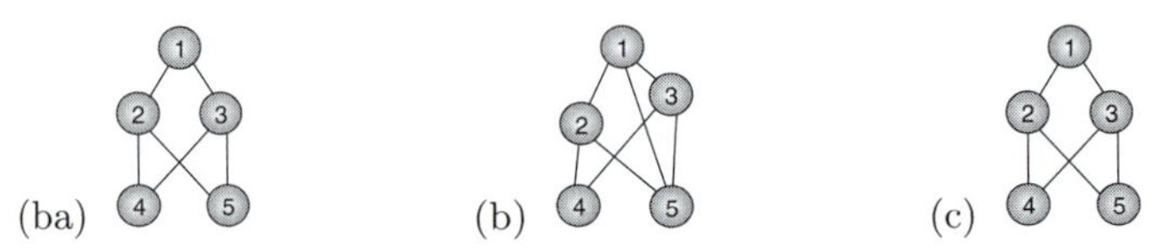

Problem 3.6 Given graphs G_1 and G_2:

Perform the following operations:(a) Union $G_1 \cup G_2$, (b) Intersection $G_1 \cap G_2$, (c) Ring sum $G_1 \oplus G_2$, and (d) Cartesian product $G_1 \otimes G_2$

Problem 3.7 Calculate the number of paths and levels in the following trees:

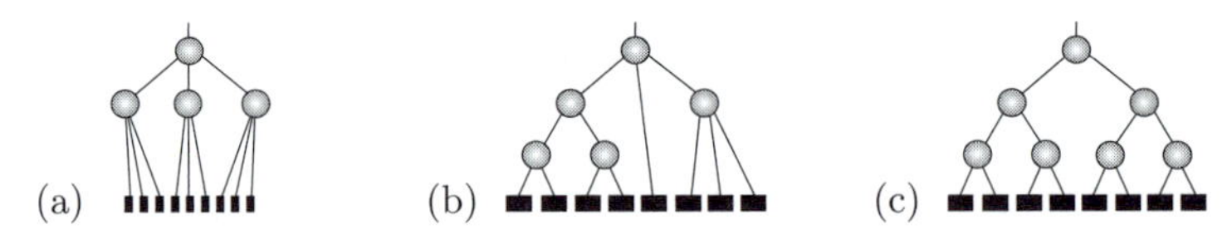

Problem 3.8 Show that tree G_1 can be embedded into graph G_2:

Problem 3.9 Derive the equation for calculating the number of intermediate nodes and terminal nodes of the complete binary tree with the following number of levels: (a) 2 levels, (b) 8 levels, (c) 64 levels, and (d) 128 levels

4

Algebra I: Boolean

Boolean algebra

- Postulates
- Duality
- Switches
- DeMorgan's law

Boolean functions

- Boolean formulas and functions
- Minterms and maxterms
- Sum-of-products (SOP) form
- Product-of-sums (POS) form
- Standard SOP and POS expressions
- Proving the validity of Boolean equations

Elementary functions

- The set of elementary functions
- Switch models for logic gates
- Local transformations

Advanced topics

- Multi-valued logic
- Probabilistic logic
- Fuzzy logic

4.1 Introduction

The theoretical foundation for digital design includes:

Theoretical foundation for digital design

- Algebra over the two-element set $\{0, 1\}$,
- Data structures for representing Boolean functions, and
- Techniques for the manipulation of these data structures, including transferring between different structures.

4.2 Definition of algebra over the set $\{0, 1\}$

A *universal* algebra consists of a set of elements and operations on this set. Boolean algebras are a particular case of universal algebra. There is an infinite number of different Boolean algebras. The basic requirements of any algebra employed to describe and manipulate Boolean functions include:

- *Functional completeness* – The algebra must be capable of describing and manipulating an arbitrary Boolean function.
- *Flexibility* – The algebra must be amenable to the manipulation of Boolean functions so that design algorithms can be set up and employed with reasonable ease.
- *Implementability* – The basic connectives and any high-level functional operations should have simple and reliable physical logic circuit counterparts.

4.2.1 Boolean algebra over the set $\{0, 1\}$

The simplest Boolean algebra is a two-valued Boolean algebra defined over the two-element set $B = \{0, 1\}$. The algebra $\{B;\ \vee, \cdot;\ ^{-};\ 0, 1\}$ is called a *Boolean algebra*, or *switching algebra*. A two-valued Boolean algebra consists of (Figure 4.1):

Two-valued Boolean algebra

- A set of two *elements*, $B = \{0, 1\}$;
- A set of *operations*: *binary* operations, Boolean sum $\vee$ and Boolean product $\cdot$ (also denoted $\wedge$); and the *unary* operation, complement, denoted by $^{-}$;
- Two distinguished elements: a unique real number called a *zero element*, 0, and a *unity element*, 1, and
- A number of *axioms* or *postulates*.

A binary operation is a rule that assigns to each pair of elements from a set of elements B a unique element from the same set B. Postulates are the basic assumptions from which it is possible to deduce the rules, theorems, and properties of the system.

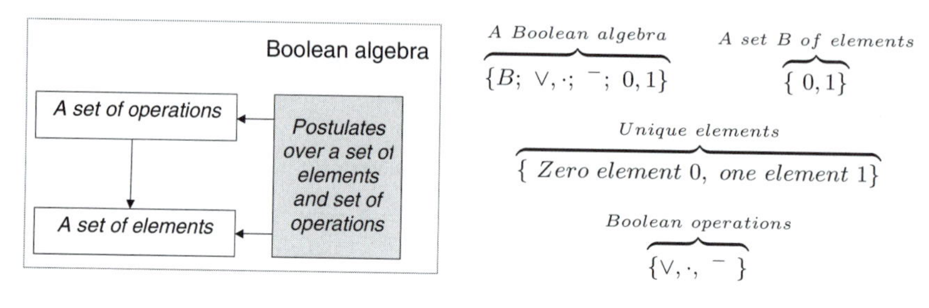

FIGURE 4.1
Boolean algebras are defined over sets of elements, operations, and postulates.

4.2.2 Postulates

The *computation rules* of Boolean algebra, known as *Huntington's postulates*, or *laws*, are defined as follows (Figure 4.2):

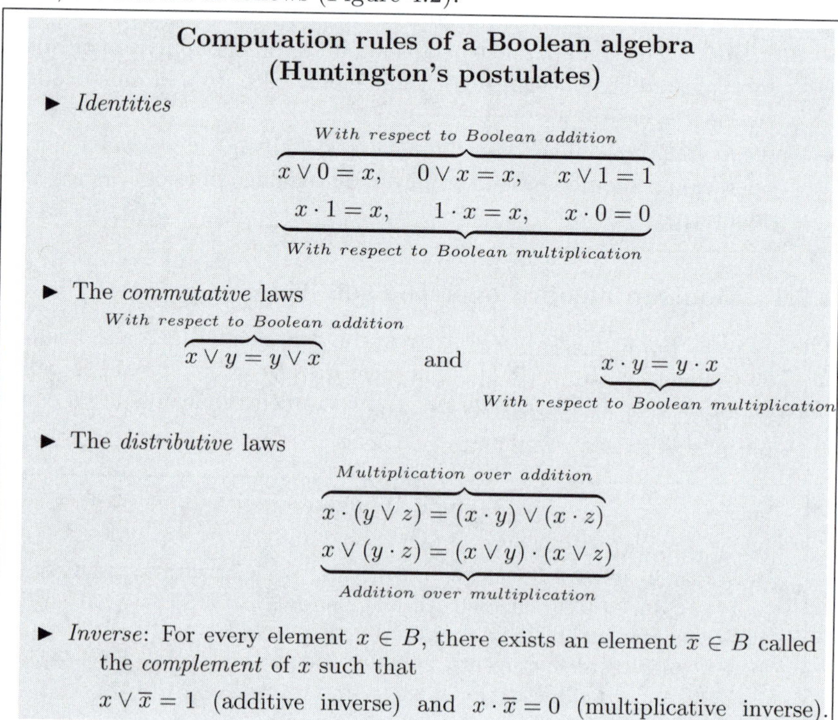

Computation rules of a Boolean algebra (Huntington's postulates)
► Identities
With respect to Boolean addition
$x \vee 0 = x, \quad 0 \vee x = x, \quad x \vee 1 = 1$
$x \cdot 1 = x, \quad 1 \cdot x = x, \quad x \cdot 0 = 0$
With respect to Boolean multiplication
► The commutative laws
With respect to Boolean addition
$x \vee y = y \vee x$ and $x \cdot y = y \cdot x$
With respect to Boolean multiplication
► The distributive laws
Multiplication over addition
$x \cdot (y \vee z) = (x \cdot y) \vee (x \cdot z)$
$x \vee (y \cdot z) = (x \vee y) \cdot (x \vee z)$
Addition over multiplication
► Inverse: For every element $x \in B$, there exists an element $\overline{x} \in B$ called the complement of x such that
$x \vee \overline{x} = 1$ (additive inverse) and $x \cdot \overline{x} = 0$ (multiplicative inverse).

Operations with constants	$x \vee 0 = x$ $x \cdot 1 = x$
Manipulation	$x \cdot (x \vee z) = (x \cdot y) \vee (x \cdot z)$ $x \vee (x \cdot z) = (x \vee y) \cdot (x \vee z)$
Elimination of variables	$x \vee \overline{x} = 1$ $x \cdot \overline{x} = 0$

The operations of Boolean algebra include the Boolean sum $\vee$, Boolean product $\cdot$, and complement $^{-}$

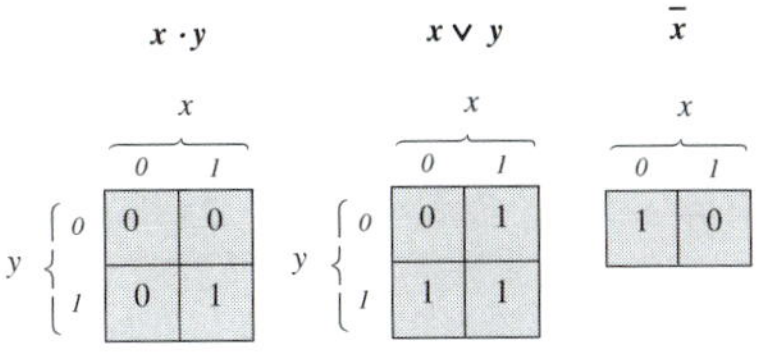

$x \cdot y$

y \ x	0	1
0	0	0
1	0	1

$x \vee y$

y \ x	0	1
0	0	1
1	1	1

$\overline{x}$

x	0	1
	1	0

FIGURE 4.2
Computation rules of Boolean algebra.

Techniques for proving some of the computational algebraic rules are given in Table 4.1.

4.2.3 The principle of duality

One can observe a consistent symmetry for each postulate: between identities with respect to the Boolean sum and product, between commutative laws with respect to the Boolean sum and product, and between distributive laws with respect to the Boolean sum and product. Symmetries can also be observed in the complement law. This fundamental interchangeability property of Boolean algebra is termed the *principle of duality.*

Principle of duality states that each theorem of a Boolean algebra has a dual, which can be obtained as follows:

Algorithm 1
for deriving a dual of a Boolean function

Given: `A Boolean function` f
Step 1: `Interchange the OR and AND operations in the expression`
Step 2: `Interchange the 0 and 1 elements of the expression.`
`The resulting function` g `is a dual of` f

Duality is used to prove that for each valid equation over a Boolean algebra, the dual equation is also valid. If we can prove, through a series of logical steps, that a given theorem is true, it immediately implies that the dual theorem is also true, since the dual of the logical steps, that proves the original theorem, proves the dual theorem.

The principle of duality is the basis for deriving many of the properties of Boolean functions, utilized in efficient computing.

TABLE 4.1
Some of the computation rules and their proofs.

Techniques for computing in two-valued Boolean algebra

Computation rule	Proof technique
Identity	$x = \overbrace{x \vee 0}^{Identity} = \overbrace{x \vee (x \cdot \overline{x})}^{Inverse} = \overbrace{(x \vee x) \cdot (x \vee \overline{x})}^{Distributivity}$ $= \overbrace{(x \vee x) \cdot 1}^{Inverse} = \overbrace{x \vee x}^{Identity}$
Property of the element 1	$x \vee 1 = \overbrace{(x \vee 1) \cdot 1)}^{Identity} = \overbrace{(x \vee 1) \cdot (x \vee \overline{x})}^{Inverse} = \overbrace{x \vee (1 \cdot \overline{x})}^{Distributivity}$ $= x \vee \overline{x} = 1$
Absorption	$x \vee (x \cdot y) = \overbrace{(x \cdot 1) \vee (x \cdot y)}^{Identity} = \overbrace{x \cdot (1 \vee y)}^{Distributivity} = x \cdot 1 = x$ $x(x \vee y) = \overbrace{(x \vee 0) \cdot (x \vee y)}^{Identity} = \overbrace{x \vee (0 \cdot y)}^{Distributivity} = x \vee 0 = x$
Adjacency	$xy \vee x\overline{y} = x\, \overbrace{(y \vee \overline{y})}^{Inverse} = \overbrace{x \cdot 1}^{Identity} = x$ $(x \vee y) \cdot (x \vee \overline{y}) = x\, \overbrace{(y \vee \overline{y})}^{Inverse} = \overbrace{x \cdot 1}^{Identity} = x$
Simplification	$x \vee \overline{x}y = \overbrace{(x \vee \overline{x})(x \vee y)}^{Distributivity} = \overbrace{1 \cdot (x \vee y)}^{Identity} = x \vee y$ $x(\overline{x} \vee y) = \overbrace{x\overline{x} \vee xy}^{Distributivity} = \overbrace{0 \vee xy}^{Identity} = xy$

Example 4.1 (Duality.) *(a) The dual to the identity postulate,* $x \vee 0 = x$*, is the equation* $x \cdot 1 = x$*:* $x \vee 0 = x \Leftrightarrow \overbrace{x \cdot 1 = x}^{Dual}$*.*
(b) The dual to $x(x \vee y) = x$ *is* $x \vee xy = x$*.*

An arbitrary Boolean function f and its dual g are not equal, $g \neq f$.

Example 4.2 (Duality.) *Consider the Boolean function* $f = \overline{x}_1\overline{x}_3 \vee \overline{x}_1 x_2 \vee x_2 x_3$*. The dual to this expression is*

$$g = (\overline{x}_1 \vee \overline{x}_3)(\overline{x}_1 \vee x_2)(x_2 \vee x_3) = \overline{x}_1 x_2 \vee \overline{x}_1 x_3 \vee x_2 \overline{x}_3$$

Note that $g \neq f$*.*

Alternatively, a dual of a Boolean function can be found using the following algorithm:

> **Algorithm 2**
> **for deriving a dual of a Boolean function**
>
> **Given:** `A Boolean function` f
> **Step 1:** `Find inverse of the Boolean function` f`:` $\overline{f}$
> **Step 2:** `Complement the variables (change` $\overline{x}$ `to` x `and` x `to` $\overline{x}$`).`
> `The resulting function` g `is a dual of` f

Example 4.3 (Duality.) *Given* $f = x_1x_2 \vee x_3$*, its complement is derived as follows:*

$$\overline{f} = \overline{x_1x_2 \vee x_3} = \overline{x_1x_2}\overline{x}_3 = (\overline{x}_1 \vee \overline{x}_2)\overline{x}_3 = \overline{x}_1\overline{x}_3 \vee \overline{x}_2\overline{x}_3$$

Complement the variables of $\overline{f}$ *and find the function* $g = x_1x_3 \vee x_2x_3$*. It is a dual to* f*, since* g *is the same function that can be obtained by finding a dual directly from* f*:*

$$g = (x_1 \vee x_3)x_3 = x_1x_2 \vee x_2x_3$$

4.2.4 Switch-based interpretation of computation rules

Boolean postulates can be interpreted using switches. This interpretation is crucial for their implementation by various physical and chemical phenomena.

In Table 4.2, several of the computation rules of Boolean algebra are interpreted using switches. For example, the Boolean product is a series, and the Boolean sum is a parallel switch arrangement. Note that while $x = 0$ is represented by an open switch, the $\overline{x} = 1$ appearing in the same problem is represented by a closed switch.

4.2.5 Boolean algebra over Boolean vectors

A vector $\mathbf{a} = (a_1, a_2, \ldots, a_n)$ with n binary elements $a_i \in \{0, 1\}$, $i = 1, 2, \ldots, n$ is called an *n-dimensional Boolean vector.* We also denote the Boolean vector as $\mathbf{a} = \begin{bmatrix} a_1 \\ a_2 \\ \cdot \\ \cdot \\ \cdot \\ a_n \end{bmatrix}$ or $\mathbf{a} = [\ a_1\ a_2\ \ldots\ a_n\ ]^T$, where T means its transposition.

Example 4.4 (Boolean vector.) *A 4-dimensional Boolean vector* $\mathbf{a} = (1, 0, 0, 1)$ *can be written as follows:*

$$\mathbf{a} = \begin{bmatrix} 1 \\ 0 \\ 0 \\ 1 \end{bmatrix} = [\ 1\ 0\ 0\ 1\ ]^T$$

FIGURE 4.2
Computation rules in terms of switches.

Techniques for computing in two-valued Boolean algebra

Computation rule	Switch-based interpretation
Idempotence $x \vee x = x$	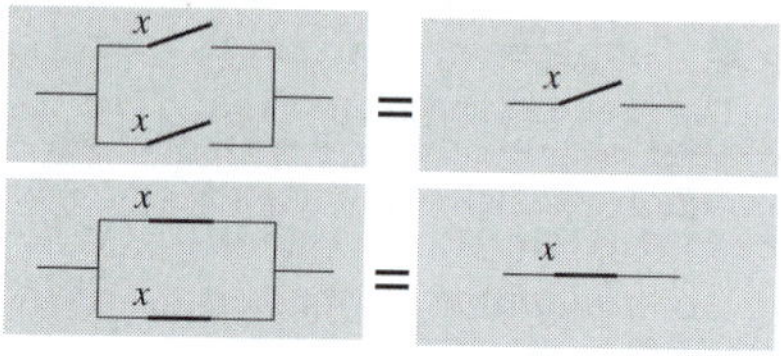
Idempotence $x \cdot x = x$	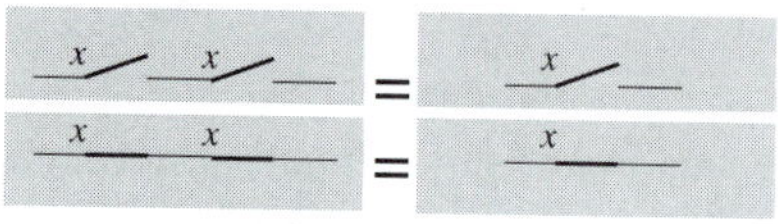
Identity $x \vee 1 = 1$	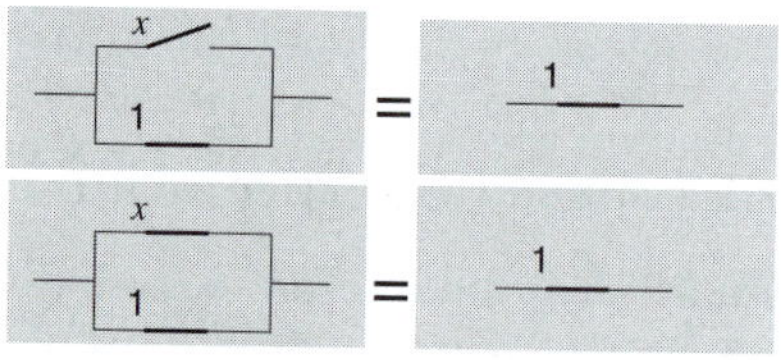
Identity $x \vee 0 = x$	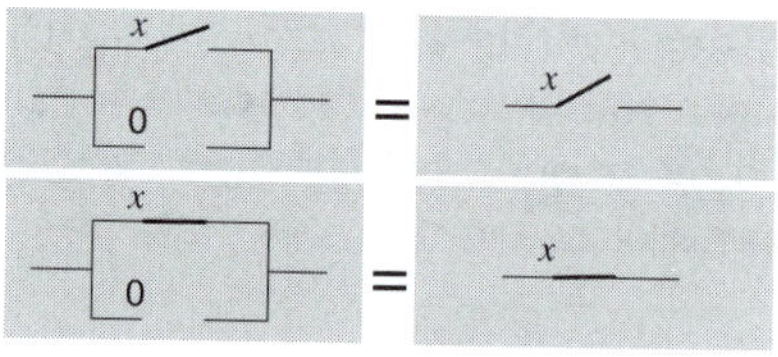
Identity $x \cdot 1 = x$	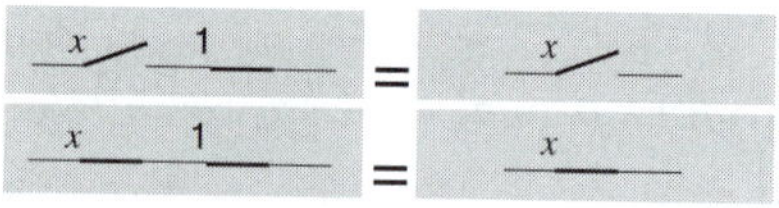
Complement $\overline{x} \vee x = 1$	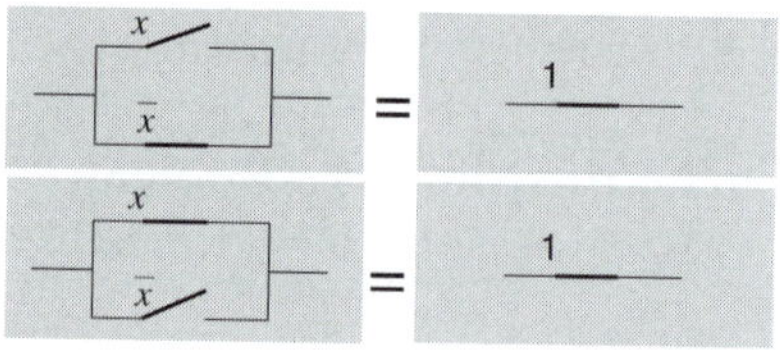
Complement $\overline{x} \cdot x = 0$	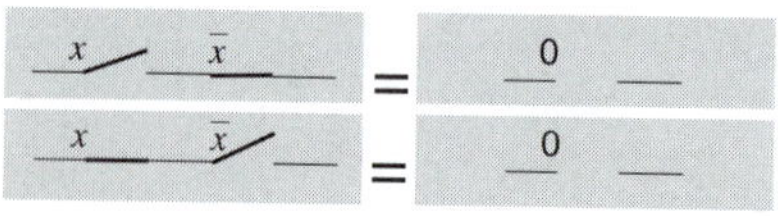

Let $\mathbf{B}^n = \{(a_1, a_2, \ldots, a_n)\}$, $a_i \in \{0, 1\}$, be the set of n-dimensional Boolean vectors. Consider two vectors, $\mathbf{a} = (a_1, a_2, \ldots, a_n)$ and $\mathbf{b} = (b_1, b_2, \ldots, b_n)$ in $\mathbf{B}^n$. The following operations with these vectors are specified:

Operations with Boolean vectors

- Boolean sum ($\vee$): $\mathbf{a} \vee \mathbf{b} = (a_1 \vee b_1,\ a_2 \vee b_2, \ldots, a_n \vee b_n)$;
- Boolean product ($\cdot$): $\mathbf{a} \cdot \mathbf{b} = (a_1 \cdot b_1,\ a_2 \vee b_2, \ldots, a_n \cdot b_n)$; and
- Complement: $\overline{\mathbf{a}} = (\overline{a}_1,\ \overline{a}_2, \ldots, \overline{a}_n)$.

If we specify the 0 and 1 elements in vector notation as $0 = \{(0, 0, \ldots, 0)\}$ and $1 = \{(1, 1, \ldots, 1)\}$, respectively, then the system $\langle \mathbf{B}^n, \vee, \cdot, 0, 1\rangle$ is a *Boolean algebra over Boolean vectors.*

Example 4.5 (Boolean vectors.) *Consider the Boolean vectors* $\mathbf{a} = \begin{bmatrix}0\\1\\1\\0\end{bmatrix}$ *and* $\mathbf{b} = \begin{bmatrix}1\\0\\0\\0\end{bmatrix}$. *The following operations hold on these vectors: Boolean sum:* $\mathbf{a} \vee \mathbf{b} = \begin{bmatrix}0\\1\\1\\0\end{bmatrix} \vee \begin{bmatrix}1\\0\\0\\0\end{bmatrix} = \begin{bmatrix}0 \vee 1\\1 \vee 0\\1 \vee 0\\0 \vee 0\end{bmatrix} = \begin{bmatrix}1\\1\\1\\0\end{bmatrix}$;

Boolean product: $\mathbf{a} \cdot \mathbf{b} = \begin{bmatrix}0\\1\\1\\0\end{bmatrix} \cdot \begin{bmatrix}1\\0\\0\\0\end{bmatrix} = \begin{bmatrix}0 \cdot 1\\1 \cdot 0\\1 \cdot 0\\0 \cdot 0\end{bmatrix} = \begin{bmatrix}0\\0\\0\\0\end{bmatrix}$;

Complement: $\overline{\mathbf{a}} = \begin{bmatrix}\overline{0}\\\overline{1}\\\overline{1}\\\overline{0}\end{bmatrix} = \begin{bmatrix}1\\0\\0\\1\end{bmatrix}$; $\quad \overline{\mathbf{b}} = \begin{bmatrix}\overline{1}\\\overline{0}\\\overline{0}\\\overline{0}\end{bmatrix} = \begin{bmatrix}0\\1\\1\\1\end{bmatrix}$.

Practice problem 4.1. (Boolean vectors.) Using the Boolean vectors $\mathbf{a}$ and $\mathbf{b}$ from Example 4.5, calculate $(\mathbf{a} \cdot \mathbf{b}) \vee \overline{\mathbf{a}}$.
Answer is given in "Solutions to practice problems."

4.2.6 DeMorgan's law

DeMorgan's law provides the possibility to manipulate complemented variables and equations. DeMorgan's law states:

DeMorgan's law

$$\overline{x_1 \cdot x_2} = \overline{x}_1 \vee \overline{x}_2; \quad \overline{x_1 \vee x_2} = \overline{x}_1 \cdot \overline{x}_2$$

These equations can be generalized for n variables $x_1, x_2, \ldots, x_n$ as follows: $\overline{x_1 \cdot x_2 \cdots x_n} = \overline{x}_1 \vee \overline{x}_2 \vee \cdots \vee \overline{x}_n$ and $\overline{x_1 \vee x_2 \vee \cdots \vee x_n} = \overline{x}_1 \cdot \overline{x}_2 \cdots \overline{x}_n$. Note that two successive complements cancel each other, that is, $\overline{\overline{x}} = x$. This rule is known as *involution*.

Example 4.6 (DeMorgan's law.) *Examples of the manipulation of Boolean equations using DeMorgan's law are as follows:*

$$(a)\ x_1 \cdot x_2 = \overbrace{\overline{\overline{x_1 \cdot x_2}}}^{Involution} = \overline{\overline{x}_1 \vee \overline{x}_2};\quad (b)\ \overline{x_1 \cdot x_2} = \overline{x}_1 \vee \overline{x}_1;$$

$$(c)\ \overline{x_1 \cdot x_2 \cdot x_3} = \overline{x}_1 \vee \overline{x}_2 \vee \overline{x}_3;\quad (d)\ x_1 \vee x_2 = \overline{\overline{x_1 \vee x_2}} = \overline{\overline{x}_1 \cdot \overline{x}_2};$$

$$(e)\ \overline{x_1 \vee x_2 \vee x_3} = \overline{x}_1 \cdot \overline{x}_2 \cdot \overline{x}_3;$$

$$(f)\ \overline{(x_1 \vee x_2)(\overline{x}_3 \vee x_4)} = (\overline{x_1 \vee x_2}) \vee (\overline{\overline{x}_3 \vee x_4}) = \overline{x}_1\overline{x}_2 \vee x_3\overline{x}_4$$

Practice problem 4.2. (DeMorgan's law.) Prove by algebraic manipulation that $\overline{x}_1\overline{x}_2 \vee \overline{x}_3\overline{x}_4 = \overline{(x_1 \vee x_2)(x_3 \vee x_4)}$.
Answer is given in "Solutions to practice problems."

4.3 Boolean functions

There are two distinguished forms of Boolean structures:

- *Boolean formulas*, and
- *Boolean functions.*

Boolean formulas are a useful form of abstraction, but they are not acceptable for computation.

4.3.1 Boolean formulas

Boolean functions are particular functions that can be described in terms of *expressions* over Boolean algebra, called *Boolean formulas.* A Boolean formula of n variables is a *string of symbols* of $x_1, x_2, \ldots, x_n$, the binary operations of Boolean sum ($\vee$), Boolean product ($\cdot$), unary operation of the complement ($^-$), and brackets ( ). A Boolean formula is the Boolean function after specification of values, given assignments of variables.

Boolean formulas are useful for the study of various Boolean algebras and their relationships, but not acceptable for computing. In order to obtain Boolean expressions suitable for efficient computation, Boolean formulas must be transformed into Boolean functions (Figure 4.3):

$$\underbrace{\texttt{Boolean formulas}}_{Abstract\ data\ structure} \xrightarrow{Specification} \underbrace{\texttt{Boolean functions}}_{Computing\ data\ structure}$$

The relationship between Boolean formulas and Boolean functions is not one-to-one: Many different formulas can represent the same Boolean function.

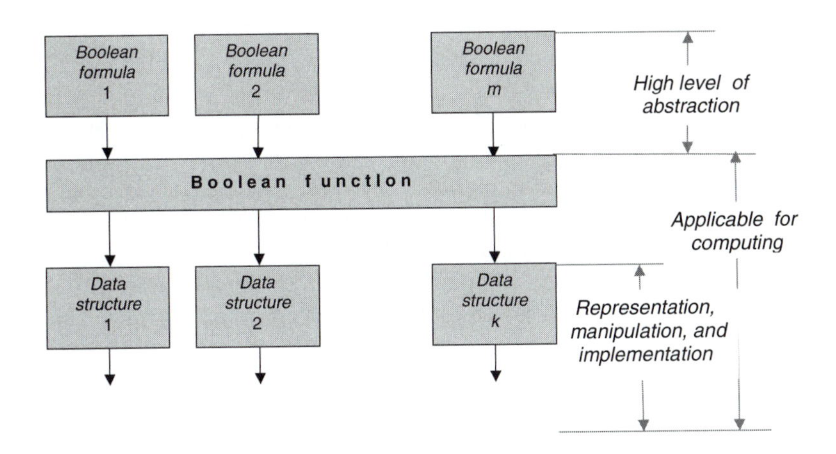

FIGURE 4.3
A Boolean function can be represented by various formulas and implemented using various data structures.

Example 4.7 (Boolean formulas.) *The Boolean formulas* $f_1 = (x_1 \vee x_2) \cdot (x_3 \vee x_2 \cdot \overline{x}_4 \vee \overline{x}_1 \cdot x_2)$ *and* $f_2 = x_3 \cdot (x_1 \vee x_2) \vee x_2 \cdot (\overline{x}_1 \vee \overline{x}_4)$ *represent the same Boolean function, because*

$$
\begin{aligned}
&\overbrace{(x_1 \vee x_2) \cdot (x_3 \vee x_2 \cdot \overline{x}_4 \vee \overline{x}_1 \cdot x_2)}^{f_1} \\
&x_1(x_3 \vee x_2 \cdot \overline{x}_4 \vee \overline{x}_1 \cdot x_2) \vee x_2(x_3 \vee x_2 \cdot \overline{x}_4 \vee \overline{x}_1 \cdot x_2) \\
= {}&x_1 \cdot x_3 \vee x_1 \cdot x_2 \cdot \overline{x}_4 \vee x_2 \cdot x_3 \vee x_2 \cdot \overline{x}_4 \vee \overline{x}_1 \cdot x_2 \\
= {}&x_1 \cdot x_3 \vee x_1 \cdot x_2 \cdot \overline{x}_4 \vee x_2 \cdot x_3 \vee x_2 \cdot \overline{x}_4 \vee \overline{x}_1 x_2 \\
= {}&x_1 \cdot x_3 \vee \cdot x_2 \cdot \overline{x}_4(x_1 \vee 1) \vee x_2 \cdot x_3 \vee x_2 \cdot \overline{x}_4 \vee \overline{x}_1 x_2 \\
= {}&x_1 \cdot x_3 \vee x_2 \cdot x_3 \vee x_2 \cdot \overline{x}_4 \vee \overline{x}_1 \cdot x_2 \\
= {}&\underbrace{x_3 \cdot (x_1 \vee x_2) \vee x_2 \cdot (\overline{x}_1 \vee \overline{x}_4)}_{f_2}
\end{aligned}
$$

4.3.2 Boolean functions

If a variable y depends on a variable x in such a way that each value of x determines exactly one value of y, then y is said to be a *function* of x. In terms of the inputs and outputs of some device, the function f is a rule that associates a unique output (exactly one) with each input. If the input is denoted by x, then the output is denoted by $f(x)$, or simply f. Thus, a function cannot assign two different outputs to the same input.

If $x_1, x_2, \ldots, x_n$ are variable inputs, there is a rule that assigns a unique value of f to given value of $x_1, x_2, \ldots, x_n$, and f is called a function of $x_1, x_2, \ldots, x_n$, $f = f(x_1, x_2, \ldots, x_n)$. The set of all possible inputs $x_1, x_2, \ldots, x_n$ is called the *domain* of f, and the set of outputs (resulting from varying $x_1, x_2, \ldots, x_n$ over the domain) is called the *range* of f.

Functions can be combined to define new functions. For example, the output of one function can be connected to the input of another. If the first function is f_1 and the second is f_2, then the composed function can be written as $f_1 \circ f_2$.

Functions can be described numerically by tables, geometrically by graphs, algebraically by formulas, or, in order to visualize the implementation of a Boolean function in terms of logic gates, logic diagrams. A logic diagram is a *visual syntax* for describing a logic network as an interconnection of logic gates, each of which emphasizes one particular input-to-output transformation of a logic signal.

A Boolean function described by an algebraic expression consists of

- Boolean (binary) variables,
- The constants 0 and 1, and
- The logic operation symbols.

Each of the variables appearing in the Boolean formula is assigned a value from the set $\{0, 1\}$ and evaluation is done using the operations of Boolean algebra.

Boolean function
When a Boolean formula is evaluated for all possible *assignments* of values to the variables, the set of pairs (assignment, value) is a *Boolean function.* For a given value of the binary variables, the function can be equal to either 1 or 0.

Example 4.8 (Boolean formulas.) *Boolean formulas can be evaluated using the following techniques:*
(a) Given the assignment $(x_1 = 1, x_2 = 0, x_3 = 1)$*, the value of the Boolean formula* $f = x_1 \vee \overline{x}_3 x_2 \vee x_2 \overline{x}_1$ *is calculated as follows:* $f = 1 \vee \overline{1} \cdot 0 \vee 0 \cdot \overline{1} = 1 \vee 0 \vee 0 = 1$.
(b) The Boolean formula $f = x_1 \vee \overline{x}_2 x_3$ *is equal to 1 if* x_1 *is equal to 1 or if both* x_2 *and* x_3 *are equal to 1. Formula* f *is equal to 0 otherwise.*

Practice problem 4.3. (Boolean formulas.) Compute the Boo-lean formula $f = \overline{x}_1 \vee \overline{x}_2 \vee \overline{x}_3$ given the assignment $x_1 = x_2 = 1$ and $x_3 = 0$.
Answer is given in "Solutions to practice problems."

Boolean functions derived from formulas are called *algebraic* representations. Operations on Boolean variables such as Boolean sum, Boolean product, and complement can also be applied to the functions.

Example 4.9 (Algebraic form.) *If* $f_1 = x_1 x_2$ *and* $f_2 = \overline{x}_1 \overline{x}_2$*, then* $f = f_1 \vee f_2 = x_1 x_2 \vee \overline{x}_1 \overline{x}_2$.

4.4 Fundamentals of computing Boolean functions

There are many forms that allow one each Boolean function to be described using algebraic representations. The efficiency of computing Boolean functions depends on the form of their representation.

4.4.1 Boolean literals and terms

The elementary groups of variables are Boolean terms:

$$\underbrace{\texttt{Boolean variables}}_{\textit{Boolean function}} \xrightarrow{\textit{Grouping}} \underbrace{\texttt{Boolean terms}}_{\textit{Boolean function}}$$

The terms are composed of literals and are characterized as follows:

Literals, product terms, and sum terms

- A Boolean *literal* is either an uncomplemented or a complemented variable x_i. The i-th literal is denoted by $x_i^{c_i}$: where $x_i^0 = \overline{x}_i$ for $c_i = 0$, and $x_i^1 = x_i$ for $c_i = 1$.
- A Boolean *product term* or product, is either a single literal, or the AND (product) of literals, that is, literals connected by the AND operation: $x_1^{c_1} x_2^{c_2} \dots x_k^{c_k}$.
- A Boolean *sum term*, or sum is either a single literal, or the OR (sum) of literals: $x_1^{c_1} \vee x_2^{c_2} \vee \dots \vee x_k^{c_k}$.

A Boolean function can be described using these groups of variables in various combinations. In determining the complexity of Boolean equations, one of the measures is the number of literals.

Example 4.10 (Literals and terms.) *The Boolean function f of five variables is described by the two terms: $f = \underbrace{(x_1 x_2 \overline{x}_3 x_4 x_5)}_{\textit{Product term}} \underbrace{(\overline{x}_1 \vee x_2 \vee \overline{x}_3 \vee x_4)}_{\textit{Sum term}}$. The product and sum terms include five and four literals, respectively.*

4.4.2 Minterms and maxterms

The Boolean variables can be grouped into a *minterm*, and a *maxterm*.

Minterms

A *minterm* of k variables $x_1, x_2, \dots, x_k$ is a Boolean *product* of m literals, in which each variable appears exactly once in either true, x_i, or complemented form, $\overline{x}_i$, but not both. A minterm has the following properties:

> **Properties of a minterm**
> - Each minterm has a value of 1 for exactly one combination of values of the variables.
> - There are 2^k minterms of k variables $x_1, x_2, \dots, x_k$

Example 4.11 (Minterms.) *(a) Four minterms can be constructed from two variables x_1 and x_2: x_1x_2, $\overline{x}_1x_2$, $x_1\overline{x}_2$, and $\overline{x}_1\overline{x}_2$. (b) For a Boolean function of four variables x_1, x_2, x_3 and x_4, the products $x_1\overline{x}_2x_3x_4$ and $x_1\overline{x}_2x_3\overline{x}_4$ are minterms, but the product $x_1\overline{x}_2x_4$ is not a minterm, since it lacks the variable x_3.*

Maxterms

A maxterm of m variables $x_1, x_2, \dots, x_m$ is a Boolean *sum* of m literals in which each variable appears exactly once in either true, x_i, or complemented form, $\overline{x}_i$, but not both. A maxterm has the following properties:

> **Properties of a maxterm**
> - Each maxterm has a value of 0 for exactly one combination of values of the variables.
> - There are 2^m maxterms of m variables $x_1, x_2, \dots, x_m$

Example 4.12 (Maxterms.) *(a) Four maxterms can be constructed from the two variables x_1 and x_2: $(x_1 \vee x_2)$, $(\overline{x}_1 \vee x_2)$, $(x_1 \vee \overline{x}_2)$, and $(\overline{x}_1 \vee \overline{x}_2)$. (b) For a Boolean function of four variables x_1, x_2, x_3 and x_4, the sums $\overline{x}_1 \vee \overline{x}_2 \vee x_3 \vee x_4$ and $x_1 \vee \overline{x}_2 \vee x_3 \vee \overline{x}_4$ are maxterms, but the sum $\overline{x}_1 \vee \overline{x}_2 \vee x_3$ is not a maxterm, since it lacks the variable x_4.*

Practice problem 4.4. (Minterms and maxterms.) Given a Boolean function of three variables, find which of the following are minterms and maxterms: $x_1 \vee x_2$, $x_1x_2\overline{x}_3$, $\text{-}x_1 \vee \overline{x}_2 \vee x_3$, $\overline{x}_2x_3$.
Answer is given in "Solutions to practice problems."

4.4.3 Canonical SOP and POS expressions

Canonical, or *standard*, *sum-of-products* (SOP) and *product-of-sums* (POS) are expressions that consist of minterms and maxterms only, respectively:

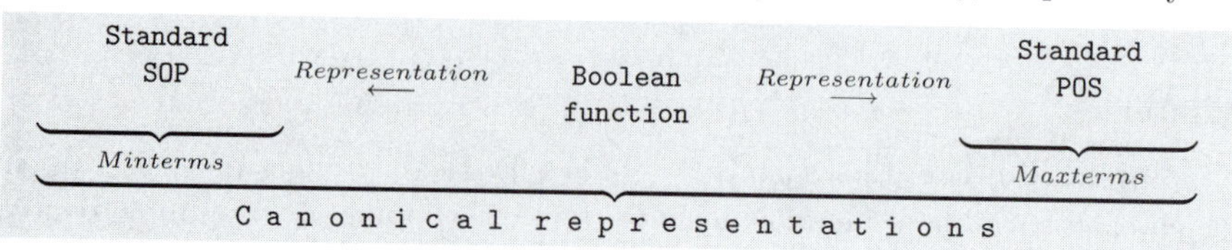

Given a Boolean function f of n variables, each product and sum term in these representations contains exactly n variables; the numbers of product and sum terms depends on the particular function.

Standard SOP representation of a Boolean function

Given: `A Boolean function of` n `variables`
Step 1: `Assume an assignment` $c_1c_2...c_n$ `to the` n `variables, and find the corresponding value of the function. Repeat for all` 2^n `possible assignments of variables.`
Step 2: `Select those assignments for which the function assumes a value of 1, and derive the product of` n `variables according to the rule` $x_1^{c_1}x_2^{c_2}\dots x_n^{c_n}$, `where` $x_i^0 = \overline{x}_i$ `for` $c_i = 0$, `and` $x_i^1 = x_i$ `for` $c_i = 1$.
Step 3: `Assemble the sum-of-products expression by combining the minterms using logical sum.`
Output: `The standard SOP expression`

Standard POS representation of a Boolean function

Given: `A Boolean function of` n `variables`
Step 1: `Assume an assignment` $c_1c_2...c_n$ `to the` n `variables, and find the corresponding value of the function. Repeat for all` 2^n `possible assignments of variables.`
Step 2: `Select those assignments for which the function assumes a value of 0, and derive the sum of` n `variables according to the rule` $x_1^{c_1} \vee x_2^{c_2} \vee \dots \vee x_k^{c_k}$, `where` $x_i^0 = \overline{x}_i$ `for` $c_i = 0$, `and` $x_i^1 = x_i$ `for` $c_i = 1$.
Step 3: `Assemble the product-of-sums expression by combining the maxterms using logical product.`
Output: `The standard POS expression`

Example 4.13 (Standard SOP and its forms.) *Given the Boolean function* $f = x_1x_2 \vee x_3$, *the standard SOP expression is derived as follows:*

$$\begin{aligned} f &= x_1x_2(\overline{x}_3 \vee x_3) \vee (\overline{x}_1 \vee x_1)(\overline{x}_2 \vee x_2)x_3 \\ &= x_1x_2\overline{x}_3 \vee x_1x_2x_3 \vee \overline{x}_1\overline{x}_2x_3 \vee \overline{x}_1x_2x_3 \vee x_1\overline{x}_2x_3 \vee x_1x_2x_3 \\ &= \overline{x}_1\overline{x}_2x_3 \vee x_1\overline{x}_2x_3 \vee \overline{x}_1x_2x_3 \vee x_1x_2\overline{x}_3 \vee x_1x_2x_3 \end{aligned}$$

The standard POS expression is derived using the following manipulations:

$$\begin{aligned} f &= (x_1 \vee x_3)(x_2 \vee x_3) = (x_1 \vee \overline{x}_2x_2 \vee x_3)(\overline{x}_1x_1 \vee x_2 \vee x_3) \\ &= (x_1 \vee \overline{x}_2 \vee x_3)(x_1 \vee x_2 \vee x_3)(\overline{x}_1 \vee x_2 \vee x_3)(x_1 \vee x_2 \vee x_3) \\ &= (x_1 \vee \overline{x}_2 \vee x_3)(\overline{x}_1 \vee x_2 \vee x_3)(x_1 \vee x_2 \vee x_3) \end{aligned}$$

The standard SOP and POS representations are dual:

$$\underbrace{\texttt{Standard SOP representation}}_{Boolean\ function} \overset{Duality}{\longleftrightarrow} \underbrace{\texttt{Standard POS representation}}_{Boolean\ function}$$

Example 4.14 (Standard SOP and POS forms.) *Given the standard POS expression* $f = (x_1 \vee \overline{x}_2 \vee x_3)(\overline{x}_1 \vee x_2 \vee x_3)(x_1 \vee x_2 \vee x_3)$*, its dual is a standard SOP* $f = x_1\overline{x}_2x_3 \vee \overline{x}_1x_2x_3 \vee x_1x_2x_3$*. Note that* $f \neq g$*.*

Any Boolean function can be expressed in standard SOP and POS form.

Example 4.15 (Majority function.) *A Boolean function* f*, called the majority function, is expressed in the standard SOP and POS forms as follows:*

$$\begin{aligned} f &= \overbrace{\overline{x}_1x_2x_3 \vee x_1\overline{x}_2x_3 \vee x_1x_2\overline{x}_3 \vee x_1x_2x_3}^{\textit{The SOP form}} \\ &= \underbrace{(x_1 \vee x_2 \vee x_3)\ (x_1 \vee x_2 \vee \overline{x}_3)\ (x_1 \vee \overline{x}_2 \vee x_3)\ (\overline{x}_1 \vee x_2 \vee x_3)}_{\textit{The POS form}} \end{aligned}$$

Let m_i and M_j be the i-th minterm and j-th maxterm, respectively, for a Boolean function f of n variables, where $i, j \in \{0, 1, 2, \ldots, 2^n - 1\}$. The short hand notation of an SOP form of the Boolean function is

SOP form of the Boolean function

$$\texttt{Minterm form} = \bigvee_i m(i)$$

where $\bigvee$ denotes the Boolean sum. The short hand notation of a POS form of the Boolean function is

POS form of the Boolean function

$$\texttt{Maxterm form} = \prod_j M(j)$$

where $\prod$ denotes the Boolean product.

Example 4.16 (SOP and POS expressions.) *A Boolean function* f *of two variables* x_1 *and* x_2 *given the minterms* $m_0 = \overline{x}_1\overline{x}_2$ *and* $m_3 = x_1x_2$*,* $i = 0, 3$*, and maxterms* $M_1 = x_1 \vee \overline{x}_2$ *and* $M_2 = \overline{x}_1 \vee x_2$*,* $j = 1, 2$*. The minterm and maxterm expressions are as follows:*

$$f = \underbrace{\bigvee_{i=0,3} m(0,3) = \overbrace{\overline{x}_1\overline{x}_2}^{m_0} \vee \overbrace{x_1x_2}^{m_3}}_{\textit{Minterm expression}} = \underbrace{\prod_{j=1,2} M(1,2) = \overbrace{(x_1 \vee \overline{x}_2)}^{M_1}\overbrace{(\overline{x}_1 \vee x_2)}^{M_2}}_{\textit{Maxterm expression}}$$

Practice problem 4.5. **(SOP and POS expressions.)** Write a Boolean function of two variables represented by (a) the standard SOP of three minterms, (b) the standard POS of three maxterms.
Answer is given in "Solutions to practice problems."

4.4.4 Algebraic construction of standard SOP and POS forms

The standard SOP expression can be derived from the SOP form of a Boolean function by multiplying by the sum $x \vee \overline{x} = 1$.

Example 4.17 (SOP expression.) *A Boolean function of three variables is given in the SOP form* $f = x_1x_3 \vee x_2x_3 \vee \overline{x}_1\overline{x}_2\overline{x}_3$. *The missing variables in the first two product terms are added as follows:* $f = x_1 \underbrace{(x_2 \vee \overline{x}_2)}_{Equal\ to\ 1} x_3 \vee \underbrace{(x_1 \vee \overline{x}_1)}_{Equal\ to\ 1} x_2x_3 \vee \overline{x}_1\overline{x}_2\overline{x}_3$.
This does not alter the equation because $x_1 \vee \overline{x}_1 = 1$ *and* $x_2 \vee \overline{x}_2 = 1$. *The standard SOP expression is obtained by using the distributive law and by deleting any duplicated terms produced:* $f = x_1\overline{x}_2x_3 \vee x_1x_2x_3 \vee \overline{x}_1x_2x_3 \vee \overline{x}_1\overline{x}_2\overline{x}_3$.

The standard POS expression can be derived from the POS form of a Boolean function by adding the product $x\overline{x} = 0$.

Example 4.18 (POS expression.) *A Boolean function of three variables is given in the POS form* $f = x_1(x_2 \vee \overline{x}_3)$. *The missing variables* x_2 *and* x_1 *in the terms are added using the rule* $x_2\overline{x}_2 = 0$ *and* $x_1\overline{x}_1 = 0$ *to get the standard POS form:*

$$
\begin{aligned}
f &= (x_1 \vee x_2\overline{x}_2 \vee x_3\overline{x}_3)(x_1\overline{x}_1 \vee x_2 \vee x_3) \\
&= (x_1 \vee x_2\overline{x}_2 \vee x_3)(x_1 \vee x_2\overline{x}_2 \vee \overline{x}_3)(x_1 \vee x_2 \vee x_3)(\overline{x}_1 \vee x_2 \vee x_3) \\
&= \underbrace{(x_1 \vee x_2 \vee x_3)}_{Maxterm\ 0}\underbrace{(x_1 \vee \overline{x}_2 \vee x_3)}_{Maxterm\ 2}\underbrace{(x_1 \vee x_2 \vee \overline{x}_3)}_{Maxterm\ 1}\underbrace{(x_1 \vee \overline{x}_2 \vee \overline{x}_3)}_{Maxterm\ 3}\underbrace{(\overline{x}_1 \vee x_2 \vee x_3)}_{Maxterm\ 4}
\end{aligned}
$$

4.5 Proving the validity of Boolean equations

Proving the validity of Boolean data structures is widely used in logic design, in particular, for verification. The *verification* of a Boolean data structure (logic network, decision diagram, network of threshold elements) is the determination of whether or not this data structure implements its specific function. Verification plays a vital role in preventing incorrect logic network designs from being manufactured and used. There are many techniques for the verification of Boolean data structures. Most of these techniques are based on finding the Boolean equations and proving their validity.

The following approaches are used, in particular, to determine if a Boolean equation is valid (Figure 4.4):

Approaches for determining if a Boolean equation is valid
(*a*) Construct a table of the values of the function (introduced in Chapter 6 as the truth table) and evaluate both sides of the equations for all combinations of values of the variables.
(*b*) Manipulate one side of the equation by applying various theorems until it is identical to the other side.
(*c*) Reduce both sides of the equation independently to the same expression.

To prove that an equation is not valid, it is sufficient to show one combination of values of the variables for which the two sides of the equation have different values.

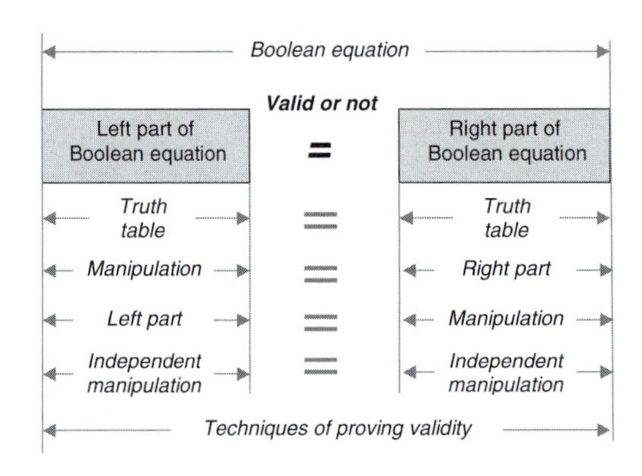

FIGURE 4.4
Techniques of proving the validity of Boolean equations.

Example 4.19 (Proving the validity.) *Prove that*

$$\underbrace{\overline{x}_1x_2\overline{x}_4 \vee x_2x_3x_4 \vee x_1x_2\overline{x}_3 \vee x_1\overline{x}_2x_4}_{Left\ part} = \underbrace{x_2\overline{x}_3\overline{x}_4 \vee x_1x_4 \vee \overline{x}_1x_2x_3}_{Right\ part}$$

One of the possible ways for proving this equation is to create truth tables for both parts. The left and right parts of the equation cover the minterms (correspond to the 1s in the table of the function's values of each part) as follows:

$$\overbrace{\overbrace{\overline{x}_1x_2\overline{x}_4}^{0100,\ 0110}\ \overbrace{x_2x_3x_4}^{0111,\ 1111}\ \overbrace{x_1x_2\overline{x}_3}^{1100,\ 1101}\ \overbrace{x_1\overline{x}_2x_4}^{1001,\ 1011}}^{Left\ part\ of\ the\ equation} = \overbrace{\overbrace{x_2\overline{x}_3\overline{x}_4}^{0100,\ 1100}\ \overbrace{x_1x_4}^{1001,\ 1011,\ 1101,\ 1111}\ \overbrace{\overline{x}_1x_2x_3}^{0110,\ 0111}}^{Right\ part\ of\ the\ equation}$$

These two sets of minterms correspond to the standard SOP expressions. These expressions are identical, and, thus, the truth tables for the left and right parts of the Boolean equation are identical. Proving this equation using algebraic manipulation is given in Example 8.11 (Chapter 8).

Practice problem 4.6. (Proving the validity.) Prove that the following Boolean equation is valid or invalid:

$$\overline{x}_1\overline{x}_2x_3 \vee x_1\overline{x}_2\overline{x}_3 \vee x_1x_2x_3 = (x_1 \vee x_3)(x_1 \vee \overline{x}_2)(\overline{x}_2 \vee x_3)(\overline{x}_1 \vee x_2 \vee \overline{x}_3)$$

Answer is given in "Solutions to practice problems."

The techniques for proving the validity of Boolean function is extended in Chapter 8 by using additional theorems for simplification and manipulation of Boolean data structures (see Table 8.1).

4.6 Gates

The fundamental principle of logic design is the *principle of assembly* of an arbitrary computing network from simple computing elements called *gates*. A gate is a module implementing a simple Boolean function such as AND, OR, etc. These gates are the basic building blocks of combinational modules (logic networks).

4.6.1 Elementary Boolean functions

The simplest Boolean functions are called *elementary*. In Figure 4.5, all $2^{2^n} = 2^{2^2} = 16$ Boolean functions of two variables are given. The following two-variable functions among these 16 are special:

Special Boolean functions of two variables

- The *constants* 0 ($f_0 = 0$) and 1 ($f_{15} = 1$) are the only two Boolean functions that do not possess any essential variables; in vector form, the Boolean constants 0 and 1 are $\mathbf{0} = [00\ldots0]^T$ and $\mathbf{1} = [11\ldots1]^T$, respectively.
- There are exactly four Boolean functions of a single variable: $f_3 = x_1$, $f_5 = x_2$, $f_{10} = \overline{x}_2$, and $f_{12} = \overline{x}_1$.
- The Boolean product is called the AND function, $f_1 = x_1x_2$. The NAND function is the complemented AND function, $f_{14} = \overline{x_1x_2}$.
- The Boolean sum is called the OR function, $f_7 = x_1 \vee x_2$. The NOR function is the complemented OR function, $f_8 = \overline{x_1 \vee x_2}$
- The exclusive OR, called the EXOR function, $f_6 = x_1 \oplus x_2$, and exclusive NOR, called the XNOR function $f_9 = \overline{x_1 \oplus x_2}$.

In Table 4.3, the symbolic representations called *logic gates*, as well as switch models for the gates, are given.

TABLE 4.3
Elementary Boolean functions of two variables and their implementation using gates and switches.

Techniques for representation of elementary Boolean functions

Function	Gate symbol	Truth table, truth vector	Switch model
AND $f = x_1 x_2$	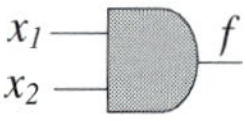 x_1, x_2, f	$x_1\ x_2 \mid f$: 0 0 | 0; 0 1 | 0; 1 0 | 0; 1 1 | 1 $\mathbf{F} = [0\ 0\ 0\ 1]^T$	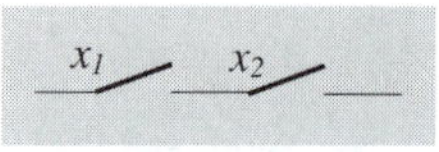 x_1 x_2
OR $f = x_1 \vee x_2$	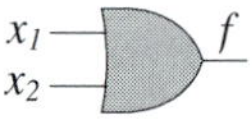 x_1, x_2, f	$x_1\ x_2 \mid f$: 0 0 | 0; 0 1 | 1; 1 0 | 1; 1 1 | 1 $\mathbf{F} = [0\ 1\ 1\ 1]^T$	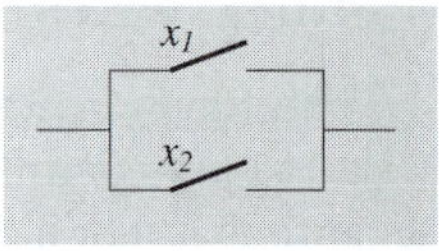 x_1 x_2
NAND $f = \overline{x_1 x_2}$	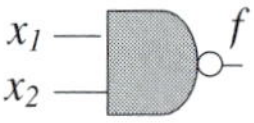 x_1, x_2, f	$x_1\ x_2 \mid f$: 0 0 | 1; 0 1 | 1; 1 0 | 1; 1 1 | 0 $\mathbf{F} = [1\ 1\ 1\ 0]^T$	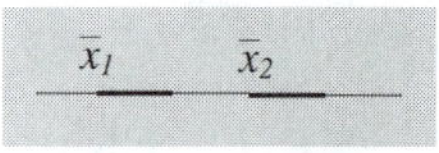 $\bar{x}_1$ $\bar{x}_2$
NOR $f = \overline{x_1 \vee x_2}$	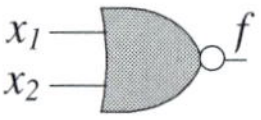 x_1, x_2, f	$x_1\ x_2 \mid f$: 0 0 | 1; 0 1 | 0; 1 0 | 0; 1 1 | 0 $\mathbf{F} = [1\ 0\ 0\ 0]^T$	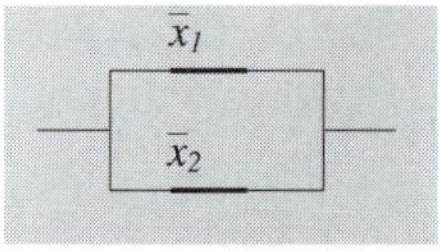 $\bar{x}_1$ $\bar{x}_2$
EXOR $f = x_1 \oplus x_2$	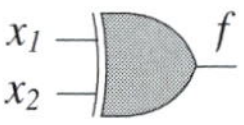 x_1, x_2, f	$x_1\ x_2 \mid f$: 0 0 | 0; 0 1 | 1; 1 0 | 1; 1 1 | 0 $\mathbf{F} = [0\ 1\ 1\ 0]^T$	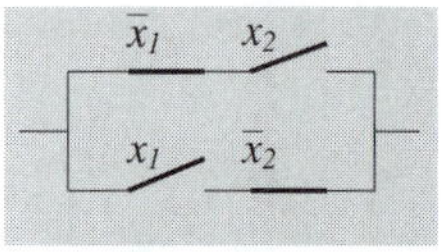 $\bar{x}_1$ x_2 x_1 $\bar{x}_2$
XOR $f = \overline{x_1 \oplus x_2}$	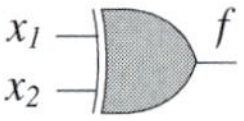 x_1, x_2, f	$x_1\ x_2 \mid f$: 0 0 | 1; 0 1 | 0; 1 0 | 0; 1 1 | 1 $\mathbf{F} = [1\ 0\ 0\ 1]^T$	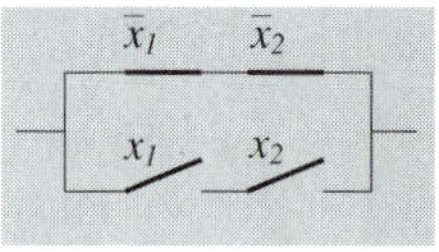 $\bar{x}_1$ $\bar{x}_2$ x_1 x_2

Design example: Boolean functions of two variables

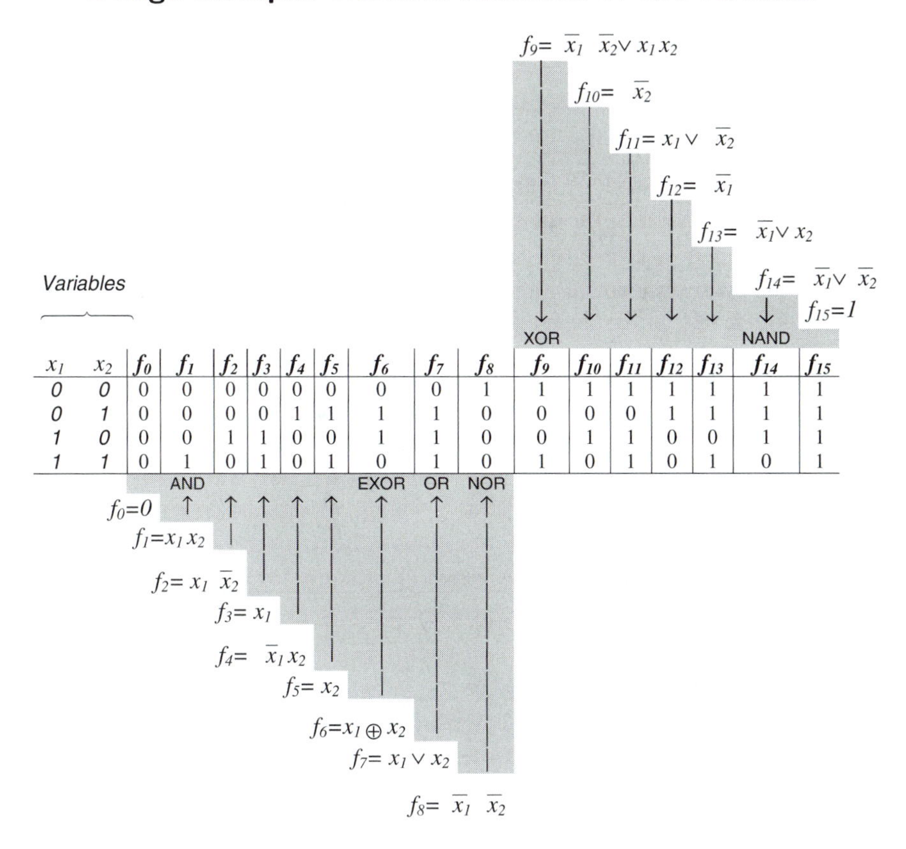

x_1	x_2	f_0	f_1	f_2	f_3	f_4	f_5	f_6	f_7	f_8	f_9	f_{10}	f_{11}	f_{12}	f_{13}	f_{14}	f_{15}
0	0	0	0	0	0	0	0	0	0	1	1	1	1	1	1	1	1
0	1	0	0	0	0	1	1	1	1	0	0	0	0	1	1	1	1
1	0	0	0	1	1	0	0	1	1	0	0	1	1	0	0	1	1
1	1	0	1	0	1	0	1	0	1	0	1	0	1	0	1	0	1

FIGURE 4.5
All 16 Boolean functions of two variables.

4.6.2 Switch models for logic gates

In logic design, combinational logic networks are compounded of switches. Switch models represent many possible physical switch mechanisms, for example, the flow of electrons or other information carriers. A switch is represented by three connection points called *terminals*, to which external signals may be applied or from which internal signals may be drawn. The lines attached to the terminals can represent wires, pipes, or other transmission media appropriate to the current technology. An *input terminal* allows a signal to enter the switch and change its state. An *output terminal* allows signals to leave the switch. The control variable x is applied to an input terminal of the switch. A terminal that can function both as an input and an output is said to be *bidirectional*. The data terminals of the switch models under consideration are assumed to be bidirectional.

The effect of the switch on the output is determined by the *state* of the switch. In Figure 4.6a, the switch has two states:

- `OPEN`, or `OFF`. This state implies that there is no closed path through the switch connecting data terminals.
- `CLOSED`, or `ON` state. This state implies that data terminals are connected via a path through the switch.

Using the switch shown in Figure 4.6a, switch-based models of elementary Boolean functions can be derived.

Example 4.20 (Switches.) *In Figure 4.6b, the switch-based model of the AND gate is shown.*

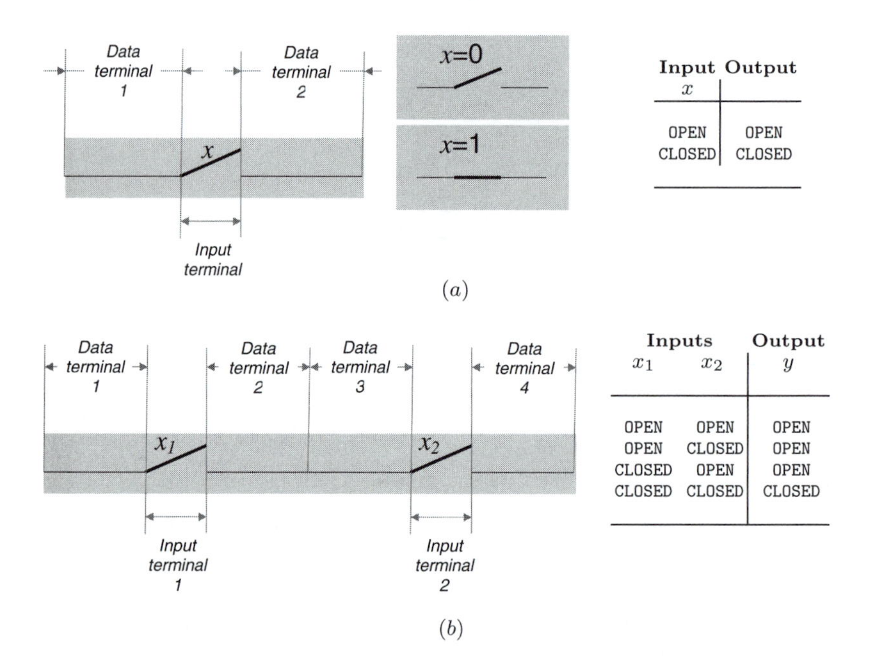

FIGURE 4.6
Model of the simplest switch (a) and switch-based models of the function AND (b) (Example 4.20).

In Table 4.2, the computation rules of Boolean algebra are interpreted using switches. These rules can be interpreted using gate notation (Table 4.4).

TABLE 4.4
The computation rules in terms of gates.

Techniques for manipulations of elementary Boolean functions

Computation rule	Gate-based interpretation
Idempotence $x \vee x = x$ $x \cdot x = x$	x x x x = x x x x = x x
Identity $x \vee 0 = x$ $x \vee 1 = 1$	x 0 x = x x x 1 1 = 1 1
Identity $x \cdot 1 = x$ $x \cdot 0 = 0$	x 1 x = x x x 0 0 = 0 0
DeMorgan's rule $\overline{x_1 \vee x_2} = \overline{x}_1 \overline{x}_2$	x_1 x_2 f = x_1 x_2 f
Distributivity $x_1(x_2 \vee x_3)$ $= x_1x_2 \vee x_1x_3$	x_2 x_3 x_1 f = x_1 x_2 x_1 x_3 f
Distributivity $x_1 \vee x_2x_3$ $= (x_1 \vee x_2)(x_1 \vee x_3)$	x_2 x_3 x_1 f = x_1 x_2 x_1 x_3 f

4.6.3 Digital waveforms and timing diagrams

The voltage used to represent a logical 1 and a logical 0 is called *voltage level.* Digital waveforms consist of voltage levels that are changing between the HIGH (logical 1) and LOW (logical 0) levels. A single *positive-going* pulse is generated when the voltage (or current) goes from the LOW level to the HIGH level and then back to the LOW level (Figure 4.7). By analogy, the *negative-going* pulse can be defined.

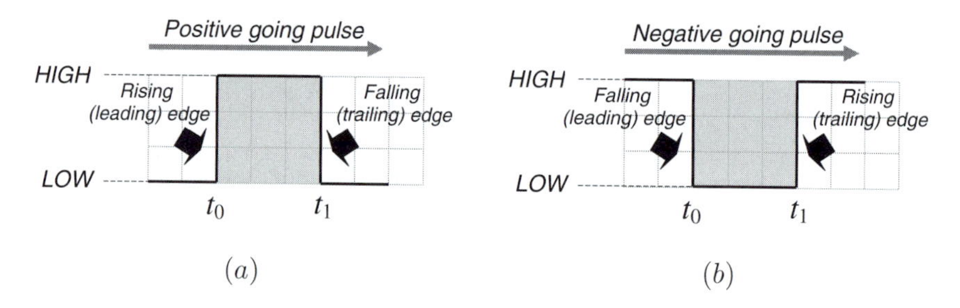

FIGURE 4.7
A positive-going (a) and negative-going (b) pulses.

The pulse has two edges: a *leading* edge and a *trailing* edge. For a positive-going pulse, the leading edge is a *rising* edge, and the trailing edge is a *falling* edge. The pulses in Figure 4.7 are ideal because the rising and falling edges are assumed to change in zero time. In practice, the rising and falling edges are characterized by the rise and fall times, respectively. The frequency f of a pulse waveform and the period T can be calculated as $f = 1/T$. The basic timing waveform in computing device is called the *clock.*

A *timing diagram* is a graphical representation of the relationship of two or more waveforms with respect to each other on a time basis. Using a timing diagram, it is possible to determine the logical level of a digital device at any specified point in time.

> **Example 4.21 (Timing diagram.)** *The inverter (NOT gate) performs the operation inversion, or complement. The inverter changes one logical level to the opposite level, that is, it changes 1 to 0 and 0 to 1. The time relationship of the output pulse to the input pulse for the inverter is shown in Figure 4.8.*

Timing diagrams for the primary logic gates are given in Table 4.5.

TABLE 4.5
Timing diagrams of the primary logic gates.

Techniques for timing description of logic gates

Function	Gate symbol	Timing diagram
AND $f = x_1 x_2$	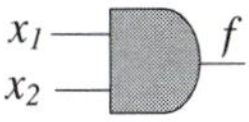	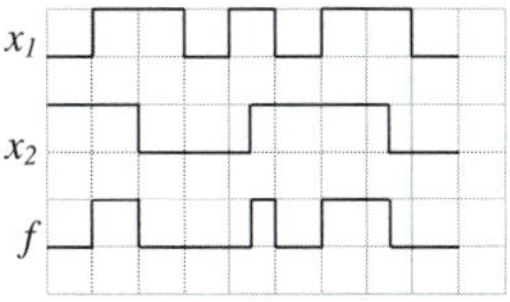
OR $f = x_1 \vee x_2$	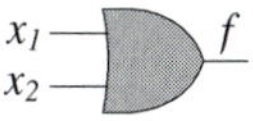	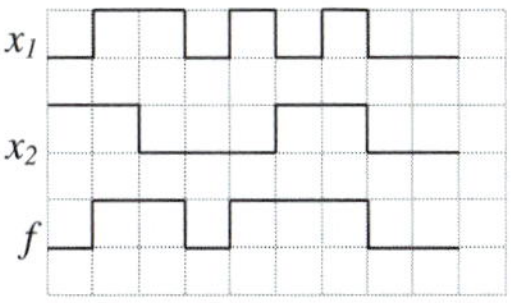
NAND $f = \overline{x_1 x_2}$	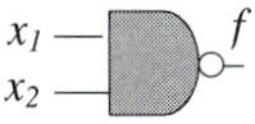	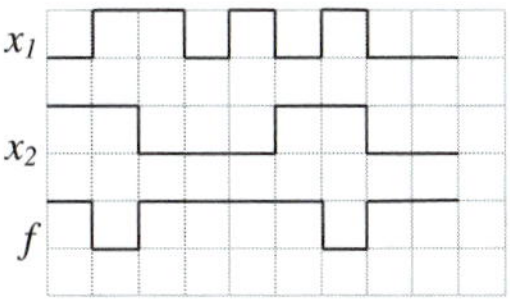
NOR $f = \overline{x_1 \vee x_2}$	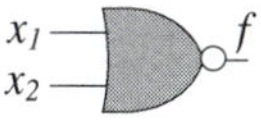	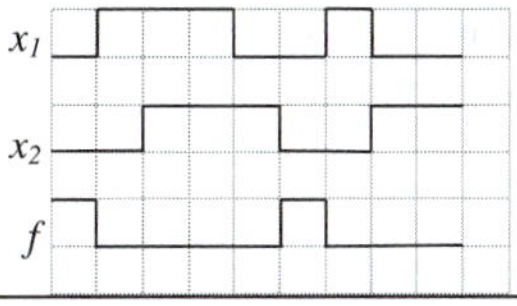
EXOR $f = x_1 \oplus x_2$	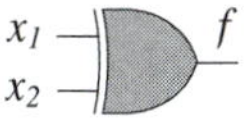	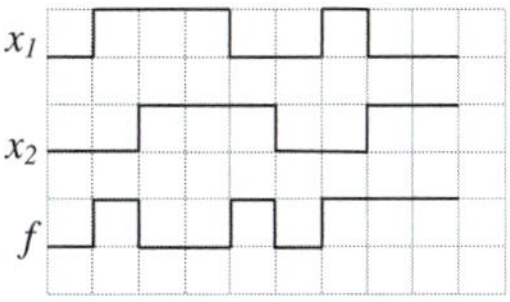
XOR $f = \overline{x_1 \oplus x_2}$	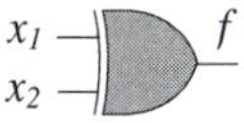	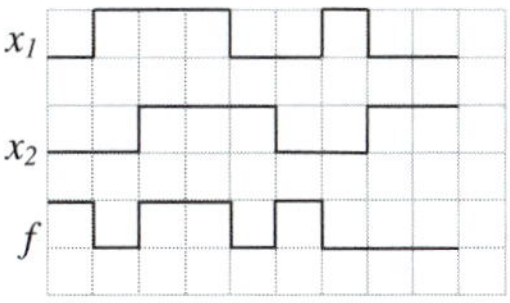

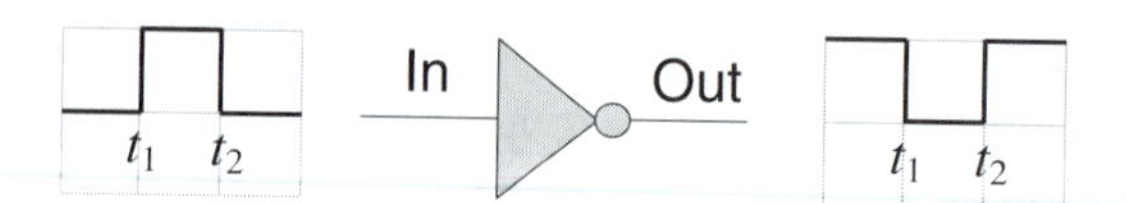

FIGURE 4.8
Timing diagram of the inverter (Example 4.21).

4.6.4 Performance parameters

The performance of a logic gates are the switching activity measured in terms of the propagation delay time and the power dissipation. The *propagation delay time* is a result of the limitation of *switching activity*. The shorter the propagation delay, the higher the speed of gate and higher the frequency at which it can operate. Propagation delay time of a logic gate is the time interval between the application of an input pulse and the occurrence of the output pulse. There are two measurements of propagation delay time (Figure 4.9): (a) the time between a specified point on the input pulse and the corresponding reference point on the output pulse, with the output changing from the HIGH level to the LOW level (t_{PHL}); and (b) the time between a specified point on the input pulse and the corresponding reference point on the output pulse, with the output changing from the LOW level to the HIGH level (t_{PLH}).

The propagation delay time is used to evaluate the speed of the logic devices, and, therefore, the maximum operating frequency. The later can not be exceeded since it can lead to violation of the timing for the surrounding logic and, therefore, incorrect operation of the devices.

Design example: Delay time measurements

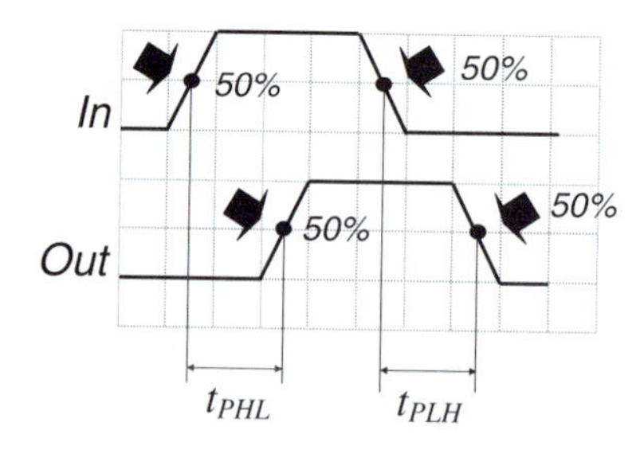

FIGURE 4.9
Measurements of propagation delay time.

4.7 Local transformations

A *local transformation* is defined as a set of rules for the simplification of a logic network. These rules are based on the theorems of Boolean algebra and polynomial algebra $GF(2)$ and applied locally. A local transformation is characterized as follows below:

Properties of a local transformations

Property 1: *Low complexity.* In contrast to the approaches that require the knowledge of the implemented Boolean function, the local transformations can be applied to the sub-functions; these local subfunctions are very simple.

Property 2: *Technological requirements.* Using a local transformations, it is easy to satisfy the requirements given technology.

Property 3: *Automation.* In automation logic design tools, the consequences of a local transformation can not be predicted, but a functionally equivalent logic networks can be generated.

Property 4: *Flexibility of optimization.* The cost of the obtained network is evaluated, in particular, in terms of the number of logic gate or connections, and a solution with lowest cost can be selected as the final design.

Property 5: *Applicable to various data structures.* Local transformations can be applied to logic networks, decision trees and diagrams.

Property 6: *Testability.* Local transformations can improve the conditions for testability.

Property 7: *Decomposability.* Local transformations can improve the conditions for decomposability.

The data structure that is obtained by applying the local transformations must be verified. For example, the functionality of logic network must be verified by comparison with an initial logic network (before application of local transformations). Local transformations for the logic networks using AND, OR, gates, and inverters, include the rules described below:

The rules for local transformations of a logic network

Rule 1: *Reduction of constants:*

- Remove a constant 1 that is connected to an AND gate
- Remove a constant 0 that is connected to an OR gate
- Replace an AND gate that has constant 0, with a constant 0
- Replace an OR gate that has constant 1, with a constant 1

Rule 2: *Reduction of duplicated gates:* if there two gates whose inputs and outputs are the same, remove one, and create a fan-out

Rule 3: *Reduction of inverters:* remove two inverters that are connected in series

Rule 4: *Deletion of unused gates:* remove gates whose outputs are not connected to other gates or terminals

Rule 5: *Merging gates:* if two AND gates (OR gates) are connected in series, then merge them into one

Example 4.22 (Local transformations.) *Figure 4.10 illustrates two types of the local transformations. Local transformation in the area A: the OR gate is replaced by the wire using the identity rule for variables and constants (Table 4.4)* $x_2 \vee 1 = 1$. *Local transformation in the area B: the AND gate is replaced by the wire based on the identity rule (Table 4.4)* $x_1 \cdot 1 = x_1$.

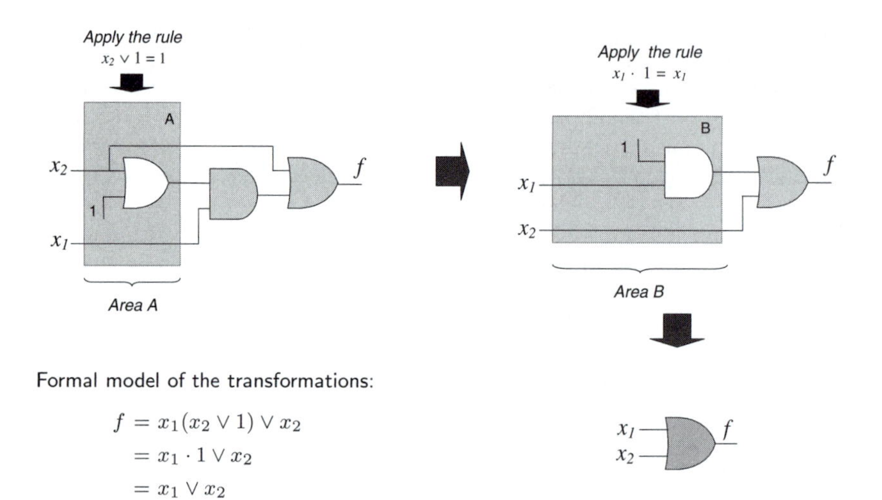

FIGURE 4.10
Local transformations in logic network (Example 4.22).

Practice problem **4.7. (Local transformations.)** Apply the local transforms to the areas A and B of a logic network given in Figure 4.11
Answer is given in "Solutions to practice problems."

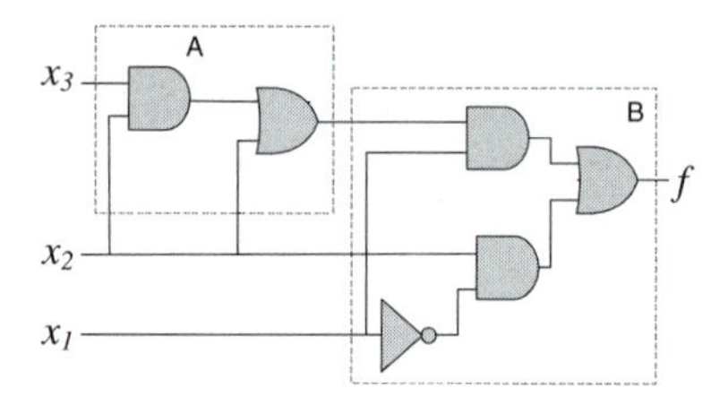

FIGURE 4.11
Logic network for the practice problem 4.7.

4.8 Summary of Boolean algebra

This chapter introduces the theoretical basis for logic design, ***Boolean algebra***. A ***universal*** algebra consists of a set of elements and operations on this set. Boolean algebras are particular cases of the universal algebra. There is an infinite number of different Boolean algebras. The simplest one is a ***two-valued*** Boolean algebra defined over the two-element set $B = \{0, 1\}$.

(*a*) Boolean algebra is the theoretical basis for ***Boolean data structures***. A ***Boolean formula*** is a ***Boolean function*** after specification of values, given assignments of variables. The relationship between Boolean formulas and Boolean functions is not one-to-one: Many different formulas can represent the same Boolean function. A Boolean function described by an algebraic expression consists of ***binary variables***, the ***constants*** 0 and 1, and the ***logic operation*** symbols.

(*b*) The fundamental representations of Boolean functions are ***sum-of-products (SOP)*** and ***product-of-sums (POS)*** expressions.

The key statements of this chapter are the following:

- The basic requirements of any algebra employed to describe and manipulate Boolean functions include:
  (*a*) ***Functional completeness*** (capability of the algebra to describe and manipulate an arbitrary Boolean function).
  (*b*) ***Flexibility*** (amenability of the algebra to manipulating Boolean functions so that design algorithms can be set up and employed with reasonable ease).
  (*c*) ***Implementability*** (the basic connectives and functional operations should have simple and reliable physical logic circuit counterparts).
- ***Axioms*** define the basic properties of Boolean algebra, from which other properties can be deduced. ***Huntingtons postulates*** provide a complete and concise set of axioms for Boolean algebra. ***Theorems*** are true statements that can be deduced from axioms.
- The fundamental interchangeability property of Boolean algebra is termed the ***principle of duality***: Whenever we derive any result, we can be sure that the result has a dual.

Summary (continuation)

- The basic description of the Boolean function includes:
  - (a) A Boolean ***literal***, or simple literal, is either an uncomplemented or a complemented variable x_i. The i-th literal is denoted by $\boxed{x_i^{c_i}}$: when $c_i = 0$, $x_i^0 = \overline{x}_i$, and when $c_i = 1$, $x_i^1 = x_i$.
  - (b) A Boolean ***product term*** or simple product is either a single literal or the AND (product) of literals: $x_1^{c_1} x_2^{c_2} \dots x_k^{c_k}$.
  - (c) A Boolean ***sum term***, or simple sum, is either a single literal or the OR (sum) of literals: $x_1^{c_1} \vee x_2^{c_2} \vee \dots \vee x_k^{c_k}$.
- The product and sum terms of n variables, given a Boolean function of n variables, are called the ***minterm*** and ***maxterm***, respectively.
- A minterm has the following properties:
  - (a) A minterm of k variables $x_1, x_2, \dots, x_k$ is a Boolean ***product*** of m literals in which each variable appears exactly once in either true x_i or complemented $\overline{x}_i$ form, but not both (a literal is a variable or its complement).
  - (b) Each minterm has a value of 1 for exactly one combination of values of the variables.
  - (c) There are 2^k minterms of k variables $x_1, x_2, \dots, x_k$.
- A maxterm of m variables $x_1, x_2, \dots, x_m$ is a Boolean ***sum*** where every variable x_i appears as x_i or $\overline{x}_i$. There are 2^m maxterms.
- A maxterm has the following properties:
  - (a) A maxterm of m variables $x_1, x_2, \dots, x_m$ is a Boolean sum of m literals in which each variable appears exactly once in either true x_i or complemented $\overline{x}_i$ form, but not both.
  - (b) Each maxterm has a value of 0 for exactly one combination of values of the variables.
  - (c) There are 2^m maxterm of m variables $x_1, x_2, \dots, x_m$.
- ***Sum-of-products (SOP)*** and ***product-of-sums (POS)*** expressions consist of minterms and maxterms, respectively. Given a Boolean function f of n variables,
  - (a) It can be represented in an SOP and/or POS form;
  - (b) Each product and sum term in this representation contains exactly n variables;
  - (c) The numbers of product and sum terms are different for a given function and depend on the particular function.

Summary (continuation)

- Proving the validity of two Boolean data structures is a typical problem of logic design. ***Verification*** of a Boolean data structure (logic network, decision diagram, network of threshold elements) is the determination of whether or not this data structure implements its specific function. Verification plays a vital role in preventing incorrect logic network designs from being manufactured and used.
- The universal logic network component is a ***switch***. In models of switches, 0 and 1 signals are used, and the physical details of real switches, which are based on various physical and chemical phenomena, are ignored. These models are ***technology-independent***.
- ***Logic gates*** are elementary logic networks that perform simple Boolean operations.
- A ***central problem*** in logic design is to obtain a circuit that realizes a given Boolean function using gates from a given set of gates, and has the lowest possible cost in terms of the number of gates.
- ***Physical constraints*** limit both the size and speed of real circuits in important ways, of which the designer must be aware. This places an upper bound on the number of input sources that may supply a gate, ***fan-in***, and places another bound on the maximum number of output devices that may be supplied with signals, ***fan-out***.
- ***Local transformation*** is defined as a set of rules for the simplification of a logic network. These rules are applied locally. Local transformations are used to simplify logic networks under technological constraints. It is also called sometimes a *Boolean matching*.
- For advances in Boolean algebra, we refer the reader to the "Further study" section.

4.9 Further study

Historical perspective

1839: Augustus DeMorgan formulated rules for expressing the negation of compound statements formed with AND and OR:

The negation of ''A AND B'' is the assertion ''Not(A) OR Not(B)''
The negation of ''A OR B'' is the assertion ''Not(A) AND Not(B)''

These laws are known as *DeMorgan's laws* in Boolean algebra. But his main contribution was in the theory of syllogisms, where he recognized that relational inferences were the core of mathematical inference and scientific reasoning. His logical papers are published in a series entitled "On the Syllogism". Augustus DeMorgan (1806–1871), together with George Boole (1815–1864), helped to make England a leading center of logic in the nineteenth century.

1847: George Boole introduced in his work "An Investigation of the Laws of Thought" a systematic treatment of logic, and developed for this purpose an algebraic system now called *Boolean algebra.* Boole's goal was essentially the development of a formalism to compute the truth or falsehood of complex compound statements from the truth values of their component statements. George Boole tied logic and mathematics together. Boole expressed this vision in an inaugural address when he was dean of the faculty:

"I speak here not of the mathematics of number and quantity alone, but a mathematics in its larger, and I believe, truer sense, as universal reasoning expressed in symbolical forms, and conducted by laws, which have their ultimate abode in the human mind. That such a science exists is simply a fact, and while it has one development in the particular science of number and quantity, it has another in a perfect logic".

1904: The formal definition of Boolean algebra is set forth by E. V. Huntington (E. V. Huntington, "Sets of Independent Postulates for the Algebra of Logic," Transactions of American Math. Soc., volume 5, pages 288–309, 1904), known as the *Huntington* postulates.

1913: H. M. Sheffer presented a set of postulates using the operation "|" that is NAND operation (H. M. Sheffer, "A set of Five Independent Postulates for Boolean Algebra with Applications to Logic Constants," *Transactions of American Math. Soc.*, volume 14, pages 481–488, 1913). The operation $x_1|x_2 = \overline{x_1 x_2}$ often called *Sheffer stroke.* For example,

$$\overline{x} = x|x, \quad x_1x_2 = \overline{x_1|x_2} = (x_1|x_2)|(x_1|x_2), \quad x_1 \vee x_2 = \overline{x}_1|\overline{x}_2 = (x_1|x_1)|(x_2|x_2)$$

1920: E. L. Post developed an approach to forming complete sets of elementary Boolean functions (primitives) by which it is possible to represent an arbitrary Boolean function. This approach is known as the *Post theorem.* E. L. Post also proposed the algebra of multiple-valued logic. The multi-valued counterparts of binary AND, OR and NOT, called the multi-valued *conjunction, disjunction* and *cycling* operators. The other basis, *modulo 2*, was extended by B. A. Bernstein in 1928 by introducing *modulo m* addition and multiplication of integers *modulo m*.

1937: Claude Shannon introduced in his master's thesis and later, in 1938 in the paper "A Symbolic Analysis of Relay and Switching Circuits," a two-valued algebra called *switching algebra*, in which he demonstrated that the properties of bistable electromechanical circuits can be represented by this algebra. Shannon recognized that the algebra developed by G. Boole to facilitate the study of the logic of the English language could be applied directly to relay switching networks.

1965: Lotfi Zadeh at the University of California, Berkeley, introduced the *fuzzy logic*. It was derived from fuzzy set and possibility theory, dealing with reasoning that is approximate rather than precisely deduced from classical predicate logic. Fuzzy logic operates with degrees of truth rather than probabilities. Fuzzy truth represents membership in vaguely defined sets. The set membership values range between 0 and 1, and have their counterparts in linguistic forms such as the imprecise statements "very," "quite," and "slightly."

Advanced topics of Boolean algebra

Topic 1: Multi-valued logic. Boolean algebra is the simplest algebra. The motivation for its implementation is that two-valued Boolean algebra is suited to the two-state physical phenomena of today's technology. However, multi-valued algebras have been the focus of interest for many years.

Multi-valued data structures are the generalization of Boolean data structures for multi-valued logic. Two-valued Boolean algebra is the dominating algebra, since it is suited to the two-state physical phenomena of today's technology. However, multi-valued Boolean algebras were the focus of interest for many years. Multi-valued algebra is based upon a set of m elements $M = \{0, 1, 2, \dots, m\}$, corresponding to multi-valued signals, and the corresponding operations. This means that multi-valued logic networks operate with multi-valued logic signals (Figure 4.12).

The primary advantage of multi-valued signals is the ability to encode more information per variable than a binary system is capable of doing. Hence, less area is used for interconnections since each interconnection carries more information. Furthermore, the reliability of systems is also relevant to the number of connections because they are sources of wiring error and weak connections. For instance, the adoption of m-valued signals enables n pins to pass q^n combinations of values rather than just the 2^n limited by the binary representation.

There are many examples of the successful implementation of 8-, 16-, and 32-valued Boolean algebra in memory devices. In particular, a 32×32 bit multiplier based on the quaternary signed-digit number system consists of only three-stage signed-digit full adder using a binary-tree addition scheme. Another application of multi-valued logic is residue arithmetic. Each residue digit can be represented by a multi-valued code. For example, by this coding,

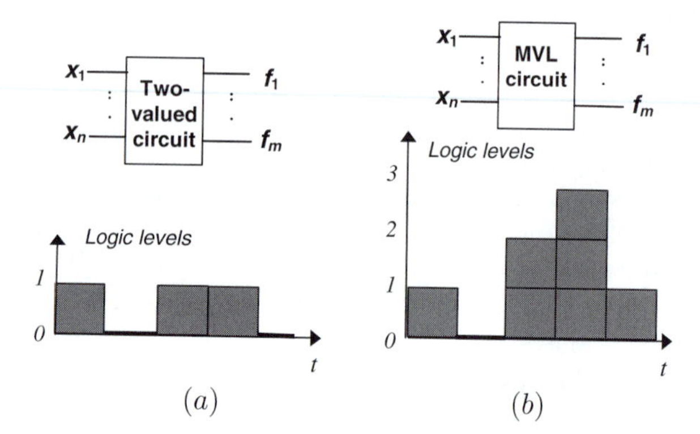

FIGURE 4.12
A binary logic network operates with two-level logic signals (a), and a quaternary logic network operates with four-valued logic signals (b).

mod m multiplication can be performed by a shift operator, and *mod m* addition can be executed using radix-arithmetic operators.

There are various universal (functionally complete) sets of operations for multi-valued algebra: *Post* algebra, *Webb* algebra, *Bernstein* algebra, and *Allen and Givone* algebra.

Multi-valued decision diagrams can be considered as an extension of decision diagram techniques for multi-valued functions. Shannon expansion has been extended for multi-valued logic as well.

It is expected that nanotechnology will provide new possibilities for implementing multi-valued logic. Details can be found in the *Proceedings of the Annual IEEE Symposium on Multiple-Valued Logic.*

Topic 2: Probabilistic logic combines the capacity of probability theory to handle uncertainty with the capacity of classical predicate, or deductive, logic. Among the applications are probabilistic models for logic relations, where the truth values of sentences are probabilities.

Topic 3: Fuzzy logic. In the case of Boolean functions, a Boolean variable can take up only one of two possible values. In fuzzy logic, there are *membership functions* which can take up infinitely many values in the interval [0, 1]. That is, Boolean algebra can handle fuzzy logic (see the book *What is Mathematics?* by R. Courant and H. Robbins, Oxford University Press, New York, 4th ed., 1947). Fuzzy logic devices provide the opportunity for adapting to variable input conditions (see, for example, the book *Fuzzy Logic with Engineering Applications* by T. J. Ross, 2nd Edition, Wiley, New York, 2004). Fuzzy logic operates with fuzzy variables that are different from logical variables.

Topic 4: Reversible logic. Contemporary computers are based on irreversible logic devices, which are being considered energy-inefficient, due to the physical effects of energy dissipation (see, for example, "Irreversibility and Heat Generation in the Computing Process" by Rolf Landauer, *IBM Journal of Research and Development*, issue 5, pages 183–191, 1961). Further study by

Charles H. Bennett ("Logical Reversibility of Computation," *IBM Journal of Research and Development*, issue 17, number 6, pages 525–532, 1973) targeted theoretical possibilities of reversible computing. With discovery of a number of energy recovering integrated circuit techniques, it became possible to exploit reversibility to reduce power consumption. An example of such circuits is the split-level charge recovery logic developed at MIT Artificial Intelligence Laboratory ("Asymptotically Zero Energy Split-Level Charge Recovery Logic" by S. G. Younis and T. F. Knight, Jr., *International Workshop on Low-Power Design*, pages 177–182, 1994). In 1982, E. F. Fredkin and T. Toffoli described a circuit implementation of a reversible element, later named the *Fredkin gate*, suggesting its possible realization on the Josephson junction-based systems ("Conservative Logic," *International Journal of Theoretical Physics*, volume 21, number 3, pages 219–253, 1982). The MIT group have further contributed to the design of asymptotically optimal computers, in particular, "adiabatic" reversible electronic logic technology for parallel reversible architectures (see, for example, "A Reversible Instruction Set Architecture and Algorithms" by J. Hall, *Physics and Computation*, pages 128–134, November 1994).

Further reading

Computer Science and Multiple-Valued Logic: Theory and Applications edited by D. C. Rine, Noth-Holland, Amsterdam, 1977.

Fundamentals of Digital Logic with VHDL Design by S. Brown and Z. Vranesic, McGraw-Hill, 2000.

"Fuzzy Logic and its Application to Switching Systems" by P. N. Marinos, *IEEE Transactions on Computers*, volume 18, pages 343–348, 1969.

Fuzzy Switching and Automata by A. Kandel and S. C. Lee, Crane, Russak, New York, 1979.

Introduction to Digital Logic Design by John P. Hayes, Addison-Wesley, 1993.

Introduction to Logic Design by Alan B. Marcovitz, McGraw-Hill, 2005.

Switching Theory for Logic Synthesis by T. Sasao, Kluwer, Dordrecht, 1999.

4.10 Solutions to practice problems

Practice problem	Solution
4.1.	$(\mathbf{a}\cdot\mathbf{b})\vee\overline{\mathbf{a}} = \left(\begin{bmatrix}0\\1\\1\\0\end{bmatrix}\cdot\begin{bmatrix}1\\0\\0\\0\end{bmatrix}\right)\vee\begin{bmatrix}\overline{0}\\\overline{1}\\\overline{1}\\\overline{0}\end{bmatrix} = \begin{bmatrix}0\cdot1\\1\cdot0\\1\cdot0\\0\cdot0\end{bmatrix}\vee\begin{bmatrix}1\\0\\0\\1\end{bmatrix} = \begin{bmatrix}1\\0\\0\\1\end{bmatrix}$
4.2.	$\overline{x}_1\overline{x}_2 \vee \overline{x}_3\overline{x}_4 = \overline{\overline{(\overline{x}_1\overline{x}_2)}\,\overline{(\overline{x}_3\overline{x}_4)}} = \overline{(x_1\vee x_2)(x_3\vee x_4)}$
4.3.	$f = \overline{1}\vee\overline{1}\vee\overline{0} = 0\vee 0\vee 1 = 1$
4.4.	$x_1x_2\overline{x}_3$ is a minterm, $x_1\vee\overline{x}_2\vee x_3$ is a maxterm.
4.5.	$f_1 = x_1x_2\vee\overline{x}_1x_2\vee\overline{x}_1\overline{x}_2$ and $f_2 = (\overline{x}_1\vee\overline{x}_2)(\overline{x}_1\vee x_2)(x_1\vee\overline{x}_2)$
4.6.	To compare both parts, let us convert them into the same form; for example, convert the left SOP form into POS form: $\overline{x}_1\overline{x}_2x_3\vee x_1\overline{x}_2\overline{x}_3\vee x_1x_2x_3 = \bigvee m(1,4,7) = \prod M(0,2,3,5,6)$ $= (x_1\vee x_2\vee x_3)(x_1\vee\overline{x}_2\vee x_3)(x_1\vee\overline{x}_2\vee\overline{x}_3)(\overline{x}_1\vee x_2\vee\overline{x}_3)(\overline{x}_1\vee\overline{x}_2\vee x_3)$ We have to prove that $\underbrace{(x_1\vee x_2\vee x_3)(x_1\vee\overline{x}_2\vee x_3)(x_1\vee\overline{x}_2\vee\overline{x}_3)(\overline{x}_1\vee x_2\vee\overline{x}_3)(\overline{x}_1\vee\overline{x}_2\vee x_3)}_{Left\ part}$ $= \underbrace{(x_1\vee x_3)(x_1\vee\overline{x}_2)(\overline{x}_2\vee x_3)(\overline{x}_1\vee x_2\vee\overline{x}_3)}_{Right\ part}$ Let us convert the right part to the POS form: $(x_1\vee x_3)(x_1\vee\overline{x}_2)(\overline{x}_2\vee x_3)(\overline{x}_1\vee x_2\vee\overline{x}_3)$ $= (\overline{x}_1\vee\underbrace{\overline{x}_2\vee x_2}_{Equal\ 0}\vee x_3)(x_1\vee\overline{x}_2\vee\underbrace{\overline{x}_3\vee x_3}_{Equal\ 0}\vee x_3)(\underbrace{\overline{x}_1\vee x_1}_{Equal\ 0}\vee\overline{x}_2\vee x_3)(\overline{x}_1\vee x_2\vee\overline{x}_3)$ $= (x_1\vee\overline{x}_2\vee x_3)(x_1\vee x_2\vee x_3)(x_1\vee\overline{x}_2\vee\overline{x}_3)(x_1\vee\overline{x}_2\vee x_3)(\overline{x}_1\vee\overline{x}_2\vee x_3)$ $\cdot\ (x_1\vee\overline{x}_2\vee x_3)(\overline{x}_1\vee x_2\vee\overline{x}_3)$ $= (x_1\vee\overline{x}_2\vee x_3)(x_1\vee x_2\vee x_3)(x_1\vee\overline{x}_2\vee\overline{x}_3)(\overline{x}_1\vee\overline{x}_2\vee x_3)(\overline{x}_1\vee x_2\vee\overline{x}_3)$ The POS expression is identical to the left part of the initial equation. Therefore, the equation is valid.

Solutions to practice problems (continuation)

Practice problem	Solution
4.7.	The condition is specified by the equation $$\overline{x_1 \vee x_2 \vee x_3} = 1.$$ The solution is $x_1 = x_2 = x_3 = 0$. Therefore, gate NOR can have the output $f = 1$ if and only if $x_1 = x_2 = x_3 = 0$.

4.11 Problems

Problem 4.1 Using the basic identities of Boolean algebra, prove the following equalities. State which postulate or theorem is applied at each step.

(*a*) $x_1 x_2 \vee x_1 \overline{x}_2 \vee \overline{x}_1 \overline{x}_2 = x_1 \vee \overline{x}_2$
(*b*) $\overline{\overline{x}_1 \overline{x}_3 \vee \overline{x}_1 x_2 \vee \overline{x}_1 x_3 \vee x_1 x_2} = x_1 \overline{x}_2$
(*c*) $(x_1 \vee x_2)(\overline{x}_1 \overline{x}_3 \vee x_3)\overline{(\overline{x}_2 \vee x_1 x_3)} = \overline{x}_1 x_2$
(*d*) $(x_1 \vee x_2)(x_2 \vee x_3)(x_1 \vee x_3) = x_1 x_2 \vee x_2 x_3 \vee x_1 x_3$

Problem 4.2 Compute the Boolean function for the given assignments of variables:

(*a*) $f = x_1 x_2 \vee \overline{x}_3$ for the assignments $x_1 x_2 x_3 = \{001, 011, 111\}$
(*b*) $f = x_1 \vee x_2 \vee x_3$ for the assignments $x_1 x_2 x_3 = \{000, 010, 111\}$
(*c*) $f = x_1 \overline{x}_2 \vee \overline{x}_3$ for the assignments $x_1 x_2 x_3 = \{000, 101, 111\}$
(*d*) $f = x_1 \oplus x_2 \oplus \overline{x}_3$ for the assignments $x_1 x_2 x_3 = \{100, 110, 111\}$

Problem 4.3 Construct truth tables for the following Boolean functions:
(a) $f = x_1 \oplus x_2 \oplus \overline{x}_3$
(b) $f = x_1 x_2 \vee x_2 x_3 \vee x_1 x_3$
(c) $f = x_1 \overline{x}_2 x_3$
(d) $f = (x_1 \vee x_2)(x_1 \vee \overline{x}_3)$

Problem 4.4 Derive the truth tables for the Boolean functions f_1, f_2, f_3, and f_4 of three variables x_1, x_2, and x_3:

x_1 x_2 x_3	$f_1 = x_1$	$f_2 = \overline{f}_1$	$f_3 = x_1 \vee x_2 \vee x_3$	$f_4 = x_1 x_2 x_3$
0 0 0				
0 0 1				
0 1 0				
0 1 1				
1 0 0				
1 0 1				
1 1 0				
1 1 1				

Problem 4.5 Derive (a) the standard SOP and (b) SOP expressions, (c) the standard POS and (d) POS expressions for the following Boolean functions of three variable given by the truth vector:

(a) $\mathbf{F} = [01101101]^T$
(b) $\mathbf{F} = [01101001]^T$
(c) $\mathbf{F} = [11101101]^T$
(d) $\mathbf{F} = [01111110]^T$

Problem 4.6 Write the standard SOP expressions for the following Boolean functions:

(a) $f = \bigvee m(1,5,7)$
(b) $f = \bigvee m(0,1,3,4,6,7)$
(c) $f = \bigvee m(2,3,4,5,6,7)$
(d) $f = \bigvee m(0,2,3,7)$

Problem 4.7 Write the standard POS expressions for the following Boolean functions:

(a) $f = \prod M(0,1,7)$
(b) $f = \prod M(1,3,4,6,7)$
(c) $f = \prod M(2,3,4,7)$
(d) $f = \prod M(1,5,7)$

Problem 4.8 Write the expression $f_1 \vee f_2$ for the Boolean functions specified as follows:

(a) $f_1 = \bigvee m(1,5,7)$ and $f_2 = \prod M(1,5,7)$
(b) $f_1 = \bigvee m(0,2)$ and $f_2 = \prod M(1,6)$
(c) $f_1 = \bigvee m(3,4,5)$ and $f_2 = \prod M(1,6,7)$
(d) $f_1 = \bigvee m(2,4,6,7)$ and $f_2 = \prod M(0,1,3,5)$

Problem 4.9 Write the expressions for the Boolean functions specified as follows:

(a) $f_1 \cdot f_2$ for $f_1 = \bigvee m(1,5,7)$ and $f_2 = \prod M(1,2,4)$
(b) $f_1 \oplus f_2$ for $f_1 = \bigvee m(0,2)$ and $f_2 = \prod M(1,3)$
(c) $f_1 \vee f_2$ for $f_1 = \bigvee m(2,4,7)$ and $f_2 = \prod M(1,5,7)$
(d) $f_1 \vee f_2$ for $f_1 = \bigvee m(0,7)$ and $f_2 = \prod M(1,2,3,4,5,6)$

Hint: If $f_1 = \bigvee m(1) = \overline{x}_1\overline{x}_2x_3$ and $f_2 = \prod M(2,4) = (x_1 \vee \overline{x}_2 \vee x_3)(\overline{x}_1 \vee x_2 \vee x_3)$, then $f_1 \cdot f_2 = \overline{x}_1\overline{x}_2x_3(x_1 \vee \overline{x}_2 \vee x_3)(\overline{x}_1 \vee x_2 \vee x_3)$

Problem 4.10 Represent each of the following Boolean functions in standard SOP form using algebraic manipulations (without first constructing a table of the function's values):

(a) $f = \overline{x}_1(\overline{x}_2 \vee x_3) \vee \overline{x}_3$
(b) $f = (\overline{x}_1 \vee \overline{x}_2)(\overline{x}_1 \vee x_3)$
(c) $f = x_1 \vee \overline{x}_2 \vee \overline{x}_3$
(d) $f = x_1 \oplus x_2x_3$

Problem 4.11 Represent each of the following Boolean functions in standard POS form using algebraic manipulations (without first constructing a table of the function's values):

(a) $f = \overline{x}_1(\overline{x}_2 \vee x_3) \vee \overline{x}_3$
(b) $f = (\overline{x}_1 \vee \overline{x}_2)(\overline{x}_1 \vee x_3)$
(c) $f = x_1\overline{x}_2\overline{x}_3$
(d) $f = x_1 \oplus x_2x_3$

Problem 4.12 Derive the standard SOP and POS and expressions for the Boolean functions f_1, f_2, f_3, and f_4 of three variables given by the table of the function's values

x_1	x_2	x_3	f_1	f_2	f_3	f_4
0	0	0	0	1	0	1
0	0	1	1	0	1	1
0	1	0	0	1	1	0
0	1	1	1	0	0	1
1	0	0	0	1	0	0
1	0	1	1	0	1	1
1	1	0	0	1	1	0
1	1	1	1	0	0	1

Problem 4.13 Given the Boolean functions f_1 and f_2, compute $f_1 \vee f_2$, $f_1 \cdot f_2$, and $f_1 \oplus f_2$ for the following functions:

(*a*) $f_1 = x_1x_2$ and $f_2 = \overline{x}_1\overline{x}_2$
(*b*) $f_1 = x_1 \vee x_2x_3$ and $f_2 = x_1x_2 \vee x_3$
(*c*) $f_1 = x_1 \oplus x_2 \oplus x_3$ and $f_2 = x_1x_2x_3$
(*d*) $f_1 = (x_1 \vee \overline{x}_2)(\overline{x}_1 \vee x_2)$ and $f_2 = (x_1 \vee x_2)(\overline{x}_1 \vee \overline{x}_2)$

Problem 4.14 Boolean functions are given by the truth vectors $\mathbf{F}_1$ and $\mathbf{F}_2$. Compute $\mathbf{F}_1 \vee \mathbf{F}_2$, $\mathbf{F}_1 \cdot \mathbf{F}_2$, and $\mathbf{F}_1 \oplus \mathbf{F}_2$ for the following functions:

(*a*) $\mathbf{F}_1 = [01101101]^T$ and $\mathbf{F}_2 = [11101101]^T$
(*b*) $\mathbf{F}_1 = [01001101]^T$ and $\mathbf{F}_2 = [01111111]^T$
(*c*) $\mathbf{F}_1 = [00001101]^T$ and $\mathbf{F}_2 = [11111101]^T$
(*d*) $\mathbf{F}_1 = [01100001]^T$ and $\mathbf{F}_2 = [01101111]^T$

Problem 4.15 Prove that

(*a*) $(x_1 \vee x_2)(x_1 \vee x_3) = \overline{x}_1(x_2 \oplus x_3)$
(*b*) $x_1 \oplus x_2x_3 \neq (x_1 \oplus x_2)(x_1 \oplus x_3)$, that is, the EXOR operation is not distributive over the AND operation
(*c*) The dual of the EXOR function is complement of the EXOR function, $x_1 \oplus x_2 \longleftrightarrow \overline{x_1 \oplus x_2}$
(*d*) $x_1 \vee x_2 = x_1 \oplus x_2 \oplus x_1x_2$

Problem 4.16 Add the extra variables using appropriate identities to the following Boolean functions:

(a) $f = x_1\overline{x}_2$, add $\overline{x}_3$
(b) $f = x_1 \oplus \overline{x}_2$, add $\overline{x}_3$
(c) $f = \overline{x}_1 \oplus \overline{x}_2$, add x_3
(d) $f = x_1 \vee \overline{x}_2$, add x_3
(e) $f = \overline{x}_1\overline{x}_2$, add $\overline{x}_3$
(f) $f = x_1x_2 \oplus x_3$, add x_4

Hint: $f = x_1x_2 = x_1x_2(x_3 \vee \overline{x}_3) = x_1x_2x_3 \vee x_1x_2\overline{x}_3$

Problem 4.17 Show that one variable can be eliminated from the following Boolean equations:

(*a*) $f = \overline{x}_2x_3 \vee x_1x_2\overline{x}_3 \vee x_1x_2$, variable x_1 can be eliminated
(*b*) $f = (x_1 \vee x_2 \vee \overline{x}_3)(x_1 \vee x_2)$, variable x_1 can be eliminated
(*c*) $f = x_1 \oplus x_2 \oplus \overline{x}_1 \oplus x_2x_3$, variable x_1 can be eliminated
(*d*) $f = (x_2 \oplus x_1x_2) \vee x_2$, variable x_1 can be eliminated

Hint: $f = x_1\overline{x}_2 \vee x_1 = x_1(1 \vee \overline{x}_2) = x_1$, variable $\overline{x}_2$ is eliminated

Problem 4.18 Derive expressions with all variables complemented for the following Boolean functions:

(a) $f = x_1 \oplus x_2$
(b) $f = x_1 \vee x_2 \vee x_3$
(c) $f = x_1x_2x_3$
(d) $f = x_1x_2 \oplus x_3$

Hint: $f = x_1 \cdot \overline{x}_2 = \overline{\overline{x_1 \cdot \overline{x}_2}} = \overline{\overline{x}_1 \vee \overline{\overline{x}}_2}$.

Problem 4.19 Simplify the following Boolean functions:

(a) $f = \overline{(x_1x_3 \vee x_2x_4)(\overline{x}_3 \vee x_4)}$

(b) $f = \overline{x_1x_2 \oplus x_3\overline{x}_4}$

(c) $f = \overline{(x_1x_2 \vee \overline{x}_1x_3)} \vee x_2$

(d) $f = \overline{x_1x_2} \vee \overline{\overline{x}_1\overline{x}_2}$

Problem 4.20 Simplify the following logic networks by applying local transformations where possible:

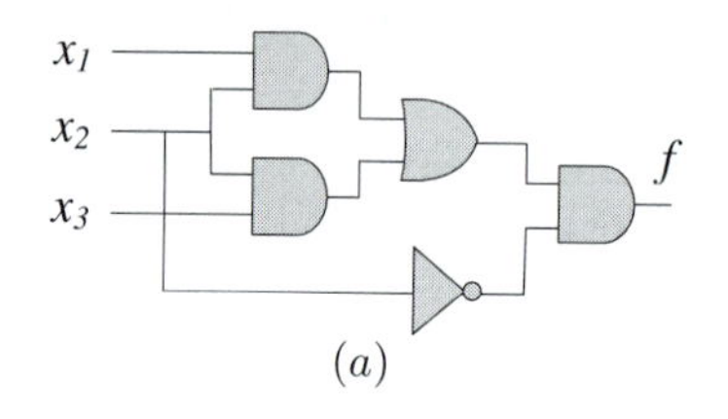

(a)

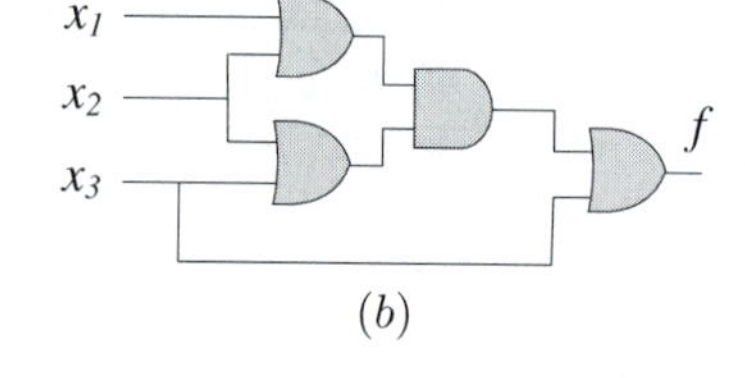

(b)

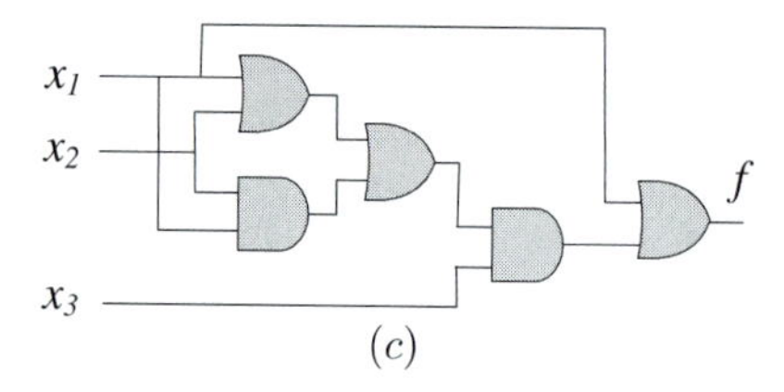

(c)

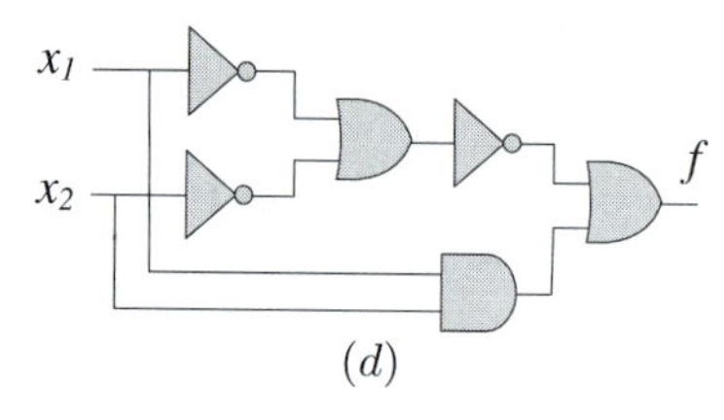

(d)

5

Fundamental Expansions

Shannon expansion for an arbitrary Boolean function

- With respect to a single variable
- With respect to a group of variables
- For all variables
- For decision trees
- Techniques for computing

Shannon expansion for symmetric Boolean functions

- Partially symmetric functions
- Totally symmetric functions
- Techniques for computing

Logic Taylor expansion

- Change in a discrete system
- Detection of change using Boolean differences
- EXOR expressions
- Expansion in terms of change
- Techniques for computing

Advanced topics

- Arithmetic analog of Shannon expansion
- Arithmetic analog of logic Taylor expansion
- Spectral techniques
- Information notation of Shannon expansion

5.1 Introduction

In this chapter, three fundamental expansions used in contemporary logic design are introduced (Figure 5.1):

- *Shannon expansion* for an arbitrary Boolean functions,
- *Shannon expansion* for symmetric Boolean functions, and
- *Logic Taylor* expansion.

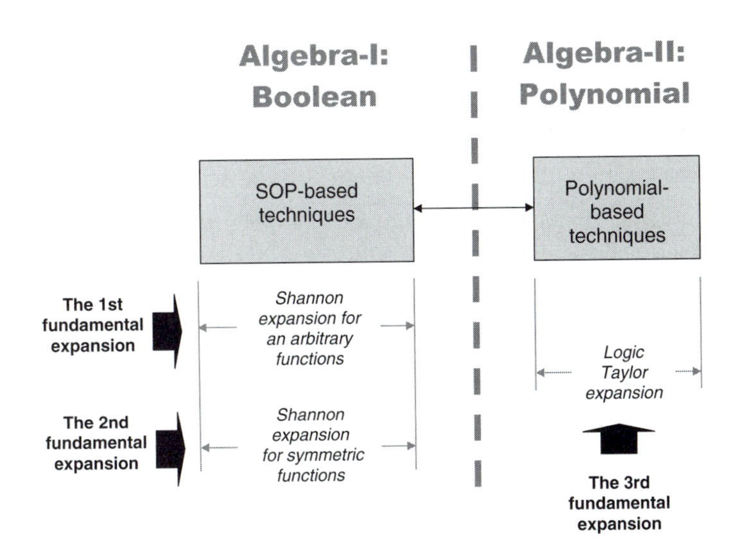

FIGURE 5.1
Fundamental expansions of Boolean functions.

The Shannon classical expansion provides the decomposition of an arbitrary Boolean function of n variables to a set of Boolean sub-functions of $m < n$ variables:

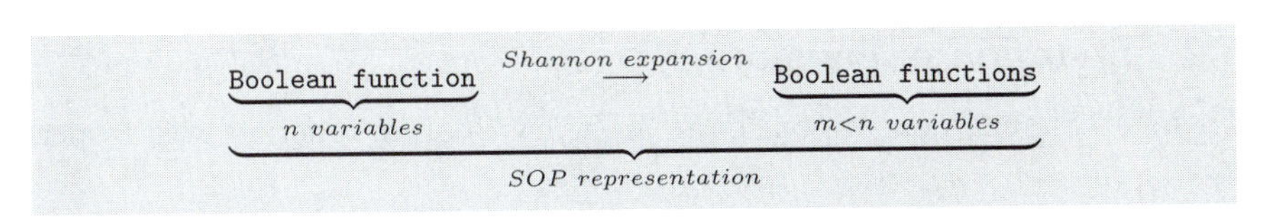

Many advanced techniques of logic analysis and synthesis are based on Shannon expansion. An arbitrary Boolean function can be expanded with respect to: a single variable, a group of variables, and all variables. This is the primary fundamental expansion of logic design, introduced by *Claude Shannon* in 1937.

In practice, one of the advantages of knowing that a given Boolean function is symmetric is the potential implementation cost reduction offered by the implementation of such function using smaller number of gates. *Shannon expansion for symmetric function* is specific for the representation and realization of function with symmetric variables. This expansion is quite distinctive from the Shannon expansion for arbitrary Boolean function; this is the second fundamental expansion of logic design.

The third fundamental expansion is called the *logic Taylor expansion*; this is an analog of the classical Taylor expansion. This expansion describes the Boolean function in terms of *change* in function values, which is formulated using a *Boolean difference*:

$$\underbrace{\texttt{Boolean function}}_{SOP\ form} \overset{Logic\ Taylor\ expansion}{\longleftrightarrow} \underbrace{\texttt{Polynomial form}}_{In\ terms\ of\ differences}$$

The logic Taylor expansion provides various polynomial representations of Boolean functions using the logical operations EXOR and AND. The classical Taylor expansion is defined as a "polynomial for the function f about the point $x = x_i$, where each coefficient is defined in the terms of derivatives." In the logic Taylor expansion, "expansion about a point" is replaced by "polarity" (details are given in Chapter 10). Given a Boolean function of n variables, the logic Taylor expansion generates 2^n polynomials of different polarities. This expansion interpreted as a representation of a Boolean function of n variables by a set of polynomials of $m < n$ variables is called the *Davio expansion*:

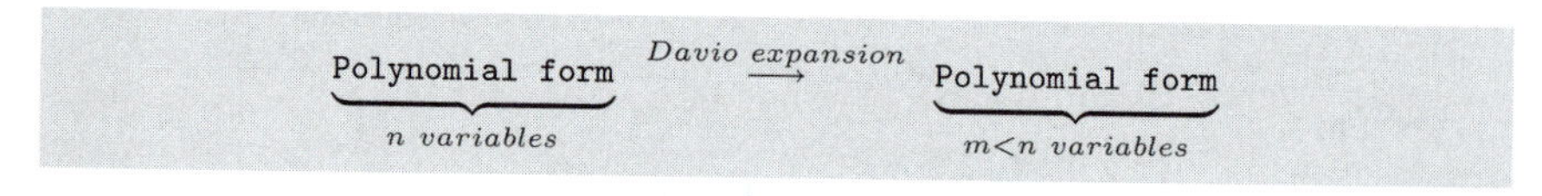

Davio expansion is a particular case of Shannon expansion.

5.2 Shannon expansion

There are Shannon expansions of an arbitrary Boolean function with respect to:

- A single variable,
- A group of variables, and
- All variables

Table 5.1 provides examples of Shannon expansion based on the above classification. In this table, Shannon expansion with respect to a single variable is defined as a *step*, or an *iteration* in the recursive procedure of Shannon expansion with respect to all variables.

5.2.1 Expansion with respect to a single variable

We denote a Boolean function f of n variables by $f = f(x_1, x_2, \ldots, x_n)$. The fact that variable x_i, $i \in \{1, 2, \ldots n\}$, takes the value 0 or 1 is represented in the form

$$f(x_1, \ldots, x_{i-1}, \overset{x_i}{\boxed{0}}, x_{i+1}, \ldots, x_n) \text{ and } f(x_1, \ldots, x_{i-1}, \overset{x_i}{\boxed{1}}, x_{i+1}, \ldots, x_n)$$

respectively.

Example 5.1 (Variables take values 0 and 1.) *Let* $f = x_1 \vee x_2 \vee x_3$, *then*

$$f = f(x_1, x_2 = 0, x_3) = f(x_1, 0, x_3) = x_1 \vee 0 \vee x_3 = x_1 \vee x_3$$
$$f = f(x_1, x_2 = 1, x_3) = f(x_1, 1, x_3) = x_1 \vee 1 \vee x_3 = 1$$

Any Boolean function $f = f(x_1, x_2, \ldots, x_n)$ of n variables can be expanded into two Boolean functions of $(n-1)$ variables:

Shannon expansion with respect to a variable

$$\overbrace{f(x_1, x_2, \ldots, x_n)}^{n \text{ variables}} = \overline{x}_i \overbrace{f(x_1, \ldots, x_{i-1}, \overset{\overline{x}_i}{\boxed{0}}, x_{i+1}, \ldots, x_n)}^{n-1 \text{ variables}}$$
$$\underset{OR}{\underset{\uparrow}{\boxed{\vee}}}\, x_i \underbrace{f(x_1, \ldots, x_{i-1}, \overset{x_i}{\boxed{1}}, x_{i+1}, \ldots, x_n)}_{n-1 \text{ variables}} \quad (5.1)$$

where $i \in \{1, 2, \ldots, n\}$. Equation 5.1 is called the *Shannon expansion*. Denoting

$$f_0 = f(x_1, \ldots, x_{i-1} \overset{\overline{x}_i}{\boxed{0}}\, x_{i+1}, \ldots, x_n) \text{ and } f_1 = f(x_1, \ldots, x_{i-1} \overset{x_i}{\boxed{1}}\, x_{i+1}, \ldots, x_n)$$

we can re-write the Shannon expansion (Equation 5.1) in the form

Shannon expansion with respect to a variable

$$f = \overline{x}_i f_0 \vee x_i f_1 \quad (5.2)$$

TABLE 5.1
Types of Shannon expansion for an arbitrary Boolean function: with respect to a single variable, a group of variables, and all variables.

Type of expansion	Formal notation and example

With respect to a single variable

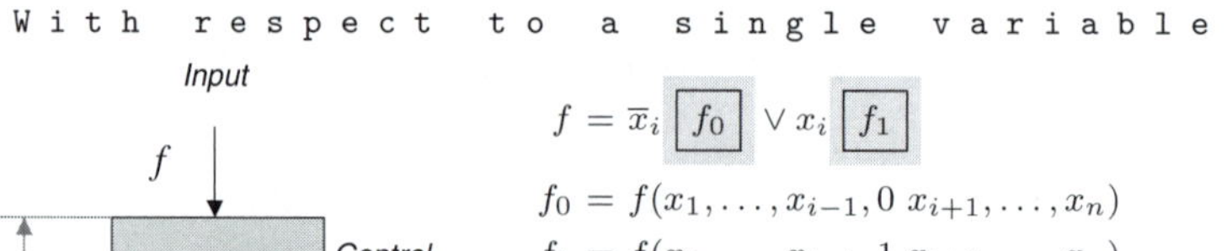

$$f = \overline{x}_i \boxed{f_0} \vee x_i \boxed{f_1}$$

$$f_0 = f(x_1, \ldots, x_{i-1}, 0\ x_{i+1}, \ldots, x_n)$$

$$f_1 = f(x_1, \ldots, x_{i-1}, 1\ x_{i+1}, \ldots, x_n)$$

Example: Given $n = 3$, $i = 1$

$$f = \overline{x}_1 f_0 \vee x_1 f_1$$

$$f_0 = f(0, x_2, x_3)$$

$$f_1 = f(1, x_2, x_3)$$

With respect to a group of variables

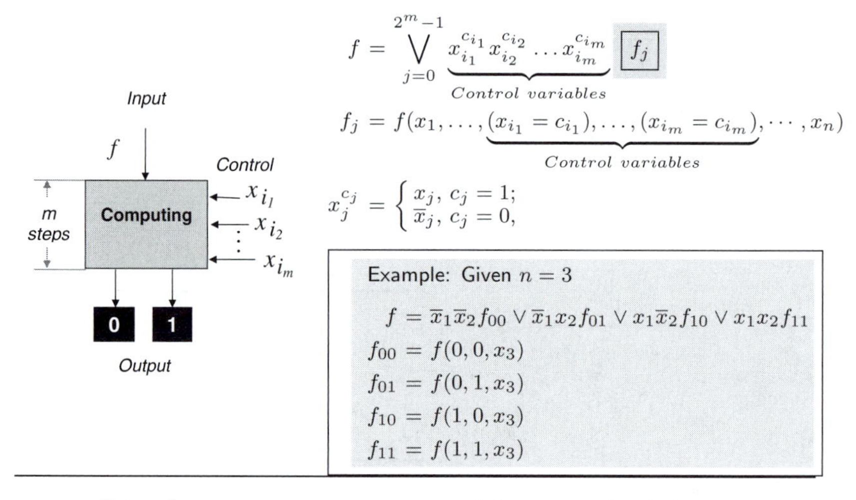

$$f = \bigvee_{j=0}^{2^m-1} \underbrace{x_{i_1}^{c_{i_1}} x_{i_2}^{c_{i_2}} \ldots x_{i_m}^{c_{i_m}}}_{Control\ variables} \boxed{f_j}$$

$$f_j = f(x_1, \ldots, \underbrace{(x_{i_1} = c_{i_1}), \ldots, (x_{i_m} = c_{i_m})}_{Control\ variables}, \cdots, x_n)$$

$$x_j^{c_j} = \begin{cases} x_j, & c_j = 1; \\ \overline{x}_j, & c_j = 0, \end{cases}$$

Example: Given $n = 3$

$$f = \overline{x}_1\overline{x}_2 f_{00} \vee \overline{x}_1 x_2 f_{01} \vee x_1\overline{x}_2 f_{10} \vee x_1 x_2 f_{11}$$

$$f_{00} = f(0, 0, x_3)$$

$$f_{01} = f(0, 1, x_3)$$

$$f_{10} = f(1, 0, x_3)$$

$$f_{11} = f(1, 1, x_3)$$

With respect to all variables

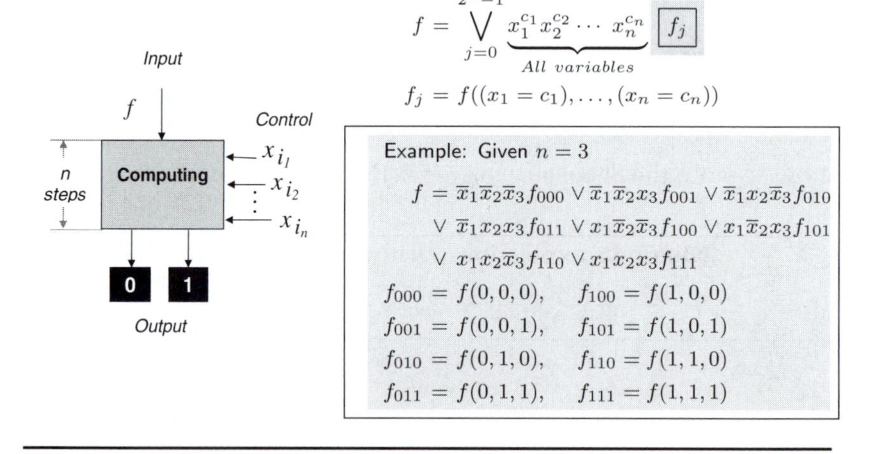

$$f = \bigvee_{j=0}^{2^n-1} \underbrace{x_1^{c_1} x_2^{c_2} \cdots x_n^{c_n}}_{All\ variables} \boxed{f_j}$$

$$f_j = f((x_1 = c_1), \ldots, (x_n = c_n))$$

Example: Given $n = 3$

$$f = \overline{x}_1\overline{x}_2\overline{x}_3 f_{000} \vee \overline{x}_1\overline{x}_2 x_3 f_{001} \vee \overline{x}_1 x_2 \overline{x}_3 f_{010} \vee \overline{x}_1 x_2 x_3 f_{011} \vee x_1\overline{x}_2\overline{x}_3 f_{100} \vee x_1\overline{x}_2 x_3 f_{101} \vee x_1 x_2 \overline{x}_3 f_{110} \vee x_1 x_2 x_3 f_{111}$$

$f_{000} = f(0,0,0)$, $f_{100} = f(1,0,0)$

$f_{001} = f(0,0,1)$, $f_{101} = f(1,0,1)$

$f_{010} = f(0,1,0)$, $f_{110} = f(1,1,0)$

$f_{011} = f(0,1,1)$, $f_{111} = f(1,1,1)$

Example 5.2 (Expansion with respect to a single variable.) *Let* $f = \overline{x}_1 x_2 \vee \overline{x}_3$ *and* $i = 2$. *Then*

$$f_0 = f(\overline{x}_1, x_2 = 0, \overline{x}_3) = f(\overline{x}_1, 0, \overline{x}_3) = \overline{x}_1 \cdot 0 \vee \overline{x}_3 = \overline{x}_3$$
$$f_1 = f(\overline{x}_1, x_2 = 1, \overline{x}_3) = f(\overline{x}_1, 1, \overline{x}_3) = \overline{x}_1 \cdot 1 \vee \overline{x}_3 = \overline{x}_1 \vee \overline{x}_3$$
$$f = \overline{x}_2 f_0 \vee x_2 f_1 = \overline{x}_2 \boxed{(\overline{x}_3)} \vee x_2 \boxed{(\overline{x}_1 \vee \overline{x}_3)}$$

Equation 5.2 describes the expanding of any Boolean function of n variables with respect to n variables as an n-step process. Reduction of this equation can be achieved by eliminating a variable x_i via replacement of this variable by the possible values 0 and 1. From Equation 5.2 follows the recursive procedure for the expansion:

Recursive application of the Shannon expansion

Step 1: `Apply Shannon expansion with respect to a given variable.`
Step 2: `Repeat Step 1 to the next variable until the further application is not required or impossible.`

Example 5.3 (Recursive expansion.) *Let* $f = \overline{x}_1 x_2 \vee \overline{x}_3$. *Find the expansion of this function: (a) with respect to variables* x_1 *and* x_3, *and (b) with respect to all variables* $x_1 \longrightarrow x_2 \longrightarrow x_3$. *Details of this computing are given in Figure 5.2.*

Practice problem 5.1. (Recursive expansion.) Given the Boolean function $f = \overline{x}_1 \overline{x}_2 \oplus \overline{x}_2$, find the Shannon expansion recursively with respect to x_1 and x_2.
Answer is given in "Solutions to practice problems."

5.2.2 Expansion with respect to a group of variables

Shannon expansion can be done in terms of $m > 1$ variables. First, consider the case of two variables. Let the variables x_i and x_j, $i, j \in \{1, 2, \ldots n\}$, $i \neq j$ both take the value 0. Thus, the expansion cofactor f_{00} is represented as follows:

$$f_{00} = f(x_1, \ldots, x_{i-1}, \boxed{0}, x_{i+1}, \ldots, x_{j-1}, \boxed{0}, x_{j+1}, \ldots .x_n)$$

Techniques for computing the Shannon expansion

Recursive expansion with respect to the variables x_1 and x_3

Step 1: Expand the function $\overline{x}_1 x_2 \vee \overline{x}_3$ with respect to the variable x_1:

$$\begin{aligned} f &= \overline{x}_1 x_2 \vee \overline{x}_3 \\ &= \overline{x}_1 f_0 \vee x_1 f_1 \\ &= \overline{x}_1(\overline{0} \cdot x_2 \vee \overline{x}_3) \vee x_1(\overline{1} \cdot x_2 \vee \overline{x}_3) \\ &= \overline{x}_1 \boxed{(x_2 \vee \overline{x}_3)} \vee x_1 \boxed{\overline{x}_3} \end{aligned}$$

Step 2: Expand the result of Step 1 with respect to the variable x_3:

$$\begin{aligned} f &= \overline{x}_3 f_{00} \vee x_3 f_{01} \\ &= \overline{x}_3(\overline{x}_1(x_2 \vee \overline{0}) \vee x_1(\overline{0})) \vee x_3(\overline{x}_1(x_2 \vee \overline{1}) \vee x_1(\overline{1})) \\ &= \overline{x}_3(\overline{x}_1 \vee x_1) \vee x_3(\overline{x}_1 x_2 \vee x_1(0)) \\ &= \overline{x}_1\overline{x}_3 \boxed{1} \vee x_1\overline{x}_3 \boxed{1} \vee \overline{x}_1 x_3 \boxed{x_2} \vee x_1 x_3 \boxed{0} \end{aligned}$$

Expansion with respect to a group of variables (x_1, x_3)

Step 1: Compute cofactors f_j, $j = 0, 1, 2, 3$, for the expansion

$$f = \overline{x}_1 x_2 \vee \overline{x}_3 = \bigvee_{j=0}^{2^2-1} x_1^{c_1} x_3^{c_3} f_j$$

as follows:

$$f_{00} = f(x_1 = 0, x_2, x_3 = 0) = f(0, x_2, 0) = \overline{0}x_2 \vee \overline{0} = \boxed{1}$$

$$f_{01} = f(x_1 = 0, x_2, x_3 = 1) = f(0, x_2, 1) = \overline{0}x_2 \vee \overline{1} = \boxed{x_2}$$

$$f_{10} = f(x_1 = 1, x_2, x_3 = 0) = f(1, x_2, 0) = \overline{1}x_2 \vee \overline{0} = \boxed{1}$$

$$f_{11} = f(x_1 = 1, x_2, x_3 = 1) = f(1, x_2, 1) = \overline{1}x_2 \vee \overline{1} = \boxed{0}$$

Step 2: Expand the function $\overline{x}_1 x_2 \vee \overline{x}_3$ with respect to a group of variables (x_1, x_3)

$$\begin{aligned} f &= \overline{x}_1\overline{x}_3 f_{00} \vee \overline{x}_1 x_3 f_{01} \vee x_1\overline{x}_3 f_{10} \vee x_1 x_3 f_{11} \\ &= \overline{x}_1\overline{x}_3(\overline{0}x_2 \vee \overline{0}) \vee \overline{x}_1 x_3(\overline{0}x_2 \vee \overline{1}) \vee x_1\overline{x}_3(\overline{1}x_2 \vee \overline{0}) \vee x_1 x_3(\overline{1}x_2 \vee \overline{1}) \\ &= \overline{x}_1\overline{x}_3 \boxed{1} \vee \overline{x}_1 x_3 \boxed{x_2} \vee x_1\overline{x}_3 \boxed{1} \vee x_1 x_3 \boxed{0} \end{aligned}$$

FIGURE 5.2
Techniques for the Shannon expansion of the Boolean function $f = \overline{x}_1 x_2 \vee \overline{x}_3$ (Examples 5.3 and 5.4).

Other cofactors are derived in a similar way:

$$f_{01} = f(x_1, \ldots, x_{i-1}, \boxed{0}, x_{i+1}, \ldots, x_{j-1}, \boxed{1}, x_{j+1}, \ldots .x_n)$$

$$f_{10} = f(x_1, \ldots, x_{i-1}, \boxed{1}, x_{i+1}, \ldots, x_{j-1}, \boxed{0}, x_{j+1}, \ldots .x_n)$$

$$f_{11} = f(x_1, \ldots, x_{i-1}, \boxed{1}, x_{i+1}, \ldots, x_{j-1}, \boxed{1}, x_{j+1}, \ldots .x_n)$$

Example 5.4 (Co-factors for groups of variables.) *Given the Boolean function of three variables $f = \overline{x}_1 x_2 \vee \overline{x}_3$, the cofactors for expansion with respect to the variables x_1 and x_3 are formed as shown in Figure 5.2.*

Applying Expansion 5.1 recursively to the remaining $(n - m)$ variables $x_{i_1}, x_{i_2}, \ldots, x_{i_m}$, we obtain the formula

Shannon expansion with respect to a group of variables

$$f = \bigvee_{j=0}^{2^m-1} \underbrace{x_{i_1}^{c_{i_1}} x_{i_2}^{c_{i_2}} \ldots x_{i_m}^{c_{i_m}}}_{Control\ variables} f_j \tag{5.3}$$

where cofactor f_j is defined as

$$f_j = f(x_1, x_2, \ldots, \underbrace{\boxed{x_{i_1} = c_{i_1}}, \boxed{x_{i_2} = c_{i_2}}, \ldots, \boxed{x_{i_m} = c_{i_m}}}_{Values\ of\ variables}, \cdots, x_{n-1}, x_n)$$

and

$$x_j^{c_j}, = \begin{cases} x_j, \, c_j = 1 \\ \overline{x}_j, \, c_j = 0 \end{cases}$$

where $j = c_{i_1} c_{i_2} \ldots c_{i_m}$ is binary representation of j. Equation 5.3 describes computing the Shannon expansion with respect to m variables as an m-step process.

Example 5.5 (Expansion with respect to a group of variables.) *Using Equation 5.3, the cofactors of the Shannon expansion of the Boolean function* $f = \overline{x}_1\overline{x}_3 \vee x_1x_2 \vee x_1x_3$ *of three variables* ($n = 3$) *with respect to the variables* x_1 *and* x_2 ($m = 2$) *are calculated as follows:*

$$\text{Co-factor } f_{00} = 1 \cdot \overline{x}_3 \vee 0 \cdot 0 \vee 0 \cdot x_3 = \boxed{\overline{x}_3}$$

$$\text{Co-factor } f_{01} = 1 \cdot \overline{x}_3 \vee 0 \cdot 1 \vee 0 \cdot x_3 = \boxed{\overline{x}_3}$$

$$\text{Co-factor } f_{10} = 0 \cdot \overline{x}_3 \vee 1 \cdot 0 \vee 1 \cdot x_3 = \boxed{x_3}$$

$$\text{Co-factor } f_{11} = 0 \cdot \overline{x}_3 \vee 1 \cdot 1 \vee 1 \cdot x_3 = \boxed{1}$$

$$f = \bigvee_{j=0}^{2^2-1} = \overline{x}_1\overline{x}_2 \boxed{f_{00}} \vee \overline{x}_1x_2 \boxed{f_{01}} \vee x_1\overline{x}_2 \boxed{f_{10}} \vee x_1x_2 \boxed{f_{11}}$$
$$= \overline{x}_1\overline{x}_2(\overline{x}_3) \vee \overline{x}_1x_2(\overline{x}_3) \vee x_1\overline{x}_2(x_3) \vee x_1x_2(1)$$

Practice problem 5.2. (Expansion with respect to a group of variables.) Derive the Shannon expansion of the Boolean function $f = x_1 \vee x_2 \vee \overline{x}_3$ with respect to the group of variables x_2 and x_3.
Answer is given in "Solutions to practice problems."

5.2.3 Expansion with respect to all variables

The boundary case of the Shannon expansion is encountered when setting $m = n$ in Equation 5.3:

Shannon expansion with respect to all variables

$$f = \bigvee_{j=0}^{2^n-1} \underbrace{x_1^{c_1}x_2^{c_2} \cdots x_n^{c_n}}_{All\ variables} f_j \tag{5.4}$$

where cofactor f_j is defined as

$$f_j = f(\underbrace{x_{i_1} = c_{i_1}, x_{i_2} = c_{i_2}, \ldots, x_{i_m} = c_{i_m}}_{Values\ of\ variables})$$

and

$$x_j^{c_j} = \begin{cases} x_j, & c_j = 1 \\ \overline{x}_j, & c_j = 0 \end{cases}$$

This equation describes the computing of any Boolean function of n variables by a single-step process.

Example 5.6 (Expansion with respect to all variables.) *Using Equation 5.4, the Shannon expansion of the Boolean function* $f = \overline{x}_1\overline{x}_3 \vee x_1x_2 \vee x_1x_3$ *with respect to all three variables* x_1, x_2*, and* x_3 *is computed as shown in Figure 5.3.*

Techniques for computing the Shannon expansion

Expansion with respect to all variables x_1, x_2, and x_3

Step 1: Compute cofactors f_j, $j = 0, 1, \ldots, 7$, for the expansion

$$f = \overline{x}_1x_2 \vee \overline{x}_3 = \bigvee_{j=0}^{2^3-1} x_1^{c_1}x_2^{c_2}x_3^{c_3} f_j$$

as follows:

$$f_{000} = f(000) = \overline{0}\cdot 0 \vee \overline{0} = \boxed{1},\quad f_{001} = f(001) = \overline{0}\cdot 0 \vee \overline{1} = \boxed{0},$$

$$f_{010} = f(010) = \overline{0}\cdot 1 \vee \overline{0} = \boxed{1},\quad f_{011} = f(011) = \overline{0}\cdot 1 \vee \overline{1} = \boxed{1},$$

$$f_{100} = f(100) = \overline{1}\cdot 0 \vee \overline{0} = \boxed{1},\quad f_{101} = f(101) = \overline{1}\cdot 0 \vee \overline{1} = \boxed{0},$$

$$f_{110} = f(110) = \overline{1}\cdot 1 \vee \overline{0} = \boxed{0},\quad f_{111} = f(111) = \overline{1}\cdot 1 \vee \overline{1} = \boxed{0}$$

Step 2: Expand the function $\overline{x}_1x_2 \vee \overline{x}_3$ with respect to a group of variables x_1, x_2, and x_3:

$$\begin{aligned} f &= \overbrace{\overline{x}_1\overline{x}_2\overline{x}_3}^{000} f_{000} \vee \overbrace{\overline{x}_1\overline{x}_2x_3}^{001} f_{001} \vee \overbrace{\overline{x}_1x_2\overline{x}_3}^{010} f_{010} \vee \overbrace{\overline{x}_1x_2x_3}^{011} f_{011} \\ &\vee \overbrace{x_1\overline{x}_2\overline{x}_3}^{100} f_{100} \vee \overbrace{x_1\overline{x}_2x_3}^{101} f_{101} \vee \overbrace{x_1x_2\overline{x}_3}^{110} f_{110} \vee \overbrace{x_1x_2x_3}^{111} f_{111} \\ &= \overline{x}_1\overline{x}_2\overline{x}_3 \boxed{1} \vee \overline{x}_1\overline{x}_2x_3 \boxed{0} \vee \overline{x}_1x_2\overline{x}_3 \boxed{1} \vee \overline{x}_1x_2x_3 \boxed{1} \\ &\vee x_1\overline{x}_2\overline{x}_3 \boxed{1} \vee x_1\overline{x}_2x_3 \boxed{0} \vee x_1x_2\overline{x}_3 \boxed{1} \vee x_1x_2x_3 \boxed{0} \end{aligned}$$

FIGURE 5.3
Techniques for the Shannon expansion of the Boolean function $f = \overline{x}_1x_2 \vee \overline{x}_3$ (Example 5.4).

Practice problem **5.3. (Expansion with respect to all variables.)** Find Shannon expansion of the Boolean function $f = x_1 \oplus x_2$ with respect to the variables x_1 and x_2.
Answer is given in "Solutions to practice problems."

5.2.4 Various forms of Shannon expansions

Shannon expansion provides a model for the multiplexing of Boolean functions, which enables representation of the Boolean function for further implementation as a logic network of universal gates such as multiplexers. Multiplexing can be implemented:

- Using an OR operation, such as in the Shannon expansion given in Equation 5.1,
- Using an AND operation, and
- Using an EXOR operation.

The dual of Shannon expansion

The dual of Shannon expansion (Equation 5.1) is the following expression:

The dual of Shannon expansion

$$\overbrace{f(x_1, x_2, \ldots, x_n)}^{n \text{ variables}} = (\overline{x}_i \vee \overbrace{f(x_1, \ldots, x_{i-1}\ \boxed{1}\ x_{i+1}, \ldots, x_n)}^{n-1 \text{ variables}}) \underset{\text{AND}}{\underset{\uparrow}{\cdot}} (x_i \vee \underbrace{f(x_1, \ldots, x_{i-1}\ \boxed{0}\ x_{i+1}, \ldots, x_n)}_{n-1 \text{ variables}}) \quad (5.5)$$

Example 5.7 (The dual of Shannon expansion.) *Let $f = x_1 \vee \overline{x}_2$. Its Shannon expansion (Equation 5.1) with respect to variable x_1 is represented as $f = \overline{x}_1(0 \vee \overline{x}_2) \vee x_1(1 \vee \overline{x}_2) = \overline{x}_1(\overline{x}_2) \vee x_1(1)$, and its dual (Equation 5.5) is $f = (\overline{x}_1 \vee (1 \vee \overline{x}_2))(x_1 \vee (0 \vee \overline{x}_2)) = (\overline{x}_1 \vee 1)(x_1 \vee \overline{x}_2)$. Both equations imply the initial function, since $f = \overline{x}_1\overline{x}_2 \vee x_1 = \overline{x}_2 \vee x_1$ and $f = (\overline{x}_1 \vee 1)(x_1 \vee \overline{x}_2) = x_1 \vee \overline{x}_2$.*

Practice problem **5.4. (The dual of Shannon expansion.)** Derive the Shannon expansion with respect to the variable x_2, and its dual, of the function $f = x_1 \oplus x_2$.
Answer is given in "Solutions to practice problems."

The EXOR form of Shannon expansion

Given a Shannon expansion in the form of Equation 5.1, the Boolean sum in this equation can be replaced by the EXOR function:

The EXOR form of Shannon expansion

$$\overbrace{f\,(x_1, x_2, \ldots, x_n)}^{n\ variables} = \overline{x}_i f\,\overbrace{(x_1, \ldots, x_{i-1}\,\boxed{0}\,x_{i+1}, \ldots, x_n)}^{n-1\ variables}$$
$$\underset{EXOR}{\oplus}\ x_i f\,\underbrace{(x_1, \ldots, x_{i-1}\,\boxed{1}\,x_{i+1}, \ldots, x_n)}_{n-1\ variables} \quad (5.6)$$

It follows from this fact that

$$\overline{x}_i\ a\ \oplus x_i\ b = \overline{x}_i\ a\ \overline{x_i\ b} \vee \overline{\overline{x}_i\ a}\ x_i\ b = \overline{x}_i\ a\ (\overline{x}_i \vee \overline{b}) \vee (x_i \vee \overline{a})x_i\ b$$
$$= \overline{x}_i\ a\ \vee \overline{x}_i\ a\ \overline{b} \vee x_i\ b \vee x_i\ \overline{a}\ b = \overline{x}_i\ a \vee x_i\ b$$

Equation 5.6 is a polynomial form of the Shannon expansion. This expansion is used in the following form

Positive Davio expansion

$$f = \overline{x}_i f_0 \oplus x_i f_1 = (1 \oplus x_i) f_0 \oplus x_i f_1$$
$$= f_0 \oplus x_i f_0 \oplus x_i f_1 = f_0 \oplus x_i (f_0 \oplus f_1) \quad (5.7)$$

Equation 5.7 is called *the positive Davio* expansion with respect to the variable x_i (details are given in Chapter 12).

Example 5.8 (The EXOR form of Shannon expansion.) *Let $f = x_1 \vee \overline{x}_2$. Then the Shannon expansion in SOP and polynomial forms is derived as follows:*

SOP form:

$$f = \overline{x}_1 f_0 \vee x_2 f_1$$
$$= \overline{x}_1(0 \vee \overline{x}_2) \vee x_1(1 \vee \overline{x}_2)$$
$$= \overline{x}_1(\overline{x}_2) \vee x_1(1)$$
$$= \overline{x}_1\overline{x}_2 \vee x_1 = \overline{x}_2 \vee x_1$$

Polynomial form:

$$f = \overline{x}_1 f_0 \oplus x_1 f_1$$
$$= \overline{x}_1(0 \vee \overline{x}_2) \oplus x_1(1 \vee \overline{x}_2)$$
$$= \overline{x}_1\overline{x}_2 \oplus x_1$$

Both forms are equivalent:

$$f = \overline{x}_1\overline{x}_2 \oplus x_1 \quad (polynomial\ form)$$
$$= \overline{\overline{x}_1\overline{x}_2} x_1 \vee \overline{x}_1\overline{x}_2\overline{x}_1$$
$$= (x_1 \vee x_2)x_1 \vee \overline{x}_1\overline{x}_2$$
$$= x_1 \vee x_1 x_2 \vee \overline{x}_1\overline{x}_2$$
$$= x_1 \vee \overline{x}_1\overline{x}_2 \vee \overline{x}_2 = x_1 \vee \overline{x}_2 \quad (SOP\ form)$$

Practice problem 5.5. (The EXOR form of Shannon expansion.) Given the Boolean function $f = \overline{x}_1 x_2 \vee x_1 \overline{x}_2$, find its EXOR Shannon expansion with respect to x_1 in polynomial form.
Answer: $f = \overline{x}_1(\overline{0}{\cdot}x_2 \vee 0{\cdot}\overline{x}_2) \oplus x_1(\overline{1}{\cdot}x_2 \vee 1{\cdot}\overline{x}_2) = \overline{x}_1(x_2) \oplus x_1(\overline{x}_2) = \overline{x}_1 x_2 \oplus x_1 \overline{x}_2$.

5.3 Shannon expansion for symmetric Boolean functions

Symmetric functions are the special class of Boolean functions. These functions are characterized by properties of symmetry and can be efficiently used in design of networks with interchangeable wires of subfunctions. A typical problem of logic design is detection of symmetric properties, since knowledge of the symmetry properties of a Boolean function can be used for more efficient implementation of functions.

Shannon theorem provides a necessary and sufficient conditions for detection of symmetry in a Boolean function f of n variables, and for the representation of this function that utilizes this symmetry. Shannon expansion for symmetric functions is the third fundamental expansion of logic design (Figure 5.1). These theorems also provide a flexibility in logic design. In particular, Shannon theorems can be applied at the *global* and *local* levels of design; that is, detection of symmetries and expansion can be used for a total Boolean function (global) or its part (local); and these theorems can be used in various Boolean data structures.

5.3.1 Symmetric functions

There are two groups of symmetric Boolean functions:

(*a*) If two variables in a Boolean function f can be exchanged without changing the function itself, it is said that this function is *partially symmetric* with respect to these two variables.

(*b*) The boundary case of symmetry is *totally symmetric* functions: in totally symmetric function, all variables can be exchanged without changing the function itself.

The symmetry properties of Boolean functions can be recognized using detection algorithms on various data structures:

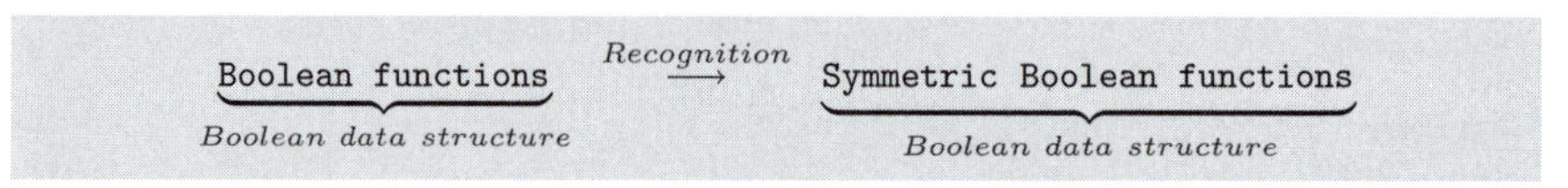

These include the algebraic forms of Boolean functions if they are available, decision diagrams, which are preferable for large-size functions, and logic networks.

> **Example 5.9 (Symmetric functions.)** *The Boolean functions of three variables (a)* $f = \overline{x}_1\overline{x}_2\overline{x}_3 \vee x_1x_2x_3$ *and (b)* $f = x_1\overline{x}_2x_3 \vee \overline{x}_1x_2x_3 \vee x_1x_2\overline{x}_3$ *are totally symmetric, since permuting the variables* $x_1, x_2,$ *and* x_3 *and* $x_1, x_2,$ *and* x_3*, respectively, leaves the function unchanged.*

Symmetric properties are specified in terms of Boolean data structure:

- In algebraic forms, symmetric properties are specified using permutation of variables in of Boolean expressions,
- In logic networks, symmetric properties are specified in terms of the permutation of inputs,
- In decision trees and diagrams, symmetric properties are specified by the assigned values of the outgoing branches of a node.

However, in all data structure types, the manipulation of their components must leaves the represented Boolean function unchanged.

> **Example 5.10 (Symmetry property.)** *The standard SOP form of the two-input OR function* $f = x_1 \vee x_2$ *is defined by three minterms:* $x_1\overline{x}_2, \overline{x}_1x_2$ *and* x_1x_2*. These minterms specify that the OR function is equal to 1 for the assignments of variables* $x_1x_2 = \{01,\ 10,\ 11\}$*. However, the OR function is symmetric in variables* x_1 *and* x_2*. That is,* f *takes value 1 for assignments* $x_1x_2 = \{01,\ 10\}$*. Observe that the number of 1s in these two assignments is the same (1).*

5.3.2 Partially symmetric functions

> A Boolean function f is *partially symmetric* in k variables $x_1, x_2, ..., x_k$, $k < n$, if and only if it is unchanged by any permutation of these k variables only.

In other words, a Boolean function f is *partially symmetric* in an arbitrary pair of variables x_i and x_j if

$$f(x_1, x_2, ..., x_{i-1}, \boxed{0}, x_{i+1}, ..., x_{j-1}, \boxed{1}, x_{j+1}, ..., x_n)$$
$$= f(x_1, x_2, ..., x_{i-1}, \boxed{1}, x_{i+1}, ..., x_{j-1}, \boxed{0}, x_{j+1}, ..., x_n)$$

Hence, a Boolean function is partially symmetric symmetric if it is partially symmetric in any pair of variables x_i and x_j, for all i, j. There are $2^{3\cdot 2^{n-2}}$ partially symmetric functions of n variables with respect to variables x_i and x_j.

Example 5.11 (Partially symmetric functions.) *The Boolean functions:*
(a) $f = x_2\overline{x}_3 \vee \overline{x}_2x_3 \vee x_1x_2x_3$ *is partially symmetric with respect to variables* x_2 *and* x_3 *(Figure 5.4).*
(b) $f = x_1x_2 \vee x_3$ *is partially symmetric with respect to variables* x_1 *and* x_2.
(b) $f = x_1\overline{x}_2x_3\overline{x}_4 \vee \overline{x}_1x_2\overline{x}_3x_4$ *is partially symmetric with respect to variables* x_1 *and* x_3 *and also with respect to the variables* x_2 *and* x_4.
There are $2^{3\cdot 2^{n-2}} = 2^{3\cdot 2^{3-2}} = 64$ *functions of three* $(n = 3)$ *variables partially symmetric with respect to two variables.*

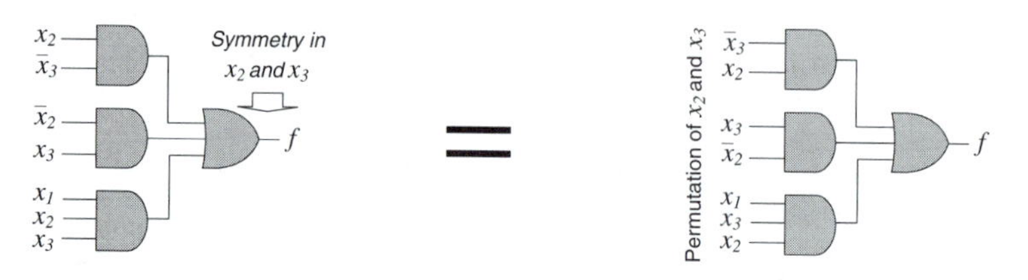

FIGURE 5.4
Logic networks for partially symmetric functions with respect variables x_2 and x_3 (Example 5.11a).

5.3.3 Totally symmetric functions

A Boolean function f is *partially symmetric symmetric* if and only if it is unchanged by any permutation of variables.

There are 2^{n+1} totally symmetric functions of n variables. However, they are only small subclass of partially symmetric functions, that is,

$$\frac{\text{Number of partially symmetric functions}}{\text{Number of totally symmetric functions}} = \frac{2^{3\cdot 2^{n-2}}}{2^{n+1}} = 2^{3\cdot 2^{n-2}-n+1}$$

Example 5.12 (Totally symmetric functions.) *There are* $2^{n+1} = 2^{2+1} = 8$ *totally symmetric Boolean functions of two variables (Figure 5.5).*

Practice problem **5.6. (Totally symmetric functions.)** Calculate the number of partially and totally symmetric functions in 2^{2^n} Boolean functions of n variables for $n = 2, 3, 4$.
Answer is given in "Solutions to practice problems."

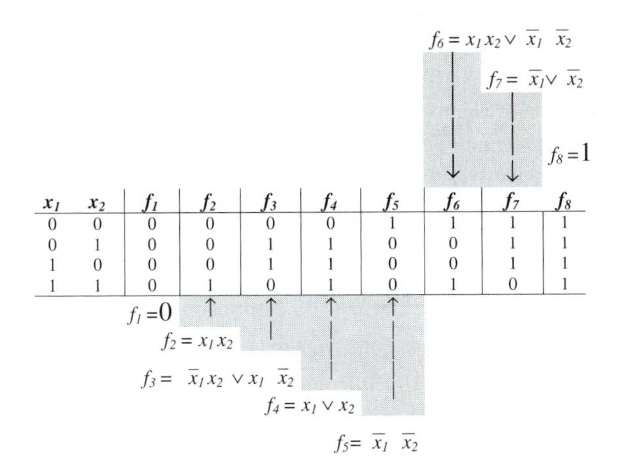

x_1	x_2	f_1	f_2	f_3	f_4	f_5	f_6	f_7	f_8
0	0	0	0	0	0	1	1	1	1
0	1	0	0	1	1	0	0	1	1
1	0	0	0	1	1	0	0	1	1
1	1	0	1	0	1	0	1	0	1

FIGURE 5.5
Totally symmetric functions of two variables (Example 5.12).

Practice problem 5.7. (Totally symmetric functions.) Determine if the logic network given in Figure 5.6 implements a totally symmetric function.
Answer: The given network implements the function $f = \overline{x}_1\overline{x}_2 \vee \overline{x}_2\overline{x}_3 \vee \overline{x}_1\overline{x}_3$. Permuting the variables x_1, x_2, and x_3 does not change the function, hence, the network implements a totally symmetric function.

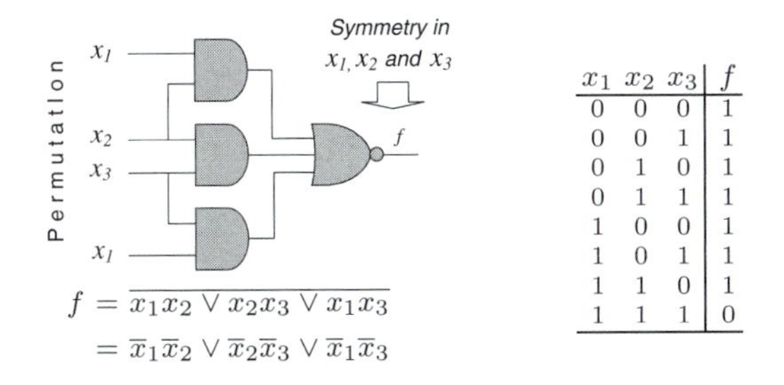

$$f = \overline{x_1x_2 \vee x_2x_3 \vee x_1x_3}$$
$$= \overline{x}_1\overline{x}_2 \vee \overline{x}_2\overline{x}_3 \vee \overline{x}_1\overline{x}_3$$

x_1	x_2	x_3	f
0	0	0	1
0	0	1	1
0	1	0	1
0	1	1	1
1	0	0	1
1	0	1	1
1	1	0	1
1	1	1	0

FIGURE 5.6
A logic network that implements a totally symmetric Boolean function (Practice problem 5.7).

5.3.4 Detection of symmetric Boolean functions

The following theorem by Shannon provide conditions for detection and computing symmetric Boolean functions:

Shannon theorem for detection of symmetric Boolean functions

A necessary and sufficient condition that a Boolean function f of n variables be symmetric is that it may be specified by starting a set of numbers $a_0, a_1, \dots, a_k$ such that if exactly a_j, $j = 1, 2, \dots a_k$, $0 \leq k \leq n$, of these variables of symmetry have the value 1, then the function has the value 1, and not otherwise

It follows from this Shannon theorem, that

(*a*) The set of numbers $a_0, a_1, \dots, a_k$ may be any set of positive integer numbers selected from the numbers zero to n inclusive.

(*b*) The values of a symmetric function are determined ***only by the number of literals which take the value 1*** (the literal is a complemented or uncomplemented variable).

This theorem is the base for development of various techniques for symmetry detection.

5.3.5 Characteristic set

A set of integers numbers a_j, $j = 0, 1, \dots k$, defined by Shannon theorem for detecting of symmetric function f of n variables is called the *characteristic set.* The structure of the complete characteristic set can be represented in terms of minterms as follows:

$$A = \overbrace{\{ \underbrace{\overset{M_{i_0}}{\downarrow}\boxed{a_0}, \overset{M_{i_1}}{\downarrow}\boxed{a_1}, \overset{M_{i_2}}{\downarrow}\boxed{a_2}, \dots, \overset{M_{i_n}}{\downarrow}\boxed{a_n}}_{Characteristic\ set} \}}^{Minterms} \quad (5.8)$$

where each number of the set corresponds to the group of minterms, except a_0 and a_n that correspond to a single minterms $M_{i_0} = M_0$ and $M_{i_n} = M_n$, respectively.

Example 5.13 (Characteristic set.) *Techniques for computing of characteristic sets for symmetric Boolean functions are given in Table 5.2. Note that the first two functions are totally symmetric, and the last function is partially symmetric with respect to the variables* x_1 *and* $\overline{x}_2$.

5.3.6 Elementary symmetric functions

Using Shannon theorem for detecting symmetric functions, the *elementary symmetric functions* $S_0^n, S_1^n, \dots, S_n^n$ of n variables can be defined as follows:

TABLE 5.2
Technique for computing the characteristic set for symmetric Boolean functions.

Technique for computing the characteristic set

Symmetric function	Characteristic set
$f = x_1 \vee x_2$ $x_1\ x_2\ f$ MINTERM 0 0 0 0 1 1 M_1 1 0 1 M_2 1 1 1 M_3	Standard SOP representation $f = M_1 \vee M_2 \vee M_3$. Characteristic set $A = \{\boxed{a_1}, \boxed{a_2}\} = \{1, 2\}$ ($M_1, M_2 \to a_1$; $M_3 \to a_2$) $a_1 \to \{M_1, M_2\} = M_{i_1}$ (Equation 5.8) $a_2 \to \{M_3\} = M_{i_2}$
$f = x_1 \oplus x_2 \oplus x_3$ $x_1\ x_2\ x_3\ f$ MINTERM 0 0 0 0 0 0 1 1 M_1 0 1 0 1 M_2 0 1 1 0 1 0 0 1 M_4 1 0 1 0 1 1 0 0 1 1 1 1 M_7	Standard SOP representation $f = M_1 \vee M_2 \vee M_4 \vee M_7$. Characteristic set $A = \{\boxed{a_1}, \boxed{a_3}\} = \{1, 3\}$ ($M_1, M_2, M_4 \to a_1$; $M_7 \to a_3$) $a_1 \to \{M_1, M_2, M_4\} = M_{i_1}$ (Equation 5.8) $a_3 \to \{M_7\} = M_{i_3}$
$f = x_1\overline{x}_2\overline{x}_3 \vee \overline{x}_1 x_2 \overline{x}_3 \vee \overline{x}_1\overline{x}_2 x_3$ $x_1\ x_2\ x_3\ f$ MINTERM 0 0 0 0 0 0 1 1 M_1 0 1 0 1 M_2 0 1 1 0 1 0 0 1 M_4 1 0 1 0 1 1 0 0 1 1 1 1	Standard SOP representation $f = M_1 \vee M_2 \vee M_4$. Characteristic set $A = \{\boxed{a_1}\} = \{1\}$ ($M_1, M_2, M_4 \to a_1$) $a_1 \to \{M_1, M_2, M_4\} = M_{i_1}$ (Equation 5.8)

$$S_0^n = \overbrace{\overline{x}_1\overline{x}_2\cdots\overline{x}_{n-1}\overline{x}_n}^{0\ inputs\ are\ equal\ to\ 1}$$

$$S_1^n = \overbrace{x_1\overline{x}_2\cdots\overline{x}_{n-1}\overline{x}_n}^{1\ input\ is\ equal\ to\ 1} \vee \overbrace{\overline{x}_1 x_2\overline{x}_3\cdots\overline{x}_{n-1}\overline{x}_n}^{1\ input\ is\ equal\ to\ 1} \vee \cdots \vee \overbrace{\overline{x}_1\overline{x}_2\cdots\overline{x}_{n-1}x_n}^{1\ input\ is\ equal\ to\ 1}$$

$$\cdots\cdots\cdots\cdots\cdots\cdots$$

$$S_n^n = \overbrace{x_1 x_2 \ldots x_{n-1}x_n}^{n\ input\ are\ equal\ to\ 1}$$

The Boolean function $S_i^n = 1$ iff exactly i out of n inputs are equal to one. That is, for all assignments of values to variables that have the same number of 1's, there is *exactly* one value of a symmetric function. The elementary symmetric functions $S_0^n, S_1^n, \ldots, S_n^n$ can be written in terms of minterms:

$$S_0^n = M_0, \; S_1^n = \underbrace{M_1 \vee M_2 \vee M_4 \vee \cdots \vee M_{2^t}}_{Group\ of\ minterms}, \ldots, S_n^n = M_n$$

In this notation, each elementary symmetric function is described by the group of minterms; the numbers of minterms can be described by the equation. For example, the elementary symmetric function S_1^n is described by the minterms with numbers $2^t, t = 0, 1, \ldots, T$, $T < 2^n$. Each minterm of this group consists of one uncomplemented variable in a product and corresponds element a_1 of the characteristic set.

Example 5.14 (Elementary symmetric functions.)
There are 3 elementary symmetric Boolean functions of two variables (Table 5.3):

$S_0^2 = M_0 = \overline{x}_1\overline{x}_2$ – *exactly* 0 *out of* 2 *inputs are equal to 1.*
$S_1^2 = M_1 \vee M_2 = x_1 \oplus x_2$ – *exactly* 1 *out of* 2 *inputs are equal to 1.*
$S_2^2 = M_3 = x_1x_2$ – *exactly* 2 *out of* 2 *inputs are equal to 1.*

TABLE 5.3
Elementary symmetric functions of two variables S_0^2, S_1^2, S_2^2, and their composition (Examples 5.14 and 5.16).

x_1	x_2	S_0^2	S_1^2	S_2^2	$S_{0,2}^2$	$S_{0,1}^2$	$S_{1,2}^2$	$S_{0,1,2}^2$
0	0	1	0	0	1	1	0	1
0	1	0	1	0	0	1	1	1
1	0	0	1	0	0	1	1	1
1	1	0	0	1	1	0	1	1
		Elementary functions			*Composed symmetric functions*			

Practice problem 5.8. (Elementary symmetric functions.)
Derive elementary symmetric Boolean functions of three variables.
Answer is given in "Solutions to practice problems."

5.3.7 Operations on elementary symmetric functions

Shannon theorem for detection of symmetric functions also provides for a considerable manipulation and simplification of symmetric Boolean functions in algebraic form. In manipulation, the characteristic sets are computed using the *set operation union* $\cup$, the *intersection* $\cap$, and *set difference* $\setminus$ operations on sets can be used in manipulation of characteristic sets.

Example 5.15 (Operations on the sets of integer numbers.) *Given symmetric Boolean functions f and g of three variables, and the characteristic sets of f and g, $A = \{1, 2, 4\}$ and $A = \{3, 5\}$, respectively. Operations on these sets are listed below:*

UNION: $A \cup B = \{1, 2, 3.4.5\}$
INTERSECTION: $A \cap B\{4\}$
SET DIFFERENCE:

$$\overline{A} = \{0, 1, 2, 3, 4, 5, 7\}\backslash\{1, 2, 4\} = \{0, 3, 5, 6, 7\}$$
$$\overline{B} = \{0, 1, 2, 3, 4, 5, 7\}\backslash\{3, 5\} = \{0, 1, 2, 4, 6, 7\}$$

Operations on elementary symmetric functions

Let the Boolean functions of n variables $f = S_A^n$ and $g = S_B^n$ be symmetric functions with respect to the set of variables A and B, $A, B \subseteq \{1, 2, \ldots n\}$, respectively; then the functions

LOGIC SUM: $f \vee g = S_{A \cup B}^n$
LOGIC MULTIPLICATION: $f \cdot g = S_{A \cap B}^n$
COMPLEMENT: $\overline{f} = S_{\overline{A}}^n$ and $\overline{g} = S_{\overline{B}}^n$

are also symmetric functions.

New symmetric functions can be formed from the elementary symmetric Boolean functions.

Example 5.16 (Operation on elementary symmetric functions.) *In Table 5.3, symmetric functions $S_{0,2}^2, S_{0,1}^2, S_{1,2}^2$ and $S_{0,1,2}^2$ are formed as follows:*

(*a*) $S_0 \vee S_2 = S_{0 \cup 2} = S_{0,2}^2 = x_1 \oplus \overline{x}_2$ – *exactly* 0 *or* 2 *out of* 2 *inputs are equal to 1.*
(*b*) $S_0 \vee S_1 = S_{0 \cup 1} = S_{0,1}^2 = \overline{x}_1 \vee \overline{x}_2$ – *exactly* 0 *or* 1 *out of* 2 *inputs are equal to 1.*
(*c*) $S_1 \vee S_2 = S_{1 \cup 2} = S_{1,2}^2 = x_1 \vee x_2$ – *exactly* 1 *or* 2 *out of* 2 *inputs are equal to 1.*
(*d*) $S_0 \vee S_1 \vee S_2 = S_{0 \cup 1 \cup 2} = S_{0,1,2}^2 = 1$ – *exactly* 0, 1, *or* 2 *out of* 2 *inputs are equal to 1.*

Example 5.17 (Operations on symmetric functions.)
The following operation can be used to generate new symmetric functions given symmetric functions of three variables:

LOGIC SUM: $S^4_{0,1,2,4} \vee S^4_{2,3,4} = S^4_{\underbrace{\{1,2,4\}\cup\{2,3,4\}}_{Union}} = S^4_{0,1,2,3,4}$

LOGIC MULTIPLICATION: $S^4_{0,1,2,4} \cdot S^4_{2,3,4} = S^4_{\underbrace{\{1,2,4\}\cap\{2,3,4\}}_{Intersection}} = S^4_{2,4}$

COMPLEMENT: $\overline{S}^7_{1,2,3,6} = S^7_{\underbrace{\{0,1,2,3,4,5,6,7\}\setminus\{1,2,3,6\}}_{Set\ difference}} = S^6_{0,4,5}$

Practice problem 5.9. **(Operations on symmetric functions.)**
Given the elementary symmetric functions S^3_1 and S^3_2 of three variables, find the new symmetric functions using the possible logic operations.
Answer is given in "Solutions to practice problems."

5.3.8 Shannon expansion with respect to a group of symmetric variables

A convenient way of the representation of symmetric function is provided by *Shannon expansion for symmetric functions*:

Shannon expansion for symmetric Boolean functions

Any Boolean function of n variables symmetric in the variables $x_1, x_2 \ldots, x_m$

$$f = f(\overbrace{x_1, x_2, \ldots, x_m}^{Symmetric\ variables}, \underbrace{x_{m+1}, \ldots, x_{n-1}, x_n}_{Non-symmetric\ variables})$$

can be represented as

$$f = \bigvee_{i=0}^{m} \overbrace{S_i^m(x_1, x_2, \ldots, x_m)}^{Elementary\ symmetric\ function} f_i(\underbrace{x_{m+1}, \ldots, x_{n-1} x_n}_{Non-symmetric\ variables}) \quad (5.9)$$

where

$$f_i(x_{m+1}, \ldots, x_{n-1} x_n) = f(\overbrace{0, \ldots, 0}^{i\ zeros}, \overbrace{1, \ldots 1}^{(m-i)\ ones}, \underbrace{x_{m+1}, \ldots, x_{n-1}, x_n}_{Non-symmetric\ variables})$$

Shannon expansion for symmetric function (Equation 5.9) assumes that the symmetry variables are already detected; that is, this expansion can be applied only to symmetric functions.

5.4 Techniques for computing symmetric functions

Techniques for computing partially and totally symmetric functions are based on the Shannon expansion for symmetric functions. The final goal is the representation of symmetric functions for efficient implementation, that is, efficient utilization of a symmetry property of Boolean function.

5.4.1 Computing partially symmetric functions

Techniques for computing partially symmetric functions are based on Shannon expansion (Equation 5.9).

Example 5.18 (Shannon expansion for symmetric functions.) *The Boolean function* $f = x_1x_2x_3x_4 \vee x_1\overline{x}_2x_3 \vee x_1x_2\overline{x}_3 \vee x_1\overline{x}_2$ *is symmetric in variables* x_2 *and* x_3*, and can be represented in the form shown in Figure 5.7.*

Practice problem 5.10. **(Shannon expansion for symmetric functions.)** Given the Boolean function $f = \overline{x}_3x_4 \vee \overline{x}_1\overline{x}_2x_4 \vee x_1x_2$ which is symmetric in variables x_1x_2, represent it using Shannon theorem on partially symmetric functions.
Answer is given in "Solutions to practice problems."

5.4.2 Computing totally symmetric functions

Techniques for computing totally symmetric functions are based on Shannon expansion (Equation 5.9) for $m = n$:

Shannon expansion for totally symmetric functions

A totally symmetric function f can be uniquely represented by elementary symmetric functions:

$$f = \bigvee_{i \in A} S_i^n \tag{5.10}$$

where $A \subseteq 0, 1, 2, \ldots n$.

Techniques for computing partially symmetric Boolean functions

Given: Boolean function of three variables symmetric in variables x_2 and x_3

$$f = x_1x_2x_3x_4 \vee x_1\overline{x}_2x_3 \vee x_1x_2\overline{x}_3 \vee x_1\overline{x}_2$$

Find: The implementation

Step 1: Shannon expansion with respect to symmetric variables x_2 and x_3

$$f = \bigvee_{i=0}^{2} S_i^2 f_i$$

(a) Assign $x_2x_3 = \{00\} \Rightarrow$ $= S_0^2(x_2,x_3)\, f_0 \overbrace{\left(x_1, \boxed{0}, \boxed{0}, x_4\right)}^{Equal\ to\ x_1x_4}$

(b) Assign $x_2x_3 = \{10, 01\} \Rightarrow$ $\vee\ S_1^2(x_2,x_3)\, f_1 \overbrace{\begin{pmatrix} x_1, \boxed{1}, \boxed{0}, x_4 \\ x_1, \boxed{0}, \boxed{1}, x_4 \end{pmatrix}}^{Equal\ to\ x_1}$

(c) Assign $x_2x_3 = \{11\} \Rightarrow$ $\vee\ S_2^2(x_2,x_3)\, f_2 \overbrace{\left(x_1, \boxed{1}, \boxed{1}, x_4\right)}^{Equal\ to\ x_1}$

$$= S_0^2(x_2,x_3)x_1x_4 \vee S_1^2(x_2,x_3)x_1 \vee S_2^2(x_2,x_3)x_1$$

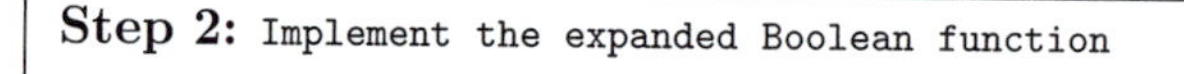

Step 2: Implement the expanded Boolean function

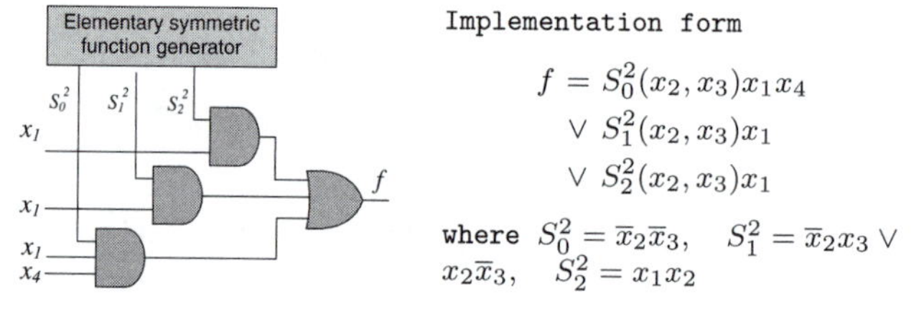

Implementation form

$$f = S_0^2(x_2,x_3)x_1x_4 \vee S_1^2(x_2,x_3)x_1 \vee S_2^2(x_2,x_3)x_1$$

where $S_0^2 = \overline{x}_2\overline{x}_3$, $S_1^2 = \overline{x}_2x_3 \vee x_2\overline{x}_3$, $S_2^2 = x_1x_2$

FIGURE 5.7
Computing partially symmetric function based on Shannon expansion for symmetric functions (Example 5.18).

Example 5.19 (Shannon expansion for symmetric functions.) *Table 5.4 shows two examples of expansion of totally symmetric functions.*

Practice problem 5.11. (Shannon expansion for symmetric functions.) Expand the function $f = x_1\overline{x}_2\overline{x}_3 \vee \overline{x}_1x_2\overline{x}_3 \vee \overline{x}_1\overline{x}_2x_3$.
Answer is given in "Solutions to practice problems."

Example 5.20 (Elementary symmetric functions.) *Given the totally symmetric Boolean functions of three variables, find its representation in terms of elementary symmetric functions:*
(a) $f = \bigvee m(1,2,4,7)$, *its representation in terms of elementary symmetric functions as follows:*

$$f = \overbrace{\overline{x}_1\overline{x}_2x_3 \vee \overline{x}_1x_2\overline{x}_3 \vee x_1\overline{x}_2\overline{x}_3}^{S_1^3} \vee \overbrace{x_1x_2x_3}^{S_3^3} = S_1^3 \vee S_3^3 = S_{1,3}^3$$

(b) $f = (x_1 \vee x_2)(x_2 \vee x_3)(x_1 \vee x_3)$. *The solution is given in Figure 5.8. Note, that this function is called a majority function.*

Practice problem 5.12. (Elementary symmetric functions.) Given the totally symmetric Boolean function $f = \overline{x}_1\overline{x}_2 \vee \overline{x}_2\overline{x}_3 \vee \overline{x}_1\overline{x}_3$ (Figure 5.6), find its representation in terms of elementary symmetric functions.
Answer is given in "Solutions to practice problems."

Practice problem 5.13. (Elementary symmetric functions.) Given the totally symmetric Boolean function $f = (x_1 \vee x_2)(x_2 \vee x_3)(x_1 \vee x_3)$, (a) find its representation in terms of elementary symmetric functions, and (b) berify that $f = (x_1 \vee x_2)(x_2 \vee x_3)(x_1 \vee x_3) = x_1S_{1,2}^2 \vee \overline{x}_1S_{2,3}^2$, where $S_{1,2}^2$ and $S_{2,3}^2$ are functions of the variables x_2 and x_3.
Answer is given in "Solutions to practice problems."

Example 5.21 (Even parity check function.) *The n variable even parity check function (n is even) and odd parity check function are totally symmetric Boolean functions*

$$\text{EVEN PARITY CHECK FUNCTION} = S_0^n \vee S_2^n \vee \cdots \vee S_n^n$$
$$\text{ODD PARITY CHECK FUNCTION} = S_1^n \vee S_3^n \vee \cdots \vee S_{n-1}^n$$

respectively. In particular, given $n = 2$, even and odd parity check functions are $S_{0,2}^2 = S_0^2 \vee S_2^2 = \overline{x}_1\overline{x}_2 \vee x_1x_2$ *and* $S_1^2 = x_1\overline{x}_2 \vee \overline{x}_1x_2$, *respectively.*

TABLE 5.4
Computing totally symmetric Boolean functions using Shannon expansion for symmetric functions.

Techniques for computing totally symmetric Boolean functions

Symmetric function	Shannon expansion for symmetric functions

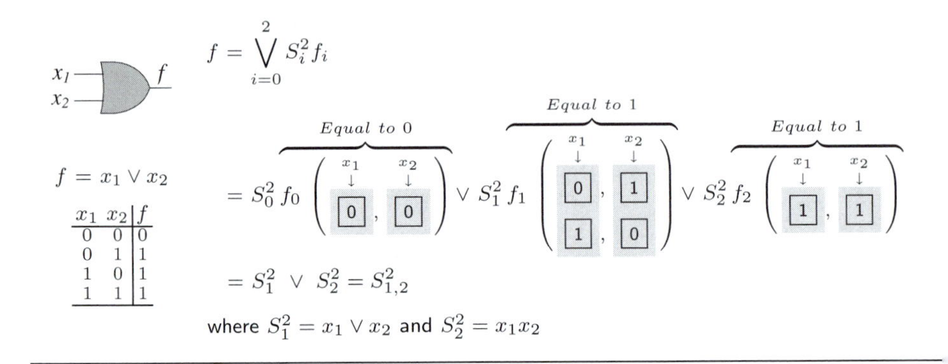

$f = \bigvee_{i=0}^{2} S_i^2 f_i$
$f = x_1 \vee x_2$
Equal to 0
Equal to 1
Equal to 1
$= S_0^2 f_0 \vee S_1^2 f_1 \vee S_2^2 f_2$
$= S_1^2 \vee S_2^2 = S_{1,2}^2$
where $S_1^2 = x_1 \vee x_2$ and $S_2^2 = x_1 x_2$

$f = x_1 \oplus x_2 \oplus x_3$

x_1	x_2	x_3	f
0	0	0	0
0	0	1	1
0	1	0	1
0	1	1	0
1	0	0	1
1	0	1	0
1	1	0	0
1	1	1	1

$$f = \bigvee_{i=0}^{3} S_i^3 f_i = S_0^3 f_0 \underbrace{\begin{pmatrix} x_1 & x_2 & x_3 \\ \downarrow & \downarrow & \downarrow \\ \boxed{0}, & \boxed{0}, & \boxed{0} \end{pmatrix}}_{\text{Equal to } 0}$$

$$\vee\ S_1^3 f_1 \underbrace{\begin{pmatrix} x_1 & x_2 & x_3 \\ \downarrow & \downarrow & \downarrow \\ \boxed{0}, & \boxed{0}, & \boxed{1} \\ \boxed{0}, & \boxed{1}, & \boxed{0} \\ \boxed{1}, & \boxed{0}, & \boxed{0} \end{pmatrix}}_{\text{Equal to } 1}$$

$$\vee\ S_2^3 f_1 \underbrace{\begin{pmatrix} x_1 & x_2 & x_3 \\ \downarrow & \downarrow & \downarrow \\ \boxed{0}, & \boxed{1}, & \boxed{1} \\ \boxed{1}, & \boxed{1}, & \boxed{0} \\ \boxed{1}, & \boxed{0}, & \boxed{1} \end{pmatrix}}_{\text{Equal to } 0}$$

$$\vee\ S_3^3 f_2 \underbrace{\begin{pmatrix} x_2 & x_2 & x_3 \\ \downarrow & \downarrow & \downarrow \\ \boxed{1}, & \boxed{1}, & \boxed{1} \end{pmatrix}}_{\text{Equal to } 1}$$

$$= S_1^3 \vee S_3^3 = S_{1,3}^3$$

where $S_1^3 = \overline{x}_1 \overline{x}_2 x_3 \vee \overline{x}_1 x_2 x_3$ and $S_3^3 = x_1 x_2 x_3$

Techniques for computing totally symmetric Boolean functions

Given: Totally symmetric Boolean function of three variables

$$f = (x_1 \vee x_2)(x_2 \vee x_3)(x_1 \vee x_3)$$

Find: The implementation

Step 1: Shannon expansion with respect to symmetric variables x_1, x_2, and x_3:

$$f = \bigvee_{i=0}^{3} S_i^2 f_i$$

(a) Assign $x_1x_2x_3 = \{000\} \Rightarrow$

$$= S_0^3\, f_0 \underbrace{\left(\begin{matrix} x_1 & x_2 & x_3 \\ \downarrow & \downarrow & \downarrow \\ 0, & 0, & 0 \end{matrix}\right)}_{Equal\ to\ 0}$$

(b) Assign $x_1x_2x_3 = \{001, 010, 100\} \Rightarrow$

$$\vee\ S_1^3\, f_1 \underbrace{\left(\begin{matrix} x_1 & x_2 & x_3 \\ \downarrow & \downarrow & \downarrow \\ 1, & 0, & 0 \\ 0, & 1, & 0 \\ 0, & 0, & 1 \end{matrix}\right)}_{Equal\ to\ 0}$$

(c) Assign $x_1x_2x_3 = \{011, 110, 101\} \Rightarrow$

$$\vee\ S_2^3\, f_2 \underbrace{\left(\begin{matrix} x_1 & x_2 & x_3 \\ \downarrow & \downarrow & \downarrow \\ 1, & 1, & 0 \\ 0, & 1, & 1 \\ 1, & 0, & 1 \end{matrix}\right)}_{Equal\ to\ 1}$$

(d) Assign $x_1x_2x_3 = \{111\} \Rightarrow$

$$\vee\ S_3^3\, f_3 \underbrace{\left(\begin{matrix} x_1 & x_2 & x_3 \\ \downarrow & \downarrow & \downarrow \\ 1, & 1, & 1 \end{matrix}\right)}_{Equal\ to\ 1}$$

$$= S_0^3 \cdot 0 \ \vee\ S_1^3 \cdot 0$$
$$\vee\ S_2^3 \cdot 1 \ \vee\ S_3^3 \cdot 1$$
$$= S_2^3 \vee S_3^3$$

Step 2: Implement the expanded Boolean function

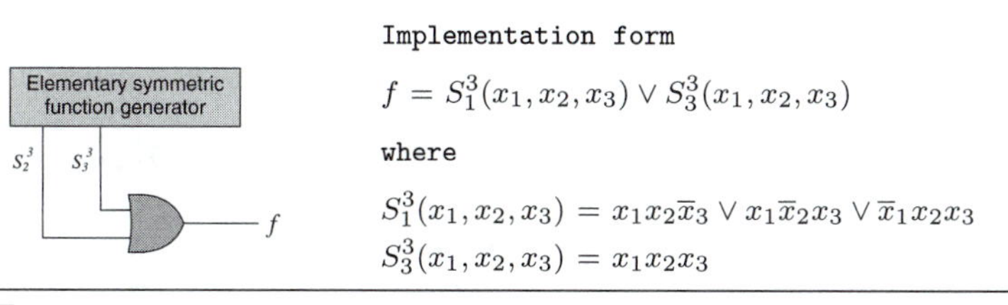

Implementation form

$$f = S_1^3(x_1, x_2, x_3) \vee S_3^3(x_1, x_2, x_3)$$

where

$$S_1^3(x_1, x_2, x_3) = x_1x_2\overline{x}_3 \vee x_1\overline{x}_2x_3 \vee \overline{x}_1x_2x_3$$
$$S_3^3(x_1, x_2, x_3) = x_1x_2x_3$$

FIGURE 5.8
Computing totally symmetric function based on Shannon expansion for symmetric functions (Example 5.20).

5.4.3 Carrier vector

Characteristic sets A can be used for the compression of a truth vectors of symmetric functions. For a totally symmetric function f symmetric in n variables that is represented by a truth vector $\mathbf{F}$, this compression is significant and can be estimate as

$$\texttt{COMPRESSION} = \frac{\texttt{Size of the truth vector } \mathbf{F}}{\texttt{Size of the carrier vector } \mathbf{F}_c} = \frac{2^n}{n}$$

where $\mathbf{F}_c$ is the compressed truth vector called the *carrier vector* of a totally symmetric function. For example, for $n = 10$, $\texttt{COMPRESSION} = 2^{10}/10 = 102.4$.

Denote f_{a_j}, $0 \leq j \leq n$, the value of Boolean function f for assignments of the variables $x_1, x_2, \ldots, x_n$ that have j 1's. Then the carrier vector $\mathbf{F}_c$ can be specified as follows:

Carrier vector

A totally symmetric function $f(x_1, x_2, \ldots, x_n)$ is completely specified by a *carrier vector*

$$\mathbf{F}_c = [f_{a_0}\ f_{a_1} \cdots\ f_{a_n}]^T$$

such that if exactly a_j, $j = 0, 1, 2, \ldots n$ of the variables $x_1, x_2, \ldots, x_n$ have the value 1, then the function has the value f_{a_j} for all assignments of values to $x_1, x_2, \ldots, x_n$ that have j 1's, where $0 \leq j \leq n$.

Example 5.22 (Carrier vector.) *The carrier vector for the totally symmetric function of three variables is formed as shown in Figure 5.9.*

Example 5.23 (Carrier vector.) *The totally symmetric function $f = \overline{x}_1\overline{x}_2\overline{x}_3 \vee x_1x_2x_3$ is specified by the carrier vector $\mathbf{F}_c = [1\ 0\ 0\ 1]^T$, since the function is 1 if and only if zero or three of the variables are 1 (Table 5.5).*

Practice problem **5.14. (Symmetric functions.)** Given:
(a) The totally symmetric function $f = x_1 \vee x_2 \vee x_3$, find its career vector.
(b) The career vector $\mathbf{F}_c = [\ 1010\ ]^T$ of a totally symmetric function, write this function in a standard SOP form.
Answer: (a) $\mathbf{F}_c = [\ 0111\ ]^T$; (b) $f = \overline{x}_1\overline{x}_2\overline{x}_3 \vee \overline{x}_1x_2x_3 \vee x_1\overline{x}_2x_3 \vee x_1x_2\overline{x}_3$.

A carrier vector of a totally symmetric Boolean function can be represented in terms of an elementary symmetric functions.

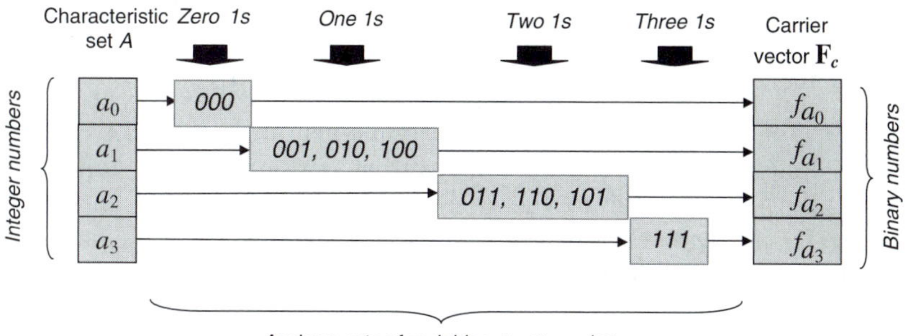

FIGURE 5.9
The carrier vector for the totally symmetric function of three variables (Example 5.22).

TABLE 5.5
Reducing the truth vector for a totally symmetric Boolean function (Example 5.23).

x_1	x_2	x_3	TRUTH VECTOR $\mathbf{F}$	$\sum 1's$	CARRIER VECTOR $\mathbf{F}_c$
0	0	0	1	0	0 ⟶ 1
0	0	1	0	1	1 ⟶ 0
0	1	0	0	1	2 ⟶ 0
0	1	1	0	2	3 ⟶ 1
1	0	0	0	1	
1	0	1	0	2	
1	1	0	0	2	
1	1	1	1	3	

Example 5.24 (Carrier vector and elementary symmetric functions.) *The Boolean function f given by the carrier vector $\mathbf{F}_c = [1001]^T$ (Table 5.5) can be represented in terms of the elementary symmetric functions as $f = S_0^3 \vee S_3^3$.*

Practice problem 5.15. (Carrier vector and elementary symmetric functions.) Represent the totally symmetric Boolean function given a carrier vector $\mathbf{F}_c = [0111]^T$ in terms of elementary symmetric functions.
Answer: $f = S_1^3 \vee S_2^3 \vee S_3^3$.

5.5 The logic Taylor expansion

The *logic Taylor expansion* is the third fundamental theorem for expanding Boolean functions (Figure 5.1). In contrast of the first two fundamental expansions that are used to the SOP representation of Boolean functions (Algebra I), the logic Taylor expansion is applied to the EXOR forms called the *polynomial* expressions (Algebra II).

5.5.1 Change in a digital system

Boolean difference is the formal model of change in a binary system. Simultaneous changes in several variables are described by Boolean difference of a higher order (second, third, and higher orders). In this chapter, Boolean differences are used for the expansion of Boolean functions using the so-called *logic Taylor expansion.*

A signal in a binary system is represented by two logical levels, 0 and 1. Let us formulate the task as detection of the change in this signal. The simplest solution is to deploy the EXOR operation (modulo 2 sum) of the signal s_{i-i} (before the "event") and the signal s_i (after the "event"), i.e., $s_{i-i} \oplus s_i$.

Example 5.25 (The signal change.) *For the signal depicted in Figure 5.10, four possible combinations of the logical values or signals 0 and 1 are analyzed.*

It follows from this example that if not change itself but direction of change is the matter, then two logical values, 0 and 1, can characterize the behavior of the logic signal $s_i \in \{0, 1\}$ in terms of change, where 0 means any change of a signal, and 1 indicates that one of two possible changes has occurred, $0 \to 1$ or $1 \to 0$.

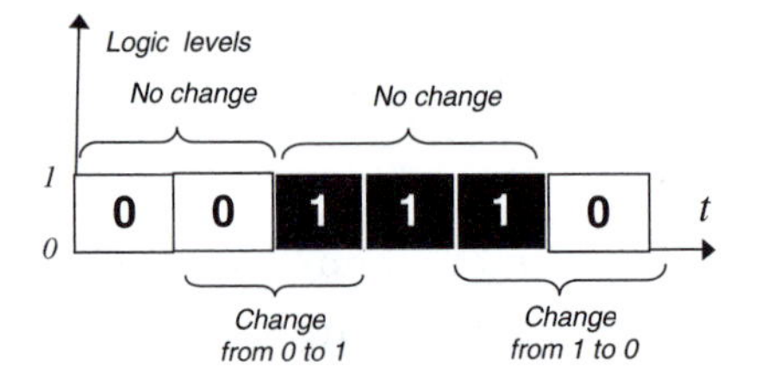

Case 1: *No change:* *Detection:* $0 \oplus 0 = 0$

Case 2: *No change:* *Detection:* $1 \oplus 1 = 0$

Case 3: *Change 0:→ 1* *Detection:* $0 \oplus 1 = 1$

Case 4: *Change 1:→ 0* *Detection:* $1 \oplus 0 = 1$

FIGURE 5.10
Change in a binary signal and its detection (Example 5.25).

5.5.2 Boolean difference

Let the i-th input of a Boolean function change from the value x_i to the opposite value, $\overline{x}_i$. This causes the circuit output to be changed from the initial value. Note that the values f_{x_i} and $f_{\overline{x}_i}$ are not necessarily different. The simplest way to recognize whether or not they are different is to find the difference between f_{x_i} and $f_{\overline{x}_i}$.

The Boolean difference of a Boolean function f of n variables with respect to a variable x_i is defined by the equation:

Boolean difference

$$\frac{\partial f}{\partial x_i} = \underbrace{f(x_1, \ldots, \boxed{x_i}, \ldots, x_n)}_{Initial\ function} \oplus \underbrace{f(x_1, \ldots, \boxed{\overline{x}_i}, \ldots, x_n)}_{Function\ with\ complemented\ x_i} \quad (5.11)$$

It follows from the Shannon expansion that

$$f(x_1, \ldots, x_i, \ldots, x_n) = \overline{x}_i f_0 \oplus x_i f_1$$
$$f(x_1, \ldots, \overline{x}_i, \ldots, x_n) = x_i f_0 \oplus \overline{x}_i f_1$$

Hence, the Boolean difference (Equation 5.11) can be represented as follows

$$\begin{aligned}\frac{\partial f}{\partial x_i} &= (x_i f_0 \oplus \overline{x}_i f_1) \oplus (\overline{x}_i f_0 \oplus x_i f_1)\\ &= f_0(x_i \oplus \overline{x}_i) \oplus f_1(\overline{x}_i \oplus x_i)\\ &= f_0 \oplus f_1\end{aligned}$$

Therefore, the second definition of the Boolean difference (Equation 5.11) is

Boolean difference

$$\frac{\partial f}{\partial x_i} = \underbrace{f(x_1, \ldots, \overset{x_i}{\boxed{0}}, \ldots, x_n)}_{x_i\ is\ replaced\ with\ 0} \oplus \underbrace{f(x_1, \ldots, \overset{x_i}{\boxed{1}}, \ldots, x_n)}_{x_i\ is\ replaced\ with\ 1}, \quad (5.12)$$

or

Boolean difference

$$\frac{\partial f}{\partial x_i} = f_{x_i=0} \oplus f_{x_i=1} = f_0 \oplus f_1 \quad (5.13)$$

Thus, the simplest (but optimal) approach to calculating the Boolean difference includes two steps:

> **Computing the Boolean difference with respect to a variable**
>
> **Given:** `A Boolean function`
> **Step 1:** `Replace` x_i `in the Boolean function with 0 to get a cofactor` $f_{x_i=0}$`; similarly, replacement of` x_i `with 1 yields` $f_{x_i=1}$
> **Step 2:** `Find the modulo 2 sum of the two cofactors`

Figure 5.11 gives an interpretation of the Boolean difference (Equation 5.11).

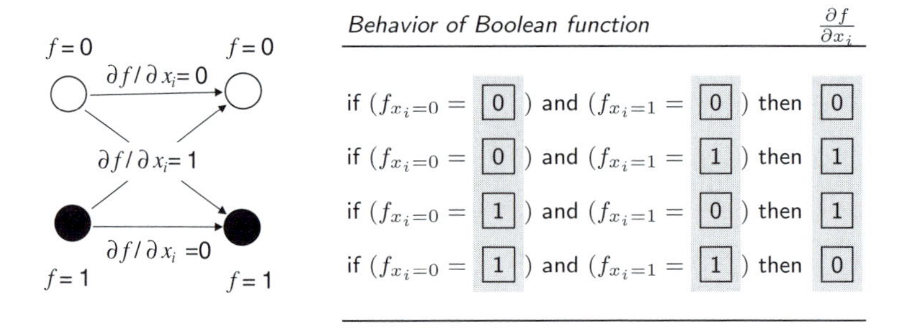

FIGURE 5.11
Formal description of change by Boolean difference.

> **Example 5.26 (Boolean difference.)** *Given an OR function of two variables, find the Boolean difference with respect to the variable x_2. The result is given in Figure 18.7.*

Practice problem **5.16. (Boolean difference.)** Find the Boolean difference of the function $f = x_1 \oplus x_2$ with respect to the variable x_1
Answer: $\frac{\partial f}{\partial x_1} = f_0 \oplus f_1 = (0 \oplus x_2) \oplus (1 \oplus x_2) = x_2 \oplus \overline{x}_2 = 1$.

5.5.3 Boolean difference and Shannon expansion

It follows from the EXOR form of the Shannon expansion (Equation 5.6) that

Techniques for computing the Boolean difference

Boolean difference for a two-input OR gate with respect to the variable x_1

$$\begin{aligned}\frac{\partial f}{\partial x_1} &= f_{x_1} \oplus f_{\overline{x}_1}\\ &= (x_1 \vee x_2) \oplus (\overline{x}_1 \vee x_2)\\ &= x_1 \oplus x_2 \oplus x_1x_2 \oplus \overline{x}_1\\ &\oplus x_2 \oplus \overline{x}_1x_2\\ &= 1 \oplus x_1x_2 \oplus \overline{x}_1x_2\\ &= 1 \oplus x_2 = \boxed{\overline{x}_2}\end{aligned}$$

or

$$\begin{aligned}\frac{\partial f}{\partial x_1} &= f_{x_1=0} \oplus f_{\overline{x}_1=1}\\ &= (0 \vee x_2) \oplus (1 \vee x_2)\\ &= x_2 \oplus 1 = \boxed{\overline{x}_2}\end{aligned}$$

Boolean difference for a two-input OR gate with respect to the variable x_2

$$\begin{aligned}\frac{\partial f}{\partial x_2} &= f_{x_2} \oplus f_{\overline{x}_2}\\ &= (x_1 \vee x_2) \oplus (x_1 \vee \overline{x}_2)\\ &= x_1 \oplus x_2 \oplus x_1x_2 \oplus x_1\\ &\oplus \overline{x}_2 \oplus x_1\overline{x}_2\\ &= 1 \oplus x_1x_2 \oplus x_1\overline{x}_2\\ &= 1 \oplus x_1 = \boxed{\overline{x}_1}\end{aligned}$$

or

$$\begin{aligned}\frac{\partial f}{\partial x_2} &= f_{x_2=0} \oplus f_{\overline{x}_2=1}\\ &= (x_1 \vee 0) \oplus (x_1 \vee 1)\\ &= x_1 \oplus 1 = \boxed{\overline{x}_1}\end{aligned}$$

FIGURE 5.12
Computing the Boolean difference with respect to the variables x_1 and x_2 for a two-input OR gate (Example 18.8).

$$\begin{aligned}f &= \overline{x}_i f_0 \oplus x_i f_1 = (1 \oplus x_i) f_0 \oplus x_i f_1\\ &= f_0 \oplus x_i f_0 \oplus x_i f_1 = f_0 \oplus x_i \underbrace{(f_0 \oplus f_1)}_{\frac{\partial f}{\partial x_i}}\end{aligned}$$

Finally

EXOR form of Shannon expansion

$$f = f_0 \oplus x_i \frac{\partial f}{\partial x_i} \tag{5.14}$$

Therefore, Boolean difference is the component of an EXOR form of Shannon expansion. The EXOR form of Shannon expansion is called *positive Davio expansion.*

Example 5.27 (Boolean difference and Shannon expansion.) *The Shannon expansion of the Boolean function* $f = x_1 \oplus x_2$ *with respect to the variable* x_1 *in terms of Boolean differences is calculated as follows:* $f = f_0 \oplus x_1 \frac{\partial f}{\partial x_1} = x_2 \oplus x_1 \cdot 1$, *where* $\frac{\partial f}{\partial x_1} = f_0 \oplus f_1 = (0 \oplus x_2) \oplus (1 \oplus x_2) = x_2 \oplus \overline{x}_2 = 1$.

Practice problem 5.17. **(Boolean difference and Shannon expansion.)** Derive an expression for the Shannon expansion of a Boolean function of two variables, with respect to the variables x_1 and x_2 in terms of Boolean differences

Answer: $f = f_0 \oplus x_2 \frac{\partial f_0}{\partial x_2} \oplus x_1 \frac{\partial f_0}{\partial x_1} \oplus x_1 x_2 \frac{\partial^2 f_0}{\partial x_1 \partial x_2}$.

5.5.4 Properties of Boolean difference

The Boolean difference (Equation 5.11) possesses the following properties:

Properties of Boolean difference

Property 1: The Boolean difference is a Boolean function of $n-1$ variables $x_1, x_2, \ldots, x_{i-1}, x_{i+1}, \ldots, x_n$; i.e., it does not depend on the variable x_i. This also implies that the Boolean difference can be represented by a 2^{n-1} truth vector which is twice as small as the original truth vector.

Property 2: The value of the Boolean difference reflects the fact of change of the Boolean function f with respect to the i-th variable x_i: the Boolean difference is equal to 0 when such change occurs, and is equal to 1 otherwise.

Example 5.28 (Properties of Boolean difference.) *The Boolean difference with respect to* x_1 *of the OR function* $f = x_1 \vee x_2$ *is equal to* $\frac{\partial f}{\partial x_1} = \overline{x}_2$. *The resulting function does not depend on the variable of differentiation,* x_1. *Similarly, the Boolean difference with respect to* x_2 *of the OR function is equal to* $\frac{\partial f}{\partial x_2} = \overline{x}_1$, *which does not depend on the variable* x_2.

Practice problem 5.18. **(Properties of Boolean difference.)** Given $f = x_1 x_2 \vee x_3$, find $\frac{\partial f}{\partial x_1}$

Answer is given in "Solutions to practice problems."

5.5.5 The logic Taylor expansion

In the *classic* Taylor series, the coefficients are calculated as *derivatives* of the initial function at a *certain point.* By analogy, the coefficients of the *logic* Taylor series are *Boolean differences* (derivatives) with respect to a variable, and multiple Boolean differences at certain points (assignments of Boolean variables). a A certain point in logic Taylor expansion is called the *polarity*. The polarity specifies which variables in Taylor expansion are complemented.

In particular, the logic Taylor series for a Boolean function f of two variables ($n = 2$) of the polarity $c \in \{0, 1, 2, 3\}$ (in binary form, $c = c_1c_2 \in \{00, 01, 10, 11\}$) is defined by the equation

$$f = \overbrace{r_0^{(c)} \underbrace{(x_1 \oplus c_1)^0 (x_2 \oplus c_2)^0}_{Equal\ to\ 1}}^{0th\ product} \oplus \overbrace{r_1^{(c)} \underbrace{(x_1 \oplus c_1)^0}_{Equal\ to\ 1} (x_2 \oplus c_2)^1}^{1st\ product}$$

$$\oplus \overbrace{r_2^{(c)} (x_1 \oplus c_1)^1 \underbrace{(x_2 \oplus c_2)^0}_{Equal\ to\ 1}}^{2nd\ product} \oplus \overbrace{r_3^{(c)} (x_1 \oplus c_1)^1 (x_2 \oplus c_2)^1}^{3rd\ product}$$

After simplification,

$$f = r_0^{(c)} \oplus r_1^{(c)} (x_2 \oplus c_2) \oplus r_2^{(c)} (x_1 \oplus c_1) \oplus r_3^{(c)} (x_1 \oplus c_1)(x_2 \oplus c_2)$$

The logic Taylor expansion for a Boolean function f of n variables of the polarity $c \in 0, 1, \ldots, 2^n - 1$, is defined by the equation

Logic Taylor expansion for a Boolean function

$$f = \bigoplus_{i=0}^{2^n-1} r_i^{(c)} \underbrace{\overbrace{(x_1 \oplus c_1)}^{c_1-polarity}{}^{i_1} \ldots \overbrace{(x_n \oplus c_n)}^{c_n-polarity}{}^{i_n}}_{i-th\ product}, \quad (5.15)$$

In Equation 5.15 $c_1, c_2, \ldots, c_n$ and $i_1, i_2, \ldots, i_n$ are the binary representations of the polarity c and index i, respectively. The value of $c_i \in \{0, 1\}$, $i = 0, 1, 2, \ldots, n$, specifies the x_i variable, that is, uncomplemented x_i if $c_i = 0$ or complemented $\overline{x}_i$ if $c_i = 1$. The i-th coefficient $r_i^{(c)}$ given polarity c is calculated as follows:

Coefficients of logic Taylor expansion (Equation 5.15)

$$r_i^{(c)} = \frac{\partial^{(j)} f(c)}{\partial x_1^{i_1} \partial x_2^{i_2} \ldots \partial x_n^{i_n}} \quad \text{and} \quad \partial x_i^{i_j} = \begin{cases} 1, & i_j = 0 \\ \partial x_j & i_j = 1 \end{cases} \quad (5.16)$$

where parameter j is the order of Boolean differences, $j = \sum_{j=1}^{n} i_j$. The ith coefficient $r_i^{(c)}$ is the value (0 or 1) of the j-order Boolean difference of the function $f(c)$ given polarity $x_1 = c_1, \ldots, x_n = c_n$.

Example 5.29 (Logic Taylor expansion.) *Given polarity $c = 2$, compute the first coefficient ($i = 1$) $r_i^{(c)}$ of the logic Taylor expansion for the Boolean function of three variables $n = 3$. Techniques for computing are introduced in Figure 5.13.*

Techniques for computing logic Taylor coefficients

Given: The polarity $c = 2$, the number of variables $n = 3$ of a Boolean function, the number of coefficient $i = 1$

Find: The coefficient $r_1^{(2)}$ of the logic Taylor expansion

Step 1: Write Equation 5.16 for $n = 3$, $i = 1$, and $c = 3$:

$$r_1^{(2)} = \frac{\partial^{(j)} f(2)}{\partial x_1^{i_1} \partial x_2^{i_2} \partial x_3^{i_3}}$$

Step 2: Compute the order j of Boolean difference for the first ($i = i_1 i_2 i_3 = 001$) coefficient :

$$j = \sum_{j=1}^{3} i_j = 0 + 0 + 1 = 1$$

Step 3: Compute $\partial x_1^{i_1}, \partial x_2^{i_2}$, and $\partial x_3^{i_3}$ using Equation 5.16:

$$\partial x_1^{i_1} = \partial x_1^0 = 1$$
$$\partial x_2^{i_2} = \partial x_2^0 = 1$$
$$\partial x_1^{i_3} = \partial x_3^1 = \partial x_3$$

Step 4: Write the first coefficient of a logic Taylor expansion given second polarity for an arbitrary Boolean function of three variables using Equation 5.16 ($n = 3$, $i = 1$, $c = 3$, and $j = 1$):

$$r_1^{(2)} = \frac{\partial^{(1)} f(2)}{\partial x_1^{i_1} \partial x_2^{i_2} \partial x_3^{i_3}} = \frac{\partial^{(1)} f(2)}{\partial x_1^0 \partial x_2^0 \partial x_3^1} = \frac{\partial^{(1)} f(2)}{\partial x_3}$$

FIGURE 5.13
Computing the coefficient of logic Taylor expansion for an arbitrary Boolean function of three variables (Example 5.29)

It follows from Equation 5.15 that

In a logic Taylor expansion:

- A variable x_j is 0-polarized if it enters into the expansion uncomplemented, and 1-polarized otherwise.
- Coefficients of the polynomial expression of a Boolean function are described in terms of a differences, formally defined as Boolean derivatives.

The logic Taylor expansion of an n-variable Boolean function f generates 2^n polynomial expressions of 2^n polarities.

Example 5.30 (Number of polynomial expressions.) *There are $2^n = 2^2 = 4$ polynomial expressions of Boolean function of two variables ($n = 2$) of polarities 0,1,2, and 3. Table 11.1 shows how each elementary function of two variables can be represented by four polynomial forms. For example, the OR gate given a polarity of $c = 3$ ($c_1c_2 = 11$) is represented by the two non-zero coefficients $r_0^{(3)}$ and $r_3^{(3)}$. That is,*

$$f = r_0^{(3)} \oplus r_3^{(3)}\overline{x}_1\overline{x}_2 = 1 \oplus \overline{x}_1\overline{x}_2$$

and corresponding truth vector $\mathbf{F} = [1001]^T$ This is an optimal representation of the OR function (optimal polarity).

While the i-th coefficient $r_i^{(c)}(d)$ is described by a Boolean expression, it can be calculated in different ways; for example, symbolic or matrix transformations, cube-based technique, decision diagram technique, and probabilistic methods (details are given in Chapter 11).

TABLE 5.6
Polynomial expressions as logic Taylor expansions of elementary Boolean functions (Example 5.30).

Techniques for computing logic Taylor expansion

Function	Vector form $r_0^{(c)}$	$\frac{\partial f}{\partial x_2}$	$\frac{\partial f}{\partial x_1}$	$\frac{\partial^2 f}{\partial x_1 \partial x_2}$	Algebraic form
$f = x_1 \wedge x_2$	0	0	0	1	$f(c=0) = x_1x_2$
	0	0	1	1	$f(c=1) = x_1 \oplus x_1\overline{x}_2$
	0	1	0	1	$f(c=2) = x_2 \oplus \overline{x}_1x_2$
	1	1	1	1	$f(c=3) = 1 \oplus \overline{x}_2 \oplus \overline{x}_1 \oplus \overline{x}_1\overline{x}_2$
$f = x_1 \vee x_2$	0	1	1	1	$f(c=0) = x_2 \oplus x_1 \oplus x_1x_2$
	1	1	0	1	$f(c=1) = 1 \oplus \overline{x}_2 \oplus x_1\overline{x}_2$
	1	0	1	1	$f(c=2) = 1 \oplus \overline{x}_1 \oplus \overline{x}_1x_2$
	1	0	0	1	$f(c=3) = 1 \oplus \overline{x}_1\overline{x}_2$
$f = x_1 \oplus x_2$	0	1	1	0	$f(c=0) = x_2 \oplus x_1$
	1	1	1	0	$f(c=1) = 1 \oplus \overline{x}_2 \oplus x_1$
	1	1	1	0	$f(c=2) = 1 \oplus x_2 \oplus \overline{x}_1$
	0	1	1	0	$f(c=3) = \overline{x}_2 \oplus \overline{x}_1$

Example below illustrates symbolic manipulation to derive algebraic polynomial forms.

Example 5.31 (Shannon and logic Taylor expansion for gates.) *Given $f = x_1 \vee x_2$, its EXOR form Shannon expansion with respect to both variables x_1 and x_2 is given in the first chart of Figure 5.14. To obtain an expression of polarity $c = 3$ ($c_1c_2 = 11$), both x_1 and x_2 should be complemented), we have to make the substitution $a = \overline{a} \oplus 1$. On the other hand, the same result can be obtained using Boolean differences (the second chart in Figure 5.14), as follows from logic Taylor expansion.*

Techniques for computing polynomial expressions

The EXOR form Shannon expansion with respect to both variables x_1 and x_2 for a two-input OR gate

$$\begin{aligned} f &= \overline{x}_1\overline{x}_2 f_{00} \oplus \overline{x}_1 x_2 f_{01} \\ &\oplus x_1\overline{x}_2 f_{10} \oplus x_1 x_2 f_{11} \\ &= \overline{x}_1\overline{x}_2(0 \vee 0) \oplus \overline{x}_1 x_2(0 \vee 1) \\ &\oplus x_1\overline{x}_2(1 \vee 0) \oplus x_1 x_2(1 \vee 1) \\ &= \overline{x}_1 x_2 \oplus x_1\overline{x}_2 \oplus x_1 x_2 \end{aligned}$$

Converting to the polarity $c = 3$ ($c_1c_2 = 11$)

$$\begin{aligned} f &= \overline{x}_1(\overline{x}_2 \oplus 1) \oplus (\overline{x}_1 \oplus 1)\overline{x}_2 \\ &\oplus (\overline{x}_1 \oplus 1)(\overline{x}_2 \oplus 1) \\ &= \overline{x}_1\overline{x}_2 \oplus \overline{x}_1 \oplus \overline{x}_1\overline{x}_2 \oplus \overline{x}_2 \\ &\oplus \overline{x}_1\overline{x}_2 \oplus \overline{x}_1 \oplus \overline{x}_2 \oplus 1 \\ &= 1 \oplus \overline{x}_1\overline{x}_2 \end{aligned}$$

Logic Taylor expansion for the polarity $c = 3$ for a two-input OR gate

$$\begin{aligned} f &= f(3) \oplus \frac{\partial f(3)}{\partial x_2}\overline{x}_2 \oplus \frac{\partial f(3)}{\partial x_1}\overline{x}_1 \oplus \frac{\partial^2 f(3)}{\partial x_1 \partial x_2}\overline{x}_1\overline{x}_2 \\ &= 1 \oplus 0 \cdot \overline{x}_2 \oplus 0 \cdot \overline{x}_1 \oplus 1 \cdot \overline{x}_1\overline{x}_2 \\ &= 1 \oplus \overline{x}_1\overline{x}_2 \end{aligned}$$

$$\begin{aligned} f(3) &= 1 \\ \frac{\partial f(3)}{\partial x_2} &= (x_1 \vee 0) \oplus (x_1 \vee 1)|_{x_1=1} \\ &= (x_1 \oplus 1)|_{x_1=1} = \overline{x}_1|_{x_1=1=\overline{1}=0} \\ \frac{\partial f(3)}{\partial x_1} &= (0 \vee x_2) \oplus (1 \vee x_2)|_{x_2=1} \\ &= \overline{x}_2|_{x_2=1} = \overline{1} = 0 \\ \frac{\partial f(3)^2}{\partial x_1 \partial x_2} &= 1 \end{aligned}$$

FIGURE 5.14
Techniques for construction of polynomial expressions of a given polarity using Shannon expansion and logic Taylor expansion (Example 5.31).

Example 5.32 (Logic Taylor expansion for a given polarity.) *Given $f = x_1\overline{x}_2 \vee x_3$, the logic Taylor expansion is derived for the polarity $c = 4$ ($c_1c_2c_3 = 100$), as shown in Figure 5.15.*

Practice problem 5.19. (Logic Taylor expansion for a given polarity.) Given $f = x_1x_2x_3$, derive the logic Taylor expansion for the polarity $c = 7$ ($c_1c_2c_3 = 111$.)
Answer is given in "Solutions to practice problems."

Techniques for computing a logic Taylor expansion

Given: A Boolean function $f = x_1\overline{x}_2 \vee x_3$

Find: A logic Taylor expansion at the point $c = 4$ $(c_1c_2c_3 = 100)$

Step 1: Apply Equation 5.15 for $n = 3$:

$$f = \bigoplus_{i=0}^{2^3-1} \frac{\partial^3 f(c)}{\partial x_1^{i_1}\partial x_2^{i_2}\partial x_3^{i_3}} \overbrace{(x_1 \oplus c_1)}^{c_1-polarity}{}^{i_1} \overbrace{(x_2 \oplus c_2)}^{c_2-polarity}{}^{i_1} \overbrace{(x_3 \oplus c_3)}^{c_3-polarity}{}^{i_3}$$

Step 2: Substitute $c = 4$ $(c_1 = 1, c_2 = 0, c = 0)$ in the equation

$$f = \bigoplus_{i=0}^{2^3-1} \frac{\partial^3 f(4)}{\partial x_1^{i_1}\partial x_2^{i_2}\partial x_3^{i_3}} \overbrace{(x_1 \oplus 1)^{i_1}(x_2 \oplus 0)^{i_1}(x_3 \oplus 0)^{i_3}}^{Polarity\ 4}$$

Step 3: Expand the equation for $i = 0, 1, 2, \ldots 7$:

$$f = \underbrace{\frac{\partial^3 f(4)}{\partial x_1^0\partial x_2^0\partial x_3^0}}_{1} \overbrace{\overline{x}_1{}^0 x_2{}^0 x_3{}^0}^{1} \oplus \underbrace{\frac{\partial^3 f(4)}{\partial x_1^0\partial x_2^0\partial x_3^1}}_{\partial x_3} \overbrace{\overline{x}_1{}^0 x_2{}^0 x_3{}^1}^{x_3} \oplus \underbrace{\frac{\partial^3 f(4)}{\partial x_1^0\partial x_2^1\partial x_3^0}}_{\partial x_2} \overbrace{\overline{x}_1{}^0 x_2{}^1 x_3{}^0}^{x_2}$$

$$\oplus \underbrace{\frac{\partial^3 f(4)}{\partial x_1^0\partial x_2^1\partial x_3^1}}_{\partial x_2\partial x_3} \overbrace{\overline{x}_1{}^0 x_2{}^1 x_3{}^1}^{x_2x_3} \oplus \underbrace{\frac{\partial^3 f(4)}{\partial x_1^1\partial x_2^0\partial x_3^0}}_{\partial x_1} \overbrace{\overline{x}_1{}^1 x_2{}^0 x_3{}^0}^{\overline{x}_1} \oplus \underbrace{\frac{\partial^3 f(4)}{\partial x_1^1\partial x_2^0\partial x_3^1}}_{\partial x_1\partial x_3} \overbrace{\overline{x}_1{}^1 x_2{}^0 x_3{}^1}^{\overline{x}_1x_3}$$

$$\oplus \underbrace{\frac{\partial^3 f(4)}{\partial x_1^1\partial x_2^1\partial x_3^0}}_{\partial x_1\partial x_2} \overbrace{\overline{x}_1{}^1 x_2{}^1 x_3{}^0}^{\overline{x}_1x_2} \oplus \underbrace{\frac{\partial^3 f(4)}{\partial x_1^1\partial x_2^1\partial x_3^1}}_{\partial x_1\partial x_2\partial x_3} \overbrace{\overline{x}_1{}^1 x_2{}^1 x_3{}^1}^{\overline{x}_1x_2x_3}$$

Step 4: Compute the products $\partial x_1{}^{i_1}\partial x_2{}^{i_2}\partial x_3{}^{i_3}$ and $x_1{}_1^i x_2{}_1^i x_3{}_3^i$:

$$f = f(4) \oplus \frac{\partial f(4)}{\partial x_3}x_3 \oplus \frac{\partial f(4)}{\partial x_2}x_2 \oplus \frac{\partial^2 f(4)}{\partial x_2\partial x_3}x_2x_3 \oplus \frac{\partial f(4)}{\partial x_1}\overline{x}_1 \oplus \frac{\partial^2 f(4)}{\partial x_1\partial x_3}\overline{x}_1x_3$$

$$\oplus \frac{\partial^2 f(4)}{\partial x_1\partial x_2}\overline{x}_1x_2 \oplus \frac{\partial^3 f(4)}{\partial x_1\partial x_2\partial x_3}\overline{x}_1x_2x_3$$

FIGURE 5.15
Logic Taylor expansion of a Boolean function (Example 5.32).

Techniques for computing a logic Taylor expansion (continuation)

Step 5: Compute the values of Boolean differences:

$$f(4) = f(1,0,0) = 1 \cdot \overline{0} \vee 0 = \boxed{1}$$

$$\frac{\partial f(4)}{\partial x_3} = \overbrace{f(1,0,0) \oplus f(1,0,1)}^{\textit{With respect to } x_3} = 1 \oplus (1 \cdot \overline{0} \vee 1) = \boxed{0}$$

$$\frac{\partial f(4)}{\partial x_2} = \overbrace{f(1,0,0) \oplus f(1,1,0)}^{\textit{With respect to } x_2} = 1 \oplus (1 \cdot \overline{1} \vee 0) = \boxed{1}$$

$$\frac{\partial^2 f(4)}{\partial x_2 \partial x_3} = \frac{\partial}{\partial x_2}\left(\frac{\partial f(4)}{\partial x_3}\right) = \frac{\partial f(1,0,0)}{\partial x_3} \oplus \frac{\partial f(1,1,0)}{\partial x_3} = 0 \oplus 1 = \boxed{1}$$

$$\frac{\partial f(4)}{\partial x_1} = \overbrace{f(1,0,0) \oplus f(0,0,0)}^{\textit{With respect to } x_1} = 1 \oplus (0 \cdot \overline{0} \vee 0) = \boxed{1}$$

$$\frac{\partial^2 f(4)}{\partial x_1 \partial x_3} = \frac{\partial}{\partial x_1}\left(\frac{\partial f(4)}{\partial x_3}\right) = \frac{\partial f(1,0,0)}{\partial x_3} \oplus \frac{\partial f(0,0,0)}{\partial x_3} = 0 \oplus 1 = \boxed{1}$$

$$\frac{\partial^2 f(4)}{\partial x_1 \partial x_2} = \frac{\partial}{\partial x_1}\left(\frac{\partial f(4)}{\partial x_2}\right) = \frac{\partial f(1,0,0)}{\partial x_2} \oplus \frac{\partial f(0,0,0)}{\partial x_2} = 1 \oplus 0 = \boxed{1}$$

$$\frac{\partial^3 f(4)}{\partial x_1 \partial x_2 \partial x_3} = \frac{\partial}{\partial x_1}\left(\frac{\partial^2 f(4)}{\partial x_2 \partial x_3}\right) = \frac{\partial^2 f(1,0,0)}{\partial x_2 \partial x_3} \oplus \frac{\partial^2 f(0,0,0)}{\partial x_2 \partial x_3} = 1 \oplus 0 = \boxed{1}$$

<u>Note:</u> The Boolean differences $\frac{\partial f(0,0,0)}{\partial x_2}$, $\frac{\partial f(0,0,0)}{\partial x_3}$, and $\frac{\partial f(0,0,0)}{\partial x_2 \partial x_3}$ are computed as follows:

$$\frac{\partial f(0,0,0)}{\partial x_2} = \overbrace{f(0,0,0) \oplus f(0,1,0)}^{\textit{With respect to } x_2} = (0 \cdot \overline{0} \vee 0) \oplus (0 \cdot \overline{1} \vee 0) = 0$$

$$\frac{\partial f(0,0,0)}{\partial x_3} = \overbrace{f(0,0,0) \oplus f(0,0,1)}^{\textit{With respect to } x_3} = (0 \cdot \overline{0} \vee 0) \oplus (0 \cdot \overline{0} \vee 1) = 1$$

$$\frac{\partial^2 f(0,0,0)}{\partial x_2 \partial x_3} = \frac{\partial f(0,0,0)}{\partial x_3} \oplus \frac{\partial f(0,1,0)}{\partial x_3} = 1 \oplus (f(0,1,0) \oplus f(0,1,1))$$
$$= 1 \oplus ((0 \cdot \overline{1} \vee 0) \oplus (0 \cdot \overline{1} \vee 1)) = 1 \oplus 1 = 0$$

Step 6: Write logic Taylor expansion using the computed Boolean differences:

$$f = \boxed{1} \oplus \boxed{0}\, x_3 \oplus \boxed{1}\, x_2 \oplus \boxed{1}\, x_2 x_3 \oplus \boxed{1}\, \overline{x}_1 \oplus \boxed{1}\, \overline{x}_1 x_3 \oplus \boxed{1}\, \overline{x}_1 x_2 \oplus \boxed{1}\, \overline{x}_1 x_2 x_3$$

FIGURE 5.16
Logic Taylor expansion of a Boolean function (continuation of Example 5.32).

5.6 Graphical representation of the fundamental expansions

A useful feature of the fundamental expansions is its relation to graphical data structures such as decision trees and decision diagrams. Equation 5.2 can be used for computing an arbitrary Boolean function in algebraic form. Another form of this equation is the graph. Figure 5.17a shows the representation of Equation 5.2 by a node of the graph.

5.6.1 Shannon expansion as a decision tree node function

Suppose that Shannon expansion, denoted by S, is implemented in this node, that is, the input is the function f, and outgoing links correspond to the parts of this equation. It is useful to interpret the function of the node as transfer-contact (Figure 5.17b).

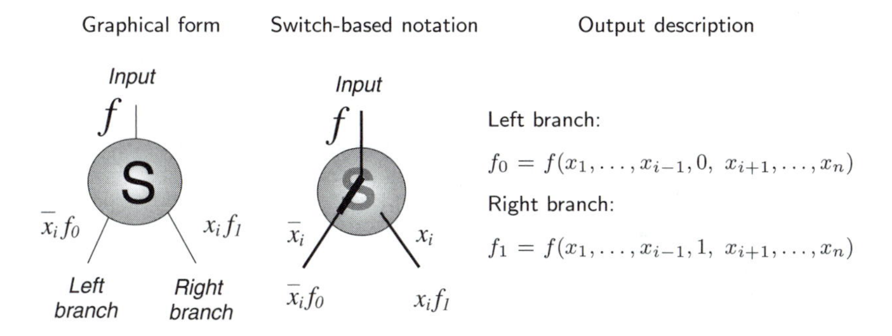

FIGURE 5.17
Graphical representation of a Shannon expansion with respect to a variable and its implementation as the transfer-contact (Equation 5.2).

Example 5.33 **(Decision tree.)** *A node of a decision tree that represents the Shannon expansion of the Boolean function $f = \overline{x}_1 x_3 \vee x_2 \overline{x}_3$ with respect to the variable x_1 is given in Figure 5.18.*

Practice problem **5.20.** **(Decision tree.)** Draw the decision tree representing the Shannon expansion of the Boolean function $f = \overline{x}_1 x_2 \vee x_1 \overline{x}_2$ with respect to the variable x_1.
Answer is given in "Solutions to practice problems."

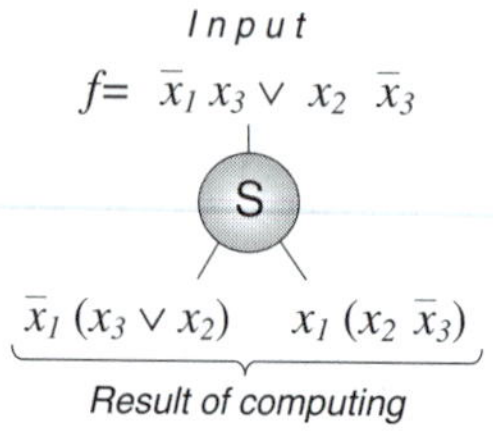

$$
\begin{aligned}
f_0 &= 1 \cdot x_3 \vee x_2\overline{x}_3 = x_3 \vee x_2 \\
f_1 &= 0 \cdot x_3 \vee x_2\overline{x}_3 = x_2\overline{x}_3 \\
f &= \overline{x}_1 f_0 \vee x_1 f_1 \\
&= \overline{x}_1(x_3 \vee x_2) \vee x_1(x_2\overline{x}_3)
\end{aligned}
$$

FIGURE 5.18
Shannon expansion with respect to a variable x_1 of the Boolean function $f = \overline{x}_1 x_3 \vee x_2\overline{x}_3$ (Example 5.33).

In the above examples, the Boolean function was replaced by two Boolean subfunctions with reduced variables. Shannon expansion can be applied to these subfunctions, too.

Figure 5.19 illustrates the computational details of the node. The node at the *intermediate* level results in two cofactors, f_0 and f_1.

Computing at the terminal level $f = \overline{x}_i f_0 \vee x_i f_1$

Output $f_0 = 0,\ f_1 = 0$:	Node computes $f = \overline{x}_i \cdot 0 \vee x_i \cdot 0 = 0$;
Output $f_0 = 0,\ f_1 = 1$:	Node computes $f = \overline{x}_i \cdot 0 \vee x_i \cdot 1 = x_i$;
Output $f_0 = 1,\ f_1 = 0$:	Node computes $f = \overline{x}_i \cdot 1 \vee x_i \cdot 0 = \overline{x}_i$;
Output $f_0 = 1,\ f_1 = 1$:	Node computes $f = \overline{x}_i \cdot 1 \vee x_i \cdot 1 = 1$;

A reduction of the decision tree is possible if $f_0 = 0,\ f_1 = 0$ (the node can be eliminated and the value $f = 0$ is assigned to the output), and if $f_0 = 1,\ f_1 = 1$ (the node can be eliminated and the value $f = 1$ is assigned to the output).

5.6.2 Matrix notation of the node function

A node in a decision tree of a Boolean function f corresponds to the Shannon expansion of a Boolean function with respect to a variable x_i: $f = \overline{x}_i f_0 \vee x_i f_1$, where $f_0 = f_{x_i=0}$ and $f_1 = f_{x_i=1}$. Here $f = f_{x_i=a}$ denotes the cofactor of f after assigning the constant a to the variable x_i.

The transformation assigned to a node of a decision tree can be represented in matrix notation. Given a function of a single variable x_i represented by the truth vector $\mathbf{F} = [\ f(0)\ f(1)\ ]^T$, its matrix notation is given as

$$
f = [\ \overline{x}_i\ x_i\ ] \begin{bmatrix} 1 & 0 \\ 0 & 1 \end{bmatrix} \begin{bmatrix} f_0 \\ f_1 \end{bmatrix} = [\ \overline{x}_i\ x_i\ ] \begin{bmatrix} f_0 \\ f_1 \end{bmatrix} = \overline{x}_i f_0 \vee x_i f_1,
$$

where $f_0 = f_{x_i=0}$, $f_1 = f_{x_i=1}$. Figure 5.20 shows the implementation of a node using a so-called *multiplexer* (details are given in Chapters 13 and 14).

Design example: A node in decision diagrams and trees

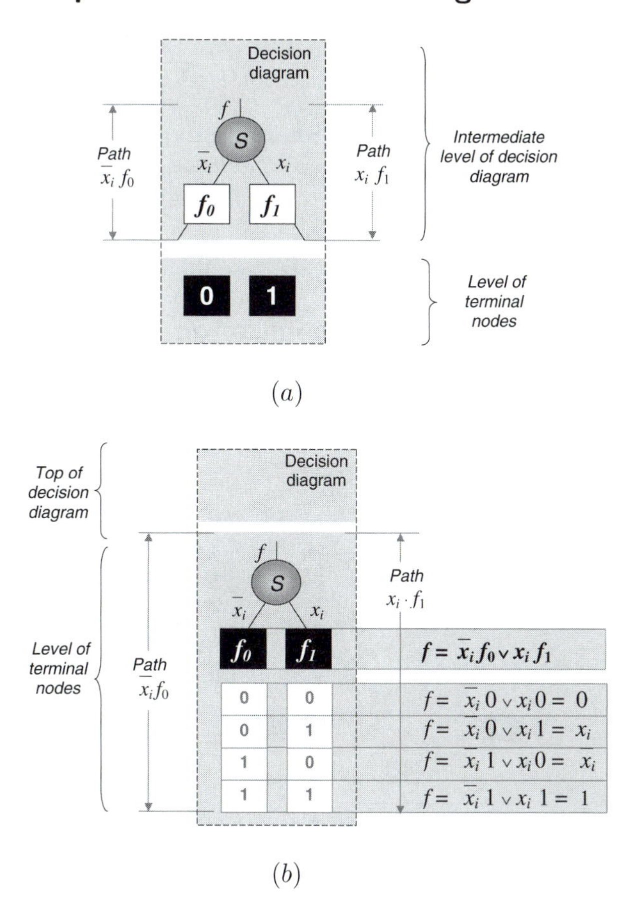

FIGURE 5.19
A node in decision diagrams and trees: computing at the intermediate level (a) and computing at the terminal level (b).

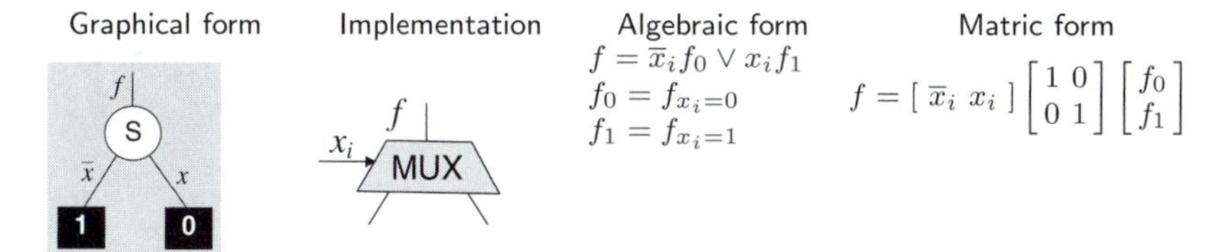

FIGURE 5.20
A node of a decision tree, its implementation by a multiplexer (MUX), algebraic, and matrix descriptions.

5.6.3 Using Shannon expansion in decision trees

Recursive application of Shannon expansion of a Boolean function with respect to all variables results in an n-level decision tree, called a *binary decision tree.*

Example 5.34 (Shannon expansion in matrix form.) *Recursive application of the Shannon expansion with respect to variables x_1 and x_2 to the Boolean function $f = \overline{x}_1x_2 \vee x_1\overline{x}_2$ is illustrated in Figure 5.21.*

Design example: Recursive application of Shannon expansion

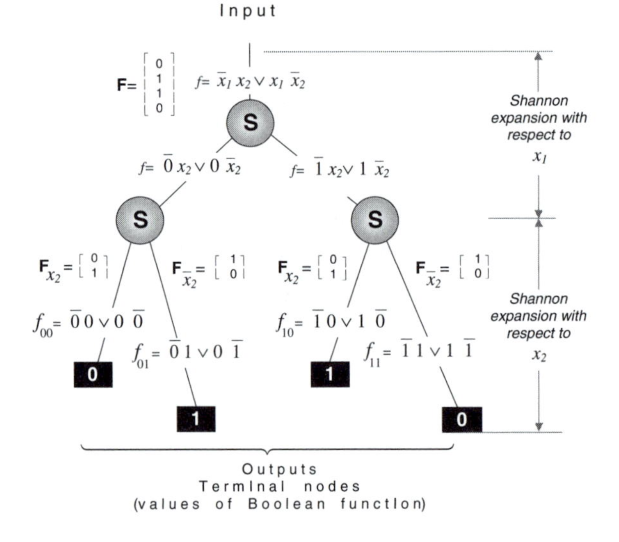

With respect to the variable x_1:

$$f_0 = f(x_1 = 0)$$
$$= \overline{0} \cdot x_2 \vee 0 \cdot \overline{x}_2 = x_2, \quad \mathbf{F}_{x_1=0} = \begin{bmatrix} 0 \\ 1 \end{bmatrix}$$
$$f_1 = f(x_1 = 1)$$
$$= \overline{1} \cdot x_2 \vee 1 \cdot \overline{x}_2 = \overline{x}_2, \quad \mathbf{F}_{x_1=1} = \begin{bmatrix} 1 \\ 0 \end{bmatrix}$$

With respect to the variable x_2:

$$f_{00} = f(x_1 = 0, x_2 = 0) = \overline{0} \cdot 0 \vee 0 \cdot \overline{0} = 0$$
$$f_{01} = f(x_1 = 0, x_2 = 1) = \overline{0} \cdot 1 \vee 0 \cdot \overline{1} = 1$$
$$f_{10} = f(x_1 = 1, x_2 = 0) = \overline{1} \cdot 0 \vee 1 \cdot \overline{0} = 1$$
$$f_{11} = f(x_1 = 1, x_2 = 1) = \overline{1} \cdot 1 \vee 1 \cdot \overline{1} = 0$$

FIGURE 5.21
Recursive application of Shannon expansion to the Boolean function $f = \overline{x}_1x_2 \vee x_1\overline{x}_2$ using a decision tree (Example 5.34).

Practice problem 5.21. (Shannon expansion in matrix form.) Given the Boolean function $f = \overline{x}_1 x_3 \vee x_2 \overline{x}_3$, draw a decision tree. Show the decomposition of the truth vector $\mathbf{F} = [\ 01110010\ ]^T$.
Answer is given in "Solutions to practice problems."

Example 5.35 (Decision diagram for symmetric functions.) *The symmetric property can be observed, for example, in decision diagrams, as illustrated in Figure 5.22.*

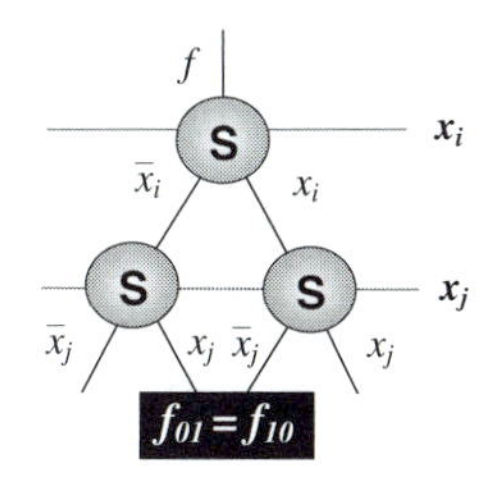

FIGURE 5.22
Symmetry conditions for adjacent variables of a decision tree (Example 5.35).

5.7 Summary of fundamental expansions of Boolean functions

This chapter introduces the fundamental expansions of Boolean functions, which play a key role in the representation, manipulation, and optimization of Boolean functions. The fundamental expansions of Boolean functions include:

1. The ***Shannon expansion***, which provides for the decomposition of an ***arbitrary*** Boolean function of n variables into a set of Boolean sub-functions of $m < n$ variables.
2. The ***Shannon expansion for symmetric*** Boolean functions, which provides for the representation using ***symmetry property*** of such a function.
3. ***Logic Taylor expansion***, which describes a Boolean function in terms of ***Boolean difference***.

Summary (continuation)

The main topics of this chapter are summarized as follows:

- Shannon expansions are used in the following design techniques:

  (*a*) In logic network design using AND, OR, and NOT gates
  (*b*) In decision trees and decision diagrams whose nodes implement Shannon expansion (details are discussed in Chapters 6 and 9)
  (*c*) In the manipulation of Boolean functions in algebraic form and matrix forms
  (*d*) In logic module design such as multiplexers (details are given in Chapter 13)

- Logic Taylor expansion is used in the following design techniques:

  (*a*) In ***functional*** decision trees and functional decision diagrams; in the nodes, a ***Davio expansion*** with respect to a single variable can be implemented using the simplest logic elements (details are discussed in Chapter 12)
  (*b*) The coefficients of expansion are the so-called ***Boolean differences***.

- Given a Boolean function f of n variables $x_1, x_2, \ldots, x_n$, the standard SOP is

$$f = \underbrace{\boxed{s_0}\ \underbrace{\overline{x}_1 \ldots \overline{x}_{n-1}\overline{x}_n}_{Minterm\ M_0} \vee \boxed{s_1}\ \underbrace{\overline{x}_1 \ldots \overline{x}_{n-1}x_n}_{Minterm\ M_1} \vee \ldots \vee \boxed{s_{2^n-1}}\ \underbrace{x_1 \ldots x_{n-1}\overline{x}_n}_{Minterm\ M_{2^n-1}}}_{Canonical\ standard\ SOP\ form}$$

  where the set of binary coefficients $s_0, s_1, \ldots s_{2^n-1}$ completely and uniquely specifies the Boolean function to be realized. These coefficients are the elements of the truth vector $\mathbf{X} = [s_0, s_1, \ldots, s_{2^n-1}]^T$.

- The simplest polynomial form (with uncomplemented variables) of a Boolean function is represented by the equation

$$f = \underbrace{\boxed{r_0} \oplus \boxed{r_1}\ x_1 \oplus \ldots \oplus \boxed{r_{2^n-1}}\ x_1 \ldots x_n}_{Canonical\ polynomial\ form}$$

  where the set of binary coefficients $r_0, r_1, \ldots r_{2^n-1}$ completely and uniquely specifies the Boolean function to be realized. These coefficients are the values of Boolean differences.

Summary (continuation)

- The formal bases for dealing with SOP and polynomial forms of Boolean functions are Boolean algebra (Algebra I) and polynomial algebra (Algebra II), respectively:

$$\underbrace{\texttt{SOP form}}_{Algebra\ I} \longleftrightarrow \underbrace{\texttt{Polynomial form}}_{Algebra\ II}$$

- Shannon expansion provides for the ***decomposition*** of an arbitrary Boolean function of n variables with respect to a variable x_i:

$$f = \overline{x}_i \boxed{f_0} \vee x_i \boxed{f_1}$$
$$f_0 = f(x_1, \ldots, x_{i-1}, 0\ x_{i+1}, \ldots, x_n)$$
$$f_1 = f(x_1, \ldots, x_{i-1}, 1\ x_{i+1}, \ldots, x_n)$$

- With respect to a group of variables,

$$f = \bigvee_{j=0}^{2^m-1} \underbrace{x_{i_1}^{c_{i_1}} x_{i_2}^{c_{i_2}} \ldots x_{i_m}^{c_{i_m}}}_{Control\ variables} \boxed{f_j}$$
$$f_j = f(x_1, \ldots, \underbrace{(x_{i_1} = c_{i_1}), \ldots, (x_{i_m} = c_{i_m})}_{Control\ variables}, \cdots, x_n)$$

- Boolean difference is the component of an EXOR form of the Shannon expansion: $f = f_0 \oplus x_i \frac{\partial f}{\partial x_i}$.
- A Boolean function f is ***totally symmetric*** if and only if it is unchanged by any permutation of variables. There are 2^{n+1} totally symmetric functions of n variables. An arbitrary totally symmetric function f can be uniquely represented by ***elementary*** symmetric functions.
- A Boolean function f is ***partially symmetric*** in k variables, if and only if it is unchanged by any permutation of these k variables only.
- ***Shannon expansion for symmetric*** in variables $x_1, x_2 \ldots, x_m$ Boolean function

$$f = f(\overbrace{x_1, x_2, \ldots, x_m}^{Symmetric\ variables}, x_{m+1}, \ldots, x_{n-1}, x_n)$$

provides the representation based on the ***elementary symmetric functions*** $S_i^m(x_1, x_2, \ldots, x_m)$

Summary (Continuation)

$$f = \bigvee_{i=0}^{2^m-1} \overbrace{S_i^m(x_1, x_2, \ldots, x_m)}^{\substack{Elementary \\ symmetric\ function}} f_i(x_{m+1}, \ldots, x_{n-1}x_n)$$

The elementary symmetric function $S_i^m(x_1, x_2, \ldots, x_m) = 1$ iff exactly i out of m inputs are equal to one. That is, for all assignments of values to variables that have the same number of 1's, there is *exactly* one value of a symmetric function.

- ***Shannon expansion for totally symmetric*** Boolean functions is the the logical sum of elementary symmetric functions S_i^n: $f = \bigvee_{i \in A} S_i^n$, where $A \subseteq 0, 1, 2, \ldots n$.
- For advances in fundamental expansions of Boolean functions, we refer the reader to the "Further study" section.

5.8 Further study

Historical perspective

1715: The English mathematician Brook Taylor (1685–1731) proved the following theorem: If a function f can be differentiated n times at point x_0, then the polynomial can be defined for f about $x = x_0$, where each coefficient is defined in terms of the derivatives. This theorem is known as the *Taylor expansion* or Taylor series.

1742: The Scottish mathematician Colin MacLaurin (1698–1746) proved the theorem of polynomial representation of a function f, known as *MacLaurin polynomials* or the MacLaurin series. This is a special case of the Taylor polynomial.

1937: Claude Shannon introduced in his master's thesis and later, in 1938, in the paper "A Symbolic Analysis of Relay and Switching Circuits" the expansion for the SOP form of Boolean functions. This expansion, known as the *Shannon expansion*, has several extensions, such as *generalized Shannon expansion* for multiple-valued logic functions, *arithmetic analog of Shannon expansion* for the representation of Boolean functions in word-level form, and *information-theoretical form of Shannon expansion* for computing Boolean functions in terms of information.

1949: Claude Shannon showed that symmetric Boolean functions may be realized with considerably fewer components than most functions (C. Shannon, "The synthesis of two-terminal switching circuits," *The Bell System Technical Journal*, volume 28, pages 59–98, 1949). This result is known as the *Shannon theorem on symmetric functions.*

1954: I. Reed and D. Muller (independently) introduced how the classical Taylor and MacLaurin expansions can be used for the representation of Boolean functions in polynomial forms known also as *Reed-Muller polynomials*. The operations associated with polynomial forms of Boolean functions were called *Boolean differences.* These differences, as in classical Taylor and MacLaurin expansions, are the coefficients of Reed-Muller polynomials, that is, the "logical form" of classical Taylor and MacLaurin expansions.

1959: Sheldon B. Akers introduced the concept of partial Boolean difference and the concept of sensitivity of a Boolean function (S. B. Akkers, Jr., "On a Theory of Boolean Functions," *Journal of the Society for Industrial and Applied Mathematics*, volume 7, number 4, 1959, pages 487–498). Much advanced research on *Boolean Differential Calculus* and its applications has been published by A. Thayse, M. Davio, J. P. Deschamps, D. Bochmann, and Ch. Posthoff.

Advanced topics of Shannon and logic Taylor expansion

Topic 1: The arithmetic analog of the Shannon expansion of a Boolean function f with respect to a variable x_i. There exist:

- *The arithmetic analog of the positive Shannon* expansion $f = f_0 + x_i f_2$, where $f_0 = f_{x_i=0}$ and $f_2 = -f_{x_i=1} + f_{x_i=0}$, and
- *The arithmetic analog of the negative Shannon* expansion $f = f_1 + \overline{x}_i f_2$, where $f_1 = f_{x_i=1}$.

For the Boolean variables x, x_1, and x_2, taking on the values of 0 or 1, the following is true, as introduced by the founder of Boolean algebra George Boole: $\overline{x} = 1 - x$, $x_1 \vee x_2 = x_1 + x_2 - x_1 x_2$, and $x_1 \wedge x_2 = x_1 x_2$. Other logic operations can be represented by arithmetic operations as well; for example, $x_1 \oplus x_2 = x_1 + x_2 - 2x_1 x_2$. The right part of the equation is called an *arithmetic expression.*

Arithmetic representations of Boolean functions are known as *word-level* forms, and are a way to describe the parallel calculation of several Boolean functions at once. Another useful property of these arithmetic representations is linearization. A multi-output Boolean function can be represented by a linear word-level arithmetic polynomial and a linear word-level decision diagram.

Topic 2: Symmetric functions. There are many forms of symmetry of Boolean functions. In practice, one of the advantages of knowing that a given Boolean function is symmetric is the potential economy offered by the implementation of such function using smaller number of gates. In addition, in verification, the effectiveness of the *input matching* procedure can be increased if the symmetric input sets of the specification are known (the goal of verification is to check whether a given implementation follows the specification for which it was designed).

Let f be a Boolean function of n variables. Shannon expansion with respect to the variables x_i and x_j yields:

$$f = \overline{x}_i\overline{x}_j f_{\overline{x}_i\overline{x}_j} \vee \overline{x}_i x_j f_{\overline{x}_i x_j} \vee x_i\overline{x}_j f_{x_i\overline{x}_j} \vee x_i x_j f_{x_i x_j}$$

From this expansion follows that

(*a*) f is partially symmetric in variables (x_i, x_j) and $(\overline{x}_i, \overline{x}_j)$ if $f_{\overline{x}_i x_j} = f_{x_i\overline{x}_j}$;
(*b*) f is partially symmetric in variables $(x_i, \overline{x}_j)$ and $(\overline{x}_i, x_j)$ if $f_{\overline{x}_i\overline{x}_j} = f_{x_i x_j}$;
(*c*) f is multiform symmetric in variables $(x_i, x_j), (\overline{x}_i, \overline{x}_j)$ and $(x_i, \overline{x}_j), (\overline{x}_i, x_j)$ if $f_{x_i\overline{x}_j} = f_{\overline{x}_i x_j}$ and $f_{\overline{x}_i\overline{x}_j} = f_{x_i x_j}$, respectively.

For example, Shannon expansion of the Boolean function $f = \overline{x}_1\overline{x}_2\overline{x}_3 \vee x_1x_2 \vee x_1x_3 \vee x_2x_3$ with respect to the variables x_1 and x_2 yields

$$f = \overline{x}_1\overline{x}_2 \overbrace{(\overline{x}_3)}^{f_{\overline{x}_i\overline{x}_j}} \vee \overline{x}_1 x_2 \overbrace{(x_3)}^{f_{\overline{x}_i x_j}} \vee x_1\overline{x}_2 \overbrace{(x_3)}^{f_{x_i\overline{x}_j}} \vee x_1 x_2 \overbrace{(x_3 \vee \overline{x}_3)}^{f_{x_i x_j}}$$

Hence, the Boolean function f is partially symmetric with respect to the variables (x_1, x_2) and $(\overline{x}_1, \overline{x}_2)$.

Shannon expansion for symmetric functions can be extended to an *arbitrary* Boolean functions; that is, an arbitrary Boolean function can be represented with respect to the elementary symmetric functions (see, for example, the pioneered paper by R. C. Born and A. K. Scidmore "Transformation of Switching Functions to Completely Symmetric Switching Functions," *IEEE Transactions on Computers*, volume C-17, number 6, 1968, pages 596–599). The result of such symmetrization is a computing structure of a regular configuration. In the implementation, a regular structure is characterized by the local and short connections between computing elements, that is attractive for technology. In addition, logic design can be combined with layout, so that no special stage of placement and routing is necessary.

Symmetrization is the underlying principle of so-called *universal computing arrays*. For example, universal Akers array is a regular and planar layout of identical computing elements (multiplexers) which is able to compute an arbitrary Boolean function (S. B. Akers, "A Rectangular Logic Array," *IEEE Transactions on Computers*, volume C-21, number 8, 1972, pages 848–857). Another effective application of the symmetry is decision diagram techniques (see, for example, the paper by M. A. Perkowski, M. Chrzanowska-Jeske, and Y. Xu, "Lattice Diagrams Using Reed-Muller Logic," *Proceedings of the 3rd International Workshop on Applications of the Reed-Muller Expansion in Circuit Design*, Oxford University, 1997, pages 85–102).

The alternative data structure for symmetric functions called transeunt triangle, is considered in Chapter 12. Figure 5.23 shows how the Pascal triangle can be used for representations of symmetric Boolean functions. The Pacal triangle can be reduced to a transeunt triangle. This data structures for the function $f = \overline{x_1 x_2}$ is shown in Figure 5.23. The carry vector $\mathbf{F}_c = [1\ 1\ 0]^T$ is formed from the initial truth vector $\mathbf{F} = [1\ 1\ 1\ 0]^T$.

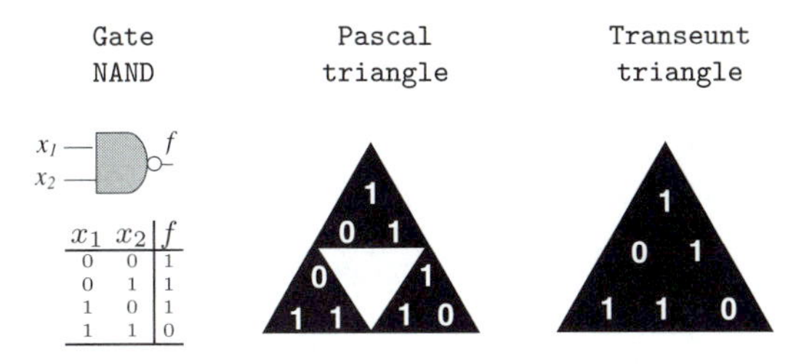

x_1	x_2	f
0	0	1
0	1	1
1	0	1
1	1	0

FIGURE 5.23
Representation of NAND gate using the Pacal triangle transeunt triangle.

Topic 3: The arithmetic analog of logic Taylor expansion represents a Boolean function in 2^n forms called *arithmetic fixed polarity forms.* An arithmetic analog of the logic Taylor expansion is expressed by the equation

$$f = \sum_{i=0}^{2^n-1} p_i^{(c)} \underbrace{\overbrace{(x_1 \oplus c_1)}^{c_1\ polarity}{}^{i_1} \ldots \overbrace{(x_n \oplus c_n)}^{c_n\ polarity}{}^{i_n}}_{i-th\ product},$$

where $c_1c_2 \ldots c_n$ and $i_1, i_2, \ldots, i_n$ are the binary representations of c (polarity) and i respectively; $p_i^{(c)}$ is a value of the arithmetic analog of a Boolean difference of f given c; i.e., $x_1 = c_1, ..., x_n = c_n$.

The arithmetic Taylor expansion produces 2^n arithmetic expressions, corresponding to 2^n polarities. Arithmetic Taylor expansions for several elementary Boolean functions are given in Table 5.7.

Topic 4: Spectral techniques are based on the fundamental expansion called *spectral transform expansion.* Many problems can be solved by transferring the problem in the original space isomorphically to some other space that reflects particular properties of the problem. Thus, in the new space, the problem is simpler or there are at least well known solution tools. Once the problem is solved, we have to be able to return to the original space. Spectral techniques offer alternative methods for solving complex tasks efficiently. The first applications of spectral techniques in logic design are related to the optimization problems and date back to the early 1950s.

Transferring a problem from the original domain into the spectral domain performs redistribution of the information content of a signal, but does not reduce it. Spectral representations are canonical, i.e., a unique spectrum corresponds to a given function, and vice versa – the function can be reconstructed from the spectrum by the inverse transform. In many cases, due to particular properties a signal may express, the main portion of the information content of the signal is encoded in a (relatively) small number of spectral coefficients. For example, the AND function can be described in terms of so-called Walsh spectrum as follows

$$f = \frac{1}{4}(1 - (1 - 2x_2) - (1 - 2x_1) + (1 - 2x_1)(1 - 2x_2))$$

TABLE 5.7
The arithmetic Taylor expansion of elementary Boolean functions.

Function	F	$\frac{\partial f}{\partial x_2}$	$\frac{\partial f}{\partial x_1}$	$\frac{\partial^2 f}{\partial x_1 \partial x_2}$	Arithmetic expression
$y = x_1 \wedge x_2$	0	0	0	1	$x_1 x_2$
	0	0	1	-1	$x_1 - x_1\overline{x}_2$
	0	1	0	-1	$x_2 - \overline{x}_1 x_2$
	1	-1	-1	1	$1 - \overline{x}_2 - \overline{x}_1 + \overline{x}_1\overline{x}_2$
$y = x_1 \vee x_2$	0	1	1	-1	$x_2 + x_1 - x_1 x_2$
	1	-1	0	1	$1 - \overline{x}_2 + x_1\overline{x}_2$
	1	0	-1	1	$1 - \overline{x}_1 + \overline{x}_1 x_2$
	1	0	0	-1	$1 - \overline{x}_1\overline{x}_2$
$y = x_1 \oplus x_2$	0	1	1	-2	$x_2 + x_1 - 2x_1x_2$
	1	-1	-1	2	$1 - \overline{x}_2 - x_1 + 2x_1\overline{x}_2$
	1	-1	-1	2	$1 - x_2 - \overline{x}_1 + 2\overline{x}_1 x_2$
	0	1	1	-2	$\overline{x}_2 + \overline{x}_1 - 2\overline{x}_1\overline{x}_2$

Decision trees are graphical representations of spectral transform expansions of Boolean function f with respect to a basis Q. In a decision tree, each path from the root node to the constant nodes corresponds to a basic function in Q. The constant nodes represent the Q-spectral coefficients.

Topic 5: Information notation of the Shannon expansion. Shannon expansion can be extended to the *integer* numbers and interpreted in terms of *Shannon information*. Let $A = \{a_1, a_2, \ldots, a_n\}$ be a complete set of events with the probability distribution $\{p(a_1), p(a_2), \ldots, p(a_n)\}$. The *entropy* of the finite field A is given by

$$H(A) = -\sum_{i=1}^{n} p(a_i) \cdot \log p(a_i)$$

where the logarithm is base 2. The entropy can never be negative; i.e., $\log p(a_i) \leq 0$, and thus $H(A) \geq 0$. The entropy is zero if and only if A contains one event only

Information notation of the Shannon expansion for a Boolean function f of n variables with respect to the variable x_i is represented by the equation

$$H^S(f|x_i) = p_{|x_i=0} \cdot H(f_{x_i=0}) + p_{|x_i=1} \cdot H(f_{x_i=1}).$$

More details can be found in the paper "Information Theoretical Approach to Minimization of AND/EXOR Expressions of Switching Functions" by V. Shmerko, S. Yanushkevich, and D.Popel, *Proceedings of the IEEE International Conference on Telecommunications*, Yugoslavia, pages 444–451,

1999, as well as in the handbook *Decision Diagram Techniques* (see "Further reading").

Further reading

"A Symbolic Analysis of Relay and Switching Circuits" by C. E. Shannon, *Transactions American Institute of Electrical Engineers*, volume 57, Part B, pages 713–723, 1938.

The Mathematical Theory of Communication by C. E. Shannon and Warren Weaver, the University of Illinois Press, Urbana, Illinois, 1949.

Handbook of Boolean Algebras edited by J. Donald Monk and R. Bonnet, North-Holland, 1989 (in 3 volumes).

Decision Diagram Techniques for Micro- and Nanoelectronic Design by S. N. Yanushkevich, D. M. Miller, V. P. Shmerko, and R. S. Stanković, CRC/Taylor & Francis Group, Boca Raton, FL, 2006.

Discrete and Switching Functions by M. Davio, J. P. Deschamps, and A. Thayse, McGraw-Hill, 1978.

"Detection of Group Invariance or Total Symmetry of a Boolean Functions" by E. J. Yr. McCluskey, *The Bell System Technical Journal*, volume 35, number 6, pages 1445–1453, 1956.

Logic Synthesis and Verification Algorithms by G. D. Hachtel and F. Somenzi, Kluwer, 1996.

Mathematical Logic: A Course with Exercises by R. Cori and D. Lascar, Oxford University Press, 2000 (Chapter 2).

Spectral Interpretation of Decision Diagrams by R. S. Stanković and J. T. Astola, Springer, 2003.

5.9 Solutions to practice problems

Practice problem	Solution
5.1.	$\begin{aligned} f &= \overline{x}_1\overline{x}_2 \oplus \overline{x}_2 \\ &= \overline{x}_1 f_{x_1=0} \vee x_1 f_{x_1=1} \\ &= \overline{x}_1(\overline{x}_2 \oplus \overline{x}_2) \vee x_1(0 \oplus \overline{x}_2) \\ &= \overline{x}_1(0) \vee x_1(\overline{x}_2) \\ &= \overline{x}_1(\overline{x}_2 f_{x_2=0} \vee x_2 f_{x_2=1}) \vee x_1(\overline{x}_2 f_{x_2=0} \vee x_2 f_{x_2=1}) \\ &= \overline{x}_1(\overline{x}_2(0) \vee x_2(0)) \vee x_1(\overline{x}_2(1) \vee x_2(0)) \\ &= \overline{x}_1\overline{x}_2(0) \vee \overline{x}_1 x_2(0) \vee x_1\overline{x}_2(1) \vee x_1 x_2(0) \end{aligned}$
5.2.	$\begin{aligned} f &= x_1 \vee x_2 \vee \overline{x}_3 \\ &= \overline{x}_2\overline{x}_3 f_{00} \vee \overline{x}_2 x_3 f_{01} \vee x_2\overline{x}_3 f_{10} \vee x_2 x_3 f_{11} \\ &= \overline{x}_2\overline{x}_3(1) \vee \overline{x}_2 x_3(x_1) \vee x_2\overline{x}_3(1) \vee x_2 x_3(1) \end{aligned}$
5.3.	$\begin{aligned} f &= x_1 \oplus x_2 \\ &= \overline{x}_1\overline{x}_2 f_{00} \vee \overline{x}_1 x_2 f_{01} \vee x_1\overline{x}_2 f_{10} \vee x_1 x_2 f_{11} \\ &= \overline{x}_1\overline{x}_2 \boxed{0} \vee \overline{x}_1 x_2 \boxed{1} \vee x_1\overline{x}_2 \boxed{1} \vee x_1 x_2 \boxed{0} \end{aligned}$
5.4.	$\begin{aligned} f &= \overline{x}_2(x_1 \oplus 0) \vee x_2(x_1 \oplus 1) \\ &= \overline{x}_2 x_1 \vee x_2\overline{x}_1 \\ f &= (\overline{x}_2 \vee (x_1 \oplus 1))(x_2 \vee (x_1 \oplus 0)) \\ &= (\overline{x}_2 \vee \overline{x}_1)(x_2 \vee x_1) \end{aligned}$

5.6.

Number of variables n	All functions 2^{2^n}	Partially symmetric functions $2^{3 \cdot 2^{n-2}}$	Totally symmetric functions 2^{n+1}
2	16	8	8
3	256	64	16
4	65536	4096	32

The numbers of partially and totally symmetric functions are equal for $n = 2$. For $n = 3$, partially and totally symmetric functions cover 25% and 6% of Boolean functions, respectively.

Solutions to practice problems (continuation)

Practice problem	Solution

5.8.

There are four elementary symmetric functions of Boolean function of three variables S_0^3, S_1^3, S_2^3, and S_3^3:

x_1	x_2	x_3	S_0^3	S_1^3	S_2^3	S_3^3
0	0	0	1	0	0	0
0	0	1	0	1	0	0
0	1	0	0	1	0	0
0	1	1	0	0	1	0
1	0	0	0	1	0	0
1	0	1	0	0	1	0
1	1	0	0	0	1	0
1	1	1	0	0	0	1

5.9.

Logic sum: $S_1^3 \vee S_2^3 = S_{\{1\}\cup\{2\}}^4 = S_{1,2}^3$

Complement: $\overline{S}_1^3 = S_{\{0,1,2,3\}-\{1\}}^3 = S_{0,2,3}^3$

Complement: $\overline{S}_2^3 = S_{\{0,1,2,3\}-\{2\}}^3 = S_{0,1,3}^3$

Using these functions, we can also generate, for example, the following symmetric functions:

Logic sum: $S_{1,2}^3 \vee S_{0,2,3}^3 = S_{\{1,2\}\cup\{0,2,3\}}^4 = S_{0,1,2,3}^3$

Complement: $\overline{S}_{1,2}^3 = S_{\{0,1,2,3\}-\{1,2\}}^3 = S_{0,3}^3$

5.10.

$$
\begin{aligned}
f &= \sum_{i=0}^{2} S_i f_i = S_0(x_1,x_2) f_0(1,1,x_3,x_4) \\
&\vee\ S_1(x_1,x_2) f_1(\frac{1,0}{0,1}, x_3, x_4) \\
&\vee\ S_2(x_1,x_2) f_2(0,0,x_3,x_4) \\
&= S_0(x_1,x_2)1 \vee S_1(x_1,x_2)\overline{x}_3 x_4 \vee S_2(x_1,x_2) x_4
\end{aligned}
$$

where $S_0 = x_1 x_2$, $S_1 = 1$, $S_2 = \overline{x}_1 \overline{x}_2$

Solutions to practice problems (continuation)

Practice problem	Solution
5.11.	Derive the table of the values for the Boolean function $f = x_1\overline{x}_2\overline{x}_3 \vee \overline{x}_1 x_2 \overline{x}_3 \vee \overline{x}_1\overline{x}_2 x_3$

x_1	x_2	x_3	f
0	0	0	0
0	0	1	1
0	1	0	1
0	1	1	0
1	0	0	1
1	0	1	0
1	1	0	0
1	1	1	1

Apply the Shannon expansion for symmetric functions:

$$f = \bigvee_{i=0}^{3} S_i^3 f_i = S_0^3(x_1,x_2,x_3)\, f_0 \overbrace{\left(\begin{matrix} x_1 & x_2 & x_3 \\ \downarrow & \downarrow & \downarrow \\ \boxed{0}, & \boxed{0}, & \boxed{0} \end{matrix}\right)}^{\textit{Equal to } 0}$$

$$\vee\, S_1^3(x_1,x_2,x_3)\, f_1 \overbrace{\left(\begin{matrix} x_1 & x_2 & x_3 \\ \downarrow & \downarrow & \downarrow \\ \boxed{1}, & \boxed{0}, & \boxed{0} \\ \boxed{0}, & \boxed{1}, & \boxed{0} \\ \boxed{0}, & \boxed{0}, & \boxed{1} \end{matrix}\right)}^{\textit{Equal to } 1}$$

$$\vee\, S_2^3(x_1,x_2,x_3)\, f_2 \overbrace{\left(\begin{matrix} x_1 & x_2 & x_3 \\ \downarrow & \downarrow & \downarrow \\ \boxed{0}, & \boxed{1}, & \boxed{1} \end{matrix}\right)}^{\textit{Equal to } 0}$$

$$\vee\, S_3^3(x_1,x_2,x_3)\, f_3 \overbrace{\left(\begin{matrix} x_1 & x_2 & x_3 \\ \downarrow & \downarrow & \downarrow \\ \boxed{1}, & \boxed{1}, & \boxed{1} \end{matrix}\right)}^{\textit{Equal to } 0}$$

$$= S_1^3(x_1,x_2,x_3)$$

where $S_1^3(x_1,x_2,x_3) = \overline{x}_1\overline{x}_2\overline{x}_3 \vee \overline{x}_1 x_2\overline{x}_3 \vee \overline{x}_1\overline{x}_2 x_3$

5.12.

$$\begin{aligned} f &= \overline{x}_1\overline{x}_2 \vee \overline{x}_2\overline{x}_3 \vee \overline{x}_1\overline{x}_3 \\ &= \overline{x}_1\overline{x}_2 \underbrace{(\overline{x}_3 \vee x_3)}_{1} \vee \underbrace{(\overline{x}_1 \vee x_1)}_{1} \overline{x}_2\overline{x}_3 \vee \overline{x}_1 \underbrace{(\overline{x}_2 \vee x_2)}_{1} \overline{x}_3 \\ &= \underbrace{\overline{x}_1\overline{x}_2\overline{x}_3}_{S_0^3} \vee \underbrace{\overline{x}_1\overline{x}_2 x_3 \vee x_1\overline{x}_2\overline{x}_3 \vee \overline{x}_1 x_2\overline{x}_3}_{S_1^3} \\ &= S_0^3 \vee S_1^3 \end{aligned}$$

Solutions to practice problems (continuation)

Practice problem	Solution
5.13.	(a) Representation in terms of elementary symmetric functions $$\begin{aligned} f &= (x_1 \vee x_2)(x_2 \vee x_3)(x_1 \vee x_3) \\ &= (x_2 \vee x_1x_3)(x_1 \vee x_3) \\ &= x_1x_2 \vee x_2x_3 \vee x_1x_3 \\ &= x_1x_2\underbrace{(\overline{x}_3 \vee x_3)}_{1} \vee \underbrace{(\overline{x}_1 \vee x_1)}_{1}x_2x_3 \vee x_1\underbrace{(\overline{x}_2 \vee x_2)}_{1}x_3 \\ &= \underbrace{x_1x_2\overline{x}_3 \vee \overline{x}_1x_2x_3 \vee x_1\overline{x}_2x_3}_{S_2^3} \vee \underbrace{x_1x_2x_3}_{S_3^3} \\ &= S_2^3 \vee S_3^3 \end{aligned}$$ (b) $$\begin{aligned} x_1S_{1,2}^2 \vee \overline{x}_1S_{2,3}^2 &= x_1(S_1^2 \vee S_2^3) \vee \overline{x}_1(S_2^2) \\ &= x_1(\underbrace{x_2\overline{x}_3 \vee \overline{x}_2x_3}_{S_1^2} \vee \underbrace{x_2x_3}_{S_2^2}) \vee \overline{x}_1\underbrace{x_2x_3}_{S_2^2} \\ &= x_1x_2\overline{x}_3 \vee x_1\overline{x}_2x_3 \vee x_1x_2x_3 \vee \overline{x}_1x_2x_3 = f \end{aligned}$$
5.18.	$$\begin{aligned} \frac{\partial f}{\partial x_1} &= f_{x_1=0} \oplus f_{x_1=1} \\ &= (0 \cdot x_2 \vee x_3) \oplus (1 \cdot x_2 \vee x_3) \\ &= x_3 \oplus (x_2 \vee x_3) \\ &= \overline{x}_3(x_2 \vee x_3) \vee x_3\overline{(x_2 \vee x_3)} \\ &= x_2\overline{x}_3 \vee \overline{x}_3x_3 \vee x_3\overline{x}_2\overline{x}_3 \\ &= x_2\overline{x}_3 \end{aligned}$$
5.19.	$$\begin{aligned} f = f(7) &\oplus \frac{\partial f(7)}{\partial x_3}\overline{x}_3 \oplus \frac{\partial f(7)}{\partial x_2}\overline{x}_2 \oplus \frac{\partial^2 f(7)}{\partial x_2 \partial x_3}\overline{x}_2\overline{x}_3 \oplus \frac{\partial f(7)}{\partial x_1}\overline{x}_1 \\ &\oplus \frac{\partial^2 f(7)}{\partial x_1 \partial x_3}\overline{x}_1\overline{x}_3 \oplus \frac{\partial^3 f(7)}{\partial x_1 \partial x_2 \partial x_3}\overline{x}_1\overline{x}_2\overline{x}_3 \\ &= 1 \oplus 1 \cdot \overline{x}_3 \oplus 1 \cdot x_2 \oplus 1 \cdot \overline{x}_2\overline{x}_3 \oplus 1 \cdot \overline{x}_1 \oplus 1 \cdot x_1x_3 \oplus 1 \cdot \overline{x}_1\overline{x}_2\overline{x}_3 \end{aligned}$$

Solutions to practice problems (continuation)

Practice problem	Solution
5.20.	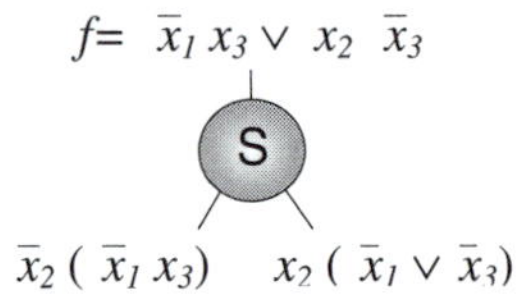 $f = \overline{x}_1 x_3 \vee x_2 \overline{x}_3$ S $\overline{x}_2(\overline{x}_1 x_3)$ $x_2(\overline{x}_1 \vee \overline{x}_3)$ $f_0 = \overline{x}_1 x_3 \vee 0\overline{x}_3 = \overline{x}_1 x_3$ $f_1 = \overline{x}_1 x_3 \vee 1\overline{x}_3$ $= \overline{x}_1 \vee \overline{x}_3$ $f = \overline{x}_2 f_0 \vee x_2 f_1$ $= \overline{x}_2(\overline{x}_1 x_3) \vee x_2(\overline{x}_1 \vee \overline{x}_3)$ 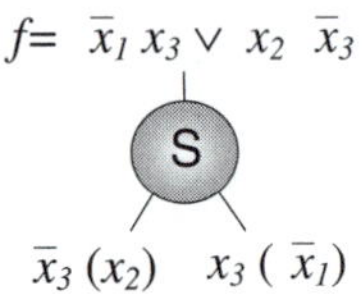 $f = \overline{x}_1 x_3 \vee x_2 \overline{x}_3$ S $\overline{x}_3(x_2)$ $x_3(\overline{x}_1)$ $f_0 = \overline{x}_1 0 \vee x_2 1 = x_2$ $f_1 = \overline{x}_1 1 \vee x_2 0 = \overline{x}_1$ $f = \overline{x}_3 f_0 \vee x_3 f_1$ $= \overline{x}_3(x_2) \vee x_3(\overline{x}_1)$
5.21.	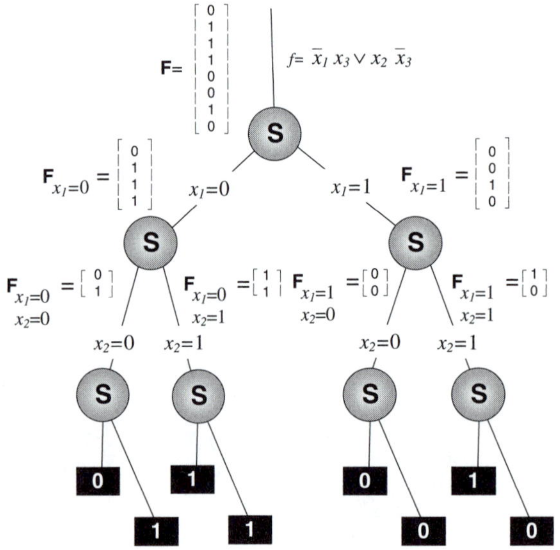 $\mathbf{F} = [0\ 1\ 1\ 1\ 0\ 0\ 1\ 0]^T$, $f = \overline{x}_1 x_3 \vee x_2 \overline{x}_3$ S: $x_1=0$, $x_1=1$ $\mathbf{F}_{x_1=0} = [0\ 1\ 1\ 1]^T$, $\mathbf{F}_{x_1=1} = [0\ 0\ 1\ 0]^T$ $\mathbf{F}_{x_1=0, x_2=0} = [0\ 1]^T$, $\mathbf{F}_{x_1=0, x_2=1} = [1\ 1]^T$, $\mathbf{F}_{x_1=1, x_2=0} = [0\ 0]^T$, $\mathbf{F}_{x_1=1, x_2=1} = [1\ 0]^T$ S: $x_2=0$, $x_2=1$; S: $x_2=0$, $x_2=1$ Terminal nodes: 0 1, 1 1, 0 0, 1 0

5.10 Problems

Problem 5.1 Compute the cofactors f_0 and f_1 of Shannon expansion of the following Boolean functions:
(a) $x_1x_2 \vee x_3,\ i=1$ (b) $x_1 \vee x_2 \vee x_3,\ i=3$ (c) $x_1x_2x_3,\ i=2$

Problem 5.2 Compute the cofactors f_{00}, f_{01}, f_{10}, and f_{11} of Shannon expansion of the following Boolean functions:
(a) $x_1x_2 \vee x_3,\ i_1=1,\ i_2=3$ (c) $x_1 \oplus x_2 \oplus x_3,\ i_1=1,\ i_2=3$
(b) $x_1 \vee x_2 \vee x_3,\ i_1=1,\ i_2=2$ (d) $x_1x_2x_3,\ i_1=1,\ i_2=3$

Problem 5.3 Compute the cofactors $f_{000}, f_{001}, f_{010}, \dots, f_{111}$ of Shannon expansion of the following Boolean functions:
(a) $x_1x_2 \vee x_3$ (b) $x_1 \oplus x_2 \oplus x_3$ (c) $x_1 \vee x_2 \vee x_3$

Problem 5.4 Derive the Shannon expansion with respect to the variable x_1 of the following Boolean functions:
(c) $x_1 \oplus x_2 \oplus x_3,\ i=1$ (b) $x_1 \vee x_2 \vee x_3,\ i=3$ (c) $x_1x_2x_3,\ i=2$

Problem 5.5 Derive the Shannon expansion with respect to the group of variables x_1 and x_3 of the following Boolean functions:
(a) $x_1 \oplus x_2 \oplus x_3,\ i=1$ (b) $x_1 \vee x_2 \vee x_3,\ i=3$ (c) $x_1x_2x_3,\ i=2$

Problem 5.6 Derive Shannon expansion with respect to all variables of the following Boolean functions:
(a) $x_1 \oplus x_2 \oplus x_3,\ i=1$ (b) $x_1 \vee x_2 \vee x_3,\ i=3$ (c) $x_1x_2x_3,\ i=2$

Problem 5.7 Recursively apply Shannon expansion with respect to the variables x_1, x_2, and x_3 to the following Boolean functions:
(a) $x_1 \oplus x_2 \oplus x_3,\ i=1$ (b) $x_1 \vee x_2 \vee x_3,\ i=3$ (c) $x_1x_2x_3,\ i=2$

Problem 5.8 Apply Shannon expansion in EXOR form with respect to the variables x_1, x_2, and x_3 to the following Boolean functions:
(a) $x_1x_2 \vee x_3$ (b) $x_1x_2 \oplus x_3$ (c) $\overline{x}_1\overline{x}_2 \vee x_3$ (d) $x_1 \oplus x_2 \oplus x_3$

Problem 5.9 Derive the Boolean difference with respect to the variable x_1 for the following Boolean functions:
(a) $x_1x_2 \vee x_3$ (b) $x_1x_2 \oplus x_3$ (c) $\overline{x}_1\overline{x}_2 \vee x_3$ (d) $x_1 \oplus x_2 \oplus x_3$

Problem 5.10 Derive the second-order Boolean difference with respect to the variables x_1 and x_2 for the following Boolean functions:
(a) $x_1x_2 \vee x_3$ (b) $x_1x_2 \oplus x_3$ (c) $\overline{x}_1\overline{x}_2 \vee x_3$ (d) $x_1 \oplus x_2 \oplus x_3$

Problem 5.11 Derive the logic Taylor expansion at point c for the following Boolean functions:
(a) $x_1x_2 \vee x_3,\ c=0$ (b) $x_1x_2 \oplus x_3,\ c=2$ (c) $x_1 \oplus x_2 \oplus x_3,\ c=3$

Problem 5.12 Derive the decision trees using Shannon expansion in the nodes for the following Boolean functions:
(a) $x_1x_2x_3$ (b) $x_1x_2 \oplus x_3$ (c) $\overline{x}_1\overline{x}_2 \vee x_3$ (d) $x_1 \oplus x_2 \oplus x_3$

Problem 5.13 Derive the decision diagrams using the Shannon expansion in the nodes and the order of variables x_3, x_1, x_2 for the following Boolean functions:
(a) $x_1x_2x_3$ (b) $x_1x_2 \vee x_3$ (c) $x_1 \vee x_2 \vee x_3$ (d) $x_1 \oplus x_2 \oplus x_3$

Problem 5.14 Which of the Boolean functions are symmetric given the following cubes:

$$f = \begin{bmatrix} 0\,\mathrm{x}\,\mathrm{x} \\ \mathrm{x}\,0\,\mathrm{x} \\ \mathrm{x}\,\mathrm{x}\,0 \end{bmatrix} \quad f = \begin{bmatrix} 0\,\mathrm{x}\,\mathrm{x} \\ \mathrm{x}\,1\,\mathrm{x} \end{bmatrix} \quad f = \begin{bmatrix} \mathrm{x}\,\mathrm{x}\,1 \\ \mathrm{x}\,1\,\mathrm{x} \\ 1\,\mathrm{x}\,\mathrm{x} \end{bmatrix} \quad f = \begin{bmatrix} \mathrm{x}\,0\,1 \\ \mathrm{x}\,1\,\mathrm{x} \end{bmatrix} \quad f = \begin{bmatrix} 1\,1\,1 \end{bmatrix}$$

(a) (b) (c) (d) (e)

Problem 5.15 Reduce the following decision trees:

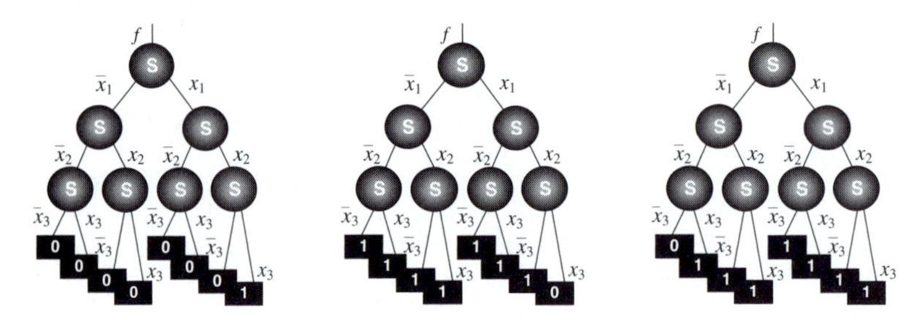

Problem 5.16 Use don't cares to provide symmetry properties for Boolean functions given the following truth tables:

(a)

x_1	x_2	x_3	f
0	0	0	**1**
0	0	1	**0**
0	1	0	**0**
0	1	1	**1**
1	0	0	**x**
1	0	1	**x**
1	1	0	**x**
1	1	1	**x**

(b)

x_1	x_2	x_3	f
0	0	0	**x**
0	0	1	**0**
0	1	0	**x**
0	1	1	**1**
1	0	0	**1**
1	0	1	**0**
1	1	0	**x**
1	1	1	**x**

(c)

x_1	x_2	x_3	f
0	0	0	**x**
0	0	1	**x**
0	1	0	**x**
0	1	1	**x**
1	0	0	**0**
1	0	1	**1**
1	1	0	**1**
1	1	1	**1**

Problem 5.17 Reduce the truth vector of the totally symmetric Boolean functions, given the following truth tables:

(a)

x_1	x_2	x_3	f
0	0	0	0
0	0	1	1
0	1	0	1
0	1	1	1
1	0	0	1
1	0	1	1
1	1	0	1
1	1	1	0

(b)

x_1	x_2	x_3	f
0	0	0	1
0	0	1	1
0	1	0	1
0	1	1	0
1	0	0	1
1	0	1	0
1	1	0	0
1	1	1	1

(c)

x_1	x_2	x_3	f
0	0	0	0
0	0	1	1
0	1	0	1
0	1	1	0
1	0	0	1
1	0	1	0
1	1	0	0
1	1	1	1

Problem 5.18 Derive Boolean function in SOP form given a carrier vector:
(a) $\mathbf{F}_c = [1010]^T$ (b) $\mathbf{F}_c = [1110]^T$ (c) $\mathbf{F}_c = [0101]^T$ (d) $\mathbf{F}_c = [0110]^T$

Problem 5.19 Apply the Shannon theorem on symmetric function given a truth table in Problem 5.17.

6

Boolean Data Structures

Data structure types

- ▶ Algebraic, tabulated, and graphical forms
- ▶ Relationships between data structures

Truth tables and K-maps

- ▶ Construction and reduction
- ▶ Incompletely specified functions

Hypercubes

- ▶ Construction and operations
- ▶ Properties

Logic and threshold networks

- ▶ Synthesis and analysis
- ▶ Properties

Decision trees and diagrams

- ▶ Construction and reduction
- ▶ Manipulation

Advanced topics

- ▶ Computing paradigms
- ▶ Multi-valued logic
- ▶ Information-theoretical measures

6.1 Introduction

Boolean *data structures* are of fundamental importance in logic design (Figure 6.1). No universal data structure exists that is efficient in all applications. Choosing an appropriate data structure is critical for the efficient representation, manipulation, and implementation of Boolean functions.

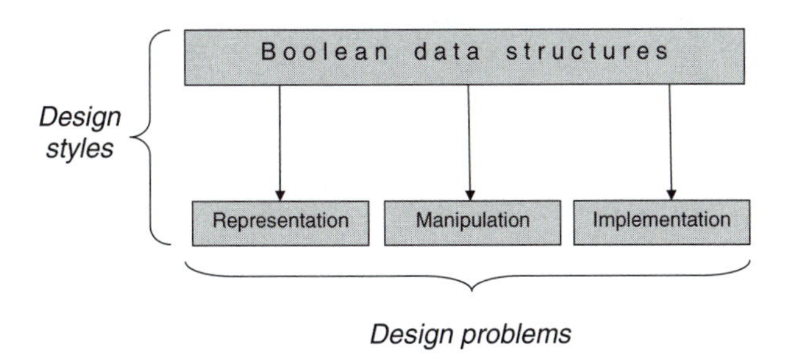

FIGURE 6.1
Design problems (representation, manipulation, and implementation) of Boolean functions can be solved using different design styles.

This chapter guides the reader through the process of the formalization, representation, and implementation of Boolean functions:

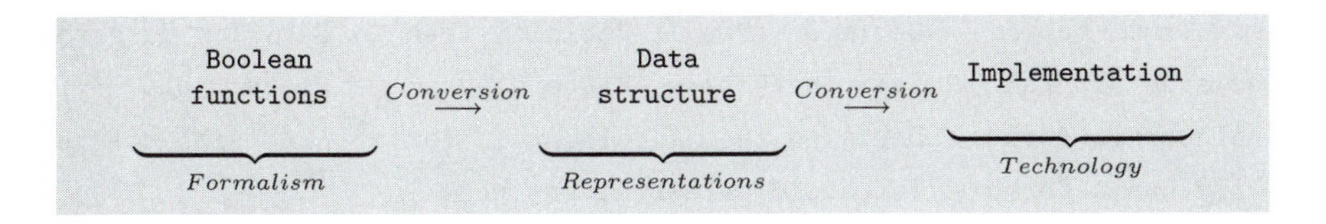

Boolean functions are implemented in computing devices using software and hardware tools. There are several steps for preparing a Boolean function for implementation, aiming at the simplification of the function and at choosing an appropriate form for implementation. No universal approach exists that is good for all situations. The effectiveness of pre-implementation depends on many factors; in particular, size (number of variables), decomposition (partitioning into subfunctions), function properties, and type of data structure.

A *data structure* representing a Boolean function is a mathematical model of the function. A *data type* specifies the properties of this mathematical model. There are, in particular, algebraic and graphical data types. Data structure types for Boolean functions are the focus of this chapter.

6.2 Data structure types

A Boolean function f of n variables $x_1, x_2, \ldots x_n$ can be represented in different mathematical *forms* or *descriptions*:

Boolean data structures

- *Graphical* forms, such as logic networks, decision trees, decision and state diagrams, and hypercubes
- *Tabulated* forms such as truth tables and state tables
- *Algebraic* and *matrix* forms such as canonical descriptions and various transforms
- *Mixed* graphical and tabulated forms such as K-maps

All of these are mathematical models or data structures of different data types.

6.3 Relationships between data structures

A Boolean function can be represented using different data structures. The relationships between data structures are widely used in contemporary logic design, in particular:

Truth table ↔ Algebraic form   Truth table ↔ K-map
K-map ↔ Algebraic form   Truth table ↔ Hypercube form
Truth table ↔ Decision tree   Decision tree ↔ Decision diagram
Decision diagram ↔ Logic network   State diagram ↔ State table
Algebraic form ↔ Decision tree   Algebraic form ↔ Decision diagram
Algebraic form ↔ Threshold gate-based network
Logic network ↔ Threshold gate-based network

The effectiveness of a given design depends on how efficiently these and other relationships between data structures are utilized.

Example 6.1 (Data structures.) *Tabulated forms such as truth tables and state tables can be used for small Boolean functions. K-maps, which combine the features of graphical and tabulated forms, are useful for the representation and optimization of Boolean functions in the so-called sum-of-product form, but are not applicable for polynomial forms. Graphical data structures, such as decision diagrams, are efficient not only for the representation and optimization of Boolean functions, but also for their implementation in hardware.*

6.4 The truth table

A Boolean function of n variables can be represented in a tabular form called a *truth table*:

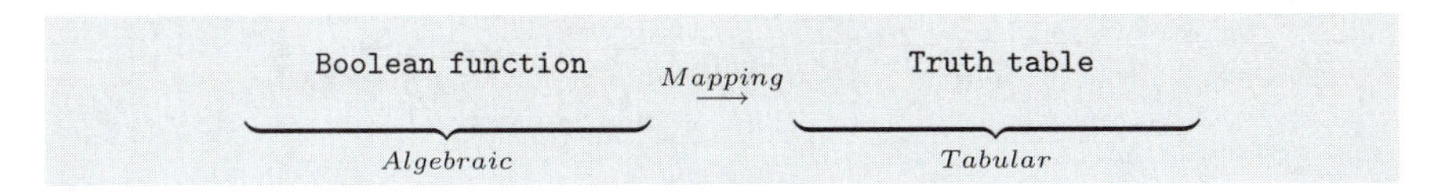

6.4.1 Construction of the truth table

A truth table includes a list of combinations of 1's and 0's assigned to the binary variables, and a column that shows the value of the function for each binary combination. The number of rows in the truth table is 2^n, where n is the number of variables of the function. The binary combinations for the truth table are obtained from the binary numbers by counting from 0 through $2^n - 1$.

Example 6.2 (Truth table.) *There are eight possible binary combinations for assigning bits to the three variables* x_1, x_2, *and* x_3 *of the Boolean function (Figure 6.2a)* $f = x_1 \vee \overline{x}_2 x_3$. *The column labeled* f *contains either 0 or 1 for each of these combinations. The table shows that the function is equal to 1 when* $x_1 = 1$ *or when* $x_2 x_3 = 01$. *It is equal to 0 otherwise.*

Practice problem 6.1. (Truth table.) Given the Boolean function $f = \overline{x}_1 \vee \overline{x}_2 \vee \overline{x}_3$, derive its truth table.
Answer is given in "Solutions to practice problems."

6.4.2 Truth tables for incompletely specified functions

Boolean algebra is flexible enough for the various situations that arise in design practice. This flexibility is demonstrated, in particular, by the utilization of Boolean algebra for the representation and manipulation of *incompletely specified* Boolean functions. As follows from the definition of Boolean algebra, Boolean formulas are not specified for assignments of variables. Boolean formulas become Boolean functions when their values are calculated. Suppose that some of these values cannot be computed; that is, they are unknown.

The $\times$ symbols in the truth table indicates that the value of the Boolean function is unknown; this is also called *don't care*; that is, it takes either the value 0 or 1 (Figure 6.3).

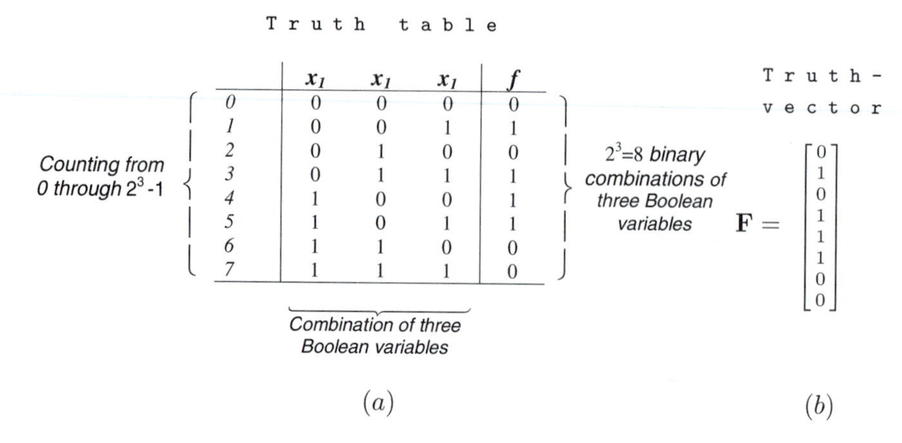

FIGURE 6.2
Representation of a Boolean function in the form of a truth table (a) and truth-vector (b) (Examples 6.2 and 6.4).

	x_1	x_1	x_1	f	
0	0	0	0	0	
1	0	0	1	d	The value is **not specified** for the assignment 001
2	0	1	0	0	
3	0	1	1	1	
4	1	0	0	d	The value is **not specified** for the assignment 100
5	1	0	1	1	
6	1	1	0	d	The value is **not specified** for the assignment 110
7	1	1	1	0	

FIGURE 6.3
Truth table for the representation of an incompletely specified Boolean function f.

Functions that contain don't cares are called *incompletely specified* Boolean functions. Incompletely specified functions, if efficiently utilized, can bring valuable benefits. In particular, don't care values can provide additional opportunities for optimization by assuming that they take values of 0 or 1, depending on the optimization strategy (details are given in Chapter 8).

Example 6.3 (Incompletely specified functions.) *The truth table of the incompletely specified Boolean function, given in Figure 6.3 can be drawn in a compact (reduced) form as follows:*

	$x_1x_2x_3$	f
0	000	0
2	010	0
3	011	1
5	101	1
7	111	0

6.4.3 Truth-vector

A *truth-vector* of a Boolean function is defined as a column f of the truth table,

$$\mathbf{F} = [f(0), f(1), \ldots, f(2^n - 1)]^T$$

The i-th element $f(i)$, $(i = 0, 1, 2, \ldots, 2^n - 1)$ of the truth vector is the value of a Boolean function given the assignment $i_1, i_2, \ldots, i_n$ (n-bit representation of i) of variables $x_1, x_2, \ldots, x_n$.

Example 6.4 (Truth-vector.) *In Figure 6.2b, the truth-vector* $\mathbf{F}$ *is derived by copying the column* f *of the truth table.*

Practice problem 6.2. (Truth-vector.) Derive the truth-vector $\mathbf{F}$ of the Boolean function f of three variables x_1, x_2, and x_3 given the truth table in Table 6.1.
Answer: $\mathbf{F} = [11010001]^T$.

6.4.4 Minterm and maxterm representations

Standard SOP and POS expressions are canonical representations of Boolean functions based on minterms and maxterms, respectively. Minterms and maxterms can be derived from a truth table. Minterms and maxterms correspond to the values 1 and 0 of the function, respectively.

Example 6.5 (Minterms and maxterms.) *The Boolean function of three variables given in Table 6.1 is represented by four minterms and four maxterms.*

Practice problem 6.3. (Minterms and maxterms.) Derive the minterms and maxterms of the Boolean function, given its truth-vector $\mathbf{F} = [\,0\,0\,0\,1\,1\,1\,1\,0\,]^T$.
Answer is given in "Solutions to practice problems."

6.4.5 Reduction of truth tables

The following manipulations of truth tables are used for small Boolean functions:

- Reduction using shorthand notation,
- Reduction using particular properties of Boolean functions such as symmetry, and
- Reduction by replacing a group of rows by variables or sub-functions.

TABLE 6.1
Minterms and maxterms for a Boolean function of three variables.

Assignment of variables x_1, x_2, x_3	Value of Boolean function	Minterm	Maxterm
000	1	$\overline{x}_1\overline{x}_2\overline{x}_3$	
001	1	$\overline{x}_1\overline{x}_2x_3$	
010	0		$x_1 \vee \overline{x}_2 \vee x_3$
011	1	$\overline{x}_1x_2x_3$	
100	0		$\overline{x}_1 \vee x_2 \vee x_3$
101	0		$\overline{x}_1 \vee x_2 \vee \overline{x}_3$
110	0		$\overline{x}_1 \vee \overline{x}_2 \vee x_3$
111	1	$x_1x_2x_3$	

Reduction using shorthand notation

A Boolean function can be represented by its minterms only, since the remaining values (0's) of the function can always be restored. This representation is called *shorthand notation.* Correspondingly, the truth table of the function can be reduced by removing the lines corresponding to maxterms.

Example 6.6 (Truth table reduction.) *Figure 6.4 shows the complete truth table and the reduced one. In a similar manner, a shorthand POS can be obtained, although the SOP form is used more often in practice.*

Reduction based on the particular properties of functions

In a *symmetric* function, for all assignments of values to variables that have the same number of 1s, there is *exactly* one value of a symmetric function (details are given in Chapter 7).

Example 6.7 (Truth table reduction using symmetries.) *Figure 6.4 shows the example of the reduction of a truth table that is symmetrical with respect to the variables x_2 and x_3.*

Practice problem 6.4. **(Truth table reduction.)** Derive a reduced truth table of the Boolean function given the truth-vector $\mathbf{F} = [\,0\,1\,0\,0\,0\,1\,1\,1\,]^T$.
Answer is given in "Solutions to practice problems."

Techniques for reduction of truth table

Reduction based on shorthand SOP notation

x_1	x_2	x_3	f
0	0	0	1
0	0	1	0
0	1	0	1
0	1	1	0
1	0	0	0
1	0	1	1
1	1	0	1
1	1	1	1

⇔

x_1	x_2	x_3	f
0	0	0	1
0	1	0	1
1	0	1	1
1	1	0	1
1	1	1	1

Reduction based on symmetry

$S = x_1 + x_2 + x_3$	x_1	x_2	x_3	f
0	0	0	0	1
1	0	0	1	0
1	0	1	0	0
2	0	1	1	1
1	1	0	0	0
2	1	0	1	1
2	1	1	0	1
3	1	1	1	0

S	x_1	x_2	x_3	f
0	0	0	0	1
1	0	0	1	0
	0	1	0	
	1	0	0	
2	0	1	1	1
	1	0	1	
	1	1	0	
3	1	1	0	0

FIGURE 6.4
Truth table reduction (Examples 6.6 and 6.7).

Reduction using variables and functions

Reduction of a truth table using variables and functions means partial replacement, together with reduction, of parts of the values of a function by its variables and sub-functions.

Example 6.8 (Truth table reduction.) *Figure 6.5 shows the results of the reduction of the complete truth tables.*

Practice problem 6.5. (Truth table reduction.) Reduce the truth table of the Boolean function given by its truth-vector $\mathbf{F} = [\,0\,0\,1\,1\,1\,1\,0\,0\,]^T$ by grouping the values of the function.
Answer is given in "Solutions to practice problems."

Practice problem 6.6. (Truth table reduction.) Reduce the truth table of the Boolean function given by its truth-vector $\mathbf{F} = [\,0\,0\,1\,1\,1\,1\,0\,0\,]^T$

Techniques for reduction of truth table

Reduction based on grouping

Initial truth table

Reduced truth table with respect to x_3

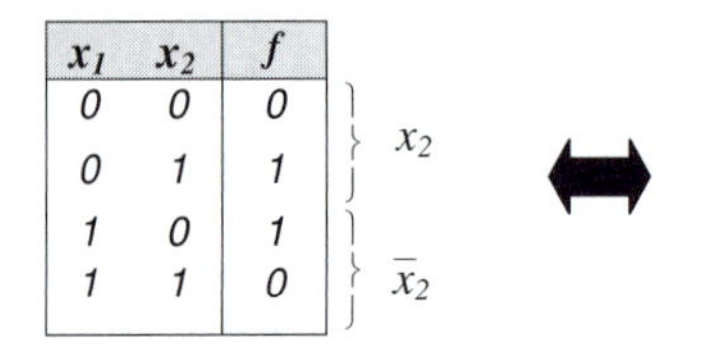

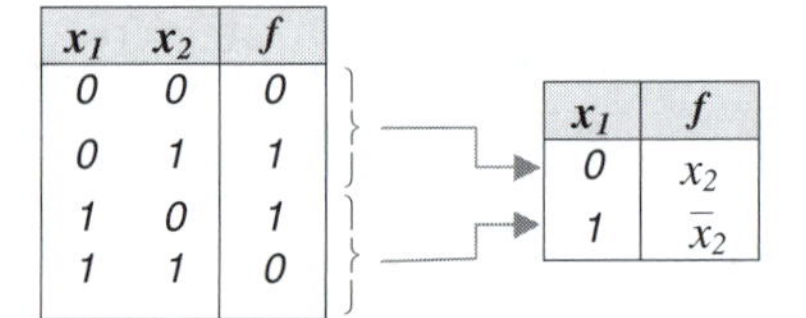

Reduced truth table with respect to x_3

Reduced truth table with respect to x_2 and x_3

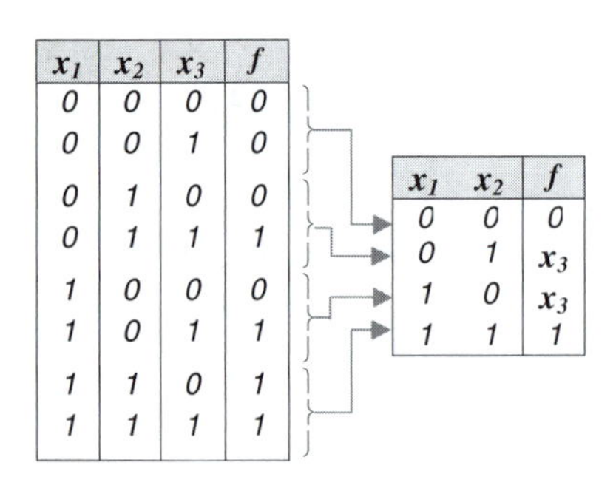

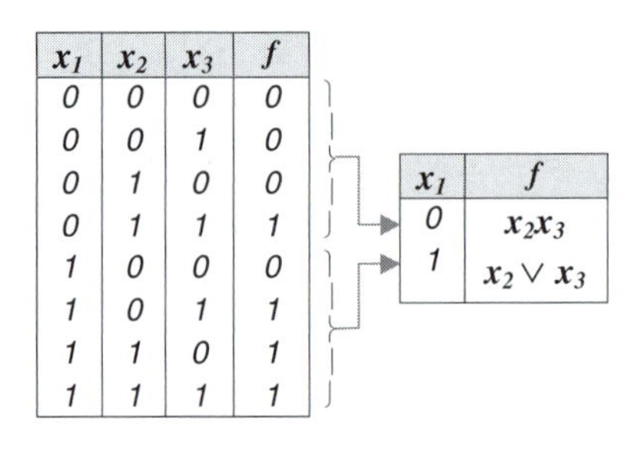

FIGURE 6.5
Truth table reduction (Example 6.8).

using grouping with respect to the variable x_2.
Answer is given in "Solutions to practice problems."

6.4.6 Properties of the truth table

Truth table-based representations of Boolean functions are characterized by the following properties:

- They allow for the observation of the values of functions and the corresponding assignments of variables;
- They allow for observation of some of the properties of functions; and
- They are ready for direct implementation.

However, truth tables grow in size quickly and therefore can be useful only for functions with a small number of variables. In tabular representation, the function's values are explicitly shown, without taking into account their possible relationships.

6.4.7 Deriving standard SOP and POS expressions from a truth table

In Figure 6.6, the procedures for deriving standard SOP and POS expressions from a truth table are given.

6.5 K-map

The *Karnaugh map* or *K-map* is a form of representation of a Boolean function that is an alternative to the truth table:

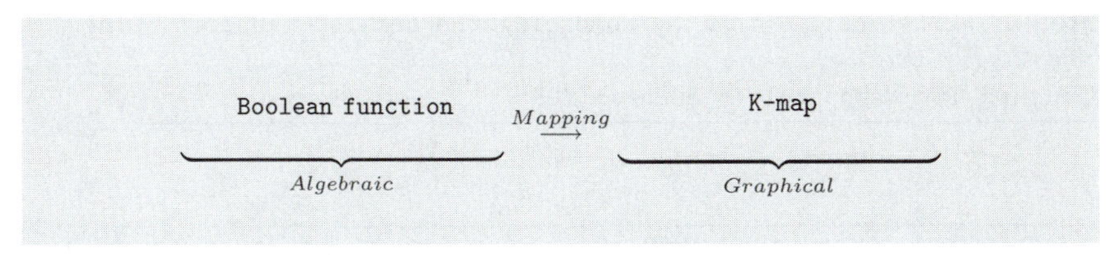

The K-map consists of *cells* that are labeled by variable assignments corresponding to the rows of the truth table (Figure 6.7). Each cell represents a minterm. In a K-map, each cell must differ from any adjacent cell by only one variable. For this, a K-map performs an algebraic factoring, such as

$$\begin{aligned}\overline{x}_1 x_2 \vee x_1 x_2 &= \overline{x}_1 \vee x_1 x_2 \\ &= 1 \cdot (x_1 x_2) = x_2\end{aligned}$$

In this way, the map displays the factorable terms in a Boolean expression.

Representation of standard SOP and POS expressions using K-maps

In Figure 6.8, the procedures for the representation of standard SOP and POS expressions using K-maps are given.

A K-map for a Boolean function of two variables

A K-map is a 2-dimensional representation of the truth table of a Boolean function. Given a Boolean function of two variables, its four-row truth table can be regrouped into a K-map consisting of four cells. The columns of the map are labeled by the values of the variable x_1, and the rows are labeled by x_2. This label leads to the locations of minterms. The columns and rows of the K-map are encoded as follows:

Deriving a standard SOP expression from a truth table

Given: `A truth table of a Boolean function`
Step 1: `Find the values of` 1 `in the truth table`
Step 2: `Identify the corresponding (to the positions of the 1s) assignments of variables`
Step 3: `Derive the minterms based on the rules:`

(*a*) `If the variable assignment is` 0 `, then the variable is included in complemented form;`

(*b*) `If the variable assignment is` 1 `, then the variable is included in the uncomplemented form.`

`Assemble variables into minterms using the AND operation`
Step 4: `Write the SOP form by combining the derived minterms using logic sum (`$\vee$`)`
Output: `The canonical (standard) SOP`

Deriving a standard POS expression from a truth table

Given: `A truth table of a Boolean function`
Step 1: `Find the values of` 0 `in the truth table`
Step 2: `Identify the corresponding (to the positions of the 1s) assignments of variables`
Step 3: `Derive the maxterms based on the rules:`

(*a*) `If the variable assignment is` 0 `, then the variable is included in uncomplemented form;`

(*b*) `If the variable assignment is` 1 `, then the variable is included in the complemented form.`

`Assemble variables into minterms using the OR operation`
Step 4: `Write the POS form by combining the derived minterms using logic product (`$\wedge$`)`
Output: `The canonical (standard) POS`

FIGURE 6.6
Procedures for deriving standard SOP and POS expressions from a truth table.

Techniques for Boolean function representation using K-maps

Boolean function of two variables

x_1	x_1	f
0	0	m_0
0	1	m_1
1	0	m_2
1	1	m_3

x_2 \ x_1	0	1
0	m_0	m_2
1	m_1	m_3

Boolean function of three variables

x_1	x_2	x_3	f
0	0	0	m_0
0	0	1	m_1
0	1	0	m_2
0	1	1	m_3
1	0	0	m_4
1	0	1	m_5
1	1	0	m_6
1	1	1	m_7

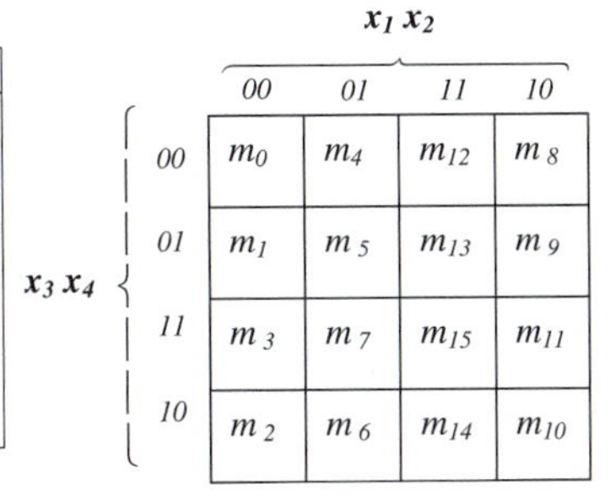

Boolean function of four variables

x_1	x_2	x_3	x_4	f	x_1	x_2	x_3	x_4	f
0	0	0	0	m_0	1	0	0	0	m_8
0	0	0	1	m_1	1	0	0	1	m_9
0	0	1	0	m_2	1	0	1	0	m_{10}
0	0	1	1	m_3	1	0	1	1	m_{11}
0	1	0	0	m_4	1	1	0	0	m_{12}
0	1	0	1	m_5	1	1	0	1	m_{13}
0	1	1	0	m_6	1	1	1	0	m_{14}
0	1	1	1	m_7	1	1	1	1	m_{15}

$x_3 x_4$ \ $x_1 x_2$	00	01	11	10
00	m_0	m_4	m_{12}	m_8
01	m_1	m_5	m_{13}	m_9
11	m_3	m_7	m_{15}	m_{11}
10	m_2	m_6	m_{14}	m_{10}

FIGURE 6.7
The K-maps are alternatives to the truth tables of Boolean functions of two, three, and four variables.

$$\texttt{2-variable K-map encoding} = \begin{cases} \overbrace{\overset{0}{\overline{x}_1},\ \overset{1}{x_1},}^{\textit{Encoding}} \text{ columns;} \\ \overset{0}{\overline{x}_2},\ \overset{1}{x_2}, \text{ rows.} \end{cases}$$

A K-map for a Boolean function of three variables

A K-map for a Boolean function of three variables is constructed by placing two two-variable maps side by side. In this case, each valuation of variables x_1 and x_2 identifies a column in the map, while the value of x_3 distinguishes

Representation of a standard SOP expression using a K-map

Given: `A K-map`
Step 1: `Identify the` `1s` `in the K-map`
Step 2: `Identify the assignments of variables corresponding to the row and line labels of the` `1s` `cells`
Step 3: `Derive the minterms using the identified assignments`
Output: `Canonical (standard) SOP`

Representation of a standard POS expression using a K-map

Given: `A K-map`
Step 1: `Identify the` `0s` `in the K-map`
Step 2: `Identify assignments of variables corresponding to the row and line labels of the` `0s` `cells`
Step 3: `Derive the maxterms using the identified assignments`
Output: `Canonical (standard) POS`

FIGURE 6.8
Procedures for the representation of standard SOP and POS expressions using K-maps.

the two rows. The columns and rows of the K-map are encoded as follows:

$$\texttt{3-variable K-map encoding} = \begin{cases} \overbrace{\overset{00}{\overline{x}_1\overline{x}_2},\ \overset{01}{\overline{x}_1 x_2},\ \overset{11}{x_1 x_2},\ \overset{10}{x_1\overline{x}_2},}^{Encoding} & \text{columns;} \\ \overset{0}{\overline{x}_3},\ \overset{1}{x_3}, & \text{rows.} \end{cases}$$

A K-map for a Boolean function of four variables

A K-map for a Boolean function of four variables is constructed by placing two three-variable maps side by side. The columns and rows of the K-map are encoded as follows:

$$\texttt{4-variable K-map encoding} = \begin{cases} \overbrace{\overset{00}{\overline{x}_1\overline{x}_2},\ \overset{01}{\overline{x}_1 x_2},\ \overset{11}{x_1 x_2},\ \overset{10}{x_1\overline{x}_2},}^{Encoding} & \text{columns;} \\ \overset{00}{\overline{x}_3\overline{x}_4},\ \overset{01}{\overline{x}_3 x_4},\ \overset{11}{x_3 x_4},\ \overset{10}{x_3\overline{x}_4}, & \text{rows.} \end{cases}$$

Example 6.9 (K-maps.) *K-maps for elementary Boolean functions of two variables and their relationships with truth-tables are given in Table 6.2.*

Practice problem 6.7 (K-maps.) Derive the K-map for the Boolean function of four variables $f = \overline{x}_1\overline{x}_2\overline{x}_3\overline{x}_4 \vee \overline{x}_1x_2\overline{x}_3x_4 \vee x_1\overline{x}_2x_3\overline{x}_4 \vee x_1x_2x_3x_4$. ***Answer*** is given in "Solutions to practice problems."

A K-map for an incompletely specified Boolean functions

An incompletely specified Boolean function can be represented by a K-map in which the unspecified values are denoted using the symbol d (the same as the one used for the representation of incompletely specified functions in Section 6.4).

Example 6.10 (Incompletely specified Boolean functions.) *Figure 6.9 illustrates the representation of the incompletely specified Boolean function given the truth-vector* $\mathbf{F} = [\,0\ d\ 0\ 1\ d\ 1\ d\ 0\,]^T$ *by the K-map and truth table, where* d *is don't care (unspecified) value.*

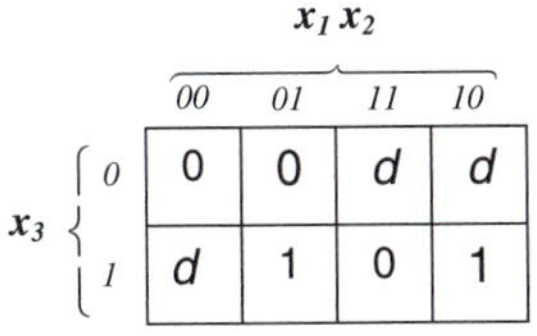

x_3 \ x_1x_2	00	01	11	10
0	0	0	d	d
1	d	1	0	1

(a)

$x_1x_2x_3$	f	Comment
000	0	
001	d	$\overline{x}_1\overline{x}_2x_3$ or $x_1 \vee x_2 \vee \overline{x}_3$
010	0	
011	1	
100	d	$x_1\overline{x}_2\overline{x}_3$ or $\overline{x}_1 \vee x_2 \vee x_3$
101	1	
110	d	$x_1x_2\overline{x}_3$ or $\overline{x}_1 \vee \overline{x}_2 \vee x_3$
111	0	

(b)

FIGURE 6.9
Representation of an incompletely specified Boolean function by a K-map and truth table (Example 6.10).

6.6 Cube data structure

Boolean function can be represented as points in n-space. The collection of 2^n possible points is said to form of an *n-cube*, or a *Boolean hypercube*.

The reduced truth table introduced in Section 6.4.5 contained only assignments of variables, corresponding to the minterms. These assignments are called *primary cubes*.

Example 6.11 (Primary cubes.) *Given a minterm* $x_1x_2\overline{x}_3$*, its assignment* $[\,1\ 1\ 0\,]$ *is a primary cube.*

TABLE 6.2
K-maps for elementary Boolean functions of two variables.

Techniques for Boolean function representation using K-maps

Elementary Boolean function	Truth table	K-map

AND

$f = x_1 x_2$

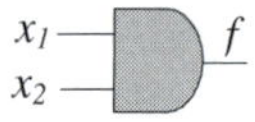

x_1	x_2	f
0	0	0
0	1	0
1	0	0
1	1	1

x_2 \ x_1	0	1
0		
1		1

OR

$f = x_1 \vee x_2$

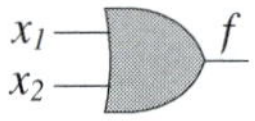

x_1	x_2	f
0	0	0
0	1	1
1	0	1
1	1	1

x_2 \ x_1	0	1
0		1
1	1	1

NAND

$f = \overline{x_1 x_2}$

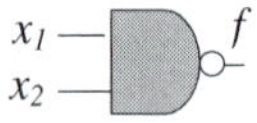

x_1	x_2	f
0	0	1
0	1	1
1	0	1
1	1	0

x_2 \ x_1	0	1
0	1	1
1	1	

NOR

$f = \overline{x_1 \vee x_2}$

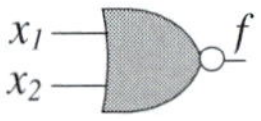

x_1	x_2	f
0	0	1
0	1	0
1	0	0
1	1	0

x_2 \ x_1	0	1
0	1	
1		

EXOR

$f = x_1 \oplus x_2$

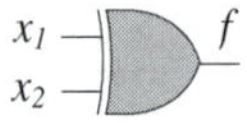

x_1	x_2	f
0	0	0
0	1	1
1	0	1
1	1	0

x_2 \ x_1	0	1
0		1
1	1	

XNOR

$f = \overline{x_1 \oplus x_2}$

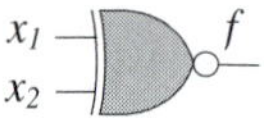

x_1	x_2	f
0	0	1
0	1	0
1	0	0
1	1	1

x_2 \ x_1	0	1
0	1	
1		1

Primary cubes can be further reduced to cubes using Boolean *adjacency* rules.

Example 6.12 (Reduction of primary cubes.) *Consider the primary cubes* $[\,1\;1\;0\,]$ *and* $[\,1\;1\;1\,]$. *These cubes are adjacent, since they are different in one element only. This means that they can form a new cube as* $[\,1\;1\;0\,] \vee [\,1\;1\;1\,] = [\,1\;1\;\times]$, *where the symbol* $\times$ *is used to represent the missing variable. This corresponds to the execution of the adjacency rule on the two minterms:*

$$x_1x_2\overline{x}_3 \vee x_1x_2x_3 = x_1x_2(\overline{x}_3 \vee x_3) = x_1x_2$$

That is, the cube $[\,1\;1\;\times]$ *corresponds to the product term* x_1x_2 *of the Boolean function of three variables.*

The rules for the OR operation on cubes are presented in Figure 6.10a. The basic operations on a cube are sum, product, and complement.

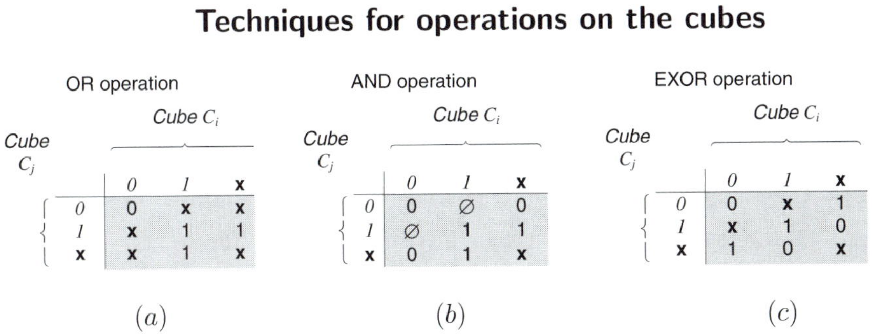

Techniques for operations on the cubes

OR operation

Cube C_j \ Cube C_i	0	1	x
0	0	x	x
1	x	1	1
x	x	1	x

(*a*)

AND operation

Cube C_j \ Cube C_i	0	1	x
0	0	∅	0
1	∅	1	1
x	0	1	x

(*b*)

EXOR operation

Cube C_j \ Cube C_i	0	1	x
0	0	x	1
1	x	1	0
x	1	0	x

(*c*)

FIGURE 6.10
OR (a), AND (b), and EXOR (c) operations on the cubes.

Practice problem 6.8 (Cube operations.) Perform AND, OR, and EXOR operations on the cubes $[\times\;0\;1\,]$ and $[\times\;0\;0\,]$.
Answer: $[\times\,0\,1\,]\wedge[\times\,0\,0\,] = \emptyset$, $[\times\,0\,1\,]\vee[\times\,0\,0\,] = [\times\,0\,\times]$, $[\times\,0\,1\,]\oplus[\times\,0\,0\,] = [\times\;0\;\times]$ (Figure 6.11).

An SOP of a Boolean function can be represented by a set of cubes, with OR operations between them, and the above rules can be used for reducing the SOP form.

Example 6.13 (Cube-based representation.) *Given a Boolean function $f = x_1 \vee x_2$, its cube representation is derived as follows:*

(a) *Using OR operations between the cubes:* $f = x_1 \vee x_2 = [\,1\ \times] \vee [\ \times\ 1\,]$.

(b) *By reducing the truth table, using primary cubes and the rules from Figure 6.11:*

x_1	x_2	f
0	0	0
0	1	1
1	0	1
1	1	1

⟺

x_1	x_2	f
0	1	1
1	0	1
1	1	1

⟺

$$f = [\,0\ 1\,] \vee [\,1\ 0\,] \vee [\,1\ 1\,] = [\,1\ \times\,] \vee [\,\times\ 1\,]$$

Note: *The cube* $[1\ \times]$ *is obtained by adjacing* $[1\ 0]$ *and* $[1\ 1]$*; the cube* $[\times\ 1]$ *is obtained by adjacing* $[0\ 1]$ *and* $[1\ 1]$.

Practice problem 6.9. (Cube-based representation.) Find the cube-based representation of the Boolean function $f = x_1x_2 \vee x_2\overline{x}_3$.
Answer is given in "Solutions to practice problems."

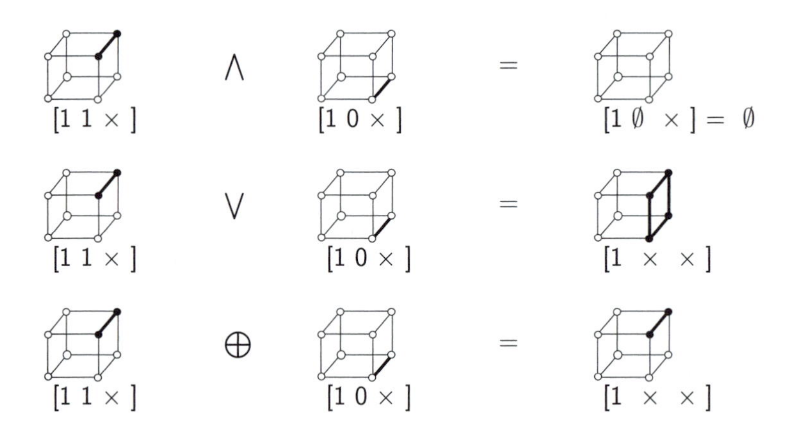

FIGURE 6.11
AND, OR, and EXOR operations on cubes (Practice problem 6.8 and Example 6.13).

6.7 Graphical data structure for cube representation

An n-bit binary number can be represented by a point in n-space.

> **Example 6.14 (Representation of binary numbers in the space.)** *The set of 1-bit binary numbers , 0 and 1, can be represented by two points in 1-space (a line). the set of 2-bit binary numbers, 00,01,10, and 11, can be represented by four points in 2-space. A 3-bit binary number can be represented by eight points in 3-space (Figure 6.12).*

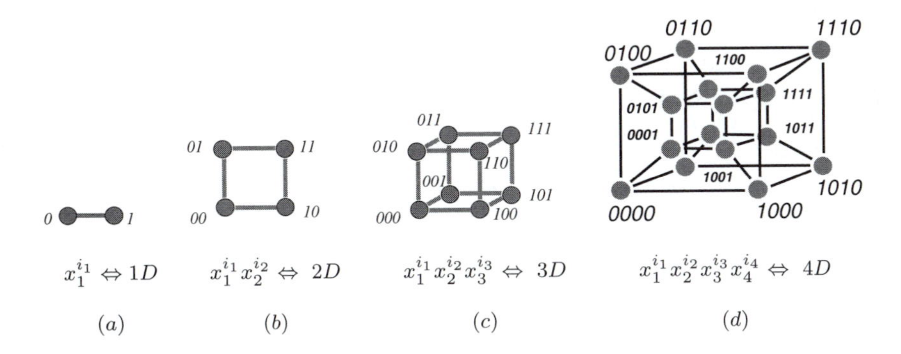

FIGURE 6.12
1-dimensional Boolean hypercube (a), 2-dimensional Boolean hypercube (b), 3-dimensional Boolean hypercube (c), and 4-dimensional Boolean hypercube (d) for representing one-, two-, three-, and four-variable Boolean functions, respectively (Example 6.14).

In general, n-bit number is represented by n-cube, or *Boolean hypercube*. A Boolean hypercube is a graphical data structure based on hypergraphs:

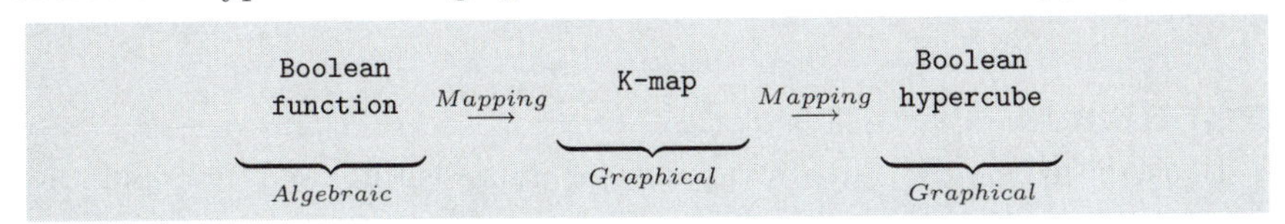

Operations on hypercubes are defined in a similar manner to operations on cubes. Thus, the manipulation of cubes involves OR, AND, and EXOR operations, applied to the appropriate literals, following the rules given in Figure 6.10. This data structure is useful for the representation, manipulation, and minimization of Boolean functions. A Boolean hypercube is defined as a collection of 2^m minterms ($m \leq n$), product terms in which each of the n variables appears once:

$$\boxed{x_1^{i_1}}\ \boxed{x_2^{i_2}} \cdots \boxed{x_{n-1}^{i_{n-1}}}\ \boxed{x_n^{i_n}}$$

where $x_j^{i_j} = 1$ for $i_j = 0$, and $x_j^{i_j} = x_j$ for $i_j = 1$.

The Boolean hypercube encodes a Boolean function assigning codes to the 2^n vertices and $n \times 2^{n-1}$ edges. Figure 6.13 shows this assigning given a function of three variables ($n = 3$). Operations between two Boolean hypercubes produce a new Boolean hypercube (product) that is useful in optimization problems. The following steps are applied for the representation of Boolean functions of n variables using Boolean hypercubes:

Representation of a Boolean function by Boolean hypercube

Given: A standard SOP form of a Boolean function

Step 1. Determine the dimension of the Boolean hypercube

Step 2. Construct the n axes corresponding to the variables. Along each axis, the variable x_i changes from 0 to 1.

Step 3. Assign the nodes of the Boolean hypercube with variable assignments, so that neighboring nodes are encoded using the Gray code rule.

Output: Boolean hypercube

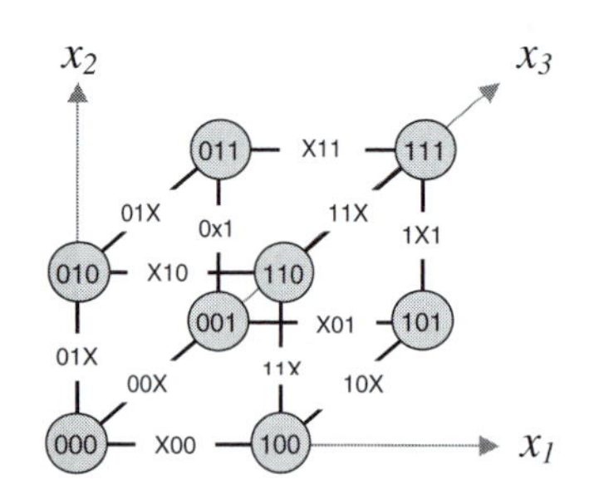

- *Each node is assigned to one of 8 codes (variable assignment)*
- *One edge out of 12 is assigned to a cube with one don't care (×)*
- *One face out of 6 is assigned to a cube with two don't cares*

FIGURE 6.13
A Boolean hypercube data structure for representing and manipulating Boolean functions of three variables.

Example 6.15 (Data structures.) *Various forms of representation of the Boolean function OR $f = x_1 \vee x_2$, and their relationships with each other are shown in Figure 6.14.*

Techniques for representation of the Boolean function

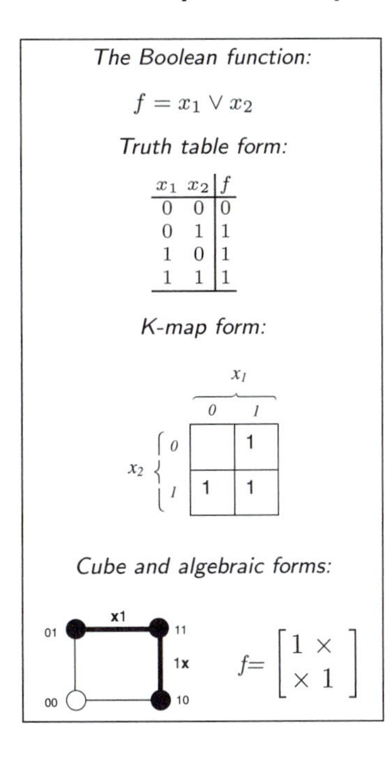

- *The 4 corners (vertices) correspond to the 4 rows of a truth table.*
- *Each vertex is identified by two coordinates.*
- *The horizontal coordinate is assumed to correspond to the variable x_1, and the vertical coordinate to x_2.*
- *The function f is equal to 1 for vertices 01, 10, and 11. The function f can be expressed as a set of vertices, $f = \{01, 10, 11\}$.*
- *The edge joins two vertices for which the labels differ in the value of only one variable.*
- *The letter **x** is used to denote the fact that the corresponding variable can be either 0 or 1.*
- *Vertices 01 and 11 are joined by the edge labeled* **x1**.
- *The edge* **1x** *means a merger of the vertices 10 and 11.*
- *The term x_1 is the sum of minterms $x_1\overline{x}_2$ and x_1x_2. It follows that $x_1\overline{x}_2 \vee x_1x_2 = x_1$.*
- *The edges* **1x** *and* **x1** *define the function f in a unique way.*

FIGURE 6.14
Representation of the Boolean function $f = x_1 \vee x_2$ in various forms (Example 6.15).

Example 6.16 (Data structure relationship.) *Various forms of representation of a Boolean function $f = \overline{x}_3 \vee x_1\overline{x}_2$, and their relationships are given in Figure 6.15.*

Practice problem 6.10. (Data structure relationship.) Given the truth-vector $\mathbf{F} = [\,0\;0\;0\;1\;1\;1\;1\;0\,]^T$, derive its K-map, Boolean hypercube, and cube forms.
Answer is given in "Solutions to practice problems."

Mapping the product into a Boolean hypercube

Let $x_j^{i_j}$ be a literal of a Boolean variable x_j such that $x_j^0 = \overline{x}_j$, $x_j^1 = x_j$ and $x_1^{i_1}x_2^{i_2}\ldots x_n^{i_n}$ is a product of literals. Topologically, this is a set of points on the plane numerated by $i = 0, 1, \ldots, n$. To map this set into a Boolean

Techniques for representation of the Boolean function

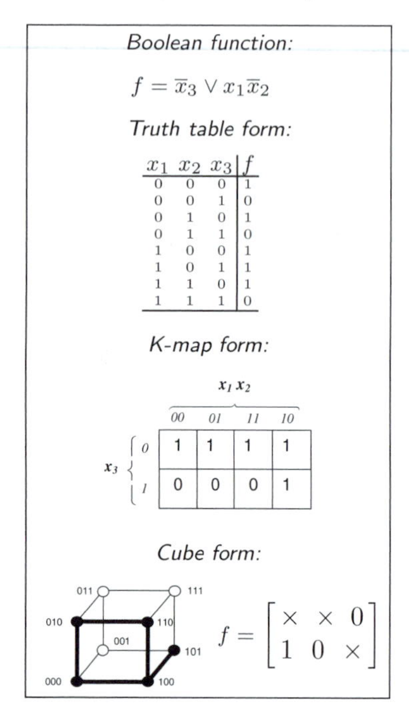

- *There are five vertices that correspond to $f = 1$:*

  000,010,100,101,110

- *Merging the vertex assignments yields*

  $\mathbf{x}00, 0\mathbf{x}0, \mathbf{x}10, 1\mathbf{x}0, 10\mathbf{x}$

- *Four of these vertices can be merged to face, or term* $\mathbf{xx}0$. *This term means that $f = 1$ if $x_3 = 0$, regardless of the values of x_1 and x_2.*

- *The function f can be represented in several ways. Some of the possibilities are*

$$
\begin{aligned}
f &= \{000, 010, 100, 101, 110\} \\
&= \{0\mathbf{x}0, 1\mathbf{x}0, 101\} \\
&= \{\mathbf{x}00, \mathbf{x}10, 101\} \\
&= \{\mathbf{x}00, \mathbf{x}10, 10\mathbf{x}\} \\
&= \{\mathbf{xx}0, 10\mathbf{x}\}
\end{aligned}
$$

FIGURE 6.15
Representation of the Boolean function $f = \overline{x}_3 \vee x_1\overline{x}_2$ by a Boolean hypercube: truth table and 3-dimensional Boolean hypercube representation (Example 6.16).

hypercube, the numbers must be encoded by the Gray code and represented by the corresponding graphs based on Hamming distance.

Example 6.17 (Boolean hypercube.) *Figure 6.12 demonstrates the representation of Boolean functions using Boolean hypercubes. The product $x_1^{i_1}$ represents two variables x_1 and $\overline{x}_1$. The product $x_1^{i_1}x_2^{i_2}$ corresponds to four minterms, x_1x_2, $\overline{x}_1x_2$, $x_1\overline{x}_2$, and $\overline{x}_1\overline{x}_2$. Products with four variables or more are represented by assembling 3-dimensional Boolean hypercubes:*

$$
\text{PRODUCT } \underbrace{x_1^{i_1}x_2^{i_2}x_3^{i_3}x_4^{i_4}}_{16\ points} \Longleftrightarrow \underbrace{\texttt{4-DIMENSIONAL HYPERCUBE}}_{n=4}
$$

Practice problem 6.11. **(Boolean hypercube.)** Construct a 5-dimensional Boolean hypercube.

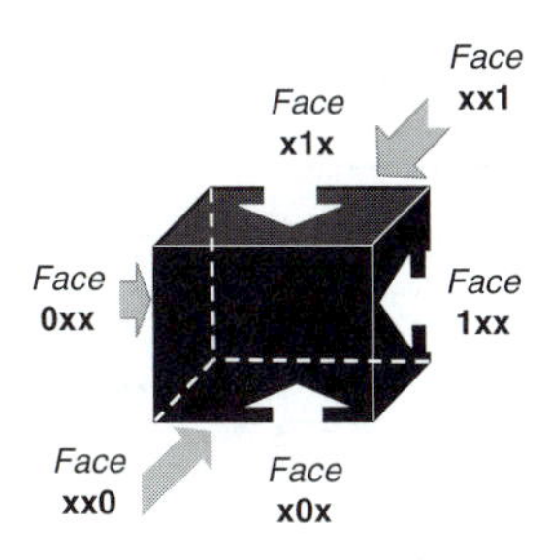

Faces of the Boolean hypercube represent several minterms of Boolean functions of three variables, grouped around one variable:
Face xx0: $\overline{x}_3(\overline{x}_1\overline{x}_2 \vee \overline{x}_1 x_2 \vee x_1\overline{x}_2 \vee x_1 x_2)$
Face xx1: $x_3(\overline{x}_1\overline{x}_2 \vee \overline{x}_1 x_2 \vee x_1\overline{x}_2 \vee x_1 x_2)$
Face 0xx: $\overline{x}_1(\overline{x}_2\overline{x}_3 \vee \overline{x}_2 x_3 \vee x_2\overline{x}_3 \vee x_2 x_3)$
Face 1xx: $x_3(\overline{x}_2\overline{x}_3 \vee \overline{x}_2 x_3 \vee x_2\overline{x}_3 \vee x_2 x_3)$
Face x0x: $\overline{x}_2(\overline{x}_1\overline{x}_3 \vee \overline{x}_1 x_3 \vee x_1\overline{x}_3 \vee x_1 x_3)$
Face x1x: $x_2(\overline{x}_1\overline{x}_3 \vee \overline{x}_1 x_3 \vee x_1\overline{x}_3 \vee x_1 x_3)$

FIGURE 6.16
Faces of the Boolean hypercube interpretation in the SOPs of a Boolean function of three variables (Example 6.18).

Answer is given in "Solutions to practice problems."

The 0-dimensional Boolean hypercube ($n = 0$) represents the constant 0. A line segment connects vertices 0 and 1, and these vertices are called the *face* of the 1-dimensional hypercube and denoted by $\times$. A 2-dimensional Boolean hypercube has four faces, denoted by the cubes $0\times$, $1\times$, $\times 0$, and $\times 1$. The total 2-dimensional Boolean hypercube can be denoted by $\times\times$.

Example 6.18 (Boolean hypercube.) *Six faces of the Boolean hypercube described by the cubes* $\times\times 0$, $\times\times 1$, $0\times\times$, $1\times\times$, $\times 1\times$, *and* $\times 0\times$ *(Figure 6.16), represent one-term products for a Boolean function of three variables.*

Example 6.19 (Boolean hypercube.) *In Table 6.3, the Boolean hypercubes for elementary Boolean functions of three variables are represented.*

Practice problem 6.12. (Boolean hypercube.) Represent the two-input Boolean functions AND, NAND, OR, EXOR using a Boolean hypercube data structure.
Answer is given in "Solutions to practice problems."

Hamming distance

Hamming distance is a useful measure of hypercube topology. The Hamming sum is defined as a bitwise operation:

$$(g_{d-1}\cdots g_0) \oplus (g'_{d-1}\cdots g'_0) = (g_{d-1} \oplus g'_{d-1}), \ldots, (g_1 \oplus g'_1), (g_0 \oplus g'_0) \quad (6.1)$$

where $\oplus$ is an exclusive OR operation. That is, the Hamming distance is the distance between two points on Boolean hypercube with n verteces defined as

TABLE 6.3
The truth tables, K-maps, and Boolean hypercubes for elementary Boolean functions of three variables.

Techniques for Boolean function representation

<table>
<tr><th>Elementary Boolean function</th><th>Truth table</th><th>K-map</th><th>Boolean hypercube</th></tr>
<tr>
<td>AND
$f = x_1x_2x_3$ 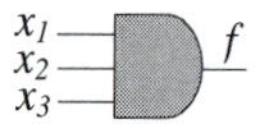
x_1 x_2 x_3 f</td>
<td><table><tr><th>x_1</th><th>x_2</th><th>x_3</th><th>f</th></tr><tr><td>0</td><td>0</td><td>0</td><td>0</td></tr><tr><td>0</td><td>0</td><td>1</td><td>0</td></tr><tr><td>0</td><td>1</td><td>0</td><td>0</td></tr><tr><td>0</td><td>1</td><td>1</td><td>0</td></tr><tr><td>1</td><td>0</td><td>0</td><td>0</td></tr><tr><td>1</td><td>0</td><td>1</td><td>0</td></tr><tr><td>1</td><td>1</td><td>0</td><td>0</td></tr><tr><td>1</td><td>1</td><td>1</td><td>1</td></tr></table></td>
<td><table><tr><td></td><td></td><td colspan="4">$x_1 x_2$</td></tr><tr><td></td><td></td><td>00</td><td>01</td><td>11</td><td>10</td></tr><tr><td rowspan="2">x_3</td><td>0</td><td></td><td></td><td></td><td></td></tr><tr><td>1</td><td></td><td></td><td>1</td><td></td></tr></table></td>
<td>011 111 010 110 001 101 000 100
$f = [1\ 1\ 1]$</td>
</tr>
<tr>
<td>OR
$f = x_1 \vee x_2 \vee x_3$
x_1 x_2 x_3 f</td>
<td><table><tr><th>x_1</th><th>x_2</th><th>x_3</th><th>f</th></tr><tr><td>0</td><td>0</td><td>0</td><td>0</td></tr><tr><td>0</td><td>0</td><td>1</td><td>1</td></tr><tr><td>0</td><td>1</td><td>0</td><td>1</td></tr><tr><td>0</td><td>1</td><td>1</td><td>1</td></tr><tr><td>1</td><td>0</td><td>0</td><td>1</td></tr><tr><td>1</td><td>0</td><td>1</td><td>1</td></tr><tr><td>1</td><td>1</td><td>0</td><td>1</td></tr><tr><td>1</td><td>1</td><td>1</td><td>1</td></tr></table></td>
<td><table><tr><td></td><td></td><td colspan="4">$x_1 x_2$</td></tr><tr><td></td><td></td><td>00</td><td>01</td><td>11</td><td>10</td></tr><tr><td rowspan="2">x_3</td><td>0</td><td></td><td>1</td><td>1</td><td>1</td></tr><tr><td>1</td><td>1</td><td>1</td><td>1</td><td>1</td></tr></table></td>
<td>011 111 010 110 001 101 000 100
$f = \begin{bmatrix} \times & \times & 1 \\ \times & 1 & \times \\ 1 & \times & \times \end{bmatrix}$</td>
</tr>
<tr>
<td>NAND
$f = \overline{x_1x_2x_3}$
x_1 x_2 x_3 f</td>
<td><table><tr><th>x_1</th><th>x_2</th><th>x_3</th><th>f</th></tr><tr><td>0</td><td>0</td><td>0</td><td>1</td></tr><tr><td>0</td><td>0</td><td>1</td><td>1</td></tr><tr><td>0</td><td>1</td><td>0</td><td>1</td></tr><tr><td>0</td><td>1</td><td>1</td><td>1</td></tr><tr><td>1</td><td>0</td><td>0</td><td>1</td></tr><tr><td>1</td><td>0</td><td>1</td><td>1</td></tr><tr><td>1</td><td>1</td><td>0</td><td>1</td></tr><tr><td>1</td><td>1</td><td>1</td><td>0</td></tr></table></td>
<td><table><tr><td></td><td></td><td colspan="4">$x_1 x_2$</td></tr><tr><td></td><td></td><td>00</td><td>01</td><td>11</td><td>10</td></tr><tr><td rowspan="2">x_3</td><td>0</td><td>1</td><td>1</td><td>1</td><td>1</td></tr><tr><td>1</td><td>1</td><td>1</td><td></td><td>1</td></tr></table></td>
<td>011 111 010 110 001 101 000 100
$f = \begin{bmatrix} 0 & \times & \times \\ \times & 0 & \times \\ \times & \times & 0 \end{bmatrix}$</td>
</tr>
<tr>
<td>NOR
$f = \overline{x_1 \vee x_2 \vee x_3}$
x_1 x_2 x_3 f</td>
<td><table><tr><th>x_1</th><th>x_2</th><th>x_3</th><th>f</th></tr><tr><td>0</td><td>0</td><td>0</td><td>1</td></tr><tr><td>0</td><td>0</td><td>1</td><td>0</td></tr><tr><td>0</td><td>1</td><td>0</td><td>0</td></tr><tr><td>0</td><td>1</td><td>1</td><td>0</td></tr><tr><td>1</td><td>0</td><td>0</td><td>0</td></tr><tr><td>1</td><td>0</td><td>1</td><td>0</td></tr><tr><td>1</td><td>1</td><td>0</td><td>0</td></tr><tr><td>1</td><td>1</td><td>1</td><td>0</td></tr></table></td>
<td><table><tr><td></td><td></td><td colspan="4">$x_1 x_2$</td></tr><tr><td></td><td></td><td>00</td><td>01</td><td>11</td><td>10</td></tr><tr><td rowspan="2">x_3</td><td>0</td><td>1</td><td></td><td></td><td></td></tr><tr><td>1</td><td></td><td></td><td></td><td></td></tr></table></td>
<td>011 111 010 110 001 101 000 100
$f = [0\ 0\ 0]$</td>
</tr>
<tr>
<td>EXOR
$f = x_1 \oplus x_2 \oplus x_3$
x_1 x_2 x_3 f</td>
<td><table><tr><th>x_1</th><th>x_2</th><th>x_3</th><th>f</th></tr><tr><td>0</td><td>0</td><td>0</td><td>0</td></tr><tr><td>0</td><td>0</td><td>1</td><td>1</td></tr><tr><td>0</td><td>1</td><td>0</td><td>1</td></tr><tr><td>0</td><td>1</td><td>1</td><td>0</td></tr><tr><td>1</td><td>0</td><td>0</td><td>1</td></tr><tr><td>1</td><td>0</td><td>1</td><td>0</td></tr><tr><td>1</td><td>1</td><td>0</td><td>0</td></tr><tr><td>1</td><td>1</td><td>1</td><td>1</td></tr></table></td>
<td><table><tr><td></td><td></td><td colspan="4">$x_1 x_2$</td></tr><tr><td></td><td></td><td>00</td><td>01</td><td>11</td><td>10</td></tr><tr><td rowspan="2">x_3</td><td>0</td><td></td><td>1</td><td></td><td>1</td></tr><tr><td>1</td><td>1</td><td></td><td>1</td><td></td></tr></table></td>
<td>011 111 010 110 001 101 000 100
$f = \begin{bmatrix} 0 & 0 & 1 \\ 0 & 1 & 0 \\ 1 & 0 & 0 \\ 1 & 1 & 1 \end{bmatrix}$</td>
</tr>
<tr>
<td>XNOR
$f = \overline{x_1 \oplus x_2 \oplus x_3}$
x_1 x_2 x_3 f</td>
<td><table><tr><th>x_1</th><th>x_2</th><th>x_3</th><th>f</th></tr><tr><td>0</td><td>0</td><td>0</td><td>1</td></tr><tr><td>0</td><td>0</td><td>1</td><td>0</td></tr><tr><td>0</td><td>1</td><td>0</td><td>0</td></tr><tr><td>0</td><td>1</td><td>1</td><td>1</td></tr><tr><td>1</td><td>0</td><td>0</td><td>0</td></tr><tr><td>1</td><td>0</td><td>1</td><td>1</td></tr><tr><td>1</td><td>1</td><td>0</td><td>1</td></tr><tr><td>1</td><td>1</td><td>1</td><td>0</td></tr></table></td>
<td><table><tr><td></td><td></td><td colspan="4">$x_1 x_2$</td></tr><tr><td></td><td></td><td>00</td><td>01</td><td>11</td><td>10</td></tr><tr><td rowspan="2">x_3</td><td>0</td><td>1</td><td></td><td>1</td><td></td></tr><tr><td>1</td><td></td><td>1</td><td></td><td>1</td></tr></table></td>
<td>011 111 010 110 001 101 000 100
$f = \begin{bmatrix} 0 & 0 & 0 \\ 0 & 1 & 1 \\ 1 & 0 & 1 \\ 1 & 1 & 0 \end{bmatrix}$</td>
</tr>
</table>

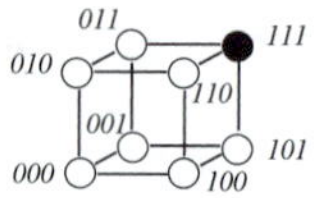

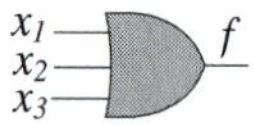

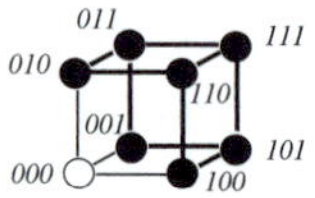

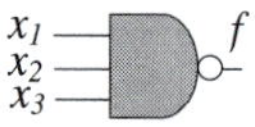

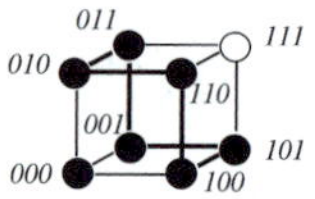

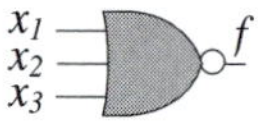

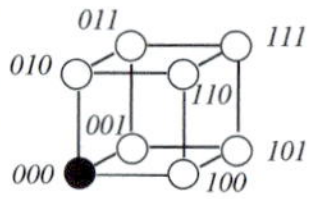

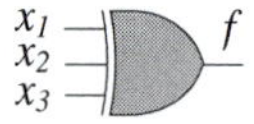

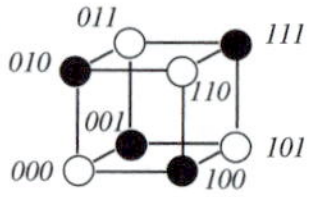

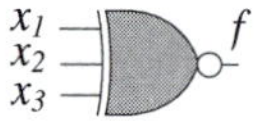

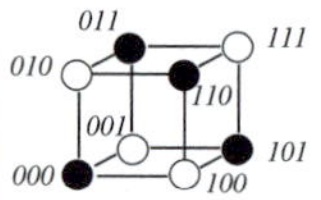

the number of coordinates (bit positions) in which the binary representations of the two points differ.

In the Boolean hypercube, two nodes are connected by a link (edge) if and only if they have labels that differ by exactly one bit. The number of bits by which the labels g_i and g_j differ is denoted by $h(g_i, g_j)$; this is the Hamming distance between the nodes.

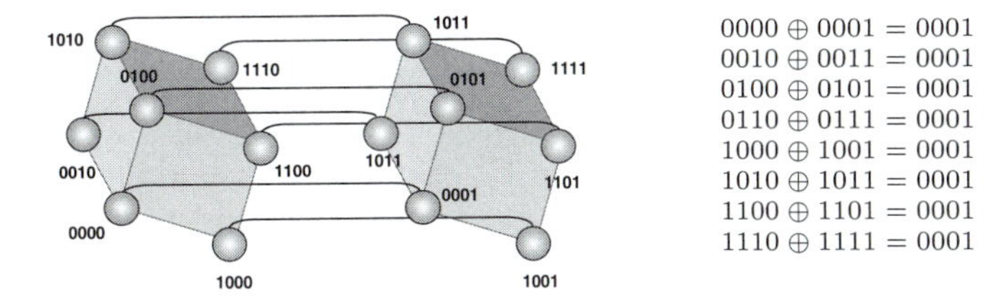

FIGURE 6.17
Hamming sum on Boolean hypercubes and the corresponding product of variables (Example 6.20).

Example 6.20 (Hamming distance.) *Performing the Hamming sum on two Boolean hypercubes for a three-variable Boolean function is illustrated in Figure 6.17. The neighboring node's assignments differ by exactly one bit (the difference is shown using the Boolean EXOR operation)*

A symmetry of the hypercube

A *symmetry* of the n-hypercube is defined to be any one-to-one translation of the binary point representations on the n-hypercube which leaves all pairwise distances the same. There only two basic translation schemes which leave pairwise distances the same:

(a) The bits of one coordinates may be interchanged with the bits of another coordinate in all vertices.

(b) The bits of one coordinate be be complemented (change 1s to 0s and 0s to 1s) in all vertices.

Since there are:

- $n!$ permutations of n coordinates,
- $n!$ translation schemes of n coordinates using the interchange (a), and
- 2^n translation schemes using the complement (b).

Thus, there are $2^n \times n!$ symmetries of the hypercube with n vertices. Details on symmetric Boolean functions are given in Chapter 5.

6.8 Logic networks

A logic network is a graphical data structure defined as an interconnection of gates that implements a combinational system. This graphical data structure can be described in various ways.

6.8.1 Design goals

A given Boolean function can be implemented in many ways using various gate types: The designer must choose the implementation that best meets the given design goals. A central problem in logic design is to obtain a logic network that realizes a given Boolean function using a given set of gates, and has the lowest possible cost:

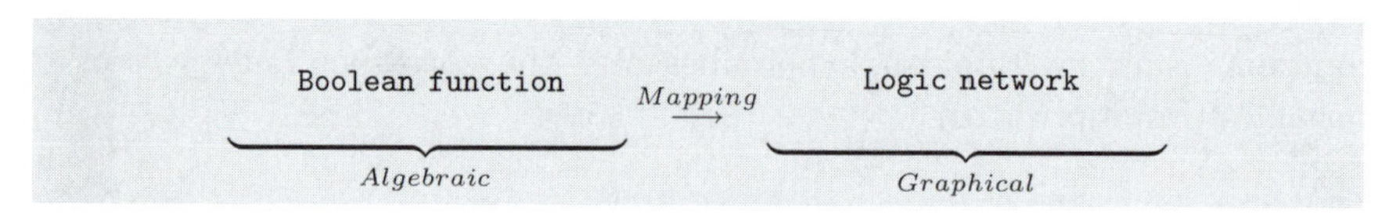

The cost of a logic network is conveniently measured in a technology-independent fashion by the total number of gates used. We assume that the logic gates and logic network can be of arbitrary size and can respond instantaneously to signal changes on their primary input lines. Physical constraints limit both the size and speed of real logic networks in important ways, of which the designer must be aware.

6.8.2 Basic components of a logic network

A logic network is composed of:

- Gates,
- External inputs and outputs, and
- Connections.

Connections are specified from external inputs and gate outputs to gate inputs and external outputs, and must meet the following requirements:

(a) Each gate input is connected to:

- A constant 0 or 1,
- A network's external input, or
- A gate output.

Only one connection per gate input is allowed.

(*b*) The output load imposed on a gate output should not be greater than its fan-out factor. This load is computed as the sum of the input load factors of all the gate inputs that are connected to the gate output.

The transfer and processing of information requires the expenditure of energy, and a physical device can absorb or produce only a certain amount of energy without failing. This places an upper bound on the number of input sources that supply a gate, and another bound on the maximum number of output devices that may be supplied with signals from the gate's output. The number of inputs of gate is termed its *fan-in*; the number of connections, which are typically inputs to other gates, that are connected to the gate's output is termed its *fan-out*.

The composition of primitive components is accomplished by physically wiring the gates together. The tabulation of gate inputs and outputs and the nets to which they are connected is called a *netlist*. The *fan-in* of a gate is its number of inputs.

A logic network is *loop-free* or *cycle-free* if there exists only one path from any point of a network through the gates in the direction from input to output.

A logic network can be described as follows:

- ▶ A *logic diagram*,
- ▶ A *netlist*, and
- ▶ A *set of language statements*.

Throughout the design cycle, all these representations are used; that is, logic networks can be synthesized simultaneously using these forms.

Example 6.21 (Logic network.) *Figure 6.18a shows the logic network implementing the Boolean function* $f = x_1x_2 \vee x_2x_3 \vee x_4$. *It can be represented by a list of gates and a list of connections (b), or by using a hardware language (c).*

A logic diagram

A logic diagram is a graphical representation of a logic network. This diagram consists of symbols of gates and connections that are drawn using the rules given above. The advantages of logic diagrams are as follows:

- ▶ A logic diagram is a directed graph, and therefore, its manipulation can be accomplished using graph-based techniques.
- ▶ A logic diagram is convenient for network analysis, including visual analysis.

A netlist description

A netlist description is a reasonable form that is required in the simplification of the manipulation of a network. A netlist description consists of

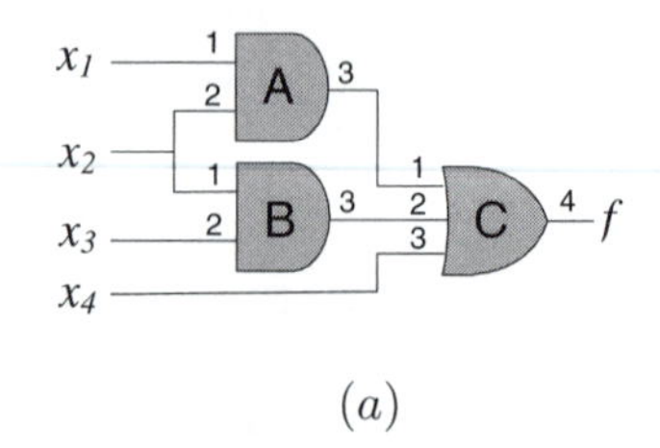

(a)

List of gates

Gate	Type	In	Out
A	AND2	A_1	A_3
		A_2	
B	AND2	B_1	B_3
		B_2	
C	OR3	C_1	C_4
		C_2	
		C_3	

List of connections

From	To
x_1	A_1
x_2	A_2
x_2	B_1
x_3	B_2
x_4	C_3
A_3	C_1
B_3	C_2
C_4	f

(b)

Language description

```
A_3 ⇐ x1 and x2
B_3 ⇐ x2 and x3
C_4 ⇐ A_3 or B_3 or x4
f ⇐ C_4
```

(c)

FIGURE 6.18
Logic network description in the form of a logic diagram (a), a netlist (b), and a hardware description language (c) (Example 6.21).

- A *list of gates*; and
- A *list of connections*

These lists can be considered as an analog of the adjustment matrix for representing directed graphs, taking into account the connection limitations.

Hardware description languages

Hardware description languages are text-based descriptions. The basic elements of these descriptions are statements aimed at translating the functional and structural properties of a logic networks into hardware equivalents.

6.8.3 Specification

The specification of a logic network is defined as the parameters and characteristics required for the representation of a network as a component of a larger logic network. According to the hierarchical design principle, this network must be suitable for assembling a large network from smaller logic networks. For this, the logic network must satisfy various requirements, including the *input load factor*, the *fan-out factor*, the *size*, the *propagation delay*, the *number of levels*, and the *dynamic* characteristics (Table 6.4).

Example 6.22 (Logic network characteristics.) *Consider a logic network (Figure 6.19). The network has 5 gates, the fan-in for every gate is 2, and the fan-out for each gate is 1, except to the AND gate with inputs x_3 and x_4. The network characteristics are as follows: the input load factor is 6, fan-out factor is -4, and the number of levels is 2.*

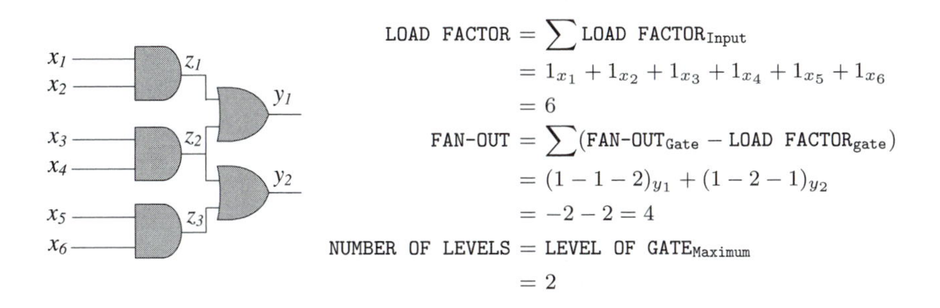

FIGURE 6.19
Logic network characteristics (Example 6.22).

6.8.4 Network verification

Each logic network design has to be verified with respect to various criteria before fabrication. In every design step, the exact input/output behavior of the system has to be specified, and functional behavior has to be realized within the bounds of the given technology. The description of system behavior is called *specification*. A realization of the specified system behavior is called an *implementation*.

If it can be *proven* by any method that the implementation satisfies the given specifications, then the chip design is functionally correct. For this proof of functional correctness, the following concepts are of practical interest:

- *Simulation*,
- *Formal* verification, and
- *Partial* verification.

TABLE 6.4
Logic network parameters and characteristics.

Characteristic	Definition
Input load factor	The sum of all the input load factors of the network inputs: $\texttt{LOAD FACTOR} = \sum \texttt{LOAD FACTOR}_{\texttt{Input}}$
Fan-out factor	The fan-out factor of the network is equal to the sum of the fan-out factors of the gate outputs minus the sum of the input load factors of those gate inputs to which that gate output is connected: $\texttt{FAN-OUT} = \sum(\texttt{FAN-OUT}_{\texttt{Gate}} - \texttt{LOAD FACTOR}_{\texttt{gate}})$
Size	The area the network occupies on the chip is estimated by the sum of the area of the gates and the area of the interconnections: $\texttt{AREA} = \texttt{AREA}_{\texttt{Gates}} + \texttt{AREA}_{\texttt{Interconnections}}$
Propagation delay	The maximum delay obtained for all paths from the input to the output of the network: $\texttt{DELAY} = \texttt{DELAY}_{\texttt{Maximum}}$
The number of levels	The level of a gate is the number of gates in the longest path from an external input to the output: $\texttt{LEVEL OF GATE} = \texttt{PATH}_{\texttt{Maximum}}$ Number of levels of a network is the maximum level of a gate: $\texttt{NUMBER OF LEVELS} = \texttt{LEVEL OF GATE}_{\texttt{Maximum}}$
Dynamic characteristics	Graphical representation of the propagation of the signals in the network

Simulation

Simulation means systematically evaluating the functional behavior of both the specification and the implementation for a large number of input vectors. If the outputs of the implementation and the specification agree for all these vectors, the design works correctly.

Formal verification

The aim of *formal verification* is a formal mathematical proof that the functional behaviors of the specification and the implementation coincides.

An appropriate data structure must be chosen for applying this approach. Decision diagrams are suitable for the efficient formal verification of logic networks.

Partial verification

Partial verification is defined as a mathematical method for proving that an implementation satisfies at least the particular important properties of the specified system behavior. To guarantee that the functionality of a logic network has not been modified through the synthesis and optimization process, a verification task has to be solved.

Example 6.23 *In Figure 6.20, the initial network (left) is interpreted as a specification and implemented using NAND gates (right). It has to be proven that both the specification and the implementation are functionally equivalent; that is, both networks compute the same Boolean function:*

$$\textit{Modified network:}\quad \overline{\overline{\overline{x_2\overline{x}_3}x_1}x_4} = \overline{x_2\overline{x}_3}x_1 \vee \overline{x}_4 = (\overline{x}_2 \vee \overline{x}_3)x_1 \vee \overline{x}_4$$
$$= x_1\overline{x}_2 \vee x_1\overline{x}_3 \vee \overline{x}_4 = f \quad \textit{Initial network}$$

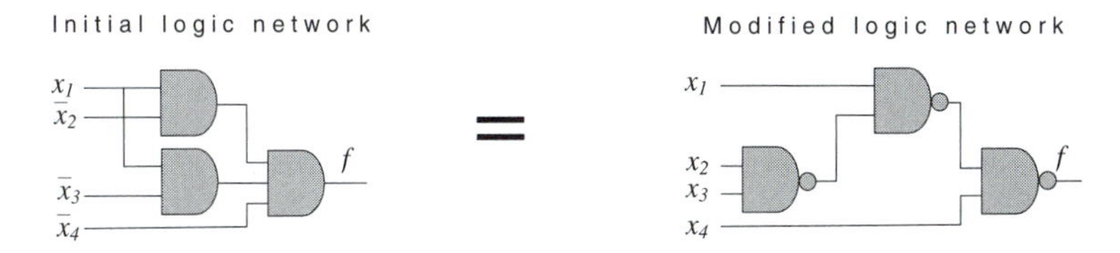

FIGURE 6.20
Logic network verification (Example 6.23).

Practice problem **6.13. (Logic network verification.)** Verify the logic networks given in Figure 6.21.
Answer is given in "Solutions to practice problems."

6.9 Networks of threshold gates

An attractive property of threshold functions is that any elementary Boolean function can be implemented by a single threshold gate using certain control

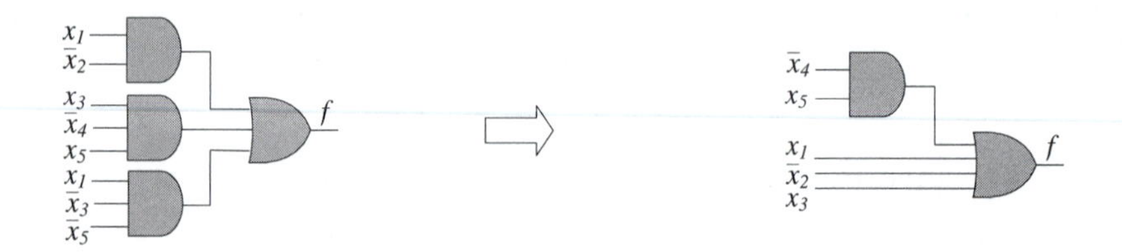

FIGURE 6.21
Verification of logic networks (Practice problem 6.13).

parameters. It follows from this property that a given threshold network can be used for computing different Boolean functions, while varying the control parameters. The ability to keep the same topology in computing various Boolean functions is an important factor in implementation. Note that a logic network is designed for the implementation of a particular Boolean function; that is, the same logic network cannot be used for computing different Boolean functions:

6.9.1 Threshold functions

The class of threshold functions is of great importance for modeling biological neurons and for constructing artificial neural networks. The neuron's cell membrane is capable of sustaining a certain electric charge. When this charge reaches or exceeds a threshold k, the neuron "fires." This effect is modeled by means of threshold functions as proposed by McCulloch and Pitts (see the "Further study section").

A Boolean function f of n variables $x_1, x_2, \ldots, x_n$ is called a *threshold* function with weights $w_1, w_2, \ldots, w_n$ (real numbers) and *threshold* k if

Threshold function

$$f = 1 \text{ if and only if } \sum_{i=1}^{n} w_i \times x_i \geq k \qquad (6.2)$$

The algorithm for designing a threshold gate to implement a given logic function is shown below:

A threshold function is implemented by a threshold gate, as shown in Figure 6.22). This threshold gate can compute various elementary Boolean functions of two variables. For this, appropriate control parameters must be chosen.

Example 6.24 (A threshold network.) *The n-input OR and AND functions have thresholds of 1 and n, respectively. That is, in this threshold gate, increasing the number of active inputs results in the output 1.*

Deriving a logic function using a threshold element

Given: A logic function and its truth table
Step 1: Derive the sum, $\sum$, of the n input signals for various combinations of x_1 and x_2.
Step 2: Find the threshold function that satisfies the equation $f = \Theta(sum)$; i.e., find the weights $w_1, w_2, \ldots w_n$.
Output: A threshold gate specification (weights and threshold function)

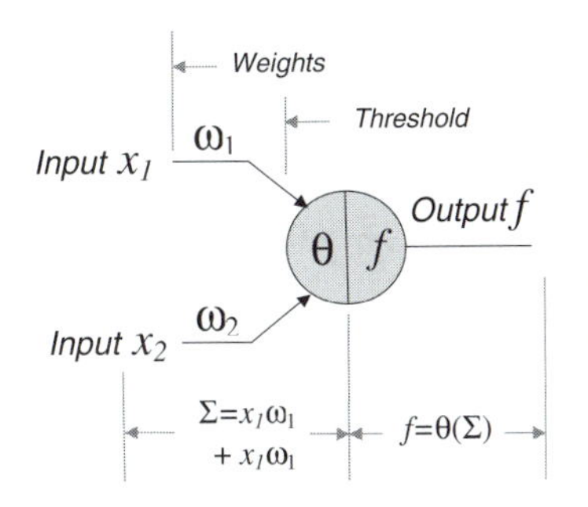

The input signal for $x_1, x_2 \in \{0,1\}$ *is*

$$\Sigma = x_1 \times w_1 + x_2 \times w_2$$

Thresholding of Σ *at the level* Θ *produces the output signal*

$$f = \Theta(\Sigma)$$

FIGURE 6.22
Threshold gate.

There are several particular implementations of threshold functions. In particular, if

$$\begin{aligned} w_1 = w_2 = \ldots = w_n = 1 \\ n = 2m + 1 \\ \Theta = m + 1 \end{aligned}$$

the threshold function implements the *majority* function. A majority function f is equal to 1 if the inputs have more 1s than 0s.

Example 6.25 (Threshold functions.) *Let* $m = 1$. *Then* $n = 2m + 1 = 2 \times 1 + 1 = 3$, *and we have the majority function of three variables:* $f = x_1x_2 \vee x_2x_3 \vee x_3x_1$. *This function is equal to 1 iff two or more inputs are equal to 1.*

Practice problem 6.14. **(Threshold functions.)** Show that the majority function in Example 6.25 is a self-dual function.
Answer is given in "Solutions to practice problem."

6.9.2 McCulloch-Pitts models of Boolean functions

Two-input AND, OR, NOR, and NAND elementary logic functions can be computed by a single neural cell known as the *McCulloch-Pitts* model. Control over the type of logic function is exercised by the threshold θ and weights $w_i \in \{1, -1\}$ of the arithmetic sum

$$w_i \times x_1 + w_i \times x_2$$

i.e., the output is $f = w_1 \times x_1 + w_2 \times x_2 - \theta$. The Boolean function $f = x_1 \oplus x_2$ cannot be represented by one neuron cell. This limitation can be explained as follows. For the weights w_1, w_2 and a threshold of one neuron θ, the following inequalities hold:

$$0 \cdot w_1 + 1 \cdot w_2 \geq k \text{ because } f(0,1) = 1$$
$$1 \cdot w_1 + 0 \cdot w_2 \geq k \text{ because } f(1,0) = 1$$
$$1 \cdot w_1 + 1 \cdot w_2 < k \text{ because } f(1,1) = 0$$

The first two inequalities can be combined into $w_1 + w_1 \geq 2k$, which contradicts the third equation. Consequently, the function f cannot be represented in terms of a weighted threshold function. A solution to this problem is given in Table 6.5, where an EXOR function is implemented using three threshold gates.

Example 6.26 (Threshold gates.) *In Table 6.5, the correspondence between the control parameters of the threshold gates and the implemented Boolean functions of two variables is illustrated.*

TABLE 6.5
The McCulloch-Pitts model for computing AND, OR, and NAND Boolean functions (Example 6.26).

Techniques for computing using the McCulloch-Pitts model

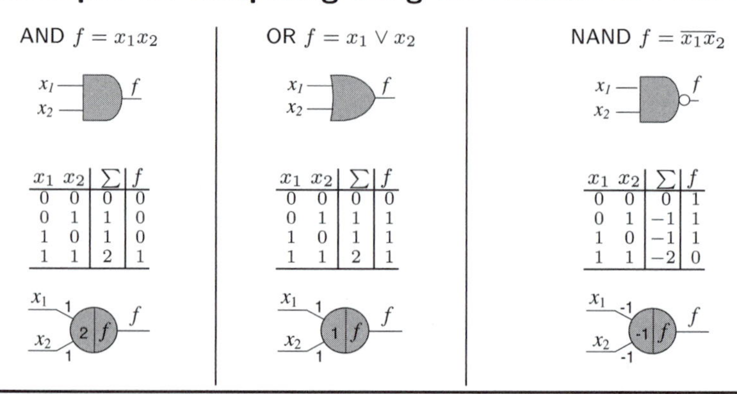

AND $f = x_1x_2$

x_1	x_2	$\sum$	f
0	0	0	0
0	1	1	0
1	0	1	0
1	1	2	1

OR $f = x_1 \vee x_2$

x_1	x_2	$\sum$	f
0	0	0	0
0	1	1	1
1	0	1	1
1	1	2	1

NAND $f = \overline{x_1x_2}$

x_1	x_2	$\sum$	f
0	0	0	1
0	1	-1	1
1	0	-1	1
1	1	-2	0

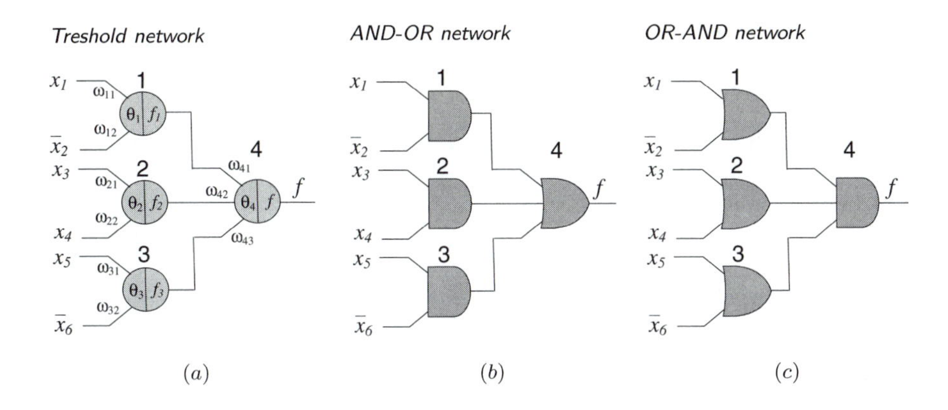

FIGURE 6.23
A network of threshold elements (a) and the corresponding AND-OR (b) and OR-AND (c) logic networks (Example 6.27).

Practice problem **6.15. (Threshold gates.)** Verify that for $w_1 = w_2 = -1$ and $\theta = -1$, the McCulloch-Pitts model implements the NAND function.
Answer is given in "Solutions to practice problems."

6.9.3 Networks of threshold gates

Example 6.27 (Threshold networks.) *A threshold network shown in Figure 6.23a is used for the implementation of AND-OR and OR-AND logic networks, which have the configurations of given threshold networks. The control parameters are taken from Table 6.5 as follows:*

(a) *An AND-OR network (Figure 6.23b) is designed from a threshold network by specifications of the weights and thresholds for an AND function, that is,* $\omega_{11} = \omega_{12} = \ldots = \omega_{43} = 1, \;\; \theta_1 = \theta_2 = \theta_3 = 2, \;\; \theta_4 = 1.$

(b) *An OR-AND network (Figure 6.23c) is designed by analogy:* $\omega_{11} = \omega_{12} = \ldots = \omega_{43} = 1, \;\; \theta_1 = \theta_2 = \theta_3 = 1, \theta_4 = 2.$

Practice problem **6.16. (Threshold network.)** Using a threshold network such as the one given in Figure 6.23a, compute the Boolean functions: (a) $x_1x_2 \vee x_3x_4 \vee x_5x_6$, (b) $(x_1 \vee x_2)(x_3 \vee x_4)(x_5 \vee x_6)$.
Answers are given in "Solutions to practice problems."

6.10 Binary decision trees

A complete binary decision tree can be constructed from the truth table of a Boolean function. The procedure for constructing a complete binary decision tree is given in Figure 6.24. This procedure is based on the iterated application of a Shannon expansion:

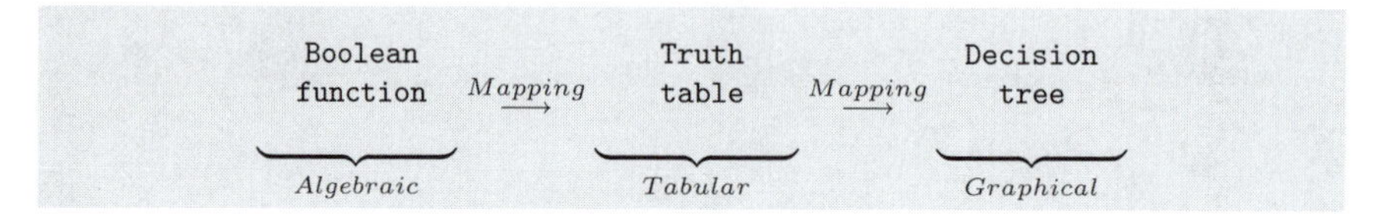

Constructing a binary decision tree

Given: A Boolean function
Step 1. Derive a truth table of the Boolean function
Step 2. Derive the first node of the decision tree as follows:

(*a*) Choose a variable to assign to the root of the tree, x_i.
(*b*) Find the Shannon expansion $f = \overline{x}_i f_{x_i=0} \vee x_i f_{x_i=1}$.
(*d*) Assign $\overline{x}_i$ to the left outgoing branch and x_i to the right one.

Step 3. Choose another variable, x_j. Construct two nodes connected to the branches of the root node:

- $f_{x_i=0}$ is the input of the left node, and
- $f_{x_i=1}$ is the input of the right node

Find the Shannon expansion for both functions:

$$f_{x_i=0} = \overline{x}_j f_{|x_i=0, |x_j=0} \vee x_j f_{|x_i=0, |x_j=1} \quad \text{and} \quad f_{x_i=1} = \overline{x}_j f_{|x_i=1, |x_j=0} \vee x_j f_{|x_i=1, |x_j=1}$$

Step 4. Repeat Step 3 $n - 2$ times, performing Shannon expansion of the factors with respect to the remaining $n-2$ variables.
Output: A complete binary decision tree
Note: The order of variables can be fixed, or chosen using minimization criteria (for further reduction of the tree to a binary decision diagram).

FIGURE 6.24
Procedure for constructing a complete binary decision tree from the truth table of a Boolean function.

Example 6.28 (Binary decision tree.) *Figure 6.25 shows a complete binary decision tree (nodes implement Shannon expansion), derived for a Boolean function of three variables ($n = 3$), and order of variables $x_1 \longrightarrow x_2 \longrightarrow x_3$.*

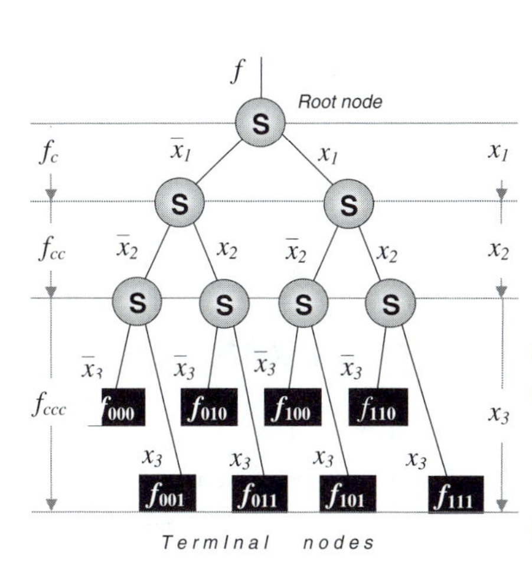

- *3 levels and 8 terminal nodes*
- *The terminal nodes are assigned with the 8 values of the function:*

$$
\begin{aligned}
f_{000} &= 1 \cdot 1 \vee 0 \cdot 0 \vee 0 \cdot 0 = 1\\
f_{001} &= 1 \cdot 0 \vee 0 \cdot 0 \vee 0 \cdot 1 = 0\\
f_{010} &= 1 \cdot 1 \vee 0 \cdot 1 \vee 0 \cdot 0 = 1\\
f_{011} &= 1 \cdot 0 \vee 0 \cdot 1 \vee 0 \cdot 1 = 0\\
f_{100} &= 0 \cdot 1 \vee 1 \cdot 0 \vee 1 \cdot 0 = 0\\
f_{101} &= 0 \cdot 0 \vee 1 \cdot 0 \vee 1 \cdot 1 = 1\\
f_{110} &= 0 \cdot 1 \vee 1 \cdot 1 \vee 1 \cdot 0 = 1\\
f_{111} &= 0 \cdot 0 \vee 1 \cdot 1 \vee 1 \cdot 1 = 1
\end{aligned}
$$

- *First level:* $f = \overline{x}_1 f_0 \vee x_1 f_1$
- *Second level:*

$$
\begin{aligned}
f = {} & \overline{x}_1\overline{x}_2 f_{00} \vee \overline{x}_1 x_2 f_{01}\\
& \vee x_1\overline{x}_2 f_{10} \vee x_1 x_2 f_{11}
\end{aligned}
$$

- *Third level:*

$$
\begin{aligned}
f = {} & \overline{x}_1\overline{x}_2\overline{x}_3 f_{000} \vee \overline{x}_1\overline{x}_2 x_3 f_{001}\\
& \vee \overline{x}_1 x_2\overline{x}_3 f_{010} \vee \overline{x}_1 x_2 x_3 f_{011}\\
& \vee x_1\overline{x}_2\overline{x}_3 f_{100} \vee x_1\overline{x}_2 x_3 f_{101}\\
& \vee x_1 x_2\overline{x}_3 f_{110} \vee x_1 x_2 x_3 f_{11}
\end{aligned}
$$

FIGURE 6.25
The complete binary tree for representing a Boolean function f of three variables x_1, x_2, and x_3 (Example 6.28).

6.10.1 Representation of elementary Boolean functions using decision trees

Example 6.29 (Decision trees.) *Decision trees for two-input Boolean functions are given in Figure 6.26.*

Practice problem **6.17. (Decision tree.)** Derive a binary decision tree for the Boolean function $f = \overline{x}_1\overline{x}_3 \vee x_1 x_3 \vee x_1 x_2$. Use the order of variables $x_1 \longrightarrow x_2 \longrightarrow x_3$.
Answer is given in "Solutions to practice problems."

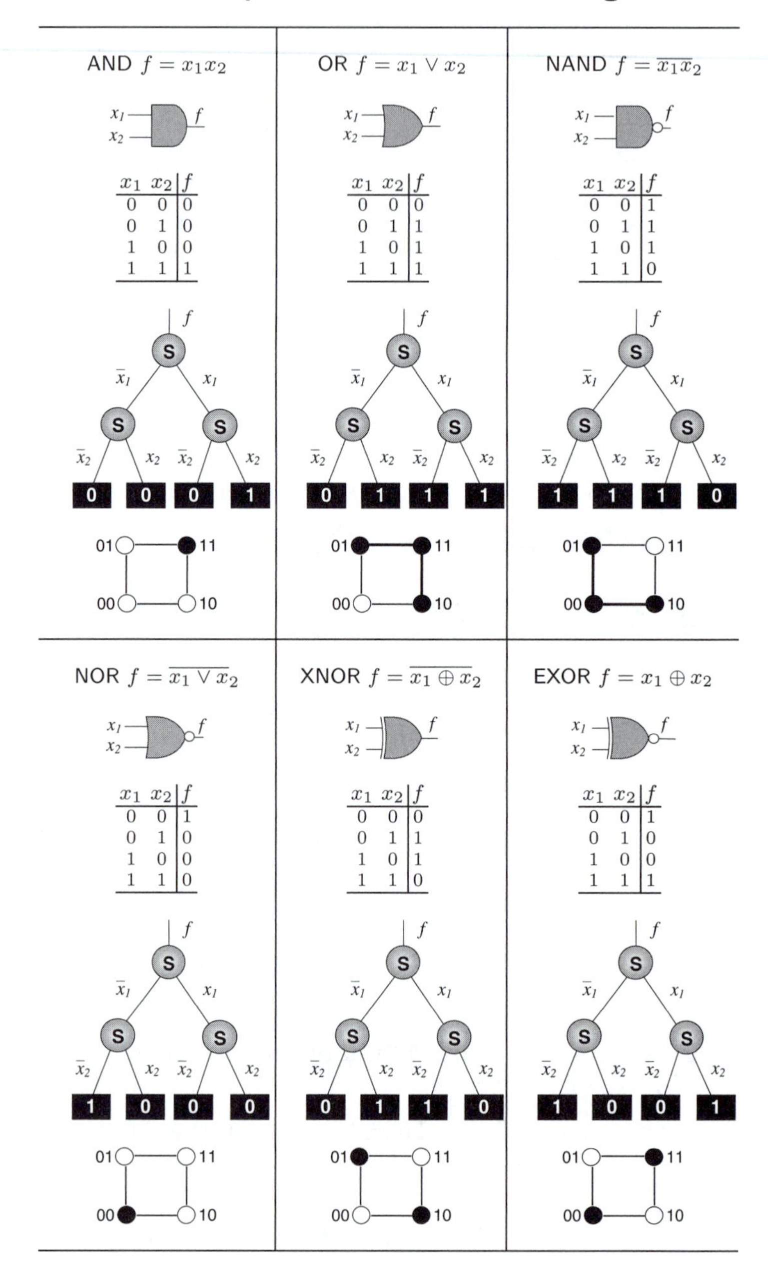

FIGURE 6.26
Complete decision trees for computing elementary Boolean functions, compared to other data structures: algebraic, tabulated (truth table), and hypercubes (Example 6.29).

6.10.2 Minterm and maxterm expression representations using decision trees

A minterm or maxterm expression can be mapped into a decision tree, and vice versa: Given a decision tree, the algebraic form of a standard SOP or POS expression can be recovered from this tree:

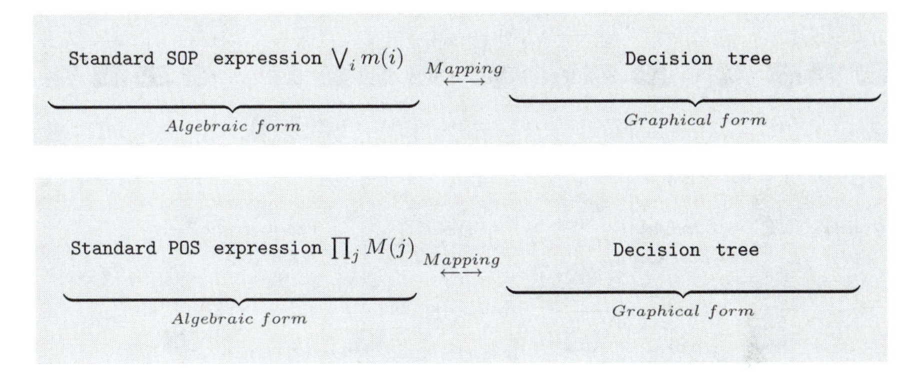

That is,

Mapping a minterm or maxterm expression into a decision tree

(*a*) The minterm in an SOP is mapped into a $\boxed{1}$ terminal node of the complete decision tree; a path in the tree from the root to a $\boxed{1}$ terminal node corresponds to a minterm in the SOP expression.

(*b*) The minterm in a POS is mapped into a $\boxed{0}$ terminal node of the complete decision tree; a path in the tree from the root to a $\boxed{0}$ terminal node corresponds to a maxterm in the POS expression.

Example 6.30 (SOP and POS expressions.) *Figure 6.27 shows the relationship of standard SOP and POS expressions and their representation using a decision trees for the Boolean function of two variables $f = x_1 \oplus x_2$.*

6.10.3 Representation of elementary Boolean functions by incomplete decision trees

In a complete decision tree, the number of nodes depends exponentially on the number of variables. Hence, this representation is not acceptable for a large number of variables. In the next chapter, techniques for decision tree reduction are considered. As a preliminary step, let us consider the properties of reduced decision trees for representing Boolean functions.

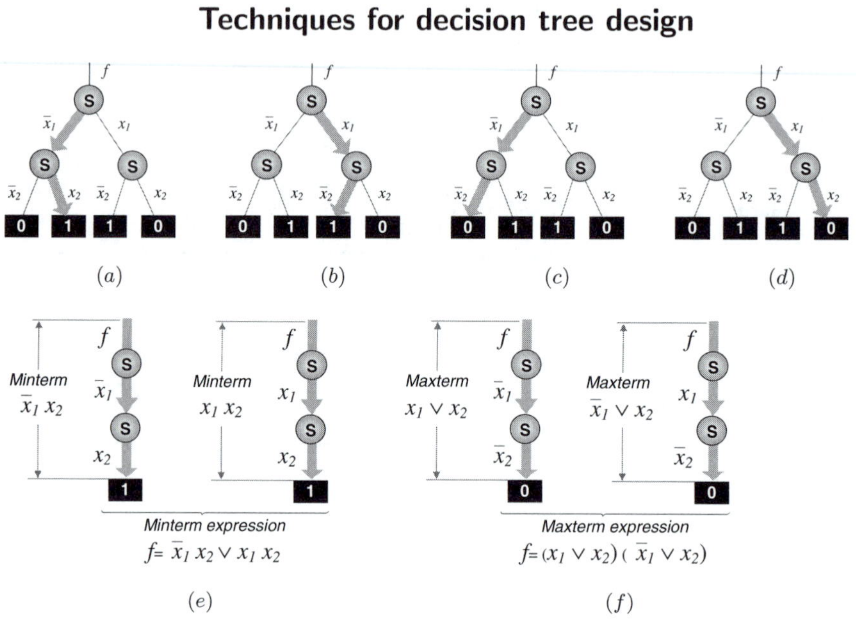

FIGURE 6.27
Minterm and maxterm representation using a decision trees: (a) minterm $\overline{x}_1 x_2$, (b) minterm $x_1 x_2$, (c) maxterm $x_1 \vee x_2$, (d) maxterm $\overline{x}_1 \vee x_2$, (e) the SOP expression $f = \overline{x}_1 x_2 \vee x_1 x_2$, (f) the POS expression $f = (x_1 \vee x_2)(\overline{x}_1 \vee x_2)$ (Example 6.30).

Example 6.31 (Reduced decision tree.) *In Figure 6.28, reduced decision trees are used for representing four Boolean functions of two variables, AND, OR, NAND, and NOR. Each of these trees is constructed from the corresponding complete tree given in Figure 6.26 by deleting the node S, the linked terminal nodes, and the corresponding link reconstruction.*

One can observe from Example 6.31 that the reduced decision trees for AND, OR, NAND, and OR functions contain repeated terminal nodes. The complete binary trees do not correspond to the minimal Boolean expression that they realize. For example, an SOP restored from an OR function is $f = \overline{x}_1 x_2 \vee x_1$. However, this SOP is not minimal; further algebraic manipulations implies

$$\begin{aligned} f &= \overline{x}_1 x_2 \vee x_1 \\ &= (x_1 \vee \overline{x}_1)(x_1 \vee x_2) = x_1 \vee x_2 \end{aligned}$$

Binary decision tree reduction

Given: A complete binary tree for a Boolean function of n variables
Step 1: Remove the nodes using the following rules:

(a) Identify the nodes at the n-th level, which are connected to the terminal nodes of the same values;
(b) Remove the nodes, and both terminal nodes, substituting them with a terminal node of the corresponding value

Step 2: Repeat Step 1 for the remaining $n-1$ levels, starting from the bottom.
Output: A reduced binary decision tree

The same can be observed for the NAND function: $f = \overline{x}_1 \vee x_1\overline{x}_2 = \overline{x}_1 \vee \overline{x}_2 = \overline{x_1x_2}$. Reduced decision trees contain less nodes than complete decision trees.

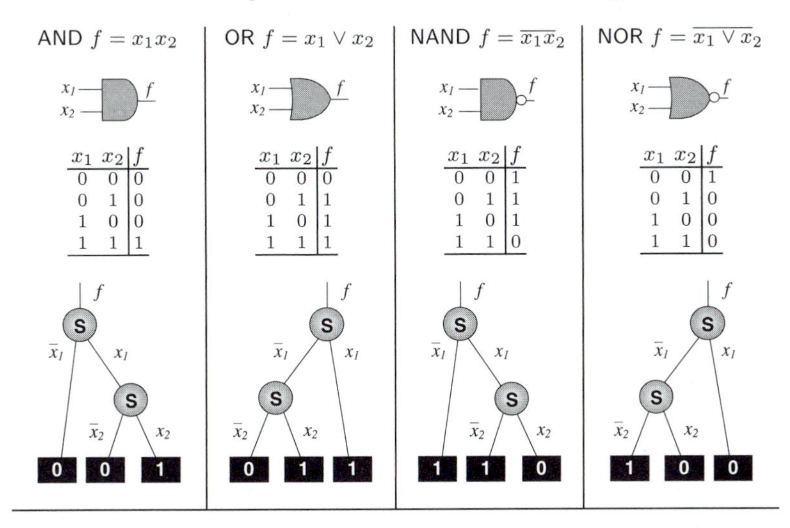

x_1	x_2	f
0	0	0
0	1	0
1	0	0
1	1	1

x_1	x_2	f
0	0	0
0	1	1
1	0	1
1	1	1

x_1	x_2	f
0	0	1
0	1	1
1	0	1
1	1	0

x_1	x_2	f
0	0	1
0	1	0
1	0	0
1	1	0

FIGURE 6.28
Representation of AND, OR, NAND, and NOR functions using reduced decision trees (Example 6.31).

Example 6.32 (Reduced decision trees.) *Decision trees in Figure 6.29 implement the function $\overline{x}_1$. The reduction implies that $f = \overline{x}_1\overline{x}_2 \vee \overline{x}_1 x_2 = \overline{x}_1$ (Figure 6.29a). This effect is caused by the redundancy of the decision tree. Shannon expansion with respect to the variable x_2 is equal to 1 because* $f = \underbrace{\overline{x}_2 f_0}_{Equal\ to\ 1} \vee \underbrace{x_2 f_1}_{Equal\ to\ 1} = 1.$
Hence, the corresponding node can be deleted and replaced with a terminal node, labeled by 1 (Figure 6.29b).

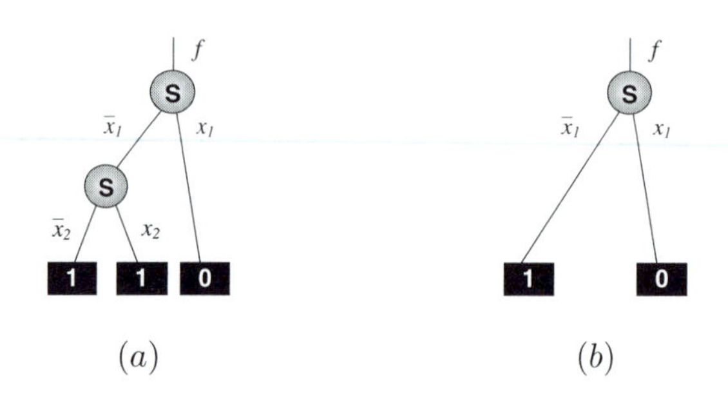

FIGURE 6.29
The redundant (a) and irredundant reduced (b) decision trees representing the function $f = \overline{x}_1$ (Example 6.32).

Practice problem **6.18.** **(Reduced decision tree.)** Construct reduced decision trees for the Boolean functions of two variables (a) $x_1 \vee \overline{x}_2$, (b) $\overline{x}_1 \vee x_2$, (c) $x_1\overline{x}_2$, (d) $\overline{x}_1 x_2$, (e) x_1, and (f) 1.
Answers are given in "Solutions to practice problems."

One can observe that reduction using the procedure described above still results in repeating terminal nodes in the reduced decision trees. Further reduction of these trees using this repetition leads to *decision diagrams.*

6.11 Decision diagrams

The manipulation of graphical data structures is especially beneficial for large-size problems. Moreover, during the logic synthesis of circuits manufactured using certain technologies, it is not necessary to convert graph data structures into logic networks. This is because the topology of the obtained graph is itself a logic network.

A *path* in a decision diagram is a sequence of edges and nodes leading from the root node to a terminal node. A path carries local information for the algebraic representation of Boolean functions. A path can be measured by a *length* t, which is the number of vertices in the path. There is always a 0-length path from node u to node u'. The *path length* is the number of non-terminal nodes in the path.

A binary decision diagram is a reduced binary decision tree:

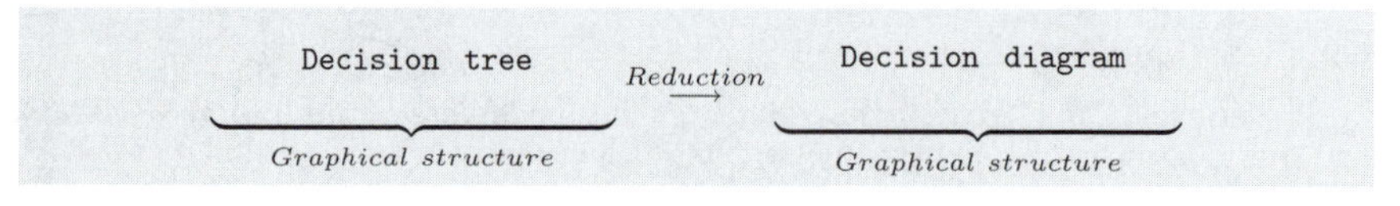

The procedures given in Figure 6.30 result in a decision diagram for the representation of a Boolean function. Note that the reduced diagram does not necessary have the minimal number of nodes. The order of variables determines the size of the reduced decision diagram. Reduced ordered binary decision diagrams (ROBDDs) are introduced in Chapters 9 and 12.

Algorithm 1: Deriving a binary decision diagram

Given: A Boolean function of n variables
Step 1: Derive a complete binary decision tree.
Step 2: Reduce the decision tree.
Step 3: Merge repeating terminal nodes into two: [0] and [1]
Output: A binary decision diagram

Algorithm 2: Deriving a binary decision diagram

Given: A Boolean function of n variables
Step 1: Find a minimal SOP or POS.
Step 2: Derive a binary decision tree by deriving the path. that corresponds to the products.
Step 3: Reduce the decision tree to a decision diagram.
Output: A binary decision diagram

FIGURE 6.30
Procedures for deriving binary decision diagrams from decision trees.

TABLE 6.6
Decision diagram parameters and characteristics.

Characteristic	Definition
Size	The sum of the numbers of nodes at all levels, excluding the terminal nodes (always two), N_N
Depth, or number of levels	The number of levels, DEPTH; this is equivalent to the number of variables plus 1, if there are no redundant variables; that is, $n+1$
Width	The maximum number of nodes at any level, WIDTH
Area	The area the decision diagram occupies is estimated as AREA=WIDTH×DEPTH
Path length	The number of edges on the path from the root to a terminal node
Weight	The number of ones in the terminal nodes
Homogeneity	The absolute difference between 1 and 0 terminal nodes

6.12 Summary of Boolean data structures

This chapter introduces the Boolean *data structures* that are the fundamentals of logic design. The key aspects of this chapter are given below:

(*a*) Data structures, such as truth tables, K-maps, SOP-based algebraic forms, cubes, decision trees and decision diagrams, and logic networks using logic gates and threshold elements.
(*b*) Relationships between data structures for Boolean functions.

The main topics of this chapter are summarized as follows:

- A data structure representing a Boolean functions is a mathematical model of a Boolean function. ***Data type*** determines the classification of this model as follows: ***algebraic*** and ***matrix*** forms such as canonical descriptions and various transforms; ***tabulated*** forms such as truth tables and state tables; ***graphical*** forms, such as logic networks, decision trees, decision and state diagrams, and hypercubes; and ***mixed*** graphical and tabulated forms such as K-maps.
- No universal data structure exists that can be efficient in all applications. Choosing an appropriate data structure based on its advantages and disadvantages, as well as the relationships between these structures, in order to satisfy design requirements, is a crucial point of logic design.
- *Algebraic* and *matrix* forms of Boolean functions are useful in the most steps of logic design. The manipulation of Boolean functions in algebraic form is used for small problems. Matrix representation is used in various algorithms for the manipulation of Boolean functions.
- In the tabular representation, ***truth tables***, all Boolean function values are explicitly shown, without taking into account their possible relationships. The size of a truth table is exponentially dependent on the number of variables. This representation of Boolean functions is unsuitable for functions of a large number of variables.
- The reduction of a truth table is the basis for its conversion to other data structures and implementations. The following reduction techniques are used in practice: using (*a*) short-hand notation, (*b*) using particular properties of Boolean functions such as symmetry, and (*c*) reduction by replacing a group of rows by variables or sub-functions.

Summary (continuation)

- A ***K-map*** is a visual (graphical) representation of a Boolean function used as an aid to simplification; a K-map is useful only for the minimization of functions with a small number of variables. A ***hypercube form*** is a graphical structure for the representation and minimization of Boolean functions. Hypercubes are useful for the manipulation of Boolean functions with a large number of variables. A ***logic network*** is a graphical representation of a network of gates. It consists of gates, external inputs and outputs, and connections, and can be described as a ***logic diagram***, a ***netlist***, and a ***set of language statements***. A ***threshold network*** is a graphical representation of a network of threshold gates useful for the manipulation and implementation of Boolean functions with a large number of variables.
- A ***decision tree*** is useful for functions with a small number of variables. A reduced decision tree with two terminal nodes (1 and 0) is a *decision diagram*. Reduction is achieved according to the particular properties of various functions. A *decision tree* is useful for functions with a small number of variables. Reduction is achieved due to the particular properties of the particular functions.
- All these data structures allow for the manipulation, representation, and validation of Boolean functions. For advances in Boolean data structures, we refer the reader to the "Further study" section.

6.13 Further study

Historical perspective

1943: Neuron cells for the computing of elementary logic functions were developed by W. S. McCulloch and W. H. Pitts, in the paper "A Logical Calculus of Ideas Immanent in Nervous Activity," *Bulletin of Mathematical Biophysics*, number 5, pages 115–137, 1943. A McCulloch-Pitts' cell was the first threshold gate. An arbitrary logic function can be computed on a network of such cells.

1948: C. Shannon suggested a measure to represent information by a numerical value, nowadays known as *Shannon entropy* ("A Mathematical theory of communication" by C. Shannon, *Bell Systems Technical Journal*, volume 27, pages 379–423, 623–656, 1948).

1958 The graphic methods used to represent binary codes were used to specify Boolean functions. Such a graphic form, n-dimensional cube, also called *algebraic topological* representation of Boolean functions, was introduced and explored by J. P. Roth ("Algebraic Topological Methods for the Synthesis of Switching Systems I," *Transactions of the American Mathematical Society*, Volume 88, July 1958).

1958: Frank Rosenblatt initiated a new phase in neural network research by introducing the idea of an auto-learning network.

1982: J. J. Hopfield developed a new computing paradigm known as a *Hopfield network* (J. J. Hopfield, "Neural Networks and Physical Systems with Emergent Collective Computational Abilities," *Proceedings of the National Academy of Sciences*, U.S.A., volume 79, pages 2554–2558, 1982).

1987: Gail Carpenter and Stephen Grossberg of Boston University introduced a network called ART 1 using binary input patterns, and another network called ART 2 using analog inputs. In 1990, the first multi-valued logic neural network, which was a simple adaptive switching network, was introduced by David Olmsted ("The Reticular Formation as a Multi-Valued Logic Neural Network," *International Joint Conference on Neural Networks*, Volume 1, pages 619–624, 1990).

Advanced topics of data structures

Topic 1: A Hopfield computing paradigm is based on a *Hopfield network*, which is defined as a logic network of interconnected, binary-valued neuron cells that implement the distributed principle of encoding and processing of information. The result (a computed Boolean function) is obtained by decoding the neuron's states. Using a Hopfield network, an arbitrary logic function can be computed via a process similar to the gradual cooling process of metal. Values of the Boolean function given an assignment of Boolean variables are computed through the relaxation of neuron cells.

Topic 2: Threshold logic theory related neural network techniques to logic design, formulating logic functions in terms of inequalities. Many existing methods perform synthesis by representing each product term in an SOP expression of a function as a threshold gate or by converting each gate in a Boolean network into a threshold gate, and this requires advanced factorization algorithms. CMOS implementations of threshold gates were proposed in the 1980s (see a survey of VLSI implementations of threshold logic in the "Further study" section). A multi-threshold logic circuit design using new technologies, such as resonant-tunneling and quantum cellular arrays have been considered just recently, upon development of efficient algorithms to factorize a multilevel network using algebraic or Boolean factorization.

Topic 3: Multi-valued logic is examined in the "Further study" section in Chapter 4.

Topic 4: Information-theoretical measures. A computing system can be seen as a process of communication between computer components. The classical concepts of information and entropy introduced by Shannon are the basis for this. The information-theoretical standpoint on computing is based of the following notations:

(*a*) *Source of information* is a stochastic process whereby an event occurs at time point i with probability p_i. That is, the source of information is defined in terms of the probability distribution for signals from this source. The problem is usually formulated in terms of sender and receiver of information and used by analogy with communication problems.

(*b*) *Information engine* is the machine that deals with information.

(*c*) *Quantity of information* is a value of a function that occurs with a probability p; this quantity is equal to $(-\log_2 p)$.

(*d*) *Entropy*, $H(f)$, is the measure of the information content of X. The greater the uncertainty in the source output, the higher is its information content. A source with zero uncertainty would have zero information content and, therefore, its entropy would be equal to zero.

Let us assume that all combinations of values of variables occur with equal probability. A value of a function that occurs with the probability p carries a quantity of information equal to $< \texttt{Quantity of information} > \; = -\log_2 p \;\; bit$, where p is the probability of that value occurring. Note that information is measured in bits. The information carried by the value of a of a random variable x_i is equal to $I(x_i)_{|x_i=a} = -\log_2 p \;\; bit$, where p is the quotient between the number of tuples whose i-th components equal a and the total number of tuples.

For example, the output of the AND function is equal to 0 with probability 0.25, and equal to 1 with probability 0.75. The entropy of the output signal is calculated as follows: $H_{out} = -0.25 \times \log_2 0.25 - 0.75 \times \log_2 0.75 = 0.81$ *bit/pattern.* (see details in "An Information Theoretic Approach to Digital Fault Testing" by V. Agraval, *IEEE Transactions on Computers*, volume 30, number 8, pages 582–587, 1981).

Further reading

"Application of Information Theory to the Construction of Efficient Decision Trees" by C. R. P. Hartmann, P. K. Varshney, K. G. Mehrotra, and C. L. Gerberich, *IEEE Transactions on Information Theory*, volume 28, number 5, pages 565–577, 1982.

Decision Diagram Techniques for Micro- and Nanoelectronic Design by S. N. Yanushkevich, D. M. Miller, V. P. Shmerko, and R. S. Stanković, CRC/Taylor & Francis Group, Boca Raton, FL, 2006.

"Depth-Size Tradeoffs for Neural Computation" by K. Y. Siu, V. P. Roychowdhury, and T. Kailath, *IEEE Transactions on Computers*, volume 40, number 12, pages 1402–1411, 1991.

Introduction to Digital Logic Design by John P. Hayes, Addison-Wesley, 1993.

Multi-Valued Logic Design by G. Epstein, Institute of Physics Publishing, London, UK, 1993.

Principles of Neurodynamics by F. Rosenblatt, Spartan, New York, 1962.

Threshold Logic by M. L. Dertouzos, Wiley, New York, 1971.

"VLSI Implementation of Threshold Logic – A Comprehensive Survey" by V. Beiu, J. M. Quintana, and M. J. Avedillo, *IEEE Neural Networks*, volume 14, number 5, pages 1217–1243, 2003.

6.14 Solutions to practice problems

Practice problem	Solution

6.1.

Let us substitute all possible combinations of variables x_1, x_2, and x_3 from 000 to 111. For instance, when $x_1 = x_2 = x_3 = 000$, then $f = \overline{0} \vee \overline{0} \vee \overline{0} = 1$. When $x_1 = x_2 = 0$ and $x_3 = 1$, then $f = \overline{0} \vee \overline{0} \vee \overline{1} = 1$. The resulting values of the function are written in the last column of the truth table

$x_1x_2x_3$	f
000	1
001	1
010	1
011	1
100	1
101	1
110	1
111	0

6.3.

Assignment	Function	Minterm	Maxterm
000	0		$x_1 \vee x_2 \vee x_3$
001	0		$x_1 \vee x_2 \vee \overline{x}_3$
010	0		$x_1 \vee \overline{x}_2 \vee x_3$
011	1	$\overline{x}_1x_2x_3$	
100	1	$x_1\overline{x}_2\overline{x}_3$	
101	1	$x_1\overline{x}_2x_3$	
110	1	$x_1x_2\overline{x}_3$	
111	0		$\overline{x}_1 \vee \overline{x}_2 \vee \overline{x}_3$

6.4.

The truth-vector $\mathbf{F} = [\ 0\ 1\ 0\ 0\ 0\ 1\ 1\ 1\ ]^T$ corresponds to the following complete and reduced truth tables:

Complete truth table

$x_1x_2x_3$	f
000	0
001	1
010	0
011	0
100	0
101	1
110	1
111	1

Reduced truth table

$x_1x_2x_3$	f
001	1
101	1
110	1
111	1

6.5, 6.6.

Practice problem 6.5

x_1	x_2	f	
0	0	0	x_2
0	1	1	
1	0	1	$\overline{x}_2$
1	1	0	

Practice problem 6.6

x_1	x_2	x_3	f
0	0	0	0
0	0	1	0
0	1	0	1
0	1	1	1
1	0	0	1
1	0	1	1
1	1	0	0
1	1	1	0

x_1	x_3	f
0	0	x_2
0	1	x_2
1	0	$\overline{x}_2$
1	1	$\overline{x}_2$

Solutions to practice problems (continuation)

Practice problem	Solution

6.7, 6.9.

Practice problem 6.7

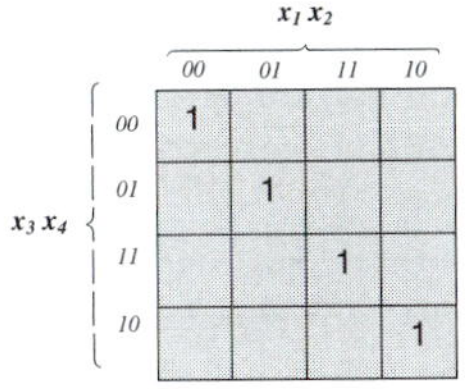

$x_1 x_2$ / $x_3 x_4$

x_3x_4 \ x_1x_2	00	01	11	10
00	1			
01		1		
11			1	
10				1

Practice problem 6.9

The first approach: $f = [11\times] \vee [\times 10]$.

The second approach:

x_1	x_2	x_3	f
0	0	0	0
0	0	1	0
0	1	0	1
0	1	1	0
1	0	0	0
1	0	1	0
1	1	0	1
1	1	1	1

⟺

x_1	x_2	x_3	f
0	1	0	1
1	1	0	1
1	1	1	1

$$f = [010] \vee [110] \vee [111] = [\times 10] \vee [11\times]$$

6.10, 6.11.

Practice problem 6.10

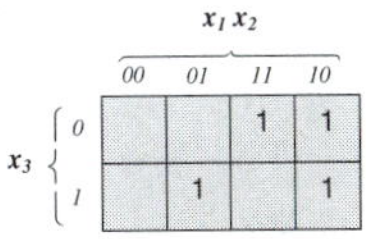

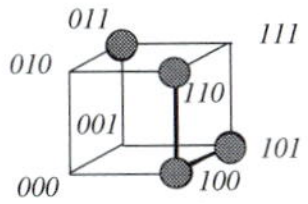

x_3 \ x_1x_2	00	01	11	10
0			1	1
1		1		1

011 111 010 110 001 101 000 100

Prime cubes

$$\begin{bmatrix} 0\,1\,1 \\ 1\,0\,0 \\ 1\,0\,1 \\ 1\,1\,0 \end{bmatrix} \Rightarrow \begin{bmatrix} 0\;1\;1 \\ 1\;0\;\times \\ 1\;\times\;0 \end{bmatrix}$$

Cubes

Practice problem 6.11

5-D hypercube

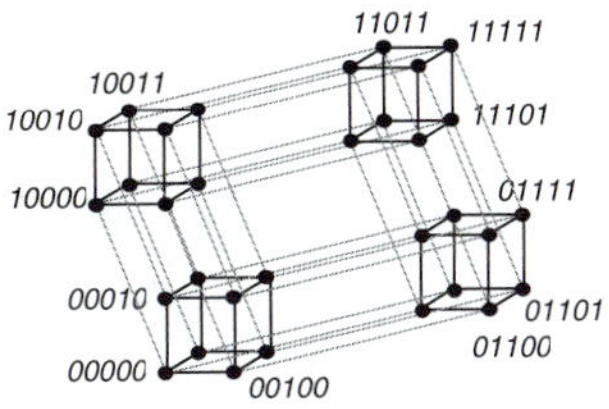

11011 11111 10011 10010 11101 10000 01111 00010 01101 01100 00000 00100

6.12.

AND $f = x_1 x_2$ — 01 11 00 10 — f= [1 1]

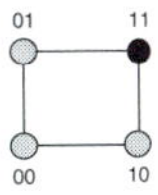

NAND $f = \overline{x_1 x_2 x}$ — 01 11 00 10 — $f = \begin{bmatrix} 0 & \times \\ \times & 0 \end{bmatrix}$

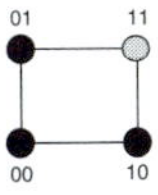

OR $f = x_1 \vee x_2$ — 01 11 00 10 — $f = \begin{bmatrix} \times & 1 \\ 1 & \times \end{bmatrix}$

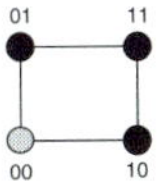

EXOR $f = x_1 \oplus x_2$ — 01 11 00 10 — $f = \begin{bmatrix} 1 & 0 \\ 0 & 1 \end{bmatrix}$

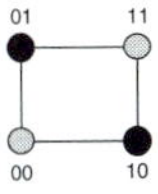

6.13.

It follows from the networks given in Figure 6.21 that

$$f = x_1\overline{x}_2 \vee x_3\overline{x}_4 x_5 \vee x_1\overline{x}_3 x_5 = x_1 \vee \overline{x}_2 \vee x_3 \vee \overline{x}_4 x_5$$

6.14.

Construct the truth table of the majority function for $n = 1$. This function satisfies the requirements of the self-dual functions.

x_1	x_2	x_3	f
0	0	0	0
0	0	1	0
0	1	0	0
0	1	1	1
1	0	0	0
1	0	1	1
1	1	0	1
1	1	1	1

Solutions to practice problems (continuation)

Practice problem	Solution

6.15.

Substitute the weights $w_1 = w_2 = -1$ and threshold $\Theta = -2$ values into Equation 6.2: $f = (\sum w_i x_i) - \Theta = w_1x_1 + w_2x_2 - \Theta = -1 \times x_1 - 1 \times x_2 - 2$. Derive the truth table:

x_1x_2	$(\sum - \Theta) \geq -1$	f
00	$-1 \times 0 - 1 \times 0 \geq -2$	1
01	$-1 \times 0 - 1 \times 1 \geq -2$	1
10	$-1 \times 1 - 1 \times 0 \geq -2$	1
11	$-1 \times 1 - 1 \times 1 \geq -2$	0

6.16.

(*a*) This is a NAND-OR network, $w_{11} = w_{12} = ... = w_{32} = -1$, $w_{41} = w_{42} = w_{43} = 1$, $\theta_1 = \theta_2 = \theta_3 = -1$, $\theta_4 = 1$.

(*b*) This is a NOR-AND network, $w_{11} = w_{12} = ... = w_{32} = -1$, $w_{41} = w_{42} = w_{43} = 0$, $\theta_1 = \theta_2 = \theta_3 = 0$, $\theta_4 = 2$.

6.17.

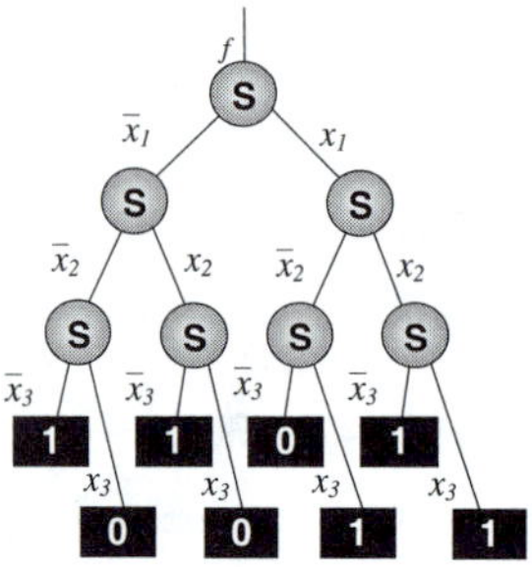

x_1	x_2	x_3	f
0	0	0	1
0	0	1	0
0	1	0	1
0	1	1	0
1	0	0	0
1	0	1	1
1	1	0	1
1	1	1	1

6.18.

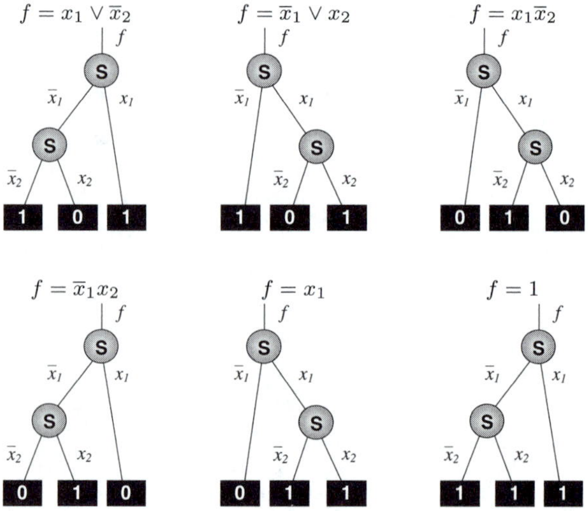

$f = x_1 \vee \overline{x}_2$

$f = \overline{x}_1 \vee x_2$

$f = x_1\overline{x}_2$

$f = \overline{x}_1 x_2$

$f = x_1$

$f = 1$

6.15 Problems

Problem 6.1 Represent the following Boolean functions using the truth table, truth-vector, and K-map:

(a) $x_1x_2 \vee \overline{x}_3$ (b) $x_1\overline{x}_2x_3$ (d) $x_1 \vee x_2 \vee \overline{x}_3$

Problem 6.2 Represent the Boolean functions from Problem 6.1 using cubes.

Problem 6.3 Convert the following cube representation of Boolean functions into a truth tables and K-maps:

$$(a)\ \begin{bmatrix} 0 \times \times \\ \times\, 0 \times \end{bmatrix} \qquad (b)\ \begin{bmatrix} 1 \times \times \\ \times\, 1 \times \\ \times \times 1 \end{bmatrix} \qquad (c)\ \begin{bmatrix} 0\, 1 \times \\ 1\, 0 \times \end{bmatrix} \qquad (d)\ \begin{bmatrix} 1 \times 1 \\ \times\, 0\, 0 \\ 0\, 1 \times \end{bmatrix}$$

Problem 6.4 Convert the following truth tables of Boolean functions into cubes:

(a)

x_1	x_2	x_3	f
0	0	0	1
0	0	1	1
0	1	0	1
0	1	1	1
1	0	0	1
1	0	1	1
1	1	0	1
1	1	1	0

(b)

x_1	x_2	x_3	f
0	0	0	0
0	0	1	1
0	1	0	1
0	1	1	1
1	0	0	1
1	0	1	1
1	1	0	1
1	1	1	0

(c)

x_1	x_2	x_3	f
0	0	0	1
0	0	1	0
0	1	0	0
0	1	1	0
1	0	0	0
1	0	1	0
1	1	0	0
1	1	1	1

(d)

x_1	x_2	x_3	f
0	0	0	1
0	0	1	0
0	1	0	0
0	1	1	1
1	0	0	1
1	0	1	1
1	1	0	1
1	1	1	0

Problem 6.5 Convert the following K-maps of Boolean functions into truth tables and cubes:

(a)

x_3 \ x_1x_2	00	01	11	10
0	1	0	1	0
1	0	1	0	1

(b)

x_3 \ x_1x_2	00	01	11	10
0	0	1	1	0
1	1	1	1	0

(c)

x_3 \ x_1x_2	00	01	11	10
0	0	1	1	0
1	1	0	0	1

Problem 6.6 Perform the following operations on the given cubes:

$(a)\ [0 \times \times] \vee [0 \times \times] \qquad (b)\ [1 \times \times] \cdot [1\, 1 \times] \qquad (c)\ [0\, 1 \times] \oplus [0 \times \times]$

Problem 6.7 Convert the following logic networks into truth tables:

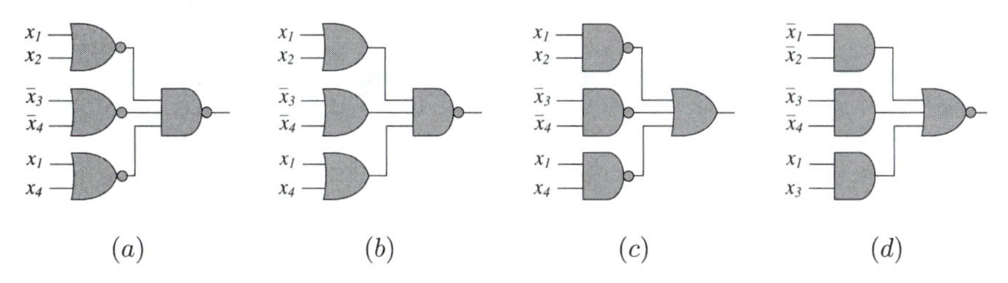

(a) (b) (c) (d)

Problem 6.8 Reduce the following truth tables of Boolean functions using (a) symmetric properties (if possible) and (b) grouping with respect to the variable x_1:

(a)

x_1	x_2	x_3	f
0	0	0	1
0	0	1	1
0	1	0	1
0	1	1	1
1	0	0	1
1	0	1	1
1	1	0	1
1	1	1	0

(b)

x_1	x_2	x_3	f
0	0	0	0
0	0	1	1
0	1	0	1
0	1	1	1
1	0	0	1
1	0	1	1
1	1	0	1
1	1	1	0

(c)

x_1	x_2	x_3	f
0	0	0	1
0	0	1	0
0	1	0	0
0	1	1	0
1	0	0	0
1	0	1	0
1	1	0	0
1	1	1	1

(d)

x_1	x_2	x_3	f
0	0	0	1
0	0	1	0
0	1	0	0
0	1	1	1
1	0	0	1
1	0	1	1
1	1	0	1
1	1	1	0

Problem 6.9 Convert the logic networks given in Problem 6.7 into a K-map.

Problem 6.10 Convert the logic networks given in Problem 6.7 into a cube form.

Problem 6.11 Convert the logic networks given in Problem 6.7 into a threshold networks.

Problem 6.12 Convert the logic networks given in Problem 6.7 into decision trees using Shannon expansion.

Problem 6.13 Convert the logic networks given in Problem 6.7 into decision diagrams using Shannon expansion.

Problem 6.14 Convert the following decision trees into truth tables:

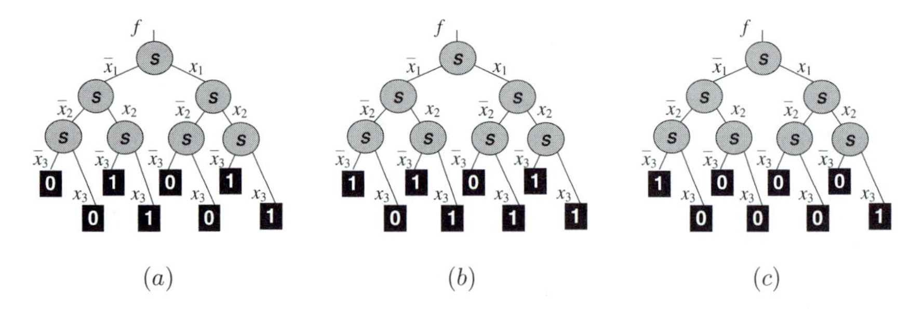

Problem 6.15 Convert the following reduced decision trees into truth tables:

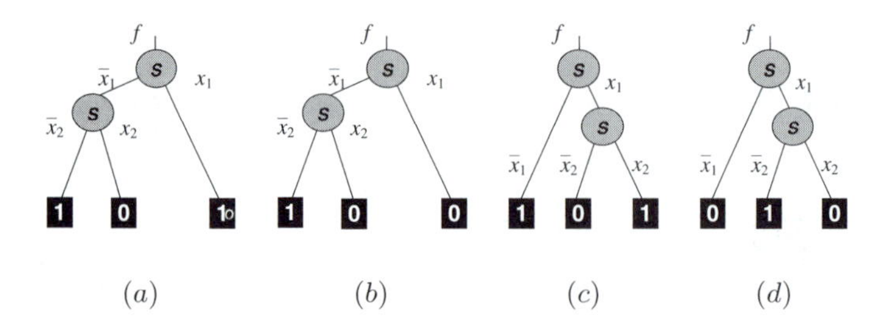

Problem 6.16 Prove using K-maps and decision diagrams that the following logic networks implement minimal Boolean functions:

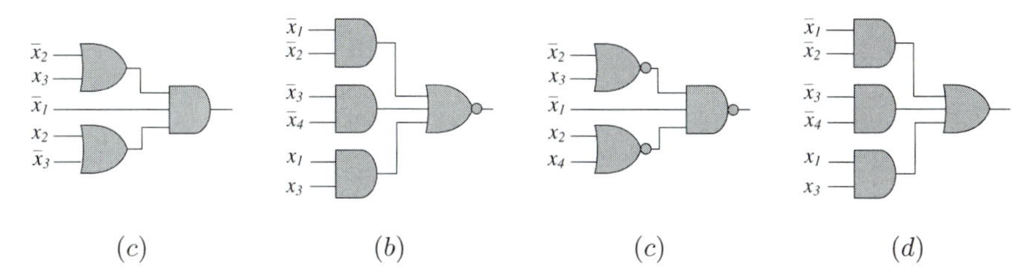

(c) (b) (c) (d)

Problem 6.17 Derive the Boolean functions in algebraic form given the following cubes:

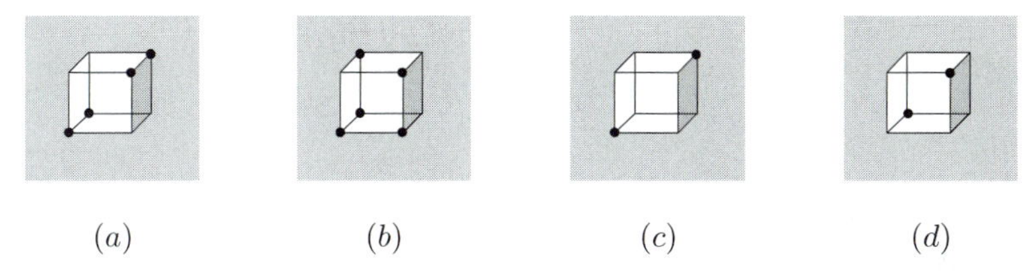

(a) (b) (c) (d)

Problem 6.18 Derive the SOP cubes for the functions given by the hypercubes in Problem 6.17.

Problem 6.19 Derive logic networks using the following cube-based SOP representations of Boolean functions:

(a) $\begin{bmatrix} 0\ 0\ \times \\ \times\ 1\ 1 \end{bmatrix}$ (b) $\begin{bmatrix} \times\ 1\ 1 \\ \times\ 0\ 0 \\ 1\ \times\ \times \end{bmatrix}$ (c) $\begin{bmatrix} 1\ \times\ 0 \\ 0\ 1\ \times \\ \times\ 0\ 1 \end{bmatrix}$ (d) $\begin{bmatrix} 1\ 0\ 0 \\ 0\ 1\ 0 \\ 0\ 0\ 1 \end{bmatrix}$

Problem 6.20 Given the Boolean function $f = x_1x_2 \vee \overline{x}_3$, build decision diagrams using the following order of variables:

(a) $x_1 \longrightarrow x_2 \longrightarrow x_3$ (b) $x_2 \longrightarrow x_1 \longrightarrow x_3$ (c) $x_1 \longrightarrow x_3 \longrightarrow x_2$

7

Properties of Boolean Functions (Optional)

Self-dual functions

- Detection
- Generation

Monotonic functions

- Detection
- Generation

Unate functions

- Detection
- Generation

Linear functions

- Detection
- Generation

Universal set of functions

- Five classes of Boolean functions
- Criterion of completeness
- Post theorem

7.1 Introduction

There are several reasons for studying the properties of Boolean functions:

(*a*) The analysis of Boolean functions of many input variables is a task of exponential complexity that requires a great amount of time and memory resources. However, the complexity can be reduced should if additional structural properties of the functions and related data structures be known.

(*b*) The properties of Boolean functions have different levels of significance in logic design (synthesis); this depends on their data structure and level of design abstraction. It is important to understand the resources available for the improvement of performance of design tools.

(*c*) Many techniques for the efficient manipulation of Boolean functions represented in various forms have been developed. These techniques utilize the properties of Boolean functions.

In this chapter, the major properties of Boolean functions are introduced.

7.2 Self-dual Boolean functions

The property of Boolean algebra called the *duality principle* has been formulated in the form of *Huntington postulates*. These postulates state that given a valid identity, another valid identity can be obtained by

- Interchanging the operations OR and AND, and
- Interchanging the constants 0 and 1, i.e., replacing 1's by 0's and 0's by 1's.

Example 7.1 (Duality principle.) *Applying the duality principle to the identity* $x_1 \cdot x_2 \vee \overline{x}_2 \cdot x_3 = x_1 \cdot x_2 \vee \overline{x}_2 \cdot x_3 \vee x_1 \cdot x_3$, *we obtain another identity,* $(x_1 \vee x_2) \cdot (\overline{x}_2 \vee x_3) = (x_1 \vee x_2) \cdot (\overline{x}_2 \vee x_3) \cdot (x_1 \vee x_3)$.

Self-dual Boolean functions

A *self-dual* Boolean function f is the function that satisfies the following requirement:

$$f = \overline{f}(\overline{x}_1, \ldots, \overline{x}_n) \tag{7.1}$$

The recognition of this and other properties can be accomplished using various data structures:

$$\underbrace{\texttt{Boolean functions}}_{\textit{Boolean data structure}} \xrightarrow{\textit{Recognition}} \underbrace{\texttt{Self-dual Boolean functions}}_{\textit{Boolean data structure}}$$

The examples below demonstrate the recognition of self-duality using algebraic data structure.

Example 7.2 (Self-dual functions.) *The Boolean function $f = (x_1 \vee x_2x_3)x_4$ is self-dual, because its dual function is reduced to the initial one:*

$$\overline{\overline{x}_1(\overline{x}_2 \vee \overline{x}_3) \vee \overline{x}_4} = (\overline{\overline{x}_1\overline{x}_2})(\overline{\overline{x}_1\overline{x}_3})x_4 = (x_1 \vee x_2)(x_1 \vee x_3)x_4$$
$$= (\underbrace{x_1 \vee x_1x_3 \vee x_1x_2}_{\textit{Equal to } x_1} \vee x_2x_3)x_4 = (x_1 \vee x_2x_3)x_4 = f$$

Practice problem 7.1 (Self-dual functions.) Determine if the function $f = x_1x_2 \vee x_2x_3 \vee x_1x_3$ is self-dual.
Answer is given in "Solutions to practice problems."

There is an inverse symmetry in the truth tables of the self-dual functions: The upper half of the truth vector is equal to the complemented value of its lower part. This property can also be detected by inspection of the function's decision diagram.

Example 7.3 (Self-dual functions.) *Figure 7.1a shows self-dual functions of two variables. This property can be used to generate the truth tables of the self-dual functions given an arbitrary upper part of the truth vector (Figure 7.1b). Since there are four possible upper part vector (00, 01, 10 and 11), there are four self-dual truth-vectors represented by the 2-variable decision tree as shown in Figure 7.1c.*

There are $2^{2^{n-1}}$ self-dual functions of n variables. For example, all Boolean functions of a single variable are self-dual, since, given $n = 1$, $2^{2^{1-1}} = 2$. Only $2^{2^{2-1}} = 4$ functions out of $2^{2^2} = 16$ are self-dual given two Boolean variables.

Practice problem 7.2 (Self-dual functions.) Prove that the class of self-dual functions of n variables includes 2^{2^n-1} out of a total number 2^{2^n} of all possible functions of n variables.
Answer: The number of elements in the truth table of the function of n variables is 2^n, of which the upper half has 2^{n-1} elements. There are $2^{2^{n-1}}$ possible combinations of 0s and 1s in this half, and, therefore, there are $2^{2^{n-1}}$ possible truth-vectors in which the lower part is a dual of the upper part.

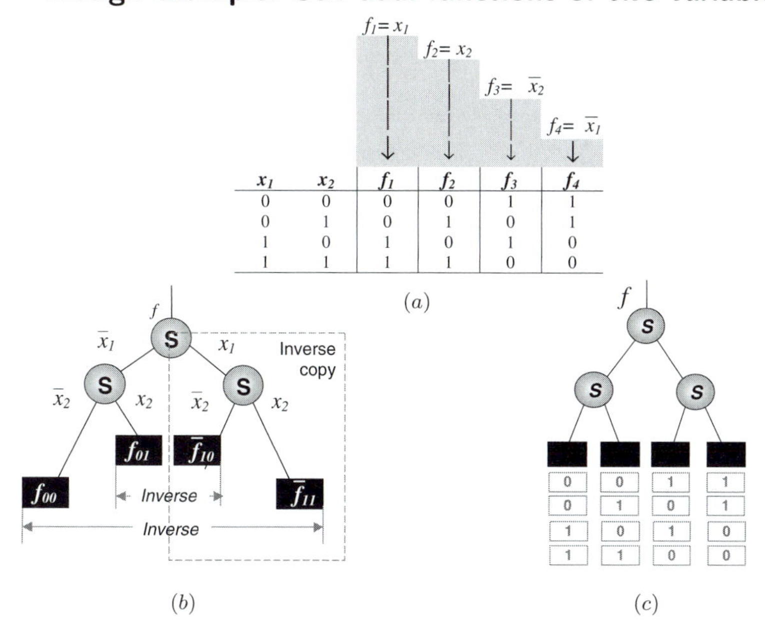

x_1	x_2	f_1	f_2	f_3	f_4
0	0	0	0	1	1
0	1	0	1	0	1
1	0	1	0	1	0
1	1	1	1	0	0

FIGURE 7.1
Self-dual functions of two variables (a), detection (b), and generation (c) of self-dual functions using decision trees (Example 7.3).

Practice problem 7.3 (Self-dual functions.) Generate the truth tables of the first eight self-dual Boolean functions of three variables.
Answer is given in "Solutions to practice problems."

A *self-anti-dual* function is the Boolean function such that

Self-anti-dual Boolean functions

$$f = f(x_1, \ldots, x_n) = f(\overline{x}_1, \ldots, \overline{x}_n) \quad (7.2)$$

Example 7.4 (Self-anti-dual functions.) *The Boolean function $f = x_1 \oplus x_2$ is self-anti-dual because $f(x_1, x_2) = (\overline{x}_1 \oplus 1) \oplus (\overline{x}_2 \oplus 1) = \overline{x}_1 \oplus \overline{x}_2 = f(\overline{x}_1, \overline{x}_2)$.*

Practice problem 7.4 (Self-anti-dual functions.) Determine if the function $f = x_1x_2 \vee \overline{x_1x_2}$ is a self-anti-dual.
Answer is given in "Solutions to practice problems."

The self-dual functions are used in *self-dual parity-checking*, a method of on-line error detection in logical networks.

7.3 Monotonic Boolean functions

A function f is *monotonically increasing*

$$\text{if } a \leq b \text{ implies } f(a) \leq f(b)$$

Similarly, it is *monotonically decreasing* function

$$\text{if } a \leq b \text{ implies } f(a) \geq f(b)$$

Recognition of the monotonicity property of a Boolean function has been found useful in fault-tolerant circuit designs:

7.3.1 Monotonically increasing Boolean functions

Let $\mathbf{a} = (a_1, a_2, \ldots, a_n)$ and $\mathbf{b} = (b_1, b_2, \ldots, b_n)$ be Boolean vectors. Consider the following relations between these vectors:

Relations between Boolean vectors

- If the elements of the vectors satisfy the condition $a_i \geq b_i$, $i = 1, 2, \ldots, n$, the vector $\mathbf{a}$ is *equal to or greater than* $\mathbf{b}$, $\mathbf{a} \geq \mathbf{b}$;
- If $a_i \leq b_i$, the vector $\mathbf{a}$ is *equal to or less than* $\mathbf{b}$, then $\mathbf{a} \geq \mathbf{b}$, and
- If $a_i \geq b_i$ and $a_i \leq b_i$, then the vectors $\mathbf{a}$ and $\mathbf{b}$ are *incomparable.*

For example, the vectors $\mathbf{a}=\begin{bmatrix}0\\1\end{bmatrix}$ and $\mathbf{b}=\begin{bmatrix}1\\0\end{bmatrix}$ are incomparable.

Example 7.5 (Monotonically increasing functions.) *Consider two cases. <u>Case 1:</u> Let the Boolean vectors be* $\mathbf{a} = (a_1a_2a_3a_4) = (1011)$ *and* $\mathbf{b} = (b_1b_2b_3b_4) = (1001)$. *The vector* $\mathbf{a}$ *is greater than the vector* $\mathbf{b}$, *that is,* $f(\mathbf{a}) \geq f(\mathbf{b})$, *since* $a_1 = b_1$ $(1 = 1)$, $a_3 > b_3$ $(1 = 0)$, $a_2 = b_2$ $(0 = 0)$, *and* $a_4 = b_4$ $(1 = 1)$.
<u>Case 2:</u> Consider the Boolean vectors $\mathbf{a} = (a_1a_2a_3a_4) = (1010)$ *and* $\mathbf{b} = (b_1b_2b_3b_4) = (0111)$. *These vectors are incomparable, because* $a_1 > b_1$ $(1 > 0)$, $a_3 = b_3$ $(1 = 1)$, $a_2 < b_2$ $(0 < 1)$, *and* $a_4 < b_4$ $(0 < 1)$.

Monotonically increasing Boolean functions

If a Boolean function f meets the requirement

$$f(\mathbf{a}) \geq f(\mathbf{b}) \tag{7.3}$$

for any vectors $\mathbf{a}$ and $\mathbf{b}$ such that $\mathbf{a}$ is equal to or greater than $\mathbf{b}$, $\mathbf{a} \geq \mathbf{b}$, then f is a *monotonically increasing* function.

Example 7.6 **(Monotonically increasing functions.)** *Figure 7.2 contains the monotonically increasing functions of two variables: 0, x_1, x_2, x_1x_2, $x_1 \vee x_2$, and 1. Note that x_2 is a monotonically increasing function as well, since $f(x_2 = 0) < f(x_2 = 1)$.*

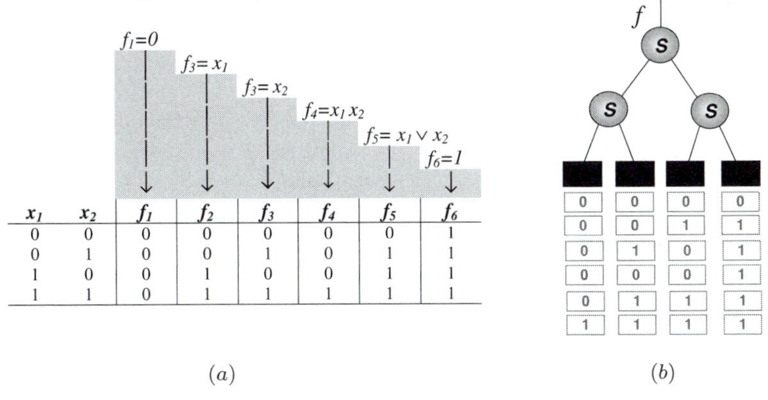

x_1	x_2	f_1	f_2	f_3	f_4	f_5	f_6
0	0	0	0	0	0	0	1
0	1	0	0	1	0	1	1
1	0	0	1	0	0	1	1
1	1	0	1	1	1	1	1

FIGURE 7.2
Monotonically increasing functions of two variables: truth table (a) and decision tree (b) (Example 7.6).

A logic network that implements a monotonic function is called *a monotonic network*.

Properties of monotonically increasing functions

- ▶ A logic product or sum of monotonically increasing function is a monotonically increasing function as well.
- ▶ A monotonically increasing function is described by an SOP *without complemented literals*. That is, they are implemented using AND and OR operations on non-complemented variables and the constant functions 0 and 1 as input variables.

Practice problem 7.5 (Monotonically increasing functions.) Prove that the logic networks consisting of AND and OR gates implement monotonically increasing functions.

Answer: The proof follows from the fact that the logic sum or product of a monotonically increasing functions is itself a monotonically increasing function.

Example 7.7 (Monotonically increasing functions.) *The logic network presented in Figure 7.3 implements a monotonically increasing function.*

Design example: Monotonically increasing functions

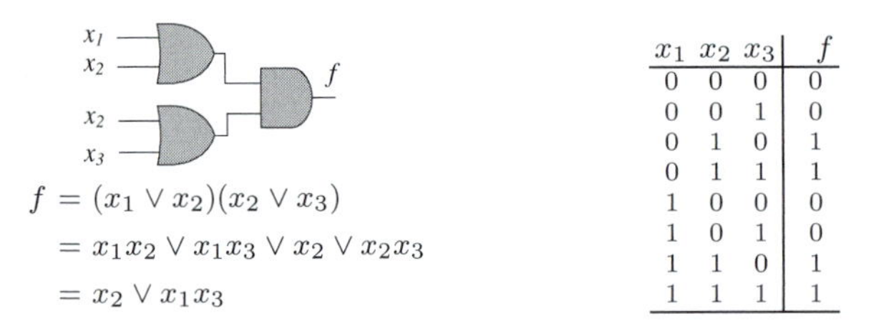

x_1	x_2	x_3	f
0	0	0	0
0	0	1	0
0	1	0	1
0	1	1	1
1	0	0	0
1	0	1	0
1	1	0	1
1	1	1	1

FIGURE 7.3
Monotonically increasing functions of two variables (Example 7.7).

7.3.2 A monotonically decreasing Boolean functions

Monotonically decreasing functions

Let **a** and **b** be Boolean vectors. If f satisfies the requirement $f(\mathbf{a}) \leq f(\mathbf{b})$ for any vectors **a** and **b** such that $\mathbf{a} \geq \mathbf{b}$, then f is a *monotonically decreasing* function.

Example 7.8 (Monotonically decreasing functions.) *Figure 7.4 represents the monotonically decreasing functions of two variables: 0, $\overline{x}_1$, $\overline{x}_2$, $\overline{x}_1\overline{x}_2$, $\overline{x}_1 \vee \overline{x}_2$, and 1.*

The useful properties of monotonically decreasing functions include the following:

Properties of monotonically decreasing functions

- A logic sum or a product of the monotonically decreasing function is a monotonically decreasing function, too.
- A monotonically decreasing function is described by an SOP *with complemented literals*. That is, they are implemented using AND and OR gates with complemented variables and the constant functions 0 and 1 as input variables.

Design example: Monotonically decreasing functions

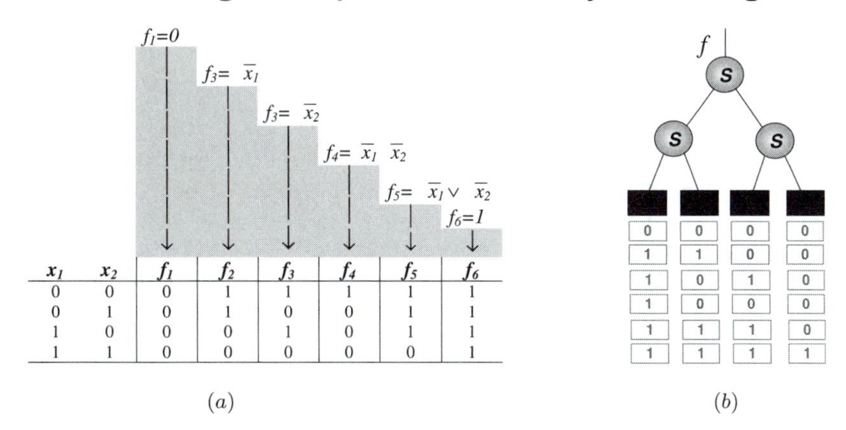

FIGURE 7.4
Monotonically decreasing functions of two variables: truth table (a) and decision tree (b) (Example 7.8).

Example 7.9 (Monotonically decreasing functions.)
Determine if the network given in Figure 7.5 (left) realizes a monotonically decreasing function. The output of the network $f = \overline{x}_1 \vee \overline{x}_2 \vee \overline{x}_3$ *and the truth table reveal that this function is the logic sum of two monotonically decreasing functions,* $\overline{x_1 x_2}$ *and* $\overline{x}_3$*. Therefore, the function, implemented by the network, is monotonically decreasing.*

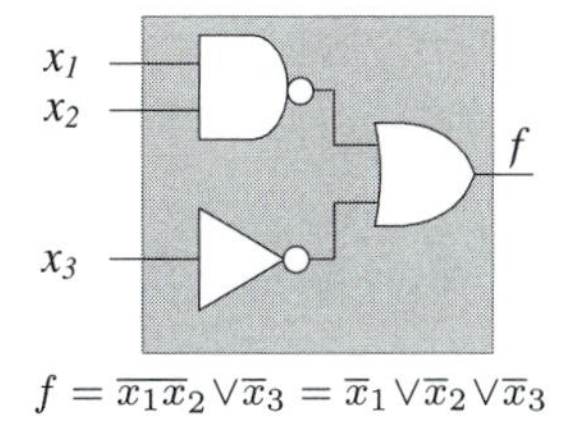

x_1	x_2	x_3	f
0	0	0	1
0	0	1	0
0	1	0	0
0	1	1	0
1	0	0	0
1	0	1	0
1	1	0	0
1	1	1	0

FIGURE 7.5
A logic network that implements a monotonically decreasing Boolean function (Example 7.9).

There are: 2^{n+1} totally symmetric functions of n variables, $n + 2$ totally symmetric functions of n variables which are monotonically increasing and

$n+2$ totally symmetric functions of n variables which are monotonically decreasing, $n+2$ totally symmetric functions of n variables which are monotonically increasing and monotonically decreasing functions.

7.3.3 Unate Boolean functions

A Boolean function f that includes both monotonically increasing and decreasing functions is called a *unate* Boolean function. A Boolean function is unate if and only if the minimal SOP expression contains either the literal x_i or $\overline{x}_i$ but not both.

Example 7.10 (Unate functions.) *The Boolean function $f = x_1x_2 \vee x_3x_4$ is unate. Function $f = x_2x_2x_3 \vee \overline{x}_2x_4$ is not unate since both x_2 and $\overline{x}_2$ appear in the minimal SOP expression.*

Example 7.11 (Unate functions.) *Determine if the network given in Figure 7.6 (left) realizes a unate function. The function $f = \overline{x}_1 \vee \overline{x}_2 \vee x_3$ is unate because the monotonically decreasing function $\overline{x_1x_2}$ and monotonically increasing function x_3 are both contained in f.*

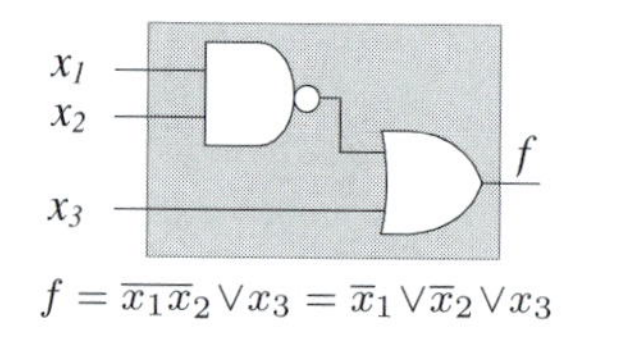

x_1	x_2	x_3	f
0	0	0	1
0	0	1	1
0	1	0	1
0	1	1	1
1	0	0	1
1	0	1	1
1	1	0	0
1	1	1	1

FIGURE 7.6
A logic network that implements a unate Boolean function (Example 7.11).

Practice problem 7.6 (Unate functions.) How many Boolean functions of two variables are unate?
Answer: There are $2^{2^n} = 2^{2^2} = 16$ Boolean functions of two variables and only two non-unate functions, $x_1 \oplus x_2$ and $\overline{x_1 \oplus x_2}$. The other 14 functions are unate functions.

7.4 Linear functions

The property of linearity of a Boolean function is important for the cost-efficient implementations of logic networks; generally, linear functions require less gates if implemented using specific gates. Linearity also greatly simplifies the manipulation and transformation of Boolean functions:

$$\underbrace{\texttt{Boolean functions}}_{\textit{Boolean data structure}} \xrightarrow{\textit{Recognition}} \underbrace{\texttt{Linear Boolean functions}}_{\textit{Boolean data structure}}$$

A linear Boolean function is represented by the equation:

Linear Boolean functions

$$f = r_0 \oplus \bigoplus_{i=1}^{n} r_i x_i = r_0 \oplus r_1 x_1 \oplus \cdots \oplus r_n x_n \tag{7.4}$$

where $r_i \in \{0, 1\}$ is the i-th coefficient, $i = 1, 2, \ldots, n$. It should be noted that Expression 7.4 is a *parity* function if $r_1 = r_2 = \cdots = r_n = 1$.

Example 7.12 (Linear functions.)

(a) *The Boolean function $f = x \vee y$ is non-linear since its polynomial form $f = x \oplus y \oplus xy$ is not linear.*

(b) *The Boolean function $f = xy \vee \overline{x}\,\overline{y}$ is linear, since its polynomial form is linear: $f = 1 \oplus x \oplus y$.*

Practice problem **7.7. (Linear functions.)** Determine whether the Boolean function $f = x_1\overline{x}_2\overline{x}_3 \vee \overline{x}_1 x_2 \overline{x}_3 \vee \overline{x}_1\overline{x}_2 x_3 \vee x_1 x_2 x_3$ is linear.
Answer: *Since $f = x_1 \oplus x_2 \oplus x_3$, the function is a linear one.*

Example 7.13 (Linear functions.) *There are eight linear functions $f(x, y)$ of two variables: $f_0 = 0$, $f_1 = 1$, $f_2 = x$, $f_3 = 1 \oplus x$, $f_4 = y$, $f_5 = 1 \oplus y$, $f_6 = x \oplus y$, and $f_7 = 1 \oplus x \oplus y$. (Figure 7.7).*

Linear expressions have several useful properties and implementations, in particular:

Design example: Linear Boolean functions

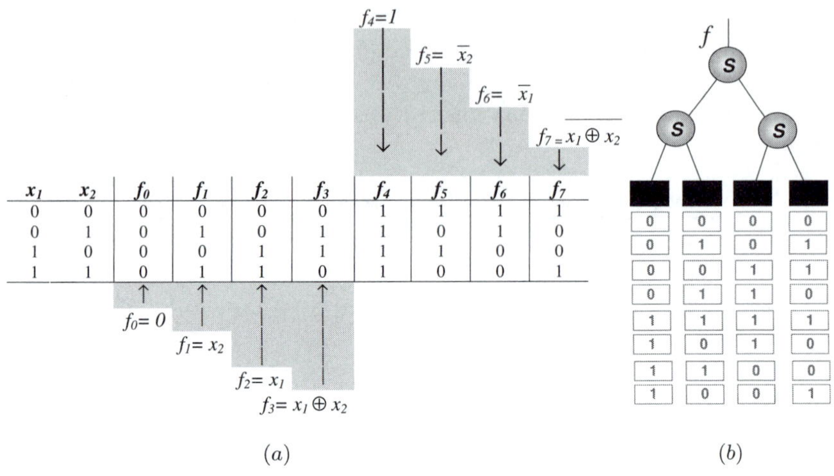

FIGURE 7.7
Linear functions of two variables (Example 7.13).

> **Properties of linear Boolean functions**
>
> ▶ There are 2^{n+1} linear expressions of n variables.
>
> ▶ A linear expression is either a self-dual or a self-anti-dual Boolean function:
>
> $$f = \begin{cases} f(\overline{x}_1, \overline{x}_2, \dots, \overline{x}_n) = \overline{f}(x_1, x_2, \dots, x_n), & \text{if } \sum_{i=0}^{n+1} r_i = 1 \\ f(\overline{x}_1, \overline{x}_2, \dots, \overline{x}_n) = f(x_1, x_2, \dots, x_n), & \text{if } \sum_{i=0}^{n+1} r_i = 0 \end{cases}$$
>
> ▶ A Boolean function obtained by the linear composition of linear expressions is also a linear expression.

7.5 Universal set of functions

If an arbitrary Boolean function is completely represented by a set F of simple Boolean functions $\{f_1, f_2, \dots, f_m\}$, this set of elementary functions is called *universal*, or *complete*. Five major classes of Boolean functions have been defined in the previous sections:

Five classes of Boolean functions

M_0, the set of 0-preserving functions,
M_1, the set of 1-preserving functions,
M_2, the set of self-dual functions,
M_3, the set of monotonically increasing functions, and
M_4, the set of linear functions.

In these classes, a *0-preserving function* is defined as $f(0,\ldots,0) = 0$, and a *1-preserving function* is specified as $f(1,\ldots,1) = 1$.

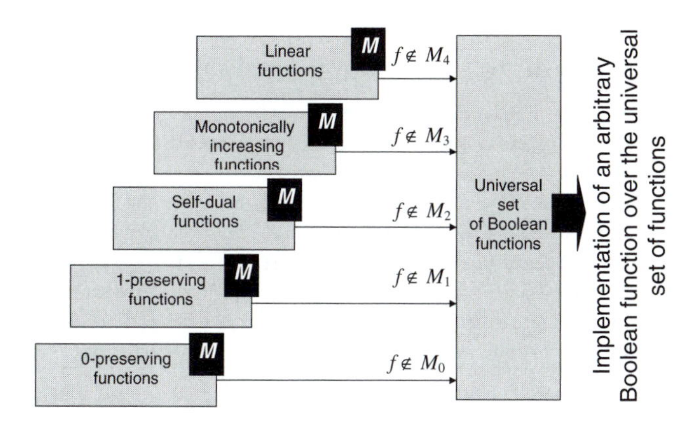

FIGURE 7.8
A universal set of Boolean functions is specified by five classes of Boolean functions.

Example 7.14 (Universal set of functions.) *Functions* $f_1 = x$, $f_2 = x_1 \vee x_2$, *and* $f_3 = x_1x_2$ *are 0- and 1-preserving functions, because* $f_1(0) = 0$, $f_1(1) = 1$, $f_2(0,0) = 0 \vee 0 = 0$, $f_2(1,1) = 1 \vee 1 = 1$, $f_3(0,0) = 0 \cdot 0 = 0$, *and* $f_3(1,1) = 1 \cdot 1 = 1$.

Post theorem

A logic network can be considered as an interconnection of simple logic networks called *logic primitives*. The function of the network can be interpreted as performing a mathematical composition with operations corresponding to the logic primitives. The problem of completeness of logical primitives in a network can be transferred to the problem of the functionally completeness of the set of operations by which it is possible to implement an arbitrary Boolean function using a finite number of of logical primitives. The *Post theorem* provides a solution to this problem:

Post theorem on the functionally complete set of Boolean functions

The set of Boolean functions F is *functionally complete*, or *universal*, if it includes (Figure 7.8):

- ▶ At least one function that is not monotonically increasing,
- ▶ At least one function that is not self-dual,
- ▶ At least one function that is not linear,
- ▶ At least one function that is not 0-preserving,
- ▶ At least one function that is not 1-preserving.

Procedure to determine functional completeness

Given: `A set of Boolean functions`
Step 1. `Find at least one function that is not monotonically increasing.`
Step 2. `Find at least one function that is not self-dual.`
Step 3. `Find at least one function that is not linear.`
Step 4. `Find at least one function that is not 0-preserving.`
Step 5. `Find at least one function that is not 1-preserving.`
Output: `Identification of functional completeness if at least one of the above functions has been found in this set`

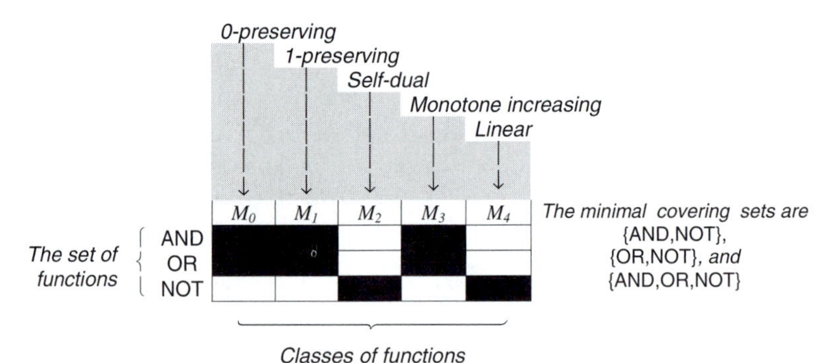

FIGURE 7.9
The minimal universal set for Boolean functions AND, OR, and NOT (Example 7.15).

Example 7.15 (Universal set of functions.) *The set of functions AND, OR, and NOT is a universal set because (Figure 7.9) the AND is a 0- and 1-preserving function, but neither a self-dual nor a linear function; the OR is a 0- and 1-preserving function, but neither a self-dual nor a linear Boolean function; the NOT is neither a 0- nor a 1-preserving function.*

Practice problem 7.8 (Universal set of functions.) Find the minimal universal sets given the following Boolean functions: $f_1 = \overline{x}_1\overline{x}_2$, $f_2 = x_1\overline{x}_2$, $f_3 = x_1 \vee \overline{x}_2$, $f_4 = x_1 \oplus \overline{x}_2, f_5 = 1$, $f_6 = 0$, $f_7 = x_1x_2 \vee x_2x_3 \vee x_1x_3, f_8 = x_1 \oplus x_2 \oplus x_3$, $f_9 = \overline{x}_1$, $f_{10} = x_1x_2$, $f_{11} = x_1$.
Answer is given in "Solutions to practice problems."

7.6 Summary of properties of Boolean functions

This chapter introduces the most important properties of Boolean functions: self-duality, monotonicity, unity, linearity, and symmetry. These properties are summarized as follows:

- A ***self-dual*** Boolean function f of n variables is based on the duality principle of Boolean algebra that satisfies the requirement $f = \overline{f}(\overline{x}_1, \ldots, \overline{x}_n)$.
- If a Boolean function f satisfies the requirement $f(\mathbf{a}) \geq f(\mathbf{b})$ for any vectors $\mathbf{a}$ and $\mathbf{b}$ such that $\mathbf{a}$ is equal to or greater than $\mathbf{b}$, $\mathbf{a} \geq \mathbf{b}$, then f is a ***monotonically increasing*** function.
- If a Boolean function f satisfies the requirement $f(\mathbf{a}) \leq f(\mathbf{b})$ for any vectors $\mathbf{a}$ and $\mathbf{b}$ such that $\mathbf{a} \geq \mathbf{b}$, then f is a ***monotonically decreasing*** function.
- A Boolean function f that includes both monotonically increasing and decreasing functions is called a ***unate*** Boolean function.
- A ***linear*** Boolean function is represented by the equation $a_0 \oplus a_1x_1 \oplus \cdots \oplus a_nx_n$.
- If an arbitrary Boolean function is represented by a set F of simple Boolean functions $\{f_1, f_2, \ldots, f_m\}$, this set of elementary functions is called ***universal***, or ***functionally complete***, if and only if it consists of at least one function that is not monotonically increasing, at least one function that is not self-dual, at least one function that is not linear, at least one function that is not 0-preserving, and at least one function that is not 1-preserving.

7.7 Further study

Historical perspective

In 1941, E. L. Post developed an approach to forming complete sets of elementary Boolean functions (primitives) by which it is possible to represent an arbitrary Boolean function (E. L. Post, "Two-Valued Iterative Systems of Mathematical Logic," Princeton University Press, Princeton, New Jersey, 1941). This approach is known as the *Post theorem.*

Further reading

Theory and Design of Switching Circuits by A. D. Friedman and P. R. Menon, Computer Science Press, Woodland Hills, California, 1975.

Modern Switching Theory and Digital Design by S. C. Lee, Prentice-Hall, Englewood Cliffs, New Jersey, 1978.

7.8 Solutions to practice problems

Practice problem	Solution

7.3.

x_1	x_1	x_1	f_1	f_2	f_3	f_4	f_5	f_6	f_7	f_8
0	0	0	0	0	0	0	0	0	0	0
0	0	1	0	0	0	0	1	1	1	1
0	1	0	0	0	1	1	0	0	1	1
0	1	1	0	1	0	1	0	1	0	1
1	0	0	1	0	0	0	0	0	0	0
1	0	1	1	0	0	0	1	1	1	1
1	1	0	1	0	1	1	0	0	1	1
1	1	1	1	1	0	1	0	1	0	1

7.4.

$$f(x_1, x_2) = x_1x_2 \vee \overline{x_1x_2} = \overline{x_1 \oplus x_2} = x_1 \oplus x_2 \oplus 1$$
$$= (\overline{x}_1 \oplus 1) \oplus (\overline{x}_2 \oplus 1) \oplus 1 = \overline{x}_1 \oplus \overline{x}_2 \oplus 1 = \overline{\overline{x}_1 \oplus \overline{x}_2} = f(\overline{x}_1, \overline{x}_2)$$

7.8.

$\{f_1\}, \{f_2, f_3\}, \{f_2, f_5\}, \{f_3, f_4\}, \{f_3, f_6\}, \{f_4, f_5, f_7\}, \{f_5, f_6, f_7, f_8\}$

	M_0	M_1	M_2	M_3	M_4
$f_1 = \overline{x}_1\ \overline{x}_2$					
$f_2 = x_1\ \overline{x}_2$					
$f_3 = x_1 \vee \overline{x}_2$					
$f_4 = x_1 \oplus x_2$					
$f_5 = 1$					
$f_6 = 0$					
$f_7 = x_1\ x_2 \vee x_2\ x_3 \vee x_1\ x_3$					
$f_8 = x_1 \oplus x_2 \oplus x_3$					
$f_9 = \overline{x}_1$					
$f_{10} = x_1\ x_2$					
$f_{11} = x_1$					

7.9 Problems

Problem 7.1 Which of the following are linear Boolean functions?

$(a)\quad f = x_1x_2$
$(b)\quad f = x_1x_2x_3 \vee \overline{x}_1x_2 \vee \overline{x}_1\overline{x}_2\overline{x}_3$
$(c)\quad f = \overline{x}_1\overline{x}_2\overline{x}_3 \vee \overline{x}_1x_2x_3 \vee x_1\overline{x}_2x_3 \vee x_1x_2\overline{x}_3$
$(d)\quad f = (x_1 \vee x_2 \vee \overline{x}_3)(x_1 \vee \overline{x}_2 \vee x_3)(\overline{x}_1 \vee x_2 \vee \overline{x}_3)(\overline{x}_1 \vee \overline{x}_2 \vee x_3)$

Problem 7.2 Which of the following set of functions are universal?

$(a)\ f_1 = x_1\overline{x}_2,\ f_2 = x_1 \oplus x_2$
$(b)\ f_1 = \overline{x}_1,\ f_2 = \overline{x}_2,\ f_3 = x_1 \oplus x_2$
$(c)\ f_1 = \overline{x}_1,\ f_2 = \overline{x}_2,\ f_3 = x_1x_2$
$(d)\ f_1 = \overline{x}_1,\ f_2 = \overline{x}_2,\ f_3 = x_1 \vee x_2$
$(e)\ f_1 = 0,\ f_2 = x_1x_2,\quad f_3 = \overline{x_1 \oplus x_2}$
$(f)\ f_1 = 0,\ f_2 = x_1 \vee x_2$
$(g)\ f_0 = x_1 \oplus x_2,\ f_2 = x_1x_2\ \ f_3 = \overline{x_1 \oplus x_2}$

Problem 7.3 Which of the Boolean functions of three variables are self-dual, monotonic, or/and symmetric, given the following standard SOP expressions:
(a) $f = \bigvee m(0,7)$ (b) $f = \bigvee m(5,6,7)$ (c) $f = \bigvee m(0,1,2)$

Problem 7.4 Which of the Boolean functions are self-dual, monotonic, or/and symmetric, given the following standard POS expressions?
(a) $f = \prod M(1,2,3,4,5)$ (b) $f = \prod M(0,1,4,5)$ (c) $f = \prod M(6,7)$

Problem 7.5 Convert (if possible) the following logic networks into logic networks using: NAND gates only, NOR gates only, OR and AND gates only, and EXOR and AND gates only. Assume, that complemented inputs are allowed.

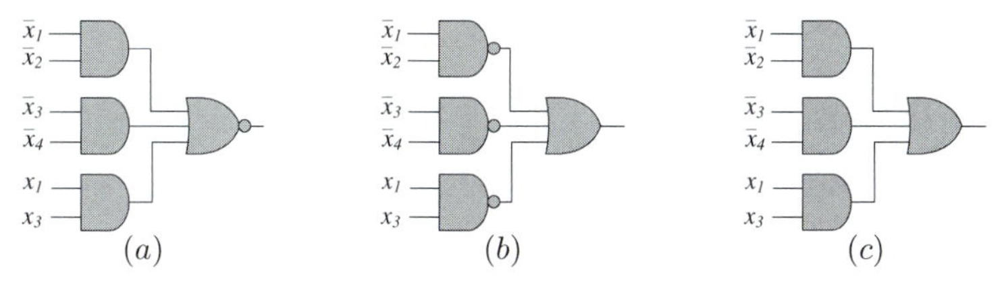

Problem 7.6 Recognize which of the Boolean functions are linear, given the following K-maps:

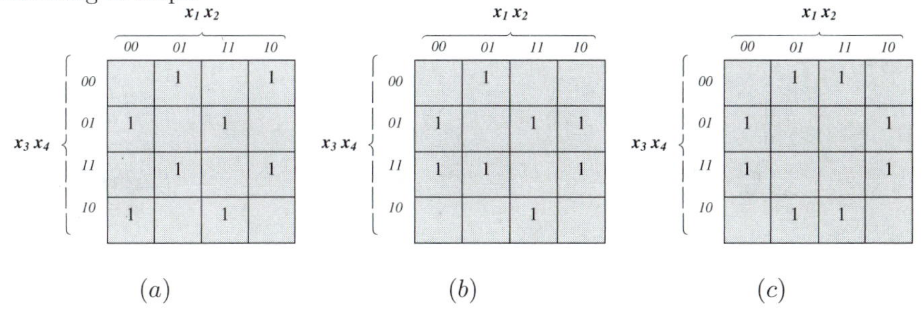

Problem 7.7 Using the K-maps given in Problem 7.6 implement the corresponding Boolean functions using AND and EXOR gates only (assume that constant "1" input is allowed).

Problem 7.8 Recognize which of the Boolean functions are self-dual, monotonic, and symmetric, given the following truth tables:

x_1	x_2	x_3	x_4	f	x_1	x_2	x_3	x_4	f
0	0	0	0	**0**	1	0	0	0	**1**
0	0	0	1	**0**	1	0	0	1	**1**
0	0	1	0	**0**	1	0	1	0	**1**
0	0	1	1	**0**	1	0	1	1	**1**
0	1	0	0	**1**	1	1	0	0	**1**
0	1	0	1	**1**	1	1	0	1	**1**
0	1	1	0	**1**	1	1	1	0	**1**
0	1	1	1	**1**	1	1	1	1	**1**

(*a*)

x_1	x_2	x_3	x_4	f	x_1	x_2	x_3	x_4	f
0	0	0	0	**0**	1	0	0	0	**1**
0	0	0	1	**0**	1	0	0	1	**1**
0	0	1	0	**1**	1	0	1	0	**0**
0	0	1	1	**1**	1	0	1	1	**0**
0	1	0	0	**0**	1	1	0	0	**1**
0	1	0	1	**1**	1	1	0	1	**0**
0	1	1	0	**1**	1	1	1	0	**0**
0	1	1	1	**0**	1	1	1	1	**1**

(*b*)

Problem 7.9 Implement the following Boolean functions, given by their truth tables, using NAND gates only:

x_1	x_2	x_3	f
0	0	0	**1**
0	0	1	**0**
0	1	0	**0**
0	1	1	**1**
1	0	0	**0**
1	0	1	**1**
1	1	0	**1**
1	1	1	**0**

(*a*)

x_1	x_2	x_3	f
0	0	0	**1**
0	0	1	**0**
0	1	0	**0**
0	1	1	**1**
1	0	0	**1**
1	0	1	**0**
1	1	0	**0**
1	1	1	**1**

(*b*)

x_1	x_2	x_3	f
0	0	0	**1**
0	0	1	**0**
0	1	0	**0**
0	1	1	**0**
1	0	0	**0**
1	0	1	**0**
1	1	0	**0**
1	1	1	**1**

(*c*)

Problem 7.10 Recognize which of the Boolean functions are self-dual, monotonic, and symmetric, given the following logic networks:

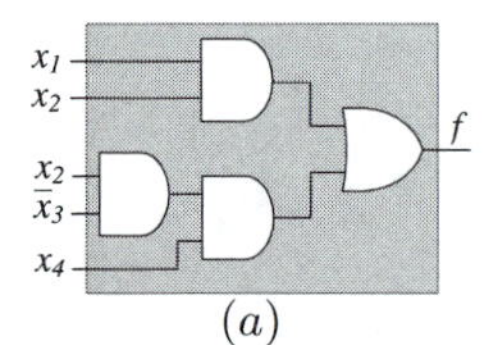

(*a*)

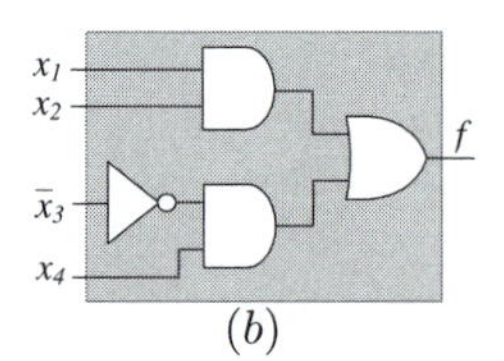

(*b*)

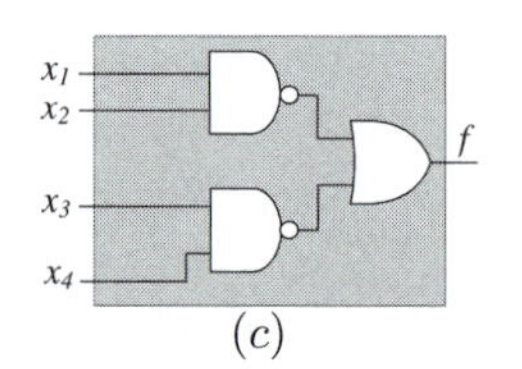

(*c*)

Problem 7.11 Recognize which of the Boolean functions are self-dual, monotonic, and symmetric, given the following decision diagrams:

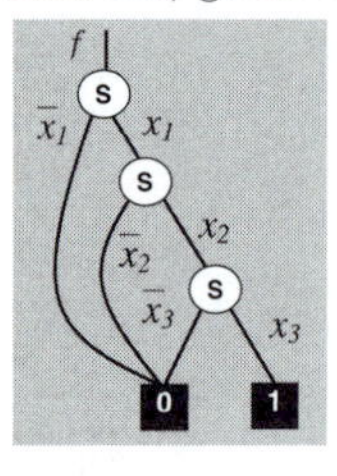

(*a*)

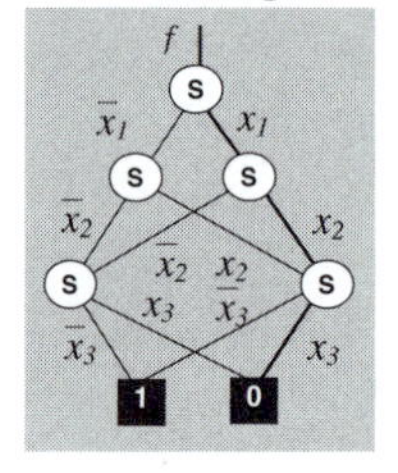

(*b*)

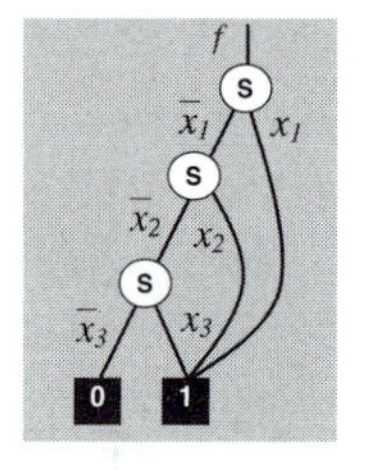

(*c*)

8

Optimization I: Algebraic and K-maps

Manipulation techniques

- Simplification techniques
- Conversion into logic networks
- Factoring of Boolean expressions

Minimization using K-maps

- K-map structure
- Using K-maps
- Techniques of minimization
- Relationships with other data structures

Quine-McCluskey algorithm

- Prime implicants
- The prime implicant chart
- Minimum covering problem

Advanced topics

- Multi-level optimization
- Heuristic optimization
- Optimization and testability

8.1 Introduction

In Chapter 6, Boolean functions were considered in various forms called *Boolean data structures*, divided into *algebraic* and *graphical* representations. Algebraic representations include algebraic forms based on Boolean formulas, matrix forms, and tabulated forms. Graphical representations include K-maps, hypercubes, logic networks over the library of logic gates, networks of threshold elements, and decision diagrams.

Optimization of Boolean data structure

Both algebraic and graphical forms are often not optimal; that is, they are redundant in their representation of a given Boolean function. The process of removing *redundancy* in Boolean data structures is called the *optimization*, or *minimization* of Boolean functions.

Since optimization at various stages of design is critical for further implementation steps, the following tools are required:

- Efficient optimization algorithms, and
- Efficient manipulation and conversion between data structures.

In this chapter, approaches to the minimization of Boolean data structures given in SOP and POS forms are introduced. These minimization techniques result, respectively, in two-level AND-OR and OR-AND logic networks. Techniques based on K-maps, hypercubes (Quine-McCluskey method), and decision diagrams are the focus of this chapter.

8.2 Minterm and maxterm expansions

In Chapters 4 and 6, SOP and POS expansions were introduced as algebraic data structures for the representation (description) of Boolean functions. In the construction of these SOP and POS expansions, minterms and maxterms, respectively, were used.

A Boolean function f, written as a sum of minterms, is referred to as a *minterm expansion*, or a *standard sum-of-products (SOP)*. Let a Boolean function f of n variables be specified by its truth-vector $\mathbf{F}$. The conversion of $\mathbf{F}$ into a minterm expansion means detecting 1s in the truth table and identifying the corresponding variable assignments.

Example 8.1 (SOP form.) *Given the truth-vector* $\mathbf{F} = [0101]^T$, *the shorthand notation of the SOP is* $f = \bigvee m(1,3)$, *since the truth-vector* $\mathbf{F}$ *contains two ones, corresponding to the assignments* $01_2 = 1_{10}$ *and* $11_2 = 3_{10}$ *of the minterms* $x_1^0 x_2^1 = \overline{x}_1 x_2$ *and* $x_1^1 x_2^1 = x_1 x_2$.

Practice problem 8.1. (SOP form.) Given the truth-vector $\mathbf{F} = [00010111]^T$, find the corresponding SOP and its shorthand notation.
Answer: The four 1s in this truth-vector correspond to the following variable assignments: 011, 101, 110 and 111. Therefore, the SOP contains four minterms: $f = \overline{x}_1 x_2 x_3 \vee x_1 \overline{x}_2 x_3 \vee x_1 x_2 \overline{x}_3 \vee x_1 x_2 x_3$. The shorthand SOP form includes the four assignments in decimal notation: $f = \bigvee m(3,5,6,7)$.

A *maxterm expansion* or *standard product of sums (POS)* is the representation of a Boolean function as a product of maxterms. Let the Boolean function f of n variables be specified by its truth-vector $\mathbf{F}$. To convert $\mathbf{F}$ into a maxterm expansion, the variable assignments corresponding to 0s in the truth-vector must be identified, and the corresponding maxterms should be generated.

Example 8.2 (POS form.) *Given the truth-vector* $\mathbf{F} = [0101]^T$, *the POS is* $f = \prod M(0,2)$, *since the truth-vector* $\mathbf{F}$ *contains two zeros, corresponding to the assignments* $00_2 = 0_{10}$ *and* $10_2 = 2_{10}$ *of the maxterms* $(x_1^{\overline{0}} \vee x_2^{\overline{0}}) = (x_1 \vee x_2)$ *and* $(x_1^{\overline{1}} \vee x_2^{\overline{0}}) = (\overline{x}_1 \vee x_2)$.

Practice problem 8.2. (POS form.) Given the truth-vector $\mathbf{F} = [00010111]^T$, find the corresponding POS expression.
Answer: The four 0s in this truth-vector correspond to the following variable assignments: 000, 001, 010 and 100. Therefore, the POS contains four maxterms: $f = (x_1 \vee x_2 \vee x_3)(x_1 \vee x_2 \vee \overline{x}_3)(x_1 \vee \overline{x}_2 \vee x_3)(\overline{x} \vee x_2 \vee x_3)$. The SOP form includes the four assignments in decimal notation: $f = \prod M(0,1,2,4)$.

The minterm expression and maxterm expression can be directly converted into logic networks, as shown in Figure 8.1.

Example 8.3 (SOP form and AND-OR network.) *Given the standard SOP expression* $f = x_1 \overline{x}_2 \vee \overline{x}_1 x_2$, *the corresponding logic network is given in Figure 8.2.*

Noncanonical (nonstandard) SOP and POS expressions can be converted to two-level AND-OR and OR-AND networks, respectively, in the same way, except that the products and sums (which may not include all the function's variables) will be implemented instead of minterms and maxterms. This mapping is the most efficient, or least costly in terms of the number of logic

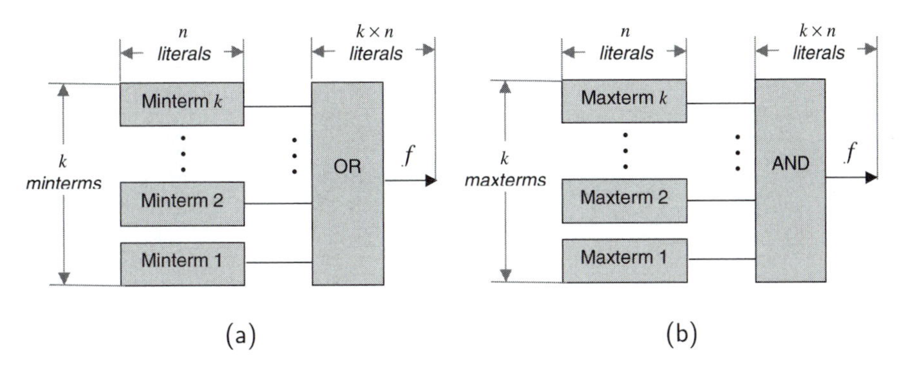

FIGURE 8.1
Converting minterm expressions (a) and maxterm expressions (b) into logic networks.

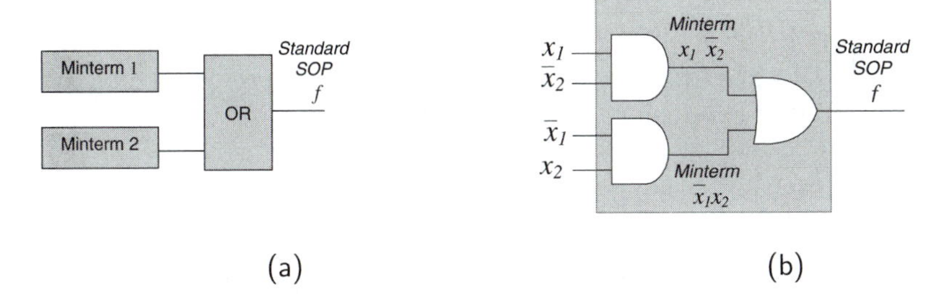

FIGURE 8.2
Mapping of a standard SOP expression into an AND-OR network (Example 8.3).

gates, if the obtained SOP and POS are minimal forms. This will be discussed later in this chapter.

Two-level AND-OR and OR-AND logic networks can be directly converted into the corresponding standard SOP and POS expressions, respectively.

Example 8.4 (Converting a network into a POS expression.) *Figure 8.3a shows an OR-AND network and its conversion into a POS expression, which includes three sums, implemented by the two-input OR gates, combined by the products (three-input OR-gate):* $f = (\overline{x}_1 \vee x_2)(\overline{x}_1 \vee x_3)(x_2 \vee x_3)$. *Note that these sums are not maxterms, since they do not include all three variables.*

Practice problem 8.3. (Converting a network into an SOP form.) Given the network shown in Figure 8.3b, derive the expression this

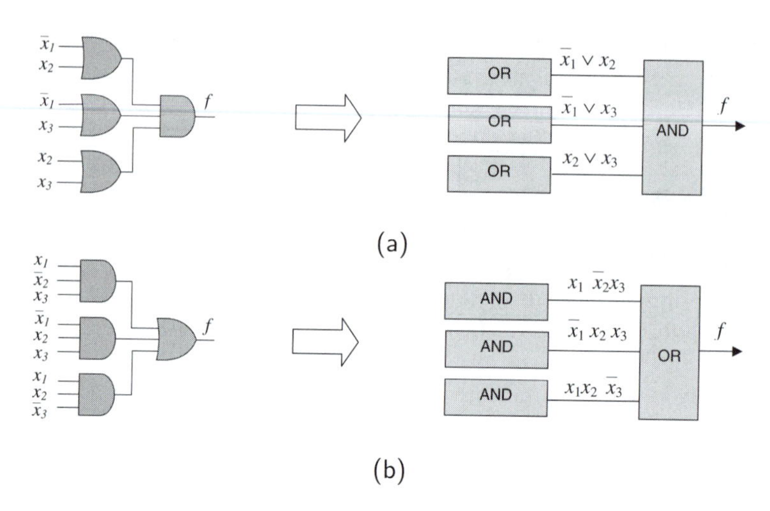

FIGURE 8.3
Converting an OR-AND logic network into (a) a POS expression (Example 8.4) and (b) an SOP expression (Practice problem 8.3).

network implements.
Answer: This is an AND-OR network, and the corresponding expression is the SOP: $f = x_1\overline{x}_2x_3 \vee \overline{x}_1x_2x_3 \vee x_1x_2\overline{x}_3$, or in short hand notation, $f = \bigvee m(2, 3, 6)$.

8.3 Optimization of Boolean functions in algebraic form

Optimization of algebraic expressions of Boolean functions means their transformation into forms that better meet the implementation criteria (Figure 8.4). The optimization criteria of the SOP and POS forms relate, in particular, to:

- The structure of the algebraic expression,
- The number of terms,
- The number of literals, and
- The fan-in.

The basic principle of minimization is *simplification*, or *reduction*. Techniques for the simplification of algebraic forms of Boolean functions are based on the following actions:

- Combining terms,

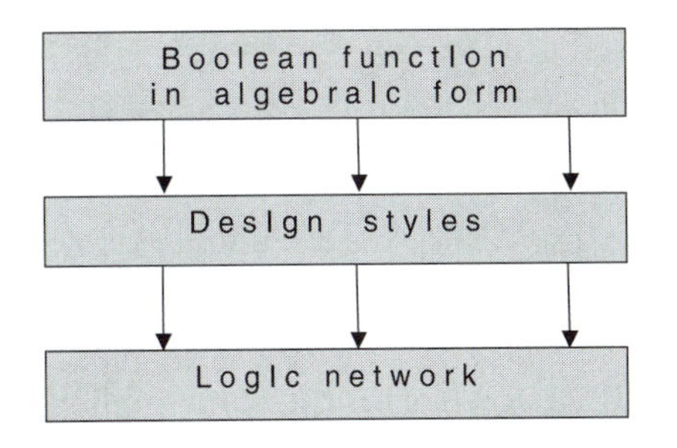

Given:

- A Boolean function in algebraic form, and
- Implementation requirements,

A cost-efficient logic network can be designed using minimization techniques

FIGURE 8.4
The conversion of Boolean functions into logic networks is based on various techniques.

- Eliminating terms,
- Eliminating literals, and
- Adding redundant terms for the efficient elimination of terms and literals.

Note that simplification does not guarantee that the resulting Boolean expression is minimal. Nevertheless, reduction is the first step to obtaining a minimal algebraic expression. The theoretical background of reduction is the consensus theorem.

8.3.1 The consensus theorem

The consensus theorem and its dual states that

The consensus theorem for a Boolean function

$$XY \vee \overline{X}Z \vee \underbrace{YZ}_{\text{Consensus term}} = XY \vee \overline{X}Z \tag{8.1}$$

$$(X \vee Y)(\overline{X} \vee Z)\underbrace{(Y \vee Z)}_{\text{Consensus term}} = (X \vee Y)(\overline{X} \vee Z) \tag{8.2}$$

where the term XZ is called the *consensus term.* This term is redundant and is eliminated. Note that the Equation 8.2 is dual of the consensus theorem.

Practice problem **8.4 (The consensus theorem.)** Prove the consensus theorem.
Answer is given in "Solutions to practice problems."

Example 8.5 (Simplification techniques.) *The Boolean function* $f = x_1x_2x_3x_4 \vee \overline{x}_2x_3x_4x_5 \vee \overline{x}_1\overline{x}_2 \vee x_2x_3\overline{x}_5$ *can be simplified by adding the consensus term* $x_1x_3x_4x_5$ *and using the consensus theorem as follows:*

$$f = x_1x_2x_3x_4 \vee \overline{x}_2x_3x_4x_5 \vee \overline{x}_1\overline{x}_2 \vee x_2x_3\overline{x}_5 \vee x_1x_3x_4x_5$$
$$= \overline{x}_1\overline{x}_2 \vee x_2x_3\overline{x}_5 \vee x_1x_3x_4x_5$$

Practice problem 8.5. (Simplification techniques.) Simplify the given Boolean expression $f = (\overline{x}_1 \vee x_2x_4)(x_1 \vee x_3)(x_2x_4 \vee x_3)$.
Answer: By the consensus theorem, $f = (\overline{x}_1 \vee x_2x_4)(x_1 \vee x_3)$.

Generalization of the consensus theorem is as follows. Let xP_1 and $\overline{x}P_2$ be the Boolean terms where P_1 and P_2 are products of literals. Then the *operator* `CONSENSUS` is defined as

The consensus operator

$$xP_1 \underbrace{\texttt{CONSENSUS}}_{Operator} \overline{x}P_2 = \underbrace{P_1P_2}_{Consensus\ term}$$

Example 8.6 (Consensus operator.) *The consensus operator exists for the following terms:*

$$(a)\quad \boxed{x_1}\ \underbrace{\overline{x}_2x_3}_{P_1}\ \texttt{CONSENSUS}\ \boxed{\overline{x}_1}\ \underset{\uparrow \atop P_2}{x_4} = \underbrace{\overline{x}_2x_3x_4}_{P_1P_2}$$

$$(b)\quad \overline{x}_1\ \boxed{\overline{x}_2}\ \overline{x}_3\ \texttt{CONSENSUS}\ \overline{x}_1\ \boxed{x_2}\ \overline{x}_3 = \overline{x}_1\overline{x}_3$$

The following property of the consensus operator is useful in reducing Boolean functions:

The consensus theorem (generalization)

$$xP_1 \vee \overline{x}P_2 \vee P_1P_2 = xP_1 \vee \overline{x}P_2 \tag{8.3}$$
$$(x \vee P_1)(\overline{x} \vee P_2)(P_1 \vee P_2) = (x \vee P_1)(\overline{x} \vee P_2) \tag{8.4}$$

The consensus theorem (Equation 8.3) and its dual (Equation 8.4) states that the consensus terms P_1P_2 and $P_1 \vee P_2$, respectively, are redundant and can be removed from a Boolean expression.

8.3.2 Combining terms

Two terms can be combined using the *adjacency* rule (see Chapter 4, Table 4.1):

The adjacency rule for a Boolean function

$$XY \vee X\overline{Y} = X$$

The adjacency rule can be used when:

- Two product terms contain the same variables, and
- Exactly one of the variables appears complemented in one term and not in the other.

When applied to an SOP expression, one product term may be duplicated, and each of the duplicates can be combined with two or more other terms. Techniques for combining terms are demonstrated in Table 8.1.

Example 8.7 (Combining terms.) *The Boolean function $f = x_1x_2x_3 \vee \overline{x}_1x_2x_3 \vee x_1\overline{x}_2x_3$ can be simplified by duplicating the product term $x_1x_2x_3$ twice and using the adjacency rule for each of the duplicates and the remaining terms: $f = x_1x_2x_3 \vee \overline{x}_1x_2x_3 \vee x_1x_2x_3 \vee x_1\overline{x}_2x_3 = x_2x_3 \vee x_1x_3$.*

Practice problem **8.6. (Combining terms.)** Simplify the given Boolean expression: $f = x_1x_2x_3x_4 \vee \overline{x}_1x_2\overline{x}_3x_4 \vee x_1x_2\overline{x}_3x_4 \vee \overline{x}_1x_2x_3x_4$.
Answer: Apply the adjacency rule twice: $f = (x_1x_2x_3x_4 \vee \overline{x}_1x_2x_3x_4) \vee (\overline{x}_1x_2\overline{x}_3x_4 \vee x_1x_2\overline{x}_3x_4) = x_2x_3x_4 \vee x_2\overline{x}_3x_4 = x_2x_4$.

8.3.3 Eliminating terms

Redundant terms can be eliminated using the *adjacency* rule (see Chapter 4, Table 4.1) $X \vee XY = X$ and the consensus theorem $XY \vee \overline{X}Z \vee YZ = XY \vee \overline{X}Z$. Techniques for eliminating terms are demonstrated in Table 8.1.

Example 8.8 (Eliminating terms.) *The Boolean function $f = \overline{x}_1x_2\overline{x}_3x_4 \vee \overline{x}_1x_2x_4 \vee x_1x_3x_4$ can be simplified by eliminating the product term $\overline{x}_1x_2\overline{x}_3x_4$: $f = (\overline{x}_1x_2\overline{x}_3x_4 \vee \overline{x}_1x_2x_4) \vee x_1x_3x_4 = \overline{x}_1x_2x_4 \vee x_1x_3x_4$.*

Practice problem **8.7. (Eliminating terms.)** Simplify the given Boolean expression $f = x_1\overline{x}_2\overline{x}_3x_4 \vee x_1\overline{x}_3x_4 \vee x_1\overline{x}_3$.
Answer: Apply elimination twice: $f = (x_1\overline{x}_2\overline{x}_3x_4 \vee x_1\overline{x}_3x_4) \vee x_1\overline{x}_3 = x_1\overline{x}_3x_4 \vee x_1\overline{x}_3 = x_1\overline{x}_3$.

8.3.4 Eliminating literals

Redundant literals can be eliminated by the simplification rule (see Chapter 4, Table 4.1):

TABLE 8.1
Techniques of Boolean function simplification.

Techniques of Boolean function simplification

The rule	Techniques
Combining terms	$(a)\ f = \overbrace{\underbrace{x_1x_2\overline{x}_3}_{X}\underbrace{\overline{x}_4}_{Y} \vee \underbrace{x_1x_2\overline{x}_3}_{X}\underbrace{x_4}_{\overline{Y}}}^{2\ terms} = \overbrace{x_1x_2\overline{x}_3}^{1\ term}$ $(b)\ f = \overbrace{x_1\overline{x}_2x_3 \vee x_1x_2x_3 \vee \overline{x}_1x_2x_3}^{3\ terms}$ $= x_1\overline{x}_2x_3 \vee \underbrace{x_1x_2x_3 \vee x_1x_2x_3}_{Duplicated} \vee \overline{x}_1x_2x_3$ $= \underbrace{x_1\overline{x}_2x_3 \vee x_1x_2x_3}_{Combining} \vee \underbrace{x_1x_2x_3 \vee \overline{x}_1x_2x_3}_{Combining} = \underbrace{x_1x_3 \vee x_2x_3}_{2\ terms}$
Eliminating terms	$(a)\ f = \overbrace{\underbrace{\overline{x}_1x_2}_{X} \vee \overline{x}_1x_2\underbrace{x_3}_{Y}}^{3\ terms} = \overbrace{\overline{x}_1x_2}^{2\ terms}$ $(b)\ f = \overbrace{\underbrace{\overline{x}_1x_2}_{Z}\underbrace{\overline{x}_3}_{\overline{X}} \vee \underbrace{x_2x_4}_{Y}\underbrace{x_3}_{X} \vee \underbrace{\overline{x}_1x_2}_{Z}x_4}^{3\ terms}$ $= \underbrace{\overline{x}_1x_2}_{Z}\underbrace{\overline{x}_3}_{\overline{X}} \vee \underbrace{x_2x_4}_{Y}\underbrace{x_3}_{X} \vee \underbrace{\overline{x}_1x_2}_{Z}\overbrace{x_2x_4}^{Y} = \underbrace{\overline{x}_1x_2\overline{x}_3 \vee x_2x_3x_4}_{2\ terms}$
Eliminating literals	$f = \overbrace{\overline{x}_1x_2 \vee \overline{x}_1\overline{x}_2\overline{x}_3\overline{x}_4 \vee x_1x_2x_3\overline{x}_4}^{10\ literals}$ $= \overline{x}_1(x_2 \vee \overline{x}_2\overline{x}_3\overline{x}_4) \vee x_1x_2x_3\overline{x}_4$ $= \overline{x}_1(x_2 \vee \overline{x}_3\overline{x}_4) \vee x_1x_2x_3\overline{x}_4$ $= x_2(\overline{x}_1 \vee x_1x_3\overline{x}_4) \vee \overline{x}_1\overline{x}_3\overline{x}_4$ $= x_2(\overline{x}_1 \vee x_3\overline{x}_4) \vee \overline{x}_1\overline{x}_3\overline{x}_4$ $= \underbrace{\overline{x}_1x_2 \vee x_2x_3\overline{x}_4 \vee \overline{x}_1\overline{x}_3\overline{x}_4}_{8\ literals}$
Adding redundant terms	$f = \overbrace{x_1x_2 \vee x_2x_3 \vee \overline{x}_2\overline{x}_4 \vee x_1\overline{x}_3\overline{x}_4}^{4\ terms}$ $= x_1x_2 \vee x_2x_3 \vee \overline{x}_2\overline{x}_4 \vee x_1\overline{x}_3\overline{x}_4 \vee \underbrace{x_1\overline{x}_4}_{Adding}$ $= x_1x_2 \vee x_2x_3 \vee \overline{x}_2\overline{x}_4 \vee \underbrace{x_1\overline{x}_3\overline{x}_4 \vee x_1\overline{x}_4}_{Eliminating}$ $= \underbrace{x_1x_2 \vee x_2x_3 \vee \overline{x}_2\overline{x}_4}_{3\ terms}$

Eliminating literals in a Boolean function

$$X \vee \overline{X}Y = X \vee Y$$

Techniques for eliminating literals are demonstrated in Table 8.1.

Example 8.9 (Eliminating literals.) *The Boolean function* $f = x_1x_2 \vee x_1\overline{x}_2x_3$ *can be simplified by eliminating the literal* $\overline{x}_2$*:* $f = x_1x_2 \vee x_1\overline{x}_2x_3 = x_1x_2 \vee x_1x_3$.

Practice problem **8.8. (Eliminating terms.)** Simplify the given Boolean function $f = x_1 \vee \overline{x}_1x_3x_4 \vee \overline{x}_1x_2$.

Answer: Apply literal elimination twice: $f = (x_1 \vee \overline{x}_1x_3x_4) \vee (x_1 \vee \overline{x}_1x_2)$ $= x_1 \vee x_3x_4 \vee x_2$.

8.3.5 Adding redundant terms

Redundant terms can be useful for the efficient combining and elimination of terms and literals. Redundant terms do not change the value of a Boolean function. They can be incorporated into Boolean expressions in various ways; in particular:

- By adding $x\overline{x}$, and
- By multiplying by $x \vee \overline{x}$.

Techniques for adding redundant terms are demonstrated in Table 8.1.

Example 8.10 (Adding redundant terms.) *Adding* x_2x_3 *to the expression* $x_1x_2 \vee \overline{x}_1x_3$ *does not change the expression:*

$$\begin{aligned} x_1x_2 \vee \overline{x}_1x_3 \vee x_2x_3 &= x_1x_2 \vee \overline{x}_1x_3 \vee (\overline{x}_1 \vee x_1)x_2x_3 \\ &= x_1x_2 \vee \overline{x}_1x_3 \vee \overline{x}_1x_2x_3 \vee x_1x_2x_3 \\ &= (x_1x_2 \vee x_1x_2x_3) \vee (\overline{x}_1x_3 \vee \overline{x}_1x_2x_3) \\ &= x_1x_2 \vee \overline{x}_1x_3 \end{aligned}$$

Adding $x_1x_2x_3$ *to the term* x_1 *does not change the term:*

$$f = x_1 \vee x_1x_2x_3 = x_1 \cdot (1 \vee x_2x_3) = x_1 \cdot 1 = x_1$$

Practice problem **8.9 (Simplification techniques.)** Simplify the following Boolean functions:

(*a*) $f = \overline{x}_1x_2x_3 \vee \overline{x}_1$

(*b*) $f = (\overline{x}_1 \vee \overline{x}_2x_3 \vee x_4 \vee x_5x_6)(x_1 \vee \overline{x}_2x_3 \vee \overline{x_3 \vee x_5x_6})$

(*c*) $f = (x_1x_2 \vee x_3)(\overline{x}_2x_4 \vee \overline{x}_3\overline{x}_5) \vee \overline{x_1x_2 \vee x_3}$

Answers are given in "Solutions to practice problems."

The formal basis of the verification is proving the validity of Boolean data structures (Chapter 4). In the example below, the techniques for proving validity of Boolean functions is extended by using consensus theorem and elimination rule.

Example 8.11 (Proving the validity.) *Prove that*

$$\underbrace{\overline{x}_1 x_2 \overline{x}_4 \vee x_2 x_3 x_4 \vee x_1 x_2 \overline{x}_3 \vee x_1 \overline{x}_2 x_4}_{Left\ part} = \underbrace{x_2 \overline{x}_3 \overline{x}_4 \vee x_1 x_4 \vee \overline{x}_1 x_2 x_3}_{Right\ part}$$

Proving using algebraic manipulations include: (a) adding three consensus terms, (b) combining terms, and (c) elimination terms to the left part as follows:

$$\text{LEFT PART} \vee \underbrace{x_2 \overline{x}_3 \overline{x}_4 \vee \overline{x}_1 x_2 x_3 \vee x_1 x_2 x_4}_{Consensus\ terms}$$
$$= x_1 x_4 \vee \underbrace{\overline{x}_1 x_2 \overline{x}_4 \vee x_2 x_3 x_4 \vee x_1 x_2 \overline{x}_3}_{Eliminated\ terms} \vee x_2 \overline{x}_3 \overline{x}_4 \vee \overline{x}_1 x_2 x_3$$
$$= x_2 \overline{x}_3 \overline{x}_4 \vee x_1 x_4 \vee \overline{x}_1 x_2 x_3$$

8.4 Implementing SOP expressions using logic gates

Let a Boolean function be represented in SOP form, and let AND, OR, NOR, and NAND gates – a *library* of gates – be available for the design of logic networks. The problem is formulated as mapping, or implementing, a function in the form of a logic network using a library of gates. In Chapter 4, a set of logic operations was called *functionally complete* if any Boolean function could be expressed in terms of the operations from this set.

Example 8.12 (Functional completeness.) *The set AND, OR, and NOT is functionally complete. Indeed, an SOP form of a Boolean function can be realized using only AND, OR, and NOT operations. This means that, given a Boolean function in its SOP form, the function can be implemented as a logic network of AND, OR and NOT gates.*

Some single Boolean operations, such as NAND or NOR, form a functionally complete set in themselves.

Example 8.13 (Functional completeness.) *An arbitrary Boolean function can be implemented by using only NAND gates or only NOR gates.*

Practice problem **8.10. (Library of gates).** Assuming that only NAND gates are available, implement the function $f = x_1x_2 \vee x_2x_3$.
Answer is given in "Solutions to practice problems."

8.4.1 Two-level logic networks

A two-level logic network composed of AND and OR gates can be converted into a network composed of NAND gates or NOR gates, or combinations of AND, OR, NAND, and NOR gates. These conversions are carried out by manipulating Boolean functions using the property $x = \overline{\overline{x}}$ and by applying DeMorgan's rule:

DeMorgan's rule

$$x_1 \vee x_2 \vee \ldots \vee x_m = \overline{\overline{x_1 \vee x_2 \vee \ldots \vee x_m}} = \overline{\overline{x_1}\overline{x_2}\ldots\overline{x_m}} \quad (8.5)$$

$$x_1x_2\ldots x_m = \overline{\overline{x_1x_2\ldots x_m}} = \overline{\overline{x}_1 \vee \overline{x}_2 \vee \ldots \vee \overline{x}_m}. \quad (8.6)$$

Example 8.14 (AND-OR networks.) *Consider a SOP* $f = x_1 \vee x_2\overline{x}_2x_3 \vee \overline{x}_2x_3x_4$ *This expression can be implemented using a two-level AND-OR network. Manipulation of this expression results in another two-level expressions, corresponding to the networks shown in Table 8.2.*

The algorithm for designing a two-level NAND-only logic network is given below.

Algorithm
for designing a minimal two-level NAND-only logic network

Step 1: `Find a minimal SOP for the Boolean function` f.
Step 2: `Draw the corresponding two-level AND-OR logic network.`
Step 3: `Replace AND and OR gates with NAND gates leaving the gate interconnections unchanged.`
Step 4: `If the output gate has any single literals as inputs, complement these literals.`

The algorithm for designing a two-level NOR-only logic network is as follows.

Algorithm
for designing a minimal two-level NOR-only logic network

Step 1: `Find a minimal POS for the Boolean function` f.
Step 2: `Draw the corresponding two-level OR-AND logic network.`
Step 3: `Replace OR and AND gates with NOR gates leaving the gate interconnections unchanged.`
Step 4: `If the output gate has any single literals as inputs, complement these literals.`

TABLE 8.2
Techniques for manipulating Boolean functions in two-level logic network design.

Design example: Two-level network design

Conversion	Two-level logic network
(a) AND-OR $f = x_1 \vee x_2\overline{x}_2x_3 \vee \overline{x}_2x_3x_4$ **(b) OR-AND** $f = (x_1 \vee x_2 \vee x_3)(x_1 \vee \overline{x}_2 \vee \overline{x}_3)(x_1 \vee \overline{x}_3 \vee x_4)$	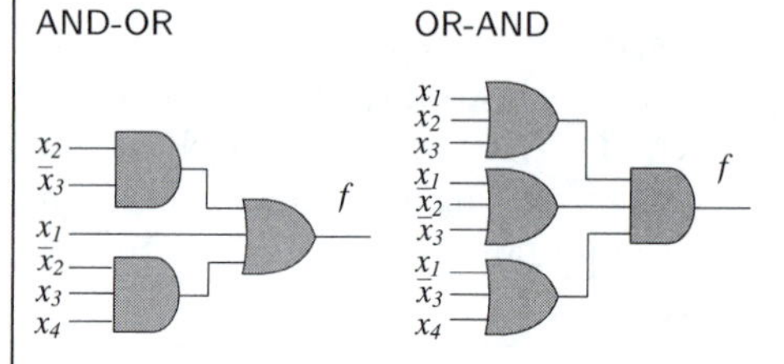
(c) NAND $f = x_1 \vee x_2\overline{x}_2x_3 \vee \overline{x}_2x_3x_4$ $= \overline{\overline{x_1 \vee x_2\overline{x}_2x_3 \vee \overline{x}_2x_3x_4}}$ $= \overline{\overline{x}_1\,\overline{x_2\overline{x}_3}\,\overline{\overline{x}_2x_3x_4}}$ **(d) NOR** $f = (x_1 \vee x_2 \vee x_3)(x_1 \vee \overline{x}_2 \vee \overline{x}_3)(x_1 \vee \overline{x}_3 \vee x_4)$ $= \overline{\overline{(x_1 \vee x_2 \vee x_3)(x_1 \vee \overline{x}_2 \vee \overline{x}_3)(x_1 \vee \overline{x}_3 \vee x_4)}}$ $= \overline{\overline{(x_1 \vee x_2 \vee x_3)} \vee \overline{(x_1 \vee \overline{x}_2 \vee \overline{x}_3)} \vee \overline{(x_1 \vee \overline{x}_3 \vee x_4)}}$	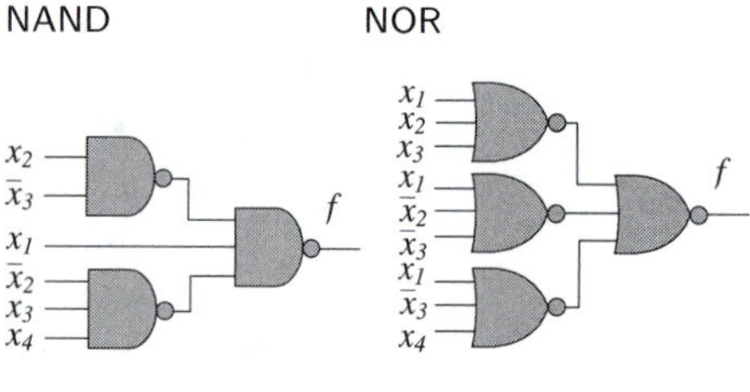
(e) OR-NAND $f = x_1 \vee x_2\overline{x}_2x_3 \vee \overline{x}_2x_3x_4$ $= \overline{\overline{x_1 \vee x_2\overline{x}_2x_3 \vee \overline{x}_2x_3x_4}}$ $= \overline{\overline{x}_1\,\overline{x_2\overline{x}_3}\,\overline{\overline{x}_2x_3x_4}}$ $= \overline{\overline{x}_1(\overline{x}_2 \vee x_3)(x_2 \vee \overline{x}_3 \vee \overline{x}_4)}$ $= x_1 \vee \overline{(\overline{x}_2 \vee x_3)} \vee \overline{x_2 \vee \overline{x}_3 \vee \overline{x}_4}$ **(f) AND-NOR** $f = \overline{\overline{(x_1 \vee x_2 \vee x_3)} \vee \overline{(x_1 \vee \overline{x}_2 \vee \overline{x}_3)} \vee \overline{(x_1 \vee \overline{x}_3 \vee x_4)}}$ $= \overline{\overline{x}_1\overline{x}_2\overline{x}_3 \vee \overline{x}_1x_2x_3 \vee \overline{x}_1x_3\overline{x}_4}$	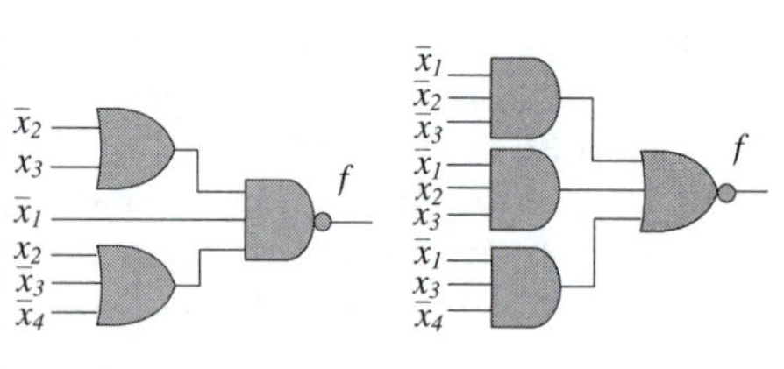
(g) OR-NAND $f = x_1 \vee x_2\overline{x}_2x_3 \vee \overline{x}_2x_3x_4$ $= \overline{\overline{x_1 \vee x_2\overline{x}_2x_3 \vee \overline{x}_2x_3x_4}}$ $= \overline{\overline{x}_1\,\overline{x_2\overline{x}_3}\,\overline{\overline{x}_2x_3x_4}}$ $= \overline{\overline{x}_1(\overline{x}_2 \vee x_3)(x_2 \vee \overline{x}_3 \vee \overline{x}_4)}$ **(h) NAND-AND** $f = \overline{\overline{\overline{x}_1\overline{x}_2\overline{x}_3 \vee \overline{x}_1x_2x_3 \vee \overline{x}_1x_3\overline{x}_4}}$ $= \overline{\overline{\overline{x}_1\overline{x}_2\overline{x}_3}\,\overline{\overline{x}_1x_2x_3}\,\overline{\overline{x}_1x_3\overline{x}_4}}$	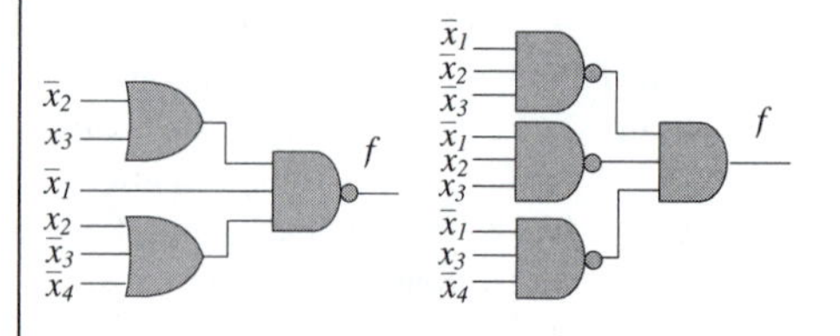

Practice problem **8.11. (Two-level networks.)** Given the AND-OR network

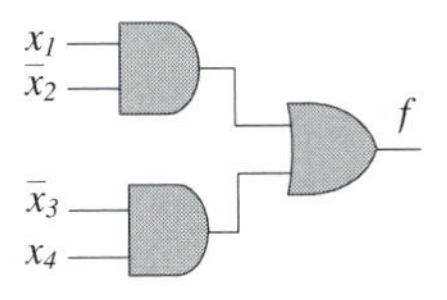

design a NOR-only circuit.
Answer is given in "Solutions to practice problems."

8.4.2 Multilevel logic networks

If the design criteria require conversion of a Boolean expression into a multilevel logic network, so-called *factorization* of the expression is used. Factoring is implemented using the SOP or POS expressions. If a SOP or POS expression is given, it has to be checked first for having a degenerate form.

Degenerate SOP and POS forms

Algebraic expressions of a Boolean function f of n variables can include products and sums of one or more (up to n, in minterms and maxterms) variables. The following forms of SOP expressions are of particular interest in function manipulation:

- SOP expressions in which products are minterms (n variables in each minterm),
- SOP expressions in which products consist of two or more variables, but less than n, and
- SOP expressions in which products may consist of a single variable; these are so-called *degenerate* SOPs.

Example 8.15 (SOP expressions.) (*a*) *The SOP form of the Boolean function f_1 of three variables, $f_1 = x_1x_2\overline{x}_3 \vee \overline{x}_1x_2x_3 \vee \overline{x}_1\overline{x}_2\overline{x}_3$, is the sum of minterms, or canonical SOP. The representation $f_2 = x_1x_2\overline{x}_3 \vee x_4$ of a function f_2 is a degenerate SOP expression.*

By analogy, in the POS form of a Boolean function, all sums are the sums of single variables. Three types of POS are distinguished:

- POS expressions in which sums are maxterms,
- POS expressions in which sums may consist of two or more variables, and
- *Degenerate* POS expressions in which sums may consist of a single variable.

Example 8.16 (POS expressions.) *The Boolean function f_1 of three variables, $f_1 = (x_1 \vee x_2 \vee \overline{x}_3)(\overline{x}_1 \vee x_2 \vee x_3)(\overline{x}_1 \vee \overline{x}_2 \vee \overline{x}_3)$, is represented by the product of maxterms. The representation of a Boolean function in the form $f_2 = (x_1 \vee x_2 \vee \overline{x}_3)x_4$ is the degenerate POS expression.*

The rationale for considering the generate form of an SOP expression is its usage for the simplification of implementation using AND and OR logic gates.

Factored expression

A *factored* expression of a Boolean function is defined as follows:

Factored expressions

- A literal is a factored expression,
- A logical sum is a factored expression, and
- A logical product of factored expressions is a factored expression.

It follows from this definition that in factored expressions, complements are permitted only for the literals. Factorization is the conversion of a degenerate SOP form into a degenerate POS form:

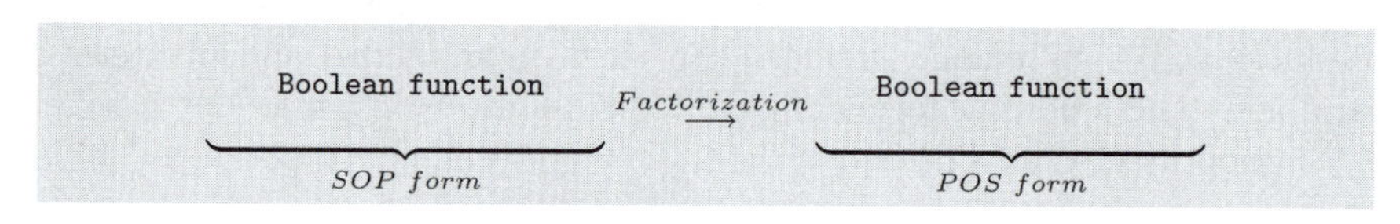

A Boolean expression is said to be:

- *Totally factored* if it is in maxterm form; if a Boolean function is in fully factored form, the function cannot be factored any further.
- *Partially factored* if it is in POS form; if a Boolean function is in partially factored form, the function can be factored further.

The distributive laws considered in Chapter 4 can be used to factor Boolean expressions.

Example 8.17 (Factorization.) *The terms $(\overline{x}_1 \vee x_2)(x_1 \vee x_3)x_2$ and $(x_1 \vee x_2)(x_3 \vee (\overline{x}_4(x_5 \vee \overline{x}_6)))$ are totally and partially factored expressions, respectively. The term $(\overline{x_1 \vee x_2})x_4$ is not a factored expression.*

Practice problem 8.12. (Factored forms.) Find whether the following expressions are totally or partially factored: (*a*) $f_1 = (x_1 \vee x_2)(\overline{x}_3)(\overline{x}_1 \vee x_2 \vee x_3)$; (*b*) $f_2 = (\overline{x}_1 \vee x_3)(\overline{x}_2\overline{x}_4)(\overline{x}_2 \vee x_3 \vee x_4)$.
Answer: The expression for f_1 is partially factored, and the expression for f_2 is totally factored.

Example 8.18 (Factorization.) *Techniques for factoring using the distributive law are illustrated below for the Boolean functions* $f_1 = x_1 \vee \overline{x}_2 x_3 x_4$ *and* $f_2 = \overline{x}_3 x_4 \vee \overline{x}_3 \overline{x}_5 \vee \overline{x}_6 x_7$*:*

$$(a)\ f_1 = \underbrace{x_1 \vee \overline{x}_2 x_3 x_4}_{\textit{Distributive law}} = (x_1 \vee \overline{x}_2)\ \underbrace{(x_1 \vee x_3 x_4)}_{\textit{Distributive law}}$$
$$= \underbrace{(x_1 \vee \overline{x}_2)(x_1 \vee x_3)(x_1 \vee x_4)}_{\textit{Totally factored form}}$$

$$(b)\ f_2 = \overline{x}_3 x_4 \vee \overline{x}_3 \overline{x}_5 \vee \overline{x}_6 x_7 = \underbrace{\overline{x}_3 (x_4 \vee \overline{x}_5) \vee \overline{x}_6 x_7}_{\textit{Distributive law}}$$
$$= \underbrace{(\overline{x}_3 \vee \overline{x}_6 x_7)}_{\textit{Distributive law}}\ \overbrace{(x_4 \vee \overline{x}_5 \vee \overline{x}_3 x_7)}^{\textit{Distributive law}}$$
$$= \underbrace{(\overline{x}_3 \vee \overline{x}_6)(\overline{x}_3 \vee x_7)(x_4 \vee \overline{x}_5 \vee \overline{x}_6)(x_4 \vee \overline{x}_5 \vee x_7)}_{\textit{Totally factored form}}$$

The factored expressions f_1 *and* f_2 *are in maxterm form; these expressions cannot be factored any further.*

Practice problem **8.13.** **(Factorization.)** Factor the Boolean function $f = x_1 \overline{x}_2 \vee \overline{x}_3 x_4$.
Answer is given in "Solutions to practice problems."

8.4.3 Conversion of factored expressions into logic networks

Factored expressions can be converted into logic networks. Totally factored expressions (the Boolean function can not be factored any further) have several useful properties; in particular, with respect to: (a) the number of literals and terms, (b) fan-out and fan-in of logic networks, and (c) the number of gates of logic networks.

Properties of totally factored expressions

Property 1: The sizes (in number of literals and terms) for totally factored expressions for a Boolean function f and its complement $\overline{f}$ are the same; this is because in a factored expression, the complement of f, $\overline{f}$, is obtained by replacing AND with OR operations, and by replacing the variables with their complements.

Property 2: A factored expression corresponds to a fan-out-free multi-level logic network.

Property 3: The amount of memory needed for factored expressions is not as large as that needed for storing SOPs or truth tables.

Property 4: Factoring

- Increases the number of gates,
- Decreases fan-in, and
- Reduces the number of literals in the expression.

An SOP expression can be implemented by one or more OR gates and by feeding a single AND gate at the logic network output. The example below illustrates how factoring can be used to deal with the fan-in problem.

Example 8.19 (Fan-in.) *Suppose that the available gates have a maximum fan-in of three, and implements the Boolean function $f = x_1\overline{x}_2\overline{x}_3\overline{x}_4 \vee x_1x_2\overline{x}_3x_4$. The direct implementation of the function f (Figure 8.5) does not meet the requirement of the maximum allowed fan-in. A solution is achieved by factoring the function $f = x_1\overline{x}_3(\overline{x}_2\overline{x}_4 \vee x_2x_4)$.*

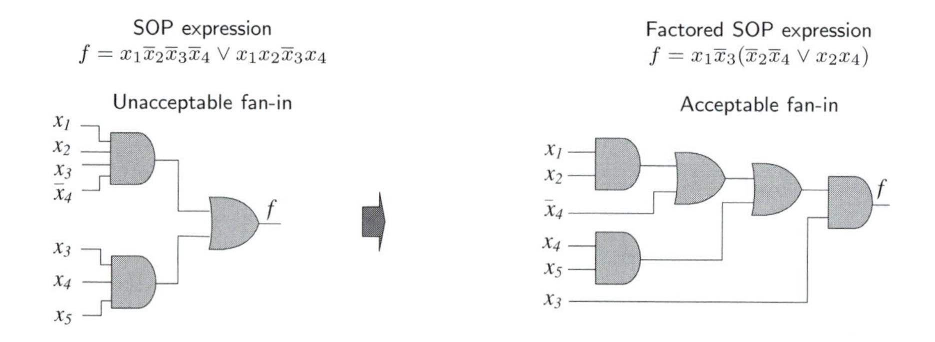

FIGURE 8.5
Implementation of the SOP form of a Boolean function and the factored form of the SOP expression (Example 8.19).

Practice problem 8.14 (Fan-in.) Implement the expression $f = x_1x_2x_3\overline{x}_4 \vee x_3x_4x_5$ using only two-input gates.
Answer is given in "Solutions to practice problems."

8.5 Minimization of Boolean functions using K-maps

In Chapter 2, the K-map was introduced as a useful data structure for the representation of Boolean functions. The K-map also provides a systematic way of performing optimization on Boolean functions:

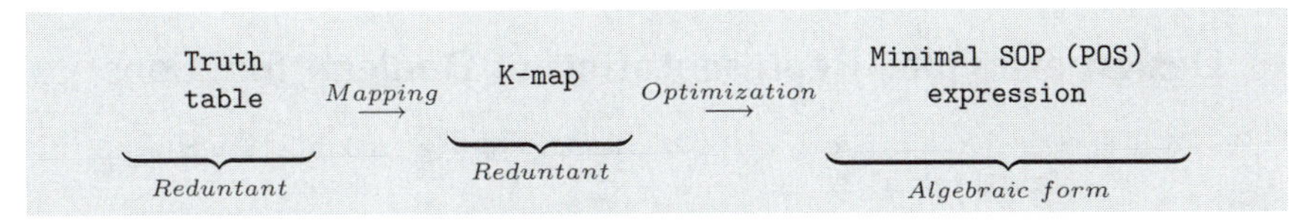

To find a minimal SOP expression, the K-map is used for the representation of Boolean function in its standard (canonical) SOP form. Each 1 in the K-map corresponds to the minterm. Reduction process is based on the use of the identities $x_1x_2 \vee \overline{x}_1x_2 = x_2$ and $\overline{x}_1 \vee \overline{x}_1x_2 = x_1 \vee x_2$ to combine and simplify the product terms. Any two adjacent squares in a K-map correspond to minterms that differ in only one variable.

Minimization of Boolean functions of two variables

A two-variable K-map is a 2-dimensional representation of a two-variable truth table. It is used for minimizing SOP or POS expressions of a Boolean function of two variables (Figure 8.6).

Example 8.20 (Minimization.) *Consider the K-map shown in Figure 8.6a.*

(a) Let $m_1 = \overline{x}_1x_2$ and $m_3 = x_1x_2$ ($m_2 = m_4 = 0$). The two minterms m_1 and m_3 differ in that the variable x_1 appears in its complimented form in m_1 and its uncomplemented form in m_3. These minterms can be combined $\overline{x}_1x_2 \vee x_1x_2 = x_2$.

(b) Let $m_0 = \overline{x}_1\overline{x}_2$, $m_1 = \overline{x}_1x_2$, and $m_2 = x_1\overline{x}_2$ ($m_3 = 0$). The minterm m_0 can be combined with both the minterm m_1 and m_2: $\overline{x}_1\overline{x}_2 \vee \overline{x}_1x_2 \vee x_1\overline{x}_2 = \overline{x}_1 \vee x_1\overline{x}_2 = \overline{x}_1 \vee x_2$.

Example 8.21 (Minimization.) *Techniques for minimization based on K-maps for Boolean functions of two variables are illustrated in Figure 8.7.*

Minimization of Boolean functions of three variables

The K-map of a function of three variables is represented as a 2×4 table, placed so that the neighbor cells are adjacent (the corresponding variable assignments of the two neighbor cells differ in exactly one bit).

Example 8.22 (Minimization.) *Minimization of various Boolean functions of three variables using their K-maps is illustrated in Figure 8.8. For example, the minimal SOP expression in the case (d) corresponds to the following manipulation*

$$f = \overline{x}_1\overline{x}_2\overline{x}_3 \vee \overline{x}_1\overline{x}_2x_3 \vee x_1\overline{x}_2x_3$$
$$= \overline{x}_1\overline{x}_2 \vee x_1\overline{x}_2 = \overline{x}_2$$

Design example: Representation of Boolean functions

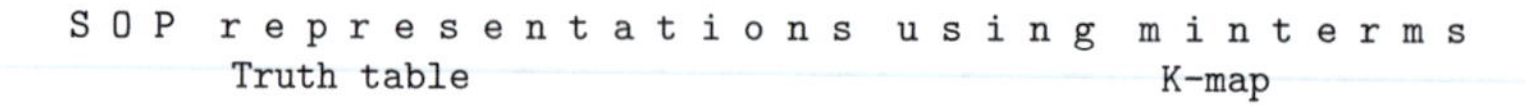

x_1	x_1	Minterm
0	0	$m_0 = \overline{x}_1 \overline{x}_2$
0	1	$m_1 = \overline{x}_1 x_2$
1	0	$m_2 = x_1 \overline{x}_2$
1	1	$m_3 = x_1 x_2$

x_2 \ x_1	0	1
0	m_0	m_2
1	m_1	m_3

(a)

P O S representations using maxterms

Truth table

K-map

x_1	x_1	Maxterm
0	0	$M_0 = x_1 \vee x_2$
0	1	$M_1 = x_1 \vee \overline{x}_2$
1	0	$M_2 = \overline{x}_1 \vee x_2$
1	1	$M_3 = \overline{x}_1 \vee \overline{x}_2$

x_2 \ x_1	0	1
0	M_0	M_2
1	M_1	M_3

(b)

FIGURE 8.6
K-maps for the representation of Boolean functions of two variables in the standard SOP form (a) and standard POS form (b).

Minimization of Boolean functions of four variables

The K-map for an SOP or POS expression of a Boolean function of four variables is a 4×4 table.

Example 8.23 **(Minimization.)** *Figure 8.9 contains various examples of K-maps of Boolean functions of four variables and the results of minimization.*

Don't care conditions in minimization

The values of a Boolean function which are not specified are called *don't care conditions*. Such a Boolean function is called *incompletely specified* function. Details are given in Chapter 6.

In the K-map, don't care conditions are marked by a d entry. In combining squares, a d entry is used whenever possible to form larger blocks of squares.

Techniques for minimization of Boolean functions

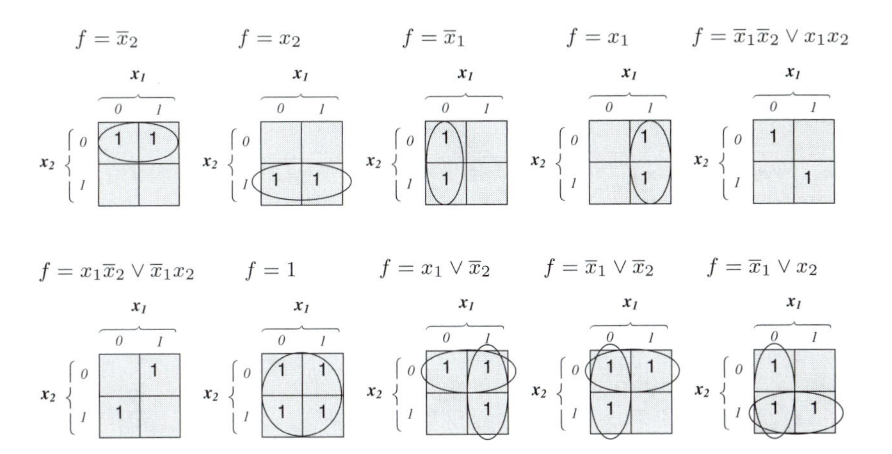

FIGURE 8.7
Technique for minimization of Boolean functions of two variables using K-maps (Example 8.21).

A d-square can be neglected if it is not needed in minimization. Incompletely specified Boolean function is described in terms of the don't care conditions d.

> **Example 8.24 (Don't care conditions.)** *Consider K-map given in Figure 8.10. In the case (a), the don't care condition d corresponding to the value of Boolean function $f(110)$ is not used in the minimization. In the case (b), the don't care condition d corresponding to $f(010)$ and $f(110)$ are used.*

Practice problem 8.15. (Don't care conditions.) Assume $d = 1$ and write the SOP expressions for Boolean function given in Figure 8.10.
Answer: $\overline{x}_1 \vee x_2 \vee x_3$.

Minimal POS expressions

The minimal POS expression can be derived using K-map. In this case, 0 entries are considered in the K-map. Each 0 entry corresponds to the maxterm. That is, a K-map represents a standard (canonical) POS. The minimal POS expression is obtained in exactly the same manner as for SOP forms except that we try to enclose the largest number of 0s into groups.

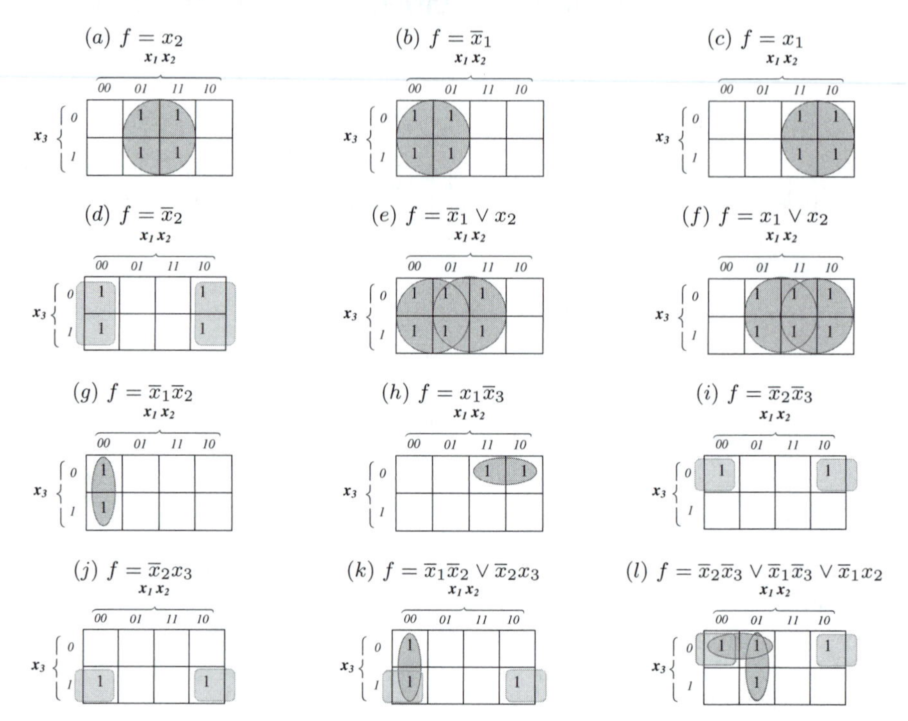

FIGURE 8.8
Techniques for the minimization of Boolean functions of three variables using K-maps.

Example 8.25 (Minimal POS expression.) *The minimization of Boolean functions in a POS form is shown in Figure 8.11.*

Practice problem 8.16. (Minimal POS expression.) Let $d = 1$, write the POS expressions for Boolean functions given in Figure 8.10.
Answer: $\overline{x}_1 \vee x_2 \vee x_3$.

Conversion of minterm expressions into maxterm expressions, and vice versa

The same K-map represents both minterms (corresponding to 1-cells in the K-map) and maxterms (corresponding to 0-cells). Minimization of the SOP form of the function is normally performed on the 1-cells by combining adjacent cells. To accomplish the minimization of a POS, 0-cells must be considered,

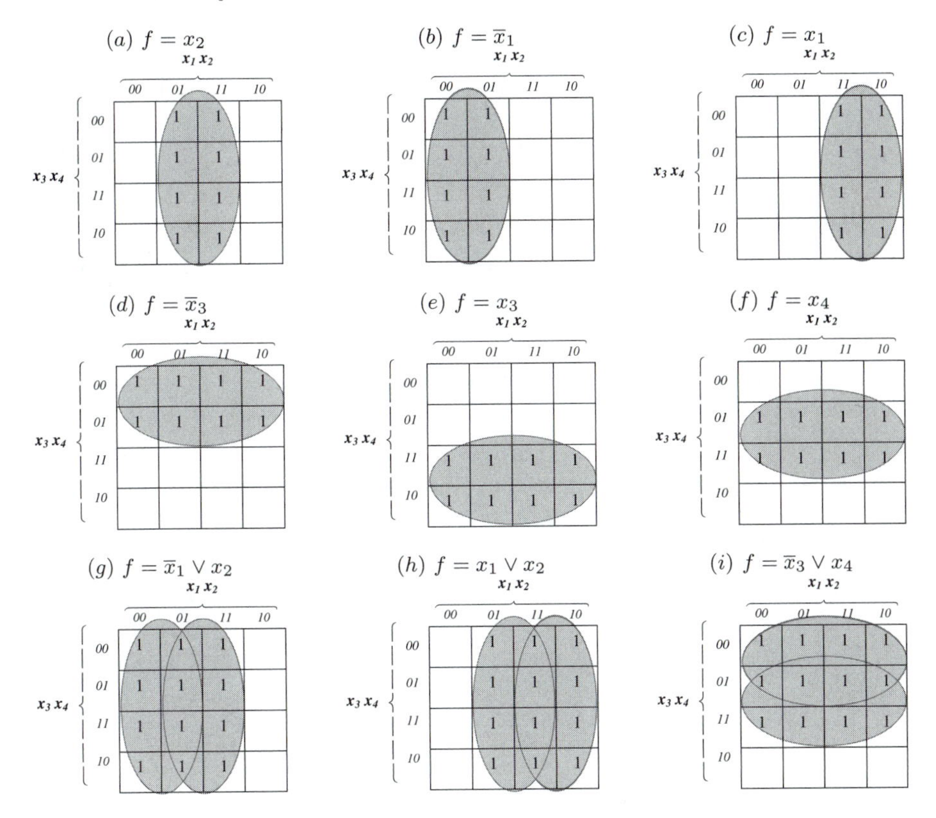

FIGURE 8.9
Techniques for the minimization of Boolean functions of four variables using K-maps.

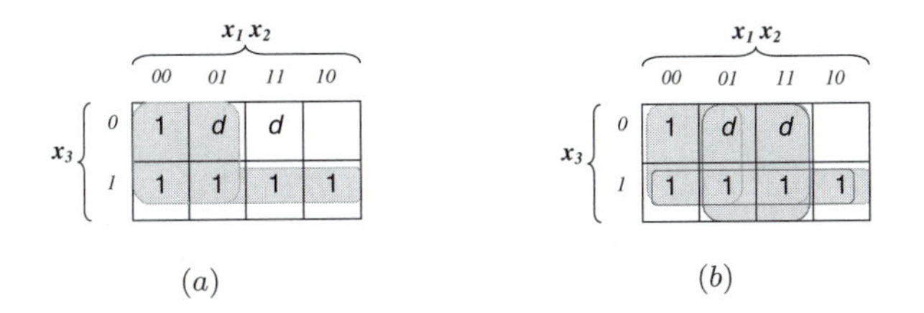

FIGURE 8.10
Don't care conditions in minimization (Example 8.24).

and the adjacent cells must be combined when possible. This operation on the 0-cells of the same map results in obtaining the POS for the same function

Standard POS expression:
$f = (x_1 \vee x_2 \vee x_3)(x_1 \vee \overline{x}_2 \vee x_3)(x_1 \vee \overline{x}_2 \vee \overline{x}_3)$.
The reduction rules: $(x \vee y)(x \vee \overline{y} = x)$ and $(x \vee y)(x \vee \overline{y} \vee z) = (x \vee y)(x \vee z)$.
The minimization using K-map corresponds to the following manipulation:

$$f = (x_1 \vee x_2 \vee x_3)(x_1 \vee \overline{x}_2 \vee x_3)(x_1 \vee \overline{x}_2 \vee \overline{x}_3)$$
$$= (x_1 \vee x_3)(x_1 \vee \overline{x}_2 \vee \overline{x}_3) = (x_1 \vee x_3)(x_1 \vee \overline{x}_2)$$

FIGURE 8.11
Minimization of Boolean function given the standard POS expression (Example 8.25).

and can be considered as a conversion of the minimal SOP to a minimal POS form of the Boolean function.

Example 8.26 (Conversion.) *Conversion of the minimal SOP expression $f = x_1 \vee x_2\overline{x}_2x_3 \vee \overline{x}_2x_3x_4$ into the minimal POS expression using K-maps is shown in Figure 8.12).*

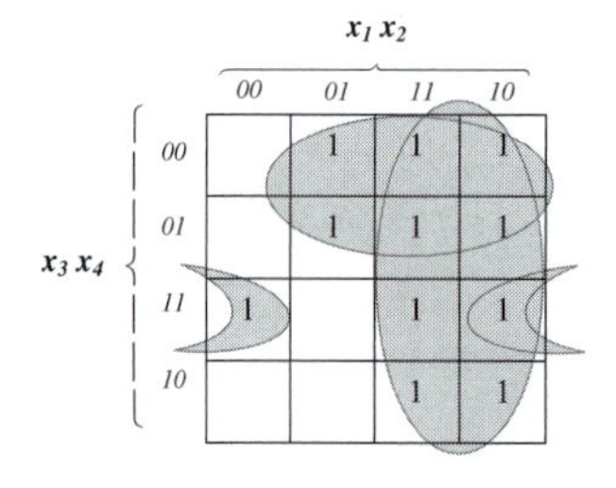

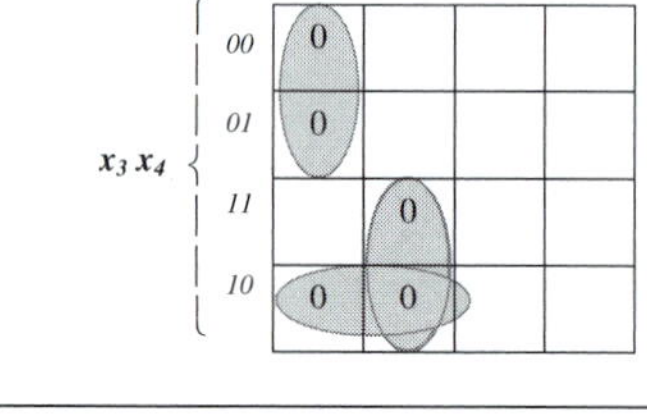

Minimumal SOP form
$f = x_1 \vee x_2\overline{x}_2x_3 \vee \overline{x}_2x_3x_4$

Minimal POS form
$f = (x_1 \vee x_2 \vee x_3)(x_1 \vee \overline{x}_2 \vee \overline{x}_3)(x_1 \vee \overline{x}_3 \vee x_4)$

FIGURE 8.12
Minimal SOP and POS expressions for the Boolean function (Example 8.26).

Practice problem 8.17 (Conversion.) Find the standard SOP and POS expressions for the functions $\overline{x}_2$, $x_1\overline{x}_2 \vee \overline{x}_1x_2$, $\overline{x}_1$, and $\overline{x}_1\overline{x}_2 \vee x_1x_2$ using K-maps for two variables.
Answers are given in "Solutions to practice problems."

Practice problem 8.18 (Conversion.) Convert the following minimal SOP expressions into minimal POS expressions using K-maps:
(a) $f = \overline{x}_1\overline{x}_2\overline{x}_3 \vee x_1x_2x_3$, (b) $f = x_1 \vee x_2 \vee x_3$.
Answers are given in "Solutions to practice problems."

Application of the consensus theorem in K-maps

The consensus theorem (Equation 8.3) and its dual (Equation 8.4) can be used in various Boolean data structures: algebraic equations, truth tables, hypercubes, K-maps, logic networks, and decision diagrams). The example below illustrates the consensus terms in K-maps.

> **Example 8.27 (Consensus theorem.)** *Figure 8.13 shows the consensus terms for Boolean functions of three and four variables. In both functions, these terms can be removed.*

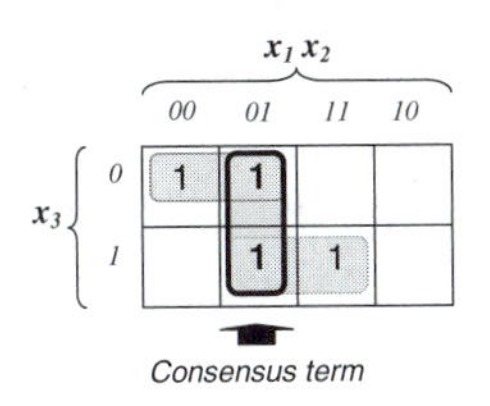

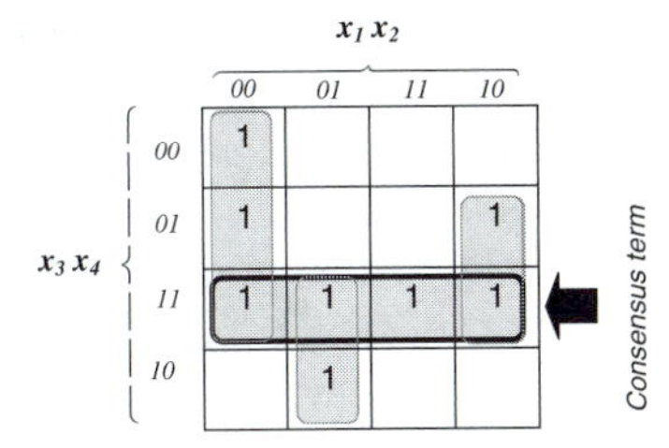

FIGURE 8.13
Consensus theorem in the form of K-maps (Example 8.27).

8.6 Quine-McCluskey algorithm

When the number of variables in a Boolean function is six or more, K-maps are not acceptable for minimization. The Quine-McCluskey algorithm can be efficiently used for a Boolean functions with $6 - 10$ variables. Again, when the number of variables exceeds 10, computation time and memory storage become unsuitable, and some heuristic algorithm that provides quasi-optimal solutions, is applied instead.

The Quine-McCluskey algorithm provides a systematic simplification procedure that has been efficiently implemented in various software packages. This algorithm reduces the minterms in an SOP representation of a Boolean function in order to obtain a minimal SOP. The algorithm consists of two main phases.

Minimal two-level NOR-only logic network design

Step 1: Derive the prime implicants.

Step 2: Construct the prime implicant chart. This chart includes a column of initial minterms and columns of cubes, some of which are prime implicants. The number of columns with cubes depends on the minimization results of every particular function.

Step 3: Select a minimal set of prime implicants from the prime implicant chart using the so-called *covering* technique.

Let a Boolean function be given in SOP form. Two minterms of this SOP expression can be combined by using the adjacency rule if they differ in exactly one variable. For example, the result of combining the two minterms x_1x_2 and $x_1\overline{x}_2 = x_1$, of a Boolean function of two variables is the following: $x_1x_2 \vee x_1\overline{x}_2 = x_1$. Here, x_1 is an implicant. If the implicant cannot be further combined with any other implicants, it is a *prime* implicant. Systematic application of this procedure eliminates as many literals as possible and results in the set of prime implicants.

> **Example 8.28 (Prime implicants.)** *Consider the Boolean function of four variables given by the canonical SOP expression below:* $f = x_1\overline{x}_2x_3\overline{x}_4 \vee x_1\overline{x}_2x_3x_4 \vee x_1x_2x_3\overline{x}_4$. *All possible implicants for this SOP expression are derived by combining the following pairs of minterms:* $x_1\overline{x}_2x_3\overline{x}_4 \vee x_1\overline{x}_2x_3x_4 = x_1\overline{x}_2x_3$ *and* $x_1\overline{x}_2x_3\overline{x}_4 \vee x_1x_2x_3\overline{x}_4 = x_1x_3\overline{x}_4$. *These implicants are also the prime implicants, since none of them can be further combined with one another.*

Practice problem 8.19. (Prime implicants.) Given the canonical SOP of the Boolean function of four variables, $f = x_1\overline{x}_2x_3\overline{x}_4 \vee x_1\overline{x}_2x_3x_4 \vee x_1x_2x_3\overline{x}_4 \vee \overline{x}_1\overline{x}_2x_3\overline{x}_4 \vee \overline{x}_1\overline{x}_2x_3x_4$, find all its prime implicants.
Answer is given in "Solutions to practice problems."

Grouping the minterms

To reduce the required number of comparisons, minterms are sorted into groups according to the number of complemented variables in the minterm. There is a group of minterms with all variables complemented, one variable complemented, two variables complemented, and so on.

> **Example 8.29 (Grouping.)** *Consider the grouping of minterms given the SOP of a Boolean function of four variables:*

$$f = \bigvee m(0,1,2,5,6,7,8,9,10,14) = \underbrace{\overset{0}{\overline{x}_1\overline{x}_2\overline{x}_3\overline{x}_4}}_{Group\ 0} \vee \underbrace{\overset{1}{\overline{x}_1\overline{x}_2\overline{x}_3x_4} \vee \overset{2}{\overline{x}_1\overline{x}_2x_3\overline{x}_4} \vee \overset{8}{x_1\overline{x}_2\overline{x}_3\overline{x}_4}}_{Group\ 1}$$

$$\vee \underbrace{\overset{5}{\overline{x}_1x_2\overline{x}_3x_4} \vee \overset{6}{\overline{x}_1x_2x_3\overline{x}_4} \vee \overset{9}{x_1\overline{x}_2\overline{x}_3x_4} \vee \overset{10}{x_1\overline{x}_2x_3\overline{x}_4}}_{Group\ 2} \vee \underbrace{\overset{7}{\overline{x}_1x_2x_3x_4} \vee \overset{14}{x_1x_2x_3\overline{x}_4}}_{Group\ 3}$$

> *This grouping results in four groups of minterms.*

Practice problem 8.20. (Grouping.) Group the minterms of the following SOP expression for a function of four variables: $f = \bigvee m(1,2,3,9,10,11,12,15)$.
Answer is given in "Solutions to practice problems."

The groups above are numbered according to the number of uncomplemented variables. The neighboring groups (for example, groups 2 and 3) are called *adjacent*.

Example 8.30 *(Continuation of Example 8.29). These groups can be encoded as follows:*

$$\underbrace{\overset{0}{0000}}_{Group\ 0}\ \underbrace{\overset{1}{0001}\ \overset{2}{0010}\ \overset{8}{1000}}_{Group\ 1}\ \underbrace{\overset{5}{0101}\ \overset{6}{0110}\ \overset{9}{1001}\ \overset{10}{1010}}_{Group\ 2}\ \underbrace{\overset{7}{0111}\ \overset{14}{1110}}_{Group\ 3}$$

In groups 0, 1, 2, and 3, the terms have 0, 1, 2, and 3 ones, respectively.

Combining minterms

After grouping, terms are combined as follows:

- Every two terms from two adjacent groups are compared, and if they differ in only one term, they are combined.
- The comparison of terms within groups is unnecessary (two terms with the same number of 1s differ in at least two variables and, therefore, cannot be combined).
- The comparison of terms in nonadjacent groups is unnecessary (the terms in these groups cannot be combined).

Example 8.31 *(Continuation of Example 8.30). The determination of prime implicants is shown in Figure 8.14. The first column "Minterms" consists of grouped and encoded minterms, the second and the third columns show the results of combining minterms. For example, the term from group 0 is compared with all terms in group 1. The result of comparing terms 0000 and 0001 (the first column) is the term 000-; that is, the elimination of the variable x_4 (second column). Three pairs of duplicates are the result of combining the same three pairs of minterm sets in different orders (third column). The detected duplicate terms are deleted.*

A cover

The set of prime implicants of a Boolean function forms an SOP, called a *cover*. In the cover, each term (prime implicant) has a minimal number of literals. However, the number of terms is not always minimal, since some prime implicants can cover the same implicant (for example, in terms of K-maps, the same 1-cells can be covered by two or more different prime implicants).

Technique for computing prime implicants

	Minterms			Cubes			Cubes		
Group 0	*0*	0000	✓	*0,1*	000-	✓	0,1,8,9	**-00-**	*Duplicate*
Group 1	*1*	0001	✓	*0,2*	00-0	✓	~~0,8,1,9~~	~~-00-~~	
	2	0010	✓	*0,8*	-000	✓	0,2,8,10	**-0-0**	*Duplicate*
	8	1000	✓	*1,5*	**0-01**		~~0,8,2,10~~	~~-0-0~~	
Group 2	*5*	0101	✓	*1,9*	-001	✓	2,6,10,14	**--10**	*Duplicate*
	6	0110	✓	*2,6*	0-10	✓	~~2,10,6,14~~	~~--10~~	
	9	1001	✓	*2,10*	-010	✓			
	10	1010	✓	*8,9*	100-	✓			
Group 3	*7*	0111	✓	*8,10*	10-0	✓			
	14	1110	✓	*5,7*	**01-1**				
				6,7	**011-**				
				6,14	-110	✓			
				10,14	1-10	✓			

FIGURE 8.14
The first phase of the Quine-McCluskey algorithm: determination of prime implicants (Example 8.31).

Example 8.32 *(Continuation of Example 8.31). The six terms that have not been checked off in the columns in Figure 8.14 are prime implicants. These prime implicants form the cover of the given Boolean function,* $f = \overline{x}_1\overline{x}_3x_4 \vee \overline{x}_1x_2x_4 \vee \overline{x}_1x_2x_3 \vee \overline{x}_2\overline{x}_3 \vee \overline{x}_2\overline{x}_4 \vee x_3\overline{x}_4$. *This cover is not the minimal SOP form of this function.*

The prime implicant chart and essential prime implicants

At the second phase of the Quine-McCluskey algorithm, the prime implicant chart is constructed to select a minimal set of prime implicants. This set includes *essential prime implicants.*

Example 8.33 (Essential implicants.) *In the expression* $f = \overline{x}_1x_2x_3 \vee x_1\overline{x}_2x_3 \vee x_1x_2\overline{x}_3 \vee x_1x_2x_3 = x_2x_3 \vee x_1x_3 \vee x_1x_2$ *none of the implicants* x_2x_3, x_1x_3, *and* x_1x_2 *can be reduced further without violating the truth table of* f*; they are prime implicants. Neither* x_2x_3, x_1x_3, *nor* x_1x_2 *can be taken away from* f *without changing its truth table; each is an essential prime implicant.*

The prime implicant chart is constructed as follows:

Prime implicant chart construction

- ▶ Classify the prime implicants according to the number of literals.
- ▶ List the prime implicants along the vertical axis, placing the implicants with the minimal number of literals on the top, followed by the prime implicants with more literals.
- ▶ List the decimal representations of minterms along the horizontal axis.
- ▶ If a prime implicant that corresponds to the i-th row contains the minterm that corresponds to the j-th column, write an $\times$ mark in the (i, j)-th cell of the chart.

Example 8.34 *(Continuation of Example 8.33). The prime implicant chart for the function* $f = \bigvee m(1, 2, 3, 9, 10, 11, 12, 15)$ *is given in Figure 8.15.*

Technique for selection of a minimal set of prime implicants

Prime implicants		MInterms										
		0	1	2	3	5	6	7	8	9	10	14
0,1,8,9	$\overline{x}_2\,\overline{x}_3$	×	×						×	×		
0,2,8,10	$\overline{x}_2\,\overline{x}_4$	×		×					×		×	
2,6,10,14	$x_3\,\overline{x}_4$			×			×				×	×
1,5	$\overline{x}_1\,\overline{x}_2\,x_4$		×			×						
5,7	$\overline{x}_1 x_2 x_4$					×		×				
6,7	$\overline{x}_1 x_2 x_3$						×	×				

FIGURE 8.15
The second phase of the Quine-McCluskey algorithm: the selection of a minimal set of prime implicants in the prime implicant chart (Example 8.34).

Minimum covering problem

If there is an $\times$ mark in the (i, j)-th cell of the prime implicant chart, then the row i *covers* the column j.

The minimum covering problem

The *minimum covering problem* is to find a set of minimal rows that covers all the columns. If a minterm is covered by only one prime implicant, then that prime implicant is called an *essential* prime implicant and must be included in the minimal SOP; that is, it must be chosen first.

Note that each row has a cost, and in the case of several rows covering the same columns, the solution is found where the sum of the costs is at a minimum. In some cases, the essential prime implicants do not cover all of the minterms; in these cases the remaining prime implicants must be examined to satisfy the covering criterion.

Example 8.35 *(Continuation of Example 8.34). There are two essential prime implicants whose rows cover all the columns in the prime implicants chart: $\overline{x}_2\overline{x}_3$ and $x_3\overline{x}_4$; so they are chosen (Figure 8.15). The remaining prime implicants are not essential. However, to cover all the minterms of the function, some non-essential implicants can be included in the minimal SOP expression. The minimal set of such prime implicants includes only one implicant $\overline{x}_1x_2x_4$, which covers the remaining two columns. The final minimal SOP expression is $f = \overline{x}_2\overline{x}_3 \vee x_3\overline{x}_4 \vee \overline{x}_1x_2x_4$.*

Some Boolean functions have two or more minimal SOP expressions, each having the same number of terms and literals. In this case, their costs are equal, and any of them can be chosen for implementation.

Practice problem **8.21 (Minimal SOP.)** Given the function of four variables, $f = \bigvee m(1,2,3,4,8,9,10,11,12,15)$, find its minimal SOP.
Answer is given in "Solution of practice problems."

8.7 Boolean function minimization using decision diagrams

Reducing a decision tree to a decision diagram representing a Boolean function corresponds to the process of minimizing the function. Detailed techniques for reducing decision trees are considered in Chapter 9. Here we will consider a brief introduction to using the decision tree to find a minimal SOP of a Boolean function.

Example 8.36 **(Reducing decision trees.)** *The complete decision trees and the reduced ordered decision diagrams for four Boolean functions are given in Figure 8.16.*

Design example: The reduction of decision trees

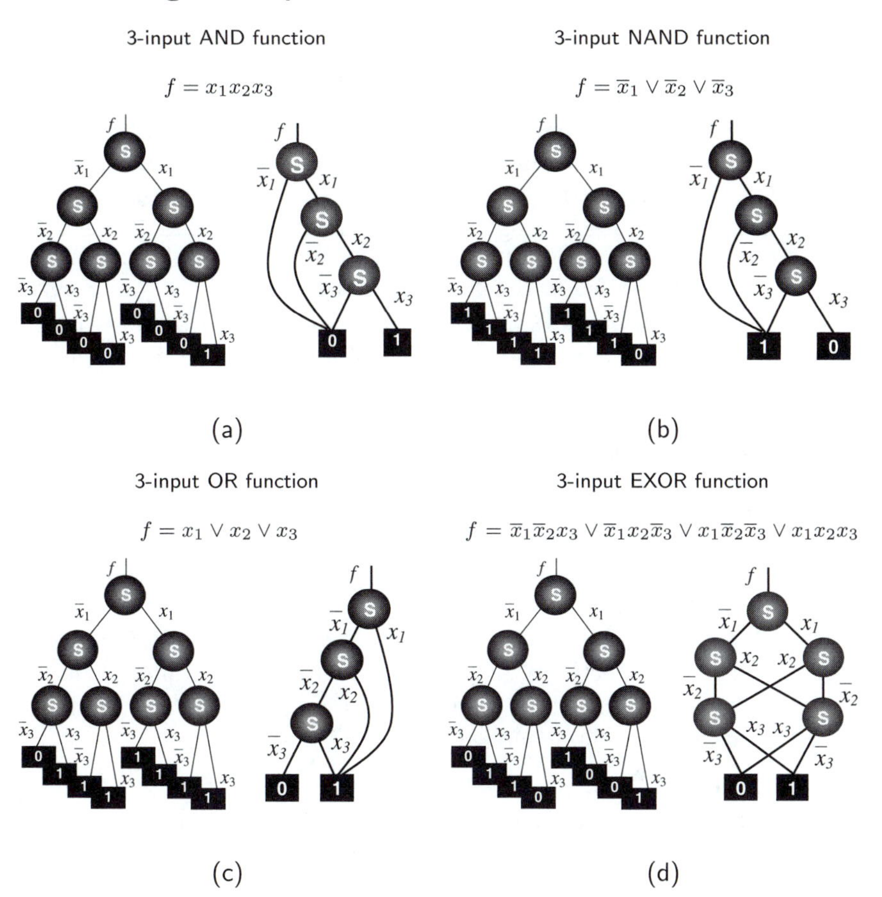

FIGURE 8.16
The complete decision trees and the reduced ordered decision diagrams for the 3-input AND (a), NAND (b), OR (c) and EXOR (d) functions (Example 8.36).

Practice problem **8.22. (Reducing decision tree.)** Derive and reduce the decision tree for the Boolean function $f = \overline{x}_1 \vee x_2$ given its truth table $\mathbf{F} = [1\ 1\ 0\ 1]^T$.
Answer is given in "Solution of practice problems."

8.8 Summary of optimization of Boolean functions

This chapter introduces the optimization of Boolean functions, which aims at removing ***redundancy*** in Boolean data structures. The final goal of logic design is the implementation of Boolean functions using hardware and software tools.

> (*a*) ***Optimization*** of Boolean functions means their transformation are better suited to the implementation criteria.
>
> (*b*) The goal of ***minimization*** is the representation of Boolean functions with a minimal number of literals and terms, given the type of representation.

The main conclusions of this chapter are as follows:

- Each Boolean data structure has specific properties and requires particular rules for manipulation and optimization. Boolean data structures often are not optimal; that is, they are redundant in the representation of a given Boolean function.
- A Boolean function f of n variables is said to be in SOP form when all products are the products of only single variables. The following forms of SOP expressions are of particular interest in Boolean function manipulation:

  (*a*) ***Canonical,*** or ***standard***, SOP expressions, in which all products are minterms (n variables in each minterm);
  (*b*) SOP expressions, in which the products consist of less than n variables; and
  (*c*) ***Degenerate*** SOP expressions, in which the products may consist of a single variable.

- The POS form of a Boolean function is a logical product of sums of single variables. Three types of POS expressions are distinguished:

  (*a*) ***Canonical,*** or ***standard*** POS expressions, in which sums are maxterms (n variables each term);
  (*b*) POS expressions, in which sums may consist of less than n variables; and
  (*c*) ***Degenerate POS*** expressions, in which sums may consist of a single variable.

- A ***factored*** expression of a Boolean function exists in the following forms: a literal, a logical sum, and a logical product of factored expressions.

Summary (continuation)

- A Boolean expression is said to be:

  (*a*) ***Totally factored*** iff it is in maxterm form (if a Boolean function is in fully factored form, this function cannot be factored further).

  (*b*) ***Partially factored*** iff it is in POS form; if a Boolean function is in partially factored form, this function can be factored further.

- Techniques of Boolean function simplification are based on combining terms, eliminating literals and terms, and adding redundant terms for the efficient elimination of terms and literals. Simplification does not guarantee that the resulting Boolean function is minimal. Finding a minimal form with respect to some criteria is based on minimization algorithms, the most popular of which are *Karnaugh* maps (K-maps) and *Quine-McCluskey* algorithm.
- A K-map can be used for the representation and optimization of Boolean functions. When the number of variables in a Boolean function is six or more, K-maps are not acceptable for minimization.
- The ***Quine-McCluskey algorithm*** provides a systematic simplification procedure. This algorithm can be efficiently used for a Boolean functions with $6-10$ variables. The Quine-McCluskey algorithm reduces the minterms in an SOP representation of a Boolean function in order to obtain a minimal SOP. The algorithm consists of two main phases:

  (*a*) Construction of the ***prime implicant chart***,

  (*b*) Selection of a ***minimal set of prime implicants*** from the prime implicant chart.

- A Boolean function can be represented by a decision tree, which can be reduced to a decision diagrams. Decision diagram are used for minimizing Boolean functions as well.
- For advances in optimization of Boolean functions, we refer the reader to the "Further study" section.

8.9 Further study

Historical perspective

1952: W. V. O. Quine and E. L. Jr. McCluskey developed the chart method of exact minimization known as the *Quine-McCluskey algorithm* ("The Problem of Simplifying Truth Functions" by W. V. Quine, *American Math Monthly*, volume 59, number 8, pages 521–531, October 1952; "Minimization of Boolean Functions" by E. J. McCluskey, *Bell Syst. Tech. Journal*, volume 35, number 5, pages 1417–1444, November 1956).

1953: M. Karnaugh proposed a map method for the minimization of Boolean functions known as the Karnaugh map or K-map (" A Map Method for Synthesis of Combinational Logic Circuits," *Trans. AIEE*, volume 73, number 9, pages 593–599, part 1, November 1953).

1970s: Computer-aided methods for large-scale Boolean minimization started to be created.

1986: The best known minimization method, ESPRESSO, was developed by Robert Brayton et.al. at the University of California, Berkeley. The ESPRESSO algorithm has been incorporated as a standard logic function minimization step into virtually any contemporary logic synthesis tool. For non-commercial purposes the logic minimization program Minilog, exploiting this ESPRESSO algorithm, can be downloaded as part of the Publicad educational design package. It is able to generate a two-level gate implementation for a combinational function block with up to 40 inputs and outputs or a synchronous state machine with up to 256 states. R. L. Rudell later published the variant ESPRESSO-MV in 1986 ("Multiple-Valued Logic Minimization for PLA Synthesis," Memorandum No. UCB/ERL M86/65 5, June 1986). For implementing a function in multi-level logic, the minimization result is optimized by factorization and mapped onto the available basic logic cells in the target technology, whether this concerns an FPGA or ASIC.

1990s: Exact minimization of complex SOP expressions become common, though often physical and geometrical constraints are now more important than pure logical ones. Minimization of Boolean expressions with depth larger than 2 became popular in the 1990s in connection with developing the binary decision diagram techniques.

Advanced topics of logic optimization

Topic 1: Implementation-driven optimization. Major optimization objectives in combinational logic design are *area* and *delay* reduction. Optimization of multi-level logic networks is related to *testability* properties. *Two-level logic optimization*, both exact and heuristic, have received much attention, since they are critical in logic synthesis. *ESPRESSO* is the most popular heuristic two-level logic optimizer. *Multi-level logic optimization* techniques aim at the synthesis logic networks with more than two levels. Multi-level networks often require fewer logic gates and fewer connections than two-level logic networks. Multi-level logic optimization techniques also provide flexibility in optimizing area and delay, and satisfying specific design constraints. The drawback to multi-level logic networks is the difficulty of modeling and optimizing. Exact optimization is not considered to be practical in contemporary multi-level synthesis. Techniques of multi-level logic synthesis have no established algorithms compared with two-level logic techniques.

Topic 2: Multi-valued logic function optimization. Multi-valued logic optimization is the generalization of a binary logic network design for multi-valued logic. The overall level of understanding of multi-level logic network design based on multi-valued logic gates is at a much less mature stage than binary multi-level networks. Details can be found in the *Proceedings of the Annual IEEE Symposium on Multiple-Valued Logic.*

Topic 3: The advent of nanoscale technologies. For earlier technologies, the relevant problems were primarily concerned with component minimization. With the development of VLSI and ULSI logic network technologies and the advent of nanoscale technologies, the digital design problems concerned with the minimization of components have become less relevant. These types of problems have been replaced by less well-defined and much more difficult problems such as physical design including partitioning, layout and routing, structural simplicity, and uniformity of modules. The last problem, uniformity, and the related data structures is of particular interest to logic design in nanoscale. Many of these problems cannot be solved based on traditional approaches. Investigation of better solutions in this area is ongoing work, which results are reported in many journals and conference publications.

Further reading

Contemporary Logic Design by R. H. Katz, The Benjamin/Cummings Publishing Company, 1994.

Design of Logic Systems by D. Lewin, Van Nostrand (UK), 1985.

Digital Principles and Design by D. D. Givone, McGraw-Hill, New York, 2003.

Fundamentals of Digital Logic with VHDL Design by S. Brown and Z. Vranesic, McGraw-Hill, 2000.

Fundamentals of Logic Design by Jr. C. H. Roth, 5th Edition, Thomson Brooks/Cole, 2004.

Introduction to Logic Design by A. B. Marcovitz, McGraw-Hill, New York, 2002.

Logic Minimization Algorithms for VLSI Synthesis by R. K. Brayton, A. Sangiovanni-Vincentelli, C. McMullen, and G. Hachtel, Kluwer, 1984.

"Multi-Level Logic Optimization" by M. Fujita, Y. Matsunaga, and M. Ciesielski, in *Logic Synthesis and Verification* edited by S. Hassoun and T. Sasao, Consulting Editor R. K. Brayton, Kluwer, 2002.

Practical Digital Logic Design and Testing by P.K. Lala, Prentice Hall, 1996.

Switching Theory for Logic Synthesis by T. Sasao, Kluwer, 1999.

Synthesis and Optimization of Digital Circuits by G. De Micheli, McGraw-Hill, 1994.

"Two-Level Logic Minimization" by O. Coudert and T. Sasao, in *Logic Synthesis and Verification* edited by S. Hassoun and T. Sasao, Consulting Editor R. K. Brayton, Kluwer, 2002.

8.10 Solutions to practice problems

Practice problem	Solution
8.4.	Apply inverse, distributive, and identity laws: $XY \vee \overline{X}Z \vee YZ = XY \vee \overline{X}Z \vee (X \vee \overline{X})YZ = (XY \vee \overline{X}ZY) \vee (\overline{X}Z \vee \overline{X})YZ$ $= XY(1 \vee Z) \vee \overline{X}Z(1 \vee Y) = XY \vee \overline{X}Z$
8.9.	$(a)\ f = \overline{x}_1x_2x_3 \vee \overline{x}_1 = \overline{x}_1 \underbrace{(x_2x_3 \vee 1)}_{Equal\ to\ 1} = \overline{x}_1$ $(b)\ f = (\underbrace{\overline{x}_1 \vee \overline{x}_2x_3}_{X} \vee \underbrace{x_4 \vee x_5x_6}_{Y})(\underbrace{x_1 \vee \overline{x}_2x_3}_{X} \vee \underbrace{\overline{x_3 \vee x_5\overline{x}_6}}_{\overline{Y}})$ $= (X \vee Y)(X \vee \overline{Y}) = X$ $= x_1 \vee \overline{x}_2x_3$ $(c)\ f = \underbrace{(x_1x_2 \vee x_3)}_{\overline{Y}}\underbrace{(\overline{x}_2x_4 \vee \overline{x}_3\overline{x}_5)}_{X} \vee \underbrace{\overline{x_1x_2 \vee x_3}}_{Y}$ $= X\overline{Y} \vee Y = X \vee Y = \overline{x}_2x_4 \vee \overline{x}_3\overline{x}_5 \vee \overline{x_1x_2 \vee x_3}$
8.10.	By applying the DeMorgan's rule, the function is transformed as follows: $f = x_1x_2 \vee x_2x_3 = \overline{\overline{x_1x_2} \vee \overline{x_2x_3}}$ This implementation requires three NAND gates: 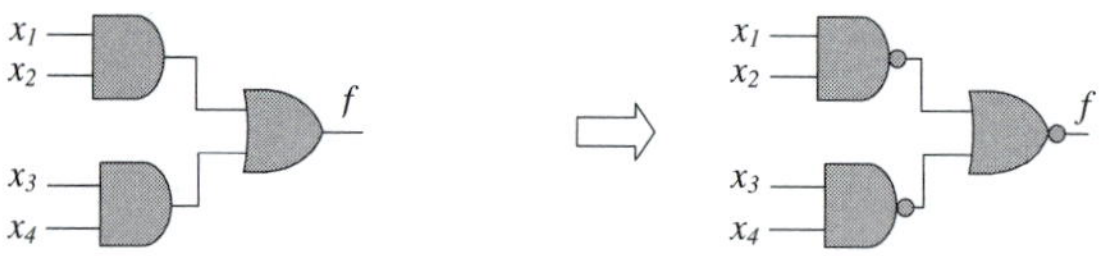
8.11.	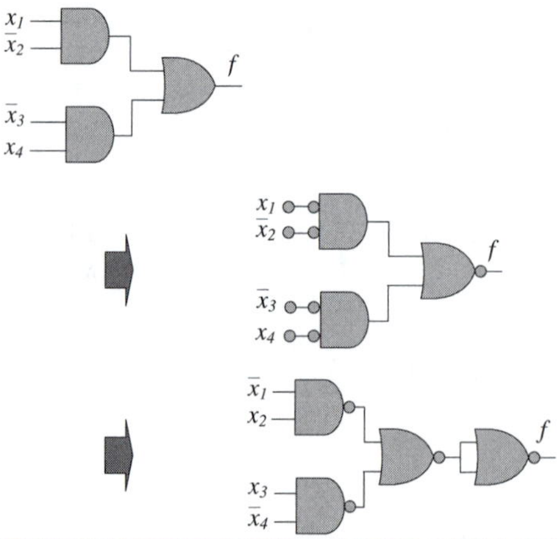

Solutions to practice problems (continuation)

Practice problem	Solution
8.13.	$f = \underbrace{x_1\overline{x}_2 \vee \overline{x}_3 x_4}_{Distributive\ law} = \underbrace{(x_1\overline{x}_2 \vee \overline{x}_3)}_{Distributive\ law} \overbrace{(x_1\overline{x}_2 \vee x_4)}^{Distributive\ law}$ $= \underbrace{(x_1 \vee \overline{x}_3)(\overline{x}_2 \vee \overline{x}_3)(x_1 \vee x_4)(\overline{x}_2 \vee x_4)}_{Totally\ factored\ form}$
8.14.	Factoring the function results in $f = x_3(x_1x_2\overline{x}_4 \vee x_4x_5) = x_3((x_1x_2)\overline{x}_4 \vee x_4x_5)$, which can be implemented on the network of two-input gates: $f = x_1x_2x_3\overline{x}_4 \vee x_3x_4x_5$ $\quad$ $f = x_3(x_1x_2\overline{x}_4 \vee x_4x_5)$ 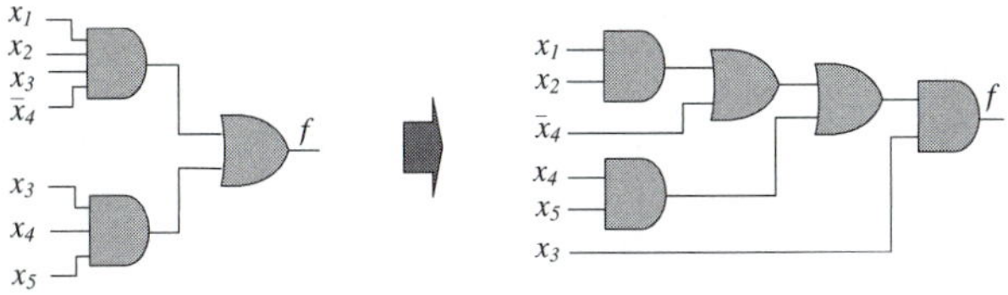 x_1 x_2 x_3 $\overline{x}_4$ x_3 x_4 x_5 f $\quad$ x_1 x_2 $\overline{x}_4$ x_4 x_5 x_3 f
8.17.	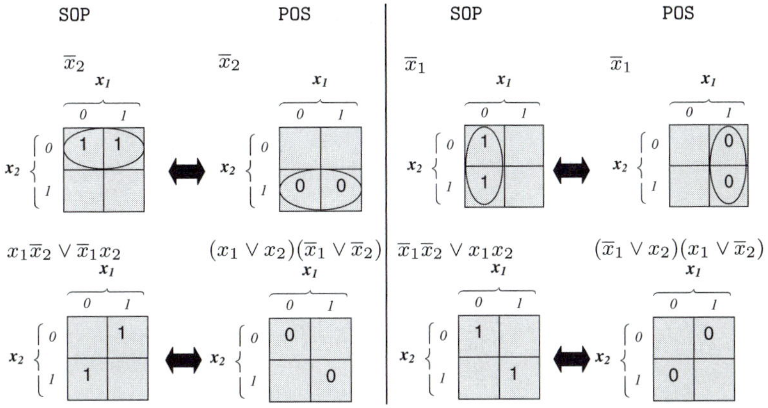 SOP: $\overline{x}_2$; K-map (x_1: 0, 1; x_2: 0, 1): row 0: 1 1; row 1: – – ⟺ POS: $\overline{x}_2$; row 0: – –; row 1: 0 0 SOP: $\overline{x}_1$; row 0: 1 –; row 1: 1 – ⟺ POS: $\overline{x}_1$; row 0: – 0; row 1: – 0 $x_1\overline{x}_2 \vee \overline{x}_1x_2$; row 0: – 1; row 1: 1 – ⟺ $(x_1 \vee x_2)(\overline{x}_1 \vee \overline{x}_2)$; row 0: 0 –; row 1: – 0 $\overline{x}_1\overline{x}_2 \vee x_1x_2$; row 0: 1 –; row 1: – 1 ⟺ $(\overline{x}_1 \vee x_2)(x_1 \vee \overline{x}_2)$; row 0: – 0; row 1: 0 –
8.18.	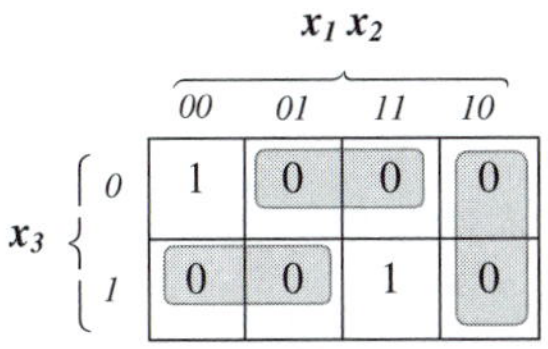 K-map x_1x_2 (00, 01, 11, 10), x_3: row 0: 1 0 0 0; row 1: 0 0 1 0 Minimal SOP: $f = \overline{x}_1\overline{x}_2\overline{x}_3 \vee x_1x_2x_3$ Minimal POS: $f = (\overline{x}_1 \vee x_2)(x_1 \vee \overline{x}_3)(\overline{x}_2 \vee x_3)$ 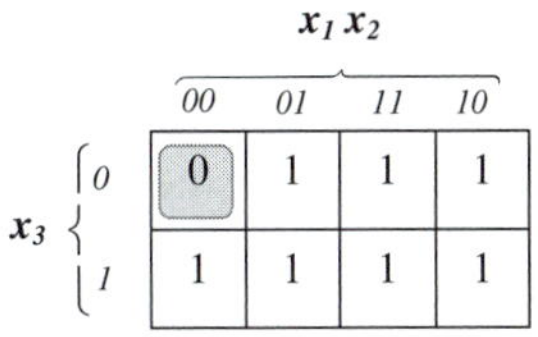 K-map x_1x_2 (00, 01, 11, 10), x_3: row 0: 0 1 1 1; row 1: 1 1 1 1 Minimal SOP: $f = x_1 \vee x_2 \vee x_3$ Minimal POS: $f = x_1 \vee x_2 \vee x_3$

Solutions to practice problems (continuation)

Practice problem	Solution

8.19.

$$x_1\overline{x}_2x_3\overline{x}_4 \vee x_1\overline{x}_2x_3x_4 = x_1\overline{x}_2x_3;\quad x_1\overline{x}_2x_3\overline{x}_4 \vee x_1x_2x_3\overline{x}_4 = x_1x_3\overline{x}_4$$

$$x_1\overline{x}_2x_3\overline{x}_4 \vee \overline{x}_1\overline{x}_2x_3\overline{x}_4 = \overline{x}_2x_3\overline{x}_4;\quad \overline{x}_1\overline{x}_2x_3\overline{x}_4 \vee \overline{x}_1\overline{x}_2x_3x_4 = \overline{x}_1\overline{x}_2x_3$$

The first and last implicants can be further reduced as follows: $x_1\overline{x}_2x_3 \vee \overline{x}_1\overline{x}_2x_3 = \overline{x}_2x_3$. The final set of prime implicants is given below: $\overline{x}_1\overline{x}_2x_3,\ x_1x_3\overline{x}_4,\ \overline{x}_2x_3\overline{x}_4$.

8.20.

$$f = \bigvee m(1,2,3,9,10,11,12,15) = \underbrace{\overset{1}{\overline{x}_1\overline{x}_2\overline{x}_3x_4} \vee \overset{2}{\overline{x}_1\overline{x}_2x_3\overline{x}_4}}_{Group\ 1}$$

$$\vee \underbrace{\overset{3}{\overline{x}_1\overline{x}_2x_3x_4} \vee \overset{9}{x_1\overline{x}_2\overline{x}_3x_4} \vee \overset{10}{x_1\overline{x}_2x_3\overline{x}_4} \vee \overset{12}{x_1x_2\overline{x}_3\overline{x}_4}}_{Group\ 2} \vee \underbrace{\overset{11}{x_1\overline{x}_2x_3x_4}}_{Group\ 3} \vee \underbrace{\overset{15}{x_1x_2x_3x_4}}_{Group\ 4}$$

8.21.

Minterms			Cubes			Cubes	
1	0001	✓	1,3	00-1	✓	1,3,9,11	-0-1
2	0010	✓	1,9	-001	✓	2,3,10,11	-01-
4	0100	✓	2,3	001-	✓	8,9,10,11	10--
8	1000	✓	2,10	-010			
3	0011	✓	4,12	-100	✓		
9	1001	✓	8,9	100-	✓		
9	1001	✓	8,10	10-0	✓		
10	1010	✓	8,9	100-	✓		
12	1100	✓	8,12	1-00	✓		
11	1011	✓	3,11	-011			
15	1111		9,11	10-1			
			10,11	101-	✓		
			11,15	1-11			

Prime implicants		1	2	3	4	8	9	10	11	12	15
1,3,9,11	$\overline{x}_2x_4$	×		×			×		×		
2,3,10,11	$\overline{x}_2x_3$		×	×				×	×		
8,9,10,11	$x_1\overline{x}_2$					×	×	×	×		
4,12	$x_2\overline{x}_3\overline{x}_4$				×					×	
8,12	$x_1\overline{x}_3\overline{x}_4$					×				×	
4,15	$x_1x_3x_4$								×		×

(Minterms span columns 1–15.)

8.22.

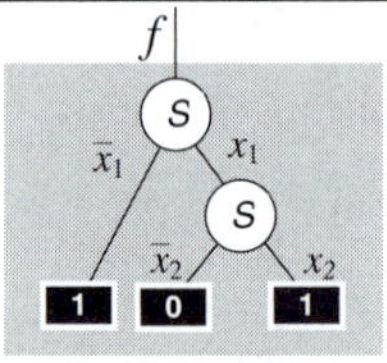

8.11 Problems

Problem 8.1 Derive the decision tree using Shannon expansion and decision diagram for the Boolean functions given below:
(a) $f = \overline{x}_1 \vee x_2 \vee x_3$ (b) $f = x_1 \vee (x_2 \oplus \overline{x}_3)$ (c) $f = x_1x_2 \vee \overline{x}_2x_3 \vee x_1\overline{x}_3$

Problem 8.2 Prove that the following gates perform the same Boolean functions:

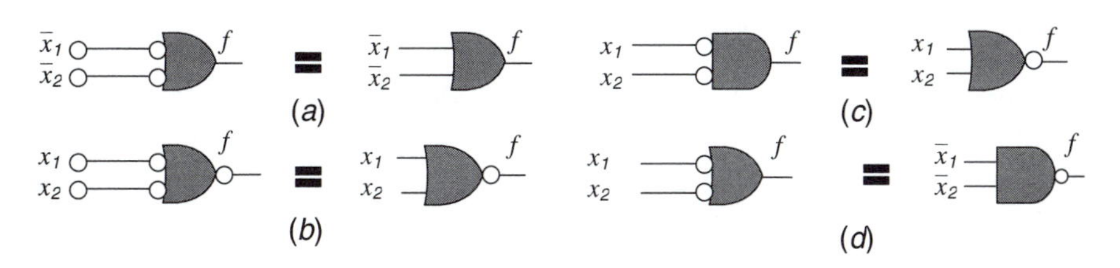

Problem 8.3 Convert the following networks into:
(a) a NOR only network
(b) a NAND only network:

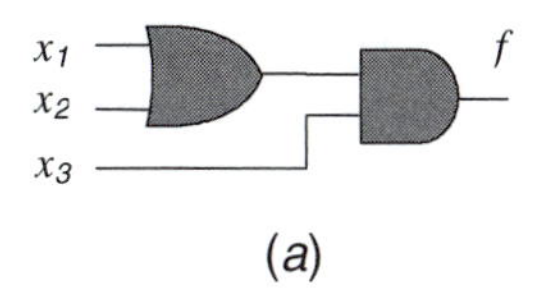

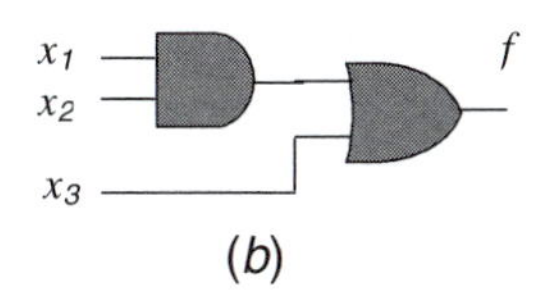

Problem 8.4 Represent on K-maps the Boolean functions given as follows:
(a) $\mathbf{F}_1 = [01100100]^T$
(b) $\mathbf{F}_2 = [\times 011 \times 010]^T$

Problem 8.5 Represent on K-maps the Boolean functions given by the following expressions:
(a) $f = \bigvee m(1, 5, 7)$
(b) $f = \overline{x}_1\overline{x}_2x_3 \vee \overline{x}_3x_4 \vee x_2\overline{x}_4$

Problem 8.6 Represent on K-maps the Boolean functions given by the following expressions:
(a) $f = \prod M(1, 3, 4, 7, 10, 13, 14, 15)$
(b) $f = \bigvee m(0, 4, 5, 9, 11, 14, 15)$ and $dc = \bigvee m(2, 8)$

Problem 8.7 For the Boolean function $f = \bigvee m(0, 1, 2, 3, 5, 7, 8, 10, 11, 15)$:
(*a*) Find all the prime implicants.
(*b*) Indicate the essential prime implicants among these prime implicants.
(*c*) Obtain a minimal SOP for f.

Problem 8.8 For the Boolean function given by the one-set $f = \bigvee m(0, 1, 2, 4, 5, 6, 7, 8, 9, 10, 15)$, determine which of the following products are implicants, prime implicants, and essential prime implicants:
(a) $\overline{x}_1\overline{x}_4$ (b) $\overline{x}_1x_3$ (c) $x_2x_3x_4$ (d) $\overline{x}_1x_2\overline{x}_3$

Problem 8.9 Using K-maps, find the minimal SOP and minimal POS expressions for the following Boolean functions:

(a) $f = \prod M(1, 3, 4, 7, 10, 13, 14, 15)$
(b) $dc = \bigvee m(2, 8)$
(c) $f = \bigvee m(0, 4, 5, 9, 11, 14, 15)$
(d) $dc = \bigvee m(0, 1, 4)$

Problem 8.10 Using K-maps, find the minimal SOP expressions for the following Boolean functions:
(a) $f = \overline{x}_1\overline{x}_2\overline{x}_3 \vee x_1\overline{x}_2\overline{x}_3 \vee x_1\overline{x}_2x_3 \vee x_1x_2\overline{x}_3$
(b) $f = \overline{x}_1\overline{x}_3\overline{x}_4 \vee \overline{x}_1\overline{x}_3x_4 \vee \overline{x}_1x_2x_3 \vee x_2x_3\overline{x}_4 \vee x_1\overline{x}_2\overline{x}_3$

Problem 8.11 Using K-maps, find the minimal POS expressions for the following Boolean functions:
(a) $f = (x_2 \vee x_3)(\overline{x}_1 \vee x_2)(x_1 \vee \overline{x}_3)(x_1 \vee x_2 \vee x_3)$
(b) $f = (x_1 \vee x_2 \vee x_3)(\overline{x}_2 \vee x_3 \vee \overline{x}_4) \vee (x_1 \vee x_2 \vee x_3)(\overline{x}_3 \vee \overline{x}_4)(x_1 \vee x_2 \vee \overline{x}_3)$

Problem 8.12 Using both algebraic manipulation and K-maps, find the minimal SOP and minimal POS expressions for the Boolean function $f = \overline{x}_1\overline{x}_3\overline{x}_4 \vee \overline{x}_1x_3x_4 \vee \overline{x}_1x_3\overline{x}_4 \vee x_3x_4 \vee x_1\overline{x}_2\overline{x}_3\overline{x}_4$.

Problem 8.13 An incompletely specified Boolean function of four variables is given by the truth-vector $\mathbf{F} = [1 \times \times \times \times \times 00 \times \times 010 \times 00]^T$. Using a K-map, assign values to the don't cares to construct three completely specified Boolean functions f_1, f_2, and f_3 with the following properties:

(*a*) f_1 is a Boolean function of exactly two variables.
(*b*) f_2 is a Boolean function with exactly five minterms.
(*c*) f_3 is a Boolean function with exactly four prime implicants.

Problem 8.14 Design a minimal two-level logic network that implements the five-input majority function using the Quine-McCluskey minimization algorithm.
Hint: In the 5-input majority function, the output is 1 whenever three or more inputs are 1.

Problem 8.15 Using K-maps,

(*a*) Find the minimal SOP and minimal POS expressions for the Boolean function $f = \bigvee m(0, 1, 2, 5, 6)$.
(*b*) Design two-level logic networks using AND and OR gates for each obtained SOP and POS.

Problem 8.16 Using K-maps, design a two-level NAND logic network that realizes the minimal SOPs of the following Boolean functions:
(a) $f = \bigvee m(0, 1, 2, 7, 8, 9)$ (b) $f = \bigvee m(0, 2, 6, 7, 8, 9, 10, 13, 15)$

Problem 8.17 Simplify algebraically each of the following Boolean functions:

(*a*) $f = x_1\overline{x}_2\overline{x}_3 \vee x_3\overline{x}_4 \vee x_2\overline{x}_3\overline{x}_4$
(*b*) $f = x_1x_2x_3\overline{x}_4 \vee \overline{x}_1\overline{x}_2x_3x_4 \vee x_3\overline{x}_4$
(*c*) $f = (x_1 \vee \overline{x}_2)(\overline{x}_1 \vee \overline{x}_2 \vee x_4)(\overline{x}_2 \vee x_3 \vee \overline{x}_4)$

Problem 8.18 Prove algebraically and using K-maps that the following equation is valid:

$$\overline{x}_1x_3\overline{x}_4x_5 \vee \overline{x}_1\overline{x}_2\overline{x}_4 \vee x_1x_2x_3x_5 \vee x_1x_2x_4 = \overline{x}_1\overline{x}_2\overline{x}_4 \vee x_1x_2x_4 \vee x_2x_3\overline{x}_4x_5$$

Problem 8.19 Factor the following Boolean functions to obtain a POS expression:

(*a*) $f = \overline{x}_1\overline{x}_2x_3 \vee x_1\overline{x}_3x_4 \vee x_1x_2x_3 \vee x_2\overline{x}_3\overline{x}_4$
(*b*) $f = x_1x_2 \vee \overline{x}_1\overline{x}_2 \vee \overline{x}_2\overline{x}_3\overline{x}_4 \vee x_2x_3\overline{x}_4$

Problem 8.20 Find the POS form for the following Boolean functions:

(*a*) $f = (x_1x_2 \oplus x_3) \vee \overline{x}_3\overline{x}_4$
(*b*) $f = \overline{x}_3(x_1 \oplus \overline{x}_4) \vee x_3x_4 \vee \overline{x}_1x_4$

Hint: Eliminate the EXOR and then factor.

Problem 8.21 Simplify the following Boolean functions using the consensus theorem (or its dual):

(*a*) $f = x_2\overline{x}_3\overline{x}_4 \vee x_1x_2\overline{x}_3 \vee x_1\overline{x}_3x_4 \vee x_1\overline{x}_2x_4 \vee \overline{x}_1x_2\overline{x}_4$
(*b*) $f = \overline{x}_1\overline{x}_2 \vee x_1x_2x_3 \vee \overline{x}_2x_3x_4 \vee x_1x_2\overline{x}_4$

Problem 8.22 Determine algebraically which of the following equations are valid:

(*a*) $\overline{x}_1x_2 \vee \overline{x}_2x_3 \vee x_1\overline{x}_3 \vee = x_1\overline{x}_2 \vee x_2\overline{x}_3 \vee \overline{x}_1x_3$
(*b*) $(x_1 \vee x_2)(x_2 \vee x_3)(x_1 \vee x_3) = (\overline{x}_1 \vee \overline{x}_2)(\overline{x}_2 \vee \overline{x}_3)(\overline{x}_1 \vee \overline{x}_3)$

Problem 8.23 Derive the minimal SOP and POS expressions for the Boolean functions given the following K-maps:

x_3x_4 \ x_1x_2	00	01	11	10
00	1	1	1	1
01	1	1	1	1
11	1	1	1	1
10	1	1	1	1

x_3x_4 \ x_1x_2	00	01	11	10
00	1	1	1	1
01	1	1	1	1
11		1	1	1
10		1	1	1

x_3x_4 \ x_1x_2	00	01	11	10
00	1	1	1	1
01	1	1	1	1
11	1	1	1	1
10	1			1

Problem 8.24 Incompletely specified Boolean functions are given in the form of K-maps:

x_3x_4 \ x_1x_2	00	01	11	10
00	1	×	×	
01	1	1	1	
11	×	×	1	
10	1	×	×	

x_3x_4 \ x_1x_2	00	01	11	10
00		×	×	1
01		1	1	×
11		×	1	1
10		×	×	1

x_3x_4 \ x_1x_2	00	01	11	10
00	1	1	1	1
01	1	×	×	1
11	1	×	×	1
10				

Derive the minimal expressions.

Problem 8.25 Assuming that the inputs $x_1x_2x_3x_4 = \{0101, 1001, 1011\}$ never occur, find a simplified expression for the Boolean function

$$f = \overline{x}_1x_2\overline{x}_3x_4 \vee \overline{x}_1\overline{x}_2x_4 \vee \overline{x}_1x_3x_4 \vee x_1x_2x_4 \vee x_1x_2x_3$$

Problem 8.26 Find all of the prime implicants for each of the following Boolean functions given by the K-maps:

x_3x_4 \ x_1x_2	00	01	11	10
00	1	1	1	1
01	1	1	1	1
11	1	1	1	1
10	1	1	1	

x_3x_4 \ x_1x_2	00	01	11	10
00		1	1	1
01		1	1	1
11		1	1	1
10	1	1	1	

x_3x_4 \ x_1x_2	00	01	11	10
00	1	1	1	1
01	1	1	1	
11	1	1	1	1
10				

9

Optimization II: Decision Diagrams

Decision trees

- Design, measurements, and implementation

Decision diagrams

- Reduction of decision trees
- Elimination and merging rules
- Efficiency

Techniques of optimization

- Single output functions
- Multi-output functions

Embedding decision trees

- Symmetric functions
- Lattice structures
- Transformation for embedding

Advanced topics

- Decomposition
- Multi-valued diagrams
- Lattice diagrams

9.1 Introduction

Decision trees, a graphical data structure for representing Boolean functions, were introduced in Chapter 2. In general, decision trees are redundant data structures, in which the redundancy of the representation of Boolean functions is expressed in graphical form. Reducing decision trees means removing redundancy in function representation. Hence, the optimization of decision trees relates to the optimization of Boolean functions. The resulting graphs are called *decision diagrams*. Besides the usage of decision diagrams as an abstract data structure at almost every stage of logic design, they are also suited for direct mapping into physical implementation for some technologies.

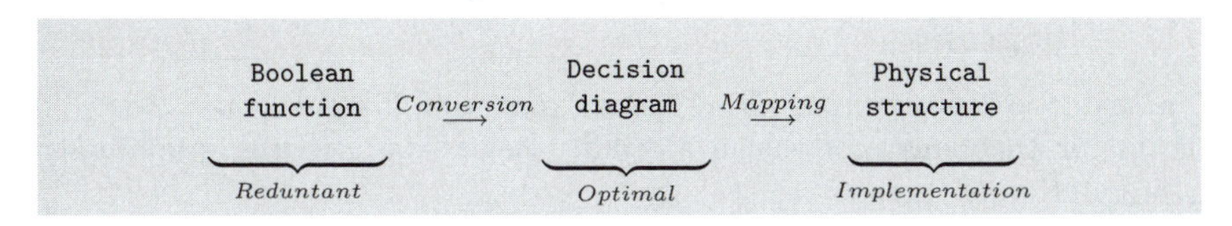

Example 9.1 (Data structure transformations.) *In Figure 9.1, the truth table of the three-input AND gate is transferred into a cube and decision diagram.*

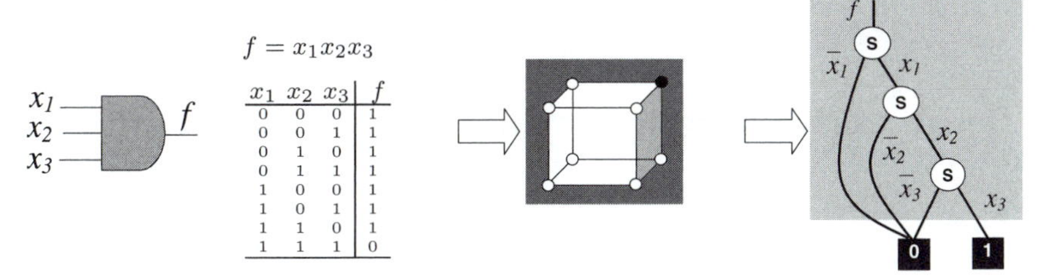

FIGURE 9.1
Relationships between truth tables, hypercubes, and decision diagrams for the representation of a Boolean function (Example 9.1).

9.2 Optimization of Boolean functions using decision trees

The conversion of a decision tree into a decision diagram is called the *optimization* of this graphical structure. This approach to optimization can be used for small- and medium-sized Boolean functions, since the number of nodes in the complete decision tree of a Boolean function of n variables is $2^{2n} - 1$; i.e., it depends exponentially on n. For instance, the complete decision tree of a function of one hundred variables has 2^{100} terminal nodes, and the total number of nodes is $2^{200} - 1$. In contemporary logic design, a decision diagram for a network is constructed and dynamically optimized directly from the circuit netlist. The disadvantage of this approach is that the structure obtained may not be canonical, since it depends heavily on the network structure and implementation. Decision trees do not have this flaw, since they are canonical, being constructed based on the truth tables of the functions. For medium-sized Boolean functions, reduced decision trees can be constructed using the properties and/or incompleteness of the data.

9.2.1 The formal basis for the reduction of decision trees and diagrams

The reduction procedures for decision trees and diagrams are based on the major theorems of Boolean algebra. Below, we give the graph-based interpretation of these rules.

> **Example 9.2 (Decision diagrams.)** *The absorption rules (see Chapter 4), $xy \vee x\overline{y} = x$ and $(x \vee y)(x \vee \overline{y}) = x$, correspond to a node reduction in binary decision trees and diagrams (Figure 9.2).*

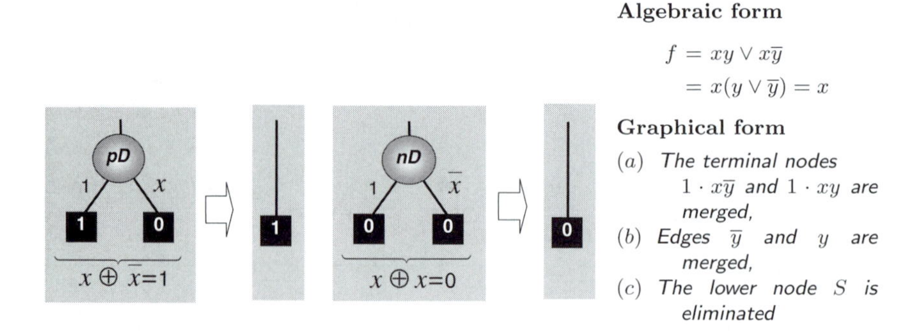

Design example: Implementation of the absorption rule
Algebraic form
$f = xy \vee x\overline{y}$
$= x(y \vee \overline{y}) = x$
Graphical form
(a) The terminal nodes $1 \cdot x\overline{y}$ and $1 \cdot xy$ are merged,
(b) Edges $\overline{y}$ and y are merged,
(c) The lower node S is eliminated
pD
1
x
1
0
1
$x \oplus \overline{x}$=1
nD
1
$\overline{x}$
0
0
0
$x \oplus x$=0

FIGURE 9.2
Absorption rules of Boolean algebra in algebraic form and in terms of decision diagrams (Example 9.2).

The consensus theorem states that $xy \vee \overline{x}z \vee yz = xy \vee \overline{x}z$ (see Chapter 8 for details). This theorem is used to eliminate the redundant terms in a Boolean expression.

Example 9.3 (Decision diagrams.) *An interpretation of the consensus theorem in terms of binary decision diagrams is given in Figure 9.3 (see also Example 8.27 in Chapter 8).*

Practice problem 9.1. (Decision diagrams.) Interpret the property $f = (x_1 \vee x_2)(\overline{x}_1 \vee x_2) = x_1$ of Boolean algebra using binary decision diagram.
Answer is given in "Solutions to practice problems."

Design example: Implementation of the consensus theorem

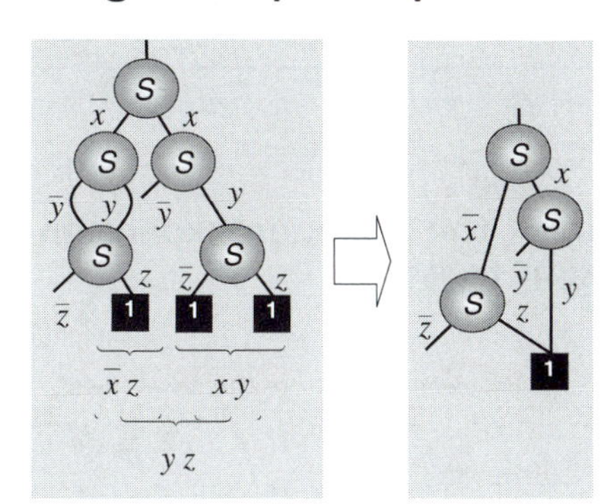

Algebraic form

$$f = xy \vee \overline{x}z \vee yz = xy \vee \overline{x}z$$

Graphical form

(a) *The terminal nodes $xy\overline{z}$ and xyz are merged to $1 \cdot xy$ and lower right S node is eliminated*

(b) *Lower left S node for variable y is eliminated*

FIGURE 9.3
Interpretation of the consensus theorem in terms of decision diagrams (Example 9.3).

Note that the resulting decision diagram is minimal with respect to the number of nodes and the order of variables. The Boolean equation, restored from the decision diagram, is not necessary minimal.

9.2.2 Decision tree reduction rules

A Boolean function can be represented by a decision tree that can be reduced and further converted into a decision diagram:

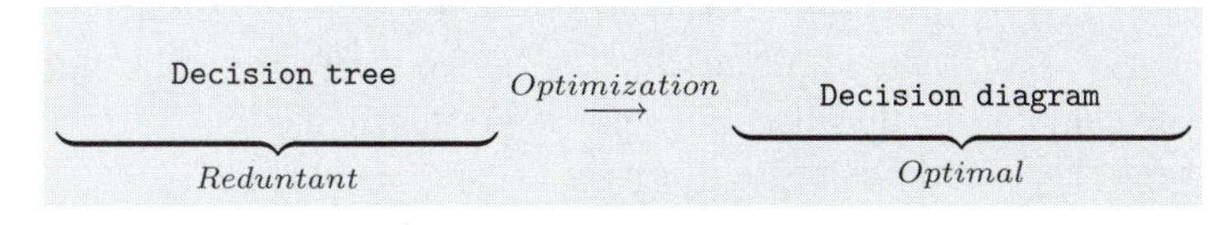

The initial binary tree or diagram has a given order of variables, that is, it is *ordered decision tree* or *diagram*. The resulting decision tree or diagram is called a *reduced ordered binary decision tree* or *diagram (ROBDD)*.

Example 9.4 (Decision tree.) *An arbitrary Boolean function f of three variables can be represented by a complete decision tree, as shown in Figure 9.4. To design this tree, Shannon expansion is used as follows:*

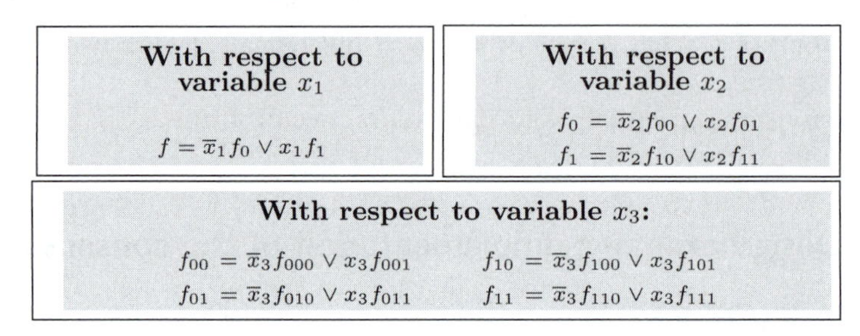

With respect to variable x_1
$f = \overline{x}_1 f_0 \vee x_1 f_1$
With respect to variable x_2
$f_0 = \overline{x}_2 f_{00} \vee x_2 f_{01}$
$f_1 = \overline{x}_2 f_{10} \vee x_2 f_{11}$
With respect to variable x_3:
$f_{00} = \overline{x}_3 f_{000} \vee x_3 f_{001}$
$f_{10} = \overline{x}_3 f_{100} \vee x_3 f_{101}$
$f_{01} = \overline{x}_3 f_{010} \vee x_3 f_{011}$
$f_{11} = \overline{x}_3 f_{110} \vee x_3 f_{111}$

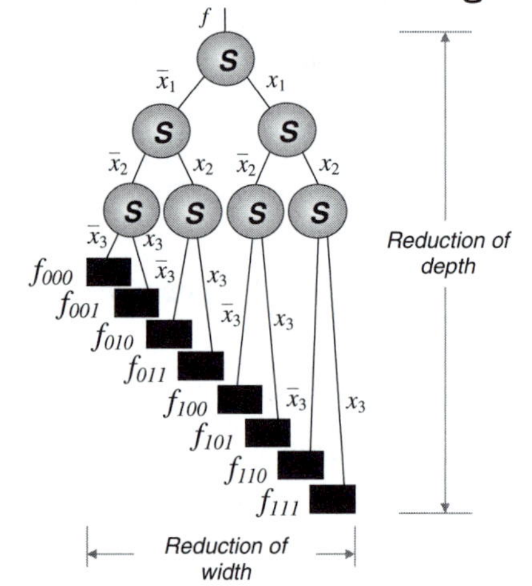

FIGURE 9.4
Decision tree for a Boolean function of three variables (Example 9.4).

A decision tree is *reduced* if it does not contain any vertex v whose successors lead to the same node, and if it does not contain any distinct vertices v and v' such that the subgraphs rooted in v and v' are isomorphic.

The reduced decision tree is still a rooted acyclic graph in which every node but the root has indegree 1. It can be further reduced to a decision diagram. In the decision diagram, the nodes can have indegree higher than 1.

> **A binary decision diagram is a directed acyclic graph that:**
>
> - Contains exactly one root.
> - Has terminal nodes labeled by the constants 1 and 0 and having indegree ≥ 2.
> - Has internal nodes with indegree ≥ 2 and outdegree equal to 2.

In an *ordered* decision diagram, a linear variable order is placed on the input variables. The variables' occurrences on each path of this diagram have to be consistent with this order.

A binary decision diagram represents a Boolean function f of n variables in the following way:

> **Boolean function representation using a decision trees**
>
> - Internal nodes represent a Shannon expansion and are labeled by S; each internal node except the root has an indegree of ≥ 1 and an outdegree equal to 2.
> - Links between the nodes are labeled by a Boolean variable x_i, $i = 1, 2, ..., n$, and each internal node has an $\overline{x}_i$-link and x_i-link.
> - Each assignment to the input variable x_i defines a unique path from the root to one of the terminal nodes.
> - The label of the terminal node gives the value of Boolean function.

Example 9.5 (Path and Boolean product.) *Figure 9.5 shows the correspondence of the path in the binary decision diagram and the Boolean product $\overline{x}_1 x_2$.*

The decision diagram for a given Boolean function f is derived from the decision tree for f by deleting redundant nodes, and by sharing equivalent subgraphs. The reduction rules are summarized in Figure 9.6 and considered in detail below.

Elimination rule

The elimination rule allows for the removal of redundant subtrees representing subfunctions. It states that if two descendent subtrees of a node are identical, then the node can be deleted, and the incoming edges of the deleted node can be connected to the corresponding successor. In terms of Shannon expansion, implemented by the node, this means that if $f_0 = f_1 = g$ then

$$f = \overline{x}_i f_0 \vee x_i f_0 = \overline{x}_i g \vee x_i g = (\overline{x}_i \vee x_i) g = g$$

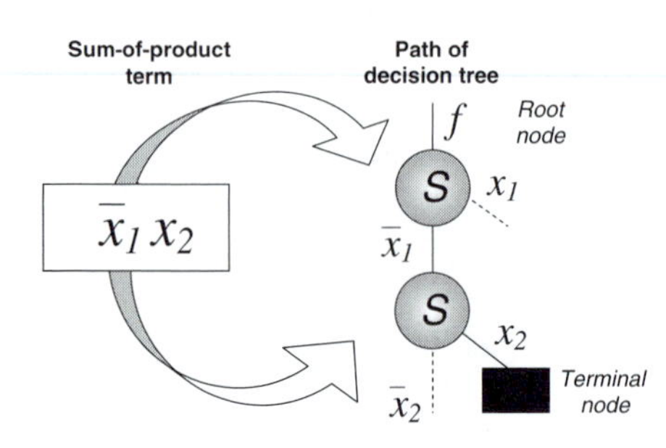

▶ *The path is a chain of two nodes (root, intermediate node(s) and one terminal node) connected by links.*
▶ *If the terminal node is marked by 0, the decision diagram represents the sum term* $(x_2 \vee \overline{x}_1) = 0$*; if the terminal node is marked by 1, the decision diagram represents the product term* $\overline{x}_1 x_2 = 1$.

FIGURE 9.5
A path in a decision diagram corresponds to the logical product or sum (Example 9.5).

Merging rules

Let the Shannon expansions for the Boolean functions α and β be

$$\alpha = \overline{x}_i \alpha_0 \vee x_i \alpha_1 \quad \text{and} \quad \beta = \overline{x}_i \beta_0 \vee x_i \beta_1$$

respectively.

Merging rule I

If $g = \alpha_0 = \beta_1$, then

$$\alpha = g \vee x_i \alpha_1 \quad \text{and} \quad \beta = \overline{x}_i \beta_0 \vee g$$

Merging rule II

If $\alpha_0 = \beta_0$ and $\alpha_1 = \beta_1$, then $\alpha = \beta$, and the nodes corresponding to α and β can be merged into one node with indegree 2.

Particular cases of elimination of internal nodes connected to the terminal nodes are shown in Figure 9.7. Elimination of the node in Figures 9.7a and 9.7b corresponds to the equations

$$f = \overline{x}_i f_0 \vee x_i f_1 = \overline{x}_i 0 \vee x_i 0 = 0 \quad \text{and}$$
$$f = \overline{x}_i f_0 \vee x_i f_1 = \overline{x}_i 1 \vee x_i 1 = 1$$

The combination of the merging and eliminating rules is illustrated in Figure 9.7c for a particular case of g. Figure 9.7d shows the reduction of the logic network to one connection (wire) based on the elimination rule:

$$f = x_i g \vee \overline{x}_i g = g(x_i \vee \overline{x}_i) = g$$

Techniques for reduction of decision trees

ELIMINATION RULE

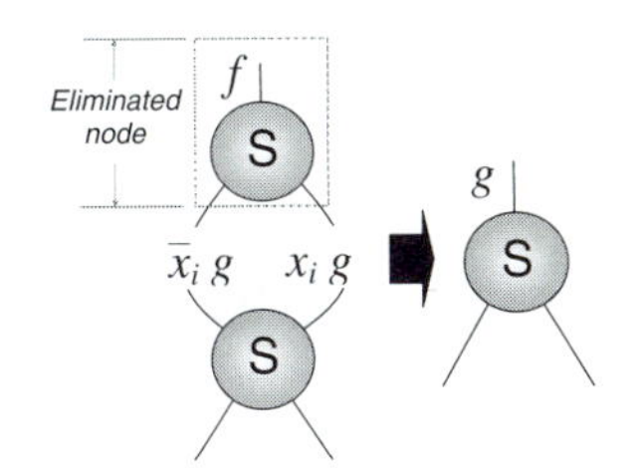

If $f_0 = f_1 = g$, then the Shannon expansion implies

$$f = \overline{x}_i g \vee x_i g = g$$

MERGING RULE I

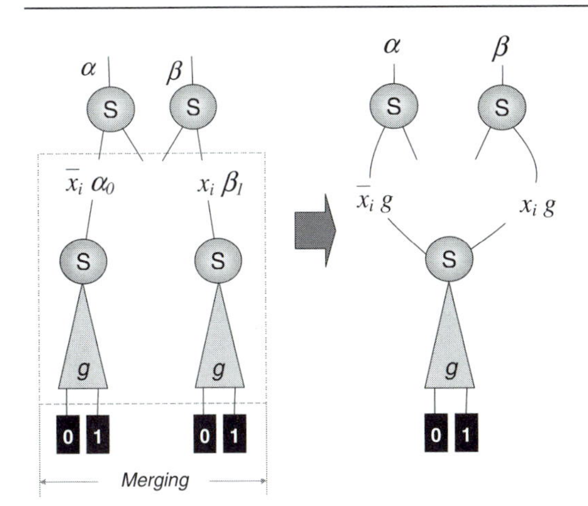

The Shannon expansions for the functions α and β are

$$\alpha = \overline{x}_i\alpha_0 \vee x_i\alpha_1$$
$$\beta = \overline{x}_i\beta_0 \vee x_i\beta_1$$

respectively. If

$$g = \alpha_0 = \beta_1$$

then

$$\alpha = g \vee x_i\alpha_1$$
$$\beta = \overline{x}_i\beta_0 \vee g$$

MERGING RULE II

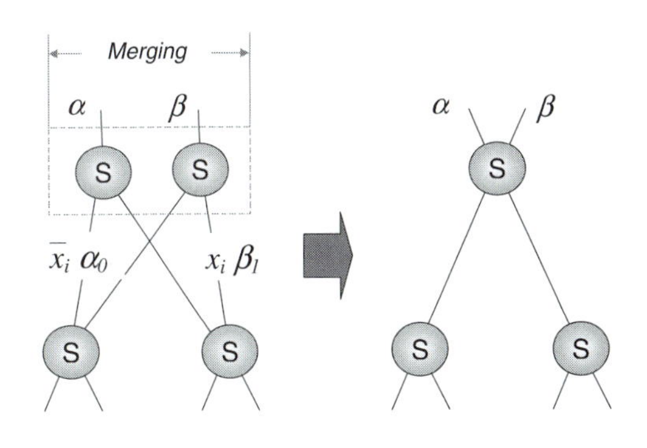

The Shannon expansions for the functions α and β are

$$\alpha = \overline{x}_i\alpha_0 \vee x_i\alpha_1$$
$$\beta = \overline{x}_i\beta_0 \vee x_i\beta_1$$

respectively. If

$$\alpha_0 = \beta_0, \quad \text{and} \quad \alpha_1 = \beta_1$$

then

$$\alpha = \beta$$

FIGURE 9.6
Reduction rules for the construction of a decision diagram from a decision tree.

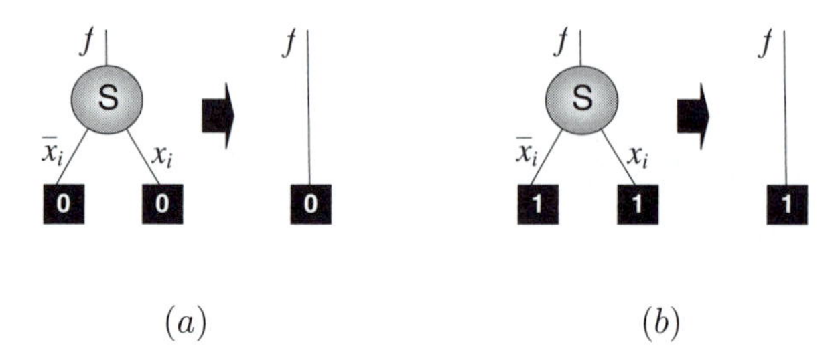

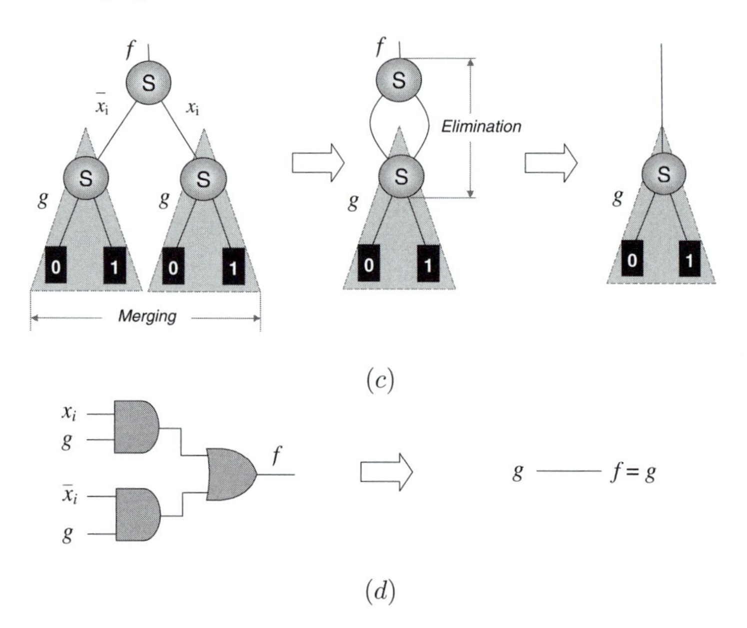

FIGURE 9.7
Application of the elimination rule (a), (b), combination of merging and elimination rules (c), and equivalent reduction using logic networks (d).

> **Deriving a reduced decision tree from a complete decision tree**
>
> **Given:** A complete decision tree for a Boolean function
> **Construct:** A reduced decision tree
> **Step 1:** Apply elimination rule where possible to remove the redundant nodes (variables in the products) of isomorphic subtrees
> **Step 2:** Repeat Step 1 as necessary until no more nodes can be eliminated

Example 9.6 (Reduced decision trees.) *Figures 9.8 and 9.9 introduce various aspects of constructing reduced decision trees from complete ones. Note that the Boolean function* $f = \overline{x}_1 \vee x_1x_2$ *derived from the tree (Figure 9.8a) is not minimal because* $f = \overline{x}_1 \vee x_1x_2 = \underbrace{(\overline{x}_1 \vee x_1)}_{Equal\ to\ 1}(\overline{x}_1 \vee x_2) = \overline{x}_1 \vee x_2$. *However, the resulting decision tree is minimal with respect to the number of nodes and the order of variables. The same situation is demonstrated by the example given in Figure 9.8c where the function obtained is not minimal:*

$$
\begin{aligned}
f &= \overline{x}_1x_2 \vee x_1x_2x_3 \vee x_1\overline{x}_2 = \overline{x}_1x_2 \vee x_1(x_2x_3 \vee \overline{x}_2) \\
&= \overline{x}_1x_2 \vee x_1 \underbrace{(x_2 \vee \overline{x}_2)}_{Equal\ to\ 1}(x_3 \vee \overline{x}_2) = \overline{x}_1x_2 \vee x_1x_3 \vee x_1\overline{x}_2
\end{aligned}
$$

The reduced decision trees are further transformed into decision diagrams. The complete algorithm for deriving the decision diagrams given a complete binary tree is given below.

Deriving a reduced decision tree from a complete decision tree

Given: A complete decision tree for a Boolean function
Construct: A reduced decision tree
Step 1: Apply elimination rule where possible, to remove the redundant nodes (variables in the products) of isomorphic subtrees
Step 2: Apply merging rules I and II, to remove the redundant nodes and isomorphic subtrees
Step 3: Repeat Steps 1 and 2 as necessary until no more nodes can be eliminated

Example 9.7 (Reduction rule.) *Application of the reduction rules to the complete decision tree of the three-input NOR function is demonstrated in Figure 9.10.*

Practice problem 9.2 (Reduction rule.) Reduce the decision tree given a three-input OR gate.
Answer is given in "Solutions to practice problems."

Practice problem 9.3. (Reducing decision tree.) Reduce the following decision trees:

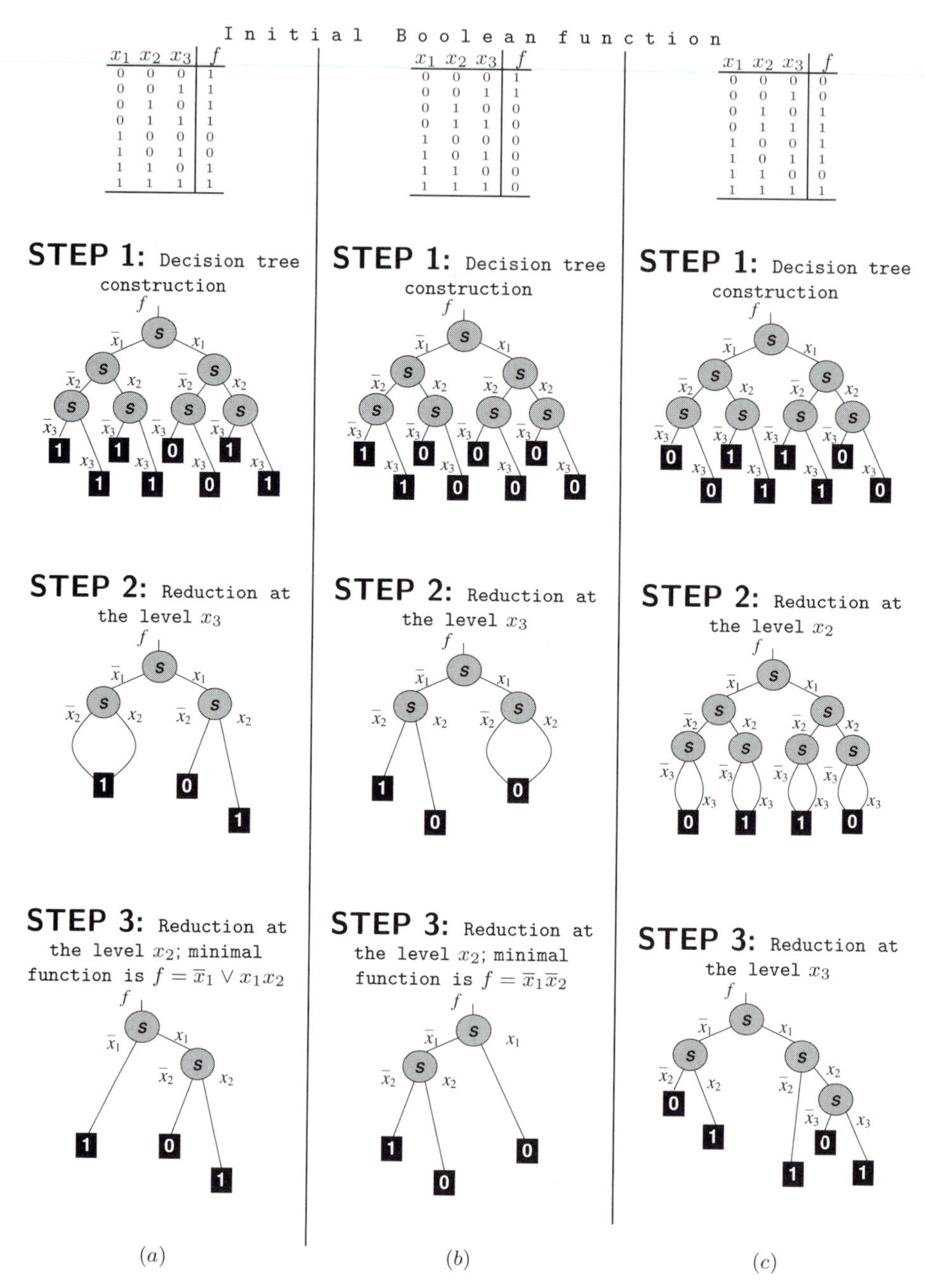
Techniques for minimization of Boolean functions
Initial Boolean function
STEP 1: Decision tree construction
STEP 2: Reduction at the level x_3
STEP 3: Reduction at the level x_2; minimal function is $f = \overline{x}_1 \vee x_1x_2$
STEP 1: Decision tree construction
STEP 2: Reduction at the level x_3
STEP 3: Reduction at the level x_2; minimal function is $f = \overline{x}_1\overline{x}_2$
STEP 1: Decision tree construction
STEP 2: Reduction at the level x_2
STEP 3: Reduction at the level x_3
(a)
(b)
(c)

FIGURE 9.8
Techniques for minimization of Boolean functions of three variables using decision trees (Example 9.6).

Techniques for minimization of Boolean functions (continuation)

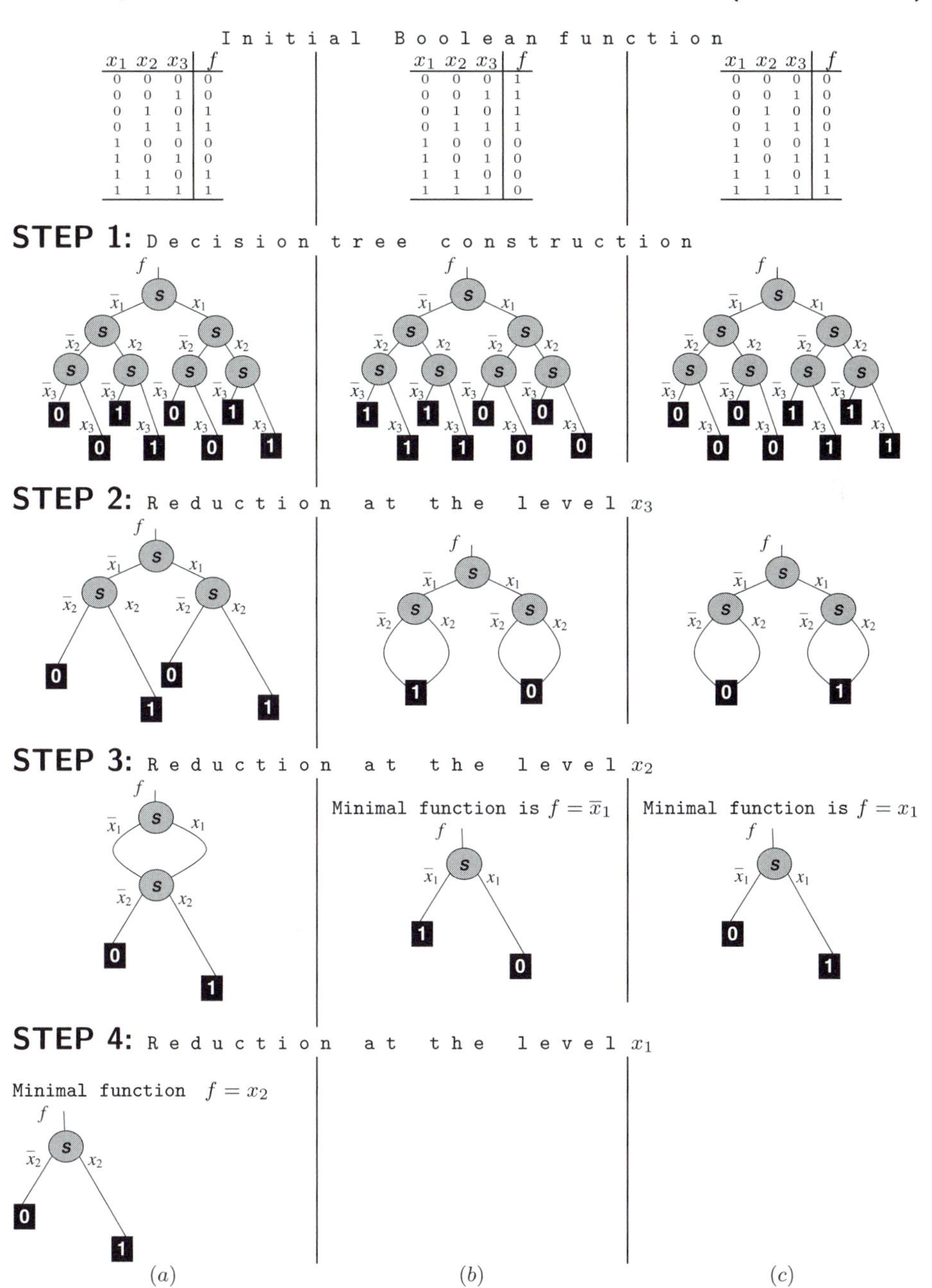

x_1	x_2	x_3	f
0	0	0	0
0	0	1	0
0	1	0	1
0	1	1	1
1	0	0	0
1	0	1	0
1	1	0	1
1	1	1	1

x_1	x_2	x_3	f
0	0	0	1
0	0	1	1
0	1	0	1
0	1	1	1
1	0	0	0
1	0	1	0
1	1	0	0
1	1	1	0

x_1	x_2	x_3	f
0	0	0	0
0	0	1	0
0	1	0	0
0	1	1	0
1	0	0	1
1	0	1	1
1	1	0	1
1	1	1	1

Initial Boolean function
STEP 1: Decision tree construction
STEP 2: Reduction at the level x_3
STEP 3: Reduction at the level x_2
Minimal function is $f = \overline{x}_1$
Minimal function is $f = x_1$
STEP 4: Reduction at the level x_1
Minimal function $f = x_2$
(a)
(b)
(c)

FIGURE 9.9
Techniques for minimization of a Boolean function of three variables using decision trees (continuation of Figure 9.8).

Design example: Decision diagram construction

GIVEN: 3-input NOR gate $f = \overline{x_1 \vee x_2 \vee x_3}$

STEP 1: Initial complete decision tree; the order of variables $x_1 \rightarrow x_2 \rightarrow x_3$

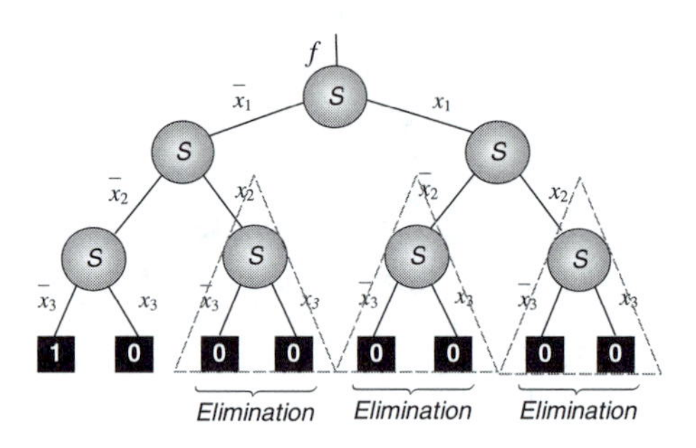

STEP 2: Reduced decision tree derived from Step 1 by using the elimination rule

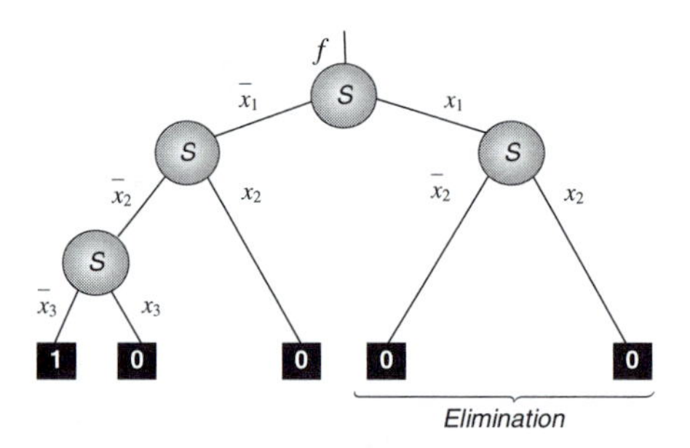

STEP 3: Reduced ordered decision diagram derived from Step 2 by using the elimination rule

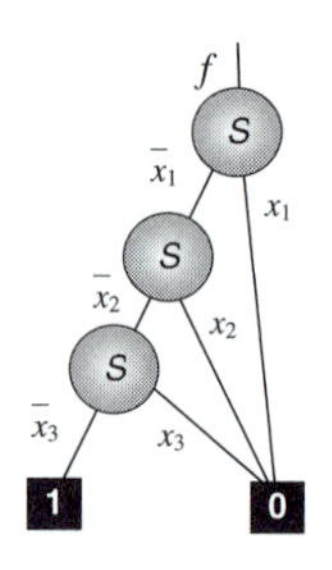

FIGURE 9.10
The three-input NOR function (a), the binary decision tree (b), and the binary decision diagram with the lexicographical order of variables (c) (Example 9.7).

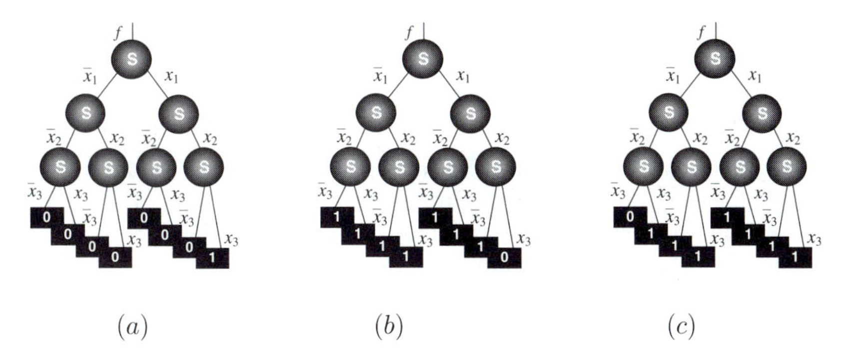

Answer is given in "Solutions to practice problems."

If an initial reduced decision tree is given, the decision diagram can be constructed from this tree using the same reduction rules.

Example 9.8 (Decision diagrams.) *A reduced decision tree that implements the Boolean function $f = x_1 \vee x_2\overline{x}_2x_3 \vee \overline{x}_2x_3x_4$ is shown in Figure 9.11. This tree can be reduced to a decision diagram, which is only different in the number of terminal nodes. The minimum SOP and POS can be derived from either the tree or the diagram.*

Design example: Decision diagram construction

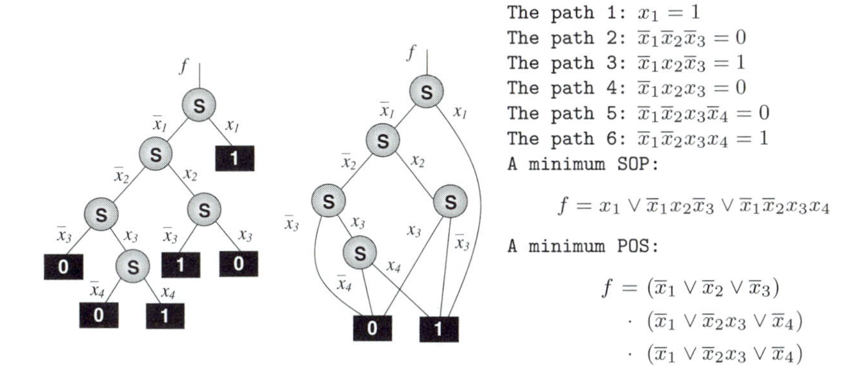

The path 1: $x_1 = 1$
The path 2: $\overline{x}_1\overline{x}_2\overline{x}_3 = 0$
The path 3: $\overline{x}_1x_2\overline{x}_3 = 1$
The path 4: $\overline{x}_1x_2x_3 = 0$
The path 5: $\overline{x}_1\overline{x}_2x_3\overline{x}_4 = 0$
The path 6: $\overline{x}_1\overline{x}_2x_3x_4 = 1$
A minimum SOP:
$f = x_1 \vee \overline{x}_1x_2\overline{x}_3 \vee \overline{x}_1\overline{x}_2x_3x_4$
A minimum POS:
$f = (\overline{x}_1 \vee \overline{x}_2 \vee \overline{x}_3)$
$\cdot\ (\overline{x}_1 \vee \overline{x}_2x_3 \vee \overline{x}_4)$
$\cdot\ (\overline{x}_1 \vee \overline{x}_2x_3 \vee \overline{x}_4)$

FIGURE 9.11
The implementation of the Boolean function $f = x_1 \vee x_2\overline{x}_2x_3 \vee \overline{x}_2x_3x_4$ (Example 9.8).

9.3 Decision diagrams for symmetric Boolean functions

The properties of Boolean functions considered in Chapter 7 can be applied for the reduction of decision trees. In particular, symmetries of the functions can be efficiently utilized in the reduction procedure. A Boolean function f is *symmetric* in variables x_i and x_j if

$$f(x_i = 1, x_j = 0) = f(x_i = 0, x_j = 1)$$

That is, a totally symmetric Boolean function is unchanged by any permutation of its variables.

Example 9.9 (Totally symmetric functions.) *Using the above definition, total symmetry can be detected:*
(a) $f = x_1 \oplus x_2$ *and* $x_1x_2 \vee x_2x_3 \vee x_1x_3$ *are totally symmetric functions.*
(b) $f = x_1x_2 \vee x_3$ *is not a totally symmetric function. It is symmetric with respect to* x_1 *and* x_2*.*

Example 9.10 (Decision diagram for a symmetric function.) *In Figure 9.12, the symmetry in variables* x_i *and* x_j *is utilized in the decision diagram construction. The Shannon expansion results in*

$$\begin{aligned} f &= \overline{x}_i\overline{x}_j f_{00} \vee \overbrace{\overline{x}_i x_j f_{01} \vee x_i\overline{x}_j f_{10}}^{Symmetry} \vee x_i x_j f_{11} \\ &= \overline{x}_i\overline{x}_j f_{00} \vee (\overline{x}_i x_j \vee x_i\overline{x}_j) f^* \vee x_i x_j f_{11} \end{aligned}$$

where $f_{01} = f_{10} = f^*$*. Instead of the four-element truth vector, this totally symmetric function can be described by the* reduced *three-element truth-vector.*

The *reduced* truth-vector of a totally symmetric Boolean function f is formed from a truth vector of f by removing the entries that are identical because of symmetry. It contains all of the information to completely specify a symmetric function. For all assignments of values to variables that have the same number of 1s, there is *exactly* one value of a symmetric function. For a totally symmetric function of n variables, the reduced truth-vector has $n + 1$ elements.

Example 9.11 (Reduced truth-vector.) *Let* $a, b, c, d \in \{0, 1\}$*. Then the reduced truth-vectors for the totally symmetric Boolean functions of two and three variables are formed as shown in Figure 9.13.*

Design example: Decision diagram for symmetric functions

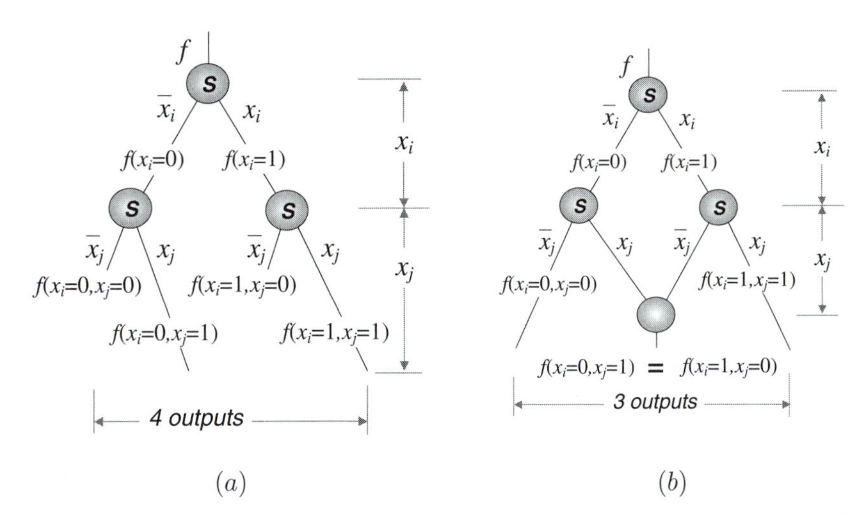

FIGURE 9.12
Fragment of a binary decision tree for a totally symmetric Boolean function (a), and the decision diagram (b) (Example 9.10).

Design example: Truth-vector for symmetric function

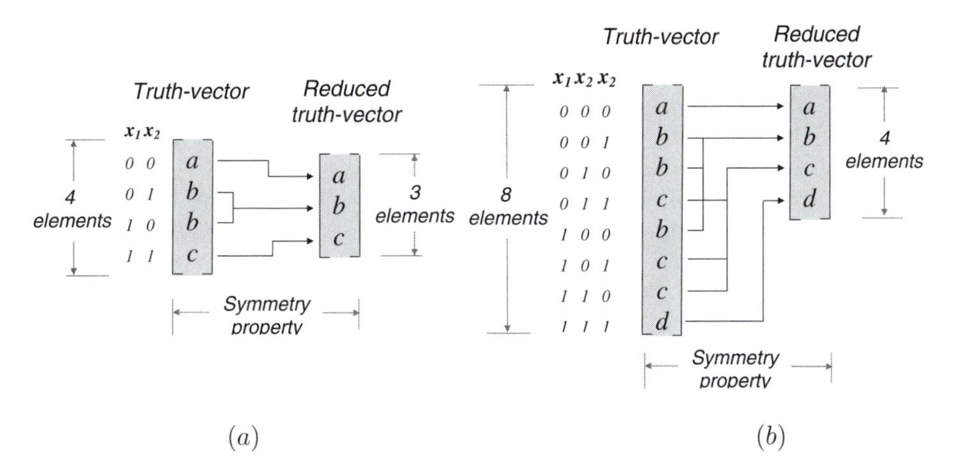

FIGURE 9.13
Forming the reduced truth-vector for a totally symmetric Boolean function of two (a) and three (b) variables (Example 9.11).

Therefore, a symmetric function is completely specified by a reduced truth vector $A = [a_0, a_1, \ldots, a_n]$, such that $f(x_1, x_2, \ldots, x_n)$ is a_i for all assignments of values to $x_1, x_2, \ldots, x_n$ that have i 1's, where $0 \leq i \leq n$.

Example 9.12 (Generation of symmetric functions.) *Figure 9.14 shows how to generate all totally symmetric Boolean functions of two variables from a reduced decision tree:*

$$f = \overline{x}_1\overline{x}_2 f_{00} \vee (\overline{x}_1 x_2 \vee x_1\overline{x}_2) f^* \vee x_1 x_2 f_{11}$$

For example, given the assignments $f_{00} = f^* = f_{11} = 1$

$$f = \overline{x}_1\overline{x}_2 \cdot 1 \vee (\overline{x}_1 x_2 \vee x_1\overline{x}_2) \cdot 1 \vee x_1 x_2 \cdot 1$$

$$= \overline{x}_1 \overbrace{(\overline{x}_2 \vee x_2)}^{Equal\ to\ 1} \vee x_1 \overbrace{(\overline{x}_2 \vee x_2)}^{Equal\ to\ 1} = \overline{x}_1 \vee x_1 = 1$$

Practice problem 9.4. (Generation of symmetric functions.) Generate the decision diagram for a totally symmetric Boolean function of three variables. Show all possible reduced truth-vectors for totally symmetric functions. Describe the first five functions in algebraic form.
Answer is given in "Solutions to practice problems."

9.4 Measurement of the efficiency of decision diagrams

A decision diagram is characterized, similarly to a decision tree, by the *size*, *depth*, *width*, *area*, and the *efficiency of reduction* of a decision tree or diagram of size SIZE_1 to a tree or diagram of size SIZE_2:

$$100 \times \frac{\text{SIZE}_1}{\text{SIZE}_2} \%$$

Example 9.13 (Reduction measures.) *In Figure 9.15, the reduction measures for the decision tree and decision diagram are given.*

Practice problem 9.5 (Decision tree and diagram.) Write the Boolean function implemented by decision tree and decision diagram given in Figure 9.15. The order of variables is $x_1 \rightarrow x_2 \rightarrow x_3$).
Answer: $f_1 = x_1 x_2 \vee x_3$.

Design example: Decision diagram for symmetric function

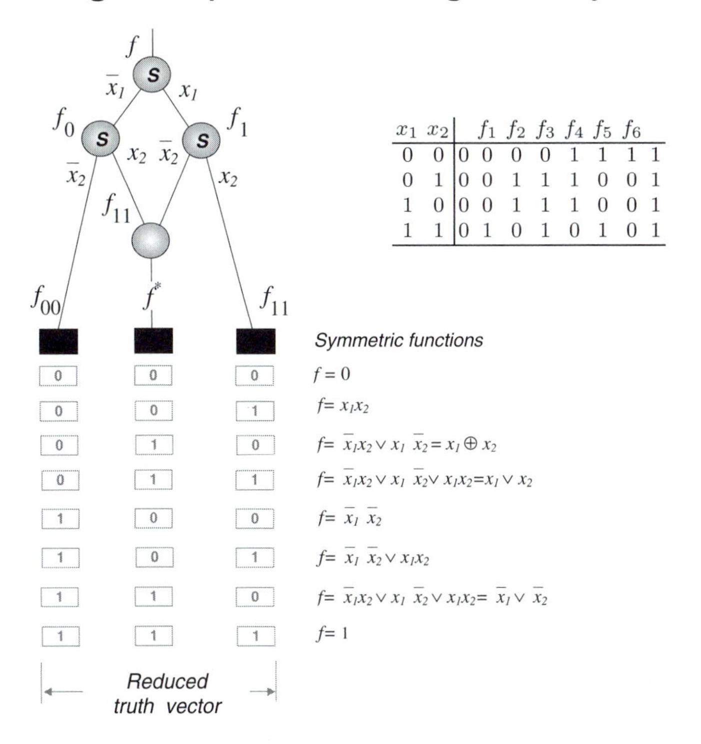

x_1	x_2	f_1	f_2	f_3	f_4	f_5	f_6
0	0	0 0	0	0	1	1	1 1
0	1	0 0	1	1	1	0	0 1
1	0	0 0	1	1	1	0	0 1
1	1	0 1	0	1	0	1	0 1

FIGURE 9.14
The decision tree for a totally symmetric Boolean function of two variables (Example 9.12).

Practice problem 9.6 (Efficiency of reduction.) The following decision tree and diagrams implement the same Boolean function:

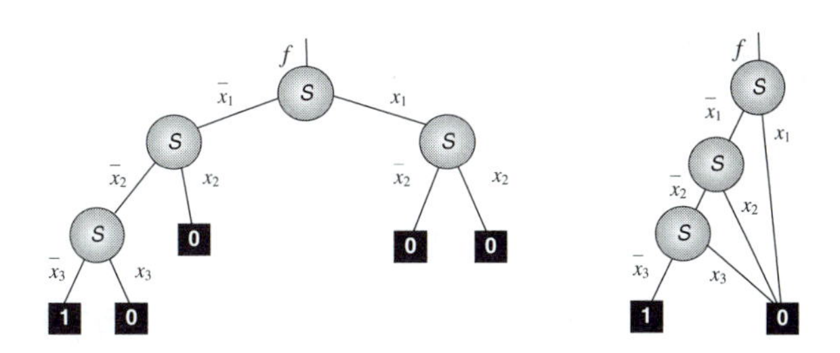

Calculate the efficiency of reduction $100 \times \frac{Size_1}{Size_2}$.
Answer is given in "Solutions to practice problems."

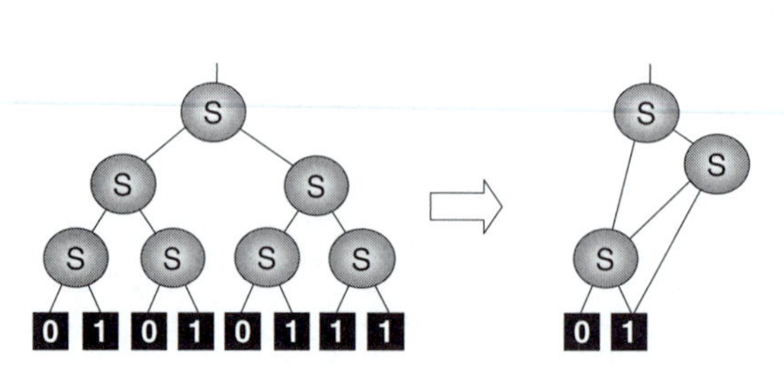

The complete decision tree:
- The size is 7
- The depth is 3
- The width is 8
- The area is 24

The decision diagram
- The size is 3
- The efficiency of reduction is

$$100 \times \frac{\text{SIZE}_1}{\text{SIZE}_2}\% = \frac{7}{3}\% = 233\%$$

FIGURE 9.15
The complete decision tree and the decision diagram for the Boolean function $f_1 = x_1x_2 \vee x_3$ (Example 9.13).

9.5 Representation of multi-output Boolean functions

A multi-output Boolean function is represented by a multi-rooted decision diagram, which is called a *shared* decision diagram.

Example 9.14 (Shared decision diagram.) *The two-output Boolean function $f_1 = x_1x_2 \vee x_3$ and $f_2 = x_1 \vee x_2 \vee x_3$ is represented by a shared decision diagram (Figure 9.16).*

Design example: Shared decision diagram

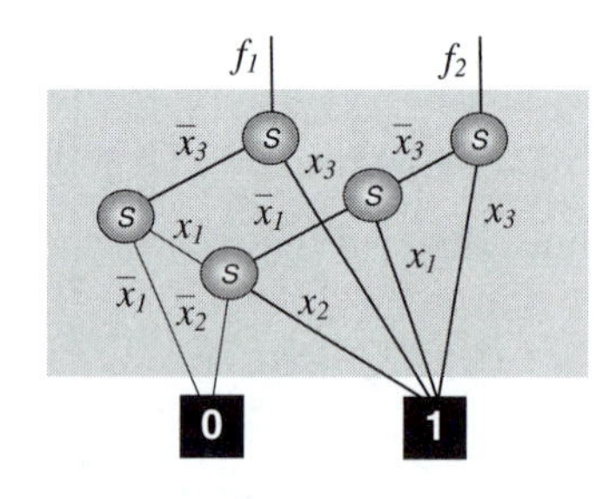

$$f_1 = x_1x_2 \vee x_3,$$
$$f_2 = x_1 \vee x_2 \vee x_3.$$

FIGURE 9.16
The shared decision diagram (Example 9.14).

Practice problem 9.7. (Shared decision diagram.) Represent the two-output Boolean function $f_1 = \overline{x}_1x_2 \vee \overline{x}_1x_3$ and $f_2 = x_1 \vee x_2 \vee x_3$ by a shared decision diagram.
Answer is given in "Solutions to practice problems."

9.6 Embedding decision diagrams into lattice structures

While decision diagrams often provide a compact representation of Boolean functions, their layout is not much simpler than that of "traditionally" designed logic networks, making placement and routing a difficult task. As an alternative, *lattice diagrams* have been proposed. The number of nodes at each level is linear, which makes the diagram fit onto a 2-dimensional structure suitable for resolution of the routing problem.

The embedding problem is formulated as follows: Embed a given decision diagram (the guest structure) into an appropriate regular lattice structure (the host structure). The particular solution to this problem is as follows:

- ▶ The host structure is specified as the lattice structure.
- ▶ The decision diagram must be transferred to a configuration that can be embedded into a lattice structure.

Example 9.15 (Embedding a decision tree into a lattice.) *An example of embedding a decision tree into a lattice is given in Figure 9.17.*

Design example: Embedding a decision tree into a lattice

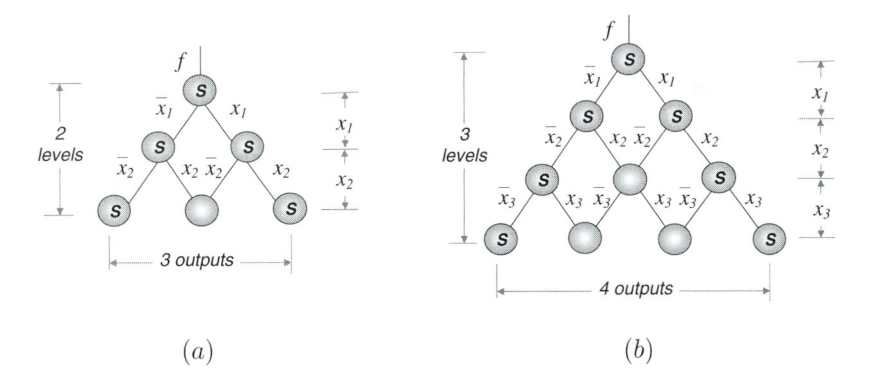

FIGURE 9.17
Embedding a decision tree into a lattice given a Boolean function of two (a) and three (b) variables (Example 9.15).

Note that an arbitrary Boolean function represented by a decision tree can be embedded into a lattice structure using the techniques known as symmetrization based on *pseudo-symmetry*.

9.7 Summary of optimization of Boolean functions using decision diagrams

This chapter introduces methods for the reduction of decision trees. In general, decision trees are redundant data structures. The results of the reduction of decision trees are called ***decision diagrams***. The key aspects of this chapter are listed below:

(*a*) Techniques for constructing decision diagrams, and
(*b*) Techniques for computing Boolean functions using decision diagrams.

The main conclusions of this chapter are summarized as follows:

- ▶ The conversion of a decision tree into a decision diagram is called the ***optimization*** of this tree.
- ▶ A decision tree is a canonical representation of f in graphical form with structural properties as follows:
  - (*a*) The nodes implement Shannon expansion.
  - (*b*) Non-terminal nodes are distributed over levels, each level corresponding to a variable x_i in f, starting from the root node corresponding to x_i, or from some other variable that was chosen first.
  - (*c*) Since the variables appear in a fixed order, such a tree is called an *ordered* decision tree.
  - (*d*) Each path from the root node to the terminal nodes corresponds to a minterm in the sum-of-products representation of the function; the minterm is determined as the product of labels at the edges.
  - (*e*) The values of terminal nodes are the values of the functions represented.
- ▶ Reduced decision trees are referred to as decision diagrams.
- ▶ A decision diagram is a directed acyclic graph that contains:
  - (*a*) Exactly one root;
  - (*b*) Terminal nodes labeled with the constants 1 and 0;
  - (*c*) Internal nodes labeled with S; Shannon expansion is implemented in these nodes;
  - (*d*) Links between nodes labeled with $\overline{x}_i$ and x_i.
- ▶ In an ***ordered*** decision diagram, a linear variable order is placed on the input variables. The variables' occurrences in each path of this diagram have to be consistent with this order.

Summary (continuation)

- A decision diagram is characterized, similarly to a decision tree, by its ***size***, ***depth***, ***width***, ***area***, and ***efficiency of reduction***.
- A decision diagram represents a Boolean function f in the following manner: (a) Each assignment to the input variables x_i defines a uniquely determined path from the root of the graph to one of its terminal nodes and (b) the label of the terminal node thus reached gives the Boolean function value of this input.
- The decision diagram for a given Boolean function f is derived from the decision tree for f by deleting redundant nodes, and by sharing equivalent subgraphs using reduction rules such as ***elimination*** and ***merging***.
- A multi-output Boolean function is represented by a multi-rooted decision diagram, which is called a ***shared*** decision diagram.
- Advanced optimization techniques using decision diagram techniques are introduced in the next section, "Further study."

9.8 Further study

Historical perspective

1959: C. Y. Lee introduced the notation of binary decision diagrams for the first time in his pioneering paper "Binary Decision Diagrams," *Bell System Technical Journal,* volume 38, number 4, pages 985–999, 1959.

1978: S. B. Akers further developed the binary decision diagram model and proposed to use it as the basis for developing tests for the Boolean function it represents ("Binary Decision Diagrams" by S. B. Akers, *IEEE Transactions on Computers,* volume 27, number 6, pages 509–516, 1978). An excellent survey on decision tree and diagram techniques and applications was written in 1982 by B. M. E. Moret ("Decision Trees and Diagrams," *Computing Surveys,* volume 14, number 4, pages 593–623, 1982).

1986: R. E. Bryant brought decision diagram techniques to a level suitable for use in practice in his revolutionary paper "Graph-Based Algorithms for Boolean Function Manipulation," *IEEE Transactions on Computers*, volume 35, number 6, pages 677–691, 1986.

1990: The first universal decision diagram package was developed by K. S. Brace, R. L. Rudell and R. E. Bryant ("Efficient Implementation of a BDD package," *Proceedings of the 27th ACM/IEEE Design Automation Conference,* Orlando, FL, pages 40–45, 1990).

Advanced topics of decision diagram techniques

Topic 1: Decision diagram-based decomposition. Details on using decision diagrams on minimization can be found, in particular, in "Multi-Level Logic Optimization" by M. Fujita, Y. Matsunaga, and M. Ciesielski, in *Logic Synthesis and Verification* Edited by S. Hassoun and T. Sasao, Consulting Editor R. K. Brayton, Kluwer, 2002.

Topic 2: Decision diagrams for multi-valued logic functions are a generalization of binary decision diagrams. Details can be found in the *Proceedings of the Annual IEEE Symposium on Multiple-Valued Logic.*

Topic 3: Lattice decision diagrams are constructed by embedding decision diagrams, representing of symmetric functions, into a *lattice array.* For this, property of symmetric Boolean functions are generalized for an arbitrary Boolean function using an approach called *pseudo-symmetry.*

An arbitrary Boolean function can be represented using the elementary symmetric functions (see, for example, "Transformation of Switching Functions to Completely Symmetric Switching Functions" by R. C. Born and A. K. Scidmore, *IEEE Transactions on Computers*, volume C-17, number 6, 1968, pages 596–599). This implies that a decision diagram, representing an arbitrary Boolean function, can be transformed into a lattice decision diagram.

Lattice decision diagrams are based on the lattice array, which is defined as a rectangular array of identical cells, each of them being a multiplexer, where every cell obtains signals from two neighbor inputs and gives them to two neighbor outputs.

Topic 4: Computing using Pascal and transeunt triangles. Table 9.1 contains the following graphical data structures for the representation of five elementary Boolean functions: both Pascal and the transeunt triangles (for symmetric functions), decision trees, and decision diagrams (see more details in Chapter 12, "Further study" section).

Further reading

"Ordered Binary Decision Diagrams" by R. E. Brayant and C. Meinel, in *Logic Synthesis and Verification* edited by S. Hassoun and T. Sasao, Consulting Editor R. K. Brayton, Kluwer, 2002.

Decision Diagram Techniques for Micro- and Nanoelectronic Design by S. N. Yanushkevich, D. M. Miller, V. P. Shmerko, and R. S. Stanković, CRC/Taylor & Francis Group, Boca Raton, FL, 2006.

Binary Decision Diagrams and Applications for VLSI CAD by S. Minato, Kluwer, 1996.

Binary Decision Diagrams. Theory and Implementation by R. Drechsler and B. Becker, Kluwer, 1998.

Spectral Interpretation of Decision Diagrams by R. S. Stanković and J. T. Astola, Springer, 2003.

Algorithms and Data Structures in VLSI Design by C. Meinel and T. Theobald, Springer, 1998.

"Boolean Division and Factorization Using Binary Decision Diagrams" by T. Stanion and C. Sechen, *IEEE Transactions on Computer-Aided Design of Integrated Circuits and Systems*, volume 13, pages 1179–1184, September 1994

TABLE 9.1
Graphical data structures: Pascal and transeunt triangles, decision tree, and decision diagram for the representation of elementary Boolean functions of two variables.

Function triangle	Pascal triangle	Transeunt tree	Decision diagram	Decision
x_1, x_2, f	1; 0 1; 0 1; 0 0 0 1	1; 0 1; 0 0 1	f; S; $\bar{x}_1$, x_1; S, S; $\bar{x}_2$, x_2, $\bar{x}_2$, x_2; 0 0 0 1	f; S; x_1, $\bar{x}_1$; S; $\bar{x}_2$, x_2; 0 1
x_1, x_2, f	1; 1 0; 1 0; 0 1 1 1	1; 1 0; 0 1 1	f; S; $\bar{x}_1$, x_1; S, S; $\bar{x}_2$, x_2, $\bar{x}_2$, x_2; 0 1 1 1	f; S; $\bar{x}_1$, x_1; S; $\bar{x}_2$, x_2; 0 1

"BDS: A BDD-Based Logic Optimization System" by C. Yang and M. Ciesielski, *IEEE Transactions on Computer-Aided Design of Integrated Circuits and Systems*, volume 21, number 7, pages 866–876, 2002.

"Spatial Interconnect Analysis for Predictable Nanotechnologies" by S. N. Yanushkevich, V. P. Shmerko, and B. Steinbach, *Journal of Computational and Theoretical Nanoscience*, American Scientific Publishers, volume 4, number 8, pages 1–14, 2007.

9.9 Solutions to practice problems

Practice problem	Solution
9.1.	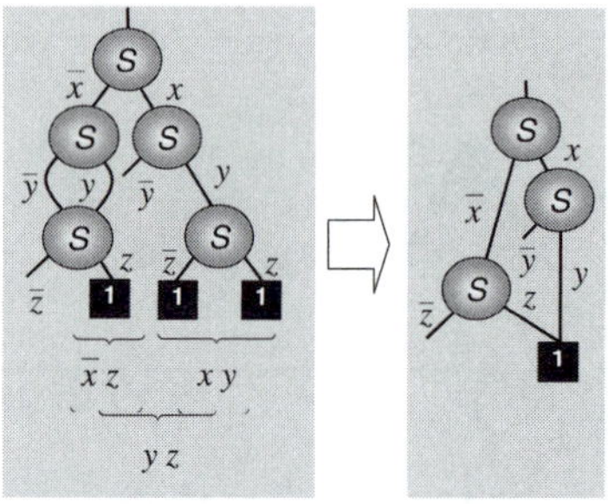
9.2.	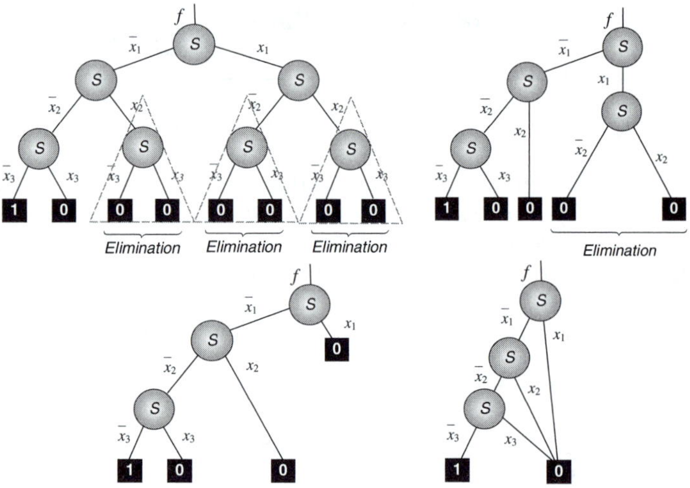
9.3.	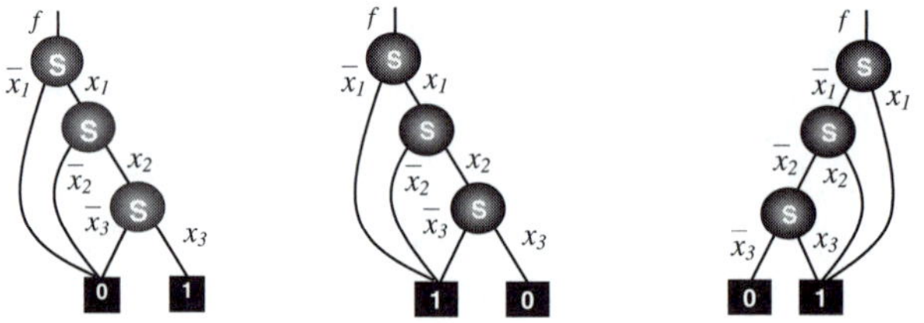 (a) (b) (c)

Solutions to practice problems (continuation)

Practice problem	Solution
9.4.	$f_1 = 0$ $f_2 = \overbrace{x_1x_2x_3}^{f_{111}}$ $f_3 = \overbrace{\overline{x}_1x_2x_3 \vee x_1\overline{x}_2x_3 \vee x_1x_2\overline{x}_3}^{f^{**}}$ $f_4 = \overbrace{\overline{x}_1x_2x_3 \vee x_1\overline{x}_2x_3 \vee x_1x_2\overline{x}_3}^{f^{**}} \vee \overbrace{x_1x_2x_3}^{f_{111}}$ $f_5 = \overbrace{\overline{x}_1\overline{x}_2x_3 \vee \overline{x}_1x_2\overline{x}_3 \vee x_1\overline{x}_2\overline{x}_3}^{f^{*}}$ $f = \overline{x}_1\overline{x}_2\overline{x}_3f_{000} \vee \overline{x}_1\overline{x}_2x_3f_{001} \vee \overline{x}_1x_2\overline{x}_3f_{010} \vee \overline{x}_1x_2x_3f_{011} \vee x_1\overline{x}_2\overline{x}_3f_{100} \vee x_1\overline{x}_2x_3f_{101} \vee x_1x_2\overline{x}_3f_{110} \vee x_1x_2x_3f_{111} = \overline{x}_1\overline{x}_2\overline{x}_3f_{000} \vee (\overline{x}_1\overline{x}_2x_3 \vee \overline{x}_1x_2\overline{x}_3 \vee x_1\overline{x}_2\overline{x}_3)f^{*} \vee (\overline{x}_1x_2x_3 \vee x_1\overline{x}_2x_3 \vee x_1x_2\overline{x}_3)f^{**} \vee x_1x_2x_3$
9.6, 9.7	Practice problem 9.6 The sizes of the first and second decision diagrams are $\text{SIZE}_1 = 4$ and $\text{SIZE}_2 = 3$. The efficiency of reduction is $100 \times \frac{Size_1}{Size_2} = 133\%$ Practice problem 9.7

9.10 Problems

Problem 9.1 Given the Boolean function $f = x_1x_2 \vee \overline{x}_3$, build its binary decision diagrams using the following order of variables:

(a) $x_1 \longrightarrow x_2 \longrightarrow x_3$ (b) $x_2 \longrightarrow x_1 \longrightarrow x_3$ (c) $x_1 \longrightarrow x_3 \longrightarrow x_2$

Problem 9.2 Reduce the following decision trees:

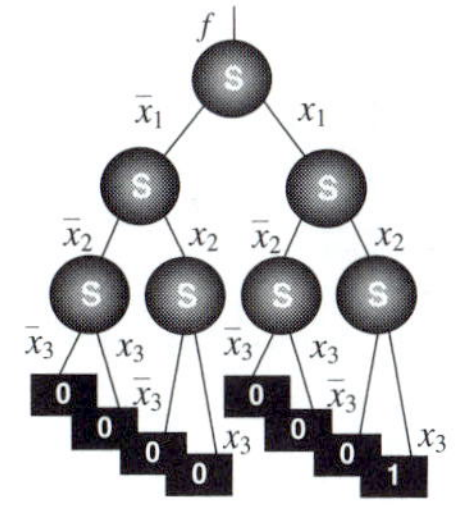

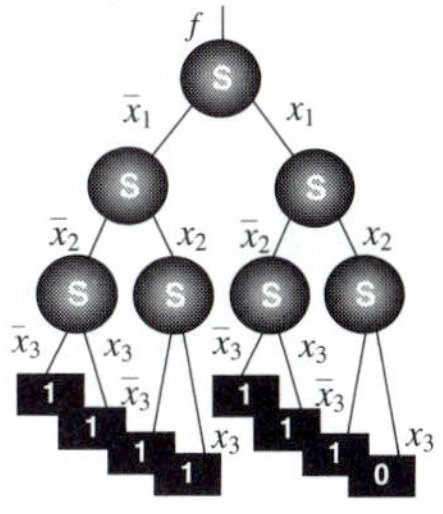

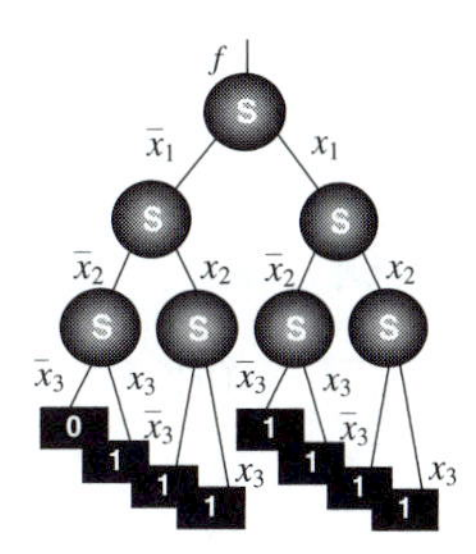

Problem 9.3 Derive decision diagrams for the Boolean functions of three variables given the following hypercubes:

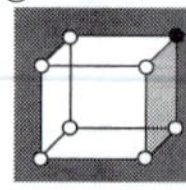
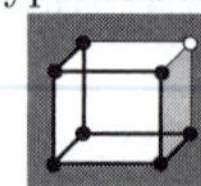
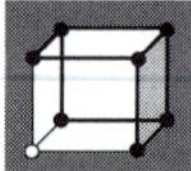
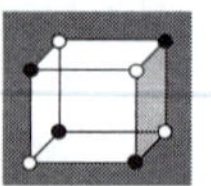

Problem 9.4 Derive decision diagrams using the following algebraic cube representation of Boolean functions:

(a) $f = [1\ 1\ 1]$ (b) $f = \begin{bmatrix} \mathbf{x}\ \mathbf{x}\ 1 \\ \mathbf{x}\ 1\ \mathbf{x} \\ 1\ \mathbf{x}\ \mathbf{x} \end{bmatrix}$ (c) $f = \begin{bmatrix} 0\ \mathbf{x}\ \mathbf{x} \\ \mathbf{x}\ 0\ \mathbf{x} \\ \mathbf{x}\ \mathbf{x}\ 0 \end{bmatrix}$ (d) $f = \begin{bmatrix} 1\ 0\ 0 \\ 0\ 1\ 0 \\ 0\ 0\ 1 \\ 1\ 1\ 1 \end{bmatrix}$

Problem 9.5 Derive the polynomial expressions given the following functional decision diagrams

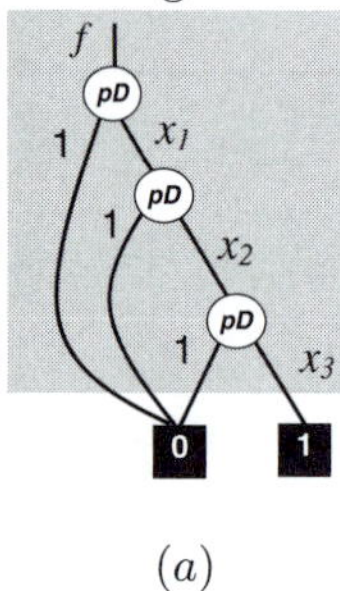

(*a*)

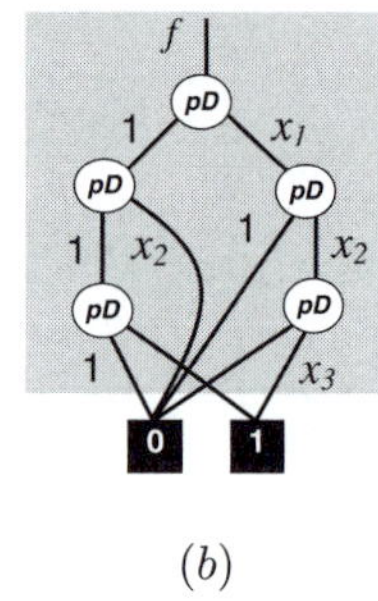

(*b*)

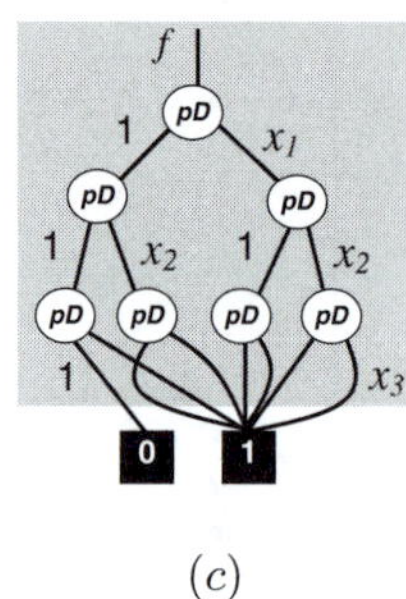

(*c*)

Problem 9.6 Derive the reduced decision trees for the following Boolean functions given by their truth tables:

(a)

x_1	x_2	x_3	f
0	0	0	1
0	0	1	1
0	1	0	1
0	1	1	1
1	0	0	1
1	0	1	1
1	1	0	1
1	1	1	0

(b)

x_1	x_2	x_3	f
0	0	0	1
0	0	1	1
0	1	0	1
0	1	1	1
1	0	0	1
1	0	1	1
1	1	0	1
1	1	1	0

(c)

x_1	x_2	x_3	f
0	0	0	1
0	0	1	1
0	1	0	1
0	1	1	1
1	0	0	1
1	0	1	1
1	1	0	1
1	1	1	0

(d)

x_1	x_2	x_3	f
0	0	0	1
0	0	1	1
0	1	0	1
0	1	1	1
1	0	0	1
1	0	1	1
1	1	0	1
1	1	1	0

Problem 9.7 Derive the algebraic SOP and POS forms of the Boolean functions from the following decision diagrams:

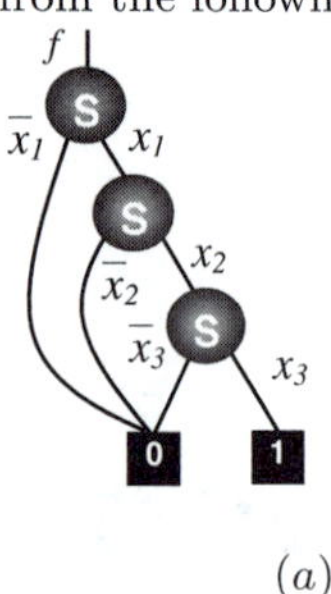

(*a*)

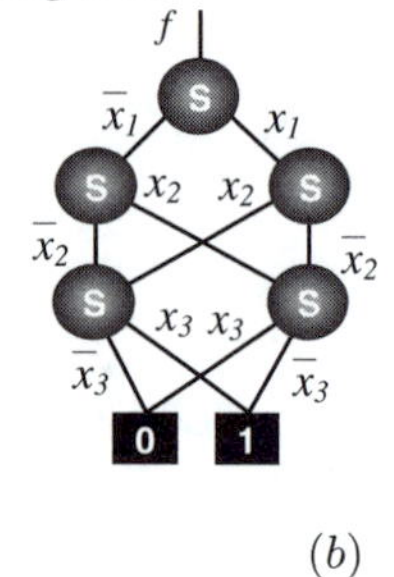

(*b*)

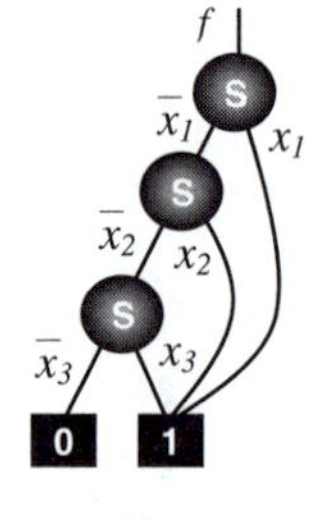

(*c*)

Problem 9.8 Convert the following logic networks into the binary decision diagrams:

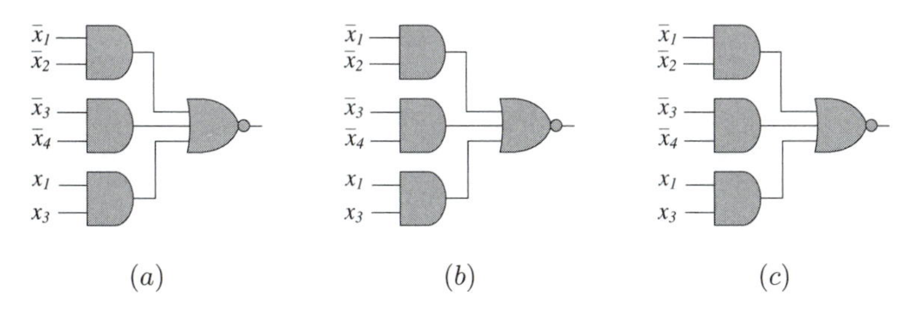

(*a*) (*b*) (*c*)

Problem 9.9 Construct the binary decision diagrams for the following Boolean function of three variable given by the truth-vectors:
(a) $\mathbf{F} = [01101101]^T$ (b) $\mathbf{F} = [01101101]^T$ (c) $\mathbf{F} = [01101101]^T$

Problem 9.10 Construct the binary decision diagrams for the following Boolean functions:
(a) $f = \bigvee m(0,1,4,5)$ (b) $f = \prod m(1,5,7)$ (c) $f = \bigvee M(0,2,3,4,7)$

Problem 9.11 Derive the binary decision diagrams for the totally symmetric Boolean functions of three variables given the following carry vectors:
(a) $\mathbf{F}_c = [011]^T$ (b) $\mathbf{F}_c = [0110]^T$ (c) $\mathbf{F}_c = [0010]^T$

Problem 9.12 Derive the binary decision diagrams using optimization techniques for the Boolean functions of four variables given the following K-maps:

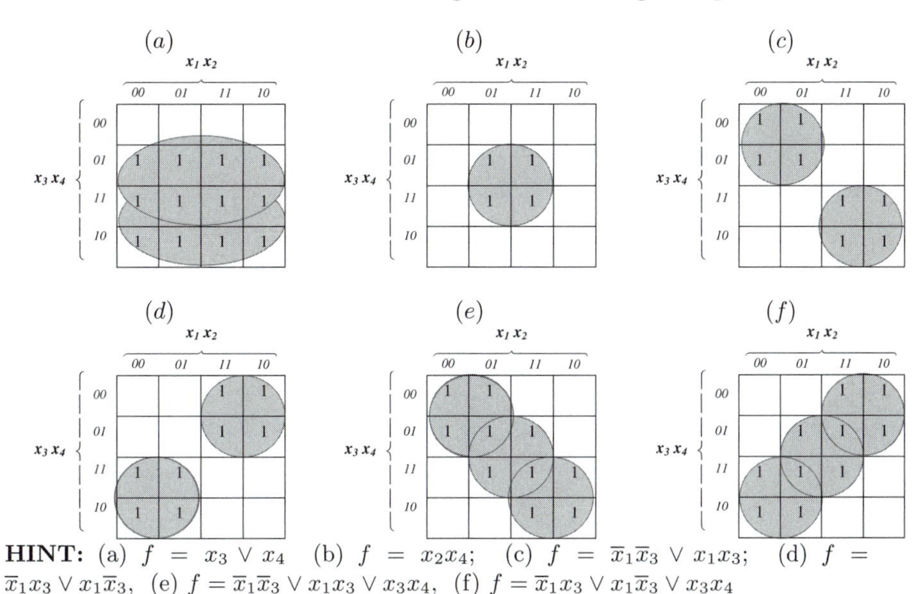

HINT: (a) $f = x_3 \vee x_4$ (b) $f = x_2x_4$; (c) $f = \overline{x}_1\overline{x}_3 \vee x_1x_3$; (d) $f = \overline{x}_1x_3 \vee x_1\overline{x}_3$, (e) $f = \overline{x}_1\overline{x}_3 \vee x_1x_3 \vee x_3x_4$, (f) $f = \overline{x}_1x_3 \vee x_1\overline{x}_3 \vee x_3x_4$

10

Algebra II: Polynomial

Domains

- Operational domain
- Functional domain

Algebra

- Types of polynomial forms
- Algebra II
- Polarized minterms
- Data structures for functional domain

Fundamental expansions

- Shannon expansion
- Taylor logic expansion

Techniques for logic networks

- Local transformations
- Factoring
- Proving validity

Advanced topics

- Arithmetic polynomials
- Polynomials and decision diagrams

10.1 Introduction

In the previous chapters, the SOP form and its dual equivalent, the POS form, of a Boolean function have been the focus of interest. The SOP forms were the data structures most often used for Boolean function representation during the 40's and 50's, when the first generation of computers was developed. Progress in computation methods dates from the beginning of the 70's, stimulated by the study of alternatives to the SOP forms (details are given in the section "Further study"). This study resulted in the development of so-called *polynomial* forms of Boolean functions (Figure 10.1). An arbitrary Boolean function can be represented in polynomial forms.

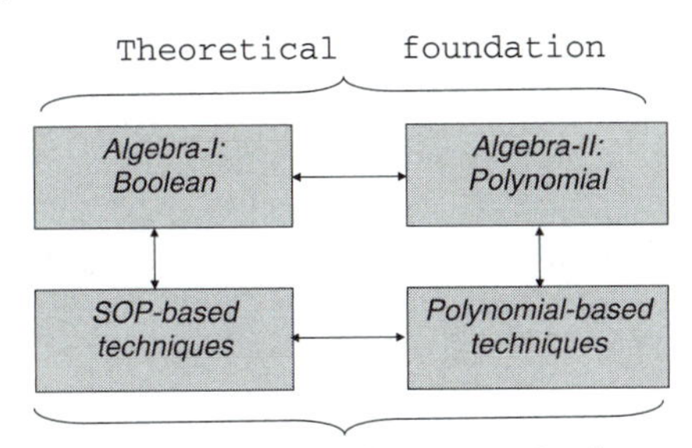

FIGURE 10.1
Techniques for the manipulation of Boolean functions are based on Boolean algebra (Algebra I) and the algebra of polynomial forms, or polynomial algebra (Algebra II).

The flexibility of polynomial forms

The motivation for studying polynomial forms of Boolean functions is as follows:

Features of the polynomial forms of Boolean functions

- ▶ Polynomial expressions provide additional flexibility in terms of choice of implementation technology. This property is efficiently utilized in logic design, especially in design of specific-area applications; in particular, encoding and encryption of information.
- ▶ There are various physical and molecular effects in predictable technology that can be interpreted as EXOR operations. Nanocomputing devices based on these effects can be used in logic network design and implementation.
- ▶ Polynomial forms are well suited to logic with more than two values – so-called *multi-valued* logic. This fact is utilized in the design of some contemporary and next-generation devices.

In this chapter, only EXOR polynomial forms are considered. The corresponding algebra is called polynomial algebra in the field GF(2). Other polynomial forms are mentioned in the "Further study" section.

Example 10.1 (EXOR and AND-OR networks.) *Figure 10.2 shows the EXOR gate and its AND-OR equivalent network. Using the criterion of the number of gates, the AND-NOR implementation is more complicated than EXOR-based networks.*

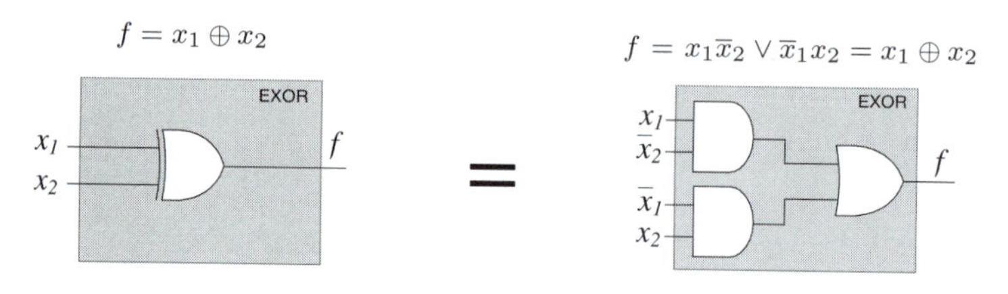

FIGURE 10.2
The EXOR function is represented by a two-level network of AND and OR gates (Example 10.1).

Practice problem 10.1. **(The EXOR realization.)** Implement the EXOR function of two variables using NAND gates.
Answer is given in "Solutions to practice problems."

Similarities between SOP and polynomial forms

Given a Boolean function, the SOP expressions can be derived from the *standard* SOP form of this function using the simplification rules. Given a complete set of minterms, standard SOP expressions are formed using the correspondence of 1's in the truth-vectors and minterms. By analogy, polynomial forms are derived from the correspondence of the polarized minterms and nonzero coefficients of the vector of coefficients.

The standard, or canonical, SOP and polynomials forms are unique given a Boolean function. The number of terms in canonical SOP and polynomial expressions are equal to 2^n. Noncanonical SOP expressions can be derived from canonical SOP forms. Similarly, canonical and noncanonical polynomial expressions can be derived given a Boolean function. Figure 10.3 shows the structural similarity of standard SOP expressions and polynomial forms of Boolean functions:

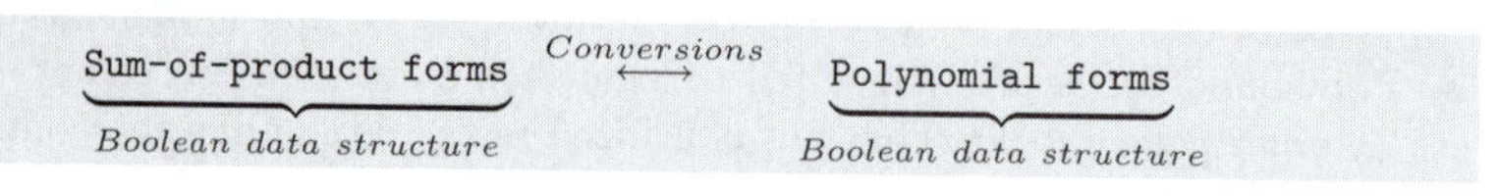

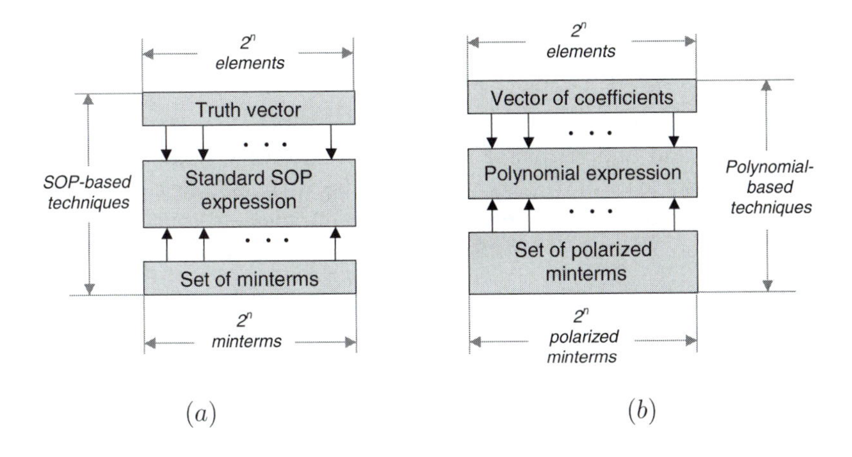

FIGURE 10.3
Structural similarity between standard SOP (a) and polynomial (b) forms.

The completeness of the set of operations used in polynomial forms

The polynomial form is a representation of a Boolean function derived from the following universal set of operations over Boolean variables:

- The constant 1,
- AND operations , and
- EXOR operations.

A formal proof of the completeness of this set of operations is given in Chapter 4.

> **Example 10.2 (Completeness of operations.)** *(a) EXOR of single variables forms a so-called "linear polynomial"* $x_1 \oplus \cdots \oplus x_n$. *(b) EXOR and AND operations form a so-called a "non-linear polynomial," such as, for example,* $x_1 \oplus x_1x_2 \oplus x_1x_2x_3$. *(c) EXOR, and AND operations and the constant 1 are used to implement an arbitrary Boolean function. For example, a complemented variable can be represented using EXOR and the constant 1:* $\overline{x} = x \oplus 1$.

Data structures for polynomial forms

Polynomial forms can be represented as follows (Figures 10.4 and 10.5):

- *Tabulated* forms, such as *functional tables* and *vectors of coefficients* (similar to truth tables and truth-vectors for SOP forms);
- *Graphical* representations, such as *functional maps* and *functional cubes*, as well as *functional decision trees and diagrams* (similar to K-maps, cubes, and decision trees and diagrams for SOP forms); and

- *Logic networks* of logic gates or threshold elements.

Operational and functional domains

The relationship between *operational* and *functional* domains is the key to the synthesis and application of the polynomial forms of Boolean functions. All satellite Boolean data structures and the corresponding techniques are aimed at providing for representation, manipulation, optimization, and implementation of Boolean functions in the functional domain, namely (Figure 10.4):

(*a*) Each data structure has particular properties and characteristics and satisfies the requirements of specific tasks of the logic design cycle. There is no "universal" data structure that can be used in all phases of logic design.

(*b*) Each data structure plays a particular role in design, and is efficient only in solving particular tasks.

(*c*) Each data structure can be converted into another one. These relationships between data structures are often used to achieve design goals.

> **Example 10.3 (Factorization of SOPs.)** *Factoring techniques aim at a multi-level implementation of Boolean functions. For example, given the Boolean function $f = x_1x_2 \vee x_2\overline{x}_3x_4$, it can be factored as follows: $f = x_2(x_1 \vee \overline{x}_3x_4)$. The corresponding logic networks are shown in Figure 10.6.*

Relationships between data structures

The relationships between various data structures in the operational and functional domains are used for the manipulation of polynomial expressions:

```
  Functional decision tree ↔ Functional decision diagram
           EXOR expression ↔ Functional decision tree
           EXOR expression ↔ Functional decision diagram
Functional decision diagram ↔ Matrix expression
            Algebraic form ↔ Matrix form
```

Analysis and synthesis of EXOR networks

The combinational networks based on SOP representations are designed, or *synthesized* and *analyzed* using Boolean algebra. Similarly, there are two classes of design problems based on polynomial descriptions: *analysis* and *synthesis*. The analysis problem is formulated as follows: Given an EXOR network, provide a tabular description of this network. The steps involved in synthesis of EXOR networks are basically the reverse of those involved in the analysis. Techniques for the analysis and synthesis of EXOR networks are based on various data structures: vector and algebraic forms, as well as functional decision trees and diagrams.

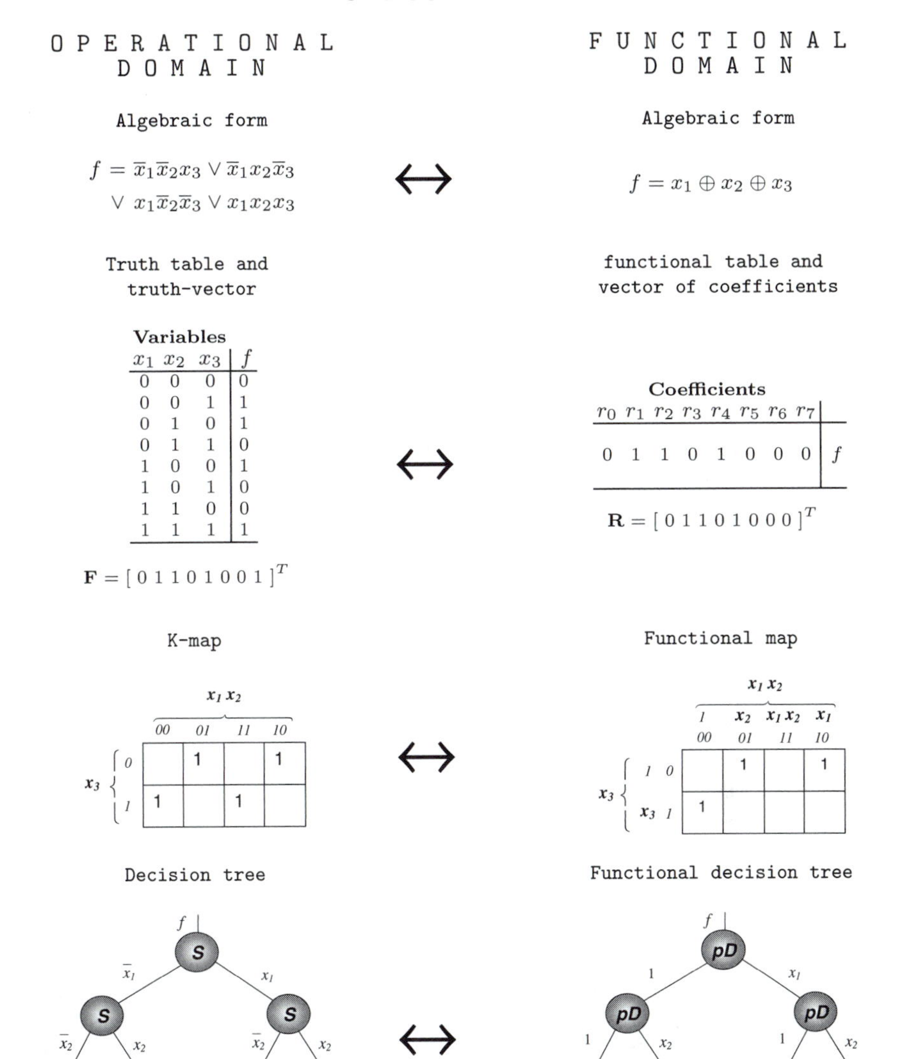

FIGURE 10.4
Data structures for representations of the EXOR function of two variables in the operational domain (SOP form) and the functional domain (polynomial form).

Techniques for polynomial representation of Boolean functions (continuation)

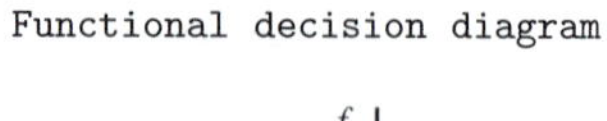

Decision diagram

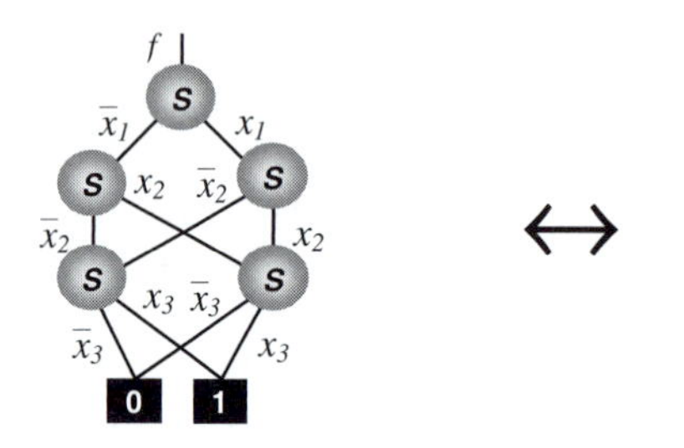

↔

Functional decision diagram

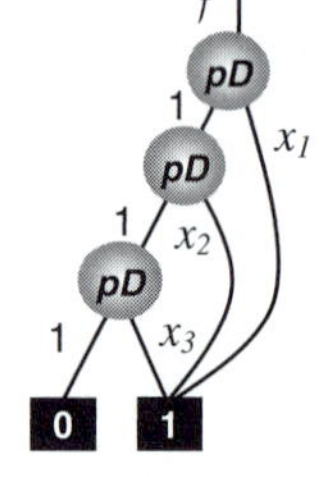

Logic network using gates

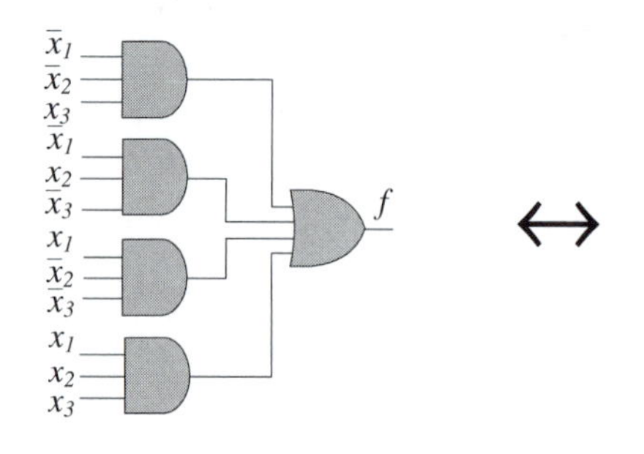

↔

Logic network using gates

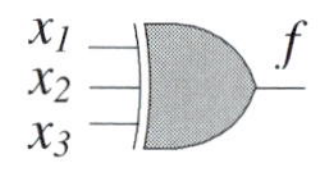

Logic network using threshold elements

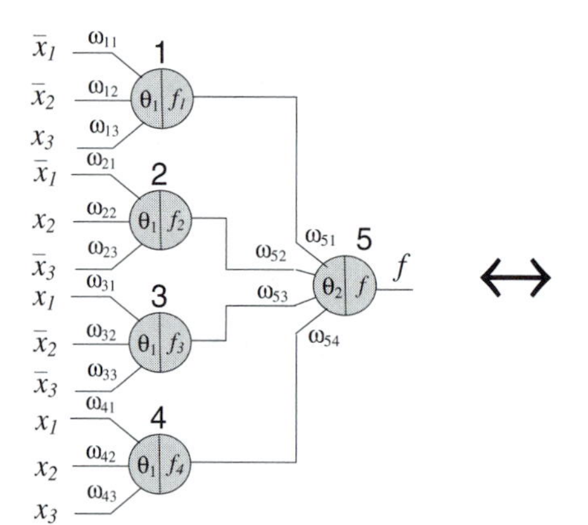

↔

Logic network using threshold elements

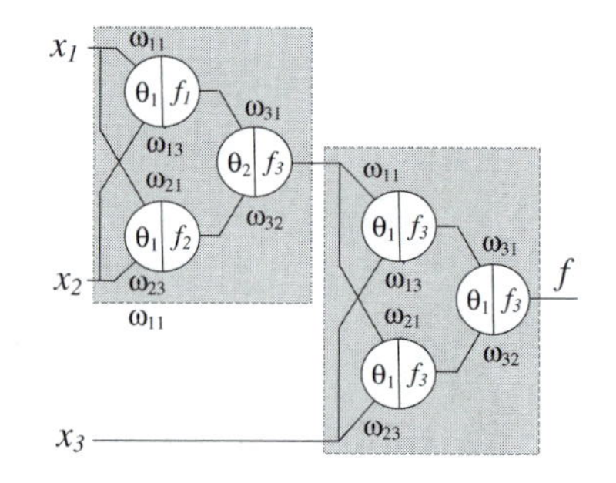

FIGURE 10.5
Data structures for representations of the EXOR function of two variables in the operational domain (SOP form) and the functional domain (polynomial form) (continuation of Figure 10.4).

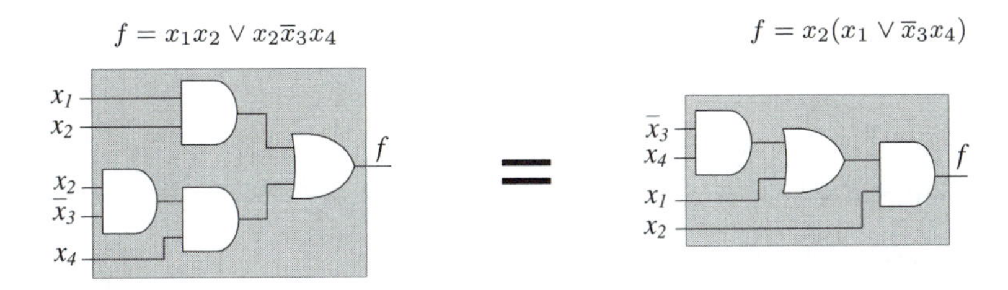

FIGURE 10.6
Implementation of original and factored SOP expressions (Example 10.3)

10.2 Algebra of the polynomial forms

In Chapter 1, Boolean algebra was defined as a set of elements, operations, and postulates. This algebraic structure is the formal basis of the SOP representation. The formal basis of the polynomial forms is the *finite field*. Finite fields are algebraic structures too, but they are characterized by the *elements*, *operations*, and *postulates* of a finite field. The theory of polynomial representations of Boolean functions has been adopted from related fields, such as digital signal processing. Details can be found in the "Further study" section.

10.2.1 Theoretical background

Fields

A *finite field* $\mathcal{F}$:

- ▶ Is an algebraic structure defined as a set of elements, together with two binary operations, each having associative, commutative, and distributive properties, closure under addition and multiplication, inverse properties, and a unique element.
- ▶ The number of elements in the field is called the **order** of the field. A field with order m exists iff m is a **prime power**, i.e., $m = p^n$ for some integer n and with p a prime integer. In this case, addition and multiplication are defined by a table composed such that the requirements for the field are true.
- ▶ In any finite field, the number of elements must be a power of a prime, p^k. This field is the Galois field GF(k).
- ▶ Every field with p^k elements is isomorphic to every other field with p^k elements. Some of these fields are useful in the representation, manipulation, analysis, and implementation of Boolean functions.

Binary operations are defined as:

- *Addition over a field $\mathcal{F}$*, and
- *Multiplication over a field $\mathcal{F}$.*

Galois field

An example of a field $\mathcal{F}$ is a *Galois field* denoted as GF(q):

- It consists of q elements $0, 1, 2, \ldots, q$.
- The number of elements in a Galois field must be equal to $p = 2^n$, where p is a *prime* number and n is a positive integer (a natural number $p \geq 2$ is called prime if and only if the only natural numbers which divide p are 1 and p).
- In cases where $p = 2$, all 2^n elements are derived using a polynomial of degree n.
- Operations in GF(q) are the modulo q sum and modulo q multiplication.

Example 10.4 (Addition and multiplication in ***GF***(4)**.)** *Galois field GF*(4) *consists of four elements* $0, 1, 2,$ *and* 3. *Addition and multiplication in GF*(4) *are defined as follows:*

Addition in GF(4)

+	*0*	*1*	*2*	*3*
0	0	1	2	3
1	1	0	3	2
2	2	3	0	1
3	3	2	1	0

Multiplication in GF(4)

·	*0*	*1*	*2*	*3*
0	0	0	0	0
1	0	1	2	3
2	0	2	3	1
3	0	3	1	2

The field of integer numbers modulo a prime number k is a field.

Polynomials

A *polynomial* in the variable x is the representation of a function f as a sum over an algebraic field $\mathcal{F}$:

$$f = \sum_{i=0}^{N-1} a_i x^i \quad \text{over the field } \mathcal{F} \tag{10.1}$$

The values $a_0, a_1, \ldots, a_{N-1}$ are called *coefficients* of the polynomial. Expression 10.1 means that there exist many polynomials, that are distinguished by the properties of the fields, namely, by the types of operation being addition and multiplication.

Example 10.5 (Fields for Boolean functions.) *The following fields are used for the representation of Boolean functions:* (*a*) Galois field *of order 2, GF(2). This field consists of two elements, 0 and 1. In GF(2), the sum and multiplication correspond to EXOR and AND operations, respectively.* (*b*) *A set of integer numbers that includes only the elements 0 and 1. In this field, traditional sum and multiplication are used.*

In addition, the sets of rational and complex numbers, together with the arithmetic operations of sum and multiplication, can be used for various representations of Boolean functions.

10.2.2 Polynomials for Boolean functions

The polynomial in Equation 10.1 is defined for a single variable x. Boolean algebra operates with a set of variables $x_1, x_2, \ldots, x_n$. To apply this polynomial equation to Boolean functions, some restrictions are needed. These restrictions are based on the fundamental theorem of arithmetic, which states that every integer $i > 2$ can be written in the form

$$i = i_1 i_2 \ldots i_n \tag{10.2}$$

for the unique *primes* $i_1 i_2 \ldots i_n$. This means that if any number is completely factored as in Expression 10.2, this expression is unique. Given a Boolean function of n variables $x_1, x_2, \ldots, x_n$,

Polynomial for Boolean function

$$f = \sum_{i=0}^{2^n-1} a_i \overbrace{(x_1^{i_1} \cdots x_n^{i_n})}^{Minterm\ over\ \mathcal{F}} \quad \text{over the field } \mathcal{F} \tag{10.3}$$

where a_i is a coefficient, i_j is the j-th bit ($j = 1, 2, \ldots, n$) in the binary representation of the index $i = i_1 i_2 \ldots i_n$, and the *literal* $x_j^{i_j}$ is defined as

$$x_j^{i_j} = \begin{cases} 1, & \text{if } i_j = 0 \\ x_j, & \text{if } i_j = 1 \end{cases} \quad \text{over the field } \mathcal{F} \tag{10.4}$$

The group of variables $x_1^{i_1} x_2^{i_2} \cdots x_n^{i_n}$ is called a *minterm over the field* $\mathcal{F}$. While the values of Boolean functions are used in SOP (operational domain), the coefficients a_i are used in polynomial forms of Boolean functions (functional domain).

Example 10.6 (Polynomial form.) *It follows from Equations 10.3 and 10.4 that*

$$\begin{aligned} f &= a_0(x_1^{i_1} x_2^{i_2}) + a_1(x_1^{i_1} x_2^{i_2}) + a_2(x_1^{i_1} x_2^{i_2}) + a_3(x_1^{i_1} x_2^{i_2}) \\ &= a_0(x_1^0 x_2^0) + a_1(x_1^0 x_2^1) + a_2(x_1^1 x_2^0) + a_3(x_1^1 x_2^1) \\ &= a_0 + a_1 x_2 + a_2 x_1 + a_3 x_1 x_2 \quad \text{over the field } \mathcal{F} \end{aligned}$$

where "+" is a sum as defined in the field $\mathcal{F}$.

Practice problem **10.2. (Polynomial form.)** Derive the polynomial for a Boolean function of three variables ($n = 3$) given its vector of coefficients $[a_0 a_1 a_2 a_3]^T = [1001^T$.
Answer is given in "Solutions to practice problems."

The polynomial form (Equation 10.3) is characterized as follows:

(a) The *operations*, of sum and multiplication, are specified by the properties of the field $\mathcal{F}$; that is, they are either *logical* or *arithmetic* operations.
(b) The *coefficients* a_i are computed for each Boolean function using the properties of the field $\mathcal{F}$.
(c) *Minterms* over the field $\mathcal{F}$ are specified by multiplication of *literals* $x_j^{i_j}$, $j = 1, 2, \ldots, n$, over $\mathcal{F}$ (Equation 10.4).

Example 10.7 (Polynomials for Boolean functi-ons.)
The following polynomials represent the Boolean function EXOR in different fields (algebras): $f_1 = x_1 \oplus x_2$ *over GF(2) and* $f_2 = x_1 + x_2 - 2x_1x_2$ *over the field of integers. In the polynomials* f_1 *and* f_2*, logical and arithmetical operations are used, respectively. In* f_1*, the coefficients are 0 or 1 because the field GF(2) consists of 0's and 1's only. The coefficients in the polynomial* f_2 *are integer numbers.*

Once the values of the Boolean variables x_1 and x_2 are assigned, the polynomials f_1 and f_2 assume the values of the initial function EXOR.

Practice problem 10.3. (Polynomials.) Given a polynomial $a_0 + a_1x_2 + a_2x_1 + a_3x_1x_2$, assume that $a_0 = a_1 = a_3 = 1$ and $a_2 = 0$, and the sum operation "+" over the field $\mathcal{F}$ is specified as the EXOR operation $\oplus$. Derive an SOP expression using truth tables and algebraic manipulation.
Answer: $\overline{x}_2 \vee x_1$; details are given in "Solutions to practice problems."

The above brief introduction to the basics of finite fields implies the following:

Features of the polynomial forms

- The polynomial forms are more complicated compared to the SOP form because special techniques are required for computing the coefficients of polynomial forms.
- The polynomial forms are an extension of Boolean algebra, and as such they are different from SOP expressions.

10.3 GF(2) algebra

The term *polynomial forms over the field GF(2)* specifies the forms of Boolean functions in which minterms are combined using the EXOR operation. An arbitrary Boolean function can be represented by polynomial expression. The

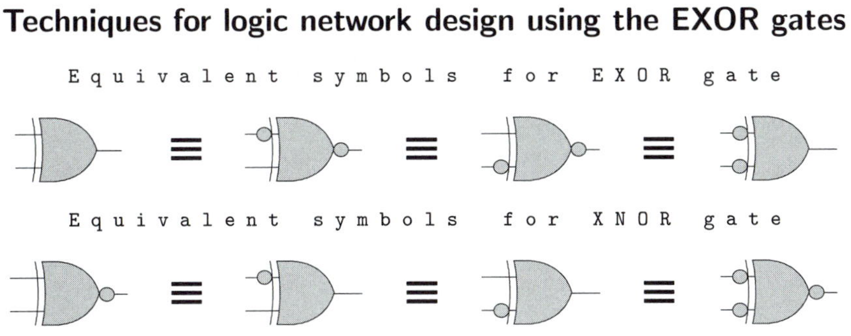

Techniques for logic network design using the EXOR gates

Equivalent symbols for EXOR gate

Equivalent symbols for XNOR gate

FIGURE 10.7
Equivalent symbols for EXOR and XNOR gates.

laws of the GF(2) algebra of polynomial forms are given in Table 10.1. In the example below, techniques for the manipulation of polynomial expressions using laws and identities are introduced.

Example 10.8 (GF(2) algebra.) *Examples of applications of the laws of algebra of polynomial forms from Table 10.1 are as follows: (a)* $x_1 \oplus x_2 \oplus x_1 x_2 = x_1 \oplus x_2(1 \oplus x_1) = x_1 \oplus x_2 \overline{x}_1$, *(b)* $1 \oplus x_1 x_2 \oplus x_1 x_3 = 1 \oplus x_1(x_2 \oplus x_3) = \overline{x_1(x_2 \oplus x_3)}$.

Practice problem 10.4. (GF(2) algebra.) Prove the following identities: *(a)* $x_1 \oplus \overline{x}_2 = \overline{x_1 \oplus x_2}$, *(b)* $\overline{x}_1 \oplus x_2 = \overline{x_1 \oplus x_2}$, *(c)* $\overline{x}_1 \oplus \overline{x}_2 = x_1 \oplus x_2$.
Answer is given in "Solutions to practice problems."

Using the theorems and rules from Table 10.1, equivalent symbols for the two input EXOR and XNOR gates can be derived (Figure 10.7). These logic symbols follow the rule: Any pair of signals (inputs or outputs) of an EXOR or XNOR gate can be complemented without changing the EXOR or XNOR function, respectively.

Practice problem 10.5. (Equivalence.) Prove the equivalence proof of symbols for the EXOR and XNOR gates given in Figure 10.7. For example, $x_1 \oplus x_2 = \overline{x}_1 \oplus 1 \oplus \overline{x}_2 \oplus 1 = \overline{x}_1 \oplus \overline{x}_1$.
Answer is given in "Solutions to practice problems."

10.3.1 Operational and functional domains

Forward and *inverse* transforms describe the relationship between *operational* and *functional* domains for Boolean data structures (Figure 10.8):

TABLE 10.1
The GF(2) algebra and identities for manipulations.

Techniques for computing the polynomial expressions

Laws and identities	Formal notation	Logic network
Associative law	$x_1 \oplus (x_2 \oplus x_3)$ $= (x_1 \oplus x_2) \oplus x_3$ $= x_1 \oplus x_2 \oplus x_3$ $x_1(x_2x_3) = (x_1x_2)x_3$ $= x_1x_2x_3$	
Distributive law	$x_1(x_2 \oplus x_3)$ $= x_1x_2 \oplus x_1x_3$	
Commutative law	$x_1 \oplus x_2 = x_2 \oplus x_1$ $x_1x_2 = x_2x_1$	
Identities for variables	$x \oplus x = 0, \quad x \cdot x = x$ $x \oplus \overline{x} = 1, \quad x \cdot \overline{x} = 0$	
Identities for variables and constants	$x \oplus 0 = x, \quad x \cdot 0 = 0$ $x \oplus 1 = \overline{x}, \quad x \cdot 1 = x$	

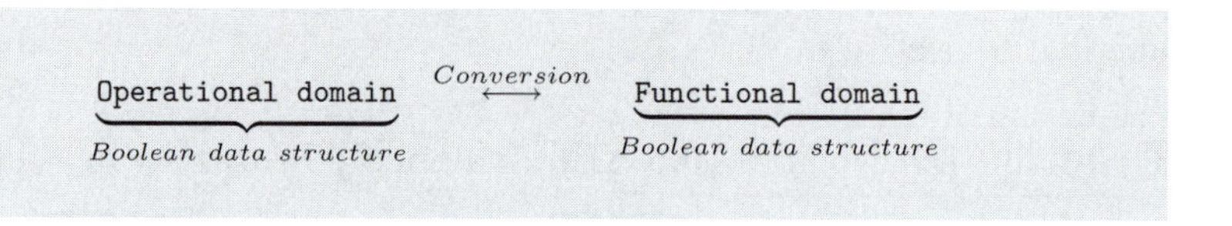

Example 10.9 (Conversions.) *An SOP expression (operational domain) can be converted into polynomial form (functional domain) using algebraic, matrix, and cube forms.*

The polarity of variables (complemented or uncomplemented) can be controlled for each transformation. Note that in SOP-based techniques, the polarities of variables can be changed using DeMorgan's rule. In the functional domain, this control can be implemented in matrix form.

TABLE 10.2
The GF(2) algebra and identities for manipulations (continuation of Table 10.1).

Techniques for computing the polynomial expressions

Rules and identities	Formal notation	Logic network
Identities for constants	$0 \oplus 0 = 0, \quad 0 \cdot 0 = 0$ $1 \oplus 1 = 0, \quad 1 \cdot 1 = 1$ $1 \oplus 0 = 1, \quad 1 \cdot 0 = 0$	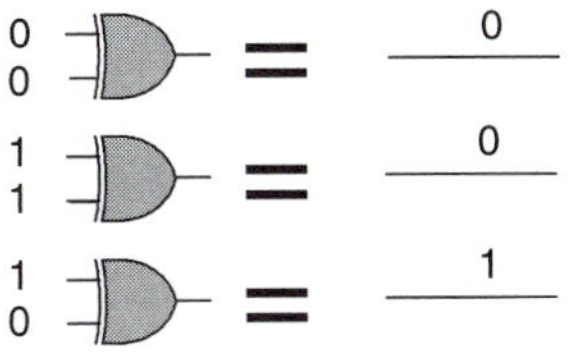 0 0 = 0 1 1 = 0 1 0 = 1
DeMorgan's rules for polynomials	$x_1 \oplus \overline{x}_2 = \overline{x_1 \oplus x_2}$ $\overline{x}_1 \oplus x_2 = \overline{x_1 \oplus x_2}$ $\overline{x}_1 \oplus \overline{x}_2 = x_1 \oplus x_2$	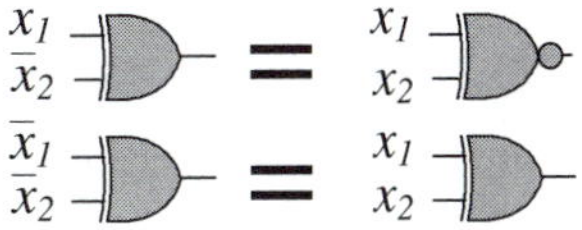 x_1 $\overline{x}_2$ = x_1 x_2 $\overline{x}_1$ $\overline{x}_2$ = x_1 x_2
Relationships with SOP form	$x_1 \oplus x_2 = x_1\overline{x}_2 \vee \overline{x}_1 x_2$ $\overline{x_1 \oplus x_2} = x_1 x_2 \vee \overline{x}_1\overline{x}_2$	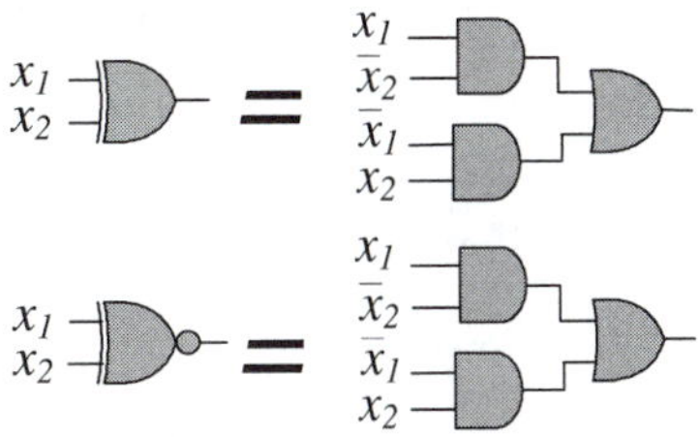 x_1 x_2 = x_1 $\overline{x}_2$ $\overline{x}_1$ x_2 x_1 x_2 = x_1 x_2 $\overline{x}_1$ $\overline{x}_2$
Simplification rules	$x_1 x_2 \oplus \overline{x}_1 x_2 = x_2$ $\overline{x}_1 x_2 \oplus x_2 = x_1 x_2$ $x_1 x_2 \oplus x_2 = \overline{x}_1 x_2$	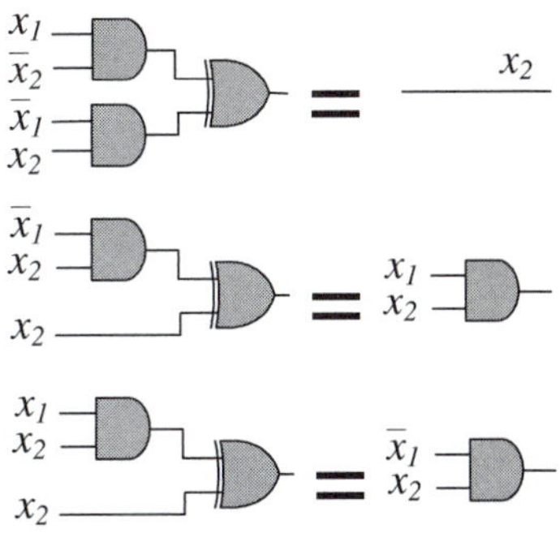 x_1 x_2 $\overline{x}_1$ x_2 = x_2 $\overline{x}_1$ x_2 x_2 = x_1 x_2 x_1 x_2 x_2 = $\overline{x}_1$ x_2

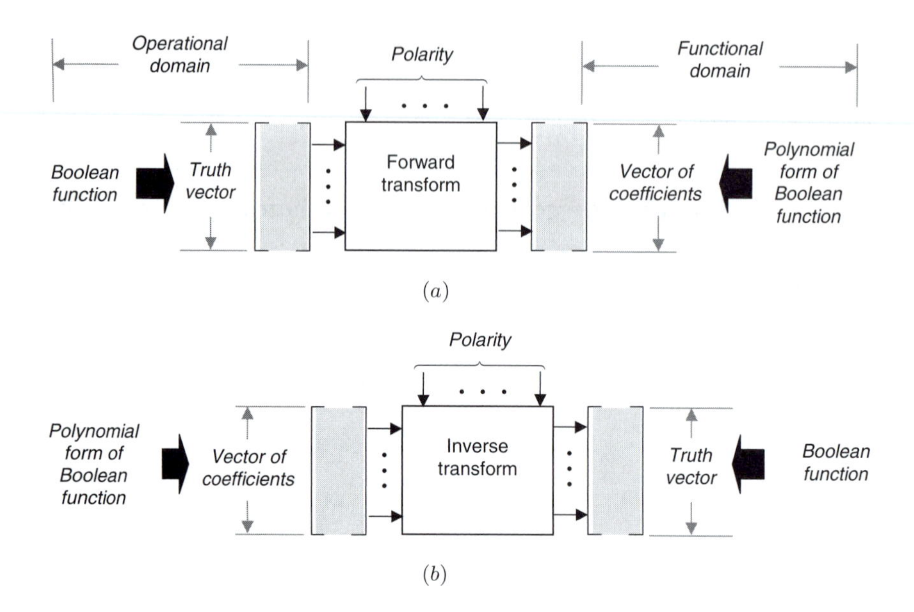

FIGURE 10.8
Forward and inverse transforms for the polynomial forms of Boolean functions.

10.3.2 The functional table

A Boolean function of n variables $f(x_i)$, $i = 1, 2, \ldots, n$, in the operational domain can be described in tabulated form using truth tables. In the functional domain, a Boolean function is represented in a polynomial form $f(x_i, a_j)\ i, j \in \{1, 2, \ldots, n\}$, which is a function of variables x_i and coefficients a_j, and can be described by a *functional table*:

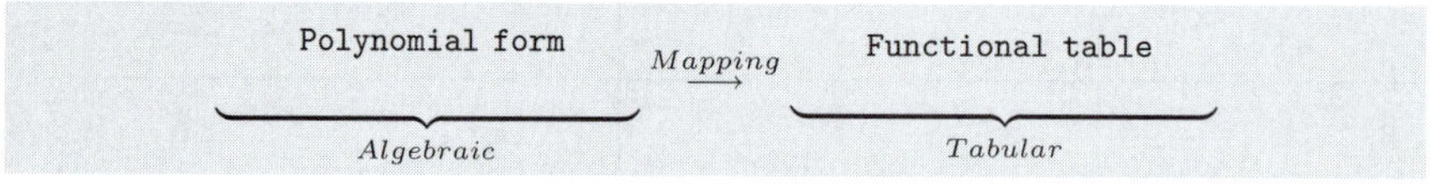

A functional table is a list of all combinations of 1's and 0's assigned to the binary coefficients $a_0, a_1, \ldots, a_{2^n-1}$ and corresponding polynomials. A functional table is characterized by the following properties:

Properties of the functional table

- Each row corresponds to a combination of the 2^n coefficients $a_0, a_1, \ldots, a_{2^n-1}$ of the polynomial. The number of rows in the table is 2^{2^n}, where n is the number of variables of the Boolean function.
- The columns are the 2^{2^n} polynomial forms of the given Boolean functions. Each polynomial represents a Boolean function assuming various 2^n polarities of its n variables.

Example 10.10 (Functional table.) *For a Boolean function f of two variables x_1 and x_2 ($n = 2$), the functional table over the field GF(2) is derived as follows. There are $2^n = 2^2 = 4$ coefficients a_0, a_1, a_2, and a_3, that specify $2^{2^n} = 2^{2^2} = 16$ polynomial forms of the Boolean function. Hence, the functional table contains 16 rows for all combinations of the coefficients a_0, a_1, a_2, a_3 and the column labeled f that contains 16 polynomials (Figure 10.9). For each polynomial in the functional table, a truth table can be derived. For example, the combination of coefficients $a_0a_1a_2a_3 = 0011$ specifies the polynomial $x_1 \oplus x_1x_2$.*

Practice problem 10.6. (Functional table.) Derive the functional table over the field GF(2) for a Boolean function of a single variable x.
Answer is given in "Solutions to practice problems."

10.3.3 The functional map

The analog of the truth-vector and K-map in the functional domain is the *vector of coefficients.* All 2^{2^n} possible vectors of coefficients for a Boolean function of n variables are represented by the *functional map.*

Example 10.11 (K-map and functional map.) *Let the Boolean function of three variables $f = \overline{x}_1\overline{x}_2\overline{x}_3 \vee \overline{x}_1x_2x_3$ be given in the form of a K-map (Figure 10.10). The functional map contains the coefficients of the polynomial in GF(2). The gates used for the implementation of the function in the operational domain are AND and OR, while the gates for polynomial implementation are AND and EXOR.*

Example 10.12 (K-map and functional map.) *In Figure 12.1, Boolean functions of two and three variables are represented in the operational and functional domains using a K-map and a functional map.*

Practice problem 10.7. (K-map and functional map.) Represent the Boolean function of four variables in the operational and functional domains.
Answer is given in "Solutions to practice problems."

10.3.4 Polarized minterms

The polynomial form of Equation 10.3 contains only uncomplemented variables. In order to achieve acceptable flexibility in a network design based

Techniques for computing using functional and truth tables of Boolean functions

Functional table

a_0	a_1	a_2	a_3	Polynomial f
0	0	0	0	0
0	0	0	1	x_1x_2
0	0	1	0	x_1
0	0	1	1	$x_1 \oplus x_1x_2$
0	1	0	0	x_2
0	1	0	1	$x_2 \oplus x_1x_2$
0	1	1	0	$x_2 \oplus x_1$
0	1	1	1	$x_2 \oplus x_1 \oplus x_1x_2$

a_0	a_1	a_2	a_3	Polynomial f
1	0	0	0	1
1	0	0	1	$1 \oplus x_1x_2$
1	0	1	0	$1 \oplus x_1$
1	0	1	1	$1 \oplus x_1 \oplus x_1x_2$
1	1	0	0	$1 \oplus x_2$
1	1	0	1	$1 \oplus x_2 \oplus x_1x_2$
1	1	1	0	$1 \oplus x_1 \oplus x_2$
1	1	1	1	$1 \oplus x_2 \oplus x_1 \oplus x_1x_2$

Corresponding truth tables

$f = 0$

x_1	x_2	f
0	0	0
0	1	0
1	0	0
1	1	0

$f = x_1x_2$

x_1	x_2	f
0	0	0
0	1	0
1	0	0
1	1	1

$f = x_1$

x_1	x_2	f
0	0	0
0	1	0
1	0	1
1	1	1

$f = x_1 \oplus x_1x_2 = x_1\overline{x}_2$

x_1	x_2	f
0	0	0
0	1	0
1	0	1
1	1	0

$f = x_2$

x_1	x_2	f
0	0	0
0	1	1
1	0	0
1	1	1

$f = x_2 \oplus x_1x_2 = \overline{x}_1x_2$

x_1	x_2	f
0	0	0
0	1	1
1	0	0
1	1	0

$f = x_2 \oplus x_1 = \overline{x}_1x_2 \vee x_1\overline{x}_2$

x_1	x_2	f
0	0	0
0	1	1
1	0	1
1	1	0

$f = x_2 \oplus x_1 \oplus x_1x_2 = x_1 \vee x_2$

x_1	x_2	f
0	0	0
0	1	1
1	0	1
1	1	1

$f = 1$

x_1	x_2	f
0	0	1
0	1	1
1	0	1
1	1	1

$f = 1 \oplus x_1x_2 = \overline{x_1x_2}$

x_1	x_2	f
0	0	1
0	1	1
1	0	1
1	1	0

$f = 1 \oplus x_1 = \overline{x}_1$

x_1	x_2	f
0	0	1
0	1	1
1	0	0
1	1	0

$f = 1 \oplus x_1 \oplus x_1x_2 = \overline{x_1\overline{x}_2}$

x_1	x_2	f
0	0	1
0	1	1
1	0	0
1	1	1

$f = 1 \oplus x_2 = \overline{x}_2$

x_1	x_2	f
0	0	0
0	1	0
1	0	1
1	1	0

$f = 1 \oplus x_2 \oplus x_1x_2 = \overline{\overline{x}_1x_2}$

x_1	x_2	f
0	0	1
0	1	0
1	0	1
1	1	1

$f = 1 \oplus x_1 \oplus x_2 = \overline{x}_1\overline{x}_2 \vee x_1x_2$

x_1	x_2	f
0	0	1
0	1	0
1	0	0
1	1	1

$f = 1 \oplus x_2 \oplus x_1 \oplus x_1x_2 = \overline{x}_1\overline{x}_2$

x_1	x_2	f
0	0	1
0	1	0
1	0	0
1	1	0

FIGURE 10.9
Representation of Boolean functions of two variables in the form of a functional table (a) and the corresponding truth tables (b) (Example 10.10).

Design example: Operational and functional domains

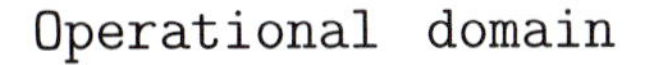

Functional domain

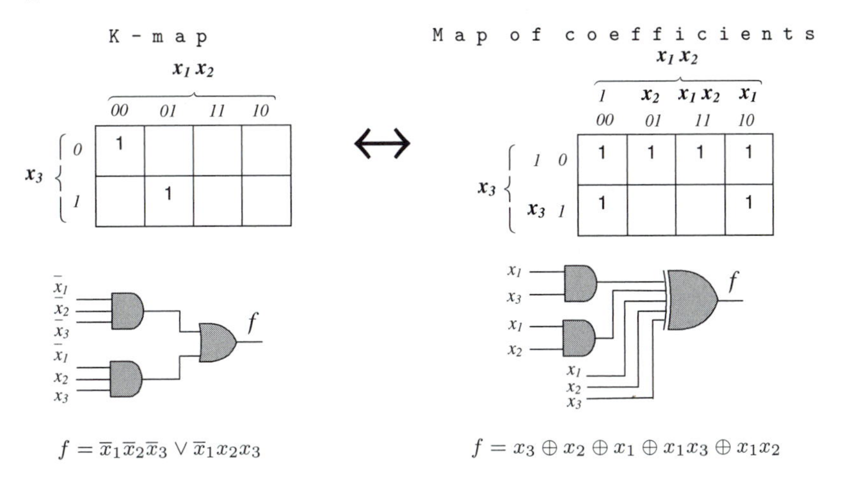

FIGURE 10.10
Representation of the Boolean function $f = \overline{x}_1\overline{x}_2\overline{x}_3 \vee \overline{x}_1 x_2 x_3$ in the operational and functional domains (Example 10.11).

on polynomial forms, so-called *polarized minterms* can be used. The polarized minterm is the product of *polarized literals.*

Literals and polarized literals

A literal is the representation of either an uncomplemented or a complemented variable:

Literal of a Boolean function

$$\texttt{Literal} = x_j^{i_j} = \begin{cases} \overline{x}_j, \text{ if } i_j = 0 \\ x_j, \text{ if } i_j = 1 \end{cases} \tag{10.5}$$

Literals in the form of Equation 10.5 are used in the standard SOP forms. The polarities of variables are specified by the particular Boolean functions. The polarity can be changed using DeMorgan's rule.

Example 10.13 (Literals.) *Let* $i = 0, 1, 2, 3$ *(*$i_1 i_2 = 00, 01, 10, 11$ *for the representation of two Boolean variables). According to Equation 10.5, literals can be generated as follows:*

$$\{x_1^0 x_2^0,\ x_1^0 x_2^1,\ x_1^1 x_2^0,\ x_1^1 x_2^1\} = \{\overline{x}_1\overline{x}_2,\ \overline{x}_1 x_2,\ x_1\overline{x}_2,\ x_1 x_2\}$$

Techniques for computing in operational and functional domains

Operational domain

Standard SOP

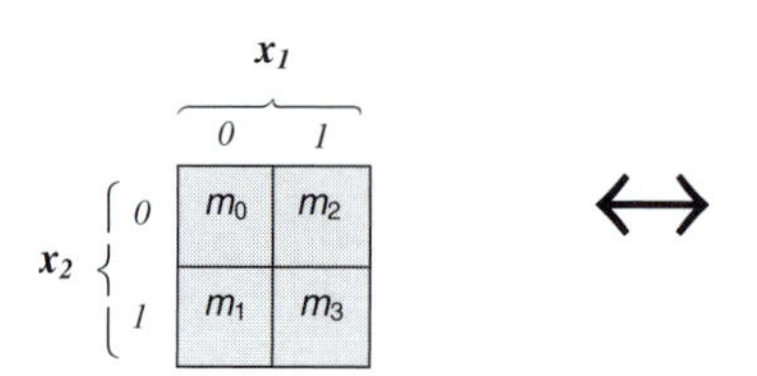

$\leftrightarrow$

Functional domain

Polynomial over field $\mathcal{F}$

x_2 \ x_1	1 (0)	x_1 (1)
1 (0)	a_0	a_2
x_2 (1)	a_1	a_3

$$f = \bigvee_{i=0}^{3} m_i$$

$$= \bigvee_{i=0}^{3} x_1^{i_1} x_2^{i_2}$$

$$= m_0\overline{x}_1\overline{x}_2 \vee m_1\overline{x}_1 x_2 \vee m_2 x_1\overline{x}_2 \vee m_3 x_1 x_2$$

$$x_j^{i_j} = \begin{cases} \overline{x}_j, & \text{if } i_j = 0; \\ x_j, & \text{if } i_j = 1. \end{cases}$$

$$f = \sum_{i=0}^{3} a_i x_1^{i_1} x_2^{i_2}$$

$$= a_0 + a_1x_2 + a_2x_1 + a_3x_1x_2$$

$$x_j^{i_j} = \begin{cases} 1, & \text{if } i_j = 0; \\ x_j, & \text{if } i_j = 1. \end{cases}$$

x_3 \ x_1x_2	00	01	11	10
0	m_0	m_2	m_6	m_4
1	m_1	m_3	m_7	m_5

$\leftrightarrow$

x_3 \ x_1x_2	1 (00)	x_2 (01)	x_1x_2 (11)	x_1 (10)
1 (0)	a_0	a_2	a_6	a_4
x_3 (1)	a_1	a_3	a_7	a_5

$$f = \bigvee_{i=0}^{7} m_i$$

$$= \bigvee_{i=0}^{7} x_1^{i_1} x_2^{i_2} x_3^{i_3}$$

$$= m_0\overline{x}_1\overline{x}_2\overline{x}_3 \vee m_1\overline{x}_1\overline{x}_2 x_3 \vee m_2\overline{x}_1 x_2\overline{x}_3$$

$$\vee\ m_3\overline{x}_1 x_2 x_3 \vee m_4 x_1\overline{x}_2\overline{x}_3 \vee m_5 x_1\overline{x}_2 x_3$$

$$\vee\ m_6 x_1 x_2\overline{x}_3 \vee m_7 x_1 x_2 x_3$$

$$x_j^{i_j} = \begin{cases} \overline{x}_j, & \text{if } i_j = 0; \\ x_j, & \text{if } i_j = 1. \end{cases}$$

$$f = \sum_{i=0}^{7} a_i x_1^{i_1} x_2^{i_2} x_3^{i_3}$$

$$= a_0 + a_1x_3 + a_2x_2 + a_3x_2x_3$$

$$+\ a_4x_1 + a_5x_1x_3 + a_6x_1x_2 + a_7x_1x_2x_3$$

$$x_j^{i_j} = \begin{cases} 1, & \text{if } i_j = 0; \\ x_j, & \text{if } i_j = 1. \end{cases} \quad \text{over the field } \mathcal{F}$$

FIGURE 10.11
Techniques for the representation of Boolean functions in the operational and functional domains using K-maps and functional maps (Example 10.12).

The *polarized* form of the literal provides an approach to *independent* control of the polarity of variables. A *polarized literal* is the representation of either an uncomplemented or a complemented variable specified by the control parameter called *polarity*, $c_1, c_2, \ldots, c_n$, $c_j \in \{0,1\}$, $j = 1, 2, \ldots, n$:

Polarized literal of a Boolean function

$$\texttt{Polarized literal} = (x_j \oplus c_j)^{i_j}$$
$$= \begin{cases} 1, & \text{if } i_j = 0 \\ (x_j \oplus c_j), & \text{if } i_j = 1 \end{cases} \quad \text{over GF(2)} \tag{10.6}$$

In Equation 10.6, the parameters i_j for the variable x_j are *separated* from the polarity c_j:

- Parameters i_j only specify the order of the minterms in the polynomial. They are an inherent property of a given form; that is, i_j are dependent parameters.
- Parameter c_j is an *independent* parameter.

The example below shows all possible combinations of the dependent and independent parameters of the literal.

Example 10.14 (Polarized literals.) *For the polarity $c_j \in \{0,1\}$ and parameter $i_j \in \{0,1\}$, the complete set of polarized literals is generated as follows:*

$$\overbrace{(x_j \oplus 0)^0}^{c_j=0,\ i_j=0} = x_j^0 = 1 \quad \text{over GF(2)}$$
$$\overbrace{(x_j \oplus 1)^0}^{c_j=1,\ i_j=0} = \overline{x}_j^0 = 1 \quad \text{over GF(2)}$$
$$\overbrace{(x_j \oplus 0)^1}^{c_j=0,\ i_j=1} = x_j^1 = x_j \quad \text{over GF(2)}$$
$$\overbrace{(x_j \oplus 1)^1}^{c_j=1,\ i_j=1} = \overline{x}_j^1 = \overline{x}_j \quad \text{over GF(2)}$$

Practice problem 10.8. (Polarized literals.) Given a polarity $c = 2$ ($c_1c_2 = 10$), and $n = 2$, find the third polarized literal for $i_1i_2 = 11$.
Answer: $(x_1 \oplus c_1)^{i_1}(x_2 \oplus c_2)^{i_2} = (x_1 \oplus 1)^1(x_2 \oplus 0)^1 = \overline{x}_1 x_2$.

Minterm structure

The minterm is defined for the assignment $i_1, i_2, \ldots, i_n$ of Boolean variables $x_1, x_2, \ldots, x_n$ for which a Boolean function is equal to 1; that is, $x_1 = i_1, x_2 = i_2, \ldots, x_n = i_n$ if $f = 1$:

Minterms of a Boolean function

$$\texttt{Minterm} = \overbrace{x_1^{i_1} x_2^{i_2} \cdots x_n^{i_n}}^{n\ literals} \quad (10.7)$$

where

$$x_j^{i_j} = \begin{cases} \overline{x}_j, \text{ if } i_j = 0 \\ x_j, \text{ if } i_j = 1 \end{cases} \quad (10.8)$$

These minterms are used in the standard SOP expressions. The simplest method for generating the minterms is to use the truth table of the Boolean function.

Polarized minterm structure

A *polarized minterm* is defined by the equation

Polarized minterm of a Boolean function

$$\texttt{Polarized minterm} = \overbrace{(x_1 \oplus c_1)^{i_1} (x_2 \oplus c_2)^{i_2} \cdots (x_n \oplus c_n)^{i_n}}^{n\ polarized\ literals} \quad \text{over GF(2)} \quad (10.9)$$

where

$$(x_j \oplus c_j)^{i_j} = \begin{cases} 1, & \text{if } i_j = 0 \\ (x_j \oplus c_j), & \text{if } i_j = 1 \end{cases} \quad \text{over GF(2)}$$

In Equation 10.9, the polarities of the variables $x_1, x_2, \ldots, x_n$ are specified by the polarity parameters $c_1, c_2, \ldots, c_n$, respectively. An arbitrary polarity $c_i \in \{0, 1\}$ can be chosen for each Boolean variable x_i, $i = 1, 2, \ldots, n$.

Example 10.15 (Polarized minterms.) *Table 10.3 includes polarized minterms,*

$$\texttt{Polarized minterm} = (x_1 \oplus c_1)^{i_1} (x_2 \oplus c_2)^{i_2} \quad \textit{over GF(2)}$$

For example, all four polarized minterms for the polarity $c = 2$ $(c_1c_2 = 10)$ can be in four forms:

$$(x_1 \oplus 1)^0 (x_2 \oplus 0)^0 = 1; \quad (x_1 \oplus 1)^0 (x_2 \oplus 0)^1 = x_2;$$
$$(x_1 \oplus 1)^1 (x_2 \oplus 0)^0 = \overline{x}_1; \quad (x_1 \oplus 1)^1 (x_2 \oplus 0)^1 = \overline{x}_1 x_2$$

Figure 10.12 shows computing with the polarized minterms.

Practice problem 10.9. (Polarized minterms.) Find all pola-rized minterms of a Boolean function of two variables $(n = 2)$ given $c = 3$ $(c_1c_2 = 11)$.
Answer is given in "Solutions to practice problems."

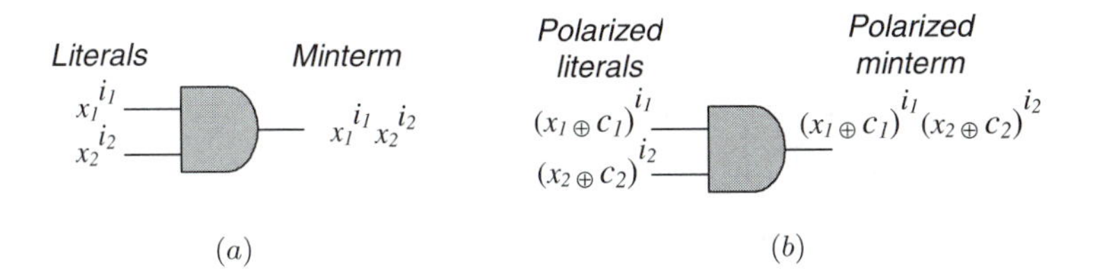

FIGURE 10.12
Computing of minterms (a) and polarized minterms (b) (Example 10.15).

TABLE 10.3
The polarized minterms $(x_1 \oplus c_1)^{i_1}(x_2 \oplus c_2)^{i_2}$ over GF(2) (Example 10.15).

Design example: Polarized minterm computing

Polarity c_1c_2	Polarized minterm $(x_1 \oplus c_1)^{i_1}(x_2 \oplus c_2)^{i_2}$	i_1i_2 00	01	10	11
00	$\overbrace{(x_1 \oplus 0)^{i_1}(x_2 \oplus 0)^{i_2}}^{c_1=0,\ c_2=0}$	1	x_2	x_1	x_1x_2
01	$\overbrace{(x_1 \oplus 0)^{i_1}(x_2 \oplus 1)^{i_2}}^{c_1=0,\ c_2=1}$	1	$\overline{x}_2$	x_1	$x_1\overline{x}_2$
10	$\overbrace{(x_1 \oplus 1)^{i_1}(x_2 \oplus 0)^{i_2}}^{c_1=1,\ c_2=0}$	1	x_2	$\overline{x}_1$	$\overline{x}_1x_2$
11	$\overbrace{(x_1 \oplus 1)^{i_1}(x_2 \oplus 1)^{i_2}}^{c_1=1,\ c_2=1}$	1	$\overline{x}_2$	$\overline{x}_1$	$\overline{x}_1\overline{x}_2$

The polarity of each variable contributes to the polarity of a polynomial over the field GF(2).

Practice problem 10.10. **(Polarity.)** Given the polynomial expression $f = x_1 \oplus x_2 \oplus x_1x_2$ of polarity 0 (variables x_1 and x_2 have polarity 0), find:

(a) The polynomial of polarity 1 (x_1 has polarity 0 and x_2 has polarity 1);
(b) The polynomial of polarity 2 (x_1 has polarity 1 and x_2 has polarity 0);
(c) The polynomial of polarity 3 (both x_1 and x_2 have polarity 1).

Answers are given in "Solutions to practice problems."

10.4 Relationship between standard SOP and polynomial forms

Polynomial form over the field GF(2) of a Boolean function can be derived directly from the SOP expression of this function using the following algorithm:

Deriving a polynomial form given a standard SOP expression

Given: `The standard SOP expression of a Boolean function`
Find: `The standard polynomial form`
Procedure: `Replace the OR operations with EXOR operation`
Result: `The standard polynomial form of the Boolean function`

This procedure results in a non-optimal polynomial representation, and additional manipulations are needed for its simplification.

Example 10.16 **(A SOP and polynomial forms.)** *The truth table of Boolean function is given in Table 10.4. A standard SOP of this function is* $f = \overline{x}_1\overline{x}_2\overline{x}_3 \vee \overline{x}_1\overline{x}_2x_3 \vee \overline{x}_1x_2x_3 \vee x_1x_2x_3$. *The polynomial expression is derived by replacing OR operations by the EXOR operations:* $f = \overline{x}_1\overline{x}_2\overline{x}_3 \oplus \overline{x}_1\overline{x}_2x_3 \oplus \overline{x}_1x_2x_3 \oplus x_1x_2x_3$. *Note that this polynomial expression consists of a the minterms with variables of different polarities.*

TABLE 10.4
Relationship between a standard SOP and polynomial forms (Example 10.16).

Assignment of variables	Value of Boolean function	Minterm
000	1	$\overline{x}_1\overline{x}_2\overline{x}_3$
001	1	$\overline{x}_1\overline{x}_2x_3$
010	0	
011	1	$\overline{x}_1x_2x_3$
100	0	
101	0	
110	0	
111	1	$x_1x_2x_3$

Practice problem **10.11. (SOP and polynomial forms.)** Simplify the polynomial form of a Boolean function given in Example 10.16.

Answer: $f = \overline{x}_1\overline{x}_2 \overbrace{(\overline{x}_3 \oplus x_3)}^{Equal\ to\ 1} \oplus x_2x_3 \overbrace{(\overline{x}_1 \oplus x_1)}^{Equal\ to\ 1} = \overline{x}_1\overline{x}_2 \oplus x_2x_3$. Note that this polynomial form has mixed polarities of variables.

10.5 Local transformations for EXOR expressions

In Chapter 4, a *local transformation* is defined as a set of rules for the simplification of a data structure. In this section, we consider a local transformations for a logic networks, which consist of various types of logic gates, including EXOR gates. These transformations are based on the theorems of Boolean algebra and polynomial algebra GF(2), and are applied locally.

Local transformations for the logic networks using AND, OR gates, and inverters, were described in Chapter 4. The following rules can be applied to logic networks with EXOR gates:

The rules for local transformations of a logic network with EXOR gates

Rule 1: *Replacing an EXOR gate with a constant:*

- Replace an EXOR gate with a corresponding constant using the rules of identities for constants if the inputs of this gates are constants
- Replace an EXOR gate with a corresponding constant using the rules of identities for variables if the inputs of this gates are literals of the same variable

Rule 2: *Replacing an EXOR gate with a variable:* An EXOR gate can be replaced with a variable using the rules of identity for variables and constants if one of the inputs is a constant

The rules for removing the duplicated gates, removing the unused gates, and merging the gates, are similar to the ones for OR and AND gates.

Example 10.17 (Local transformations.) *Figure 10.13 illustrates two types of the local transformations:*

Local transformation in area A: the EXOR gate is replaced with the inverter using the identity rule for variables and constants (Table 10.1) $x_2 \oplus 1 = \overline{x}_2$.

Local transformation in area B: the inverter and EXOR gates are replaced by the wire using the simplification rule (Table 10.2) $x_1\overline{x}_2 \oplus x_1 = x_1(\overline{x}_2 \oplus 1) = x_1x_2$.

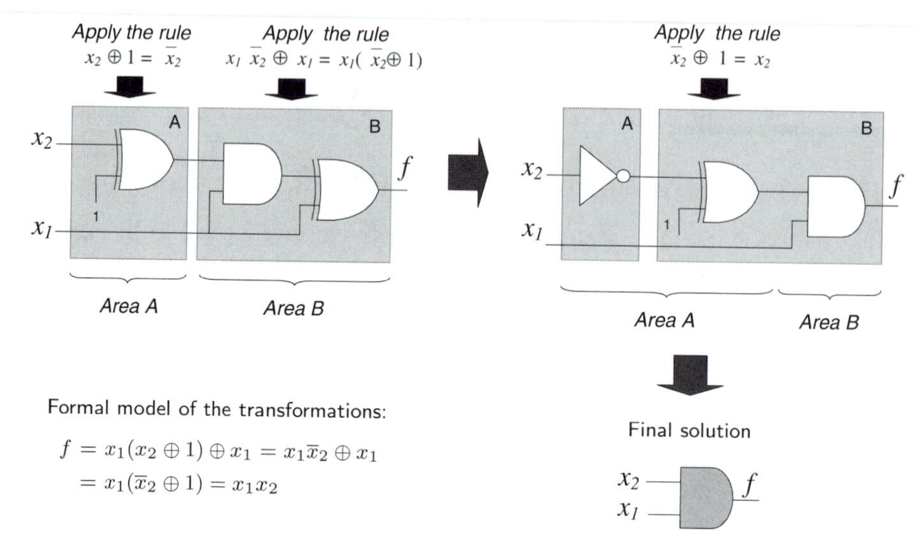

FIGURE 10.13
Local transformations in logic network (Example 10.17).

Practice problem **10.12. (Local transformations.)** Apply the local transforms to the areas A, B, and C of the logic network given in Figure 10.14a.
Answer is given in "Solutions to practice problems."

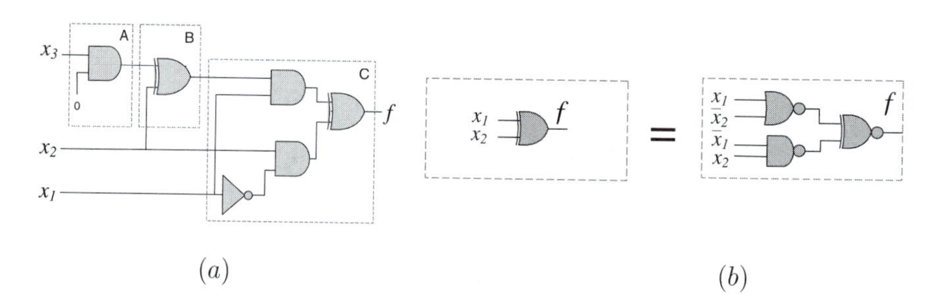

FIGURE 10.14
Logic networks for Practice problems 10.12 (a) and 10.13 (b).

10.6 Factorization of polynomials

Factoring of polynomial expressions is used, in particular, for dealing with the fan-in problem, and also if a logic network is designed using limited numbers of gate inputs. However, techniques for factoring SOP forms are not acceptable for polynomial forms. Factorization of polynomial expressions is based on the laws and identities given in Table 10.1. Similarly to SOP-based techniques, the factoring of polynomial expressions is based on the designer's experience and may be built into CAD tools to a limited extent. The application of various identities does not guarantee satisfactory results from factoring. In particular, an arbitrary Boolean variable can be replaced by its complement as follows:

$$x_1 \oplus x_2 = \overline{\overline{x}_1 \oplus x_2} = \overline{x_1 \oplus \overline{x}_2} = (1 \oplus x_1)x_2 \oplus 1 = x_1 \oplus (1 \oplus x_2) \oplus 1$$

Extra variables can be included in an equation using the following properties: $x \oplus x \oplus x = x$ and $x \oplus x = 0$.

Example 10.18 (Factorization .) *The polynomial expression*

$$f = \underbrace{1 \oplus x_4 \oplus x_3 \oplus x_2 \oplus x_2x_3 \oplus x_1x_3 \oplus x_1x_2 \oplus x_1x_2x_3}_{2\ level\ logic\ network}$$

can be directly implemented by the two-level logic network as shown in Figure 10.15a. The fan-in of the EXOR gate is equal to 7, it is often not acceptable. Factoring results in the expression $f = \underbrace{1 \oplus x_4 \oplus (x_3 \oplus x_2 \oplus x_2x_3)x_1}_{4\ level\ logic\ network}$. *This polynomial expression is implemented by a four-level logic network (Figure 10.15b) using three-input EXOR gates.*

10.7 Validity check for EXOR networks

Similarly to proving validity in SOP forms (Chapter 4), two Boolean functions in polynomial form can be said to be equivalent or non-equivalent. This check is based on algebra in GF(2).

The validity of two logic networks is referred to as an equivalence proof. The algorithm for the equivalence proof is given below.

Design example: Factoring

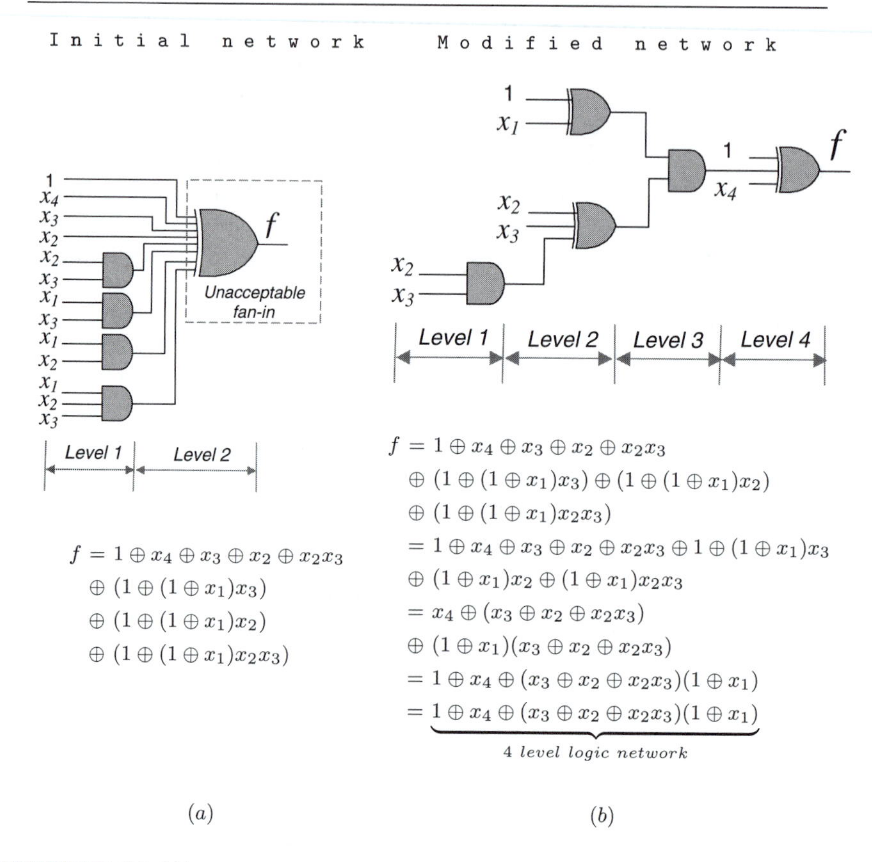

FIGURE 10.15
Two-level logic network implementation of a non-factored polynomial expression (a) and four-level logic network implementation of a factored polynomial expression (Example 10.18).

Equivalence proof problem
Given: `Two logic networks`
Prove: `The equivalence of these networks`
Step 1: `Describe the outputs for each network`
Step 2: `Form the logic equation for the outputs`
Step 3: `Prove this logic equation`

Proving the validity of Boolean expressions that combine SOP and polynomial terms is performed by transforming both sides of the equation to one domain (operational or functional).

Example 10.19 (Equivalence.) *The equivalence of the logic networks given in Figure 10.16 can be determined by proving the validity of the Boolean equations:* $\underbrace{x_1 \oplus x_2}_{Left\ network} = \underbrace{(x_1 \vee x_2) \oplus x_1x_2}_{Right\ network}$. *Since* $x_1 \vee x_2 = x_1 \oplus x_2 \oplus x_1x_2$, *then* $x_1 \oplus x_2 = (x_1 \oplus x_2 \oplus x_1x_2) \oplus x_1x_2$ *and* $x_1 \oplus x_2 = x_1 \oplus x_2$.

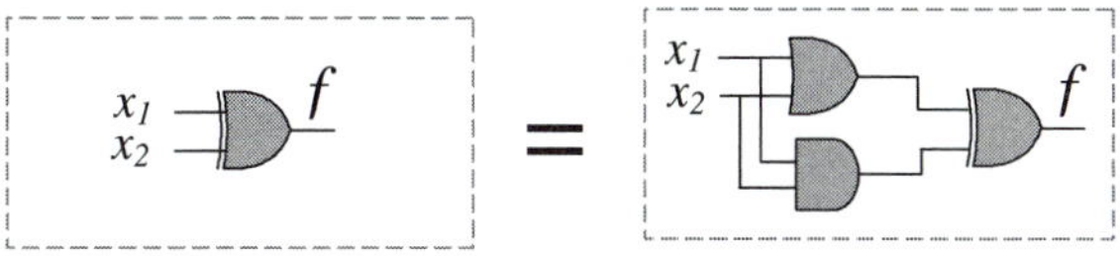

FIGURE 10.16
Proving the validity of two networks using manipulation of the Boolean functions in polynomial form (Example 10.19).

Practice problem **10.13. (Equivalence.)** Prove the equivalence of two networks given in Figure 10.14b.
Answer is given in "Solutions to practice problems."

10.8 Summary of polynomial algebra

This chapter introduces an alternative to SOP forms: ***polynomial*** forms of Boolean functions. Some polynomial expressions provide flexibility in the representation of Boolean functions and optimality of solutions compared to SOP expressions. This property is efficiently utilized in logic design, especially in the design of specific-area application discrete devices (for encoding and encryption of information, in particular). The key aspects of this chapter are listed below:

(*a*) Transformations between the SOP and polynomial forms using the notation of polarized literals, polarized minterms, coefficients, and operational and functional domains.

(*b*) Manipulation of polynomial forms such as factoring and simplification.

(*c*) Data structures for the representation and manipulation of Boolean functions in operational and functional domains.

Summary (continuation)

The following conclusions can be made based on this chapter:

- The formal basis of the polynomial forms is algebraic structures called ***fields***. The smallest finite field has an order of 2 and is called a ***Galois field*** of order 2; it is denoted as GF(2). This field consists of two elements, 0 and 1. In GF(2), sum and multiplication correspond to EXOR and AND operations, respectively.
- The polynomial expression is unique to a given Boolean function; that is, the polynomial form is the canonical form. The polynomial form is a universal basis that includes: (*a*) the constant 1, (*b*) AND operations, and (*c*) EXOR operations over Boolean variables. An arbitrary Boolean function can be represented in polynomial form.
- Polynomial forms can be represented using: (*a*) Tabulated forms, such as truth tables and ***functional tables***, truth-vectors, and ***vectors of coefficients***; (*b*) Graphical representations such as ***functional maps***; ***functional cubes***, and ***functional decision trees*** and ***functional decision diagrams***; (*c*) Logic networks of logic gates or threshold elements.
- The relationship between ***operational*** and ***functional*** domains is a key to the synthesis and application of the polynomial forms of Boolean functions.
- The standard, or canonical SOP expressions and polynomial forms have the following structure:

$$\texttt{Standard SOP form:} \quad f = \bigvee_{i=0}^{2^n-1} s_i \cdot \underbrace{x_1^{i_1} \cdots x_n^{i_n}}_{i-th\ minterm}$$

$$\texttt{Polynomial form:} \quad = \bigoplus_{i=0}^{2^n-1} r_i \cdot \underbrace{(x_1 \oplus c_1)^{i_1} \cdots (x_n \oplus c_n)^{i_n}}_{i-th\ minterm}$$

  The number of terms in canonical SOP and polynomial expressions are equal to 2^n.
- The ***polarized literal*** is the representation of either an uncomplemented or a complemented variable together with the control parameter, called ***polarity***, $c_1, c_2, \ldots, c_n$, $c_j \in \{0,1\}$, $j = 1, 2, \ldots, n$:

$$\texttt{Polarized literal} = (x_j \oplus c_j)^{i_j} = \begin{cases} 1, & \text{if } i_j = 0 \\ (x_j \oplus c_j), & \text{if } i_j = 1 \end{cases} \quad \text{over GF(2)}$$

Summary (continuation)

- The ***polarized minterm*** is defined by the equation

Polarized minterm

$$= \overbrace{(x_1 \oplus c_1)^{i_1}(x_2 \oplus c_2)^{i_2} \cdots (x_n \oplus c_n)^{i_n}}^{n \text{ polarized literals}} \quad \text{over GF}(2)$$

where

$$(x_j \oplus c_j)^{i_j} = \begin{cases} 1, & \text{if } i_j = 0 \\ (x_j \oplus c_j), & \text{if } i_j = 1 \end{cases} \quad \text{over GF}(2)$$

The polarities of the variables $x_1, x_2, \ldots, x_n$ are specified by polarity parameters $c_1, c_2, \ldots, c_n$, respectively. An arbitrary polarity $c_i \in \{0, 1\}$ can be chosen for each Boolean variable x_i, $i = 1, 2, \ldots, n$.

- If all the variables in the polynomial expression are uncomplemented only, the expression is of positive polarity. In another form of polynomial expressions, each variable can appear in the expression in complemented or uncomplemented form, i.e., the polarity of each variable is ***fixed*** through the expression. Finally, the polarity of a variable can be different in various products of the expression, i.e., the polarity of each variable is ***mixed*** through the expression.
- The data structure for polynomial forms must satisfy the formal requirements of the functional domain. In the functional domain, the following data structures are used: a ***functional table*** instead of a truth table, modified cubes, and ***functional decision diagrams***.
- ***Proving validity*** in algebraic forms is the formal basis for verification algorithms. Similarly to proving validity in SOP forms, polynomial forms can be checked for equivalence. However, proving validity for polynomial equations is based on algebra in GF(2).
- A ***local transformation*** is defined as a set of rules for the simplification of a data structure. These rules are based on the theorems of Boolean algebra and polynomial algebra GF(2), and are applied locally.
- ***Factorization*** of polynomial expressions is used, in particular, for dealing with the fan-in problem. Factoring is based on the laws and identities of a polynomial forms.
- For advances in polynomial algebra, we refer the reader to the "Further study" section.

10.9 Further study

Historical perspective

1830: Evariste Galois used finite fields to show the conditions under which algebraic equations have solutions in radicals (roots). A field with q elements is called *Galois field GF(q)*. Galois field GF(q) exists whenever q is prime or the integer power of a single prime; that is, $q = p^k$. There are many applications of GF(p^k) fields such as polynomial forms of Boolean functions, error-correcting codes, encryption, digital and signal processing.

1927: I. I. Zhegalkin first introduced the polynomial form of Boolean functions using the EXOR operations (I. I. Zhegalkin, "The Technique of Calculation of Statements in Symbolic Logic," Math. Journal, volume 34, pages 9–28, 1927, in Russian).

1954: I. Reed and D. Muller (independently) showed how classic Taylor and Maclaurin expansions can be used for the representations of Boolean functions in polynomial form, also known as *Reed-Muller polynomials*. The operations associated with the polynomial forms of Boolean functions were called *Boolean differences.* These differences, as in classical Taylor and Maclaurin expansions, are the coefficients of Reed-Muller polynomials; that is, the "logical form" of classical Taylor and Maclaurin expansions.

Advanced topics of polynomial algebra for Boolean functions

Topic 1: Arithmetic polynomials. The fields that are used for the representation of Boolean functions are given in Figure 10.17.

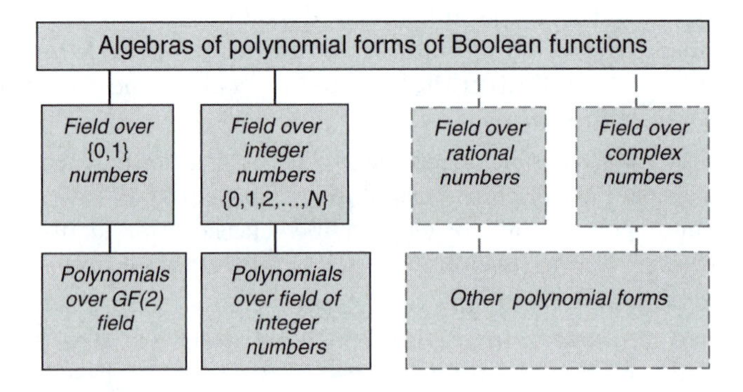

FIGURE 10.17
Algebras of polynomial forms of Boolean functions.

It follows from the above that fields over the numbers $\{0, 1\}$ and over the integer numbers $\{0, 1, 2, \ldots, N\}$ provide more possibilities for the representation of Boolean functions compared to SOP form.

Apparently, the first attempts to represent logic operations by arithmetical ones were taken by the founder of Boolean algebra, Boole (1854). He did not use the Boolean operators known today. Rather, he used arithmetic expressions. It is interesting to note that H. Aiken first found that arithmetic expressions can be useful in designing logic circuits and used them in the Harvard MARK 3 and MARK 4 computers (H. H. Aiken, "Synthesis of Electronic Computing and Control Circuits," *Ann. Computation Laboratory of Harvard University*, XXVII, Harvard University, Cambridge, MA, 1951).

Arithmetic expressions are closely related to polynomial expressions over GF(2), but with variables and function values interpreted as the integers 0 and 1 instead of logic values. In this way, arithmetic expressions can be considered as integer counterparts of polynomial expressions over GF(2).

For two Boolean variables x_1 and x_2 the following are true:

$$\overline{x} = 1 - x, \qquad x_1 \vee x_2 = x_1 + x_2 - x_1x_2,$$
$$x_1 \wedge x_2 = x_1x_2, \qquad x_1 \oplus x_2 = x_1 + x_2 - 2x_1x_2.$$

The right parts of the equations are called the *arithmetic expressions.* A Boolean function of n variables is the mapping $\{0,1\}^n \rightarrow \{0,1\}$, while an integer-valued function in arithmetical logic denotes the mapping $\{0,1\}^n \rightarrow \{0,1,\ldots,p-1\}$, where $p > 2$. For a Boolean function f of n variables, the arithmetic expression is given by

$$f = \sum_{i=0}^{2^n-1} a_i \cdot \underbrace{(x_1^{i_1} \cdots \; x_n^{i_n})}_{i-th\ product}$$

where a_i is a coefficient (integer number); i_j is the j-th bit $1, 2, \ldots, n$, in the binary representation of the index $i = i_1i_2 \ldots i_n$; and $x_j^{i_j}$ is defined as

$$x_j^{i_j} = \begin{cases} 1, & i_j = 0 \\ x_j, & i_j = 1 \end{cases}$$

Note that $\sum$ is arithmetic addition. For example, an arbitrary Boolean function of three variables is represented in the arithmetic expression

$$f = a_0 + a_1x_3 + a_2x_2 + a_3x_2x_3 + a_4x_1 + a_5x_1x_3 + a_6x_1x_2 + a_7x_1x_2x_3$$

The coefficients a_i, $i = 0, 1, \ldots, 7$, are the integer numbers. These coefficients are derived from the truth table:

x_1	x_2	x_3	f
0	0	0	a_0
0	0	1	a_1
0	1	0	a_2
0	1	1	a_3

x_1	x_2	x_3	f
1	0	0	a_4
1	0	1	a_5
1	1	0	a_6
1	1	1	a_7

However, not arbitrary chosen integer numbers specify a Boolean function. Special rules are used to determine the values of the polynomial coefficients. Arithmetic logic has many applications in contemporary logic design; for example, in the computation of signal probabilities for test generations, and switching activities for power and noise analysis.

Topic 2: Decision diagrams for polynomial forms. Various decision diagrams for polynomial forms have been developed. They can be classified with respect to functions of non-terminal nodes, number of outgoing edges, decomposition rules, terminal nodes, and edges and their labels. In the case of GF(p) field, the corresponding decision diagrams are called *bit-level* diagrams, allowing p-valued bits in the case of multi-valued logic functions. Otherwise, decision diagrams are called *word-level* diagrams, since the value of a terminal node is represented by a word. The terminal nodes of decision diagrams can be fuzzy values.

Further reading

Binary Decision Diagrams and Applications for VLSI CAD by S. Minato, Kluwer, Dordrecht, 1996.

Decision Diagram Techniques for Micro- and Nanoelectronic Design, by S. N. Yanushkevich, D. M. Miller, V. P. Shmerko, and R. S. Stanković, CRC/Taylor & Francis Group, Boca Raton, FL, 2006.

Modern Logic Design by D. Green, Addison-Wesley, 1986.

Spectral Interpretation of Decision Diagrams by R. S. Stanković and J. T. Astola, Springer, 2003.

10.10 Solutions to practice problems

Practice problem	Solution
10.1.	Let us express the EXOR operation in terms of AND and NOT operations. $f = x_1 \oplus x_2 = x_1\overline{x}_2 \vee \overline{x}_1 x_2 = \overline{\overline{x_1\overline{x}_2 \vee \overline{x}_1 x_2}} = \overline{(\overline{x_1\overline{x}_2})\,(\overline{\overline{x}_1 x_2})}$ This requires three NAND gates, assuming that complemented inputs are allowed. 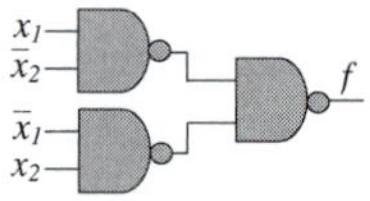
10.2.	$f = a_0 + a_1x_2 + a_2x_1 + a_3x_1x_2 = 1 + x_1x_2.$ The value of the Boolean function given the assignment of variables is computed as follows: $x_1x_2 = 00: \; f = 1 + 0 \times 0 = 1; \; x_1x_2 = 01: \; f = 1 + 0 \times 1 = 1$ $x_1x_2 = 10: \; f = 1 + 1 \times 0 = 1; \; x_1x_2 = 11: \; f = 1 + 1 \times 1 = 1 + 1 = 2$ The truth-vector of f is $\mathbf{F} = [1112]^T$.

Solutions to practice problems (continuation)

Practice problem	Solution

10.3.

Manipulation in algebraic form results in the SOP expression

$$f = 1 \oplus x_2 \oplus x_1x_2 = 1 \oplus x_2(1 \oplus x_1) = 1 \oplus x_2\overline{x}_1$$
$$= \overline{x_2\overline{x}_1} = x_1 \vee \overline{x}_2$$

The same result can be obtained from the truth table

Assignment of variables	Value of Boolean function	Minterm
00	1	
01	0	$x_1 \vee \overline{x}_2$
10	1	
11	1	

10.4.

$$(a)\quad x_1 \oplus \overline{x}_2 = x_1 \oplus 1 \oplus x_2 = 1 \oplus (x_1 \oplus x_2) = \overline{x_1 \oplus x_2}$$
$$(b)\quad \overline{x}_1 \oplus x_2 = 1 \oplus x_1 \oplus x_2 = \overline{x_1 \oplus x_2}$$
$$(c)\quad \overline{x}_1 \oplus \overline{x}_2 = 1 \oplus x_1 \oplus 1 \oplus x_2 = x_1 \oplus x_2$$

10.5.

$$\overline{x_1 \oplus x_2} = 1 \oplus x_1 \oplus x_2 = \overline{x}_1 \oplus x_2$$
$$\overline{x_1 \oplus x_2} = \overline{1 \oplus x_1 \oplus 1 \oplus x_2} = \overline{\overline{x}_1 \oplus \overline{x}_2}$$

10.6.

Coefficients a_0	Coefficients a_1	Polynomial
0	0	0
0	1	x_1
1	0	1
1	1	$1 \oplus x_1$

10.7.

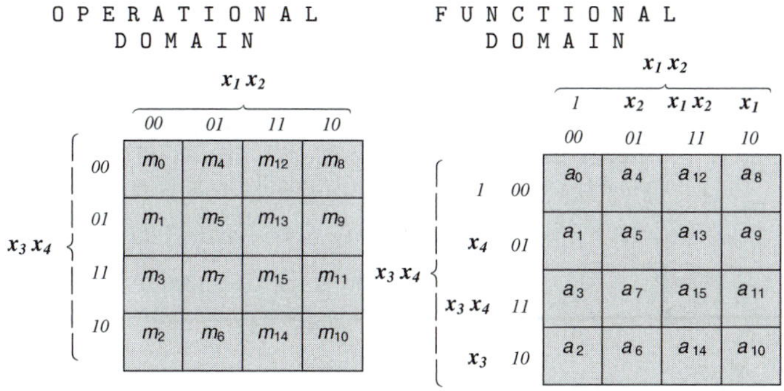

OPERATIONAL DOMAIN

x_3x_4 \ x_1x_2	00	01	11	10
00	m_0	m_4	m_{12}	m_8
01	m_1	m_5	m_{13}	m_9
11	m_3	m_7	m_{15}	m_{11}
10	m_2	m_6	m_{14}	m_{10}

FUNCTIONAL DOMAIN

x_3x_4 \ x_1x_2	1 (00)	x_2 (01)	x_1x_2 (11)	x_1 (10)
1 (00)	a_0	a_4	a_{12}	a_8
x_4 (01)	a_1	a_5	a_{13}	a_9
x_3x_4 (11)	a_3	a_7	a_{15}	a_{11}
x_3 (10)	a_2	a_6	a_{14}	a_{10}

Solutions to practice problems (continuation)

Practice problem	Solution
10.7.	(Continuation) OPERATIONAL DOMAIN $f = \bigvee_{i=0}^{15} m_i = \bigvee_{i=0}^{15} x_1^{i_1} x_2^{i_2} x_3^{i_3} x_4^{i_4}$ $= m_0\overline{x}_1\overline{x}_2\overline{x}_3\overline{x}_4 \vee m_1\overline{x}_1\overline{x}_2\overline{x}_3 x_4$ $\vee\ m_2\overline{x}_1\overline{x}_2 x_3\overline{x}_4 \vee m_3\overline{x}_1\overline{x}_2 x_3 x_4$ $\vee\ m_4\overline{x}_1 x_2\overline{x}_3\overline{x}_4 \vee m_5\overline{x}_1 x_2\overline{x}_3 x_4$ $\vee\ m_6\overline{x}_1 x_2 x_3\overline{x}_4 \vee m_7\overline{x}_1 x_2 x_3 x_4$ $\vee\ m_8 x_1\overline{x}_2\overline{x}_3\overline{x}_4 \vee m_9 x_1\overline{x}_2\overline{x}_3 x_4$ $\vee\ m_{10} x_1\overline{x}_2 x_3\overline{x}_4 \vee m_{11} x_1\overline{x}_2 x_3 x_4$ $\vee\ m_{12} x_1 x_2\overline{x}_3\overline{x}_4 \vee m_{13} x_1\overline{x}_2\overline{x}_3 x_4$ $\vee\ m_{14} x_1\overline{x}_2 x_3\overline{x}_4 \vee m_{15} x_1 x_2 x_3 x_4$ $x_j^{i_j} = \begin{cases} \overline{x}_j, & \text{if } i_j = 0; \\ x_j, & \text{if } i_j = 1. \end{cases}$ FUNCTIONAL DOMAIN $f = \sum_{i=0}^{15} a_i x_1^{i_1} x_2^{i_2} x_3^{i_3} x_4^{i_4}$ $= a_0 + a_1 x_4 + a_2 x_3 + a_3 x_3 x_4$ $+\ a_4 x_2 + a_5 x_2 x_4 + a_6 x_2 x_3$ $+\ a_7 x_2 x_3 x_4 + a_8 x_1 + a_9 x_1 x_4$ $+\ a_{10} x_1 x_3 + a_{11} x_1 x_2 x_4 + a_{12} x_1 x_2$ $+\ a_{13} x_1 x_2 x_4 + a_{14} x_1 x_2 x_3$ $+\ a_{15} x_1 x_2 x_3 x_4$ $x_j^{i_j} = \begin{cases} 1, & \text{if } i_j = 0; \\ x_j, & \text{if } i_j = 1. \end{cases}$ over the field $\mathcal{F}$
10.9.	There are four polarized minterms $(x_1 \oplus 1)^{i_1}(x_2 \oplus 1)^{i_2} = \overline{x}_1^{i_1}\overline{x}_2^{i_2}$; that is $(x_1 \oplus 1)^0(x_2 \oplus 1)^0 = 1;\ (x_1 \oplus 1)^0(x_2 \oplus 1)^1 = \overline{x}_2;$ $(x_1 \oplus 1)^1(x_2 \oplus 1)^0 = \overline{x}_1;\ (x_1 \oplus 1)^1(x_2 \oplus 1)^1 = \overline{x}_1\overline{x}_2$
10.10.	The EXOR expression in 1-polarity can be obtained from a 0-polarity expression as follows: $f = \underbrace{x_2 \oplus x_1 \oplus x_1 x_2}_{Polarity\ 0} = \underbrace{1 \oplus 1}_{Equal\ to\ 0} \oplus x_2 \oplus x_1 \oplus x_1 x_2$ $= 1 \oplus \underbrace{(1 \oplus x_2)}_{Equal\ to\ \overline{x}_2} \oplus x_1 \underbrace{(1 \oplus x_2)}_{Equal\ to\ \overline{x}_2} = \underbrace{1 \oplus \overline{x}_2 \oplus x_1\overline{x}_2}_{Polarity\ 1}$ The EXOR expression in 2-polarity can be obtained from a0-polarity expression by analogy: $f = 1 \oplus \overline{x}_1 \oplus \overline{x}_1 x_2$. The EXOR expression in 3-polarity can be derived from 1-polarity as follows: $f = \underbrace{1 \oplus \overline{x}_2 \oplus x_1\overline{x}_2}_{Polarity\ 1} = 1 \oplus \overline{x}_2 \underbrace{(1 \oplus x_1)}_{Equal\ to\ \overline{x}_1} = \underbrace{1 \oplus \overline{x}_1\overline{x}_2}_{Polarity\ 3}$

Solutions to practice problems (continuation)

Practice problem	Solution
10.12.	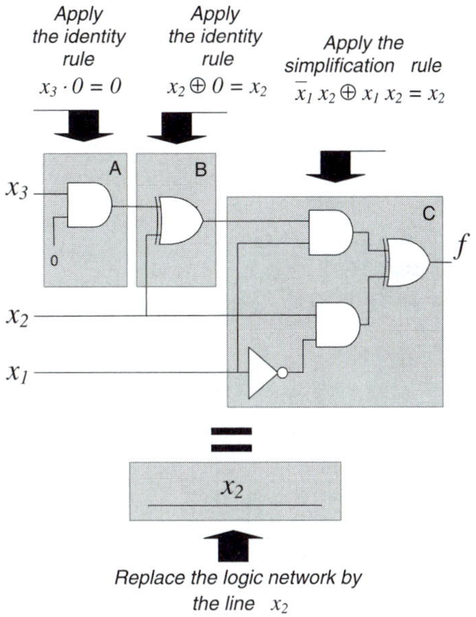 *Apply the identity rule* $x_3 \cdot 0 = 0$ *Apply the identity rule* $x_2 \oplus 0 = x_2$ *Apply the simplification rule* $\overline{x}_1 x_2 \oplus x_1 x_2 = x_2$ A B C x_3 0 x_2 x_1 f x_2 *Replace the logic network by the line* x_2
10.13.	**The first network:** $f = x_1 \oplus x_2$. **The second network:** $f = \overline{\overline{x_1 \vee x_2} \oplus \overline{\overline{x}_1 x_2}} = \overline{\overline{x}_1\overline{x}_2 \oplus (x_1 \vee \overline{x}_2)}$ $= \overline{x}_1\overline{x}_2 \oplus (x_1 \oplus \overline{x}_2 \oplus x_1\overline{x}_2) \oplus 1$ $= x_1 \oplus \overline{x}_2 \oplus \overbrace{(\overline{x}_1 \oplus x_1)}^{Equal\ to\ 1}\, \overline{x}_2 \oplus 1$ $= x_1 \oplus \overbrace{\overline{x}_2 \oplus \overline{x}_2}^{Equal\ to\ 0} \oplus 1 = x_1 \oplus 1 = \overline{x}_1$ **Conclusion:** These logic networks are not equivalent

10.11 Problems

Problem 10.1 Represent the Boolean functions given below by polynomial expressions of polarities $c = 1$ and $c = 2$:
(a) $f = \overline{x}_1 x_2 \vee x_1 x_2 \overline{x}_3 \vee x_3$ (b) $f = x_1 \vee x_1 x_2 x_3 \vee x_1 x_3$ (c) $f = x_1 \vee x_2 \vee \overline{x}_1 x_3$

Problem 10.2 Represent the following Boolean functions in a polynomial form of polarity $c = 0$:
(a) $f = \overline{x}_1 x_2 \oplus x_1 x_2 \overline{x}_3 \oplus x_3$
(b) $f = x_1 \oplus x_2 \oplus x_3$
(c) $f = x_1 \oplus x_1 x_2 x_3 \oplus x_1 x_3$
(d) $f = 1 \oplus x_1 \oplus x_2 \oplus \overline{x}_1 x_3$

Problem 10.3 Prove that the following polynomials represent the same Boolean function, $f_1 = x_1 x_3 \vee x_2 \overline{x}_3 \vee x_2 \overline{x}_4$:

(*a*) $f_1 = 1 \oplus \overline{x}_2 \overline{x}_3 x_4 x_5 \oplus \overline{x}_2 \overline{x}_4 \overline{x}_5 \oplus \overline{x}_2 \overline{x}_4 \oplus x_2 x_3 x_4$
(*b*) $f_2 = x_2 \oplus x_5 \oplus x_2 x_5 \oplus x_2 x_3 x_4 \oplus x_3 x_4 x_5 \oplus x_2 x_3 x_4 x_5$
(*c*) $f_3 = 1 \oplus \overline{x}_2 \overline{x}_5 \oplus x_3 x_4 \oplus \overline{x}_2 x_3 x_4 \overline{x}_5$
(*d*) $f_4 = x_2 \oplus \overline{x}_2 x_5 \oplus x_2 x_3 x_4 \oplus x_2 x_3 x_4 \overline{x}_5$
(*e*) $f_5 = \overline{x}_2 \overline{x}_5 \oplus \overline{x}_2 x_3 x_4 x_5 \oplus x_2 x_3 x_4$
(*f*) $f_6 = x_1 x_3 \oplus x_2 \overline{x}_3 \oplus x_3 \overline{x}_4 \oplus \overline{x}_1 \overline{x}_2 x_3 \overline{x}_4$

Problem 10.4 Derive the polynomial expressions given functional maps as follows:

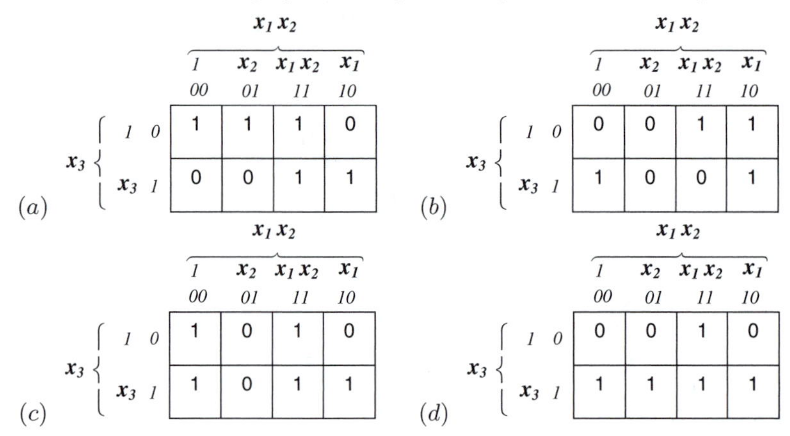

Problem 10.5 Derive the vector of coefficients for the following polynomial form:
(a) $f = \overline{x}_1 x_2 \oplus x_1 x_2 \overline{x}_3 \oplus x_3$
(b) $f = x_1 \oplus x_2 \oplus x_3$
(c) $f = x_1 \oplus x_1 x_2 x_3 \oplus x_1 x_3$
(d) $f = 1 \oplus x_1 \oplus x_2 \oplus \overline{x}_1 x_3$

Problem 10.6 Derive the polarized minterms for the following Boolean functions:
(a) $f = \overline{x}_1 x_2 \vee x_1 x_2 \overline{x}_3 \vee x_3$
(b) $f = x_1 \vee x_2 \vee \overline{x}_1 x_3$
(c) $f = x_1 \vee x_1 x_2 x_3 \vee x_1 x_3$
(d) $f = x_1 \vee (x_2 \oplus x_3)$

Problem 10.7 Simplify the following polynomial expressions:
(a) $f = \overline{x}_1 x_2 \oplus x_1 x_2 \overline{x}_3 \oplus x_3$
(b) $f = x_1 \oplus x_2 \oplus x_3$
(c) $f = x_1 \oplus x_1 x_2 x_3 \oplus x_1 x_3$
(d) $f = 1 \oplus x_1 \oplus x_2 \oplus \overline{x}_1 x_3$

Problem 10.8 Factor the following polynomial expressions:
(a) $f = \overline{x}_1 x_2 \oplus x_1 x_2 \overline{x}_3 \oplus x_3$ (c) $f = x_1 \oplus x_1 x_2 x_3 \oplus x_1 x_3$
(b) $f = x_1 \oplus x_2 \oplus x_3$ (d) $f = 1 \oplus x_1 \oplus x_2 \oplus \overline{x}_1 x_3$

Problem 10.9 Represent each of the following polynomials in all polarities:
(a) $f = \overline{x}_1 x_2 \oplus x_1 x_2 \overline{x}_3 \oplus x_3$ (b) $f = x_1 \oplus x_1 x_2 x_3 \oplus x_1 x_3$ (c) $f = x_1 \oplus x_2 \oplus x_3$

Problem 10.10 Given the below network, derive the equivalent EXOR-AND network:

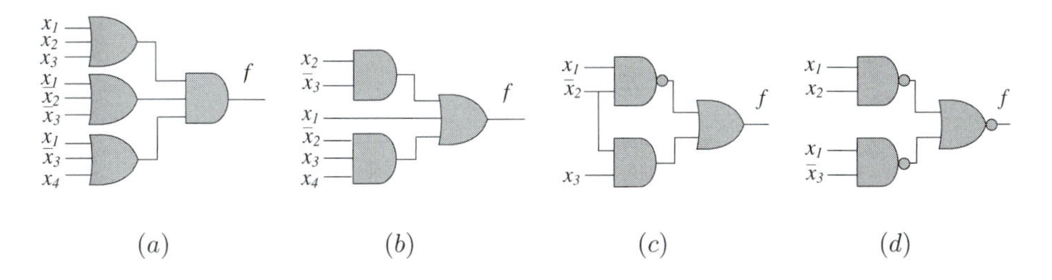

Problem 10.11 Prove that the following logic networks perform the same Boolean functions:

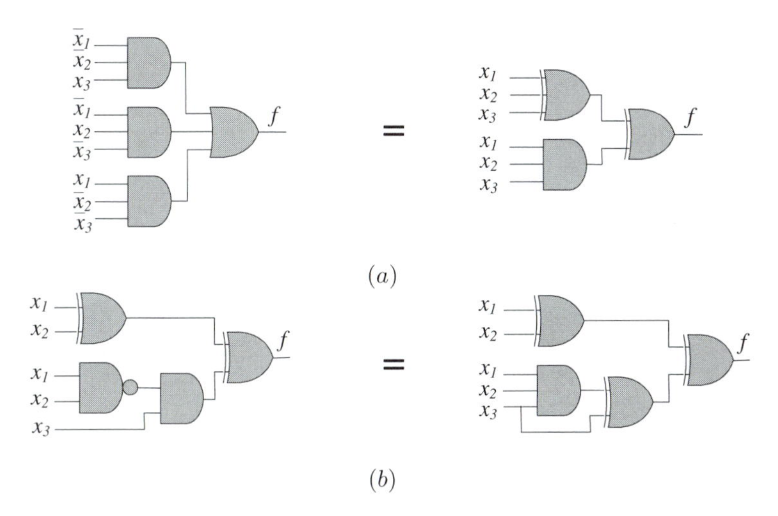

Problem 10.12 Prove the validity of the following expressions:
(a) $\overline{x}_1 x_2 \oplus x_1 x_2 \overline{x}_3 \oplus x_3$ and $\overline{x}_1 x_2 \oplus (x_1 x_2 \vee x_3)$
(c) $x_1 \oplus x_2 \oplus x_3$ and $x_1 \overline{x}_2 \overline{x}_3 \vee \overline{x}_1 \overline{x}_2 x_3 \vee \overline{x}_1 x_2 \overline{x}_3 \vee x_1 x_2 x_3$
(b) $x_1 \oplus x_1 x_2 x_3 \oplus x_1 x_3$ and $x_1 \overline{x}_3 \oplus x_1 x_2 x_3$
(d) $1 \oplus x_1 \oplus x_2 \oplus \overline{x}_1 x_3$ and $x_2 \oplus \overline{x}_1 \overline{x}_3$

Problem 10.13 Find the conditions for observability of $f = 1$ for the following logic networks:

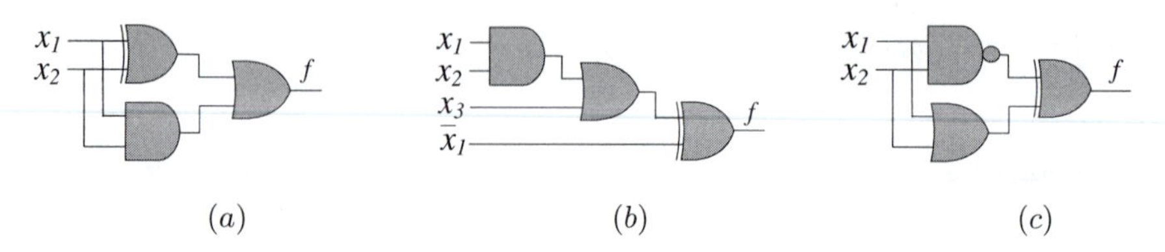

Problem 10.14 Apply the local transformations to reduce the following logic networks:

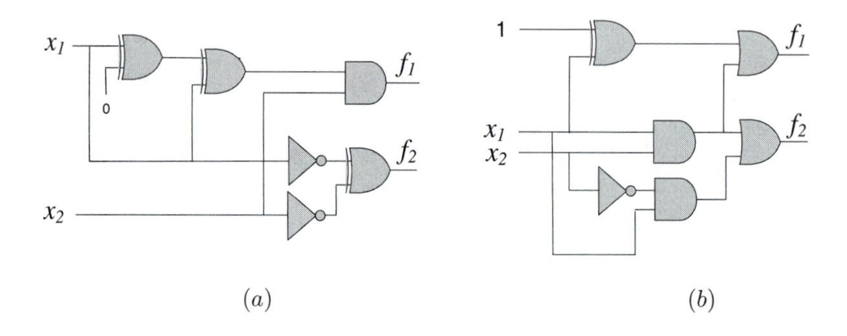

11

Manipulation of Polynomial Expressions

Polynomial representations

- Fixed polarity
- Mixed polarity
- SOP and polynomial expressions
- K-maps and polynomial expressions
- Simplification techniques
- Techniques for logic network design

Computing in matrix form

- Matrix operations
- Polarized literals
- Polarized minterms
- Forward transform
- Inverse transform

Advanced topics

- Classification of polynomial forms
- Advanced computing techniques

11.1 Introduction

Polynomial forms can be represented using various canonical expansions such as logic Taylor expansion. A logic Taylor expansion at a the point c is also called a *fixed polarity Reed-Muller expansion*. The value c is the so-called *fixed polarity* c. The binary representation of c, the n-tuple $(c_1, c_2, \ldots, c_n)$, determines the polarity of variables in the polynomial.

11.2 Fixed and mixed polarity polynomial forms

In terms of polynomial forms, two types of polarity are distinguished:

- *Fixed* polarity and
- *Mixed* polarity.

In a fixed polarity polynomial expression of a Boolean function f, every variable appears either complemented ($\overline{x}_i$) or uncomplemented (x_i), and never in both forms. There are 2^n fixed polarity forms given a function of n variables. In a mixed polarity form, a variable can appear in one or both polarities. There are 3^n mixed polarity forms given a function of n variables.

Example 11.1 (Fixed and mixed polarity.) *In Figure 11.1, the polynomial expressions*

$$\overbrace{\overline{x}_1 x_2 \overline{x}_3 \oplus \overline{x}_1 x_2 \oplus \overline{x}_1 \overline{x}_3}^{\textit{Fixed polarities of variables}} \quad \text{and} \quad \overbrace{\overline{x}_1 x_2 \overline{x}_3 \oplus x_1 \overline{x}_2 \oplus \overline{x}_1 x_3}^{\textit{Mixed polarities of variables}}$$

are a fixed and a mixed polarity form, respectively.

Practice problem 11.1. (Fixed and mixed polarity.) Prove that for a Boolean function of two variables there are 2^2 fixed polarity expressions and 3^2 mixed polarity expressions.
Answer is given in "Solutions to practice problems."

11.2.1 Fixed polarity polynomial forms

A fixed polarity polynomial expression of a Boolean function f of n variables is *unique*; that is, only one representation exists given a polarity c $(c_1, c_2, \ldots, c_n)$.

Techniques for computing the fixed and mixed polarity polynomial forms

Fixed polarity polynomial expressions

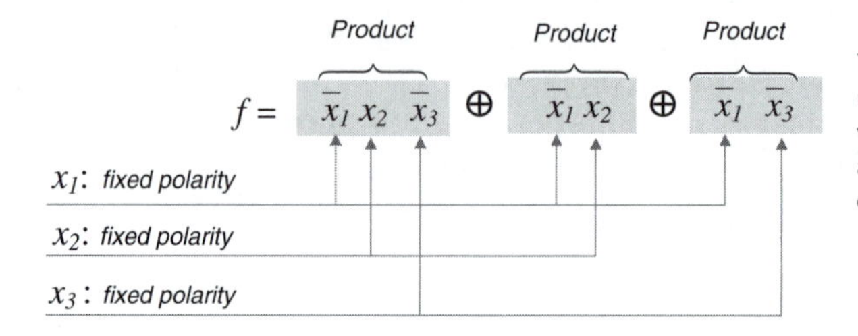

The variable x_2 appears uncomplemented; the variables x_1 and x_3 appear complemented only

Mixed polarity polynomial expressions

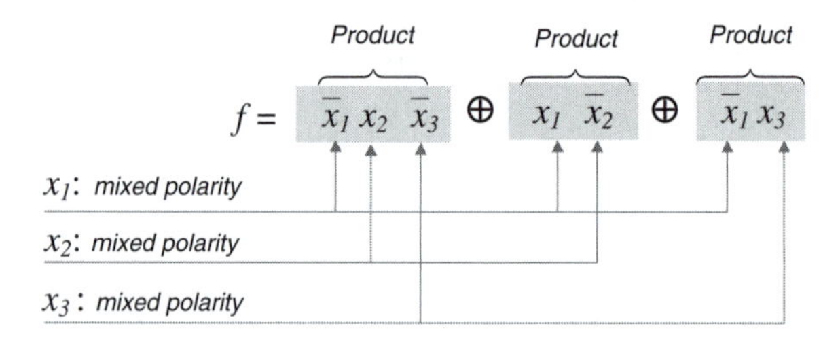

The variables x_1, x_2, and x_3 appear in both uncomplemented and complemented forms

FIGURE 11.1
Illustration of fixed and mixed polarity polynomial forms (Example 11.1).

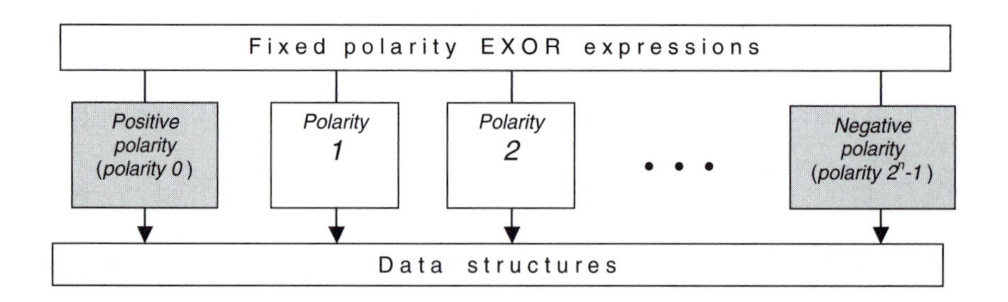

FIGURE 11.2
An arbitrary Boolean function can be represented in fixed polarity polynomial form.

There are 2^n various polarities, and two boundary cases among them: a *positive polarity* form in which all variables are uncomplemented, and a *negative polarity* form, in which all variables are complemented (Figure 11.2).

Example 11.2 (Fixed polarity.) *Polynomial expressions in positive polarity $c = 0$ ($c_1c_2 = 00$) (uncomp-lemented variables) and the polarity $c = 2$, $c_1c_2 = 10$ (only the variable x_1 is complemented) are as follows:*

$$
\begin{aligned}
&\textbf{\textit{Polarity}}\ \mathbf{c = 0}\ (c_1 = 0, c_2 = 0):\\
f &= r_0(x_1 \oplus c_1)^{i_1}(x_2 \oplus c_2)^{i_2} \oplus r_1(x_1 \oplus c_1)^{i_1}(x_2 \oplus c_2)^{i_2}\\
&\oplus r_2(x_1 \oplus c_1)^{i_1}(x_2 \oplus c_2)^{i_2} \oplus r_3(x_1 \oplus c_1)^{i_1}(x_2 \oplus c_2)^{i_2}\\
&= r_0(x_1^0x_2^0) \oplus r_1(x_1^0x_2^1) \oplus r_2(x_1^1x_2^0) \oplus r_3(x_1^1x_2^1)\\
&= r_0 \oplus r_1x_2 \oplus r_2x_1 \oplus r_3x_1x_2\\
&\textbf{\textit{Polarity}}\ \mathbf{c = 2}\ (c_1 = 1, c_2 = 0):\\
f &= r_0(x_1 \oplus 1)^0(x_2 \oplus 0)^0 \oplus r_1(x_1 \oplus 1)^0(x_2 \oplus 0)^1\\
&\oplus r_2(x_1 \oplus 1)^1(x_2 \oplus 0)^0 \oplus r_3(x_1 \oplus 1)^1(x_2 \oplus 0)^1\\
&= r_0 \oplus r_1x_2 \oplus r_2\overline{x}_1 \oplus r_3\overline{x}_1x_2.
\end{aligned}
$$

Let $f = x \vee y$. Then four fixed polarity polynomial expressions can be derived as shown in Figure 11.3.

Design example: Polynomial forms of the OR gate

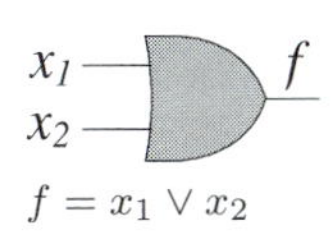

0-polarity:	$f = x_1 \oplus x_2 \oplus x_1x_2$,	*no complemented variables*
1-polarity:	$f = 1 \oplus \overline{x}_2 \oplus x_1\overline{x}_2$,	x_2 *is complemented*
2-polarity:	$f = 1 \oplus \overline{x}_1 \oplus \overline{x}_1x_2$,	x_1 *is complemented*
3-polarity:	$f = 1 \oplus \overline{x}_1\overline{x}_2$,	x_1 *and* x_2 *are complemented*

FIGURE 11.3
Representation of the two-input OR gate by polynomial forms of $2^2 = 4$ different fixed polarities (Example 11.2).

Practice problem 11.2. (Negative polarity.) Derive a common form for the negative polarity polynomial expression given a Boolean function of two variables.
Answer is given in "Solutions to practice problems."

Figure 11.4 shows how a polynomial expression is constructed using the polynomial coefficients and minterms over GF(2).

Deriving a fixed polarity polynomial expansion given a Boolean function is a necessary step in the process of implementation of the function given a library of logic gates AND and EXOR together with a constant 1 signal. It forms a universal basis of operations for implementing an arbitrary Boolean function.

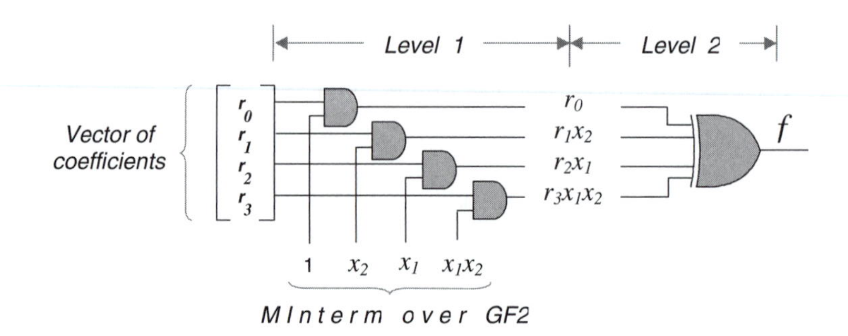

FIGURE 11.4
A two-level logic network for computing the EXOR expression of two variables (Example 11.2).

Example 11.3 (Fixed polarities for gates.) *In Table 11.1, the expressions of the AND, OR, and EXOR Boolean functions of two variables are given in fixed polarities. For example, the polynomial in polarity* $c = 3$ $(c_1c_2 = 11)$ *for the OR function is represented by two nonzero coefficients* $f = 1 \oplus \overline{x}_1\overline{x}_2$. *This is an optimal polynomial form of the OR function with respect to the criterion of the minimal number of literals.*

Practice problem 11.3. (Fixed polarities for gates.) Represent the AND function $f = x_1x_2$ in polarities $c_1c_2 = 01, 10, 11$ using the algebraic manipulation.
Answer is given in "Solutions to practice problems."

11.2.2 Deriving polynomial expressions from SOP forms

Polynomial expressions can be derived from SOP forms using algebraic transformations. Given a canonical, or standard, SOP form, a polynomial form of mixed polarity is derived by replacing (a) the logical sum "$\vee$" with the EXOR operation "$\oplus$", and (b) the complement of a variable x with the $1 \oplus x$.

Example 11.4 (Polynomial and SOP expressions.) *A 0-polarity polynomial form of the function* $f = x_1 \vee x_2$ *is derived as follows:* $x_1 \vee x_2 = \overline{\overline{x_1 \vee x_2}} = \overline{\overline{x}_1\overline{x}_2} = (1 \oplus x_1)(1 \oplus x_2) \oplus 1 = x_1 \oplus x_2 \oplus x_1x_2$.

Practice problem 11.4. (Polynomial and SOP expressions.) Derive a 0-polarity polynomial form for the OR function of three variables.
Answer: $f = x_1 \vee x_2 \vee x_3 = \overline{\overline{x_1 \vee x_2 \vee x_3}} = \overline{\overline{x}_1\overline{x}_2\overline{x}_3} = (1 \oplus x_1)(1 \oplus x_2)(1 \oplus x_3) \oplus 1 = (1 \oplus x_1 \oplus x_2 \oplus x_1x_2)(1 \oplus x_3) \oplus 1 = x_1 \oplus x_2 \oplus x_3 \oplus x_1x_2 \oplus x_1x_3 \oplus x_2x_3$.

TABLE 11.1
Polynomial expressions of fixed polarities of elementary Boolean functions (Example 11.3).

Design example: Fixed polarity polynomial forms of the gates

Gate	r_0	r_1	r_2	r_3	Polynomial expressions
$f = x_1 x_2$	0	0	0	1	$x_1 x_2$
	0	0	1	1	$x_1 \oplus x_1 \overline{x}_2$
	0	1	0	1	$x_2 \oplus \overline{x}_1 x_2$
	1	1	1	1	$1 \oplus \overline{x}_2 \oplus \overline{x}_1 \oplus \overline{x}_1 \overline{x}_2$
$f = x_1 \vee x_2$	0	1	1	1	$x_2 \oplus x_1 \oplus x_1 x_2$
	1	1	0	1	$1 \oplus \overline{x}_2 \oplus x_1 \overline{x}_2$
	1	0	1	1	$1 \oplus \overline{x}_1 \oplus \overline{x}_1 x_2$
	1	0	0	1	$1 \oplus \overline{x}_1 \overline{x}_2$
$f = x_1 \oplus x_2$	0	1	1	0	$x_2 \oplus x_1$
	1	1	1	0	$1 \oplus \overline{x}_2 \oplus x_1$
	1	1	1	0	$1 \oplus x_2 \oplus \overline{x}_1$
	0	1	1	0	$\overline{x}_2 \oplus \overline{x}_1$

11.2.3 Conversion between polarities

Given one polarity of a polynomial expression of a Boolean function, one can convert it to another polarity expression by algebraic manipulations.

Example 11.5 (Conversion between polarities.) *A mixed polarity polynomial expression can be transformed into a polynomial form of polarity $c = 2$ $(c_1 = 1, c_2 = 0)$:*

$$\begin{aligned} f &= \overbrace{\overline{x}_1\overline{x}_2 \oplus \overline{x}_1 x_2 \oplus x_1 \overline{x}_2}^{Mixed\ polarity} = \overline{x}_1(1 \oplus x_2) \oplus \overline{x}_1 x_2 \oplus (\overbrace{x_1 \oplus 1}^{\overline{x}_1} \oplus 1) x_2 \\ &= \overline{x}_1 \oplus \underbrace{\overline{x}_1 x_2 \oplus \overline{x}_1 x_2}_{Equal\ to\ 0} \oplus \overline{x}_1 x_2 \oplus x_2 = \underbrace{\overline{x}_1 \oplus \overline{x}_1 x_2 \oplus x_2}_{Fixed\ 2-polarity} \end{aligned}$$

Figure 11.5 illustrates the implementation of the initial SOP form (AND-OR logic network), the conversion of a mixed polarity polynomial form (AND-EXOR logic network), and the fixed polarity polynomial form, given $c = 2$. The polynomial form of the fixed polarity $c = 1(c_1 = 0, c_2 = 1)$ is obtained as follows:

$$\begin{aligned} f &= \overline{x}_1\overline{x}_2 \oplus \overline{x}_1 x_2 \oplus x_1 x_2 \\ &= (x_1 \oplus 1)\overline{x}_2 \oplus (x_1 \oplus 1)(\overline{x}_2 \oplus 1) \oplus x_1(\overline{x}_2 \oplus 1) \\ &= x_1\overline{x}_2 \oplus \overline{x}_2 \oplus x_1\overline{x}_2 \oplus x_1 \oplus \overline{x}_2 \oplus 1 \oplus x_1\overline{x}_2 \oplus x_1 = x_1\overline{x}_2 \oplus 1 \end{aligned}$$

Design example: Converting logic network

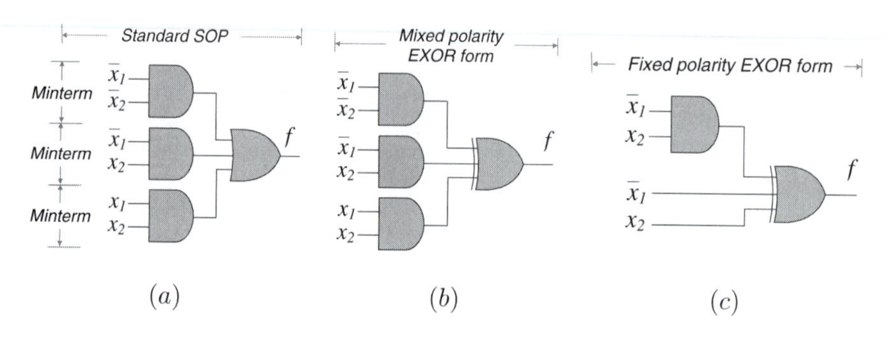

FIGURE 11.5
Interpretation of converting an SOP form into a mixed polarity and then into a fixed polarity expression (Example 11.5).

Practice problem **11.5. (Conversion between polarities.)** Derive a negative polarity polynomial expression for the OR function from a positive polarity.
Answer is given in "Solution to practice problems."

11.2.4 Deriving polynomial expressions from K-maps

K-maps are known as a graphical form for representing Boolean functions. Since canonical SOP forms can be derived directly from K-maps, the latter can be easily converted to mixed polarity polynomial forms, which can be further simplified to EXOR expressions of other polarities:

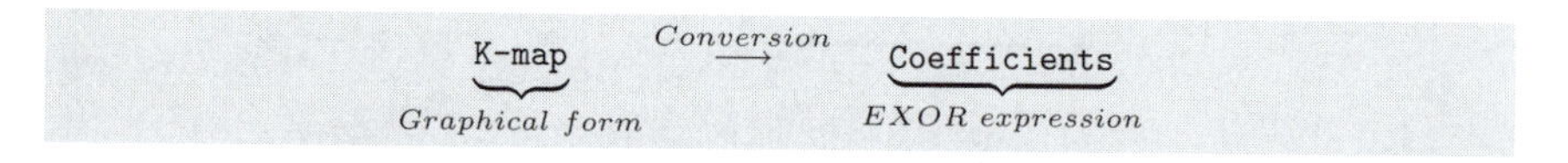

Example 11.6 (Polynomials and K-maps.) *In Figure 11.6, mixed polarity polynomial expressions are derived from K-maps for two Boolean functions of two and three variables. This is accomplished by deriving the canonical SOP forms first, and then replacing "$\vee$" with "$\oplus$". The obtained polynomial expressions of mixed polarity are not minimal and can be further simplified to the expressions* $x_1 \oplus x_2$ *and* $x_1 \oplus x_2 \oplus x_3$*, respectively.*

Design example: Deriving polynomial forms

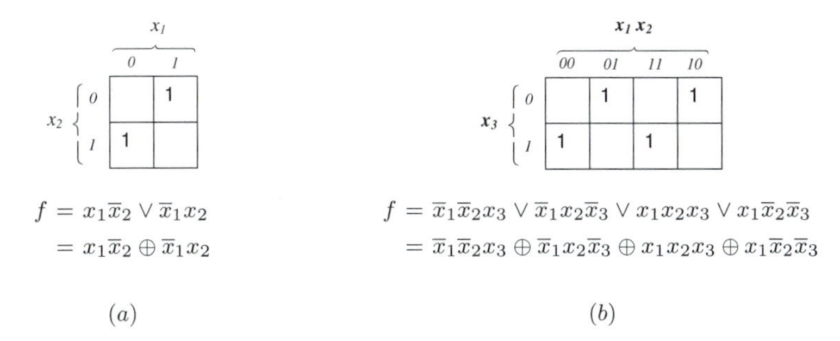

FIGURE 11.6
Deriving polynomial expressions from K-maps for Boolean functions of two variables (a) and three variables (b) (Example 11.6).

Practice problem **11.6. (Polynomials and K-maps.)** Derive a polynomial expression from the following K-map:

x_3 \ $x_1 x_2$	00	01	11	10
0	1			
1		1		

Answer: $f = \overline{x}_1\overline{x}_2\overline{x}_3 \oplus \overline{x}_1 x_2 x_3$.

It should be noted that K-maps cannot be used for the minimization of polynomial forms of Boolean functions. Special techniques are required for minimizing polynomial expressions.

11.2.5 Simplification of polynomial expressions

Given a mixed polarity polynomial form derived from a canonical SOP, further algebraic transformations can lead to a minimized polynomial representation. Minimization on a K-map cannot be directly applied to polynomial forms, since the rules for reducing are different in GF(2). Some rules are given below:

Rules for simplification of polynomial expressions

$$x \oplus \overline{x} = 1 \qquad x \oplus x = 0$$

$$x \oplus 1 = \overline{x} \qquad x \oplus 0 = x$$

Since the EXOR operation is commutative and associative, the following rules hold true as well:

Rules for simplification of polynomial expressions

$$x_1x_2 \oplus \overline{x}_1x_2 = x_2$$
$$x_1\overline{x}_2 \oplus x_1 = x_1\overline{x}_2$$
$$(x_1 \oplus x_2)x_1 = x_1\overline{x}_2$$

The details of such minimizations are not the subject of this textbook. Exhaustive search techniques; that is, the generation of all possible mixed polarity forms, can be applied to find the minimal one. Obviously, such techniques can only be applied to small functions, and for larger functions some heuristic approaches have been developed.

Example 11.7 (Minimal expressions.) *The expression* $\overline{x}_1 \oplus \overline{x}_1x_2 \oplus x_2$ *obtained in Example 11.5 can be further simplified as follows:*

$$f = \overline{x}_1 \oplus \overline{x}_1x_2 \oplus x_2 = \overline{x}_1(1 \oplus x_2) \oplus x_2$$
$$= (\overline{x}_1 \oplus x_2)(\overline{x}_2 \oplus x_2) = \overline{x}_1 \oplus x_2$$

The resulting expression is a minimal one.

Practice problem **11.7. (Minimal expressions.)** Prove that the minimal polynomial expressions for the Boolean functions of two and three variables given in Example 11.6 are $x_1 \oplus x_2$ and $x_1 \oplus x_2 \oplus x_3$, respectively.
Answer is given in "Solutions to practice problems."

11.3 Computing the coefficients of polynomial forms

The coefficients of polynomial forms can be calculated using matrix representations. In this section, the basics of matrix computations for deriving vectors of coefficients of fixed polarity polynomial expressions are given. Matrix-based computations are well suited, in particular, for programming and fast transforms.

11.3.1 Matrix operations over GF(2)

In matrix operations in GF(2), *binary* matrices, consisting of 0's and 1's, are used.

Example 11.8 (Binary matrices.) *A rectangular array of binary numbers,* $\mathbf{A} = \begin{bmatrix} 1 & 0 \\ 1 & 1 \end{bmatrix}$ *is a binary matrix with two columns and two rows. We thus say that* $\mathbf{A}$ *is a* 2×2 *(two by two) matrix. The first and second rows of* $\mathbf{A}$ *are* $[1\ 0]$ *and* $[1\ 1]$*, respectively.*

A matrix is a *square* matrix if it has the same number of rows and columns. A square matrix may be multiplied by itself. The *identity* matrix $\mathbf{I}$ is a matrix that has 1s along the diagonal and 0s everywhere else. If a matrix $\mathbf{A}$ has an inverse, then that inverse is denoted $\mathbf{A}^{-1}$, and $\mathbf{A}\mathbf{A}^{-1} = \mathbf{A}^{-1}\mathbf{A} = \mathbf{I}$. Matrix multiplication also works if one of the matrices is a vector, which is a matrix containing a single column or a single row. The following matrix operations are used for deriving vectors representing the polynomial forms of Boolean functions:

Matrix operations over GF(2)

Addition over GF(2): If a_{ij} and b_{ij} are the elements in the i-th row and j-th column of the binary matrices $\mathbf{A}$ and $\mathbf{B}$, respectively, then the matrix $\mathbf{C} = \mathbf{A} \oplus \mathbf{B}$ has elements $c_{ij} = a_{ij} \oplus b_{ij}$.

Multiplication over GF(2): If $a_{ik} \in \{0,1\}$ is the element in the i-th row and k-th column of a $2^n \times 2^m$ matrix $\mathbf{A}$ and $b_{kj} \in \{0,1\}$ is the element in the k-th row and j-th column of a $2^n \times 2^m$ matrix $\mathbf{B}$, then the $2^n \times 2^m$ matrix $\mathbf{C} = \mathbf{AB}$ has elements $c_{ij} \in \{0,1\}$, $c_{ij} = \bigoplus_{k=1}^{2^n} a_{ik} b_{kj}$.

Inversion: The inverse of a $2^n \times 2^n$ matrix $\mathbf{A}$ is denoted as $\widehat{\mathbf{A}}$ or $\mathbf{A}^{-1}$ and satisfies $\mathbf{A}\mathbf{A}^{-1} = \mathbf{A}^{-1}\mathbf{A} = \mathbf{I}$, where $\mathbf{I}$ is the $2^n \times 2^n$ *identity* matrix containing unity entries on the diagonal and zeros elsewhere.

Example 11.9 (Matrix-vector operations.) *An example of multiplication of a binary matrix by a binary vector over GF(2) is given in Figure 11.7.*

Example 11.10 (Identity matrices.) *The following multiplications of $2^1 \times 2^1$ matrices result in identity matrices:*

$$(a)\ \ \mathbf{A}\widehat{\mathbf{A}} = \begin{bmatrix} 1 & 0 \\ 1 & 1 \end{bmatrix} \begin{bmatrix} 1 & 0 \\ 1 & 1 \end{bmatrix} = \begin{bmatrix} 1 & 0 \\ 0 & 1 \end{bmatrix} = \mathbf{I} \ \ \textit{over GF(2)}$$

$$(b)\ \ \mathbf{B}\widehat{\mathbf{B}} = \begin{bmatrix} 0 & 1 \\ 1 & 1 \end{bmatrix} \begin{bmatrix} 1 & 1 \\ 1 & 0 \end{bmatrix} = \begin{bmatrix} 1 & 0 \\ 0 & 1 \end{bmatrix} = \mathbf{I} \ \ \textit{over GF(2)}$$

$$(c)\ \ \mathbf{C}\widehat{\mathbf{C}} = \begin{bmatrix} 1 & 0 \\ -1 & 1 \end{bmatrix} \begin{bmatrix} 1 & 0 \\ 1 & 1 \end{bmatrix} = \begin{bmatrix} 1 & 0 \\ 0 & 1 \end{bmatrix} = \mathbf{I}$$

$$(d)\ \ \mathbf{D}\widehat{\mathbf{D}} = \begin{bmatrix} 0 & 1 \\ 1 & -1 \end{bmatrix} \begin{bmatrix} 1 & 1 \\ 1 & 0 \end{bmatrix} = \begin{bmatrix} 1 & 0 \\ 0 & 1 \end{bmatrix} = \mathbf{I}$$

Cases (a) and (b) concern operations over GF(2); in cases (c) and (d), identity matrices are obtained by operations over the field of integer numbers.

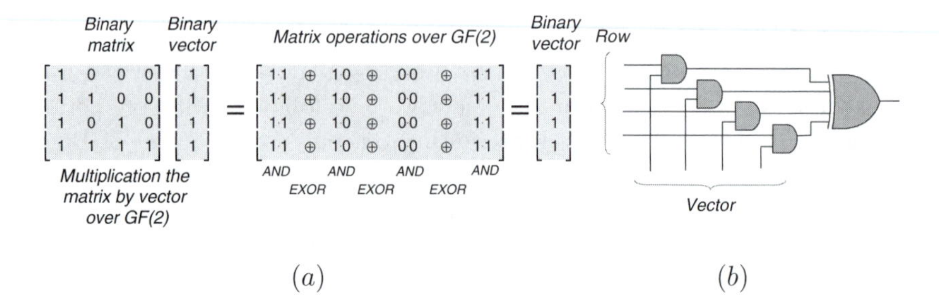

FIGURE 11.7
Multiplication of a binary matrix by a binary vector over GF(2) (a) and corresponding logic network b (Example 11.8).

Practice problem **11.8. (Binary matrices.)** Perform multiplication of the following matrices over GF(2): $\begin{bmatrix} 1 & 0 \\ 1 & 1 \end{bmatrix}$ and $\begin{bmatrix} 0 & 1 \\ 1 & 1 \end{bmatrix}$.
Answer is given in "Solutions to practice problems."

11.3.2 Polarized literals and minterms in matrix form

Polarized minterms are the basic components of the polynomial forms. Each polarized minterm is formed from n polarized literals, where n is the number of variables in a Boolean function.

In order to describe the polarized minterm in matrix form, its algebraic description must be converted into a description in matrix terms:

$$\underbrace{\texttt{Polarized literal}}_{Algebraic\ form} \overset{Conversion}{\longleftrightarrow} \underbrace{\texttt{Polarized literal}}_{Matrix\ form}$$

The function of the variable x_j, its polarity parameter c_j, and the parameter i_j which indicates the "presence" or "absence" of this variable in a logic expression, is called the *polarized literal*; its algebraic form is as follows:

$$\texttt{Polarized literal} = (x_j \oplus c_j)^{i_j} = \begin{cases} 1, & \text{if } i_j = 0 \\ (x_j \oplus c_j), & \text{if } i_j = 1 \end{cases} \quad \text{over GF}(2)$$

The matrix form of a polarized literal is based on the assumption that

- All operations are performed over GF(2);

► Multiplication of the *elementary* transform matrix and the truth-vector of the variable x_j results in a vector of coefficients which corresponds to the simplest polynomial expressions, literals $x_j \oplus 0$ or $x_j \oplus 1$.

The elementary $2^1 \times 2^1$ transform matrix is denoted as $\mathbf{R}_{2^1}^{(c_j)}$, where $c_j \in \{0, 1\}$ is the polarity of the variable x_j. The polarized literal corresponds to the elementary transform matrix for $c_j = 0$ and $c_j = 1$ as follows:

Polarized literal in matrix form

$$\texttt{Polarized literal} = \mathbf{R}_{2^1}^{(c_j)}$$

$$= \begin{cases} \mathbf{R}_{2^1}^{(0)} = \begin{bmatrix} 1 & 0 \\ 1 & 1 \end{bmatrix}, & \text{if } c_j = 0 \\ \\ \mathbf{R}_{2^1}^{(1)} = \begin{bmatrix} 1 & 0 \\ 1 & 1 \end{bmatrix}, & \text{if } c_j = 1 \end{cases} \quad \text{over GF(2)} \qquad (11.1)$$

Computing the polarized literal means multiplying the elementary transform matrix $\mathbf{R}_{2^1}^{(c_j)}$ by the truth-vector of a single variable x_j, $\mathbf{F} = [0\ 1]^T$.

Example 11.11 (Polarized literal.) *In Figure 11.8, the polarized literal is computed by the multiplication of the elementary matrix and the truth-vector for the variable x_j,* $\mathbf{F} = \begin{bmatrix} 0 \\ 1 \end{bmatrix}$*, over GF(2):*

$$\mathbf{R}_{2^1}^{(0)}\mathbf{F} = \begin{bmatrix} 1 & 0 \\ 1 & 1 \end{bmatrix} \begin{bmatrix} 0 \\ 1 \end{bmatrix} = \begin{bmatrix} 0 \\ 1 \end{bmatrix} \longrightarrow r_0 \oplus r_1 x_j = x_j$$

$$\mathbf{R}_{2^1}^{(1)}\mathbf{F} = \begin{bmatrix} 0 & 1 \\ 1 & 1 \end{bmatrix} \begin{bmatrix} 0 \\ 1 \end{bmatrix} = \begin{bmatrix} 1 \\ 1 \end{bmatrix} \longrightarrow r_0 \oplus r_1 x_j = 1 \oplus x_j$$

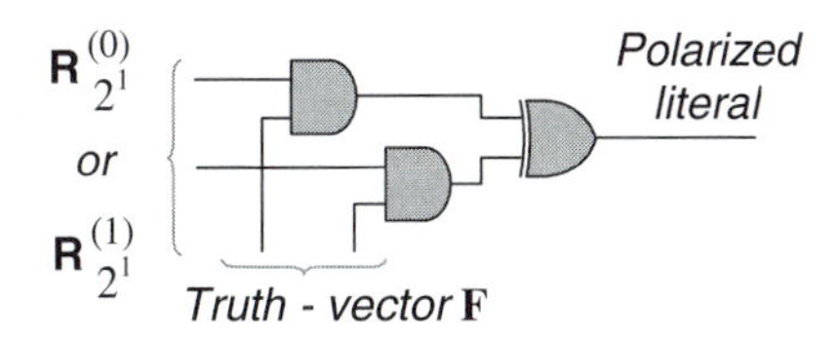

FIGURE 11.8
Computing a polarized literal in matrix form (Example 11.11).

Polarized minterms in matrix form

A polarized minterm can be described in a matrix equation as a conversion of its algebraic form into matrix form

$$\underbrace{\texttt{Polarized minterm}}_{Algebraic\ form} \overset{Conversion}{\longleftrightarrow} \underbrace{\texttt{Polarized minterm}}_{Matrix\ form}$$

For this, a matrix description of polarized literals is used (Equation 11.1), that is, n polarized literals in algebraic form are replaced by their n matrix equivalents. In forming vectors and matrices of sizes $2^n \times 1$ and $2^n \times 2^n$, respectively, the operation of Kronecker product must be used. Recall that the Kronecker product of the matrices

$$\mathbf{A}_2 = \begin{bmatrix} a_{11} & a_{12} \\ a_{21} & a_{22} \end{bmatrix} \quad \text{and} \quad \mathbf{B}_2 = \begin{bmatrix} b_{11} & b_{12} \\ b_{21} & b_{22} \end{bmatrix}$$

results in the 4×4 matrix

$$\mathbf{A}_2 \otimes \mathbf{B}_2 = \left[\begin{array}{c|c} a_{11}\mathbf{B}_2 & a_{12}\mathbf{B}_2 \\ \hline a_{21}\mathbf{B}_2 & a_{22}\mathbf{B}_2, \end{array}\right] = \left[\begin{array}{cc|cc} a_{11}b_{11} & a_{11}b_{12} & a_{12}b_{13} & a_{12}b_{14} \\ a_{11}b_{21} & a_{11}b_{22} & a_{12}b_{23} & a_{12}b_{24} \\ \hline a_{21}b_{31} & a_{21}b_{32} & a_{22}b_{33} & a_{22}b_{34} \\ a_{21}b_{41} & a_{21}b_{42} & a_{22}b_{43} & a_{22}b_{44} \end{array}\right]$$

The polarized minterm is formed using the Kronecker product between n elementary matrices $\mathbf{R}_{2^1}^{(c_j)}$ as follows:

The polarized minterm in matrix form

$$\mathbf{R}_{2^n}^{(c)} = \bigotimes_{j=1}^{n} \mathbf{R}_{2^1}^{(c_j)} \tag{11.2}$$

where $\mathbf{R}_{2^1}^{(c_j)}$ is defined by Equation 11.1.

The resulting $2^n \times 2^n$ matrix $\mathbf{R}_{2^n}^{(c)}$ represents a minterm of polarity $c = c_1c_2 \ldots c_n$.

Example 11.12 (Polarized minterms.) *For the polarity $c = 1$ ($c_1 = 1,\ c_2 = 0$) and parameter $i_j \in \{0,1\}$, polarized literals are generated in matrix form as follows:*

$$(x_j \oplus 0) \longleftrightarrow \mathbf{R}_{2^1}^{(0)} = \begin{bmatrix} 1 & 0 \\ 1 & 1 \end{bmatrix}$$

$$(x_j \oplus 1) \longleftrightarrow \mathbf{R}_{2^1}^{(1)} = \begin{bmatrix} 0 & 1 \\ 1 & 1 \end{bmatrix}$$

$$(x_j \oplus 0)(x_t \oplus 1) \longleftrightarrow \mathbf{R}_{2^1}^{(0)} \otimes \mathbf{R}_{2^1}^{(1)} = \begin{bmatrix} 1 & 0 \\ 1 & 1 \end{bmatrix} \otimes \begin{bmatrix} 0 & 1 \\ 1 & 1 \end{bmatrix}$$

Example 11.13 (Kronecker product.) *The Kronecker product of matrices* $\mathbf{R}_{2^1}^{(0)}$ *is computed as follows*

$$\mathbf{R}_{2^2}^{(0)} = \mathbf{R}_{2^1}^{(0)} \otimes \mathbf{R}_{2^1}^{(0)} = \begin{bmatrix} 1 & 0 \\ 1 & 1 \end{bmatrix} \otimes \begin{bmatrix} 1 & 0 \\ 1 & 1 \end{bmatrix} = \left[\begin{array}{cc|cc} 1 & 0 & 0 & 0 \\ 1 & 1 & 0 & 0 \\ \hline 1 & 0 & 1 & 0 \\ 1 & 1 & 1 & 1 \end{array}\right]$$

Practice problem 11.9. (Polarized minterms.) Derive the matrix $\mathbf{R}_{2^2}^{(01)} = \mathbf{R}_{2^1}^{(0)} \otimes \mathbf{R}_{2^1}^{(1)}$.

Answer: $\mathbf{R}_{2^1}^{(0)} \otimes \mathbf{R}_{2^1}^{(1)} = \begin{bmatrix} 1 & 0 \\ 1 & 1 \end{bmatrix} \bigotimes \begin{bmatrix} 0 & 1 \\ 1 & 1 \end{bmatrix} = \left[\begin{array}{cc|cc} 0 & 1 & 0 & 0 \\ 1 & 1 & 0 & 0 \\ \hline 0 & 1 & 0 & 1 \\ 1 & 1 & 1 & 1 \end{array}\right]$

Example 11.14 (Polarized minterms.) *A polarized minterm of fifth polarity ($c = 5$, $n = 3$) is constructed in matrix form as shown in Figure 11.9.*

11.3.3 Computing the coefficients in fixed polarity polynomial forms

Using Equation 11.2, the polarized minterms can be generated for a given polarity of an polynomial expression. In algebraic form, the polynomial expression is a sum of polarized minterms over GF(2):

$$f = \bigoplus_{i=0}^{2^n-1} (x_1 \oplus c_1)^{i_1} \cdots (x_n \oplus c_1)^{i_n} \quad (11.3)$$

The particular properties of a Boolean function in algebraic polynomial representation (Equation 11.3) are specified by the index $i = i_1 i_2 \dots i_n$as follows:

$$(x_j \oplus c_j)^{i_j} = \begin{cases} 1, & \text{if } i_j = 0 \\ (x_j \oplus c_j), & \text{if } i_j = 1 \end{cases} \quad \text{over GF(2)}$$

In matrix form, this specification is realized by the multiplication of the matrix $\mathbf{R}_{2^n}^{(c_j)}$ and the truth-vector $\mathbf{F}$ of the Boolean function f over GF(2).

Forward transform

A forward transform is used for the representation of the truth vector of a Boolean function (operational domain) in the form of a vector of coefficients in polynomial form (functional domain):

$$\underbrace{\texttt{Truth-vector}}_{\textit{Operational domain}} \quad \underbrace{\overset{\textit{Transform}}{- - - - \longrightarrow}}_{\textit{over } GF(2)} \quad \underbrace{\texttt{Vector of coefficients}}_{\textit{Functional domain}}$$

Techniques for deriving a polarized minterms in matrix form

Step 1: Find the corresponding elementary matrix for each literal:

$$\text{Algebraic form} \longrightarrow \underbrace{\underbrace{(x_1 \oplus 1)^{i_1}}\ \underbrace{(x_2 \oplus 0)^{i_2}}\ \underbrace{(x_3 \oplus 1)^{i_3}}}$$

$$\underbrace{\begin{bmatrix} 0 & 1 \\ 1 & 1 \end{bmatrix}}_{\mathbf{R}_{2^1}^{(1)}} \quad \underbrace{\begin{bmatrix} 1 & 0 \\ 1 & 1 \end{bmatrix}}_{\mathbf{R}_{2^1}^{(0)}} \quad \underbrace{\begin{bmatrix} 0 & 1 \\ 1 & 1 \end{bmatrix}}_{\mathbf{R}_{2^1}^{(1)}}$$

Step 2: Form the $2^3 \times 2^3$ transform matrix $\mathbf{R}_{2^3}^{(5)}$ for the fifth polarity as the Kronecker product of the elementary matrices:

$$\overbrace{\mathbf{R}_{2^1}^{(1)} \otimes \overbrace{\mathbf{R}_{2^1}^{(0)} \otimes \mathbf{R}_{2^1}^{(1)}}^{\textit{The 1st step}}}^{\textit{The 2nd step}} = \overbrace{\begin{bmatrix} 0 & 1 \\ 1 & 1 \end{bmatrix} \otimes \overbrace{\begin{bmatrix} 1 & 0 \\ 1 & 1 \end{bmatrix} \otimes \begin{bmatrix} 0 & 1 \\ 1 & 1 \end{bmatrix}}^{\textit{The 1st step}}}^{\textit{The 2nd step}}$$

$$= \overbrace{\begin{bmatrix} 0 & 1 \\ 1 & 1 \end{bmatrix} \otimes \left[\begin{array}{cc|cc} 0 & 1 & 0 & 0 \\ 1 & 1 & 0 & 0 \\ \hline 0 & 1 & 0 & 1 \\ 1 & 1 & 1 & 1 \end{array}\right]}^{\textit{The 2nd step}} = \left[\begin{array}{cccc|cccc} 0 & 0 & 0 & 0 & 0 & 1 & 0 & 0 \\ 0 & 0 & 0 & 0 & 1 & 1 & 0 & 0 \\ 0 & 0 & 0 & 0 & 0 & 1 & 0 & 1 \\ 0 & 0 & 0 & 0 & 1 & 1 & 1 & 1 \\ \hline 0 & 1 & 0 & 0 & 0 & 1 & 0 & 0 \\ 1 & 1 & 0 & 0 & 1 & 1 & 0 & 0 \\ 0 & 1 & 0 & 1 & 0 & 1 & 0 & 1 \\ 1 & 1 & 1 & 1 & 1 & 1 & 1 & 1 \end{array}\right] = \mathbf{R}_{2^3}^{(5)}$$

Step 3: Use the matrix $\mathbf{R}_{2^3}^{(5)}$ for the matrix transform of the vector $\mathbf{F}$ to a vector of coefficients in the fifth polarity. In Boolean expressions, the variables x_1, x_2, and x_3 are used as $\overline{x}_1$, x_2, and $\overline{x}_3$, respectively; that is, the polarities of variables are fixed.

FIGURE 11.9
Deriving a polarized minterm in matrix form (Example 11.14).

Specifically, given the truth-vector $\mathbf{F} = [f(0)\ f(1) \ldots f(2^n - 1)]^T$, the vector of coefficients in polarity c, $\mathbf{R}^{(\mathbf{c})} = [r_0^{(c)}\ r_1^{(c)} \ldots r_{2^n-1}^{(c)}]^T$ is derived by the matrix equation as follows:

Techniques for computing the polynomials

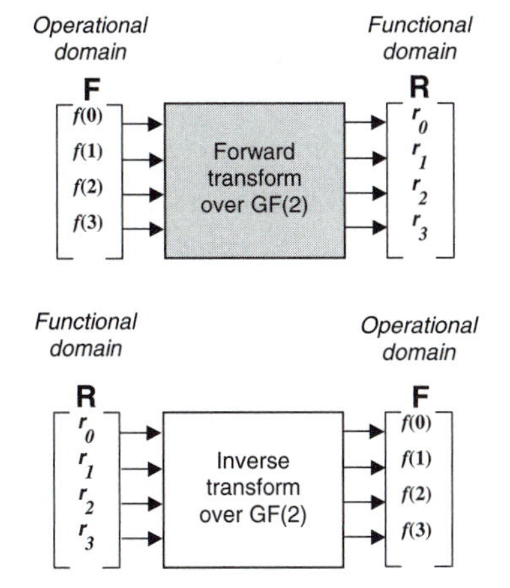

The pair of forward and inverse transforms

$$\mathbf{R} = \mathbf{R}_{2^2} \cdot \mathbf{F} \quad \text{over GF(2)}$$
$$\mathbf{F} = \mathbf{R}_{2^2}^{-1} \cdot \mathbf{R} \quad \text{over GF(2)}$$

where

$$\mathbf{R}_{2^2} = \mathbf{R}_{2^2}^{-1}$$
$$\mathbf{R}_{2^2} = \bigotimes_{j=1}^{2} \mathbf{R}_{2^1}$$
$$\mathbf{R}_{2^1} = \begin{bmatrix} 1 & 0 \\ 1 & 1 \end{bmatrix}$$

FIGURE 11.10
Forward and inverse transforms for a Boolean function over GF(2) of two variables.

Direct transform for polynomial expressions

$$\mathbf{R}^{(\mathbf{c})} = \mathbf{R}_{2^n}^{(c)} \cdot \mathbf{F} \quad \text{over GF}(2) \tag{11.4}$$

where the $2^n \times 2^n$ matrix $\mathbf{R}_{2^n}^{(c)}$ is generated by the Kronecker product:

$$\mathbf{R}_{2^n}^{(c)} = \bigotimes_{j=1}^{n} \mathbf{R}_{2^1}^{(c_j)}, \qquad \mathbf{R}_{2^1}^{(c)} = \begin{cases} \begin{bmatrix} 1 & 0 \\ 1 & 1 \end{bmatrix}, & c_j = 0 \\ \begin{bmatrix} 0 & 1 \\ 1 & 1 \end{bmatrix}, & c_j = 1 \end{cases}$$

Example 11.15 (Forward transform.) *Given a Boo-lean function of two variables in the form of a truth-vector* $\mathbf{F} = [1011]^T$, *the vector of coefficients is computed as follows:*

$$\mathbf{R}^{(2)} = \mathbf{R}_{2^2}^{(2)} \cdot \mathbf{F} = \left[\begin{array}{cc|cc} 0 & 0 & 0 & 1 \\ 0 & 0 & 1 & 1 \\ \hline 0 & 1 & 0 & 1 \\ 1 & 1 & 1 & 1 \end{array}\right] \begin{bmatrix} 1 \\ 0 \\ 1 \\ 1 \end{bmatrix} = \begin{bmatrix} 1 \\ 0 \\ 1 \\ 1 \end{bmatrix} \quad \text{over GF}(2)$$

where the matrix $\mathbf{R}_{2^2}^{(2)}$ *given* $c = 2$ *is generated using the Kronecker product* $\mathbf{R}_{2^2}^{(2)} = \mathbf{R}_{2^1}^{(1)} \otimes \mathbf{R}_{2^1}^{(0)} = \begin{bmatrix} 0 & 1 \\ 1 & 1 \end{bmatrix} \otimes \begin{bmatrix} 1 & 0 \\ 1 & 1 \end{bmatrix}$.
The vector of coefficients $\mathbf{R}^{(2)} = [1\ 0\ 1\ 1]^T$ *corresponds to the expression* $f = 1 \oplus \overline{x}_1 \oplus \overline{x}_1 x_2$.

Practice problem 11.10. (Forward transform.) Given the truth-vector $\mathbf{F} = [00100000]^T$ of a Boolean function of three variables, find its polynomial representation in the 5th polarity using the matrix $\mathbf{R}_{2^3}^{(5)}$ from Example 11.14.
Answer is given in "Solutions to practice problems."

Example 11.16 (Forward transform.) *Given a logic network that implements a standard SOP expression (Figure 11.11), this logic network can be converted into an AND-EXOR network as follows:*

Step 1: *Compute the truth-vector* $\mathbf{F} = [1\ 0\ 1\ 1]^T$

Step 2: *Compute the vector of coefficients. Use the forward transform with the given polarity. Let a positive polarity be required:* $\mathbf{R} = \mathbf{R}_{2^3} \cdot \mathbf{F} = [1\ 1\ 0\ 1]^T$. *The vector of coefficients corresponds to the algebraic form* $f = 1 \oplus x_2 \oplus x_1x_2$.

Step 3: *Design the AND-EXOR network.*

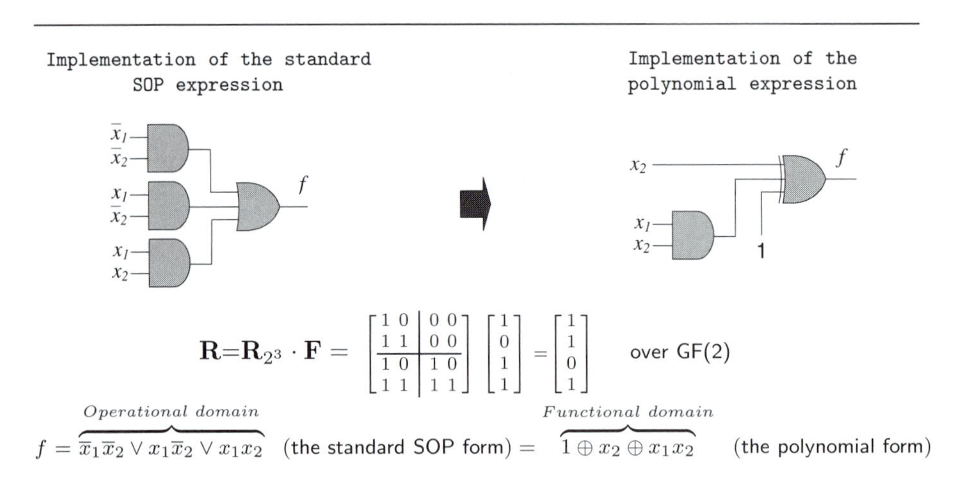

FIGURE 11.11
Conversion of the AND-OR logic network into the AND-EXOR network using forward transform (Example 11.16).

Practice problem 11.11. (Forward transform.) Convert the AND-OR logic network shown in Figure 11.12a into an AND-EXOR network using a forward transform.
Answer is given in "Solutions to practice problems."

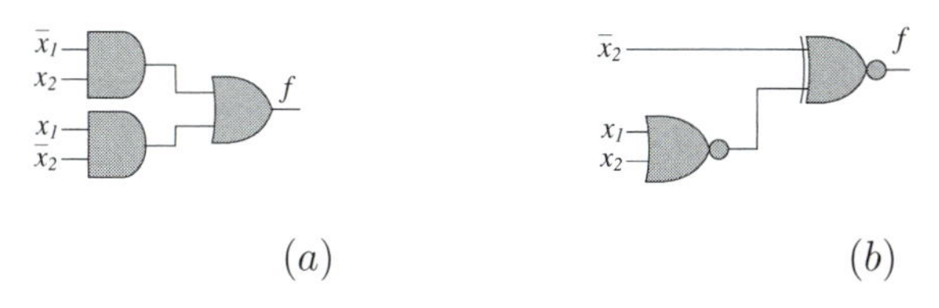

FIGURE 11.12
The logic network for Practice problems 11.11 (a) and 11.12 (b).

Inverse transform

The inverse transform is used for the conversion of a vector of coefficients of the polynomial form of a Boolean function (functional domain) into its truth-vector (operational domain):

$$\underbrace{\texttt{Vector of coefficients}}_{\textit{Functional domain}} \quad \underbrace{\xrightarrow{\textit{Transform}}}_{\textit{over } GF(2)} \quad \underbrace{\texttt{Truth-vector}}_{\textit{Operational domain}}$$

Given a vector of positive polarity polynomial coefficients $\mathbf{R} = [r_0\ r_1 \ldots r_{2^n-1}]^T$, the truth-vector $\mathbf{F} = [f(0)\ f(1) \ldots f(2^n-1)]^T$ of a Boolean function f is derived as follows (Figure 11.10):

Inverse transform for polynomial expressions

$$\mathbf{F} = \mathbf{R}_{2^n}^{-1} \cdot \mathbf{R} \quad \text{over GF(2)} \tag{11.5}$$

where $\mathbf{R}_{2^1}^{-1} = \mathbf{R}_{2^1}$.

Notice that the matrix $\mathbf{R}_{2^1}$ is a self-inverse matrix over AND and polynomial operations.

Example 11.17 (Inverse transform.) *Given the an AND-EXOR network (Figure 11.13), this network can be converted into an AND-OR network as follows:*

Step 1: *Compute the vector of coefficients* $\mathbf{R} = [0\ 1\ 0\ 1]^T$.
Step 2: *Compute the truth-vector. Use the inverse transform given a certain polarity. Let the positive polarity is required:* $\mathbf{F} = \mathbf{R}_{2^3}^{(-1)} \cdot \mathbf{R} = [0\ \ 1\ \ 0\ \ 0]^T$. *The truth-vector corresponds to the standard SOP expression in algebraic form* $f = \overline{x}_1 x_2 \vee x_1 \overline{x}_2$.
Step 3: *Design the AND-OR network.*

Practice problem **11.12. (Inverse transform.)** Convert the AND-EXOR logic network shown in Figure 11.12b into an AND-OR network using the forward transform.
Answer is given in "Solutions to practice problems."

Design example:
Conversion from functional to operational domain

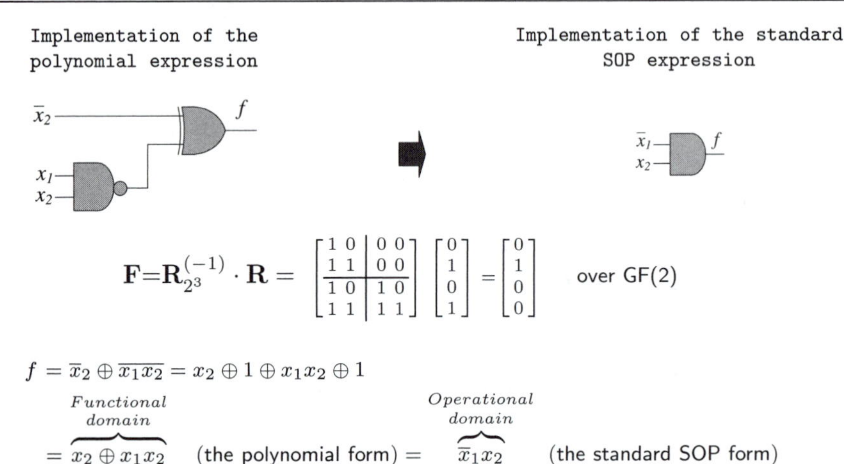

Implementation of the
polynomial expression
Implementation of the standard
SOP expression
$\overline{x}_2$
f
x_1
x_2
$\overline{x}_1$
x_2
f
$\mathbf{F}=\mathbf{R}_{2^3}^{(-1)} \cdot \mathbf{R} = \left[\begin{array}{cc|cc} 1 & 0 & 0 & 0 \\ 1 & 1 & 0 & 0 \\ \hline 1 & 0 & 1 & 0 \\ 1 & 1 & 1 & 1 \end{array}\right] \begin{bmatrix} 0 \\ 1 \\ 0 \\ 1 \end{bmatrix} = \begin{bmatrix} 0 \\ 1 \\ 0 \\ 0 \end{bmatrix}$ over GF(2)
$f = \overline{x}_2 \oplus \overline{x_1 x_2} = x_2 \oplus 1 \oplus x_1 x_2 \oplus 1$
Functional domain
Operational domain
$= x_2 \oplus x_1 x_2$ (the polynomial form) $= \overline{x}_1 x_2$ (the standard SOP form)

FIGURE 11.13
Conversion of the AND-EXOR logic network into the AND-OR network using inverse transform (Example 11.17).

11.4 Summary of the polynomial expressions

This chapter introduces the techniques for computing polynomial expressions. Polynomial forms such as EXOR expressions can be represented using various canonical expansions, also known as ***Reed-Muller*** forms or ***logic Taylor*** expansions. The main aspects of this chapter cover:

(*a*) Computing of fixed and mixed polarity polynomials in algebraic form and

(*b*) Techniques of computing using matrix transformations.

The main topics of this chapter are summarized as follows:

- In terms of polynomial forms, two types of polarity are distinguished: ***fixed*** and ***mixed*** polarity.

Summary (continuation)

- In a fixed polarity polynomial expression of a Boolean function f, every variable appears either complemented ($\overline{x}_i$) or uncomplemented (x_i); but never in both forms. There are 2^n fixed polarity forms. In a mixed polarity, a variable can appear in one or both polarities. There are 3^n mixed polarity forms. Fixed and mixed polarity polynomial expressions are unique. Thus, only one representation exists for a given polarity c ($c_1, c_2, \ldots, c_n$ of variables of f).
- If all variables are uncomplemented, this polynomial expression is called ***positive*** polarity expression. If all variables are complemented, this polynomial expression is called ***negative*** polarity expression.
- The logic Taylor expansion in the point c is also called a ***fixed polarity Reed-Muller expansion***. The binary representation of fixed polarity c, the n-tuple $(c_1, c_2, \ldots, c_n)$, determines the polarity of variables in the polynomial.
- Polynomial expressions can be derived from SOP forms using algebraic transformations. If the SOP form is canonical (standard), the polynomial form of mixed polarity is derived by replacing the logical sum "$\vee$" with the EXOR operation "$\oplus$."
- Matrix-based computing of the polynomial forms of Boolean functions is based on matrix-matrix, matrix-vector, and vector-vector operations over GF(2). In a binary matrix, the binary numbers 0 and 1 are used.

  The matrix form of a polarized literal is based on the assumption that operations are performed over GF(2). The ***elementary*** $2^1 \times 2^1$ transform matrix $\mathbf{R}_{2^1}^{(c_j)}$ ($c_j = 0$ or $c_j = 1$) and truth-vector of the variable x_j of the polarity $c_j \in \{0, 1\}$ results in a vector of coefficients that corresponds to the simplest polynomial expressions; that is, literals $x_j \oplus 0$ or $x_j \oplus 1$:

$$\texttt{Polarized literal} = \mathbf{R}_{2^1}^{(c_j)}$$

$$= \begin{cases} \mathbf{R}_{2^1}^{(0)} = \begin{bmatrix} 1 & 0 \\ 1 & 1 \end{bmatrix}, & \text{if } c_j = 0 \\ \mathbf{R}_{2^1}^{(1)} = \begin{bmatrix} 1 & 0 \\ 1 & 1 \end{bmatrix}, & \text{if } c_j = 1 \end{cases} \quad \text{over GF(2)}$$

Summary (continuation)

- The computing of a polarized literal is accomplished by multiplying the elementary transform matrix $\mathbf{R}_{2^1}^{(c_j)}$ by the truth-vector of a single variable x_j, $\mathbf{F} = [0\ 1]^T$:

$$\mathbf{R}_{2^1}^{(0)}\mathbf{F} = \begin{bmatrix} 1 & 0 \\ 1 & 1 \end{bmatrix} \begin{bmatrix} 0 \\ 1 \end{bmatrix} = \begin{bmatrix} 0 \\ 1 \end{bmatrix} \longrightarrow r_0 \oplus r_1 x_j = x_j$$

$$\mathbf{R}_{2^1}^{(1)}\mathbf{F} = \begin{bmatrix} 0 & 1 \\ 1 & 1 \end{bmatrix} \begin{bmatrix} 0 \\ 1 \end{bmatrix} = \begin{bmatrix} 1 \\ 1 \end{bmatrix} \longrightarrow r_0 \oplus r_1 x_j = 1 \oplus x_j$$

- The polarized minterm is formed using the ***Kronecker product*** operation of n elementary matrices $\mathbf{R}_{2^1}^{(c_j)}$ as $\mathbf{R}_{2^n}^{(c)} = \bigotimes_{j=1}^{n} \mathbf{R}_{2^1}^{(c_j)}$, where $\mathbf{R}_{2^1}^{(c_j)}$ corresponds to polarized literal (the variable x_j of the polarity $c_j \in \{0, 1\}$). The result is the $2^n \times 2^n$ matrix $\mathbf{R}_{2^n}^{(c)}$ that represents the minterm of polarity $c = c_1 c_2 \ldots c_n$.
- For advances in computing Boolean functions in polynomial form, we refer the reader to the "Further study" section.

11.5 Further study

Advanced topics of polynomial techniques for Boolean functions

Advanced topics of polynomial techniques, such as word-level representations, functional decision diagrams, probabilistic computing, and new computing paradigms can be found, in particular, in *Proceedings of the International Workshop on Applications of the Reed-Muller Expansion in Circuit Design.*

Further reading

Binary Decision Diagrams and Applications for VLSI CAD by S. Minato, Kluwer, Dordrecht, 1996.

Decision Diagram Techniques for Micro- and Nanoelectronic Design, by S. N. Yanushkevich, D. M. Miller, V. P. Shmerko, and R. S. Stanković, CRC/Taylor & Francis Group, Boca Raton, FL, 2006.

Modern Logic Design by D. Green, Addison-Wesley, 1986.

Spectral Interpretation of Decision Diagrams by R. S. Stanković and J. T. Astola, Springer, 2003.

Switching Theory for Logic Synthesis by T. Sasao, Kluwer, 1999.

11.6 Solutions to practice problems

<table>
<tr><th>Practice problem</th><th>Solution</th></tr>
<tr><td>11.1.</td><td>Given $n = 2$, there are 2^2 various fixed polarities c, $c = c_1c_2 = \{00, 01, 10, 11\}$. In fixed polarity form, each variable can appear as complemented, uncomplemented, or in both forms. That is, for two variables, there are 3^2 combinations $\{00, 01, 02, 10, 11, 12, 20, 21, 22\}$, where "2" means that the variable can appear both complemented and uncomplemented.</td></tr>
<tr><td>11.2.</td><td>Since $n = 2, c = 3$ $(c_1 = 1, c_2 = 1)$, then

$$\begin{aligned} f &= r_0(x_1 \oplus c_1)^{i_1}(x_2 \oplus c_2)^{i_2} \oplus r_1(x_1 \oplus c_1)^{i_1}(x_2 \oplus c_2)^{i_2} \\ &\oplus r_2(x_1 \oplus c_1)^{i_1}(x_2 \oplus c_2)^{i_2} \oplus r_3(x_1 \oplus c_1)^{i_1}(x_2 \oplus c_2)^{i_2} \\ &= r_0 \underbrace{(x_1 \oplus 1)^0}_{1} \underbrace{(x_2 \oplus 1)^0}_{1} \oplus r_1 \underbrace{(x_1 \oplus 1)^0}_{1} \underbrace{(x_2 \oplus 1)^1}_{\overline{x}_2} \\ &\oplus r_2 \underbrace{(x_1 \oplus 1)^1}_{\overline{x}_1} \underbrace{(x_2 \oplus 1)^0}_{1} \oplus r_3 \underbrace{(x_1 \oplus 1)^1}_{\overline{x}_1} \underbrace{(x_2 \oplus 1)^1}_{\overline{x}_2} \\ &= r_0 \oplus r_1\overline{x}_2 \oplus r_2\overline{x}_1 \oplus r_3\overline{x}_1\overline{x}_2 \end{aligned}$$</td></tr>
<tr><td>11.3.</td><td>

$$\begin{aligned} f &= x_1x_2 = x_1(1 \oplus 1 \oplus x_2) \\ &= x_1(1 \oplus \overline{x}_2) = x_1 \oplus x_1\overline{x}_2 \text{ (polarity } c_1c_2 = 01) \\ f &= x_1x_2 = (1 \oplus 1 \oplus x_1)x_2 \\ &= x_2 \oplus x_2\overline{x}_1 \text{ (polarity } c_1c_2 = 10) \\ &= (1 \oplus \overline{x}_2) \oplus (1 \oplus \overline{x}_2)\overline{x}_1 = 1 \oplus \overline{x}_2 \oplus \overline{x}_1 \oplus \overline{x}_1\overline{x}_2 \end{aligned}$$</td></tr>
<tr><td>11.5.</td><td>Given $f = x_1 \oplus x_2 \oplus x_1x_2$ (see Figure 11.3), we substitute $x_1 = \overline{x}_1 \oplus 1$ and $x_2 = \overline{x}_2 \oplus 1$:

$$\begin{aligned} f &= (\overline{x}_1 \oplus 1) \oplus (\overline{x}_2 \oplus 1) \oplus (\overline{x}_1 \oplus 1)(\overline{x}_2 \oplus 2) \\ &= \overline{x}_1 \oplus 1 \oplus \overline{x}_2 \oplus 1 \oplus \overline{x}_1\overline{x}_2 \oplus \overline{x}_1 \oplus \overline{x}_2 \oplus 1 = x_1x_2 \oplus 1 \end{aligned}$$</td></tr>
<tr><td>11.7.</td><td>The proof is based on the derivation of all possible polarities of the polynomial expressions and a comparison of their costs (the number of literals). The following fixed and mixed polarities have been generated:

$$f = \overline{x}_1\overline{x}_2\overline{x}_3 \oplus \overline{x}_1\overline{x}_2x_3 \oplus \overline{x}_1x_2x_3 \oplus x_1x_2x_3$$

This can be further simplified to the expression

$$f = \overline{x}_1\overline{x}_2(\overline{x}_3 \oplus x_3) \oplus (\overline{x}_1 \oplus x_1)x_2x_3 = \overline{x}_1\overline{x}_2 \oplus x_2x_3$$

Thus, only the polynomials of polarities 1 (00) and 3 (11) have the minimal form.</td></tr>
<tr><td>11.8.</td><td>over GF(2) ... over GF(2)

$$\begin{bmatrix} 1 & 0 \\ 1 & 1 \end{bmatrix} \times \begin{bmatrix} 0 & 1 \\ 1 & 1 \end{bmatrix} = \begin{bmatrix} 1{\cdot}0 \oplus 0{\cdot}1 & 1{\cdot}1 \oplus 0{\cdot}1 \\ 1{\cdot}0 \oplus 1{\cdot}1 & 1{\cdot}1 \oplus 1{\cdot}1 \end{bmatrix} = \begin{bmatrix} 0 & 1 \\ 1 & 0 \end{bmatrix}$$</td></tr>
</table>

Solutions to practice problems (continuation)

<table>
<tr><th>Practice problem</th><th>Solution</th></tr>
<tr><td>11.10.</td><td>

$$\mathbf{R}^{(5)} = \mathbf{R}_{2^3}^{(5)}\mathbf{F} = \left[\begin{array}{cccc|cccc} 0&0&0&0&0&1&0&0\\ 0&0&0&0&1&1&0&0\\ 0&0&0&0&0&1&0&1\\ 0&0&0&0&1&1&1&1\\ \hline 0&1&0&0&0&1&0&0\\ 1&1&0&0&1&1&0&0\\ 0&1&0&1&0&1&0&1\\ 1&1&1&1&1&1&1&1 \end{array}\right] \left[\begin{array}{c} 0\\0\\1\\0\\ \hline 0\\0\\0\\0 \end{array}\right] = \overline{x}_1 x_2 \overline{x}_3$$

$$(x_1 \oplus 1)^{i_1}(x_1 \oplus 0)^{i_1}(x_1 \oplus 1)^{i_1} = \overline{x}_1 x_2 \overline{x}_3$$

</td></tr>
<tr><td>11.11.</td><td>

$$\mathbf{R}{=}\mathbf{R}_{2^3} \cdot \mathbf{F} = \left[\begin{array}{cc|cc} 1&0&0&0\\ 1&1&0&0\\ \hline 1&0&1&0\\ 1&1&1&1 \end{array}\right] \begin{bmatrix} 0\\1\\1\\0 \end{bmatrix} = \begin{bmatrix} 0\\1\\1\\0 \end{bmatrix} \quad \text{over GF(2)}$$

$f = \overline{x}_1 x_2 \vee x_1 \overline{x}_2$ (standard SOP form) $= x_1 \oplus x_2$ (polynomial form)

</td></tr>
<tr><td>11.12.</td><td>

$$\begin{aligned} f &= \overline{\overline{x}_2 \oplus (\overline{x_1 \vee x_2})} \\ &= x_2 \oplus 1 \oplus \overline{x}_1 \overline{x}_2 \oplus 1 \\ &= x_2 \oplus (x_1 \oplus 1)(x_2 \oplus 1) \\ &= 1 \oplus x_1 \oplus x_1 x_2 \end{aligned}$$

$$\mathbf{F} = \mathbf{R}_{2^3}^{(-1)} \cdot \mathbf{R} = \left[\begin{array}{cc|cc} 1&0&0&0\\ 1&1&0&0\\ \hline 1&0&1&0\\ 1&1&1&1 \end{array}\right] \begin{bmatrix} 1\\0\\1\\1 \end{bmatrix} = \begin{bmatrix} 1\\0\\0\\1 \end{bmatrix} \quad \text{over GF(2)}$$

The vector of coefficients is $\mathbf{R} = [1011]^T$. The truth vector $\mathbf{F} = [1001]$ corresponds to the standard SOP expression $f = \overline{x}_1 \overline{x}_2 \vee x_1 x_2$.

</td></tr>
</table>

11.7 Problems

Problem 11.1 Derive the following polarities of the polynomial expansion of the function $f = x_1 \overline{x}_2 \vee \overline{x}_1 \overline{x}_3$ using matrix and algebraic manipulations: $c = 0$, $c = 2$, $c = 6$, and $c = 7$.

Problem 11.2 Derive the mixed polarity forms given the following decision diagrams, and then convert them into positive polarity polynomials:

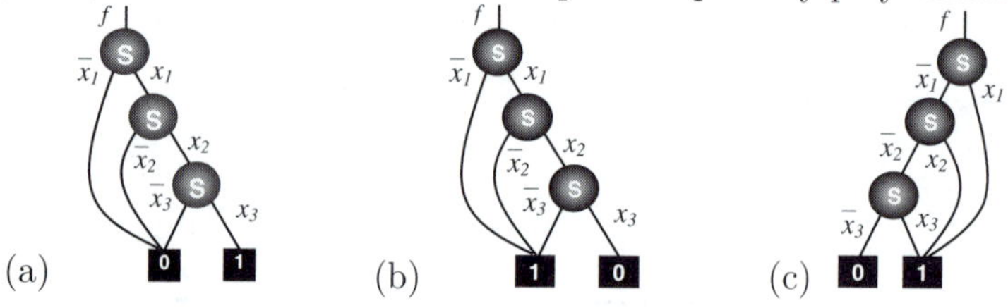

Problem 11.3 Derive positive polarity expressions given the following functional decision diagrams:

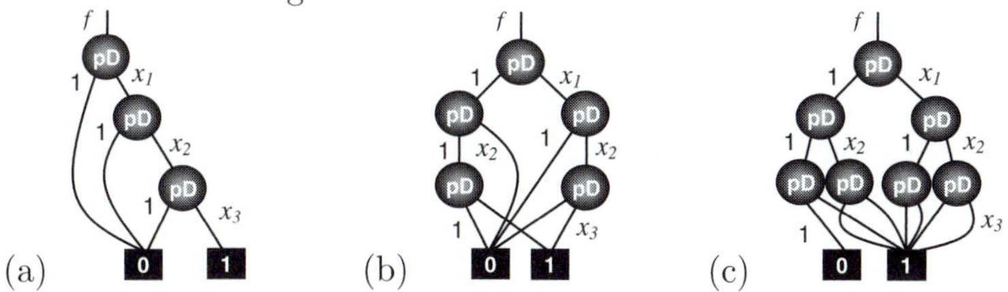

Problem 11.4 Derive negative polarity expressions given the functional decision diagrams from Problem 11.3.

Problem 11.5 Derive positive and negative polarity polynomial forms given the following functions using their cube representation:

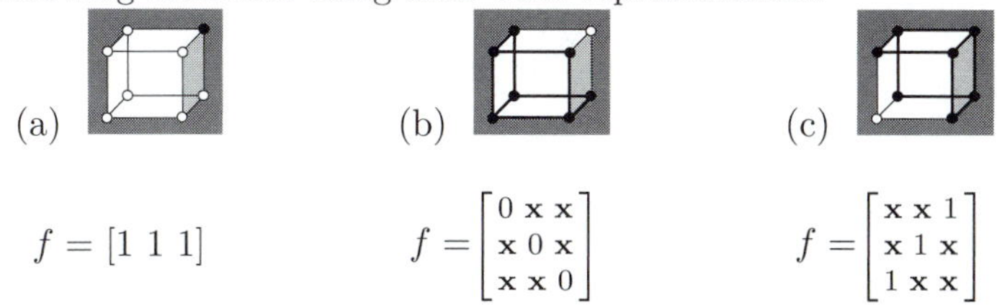

(a) $f = [1\ 1\ 1]$ (b) $f = \begin{bmatrix} 0 & \mathrm{x} & \mathrm{x} \\ \mathrm{x} & 0 & \mathrm{x} \\ \mathrm{x} & \mathrm{x} & 0 \end{bmatrix}$ (c) $f = \begin{bmatrix} \mathrm{x} & \mathrm{x} & 1 \\ \mathrm{x} & 1 & \mathrm{x} \\ 1 & \mathrm{x} & \mathrm{x} \end{bmatrix}$

Problem 11.6 Given the following K-maps, derive mixed polarity forms, and then find the negative polarity polynomial expressions:

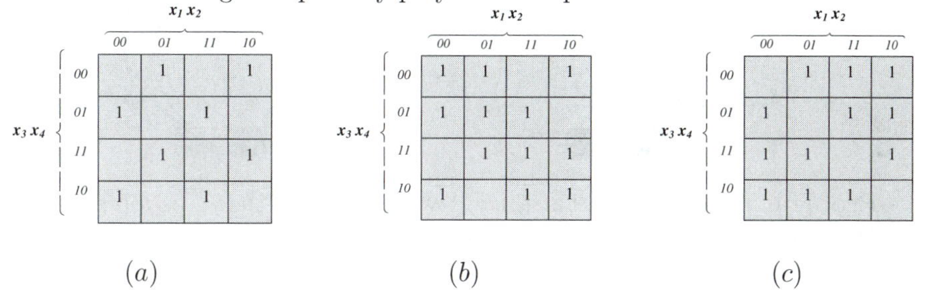

Problem 11.7 Convert the logic networks given below into AND-EXOR networks:

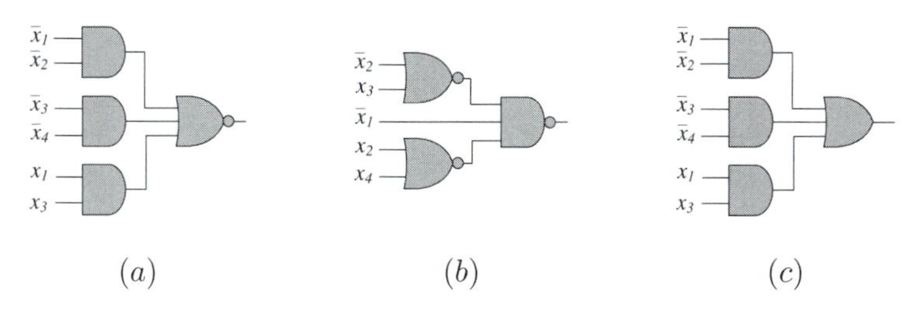

Problem 11.8 Convert the logic networks given in Problem 16.3 into tree-level networks, using matrix transforms.

Problem 11.9 Derive the 4×4 matrices for the Reed-Muller transforms of the following polarities:

(a) negative for the first variable, and positive for the second variable
(c) positive for the first variable and negative for the second variable
(b) mixed for the first variable and positive for the second variable
(d) mixed for the first variable and negative for the second variable

Problem 11.10 Verify that the following logic networks perform the same Boolean function:

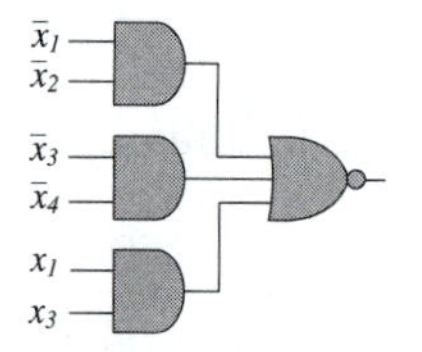

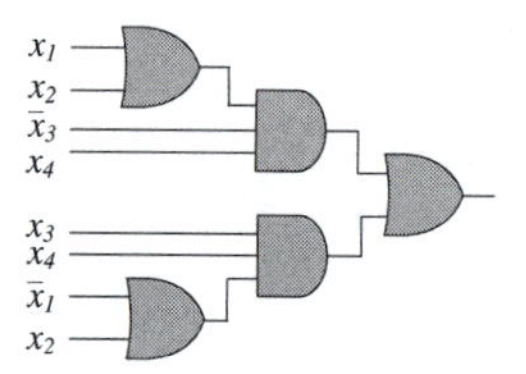

Problem 11.11 Verify that the the following decision diagrams perform the same Boolean function:

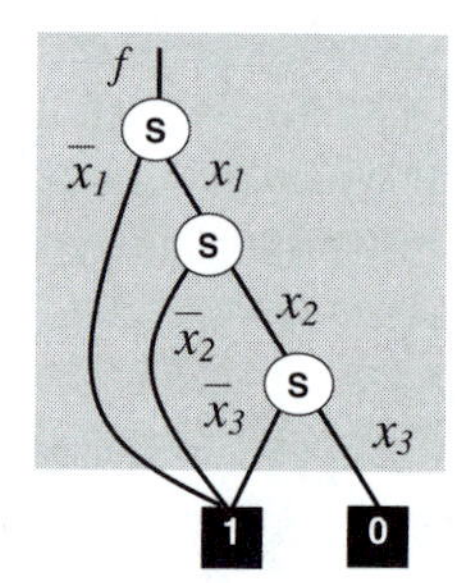

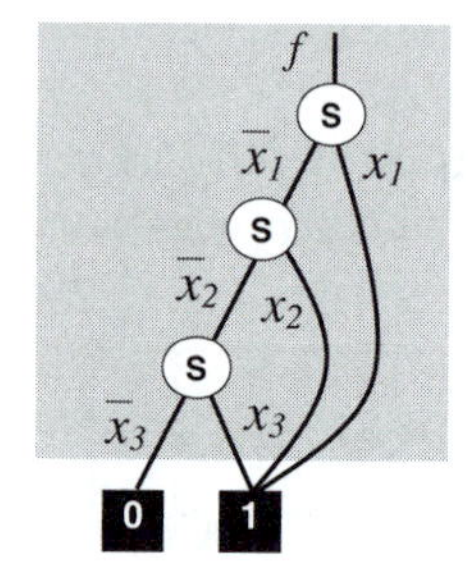

Problem 11.12 Prove that the following holds for the three-input majority function: $\texttt{MAJORITY} = x_1x_2 \vee x_2x_3 \vee x_1x_3 = (x_1 \vee x_2)(x_2 \vee x_3)(x_1 \vee x_3) = x_1x_2 \oplus x_2x_3 \oplus x_1x_3$.

12

Decision Diagrams for Polynomial Forms (Optional)

Davio expansion

- Positive Davio expansion
- Negative Davio expansion
- Gate level implementation

Functional decision trees

- For positive polarity polynomials
- For negative polarity polynomials
- For mixed polarity polynomials

Functional decision diagrams

- Elimination rule
- Merging rule
- Particular cases

Advanced topics

- Computing using Pascal and transeunt triangles

12.1 Introduction

In Chapter 2, Shannon expansion was represented by decision trees for processing Boolean functions in an *operational domain*. In this domain, Boolean functions are computed using SOP representation. The EXOR analog of Shannon expansion, known as *Davio expansion*, operates in the *functional domain*. In a functional domain, Boolean functions are computed using polynomial representations.

Figure 12.1 shows the relationships between data structures in operational and functional domains that are used in decision diagram techniques. Techniques for decision diagram construction in an operational domain using Shannon expansion are based on algebraic and matrix notations of node functions. The same is true for a functional domain, where Davio expansion is used in the nodes. For example, a Boolean functions in algebraic and matrix forms can be converted from an operational into a functional domain, and vice versa. This is the basis for decision diagram construction and computing.

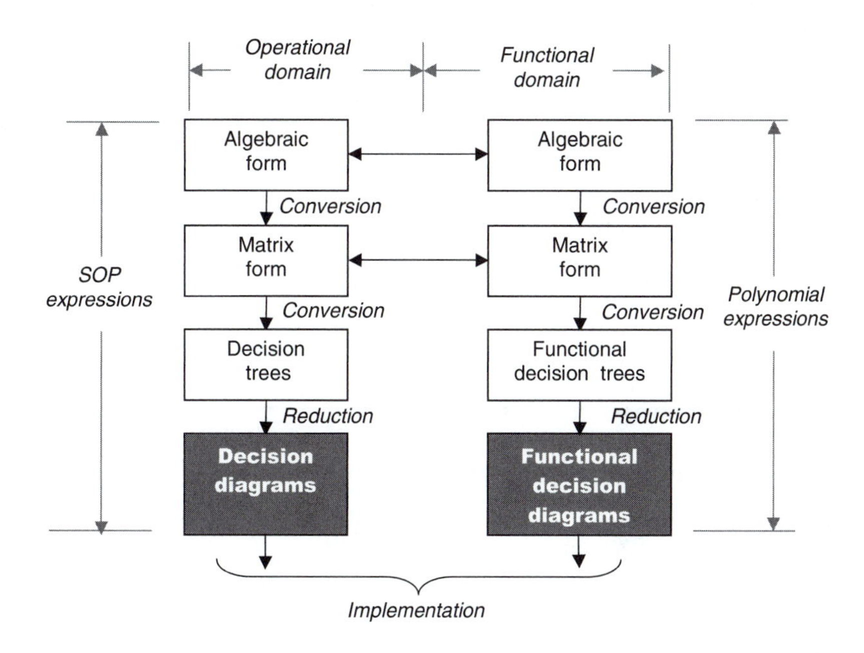

FIGURE 12.1
Data structures for the operational and functional domains of Boolean functions.

A decision tree using Davio expansion is called a *functional* decision tree. Note that in conversions of decision diagrams into functional decision diagrams and vice versa, algebraic and matrix forms are used as intermediate data structures:

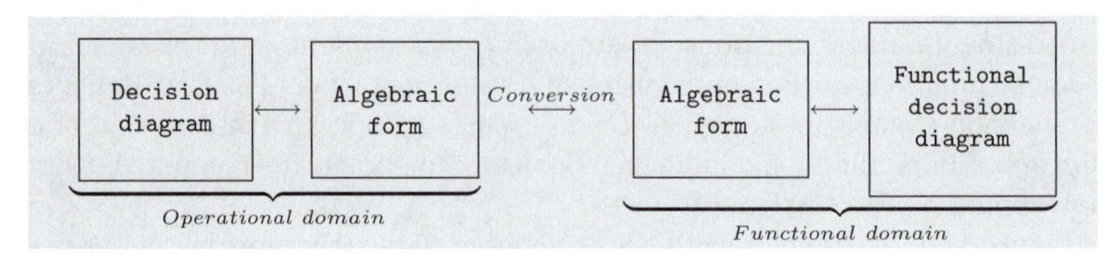

Functional decision tree

The differences between decision trees and functional decision trees are specified by the properties of SOP and polynomial expressions of Boolean functions. That is, decision trees using Shannon expansion in the nodes operate in SOP form, and a functional decision trees using Davio expansion in the nodes operate in the polynomial form of a Boolean function.

After reduction, the decision tree becomes a decision diagram. By analogy, a functional decision tree is called a *functional decision diagram* after applying the reduction procedure. The reduction procedure for functional decision diagrams is different from the reduction of decision diagrams using Shannon expansion. This difference is specified by different techniques for simplification of SOP and polynomial expressions.

Similarly to decision diagrams based on Shannon expansion, functional decision diagrams are used for various design tasks such as the representation, manipulation, optimization, and implementation of Boolean functions. The difference is that solutions of these tasks are in the functional domain.

12.2 Function of the nodes

A node in a functional decision tree of a Boolean function f corresponds to the EXOR analog of Shannon expansion with respect to the variable x_i. It is called *Davio* expansion in honor of Professor *M. Davio*, who made pioneering contributions in the development of the theory of polynomial forms of Boolean functions.

In decision tree and diagram construction using an SOP form, only one type of node is used; that is, nodes that implement Shannon expansion. There are two expansions in the functional domain: *positive Davio* expansion and *negative Davio* expansion. This is because the polynomial form of a Boolean function is characterized by polarity. These two expansions provide the construction of polynomial forms as follows:

The EXOR analog of Shannon expansion (Davio expansion)

Positive polarity polynomials: If only the positive Davio expansion is applied, the resulting polynomial is of zero polarity; that is, all variables in the polynomial are uncomplemented.

Negative polarity polynomials: If only the negative Davio expansion is applied, the resulting polynomial is of $2^n - 1$ polarity; that is, all variables in the polynomial are complemented.

Fixed polarity polynomials: Application of both positive and negative Davio expansions results in fixed polarity; that is, polarity from 1 to $2^n - 2$ of the polynomial.

12.2.1 Algebraic form of the positive Davio expansions

Given a Boolean function f of n variables $x_1, x_2, \dots, x_{i-1}, x_i, x_{i+1}, \dots, x_n$,

$$f = f(x_1, x_2, \dots, x_{i-1}, \boxed{x_i}, x_{i+1}, \dots, x_n)$$

The positive Davio expansion with respect to the variable x_i is defined by the equation

The positive Davio expansion

$$f = f_0 \oplus x_i f_2 \tag{12.1}$$

where $f_2 = f_0 \oplus f_1$.

Equation 12.1 is derived as follows. Shannon expansion of a Boolean function f with respect to the variable x_i results in the expression

$$\begin{aligned} f &= \overline{x}_i f_0 \oplus x_i f_1 = (1 \oplus x_i) f_0 \oplus x_i f_1 \\ &= f_0 \oplus x_i f_0 \oplus x_i f_1 = f_0 \oplus x_i \underbrace{(f_0 \oplus f_1)}_{f_2} \end{aligned}$$

Given $f_2 = f_0 \oplus f_1$, Equation 12.1 follows straightforwardly. From Equation 12.1 it follows that an arbitrary Boolean function f of n variables can be represented in expanded form with respect to the i-th variable x_i, $i \in 1, 2, \dots, n$. Hence, positive Davio expansion given by Equation 12.1 is specified by the parameters f_0, f_1, and f_2:

Specification of the positive Davio expansion

Factor f_0*:* This is the function that is obtained from the function f by replacing the variable x_i by the logic value 0:

$$f_0 = f_{x_i=0} = f(x_1, \cdots, x_{i-1}, \boxed{x_i = 0}, x_{i+1}, \cdots, x_n)$$

Factor f_1*:* This is the function that is obtained from the function f by replacing the variable x_i by the logic value 1:

$$f_1 = f_{x_i=1} = f(x_1, \cdots, x_{i-1}, \boxed{x_i = 1}, x_{i+1}, \cdots, x_n)$$

Factor f_2*:* This is the function that is obtained by the EXOR sum of factors f_0 and f_1; that is,

$$f_2 = f_0 \oplus f_1 = f_{x_i=0} \oplus f_{x_i=1}$$

Factor $x_i f_2$*:* This is the function that is obtained by the AND multiplication of the variable x_i by the factor f_2, $x_i f_2$.

Computing the positive Davio expansion

- The node that implements the positive Davio expansion, denoted by pD, has two outputs:

  The left branch corresponds to the factor $1 \cdot f_0$ and

  The right branch corresponds to the factor $x_i \cdot f_2$

- Four possible combinations of the outputs f_0 and f_2 can be observed in computing:

  $\{f_0, f_2\} = \{0, 0\}$*:* Outputs of the left and right branches are both zero, hence, the input is $f = 0$;

  $\{f_0, f_2\} = \{0, 1\}$*:* The output of the right branch is 1, hence, the input is $f = x_i$;

  $\{f_0, f_2\} = \{1, 0\}$*:* Outputs of the left and right branches are both 1, hence, the input is $f = 1$;

  $\{f_0, f_2\} = \{1, 1\}$*:* The output of the left branch is 1, hence, the input is $f = \overline{x}_i$.

Example 12.1 **(Positive Davio expansion.)** *Let* $f = x_1 \oplus x_2 \oplus x_1 x_3$*. The positive Davio expansion of the Boolean function* f *with respect to the variable* x_2 *is defined as follows:*

$$f_0 = x_1 \oplus (x_2 = 0) \oplus x_1 x_3 = x_1 \oplus x_1 x_3$$

$$f_1 = x_1 \oplus (x_2 = 1) \oplus x_1 x_3 = 1 \oplus x_1 \oplus x_1 x_3$$

$$f_2 = f_0 \oplus f_1 = \underbrace{x_1 \oplus x_1 x_3}_{f_0} \oplus \underbrace{1 \oplus x_1 \oplus x_1 x_3}_{f_1} = 1$$

$$f = f_0 \oplus x_2 f_2 = x_1 \oplus x_1 x_3 \oplus x_2$$

Practice problem **12.1.** **(Positive Davio expansion.)** Apply positive Davio expansion with respect to the variable x_1 of the function $f = x_1x_2 \oplus x_3x_4x_5$.
Answer is given in "Solutions to practice problems."

Figure 12.2 illustrates the computational aspects of the positive Davio expansion.

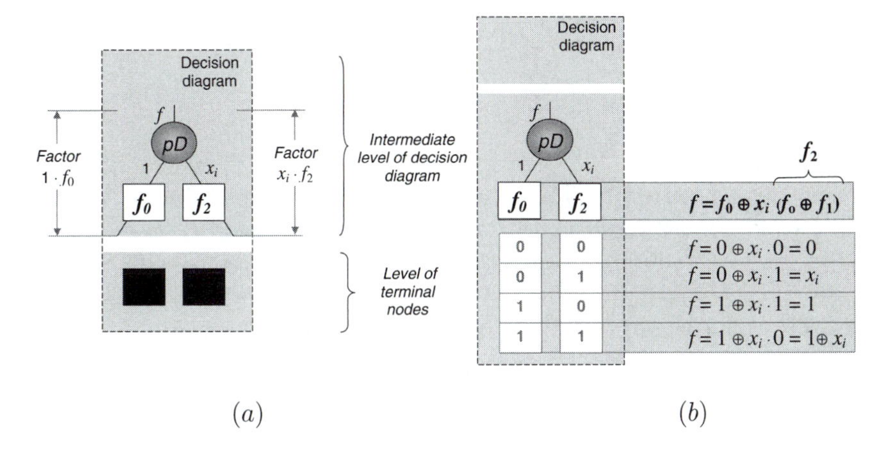

FIGURE 12.2
Nodes in functional decision diagrams and trees that implement the positive Davio expansion pD: function of the node (a) and computing of the node (b).

Practice problem **12.2.** **(Polynomial forms and functional decision trees.)** Given the following decision trees:

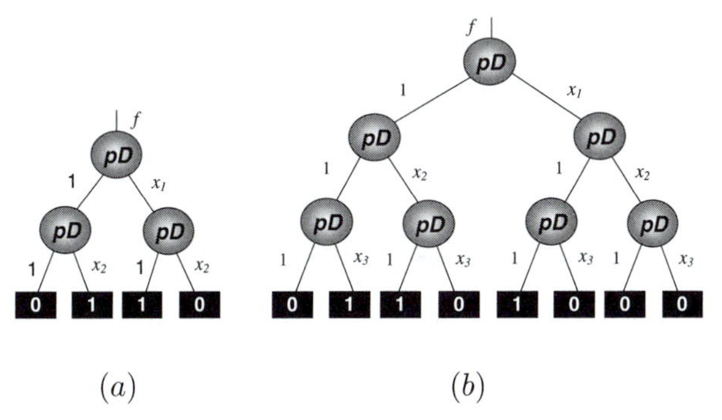

Derive the polynomial forms of the Boolean functions.
Answer is given in "Solutions to practice problems."

12.2.2 Algebraic form of the negative Davio expansion

Given the Boolean function f of n variables $x_1, x_2, \ldots, x_{i-1}, x_i, x_{i+1}, \ldots, x_n$, *the negative Davio* expansion with respect to the variable x_i is expressed by the equation

The negative Davio expansion

$$f = f_1 \oplus \overline{x}_i f_2 \tag{12.2}$$

By analogy with positive Davio expansion, negative Davio expansion is specified by the factors f_0, f_1, f_2, and $\overline{x}_i f_2$. Negative Davio expansion (Equation 12.2) with respect to the variable x_i is defined by analogy with positive Davio expansion:

$$\begin{aligned} f &= \overline{x}_i f_0 \oplus x_i f_1 = \overline{x}_i f_0 \oplus (1 \oplus \overline{x}_i) f_1 \\ &= \overline{x}_i f_0 \oplus f_1 \oplus \overline{x}_i f_1 = f_1 \oplus \overline{x}_i (f_0 \oplus f_1) = f_1 \oplus \overline{x}_i f_2 \end{aligned}$$

Example 12.2 **(Negative Davio expansion.)** *Negative Davio expansion with respect to the variable* x_1 *is defined as follows:* $f = f_1 \oplus x_2 f_2 = 1 \oplus x_1 \oplus x_1 x_3 \oplus \overline{x}_2$.

Practice problem **12.3.** **(Negative Davio expansion.)** Derive negative Davio expansion with respect to the variable x_1 of the Boolean function $f = 1 \oplus x_1 x_2 \oplus x_3 x_4$.
Answer is given in "Solutions to practice problems."

Figure 12.3 illustrates the computational aspects of negative Davio expansion.

Computing negative Davio expansion

- The node that implements the negative Davio expansion, denoted by nD, has two outputs:

  The left branch corresponds to the factor $1 \cdot f_1$ and
  The right branch corresponds to the factor $\overline{x}_i \cdot f_2$

- Four possible combinations of the outputs f_1 and f_2 can be observed in computing:

  $\{f_1, f_2\} = \{0, 0\}$: Outputs of the left and right branches are both zero, hence, the input is $f = 0$;

  $\{f_1, f_2\} = \{0, 1\}$: The output of the right branch is 1, hence, the input is $f = \overline{x}_i$;

  $\{f_1, f_2\} = \{1, 0\}$: Outputs of the left and right branches are both 1, hence, the input is $f = 1$;

  $\{f_1, f_2\} = \{1, 1\}$: The output of the left branch is 1, hence, the input is $f = x_i$.

Techniques for computing the negative Davio expansion

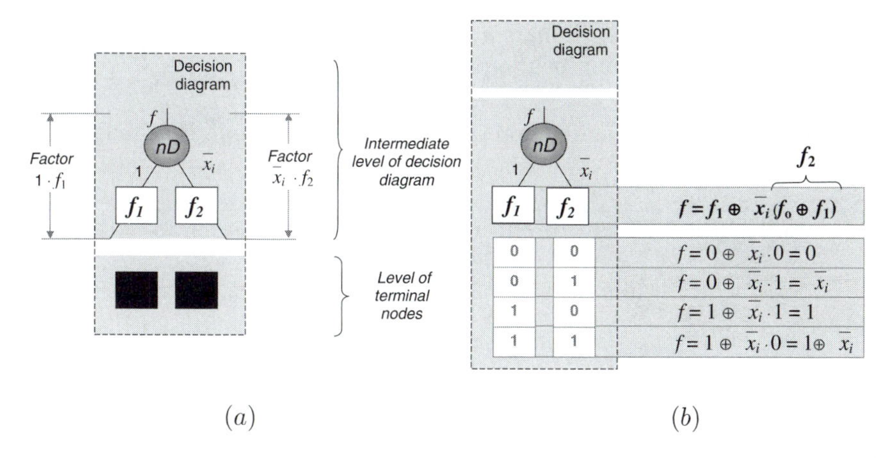

FIGURE 12.3
Nodes in the functional decision diagrams and trees that implement the negative Davio expansion nD: function of the node (a) and computing of the node (b).

12.2.3 Matrix forms of positive and negative Davio expansions

In matrix notation, the function f of the node is a Boolean function of a single variable x_i given by the truth-vector $\mathbf{F} = [\ f(0)\ f(1)\ ]^T$. Hence, the node can be described by the simplest matrix equation. The function of the node using a positive Davio expansion is as follows:

Matrix form of the positive Davio expansion

$$f = [\ 1\ x_i\ ] \begin{bmatrix} 1 & 0 \\ 1 & 1 \end{bmatrix} \begin{bmatrix} f_0 \\ f_1 \end{bmatrix} = [\ 1\ x_i\ ] \begin{bmatrix} f_0 \\ f_2 \end{bmatrix}$$
$$= f_0 \oplus x_i f_2 \qquad (12.3)$$

Example 12.3 (Matrix form of *pD* expansion). *Let the input variable be given by the truth-vector* $\mathbf{F} = [\ f(0)\ f(1)\ ]^T = [\ 1\ 0\ ]^T$*; that is,* $f_0 = 1$ *and* $f_1 = 0$*. The positive Davio transform (Equation 12.3) is*

$$f = [\ 1\ x_i\ ] \begin{bmatrix} 1 & 0 \\ 1 & 1 \end{bmatrix} \begin{bmatrix} 1 \\ 0 \end{bmatrix} = [\ 1\ x_i\ ] \begin{bmatrix} 1 \\ 1 \end{bmatrix} = 1 \oplus x_i$$

Practice problem **12.4. (Matrix form of *pD* expansion.)** Given the truth-vector $\mathbf{F} = [\ 0\ 1\ ]^T$, compute the positive Davio expansion with respect to the variable x_2 in matrix form.
Answer is given in "Solutions to practice problems."

In matrix notation, the function of a negative Davio node is as follows:

Matrix form of the negative Davio expansion

$$f = [\ 1\ \overline{x}_i\ ] \begin{bmatrix} 0 & 1 \\ 1 & 1 \end{bmatrix} \begin{bmatrix} f_0 \\ f_1 \end{bmatrix} = [\ 1\ \overline{x}_i\ ] \begin{bmatrix} f_1 \\ f_2 \end{bmatrix}$$
$$= f_1 \oplus \overline{x}_i f_2 \qquad (12.4)$$

Example 12.4 *(Continuation of Example 12.3). The negative Davio transform (Equation 12.4) is*

$$f = [\ 1\ \overline{x}_i\ ] \begin{bmatrix} 0 & 1 \\ 1 & 1 \end{bmatrix} \begin{bmatrix} 1 \\ 0 \end{bmatrix} = [\ 1\ \overline{x}_i\ ] \begin{bmatrix} 1 \\ 1 \end{bmatrix} = 1 \oplus \overline{x}_i$$

12.2.4 Gate level implementation of Shannon and Davio expansions

Consider the gate level implementation of Shannon and Davio expansions. The logic networks are given in Figure 12.4 in comparison with a network for Shannon expansion.

Example 12.5 (Implementation of Davio expansion.) *Given the Boolean function $f = x_1 \vee x_2$, its positive (pD) and negative (nD) Davio expansions with respect to the variable x_1 result in the polynomial expressions*

$$\textbf{pD: } f = f_0 \oplus x_1(f_0 \oplus f_1) = \underbrace{x_2}_{f_0} \oplus x_1 \underbrace{(x_2 \oplus 1)}_{f_0 \oplus f_1} = x_2 \oplus x_1 \oplus x_1 x_2$$

$$\textbf{nD: } f = f_1 \oplus \overline{x}_1(f_0 \oplus f_1) = \underbrace{x_2}_{f_1} \oplus x_1 \underbrace{(x_2 \oplus 1)}_{f_0 \oplus f_1} = 1 \oplus \overline{x}_1 \oplus \overline{x}_1 x_2$$

The logic networks for the Davio expansion given in Figure 12.4 can be used for computing by specification of the inputs; that is, $f_0 = x_2$ and $f_1 = 1$.

Table 12.1 summarizes the functions of the nodes for positive Davio and negative Davio expansions, labeled as pD and nD respectively. For simplification, realization of the nodes is given using a single EXOR gate.

Design example: Shannon and Davio expansions

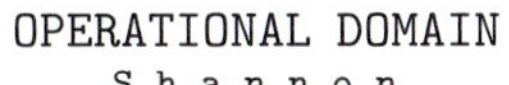

OPERATIONAL DOMAIN

Shannon expansion

$f = \overline{x}_i f_0 \vee x_i f_1$

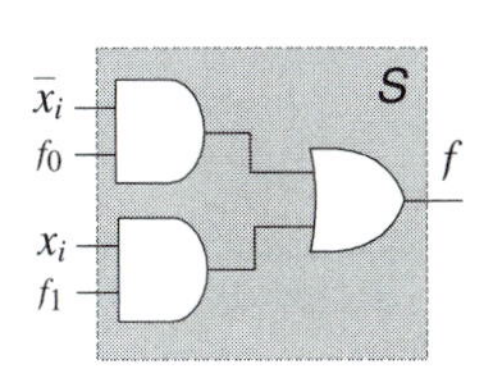

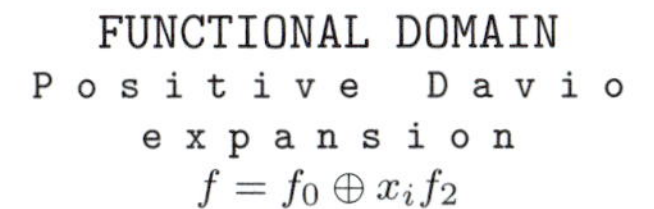

FUNCTIONAL DOMAIN

Positive Davio expansion

$f = f_0 \oplus x_i f_2$

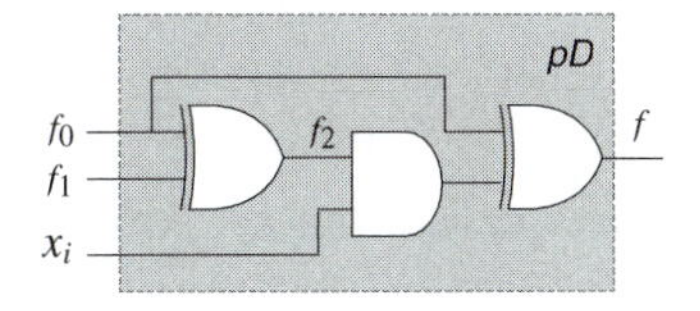

Negative Davio expansion

$f = f_1 \oplus \overline{x}_i f_2$

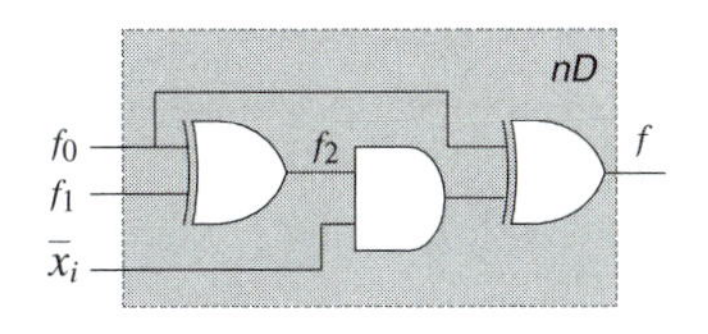

FIGURE 12.4

Gate level representation of the nodes of decision trees and diagrams using Shannon and Davio expansions.

TABLE 12.1

The nodes of the functional decision tree that implement Davio expansion at the gate level, and their description in algebraic and matrix forms.

Techniques for computing Davio expansions

Node	Realization	Algebraic form	Matrix form
Positive Davio node (pD; f, 1, x_i)	f_0, $x_i f_2$ → f	$f = f_0 \oplus x_i f_2$ $f_0 = f\vert_{x_i=0}$ $f_2 = f\vert_{x_i=1} \oplus f\vert_{x_i=0}$	$f = [\, 1 \; x_i \,] \begin{bmatrix} 1 & 0 \\ 1 & 1 \end{bmatrix} \begin{bmatrix} f_0 \\ f_1 \end{bmatrix}$
Negative Davio node (nD; f, 1, $\overline{x}_i$)	f_1, $\overline{x}_i f_2$ → f	$f = f_1 \oplus \overline{x}_1 f_2$ $f_1 = f\vert_{x_i=1}$ $f_2 = f\vert_{x_i=0} \oplus f\vert_{x_i=1}$	$f = [\, 1 \; \overline{x}_i \,] \begin{bmatrix} 0 & 1 \\ 1 & 1 \end{bmatrix} \begin{bmatrix} f_1 \\ f_2 \end{bmatrix}$

12.3 Techniques for functional decision tree construction

Techniques for functional decision tree construction consist of techniques for the reduction of functional decision trees, matrix-based designs, and manipulation of pD and nD nodes for conversion between polarities.

12.3.1 The structure of functional decision trees

The most important structural properties of the functional decision tree with positive Davio nodes are as follows:

Structural properties of functional decision trees

- ▶ A Boolean function of n variables is represented by an n-level functional decision tree. The i-th level of the functional decision tree, $i = 1, \ldots, n$, includes 2^{i-1} nodes.
- ▶ Nodes at the n-th level are connected to 2^n terminal nodes, which take values 0 or 1. The nodes, corresponding to the i-th variable, form the i-th level in the functional decision tree.
- ▶ In every path from the root node to a terminal node, the variables appear in a fixed order; the tree is thus sa id to be ordered.
- ▶ The values of constant nodes are the values of the coefficients of the polynomial expression for the Boolean function represented.

12.3.2 Design example: Manipulation of pD and nD nodes

This design example introduces techniques for the design of functional decision diagrams for computing polynomial expressions of various polarities. This computing ability is provided by the distribution of pD and nD nodes in the levels of a decision tree. There are 2^n various combinations of the pD and nD nodes in the levels of a decision tree. Each combination corresponds to one polarity of a polynomial. There are two trivial cases in these 2^n combinations:

(a) The tree consisting of only pD nodes; it computes only the positive polarity polynomial (all variables are noncomplemented).

(b) The tree consisting of only nD nodes; it computes only the negative polarity polynomial (all variables are complemented).

Example 12.6 (Manipulation of pD and nD nodes.) *Design functional decision trees for computing all positive fixed polarity polynomial expressions of Boolean functions of two variables. Figure 12.5 shows all four possible decision trees. The functional decision tree that represents the polynomial of polarity $c = 1$ is shown in Figure 12.6. All possible coefficients of the polynomial expression are also shown.*

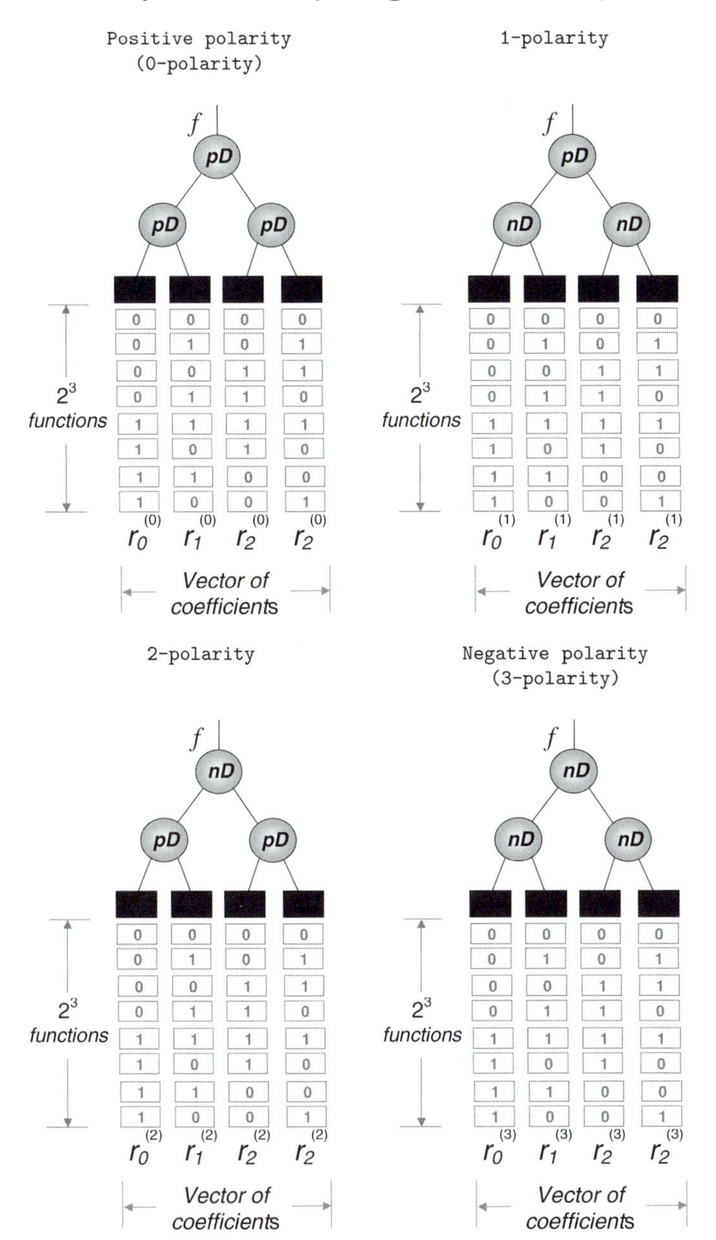

FIGURE 12.5
Functional decision trees for computing polynomial expressions in positive, negative, and fixed polarities of Boolean functions of two variables (Example 12.6).

Practice problem **12.5. (Decision trees for EXOR functions.)** Derive the decision trees for computing the EXOR function of two variables using (a) pD and (b) nD expansions in the nodes.
Answer is given in "Solutions to practice problems."

Techniques for computing the Davio expansions

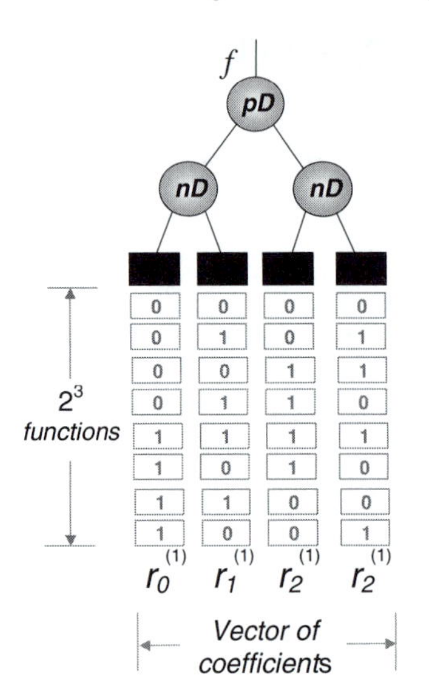

The polynomial expression of an arbitrary Boolean function of two variables ($n = 2$) and the polarity

$$c = 1 \; (c_1 c_2 = 01):$$

$$f = \bigoplus_{i=0}^{3} r_i^{(1)} \overbrace{(x_1 \oplus 0)}^{0-polarity}{}^{i_1} \overbrace{(x_2 \oplus 1)}^{1-polarity}{}^{i_2}$$

$$= \bigoplus_{i=0}^{3} r_i^{(1)} x_1{}^{i_1} \overline{x}_2{}^{i_2}$$

$$= r_0^{(1)} \oplus r_1^{(1)} \overline{x}_2 r_3^{(1)} x_1 \oplus r_3^{(1)} x_1 \overline{x}_2$$

Vector of coefficients:

$$\mathbf{R}^1 = [r_0^{(1)} r_1^{(1)} r_2^{(1)} r_3^{(1)}]^T$$

FIGURE 12.6
The functional decision tree for a Boolean function of two variables (Example 12.6).

Practice problem **12.6. (Decision tree.)** Derive the decision tree for the polynomial expression of the Boolean function $f = x_1\overline{x}_2 \vee \overline{x}_3$.
Answer is given in "Solutions to practice problems."

12.3.3 Design example: Application of matrix transforms

This design example introduces techniques of functional decision tree construction using a matrix description of computing. Given a truth-vector $\mathbf{F} = [f(0)\ f(1) \ldots f(2^n - 1)]^T$ of the Boolean function f, the positive polarity polynomial expression in algebraic form is defined as the following matrix transforms:

Application of matrix transforms

$$f = \widehat{\mathbf{X}}\ \mathbf{R}_{2^n}\ \mathbf{F} \quad \text{over GF}(2) \tag{12.5}$$

where the vector $\widehat{\mathbf{X}}$ and matrix $\mathbf{R}_{2^n}$ are constructed using the Kronecker product, denoted as $\otimes$:

$$\widehat{\mathbf{X}} = \bigotimes_{i=1}^{n} [\ 1\ x_i\ ], \quad \mathbf{R}_{2^n} = \bigotimes_{i=1}^{n} \mathbf{R}_2,$$

and the elementary matrix $\mathbf{R}_2 = \begin{bmatrix} 1 & 0 \\ 1 & 1 \end{bmatrix}$

Example 12.7 (The Kronecker product.) *Given the Boolean functions of (a) a single variable x_i $(n = 1)$ and (b) two variables $x_1 \vee x_2$ $(n = 2)$, using Equation 12.5, the positive polarity polynomial expressions are as follows:*

$$(a)\ n = 1:\ f = \widehat{\mathbf{X}}\ \mathbf{R}_{2^1}\ \mathbf{F} = [\ 1\ x_i\ ] \begin{bmatrix} 1 & 0 \\ 1 & 1 \end{bmatrix} \begin{bmatrix} 0 \\ 1 \end{bmatrix} = [\ 1\ x_i\ ] \begin{bmatrix} 0 \\ 1 \end{bmatrix} = x_i$$

$$(b)\ n = 2:\ f = \widehat{\mathbf{X}}\ \mathbf{R}_{2^2}\ \mathbf{F} = [\ 1\ x_2\ x_1\ x_1x_2] \left[\begin{array}{cc|cc} 1 & 0 & 0 & 0 \\ 1 & 1 & 0 & 0 \\ \hline 1 & 0 & 1 & 0 \\ 1 & 1 & 1 & 1 \end{array}\right] \begin{bmatrix} 0 \\ 1 \\ 1 \\ 1 \end{bmatrix}$$

$$= [\ 1\ x_2\ x_1\ x_1x_2] \begin{bmatrix} 0 \\ 1 \\ 1 \\ 1 \end{bmatrix} = x_2 \oplus x_1 \oplus x_1x_2$$

Practice problem **12.7. (The Kronecker product.)** Using a matrix transform, find the positive polarity polynomial representation of the Boolean function given by the truth-vector $\mathbf{F} = [1\ 0\ 0\ 1]^T$.
Answer is given in "Solutions to practice problems."

Example 12.8 (Matrix-based design.) *Let the Boolean function $f = \overline{x}_1 \vee x_2$ be given by its truth-vector $\mathbf{F} = [1\ 1\ 0\ 1]^T$. The functional decision tree with all possible Davio expansion nodes is shown in Figure 12.9. The values of this tree's terminal nodes are used to derive the polynomial expression, in which the product terms are generated by the Kronecker product $\widehat{\mathbf{X}}$. The 4×4 transform matrix $\mathbf{R}$ is generated by the Kronecker product of the basic matrix $\mathbf{R}_{2^1}$.*

The example below shows functional decision tree design using the truth-vector of a Boolean function.

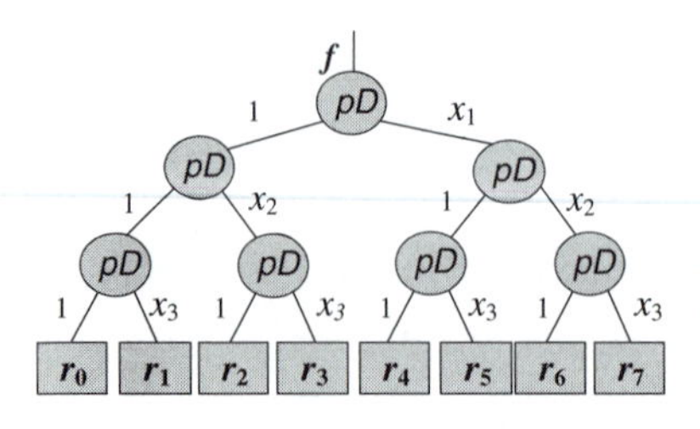

There are 8 paths from f to the terminal nodes:

Path 1:	$t_1 = 1$	*Path 2:*	$t_2 = x_3$
Path 3:	$t_3 = x_2$	*Path 4:*	$t_4 = x_2x_3$
Path 5:	$t_5 = x_1$	*Path 6:*	$t_6 = x_1x_3$
Path 7:	$t_7 = x_1x_2$	*Path 8:*	$t_8 = x_1x_2x_3$

Polynomial expression
$f = f_{000}t_1 \oplus f_{002}t_2 \oplus ... \oplus f_{222}t_8$

FIGURE 12.7
Positive polarity polynomial representation of a Boolean function of three variables by a functional decision tree (Example 12.10).

Example 12.9 (Tree design in terms of truth-vectors.) *A Boolean function is given by the truth-vector* $\mathbf{F} = [\,0\ 1\ 1\ 0\,]^T$ *(Figure 12.8).*

Step 1: *Root pD node with input* $\mathbf{F} = [\,0\ 1\ 1\ 0\,]^T$. *Positive Davio expansion results in left branch,* $\mathbf{F}_{x_1=0} = [0\ 1]^T$, *and right branch,* $\mathbf{F}_{x_1=0} \oplus \mathbf{F}_{x_1=1} = [0\ 1]^T \oplus [1\ 0]^T = [\,1\ 1\,]^T$.
Both outputs results in functions and require further application of a Davio expansion:
Step 2: *Left node, left branch:* $\mathbf{F}_{\substack{x_1=0\\x_2=0}} = [\,0\,]$. *Left node,right branch:* $\mathbf{F}_{\substack{x_1=0\\x_2=0}} \oplus \mathbf{F}_{\substack{x_1=0\\x_2=1}} = [\,0\,] \oplus [\,1\,] = [\,1\,]$.
Step 3: *Right node, left branch:* $\mathbf{F}_{\substack{x_1=1\\x_2=0}} = [\,1\,]$. *Right node, left branch:* $\mathbf{F}_{\substack{x_1=1\\x_2=0}} \oplus \mathbf{F}_{\substack{x_1=1\\x_2=1}} = [\,1\,] \oplus [\,1\,] = [\,0\,]$.

Example 12.10 (Decision trees.) *An arbitrary Boolean function f of three variables can be represented by the decision tree shown in Figure 12.7 (3 levels, 7 nodes, 8 terminal nodes). To design this tree, the positive polynomial expansion (Equation 12.1) is used as follows:*

Step 1: *With respect to variable* x_1: $f = f_0 \oplus x_1 f_2$
Step 2: *With respect to variable* x_2: $f_0 = f_{00} \oplus x_2 f_{02}$, $f_1 = f_{10} \oplus x_2 f_{22}$.
Step 3: *With respect to variable* x_3:

$$f_{00} = f_{000} \oplus x_3 f_{002}, \quad f_{02} = f_{020} \oplus x_3 f_{022},$$
$$f_{20} = f_{200} \oplus x_3 f_{202}, \quad f_{22} = f_{220} \oplus x_3 f_{222}$$

Step 4: *Representation of Boolean function f in t polynomial form* $f = f_{000} = f_{002}x_3 \oplus f_{020}x_2 \oplus f_{022}x_2x_3 \oplus f_{200}x_1 \oplus f_{202}x_1x_3 \oplus f_{220}x_1x_2 \oplus f_{222}x_1x_2x_3$.

Design example: Functional decision tree

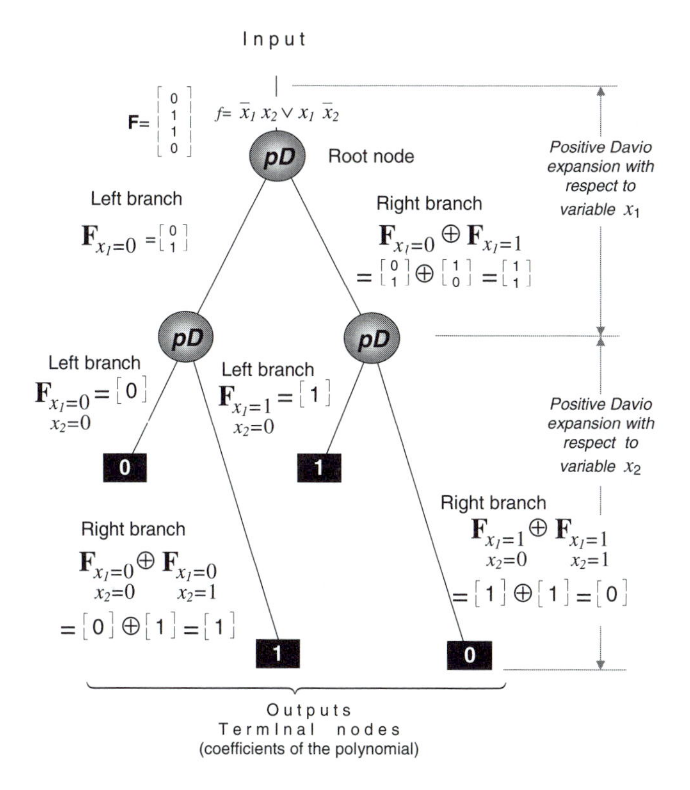

FIGURE 12.8
Functional decision tree design with positive Davio expansion in the nodes using a truth-vector of the Boolean function $f = \overline{x}_1 x_2 \vee x_2 \overline{x}_2$ (Example 12.9).

12.3.4 Design example: Minterm computing

Example 12.11 (Decision trees of polarized minterms.) *Figure 12.10 shows a representation of polarized minterms using a decision tree for the Boolean function of two variables.*

Practice problem 12.8. (Decision trees of polarized minterms.) Derive all polarized minterms for the Boolean functions of two variables.
Answer is given in "Solutions to practice problems."

Design example: Functional decision tree

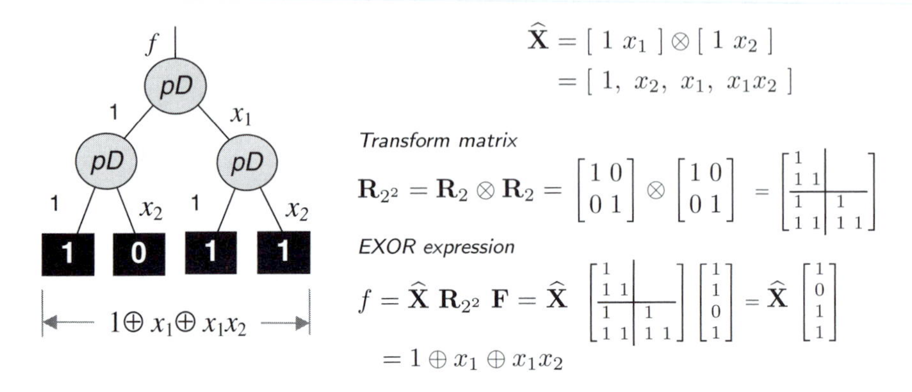

FIGURE 12.9
The functional decision tree for the Boolean function $f = \overline{x}_1 \vee x_2$ (Example 12.8).

Design example: Polarized minterms

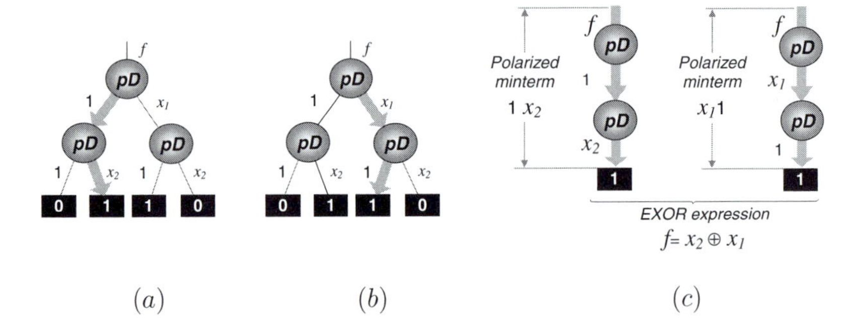

FIGURE 12.10
Polarized minterm representation using a decision tree: (a) polarized minterm $1x_2$, (b) polarized minterm x_21, (c) polynomial expression $f = x_2 \oplus x_1$ (Example 12.11).

12.4 Functional decision tree reduction

The functional decision diagram for polynomial forms is derived from the functional decision tree by deleting redundant nodes, and by sharing equivalent subgraphs. The rules below produce reduced functional decision diagrams.

12.4.1 Elimination rule

If the outgoing edge of a node labeled with x_i and $\overline{x}_i$ points to the constant zero, then delete the node and connect the edge to the other subgraph directly. The formal basis of this rule is as follows (Figure 12.11):

Elimination rule

$$\varphi = \varphi_0 \oplus x_i \varphi_2$$

If $\varphi_2 = 0$, then $\varphi = \varphi_0$.

12.4.2 Merging rule

In a tree, edges longer than one; i.e., connecting nodes at non-successive levels, can appear. For example, the length of an edge connecting a node at the $(i-1)$-th level with a node at the $(i+1)$-th level is two.

Example 12.12 (Merging nodes.) *Provide a formal explanation of the reduction of the following functional decision trees:*

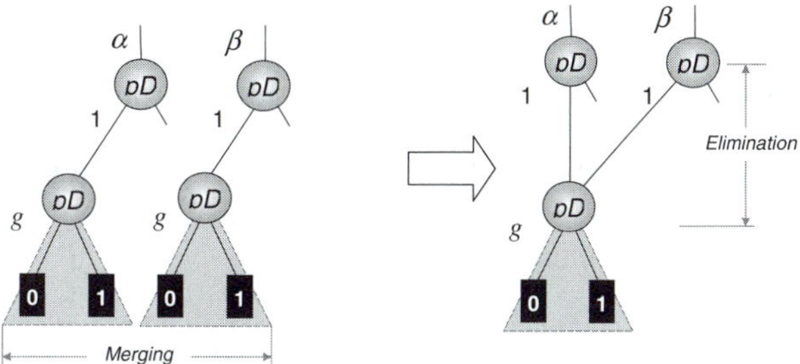

Example 12.13 (Reduction rules.) *Application of reduction rules to the three-variable NAND function is demonstrated in Figure 12.12.*

Practice problem **12.9. (Reduction rules.)** Derive the functional decision tree equivalent for the following notations: $f_0 \oplus x_i f_2 = 1$ if $f_0 = 0$, $f_1 = 0$ and $f_1 \oplus \overline{x}_i f_2 = 0$ if $f_0 = 0$, $f_1 = 0$.
Answer is given in "Solutions to practice problems."

The functional decision diagram is derived from the functional decision tree by deleting redundant nodes, and by sharing equivalent subgraphs. The rules below produce the reduced Davio diagram.

Example 12.14 (Functional decision diagram.) *Figure 12.13 shows the derived functional decision diagrams for some Boolean functions of three variables.*

Practice problem 12.10. (Equivalent decision diagrams.) Given the following decision diagrams:

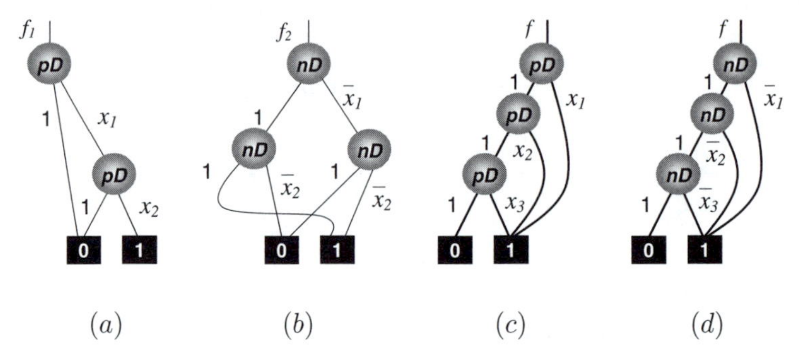

(*a*) (*b*) (*c*) (*d*)

Prove that they represent the same Boolean function.
Answers are given in "Solutions to practice problems."

Practice problem 12.11. (Decision diagrams for EXOR functions.) Derive decision diagrams for computing the EXOR function of two variables using (a) pD and (b) nD expansions in the nodes
Answer is given in "Solutions to practice problems."

Example 12.15 (Decision diagrams.) *Derive the polynomial forms of the Boolean functions from the following functional decision diagrams:*

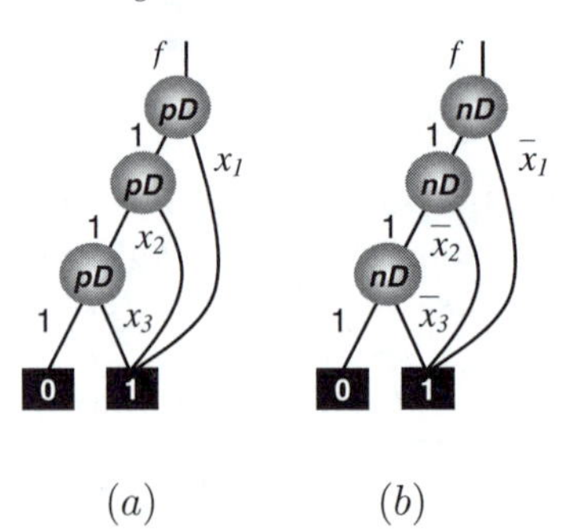

(*a*) (*b*)

Diagrams (a) and (b) represent the positive polarity and negative polarity polynomial expressions $f = x_1 \oplus x_2 \oplus x_3$ *and* $f = \overline{x}_1 \oplus \overline{x}_2 \oplus \overline{x}_3$*, respectively.*

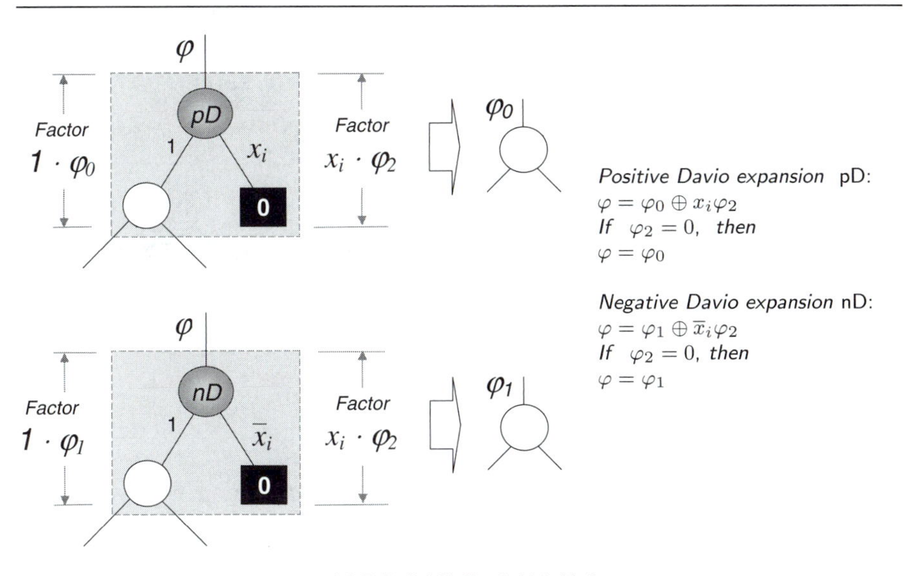

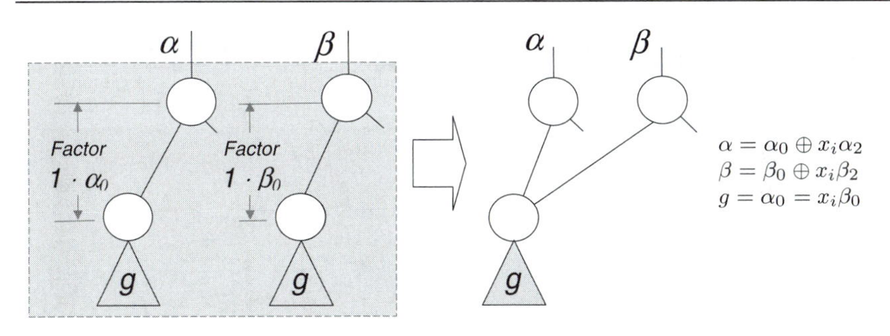

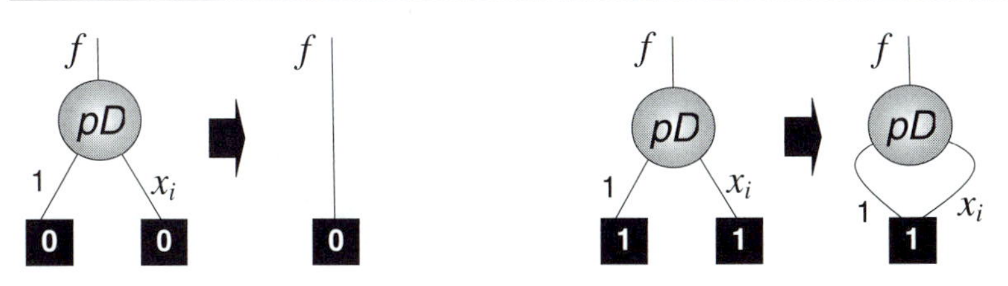

FIGURE 12.11
Reduction rules for functional decision diagram construction.

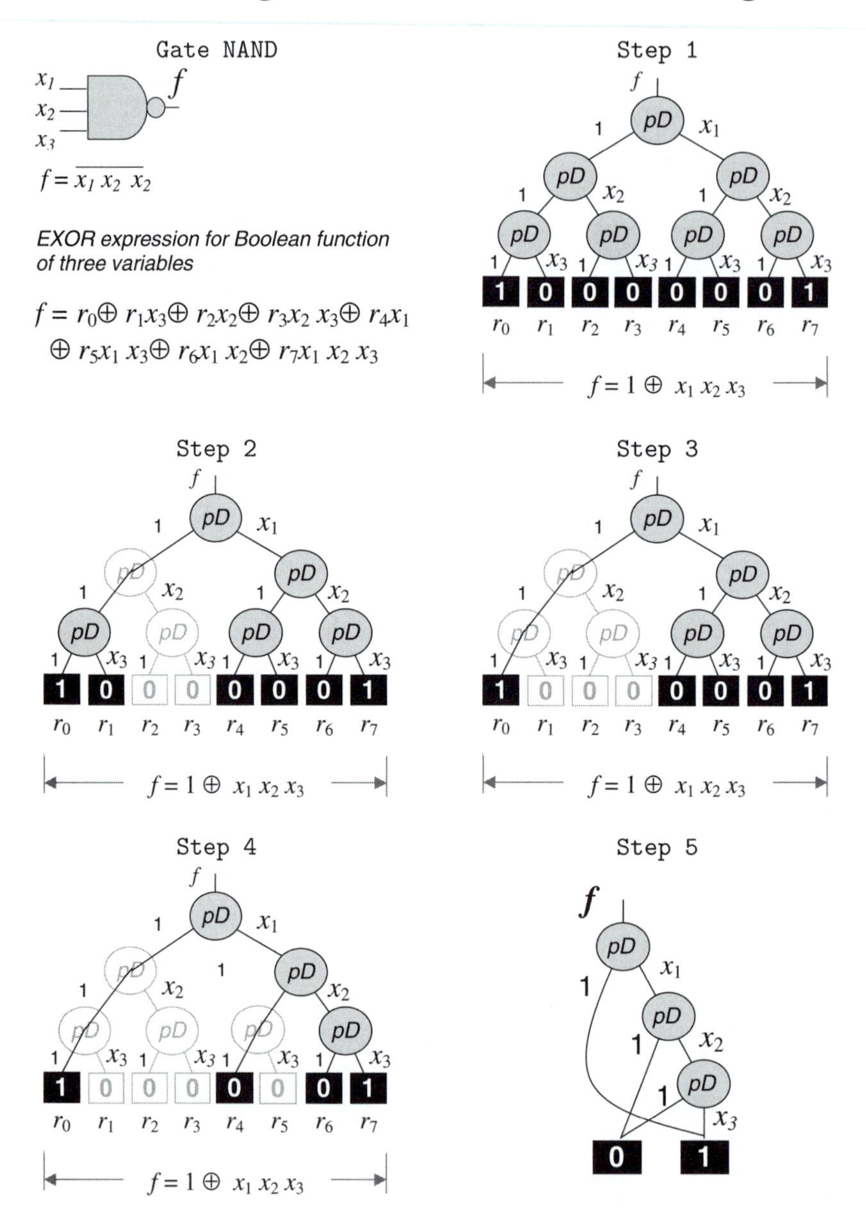

FIGURE 12.12
Functional decision diagram design using pD nodes for the three-variable NAND function (Example 12.13).

Design techniques:
Computing the elementary functions

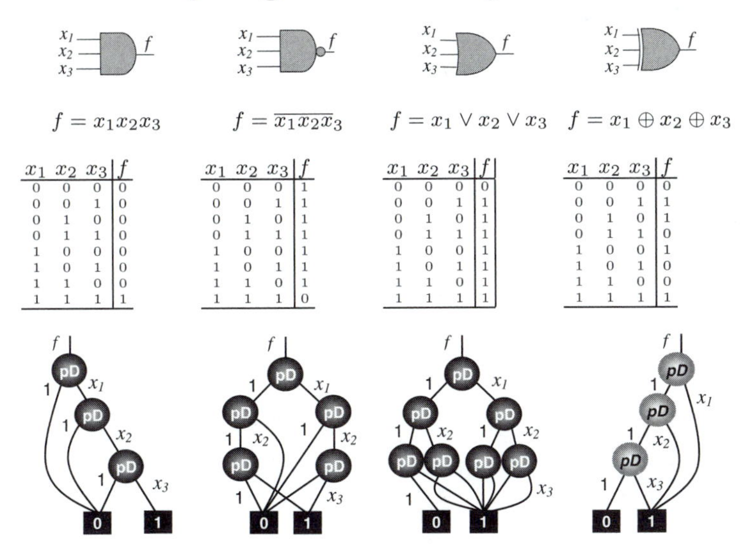

x_1	x_2	x_3	f
0	0	0	0
0	0	1	0
0	1	0	0
0	1	1	0
1	0	0	0
1	0	1	0
1	1	0	0
1	1	1	1

x_1	x_2	x_3	f
0	0	0	1
0	0	1	1
0	1	0	1
0	1	1	1
1	0	0	1
1	0	1	1
1	1	0	1
1	1	1	0

x_1	x_2	x_3	f
0	0	0	0
0	0	1	1
0	1	0	1
0	1	1	1
1	0	0	1
1	0	1	1
1	1	0	1
1	1	1	1

x_1	x_2	x_3	f
0	0	0	0
0	0	1	1
0	1	0	1
0	1	1	0
1	0	0	1
1	0	1	0
1	1	0	0
1	1	1	1

FIGURE 12.13
Functional decision diagrams for the Boolean functions AND, NAND, OR, and EXOR of three variables (Example 12.14).

12.5 Summary of functional decision diagrams

Decision trees and diagrams are graphical data structures. These data structures are useful for the representation, manipulation, optimization, and implementation of Boolean functions. There are two main types of decision trees and diagrams:

(*a*) In the first type of decision trees and diagrams, Shannon expansion is utilized in the nodes. These data structures are used for processing Boolean functions in SOP form; that is, in the *operational domain.*

(*b*) In the second type of decision trees and diagrams, modified Shannon expansion, known as *Davio* expansion, is utilized in the nodes. These data structures are used for processing Boolean functions in polynomial form; that is, in the *functional domain.*

In every path from the root node to a terminal node, the variables appear in a fixed order.

Summary (continuation)

- There are two types of Davio expansion: *positive Davio* expansion and *negative Davio* expansion. Application of these expansions results in polynomials of various polarities.
- Positive and negative Davio expansion are defined in algebraic form as follows:

`Positive Davio:` $f = f_0 \oplus x_i f_2$

`Negative Davio:` $f = f_1 \oplus \overline{x}_i f_2$

- Positive and negative Davio expansions are defined in matrix form as follows:

`Positive Davio:` $f = [\,1\ x_i\,] \begin{bmatrix} 1 & 0 \\ 1 & 1 \end{bmatrix} \begin{bmatrix} f_0 \\ f_1 \end{bmatrix} = [\,1\ x_i\,] \begin{bmatrix} f_0 \\ f_2 \end{bmatrix}$

`Negative Davio:` $f = [\,1\ \overline{x}_i\,] \begin{bmatrix} 1 & 0 \\ 1 & 1 \end{bmatrix} \begin{bmatrix} f_0 \\ f_1 \end{bmatrix} = [\,1\ \overline{x}_i\,] \begin{bmatrix} f_1 \\ f_2 \end{bmatrix}$

- If positive and negative Davio expansion at the node are implemented at the gate level, an arbitrary decision diagram can be implemented.
- The decision diagram for polynomial forms is derived from the decision tree by deleting redundant nodes, and by sharing equivalent subgraphs. There are two types of reducing rules: the *elimination* rule and the *merging* rule.

12.6 Further study

Historical perspective

1993: The first International Workshop on "Applications of the Read-Muller Expansions in Circuit Design" was held in Hamburg, Germany. This initiative was continued by the 2nd (Makuhari, Chiba, Japan, 1995), 3rd (Oxford, UK, 1997), 4th (University of Victoria, Canada, 1999), 5th (Mississippi State University, USA, 2001), and 6th Workshops (Oslo, Norway, 2007).

Advanced topics of the functional decision diagrams

Computing polynomials using Pascal and transeunt triangles. The Pascal triangle for a Boolean function $f(x_1, x_2, ..., x_n)$ is a fractal structure formed by modulo 2 addition of 0's and 1's starting from the bottom row, which is the truth-vector of f. The Pascal triangle for an n-variable Boolean function has a width of 2^n and a height of 2^n. Since the truth-vector for functions with one or more variables has an even number of entries, it can be divided evenly into two parts. Each part produces, on its own, two sub-triangles. In general,

- ▶ The bits along the triangle's left side are coefficients of zero polarity polynomial form RM_0,
- ▶ The bits along the right side of each subtriangle on the left side of the triangle are the coefficients of the polynomial form of polarity 1, RM_1,
- ▶ The bits along the left or right side of each following row of subtriangles, such that the bottom element is the jth element of the truth-vector $\mathbf{F}$, are the coefficients of polynomial forms of polarity j, RM_j, $j = 1, 2, \ldots, 2^n - 2$.
- ▶ The bits along the left side of the triangle are the coefficients of the polynomial form of polarity $2^n - 1$, RM_{2^n-1}.

Given the Boolean function $f = x_1\overline{x}_2$ and its truth-vector $\mathbf{F} = [0010]^T$, its Pascal triangles are shown in Figure 12.14. The location of coefficients of the fixed polynomial expressions of polarities 0,1,2, and 3,

$$RM(0) = [0011]^T, \quad RM(1) = [0001]^T, \quad RM(2) = [1111]^T, \quad RM(3) = [0101]^T$$

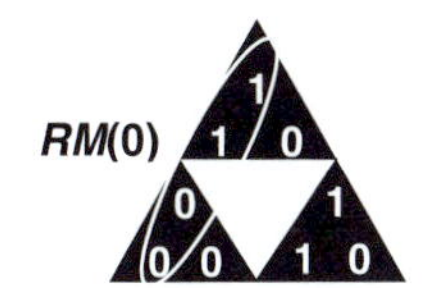

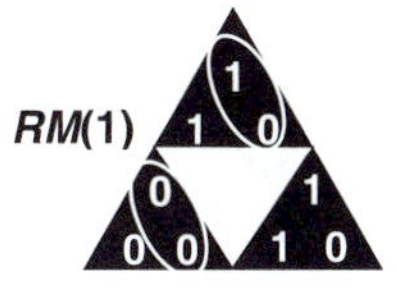

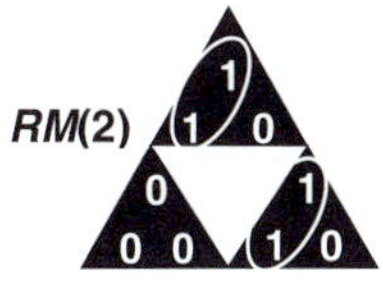

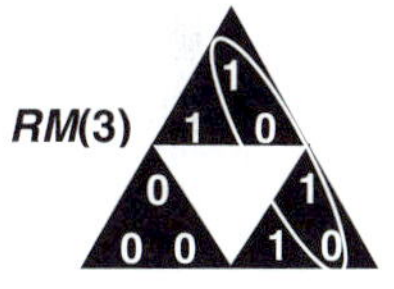

FIGURE 12.14
Encoding the Pascal triangles for the polynomial representation of the Boolean function $f = x_1\overline{x}_2$.

The Pascal triangle for totally symmetric function called transeunt triangle is formed as follows (J. Butler, G. Dueck, V. Shmerko, S. Yanushkevich, "On the Number of Generators of Transeunt Triangles", *Discrete Applied Mathematics*, number 108, pages 309–316, 2001):

- ▶ The carry vector is located at the base of the triangle.
- ▶ A vector of n 1's and 0's is formed by the exclusive OR of adjacent bits in the carry vector.
- ▶ A vector of $(n-1)$ 1's and 0's is formed by the exclusive OR of adjacent bits in the previous vector, etc.

▶ At the apex of the triangle is a single 0 or 1.

The transeunt triangle can be generated from the carry vector of coefficients in RM_0 or in RM_{2^n-1} by taking the exclusive OR of adjacent bits repeatedly until a single bit is obtained. This is because the exclusive OR is *self-invertible.* That is, given $A \oplus B$ and the value of A, we can find the value of B $(= A \oplus (A \oplus B))$. Figure 12.15 shows the transeunt triangle given the symmetric Boolean function $f = \overline{x}_1\overline{x}_2x_3 \vee x_1x_2x_3 \vee x_1\overline{x}_2\overline{x}_3 \vee \overline{x}_1x_2\overline{x}_3$. In this triangle, only the elements that correspond to a truncated truth-vector of the symmetric function, are given. The corresponding functional trees based on Davio expansion for $RM(0) = x_1 \oplus x_2 \oplus x_3$ and $RM(7) = 1 \oplus \overline{x}_1 \oplus \overline{x}_2 \oplus \overline{x}_3$, and the decision tree based on Shannon expansion are given in Figure 12.15 as well.

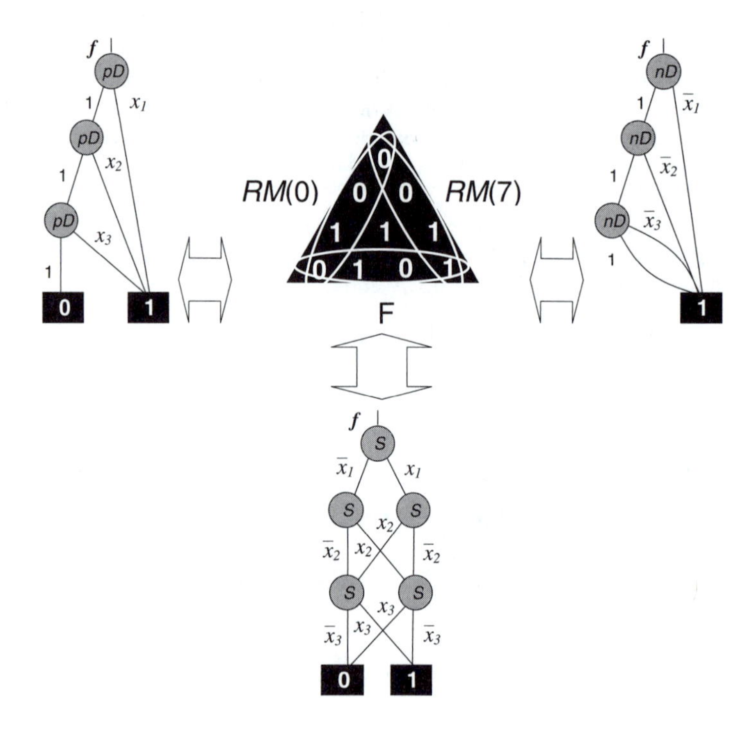

FIGURE 12.15
The relationship between the transeunt triangle and the corresponding decision diagrams for the totally symmetric Boolean function $f = x_1 \oplus x_2 \oplus x_3$.

12.7 Solutions to practice problems

Practice problem	Solution
12.1.	$f_0 = x_3x_4x_5;\ f_1 = x_2 \oplus x_3x_4x_5;\ f_2 = f_0 \oplus f_1 = x_2;$ $f = f_0 \oplus x_1f_1 = x_3x_4x_5 \oplus x_1(x_2)$
12.2.	$(a)\ \ f = x_1 \oplus x_2;\ (b)\ \ f = x_1 \oplus x_2 \oplus x_3$
12.3.	$f_0 = 1 \oplus x_3x_4;\ f_1 = 1 \oplus x_2 \oplus x_3x_4;\ f_2 = f_0 \oplus f_1 = x_2;$ $f = (1 \oplus x_2 \oplus x_3x_4) \oplus \overline{x}_1(x_2)$
12.4.	$f = [\ 1\ x\ ]\begin{bmatrix}1 & 0\\ 1 & 1\end{bmatrix}\begin{bmatrix}0\\ 1\end{bmatrix} = [\ 1\ x\ ]\begin{bmatrix}0\\ 1\end{bmatrix} = x$
12.5.	$f = x_1 \oplus x_2$ 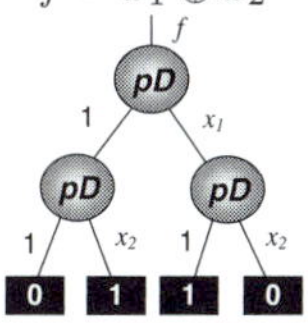 $f = \overline{x}_1 \oplus \overline{x}_2 = x_1 \oplus x_2$ 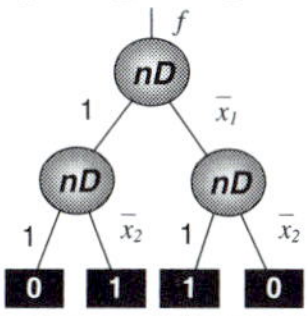
12.6.	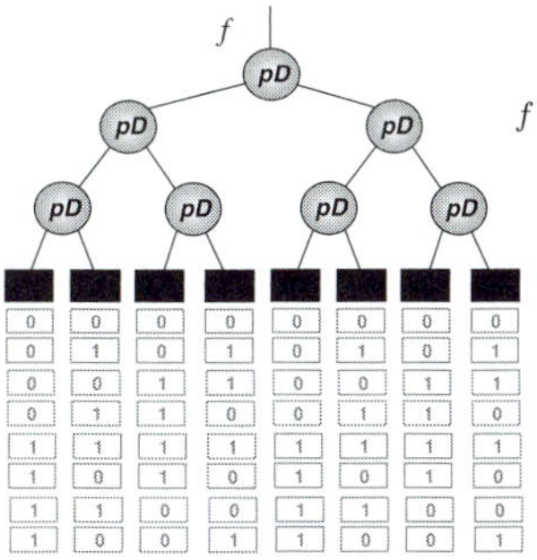 $f = x_1\overline{x}_2 \vee \overline{x}_3$ $= \begin{cases} 1 \oplus x_3 \oplus x_1x_3 \oplus x_1x_2x_3, & c=0\\ \overline{x}_3 \oplus x_1 \oplus x_1\overline{x}_3 \oplus x_1x_2 \oplus x_1x_2\overline{x}_3, & c=1\\ 1 \oplus x_3 \oplus x_1\overline{x}_2x_3, & c=2\\ \overline{x}_3 \oplus x_1\overline{x}_2 \oplus x_1\overline{x}_2\overline{x}_3, & c=3\\ 1 \oplus x_2x_3 \oplus \overline{x}_1x_3 \oplus \overline{x}_1x_2x_3, & c=4\\ 1 \oplus x_2 \oplus x_2\overline{x}_3 \oplus \overline{x}_1 \oplus \overline{x}_1\overline{x}_3 & \\ \oplus\overline{x}_1x_2 \oplus \overline{x}_1x_2\overline{x}_3, & c=5\\ 1 \oplus x_3 \oplus \overline{x}_2x_3 \oplus \overline{x}_1\overline{x}_2x_3, & c=6\\ \overline{x}_3 \oplus \overline{x}_2 \oplus \overline{x}_2\overline{x}_3 \oplus \overline{x}_1\overline{x}_2 \oplus \overline{x}_1\overline{x}_2\overline{x}_3, & c=7 \end{cases}$
12.7.	$f = \widehat{\mathbf{X}}\ \mathbf{R}_{2^2}\ \mathbf{F} = [\ 1\ x_2\ x_1\ x_1x_2\ ]\left[\begin{array}{cc\|cc}1 & & & \\ 1 & 1 & & \\ \hline 1 & & 1 & \\ 1 & 1 & 1 & 1\end{array}\right]\begin{bmatrix}1\\ 0\\ 0\\ 1\end{bmatrix} = [\ 1\ x_2\ x_1\ x_1x_2\ ]\begin{bmatrix}1\\ 1\\ 0\\ 1\end{bmatrix}$ $= (1 \oplus x_2) \oplus x_1(x_2)$
12.8.	Solutions are given in Figure 12.16.
12.9.	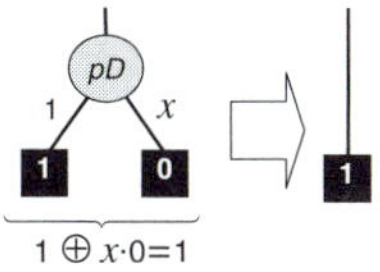 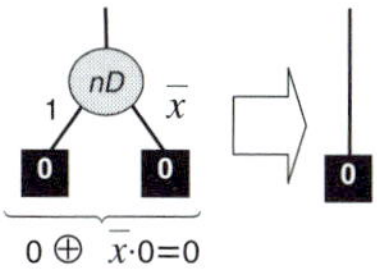 $1 \oplus x{\cdot}0 = 1$ $0 \oplus \overline{x}{\cdot}0 = 0$

Solutions to practice problems (continuation)

Practice problem	Solution
12.10.	$f_1 = x_1 \oplus x_2 \oplus x_1x_2$ $f_2 = 1 \oplus \overline{x}_1\overline{x}_2 = 1 \oplus (1 \oplus x_1)(1 \oplus x_2) = 1 \oplus 1 \oplus x_1 \oplus x_2 \oplus x_1x_2$ $= x_1 \oplus x_2 \oplus x_1x_2$ Hence, $f_1 = f_2$
12.11.	$f = x_1 \oplus x_2 \oplus x_3$ $f = 1 \oplus \overline{x}_1 \oplus \overline{x}_2 \oplus \overline{x}_3$ $= 1 \oplus 1 \oplus x_1 \oplus 1$ $\oplus\, x_2 \oplus 1 \oplus x_3$ $= x_1 \oplus x_2 \oplus x_3$

12.8 Problems

Problem 12.1 Prove that the following decision diagrams represent the same Boolean function:

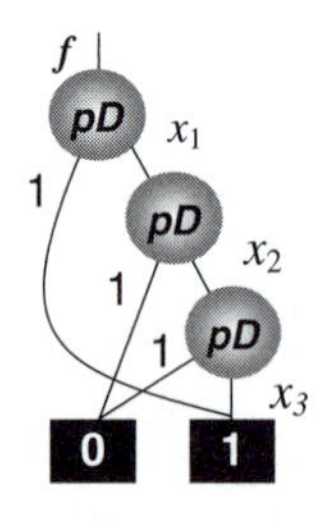

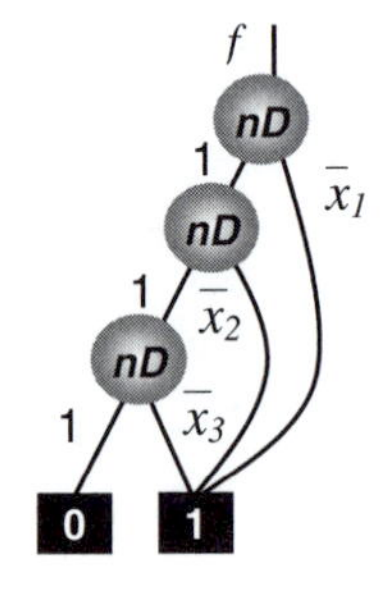

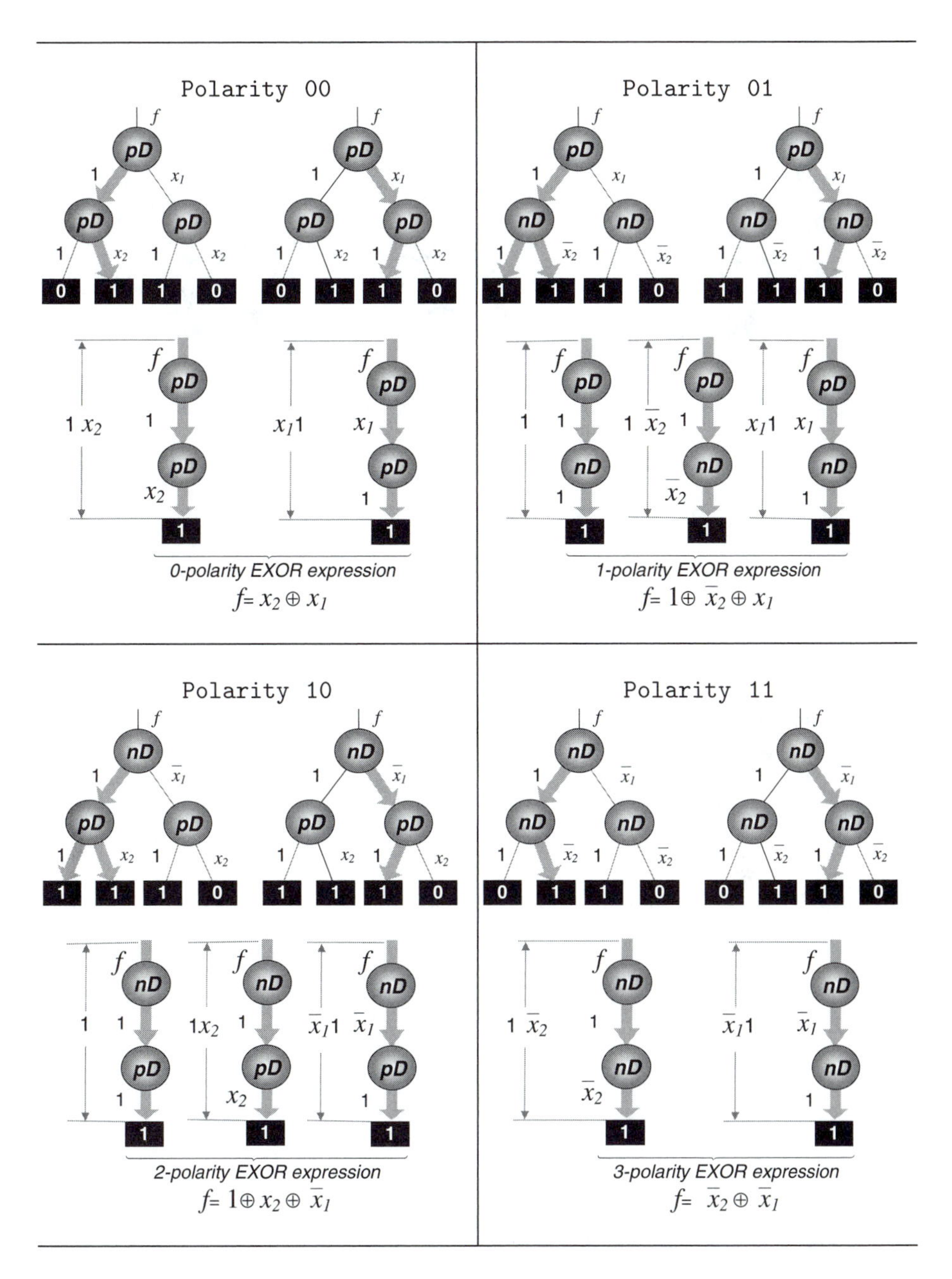

FIGURE 12.16
Polarized minterms and corresponding polynomial expressions in various polarities for Boolean functions of two variables (Practice problem 12.8).

Problem 12.2 Derive logic networks given the following functional decision diagrams:

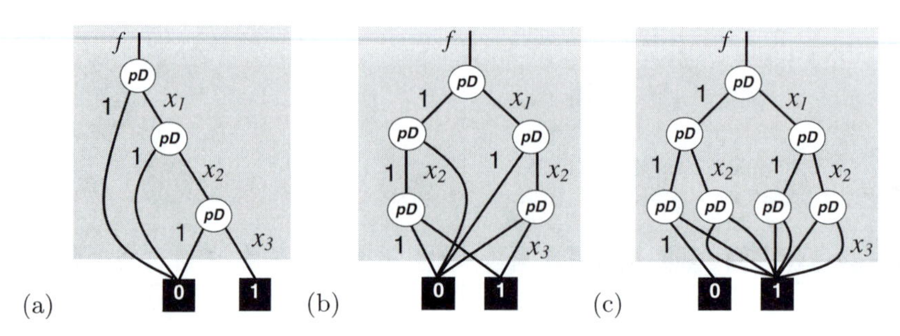

Problem 12.3 Derive functional decision diagrams for the following logic networks:

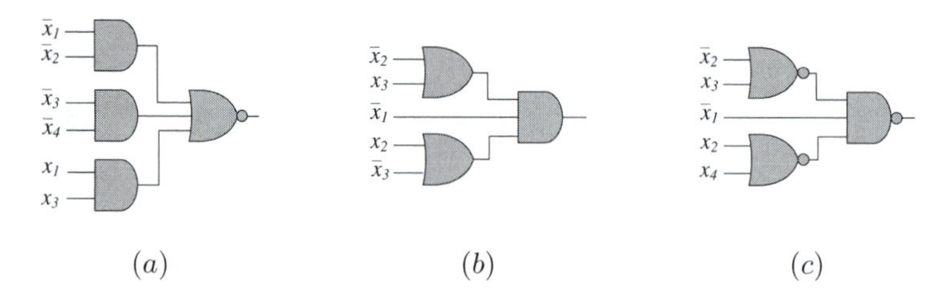

Problem 12.4 Derive functional decision diagrams for the following Boolean functions:

(a) $f = \overline{x}_1 \vee \overline{x}_2\overline{x}_3$

(b) $f = x_1 \vee x_2 \vee x_3$

(c) $f = x_1x_2 \vee x_1x_3 \vee x_2x_3$

(d) $f = x_1x_2x_3$

Problem 12.5 Compare decision diagrams based on Shannon and Davio expansions for the following logic networks given in Problem 12.3.

Problem 12.6 Simplify the the following functional decision diagrams:

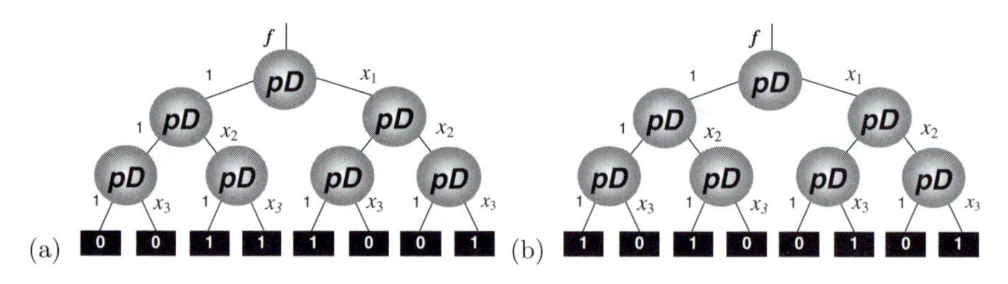

Problem 12.7 Derive functional decision diagrams given the following functional maps:

x_3x_4 \ x_1x_2	00	01	11	10
00		1		1
01	1		1	
11		1		1
10	1		1	

x_3x_4 \ x_1x_2	00	01	11	10
00	1	1		1
01	1	1	1	
11		1	1	1
10	1		1	1

Problem 12.8 Compare decision diagrams based on Shannon and Davio expansion for Boolean functions given the following K-maps:

x_3x_4 \ x_1x_2	00	01	11	10
00		1	1	1
01	1	1	1	
11	1	1		1
10	1		1	

x_3x_4 \ x_1x_2	00	01	11	10
00	1	1		1
01	1		1	1
11		1		1
10	1	1	1	

Problem 12.9 Optimize the following multiplexer-based logic networks:

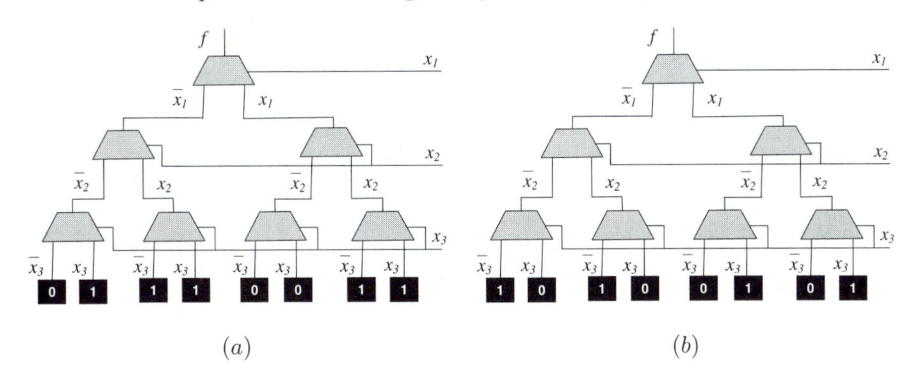

(a) (b)

Problem 12.10 Simplify the following logic networks using algebraic manipulations:

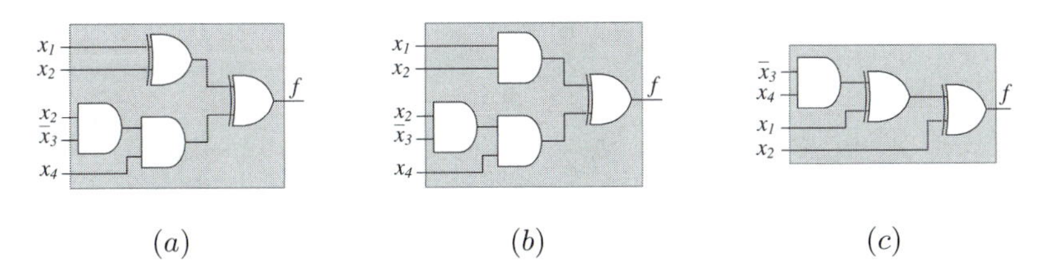

(a) (b) (c)

Problem 12.11 Derive functional decision diagrams using various orders of variables given the following truth table of a Boolean function:

x_1	x_2	x_3	x_4	f	x_1	x_2	x_3	x_4	f
0	0	0	0	**0**	1	0	0	0	**0**
0	0	0	1	**0**	1	0	0	1	**1**
0	0	1	0	**1**	1	0	1	0	**0**
0	0	1	1	**1**	1	0	1	1	**1**
0	1	0	0	**0**	1	1	0	0	**0**
0	1	0	1	**0**	1	1	0	1	**0**
0	1	1	0	**1**	1	1	1	0	**0**
0	1	1	1	**0**	1	1	1	1	**1**

Problem 12.12 Derive functional decision diagrams for the totally symmetric Boolean functions given the following truth tables:

(a)

x_1	x_2	x_3	f
0	0	0	**1**
0	0	1	**0**
0	1	0	**0**
0	1	1	**1**
1	0	0	**0**
1	0	1	**1**
1	1	0	**1**
1	1	1	**0**

(b)

x_1	x_2	x_3	f
0	0	0	**1**
0	0	1	**1**
0	1	0	**1**
0	1	1	**0**
1	0	0	**1**
1	0	1	**0**
1	1	0	**0**
1	1	1	**0**

(c)

x_1	x_2	x_3	f
0	0	0	**1**
0	0	1	**0**
0	1	0	**0**
0	1	1	**0**
1	0	0	**0**
1	0	1	**0**
1	1	0	**0**
1	1	1	**1**

Problem 12.13 Represent the logic networks given in Problem 12.10 using functional decision diagrams.

Problem 12.14 Derive the functional decision diagrams given the standard SOP and POS expressions ($n = 3$):

(a) $f = \bigvee m(0, 1, 2, 6, 7)$
(b) $f = \bigvee m(3, 5, 6)$
(c) $f = \prod M(1, 4, 6)$
(d) $f = \prod M(2, 3, 5, 6, 7)$

Problem 12.15 Construct functional decision diagrams for the following Boolean functions of three variables:
(a) AND function $x_1x_2x_3$
(b) NAND function $\overline{x_1x_2x_3}$
(c) OR function $\overline{x}_1 \vee x_2 \vee x_3$
(d) NOR function $\overline{x_1 \vee x_2 \vee x_3}$
(e) EXOR function $x_1 \oplus x_2 \oplus x_3$

13

Standard Modules of Combinational Networks

Data transfer logic

- Shared data paths
- Multiplexers and demultiplexers
- Word-level multiplexers
- Design using Shannon expansion theorem
- Multiplexer and demultiplexer trees

Boolean function computing

- Computing using multiplexers
- Computing using demultiplexers
- Implementation of decision trees and diagrams

Encoders, decoders, and code converters

- Design
- Applications

Advanced topics

- Threshold elements
- Multi-valued logic elements

13.1 Introduction

Complex digital system design is based on a modular principle, that is, systems are built using *standard* modules. These modules (basic blocks, components) correspond to subfunctions (subtasks) of the system's functionality. The set of subfunctions has been identified as useful for a large variety of logic networks, and its elements are available as *library* components. The design of a system using standard modules consists of decomposition of the overall function of the system into subfunctions of standard components, and the interconnection of these components.

The concept of the *assembly* of complex computing logic networks from basic elements is well developed in contemporary logic design and widely implemented based on advanced technology. This concept is also intensively developed for predictable technologies. In this chapter, the following standard modules are introduced: multiplexers and demultiplexers, encoders and decoders, code converters, comparators, and shifters.

13.2 Data transfer logic

Data transfer logic aims at controlling the transfer of binary data between several points in a logic network. This is a *moving* or *copying* operation, in which no changes are made to the logic values of signals; that is, no data processing takes place. Data transfer logic is useful for proving data-paths between logic networks and for implementing designs for Boolean functions.

13.2.1 Shared data path

In data transfer logic, inputs are associated with *data sources* and outputs are associated with *data destinations.* There are two typical tasks of data transfer (Figure 13.1):

- Several data sources have a common destination, or
- One source must send data to several destinations.

Because a particular destination can take signals from only one source at a time, the source-to-destination data transfer paths must be shared to some degree. Shared data paths are also employed primarily to reduce the number of lines needed for communication purposes. The devices that enable the communication of data are the *multiplexers* and the *demultiplexers.*

The multiplexer allows only one of many sources to be logically connected to a given common destination at any time (Figure 13.1a). Multiplexers are

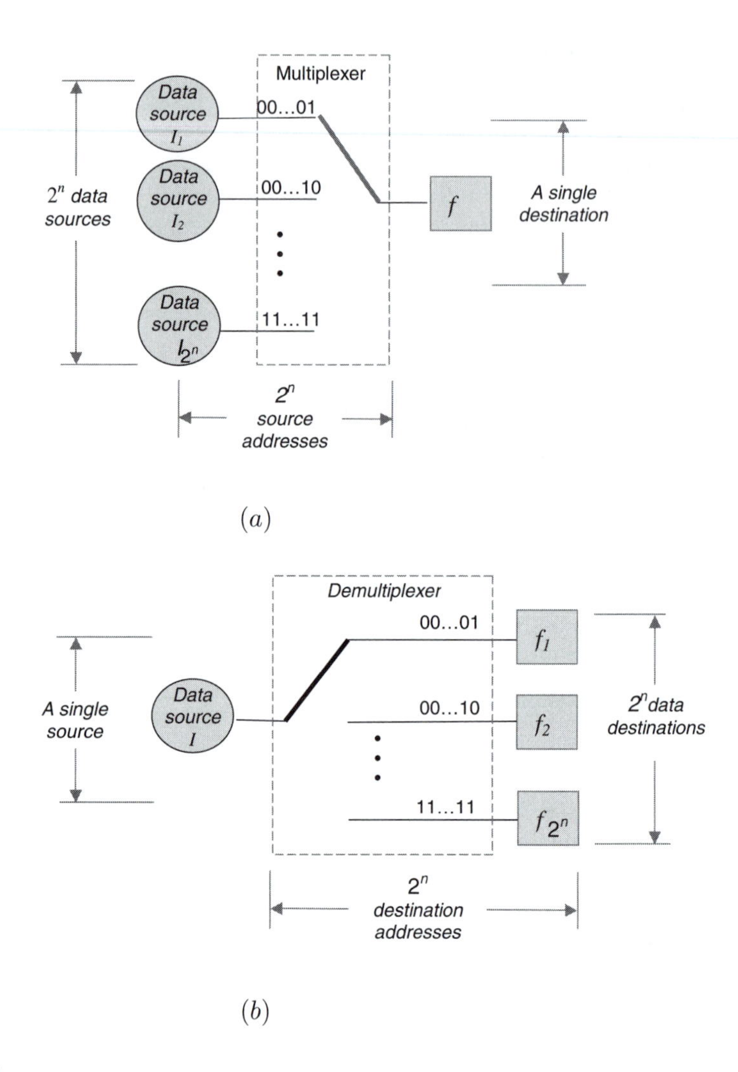

FIGURE 13.1
Data transfer connections: a switch model of the multiplexer for transferring data from one source to many simultaneous destinations (a), and a model of the demultiplexer, that transfers data from one source to one of many destinations (b).

used to implement designs for Boolean functions and to provide data-flow paths between logic networks. An arbitrary Boolean function of n variables can be implemented by using at most $2^n - 1$ 2-to-1 multiplexers. The inverse of a multiplexer is a 1-to-2^n *demultiplexer* (Figure 13.1b), which serves to connect a common source I to one of 2^n destinations:

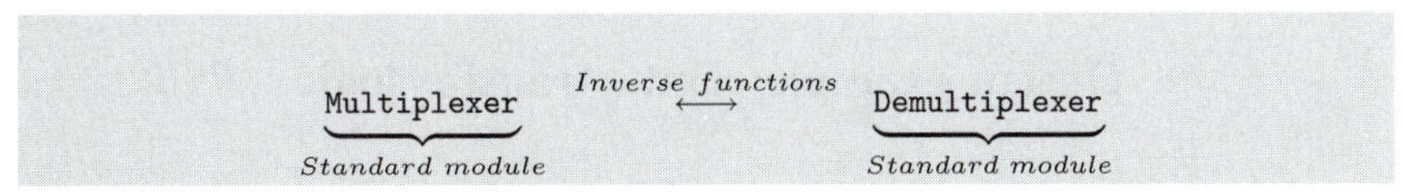

A set of selected bits (addresses) specifies which destination is to be selected at any time. The simplest demultiplexer, which connects the common source to one of two destinations, and is called a *single-bit* or *1-to-2* demultiplexer. Also, a *decoder* with an enable input is called a demultiplexer. Decoder is a logic network that converts a binary code applied to n input lines to one of 2^n different outputs lines.

Logic symbols for the multiplexer (MUX) and demultiplexer (DMUX) are shown in Figure 13.2a,b. In Figure 13.2c, the multiplexer's output is fed into the demultiplexer input and creates a router network. Depending on the settings of the control signals, any one input can be routed to any one output. This router network is not a full crossbar switch because at any time only one output is active. This is because a 1-bit bus is used for the transmission of 2^n logic signals.

13.2.2 Multiplexer

The multiplexer is specified as follows:

Multiplexer design

- A logic network called a 2^n-to-1 *multiplexer* or *data selector* allows one and only one of 2^n sources to be logically connected to a common destination f at a time.
- The source I_i, $i = 1, 2, ..., \ldots 2^n$, is transferred to the output if the address input is i.
- The address signal is a binary integer i, which controls the selection of the desired source I_i.
- By changing the address from i to j, $i, j \in \{1, 2, ..., \ldots 2^n\}$, the input source for the multiplexer is changed from I_i to I_j.

Let the number of sources be $k = 2^n, k \in \{1, 2, \ldots, K\}$. Then, n address bits allow 2^n distinct addresses to be specified. The output of the multiplexer is defined by the following relationship of its data input and the address:

Multiplexer

$$f = \bigvee_{i=0}^{2^n-1} \underset{\substack{\uparrow \\ In}}{I_{i+1}} \cdot \underbrace{S_1^{i_1} S_2^{i_2} \cdots S_n^{i_n}}_{Address} \tag{13.1}$$

where

$$S_j^{i_j} = \begin{cases} \overline{S}_j, \text{ if } i_j = 0 \\ S_j, \text{ if } i_j = 1 \end{cases}$$

Design example: Data transfer logic

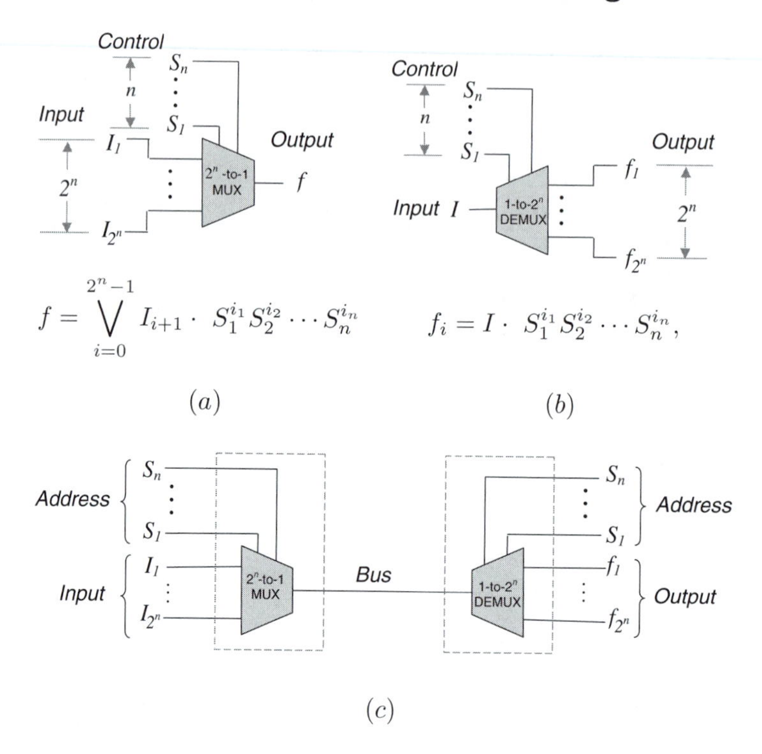

FIGURE 13.2
Data transfer logic and its formal description: 2^n-to-1 multiplexer (a), 1-to-2^n demultiplexer (b), and a router network (c).

The address $S_1^{i_1} S_2^{i_2} \cdots S_n^{i_n}$ in Equation 13.1 is a minterm, or product of n select variables, S_j, $j = 1, 2, \ldots, n$, and the corresponding data input I_{i+1}, $i = 0, 1, \ldots, 2^n - 1$.

Example 13.1 (Multiplexer.) *It follows from Equation 13.1 that 2^n-to-1 multiplexer have 2^n sources of data and $n = \log_2 2^n$ selected lines. 2-to-1, 4-to-1, and 8-to-1 multiplexers transmit 2, 4, and 8 inputs to one output, respectively (Table 13.1).*

Practice problem **13.1. (Multiplexer.)** Calculate the number of selected lines for 16-to-1 and 32-to-1 multiplexers.
Answer is given in "Solutions to practice problems."

TABLE 13.1
Multiplexers and their formal descriptions (Example 13.1).

Design techniques for multiplexers

Graphical representation | **Formal description**

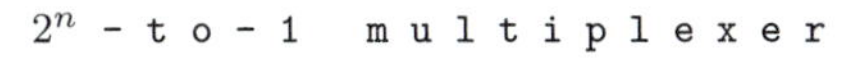

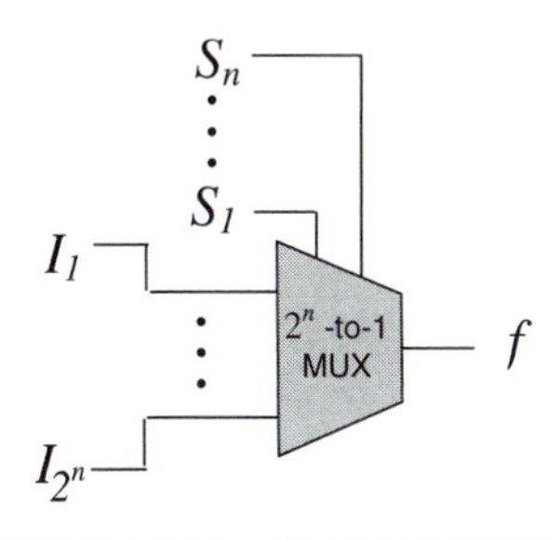

$$f = \bigvee_{i=0}^{2^n-1} I_{i+1} \cdot \overbrace{S_1^{i_1} S_2^{i_2} \cdots S_n^{i_n}}^{n-bit\ address}$$

---

2 - to - 1 multiplexer

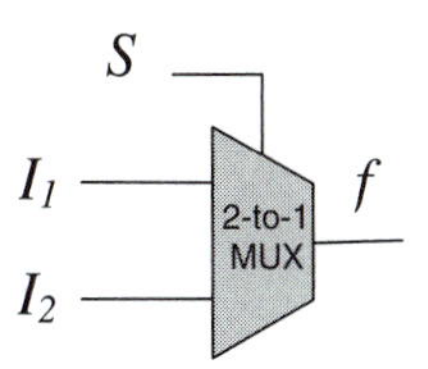

$$f = \bigvee_{i=0}^{2^1-1} I_{i+1} S$$

$$= I_1 \overline{S} \vee I_2 S = \begin{cases} I_1, \text{ if } S = 0; \\ I_2, \text{ if } S = 1. \end{cases}$$

---

4 - to - 1 multiplexer

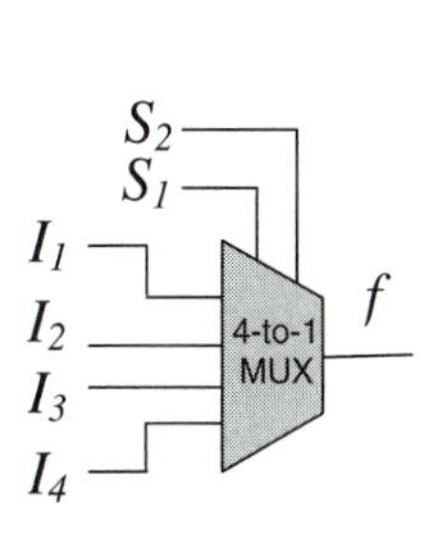

$$f = \bigvee_{i=0}^{2^2-1} I_{i+1} \cdot S_1^{i_1} S_2^{i_2}$$

$$= I_1 \overline{S}_1 \overline{S}_2 \vee I_2 \overline{S}_1 S_2 \vee I_3 S_1 \overline{S}_2 \vee I_4 S_1 S_2$$

$$= \begin{cases} I_1, \text{ if } S_1 S_2 = 00 \\ I_2, \text{ if } S_1 S_2 = 01 \\ I_3, \text{ if } S_1 S_2 = 10 \\ I_4, \text{ if } S_1 S_2 = 11 \end{cases}$$

---

8 - to - 1 multiplexer

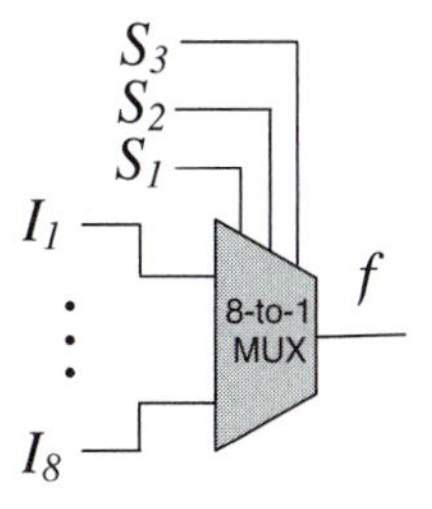

$$f = \bigvee_{i=0}^{2^3-1} I_{i+1} S_1^{i_1} S_2^{i_2} S_3^{i_3}$$

$$= I_1 \overline{S}_1 \overline{S}_2 \overline{S}_3 \vee I_2 \overline{S}_1 \overline{S}_2 S_3 \vee I_3 \overline{S}_1 S_2 \overline{S}_3$$

$$\vee\ I_4 \overline{S}_1 S_2 S_3 \vee I_5 S_1 \overline{S}_2 \overline{S}_3 \vee I_6 S_1 \overline{S}_2 S_3$$

$$\vee\ I_7 S_1 S_2 \overline{S}_3 \vee I_8 S_1 S_2 S_3$$

13.2.3 Multiplexers and Shannon expansion theorem

Consider the Shannon expansion $f = \overline{x}_i f_0 \vee x_i f_1$. This expansion means that if $x_i = 0$, then f_0 is selected; and if $x_i = 1$, then f_1 is selected. Thus, the variable x_i in a Shannon expansion can be interpreted as the select input to a 2-to-1 multiplexer (Figure 13.3) that has f_0 and f_1 as the data inputs:

$$f_0 = f(x_1, \ldots, x_{i-1} \underset{\text{Select}}{\boxed{0}} x_{i+1}, \ldots, x_n)$$

$$f_1 = f(x_1, \ldots, x_{i-1} \underset{\text{Select}}{\boxed{1}} x_{i+1}, \ldots, x_n)$$

Consider the Shannon expansion with respect to two variables:

$$f = \overline{x}_1\overline{x}_2 f_{00} \vee \overline{x}_1 x_2 f_{01} \vee x_1\overline{x}_2 f_{10} \vee x_1 x_2 f_{11}$$

If $x_1, x_2 = 0, 0$, then f_{00} is chosen, and for the remaining selections of x_1, x_2 (01,10 and 11), the values f_{01}, f_{10}, and f_{11} are chosen, respectively. This corresponds to the function of the 4-to-1 multiplexer (Figure 13.3).

13.2.4 Single-bit (2-to-1) multiplexer

For a 2-to-1 multiplexer, only one address bit S is needed to identify the selected source: $S = 0$ selects I_0 to connect to output, and $S = 1$ selects I_1. The single-bit multiplexer is defined by the equation

2-to-1 multiplexer

$$f = I_0\overline{S} \vee I_1 S \tag{13.2}$$

In this multiplexer, the output is equal I_0 when S is 1, and it is equal to I_1 when S is 0.

Example 13.2 (Multiplexer.) *The single-bit multiplexer is shown in Figure 13.4. In this circuit, two AND gates determine whether to pass their respective data inputs to the OR gate. The upper AND gate passes signal I_0 when S is 0 (since the other input to the gate is $\overline{S}$). The lower AND gate passes signal I_1 when S is 1.*

Practice problem 13.2. (Multiplexer.) Design a single-bit multiplexer given the input signals x_1, x_2, and x_3 using one of them as a selected signal.
Answer is given in "Solutions to practice problems."

Design techniques for multiplexers

Shannon expansion
with respect to the variable x_i

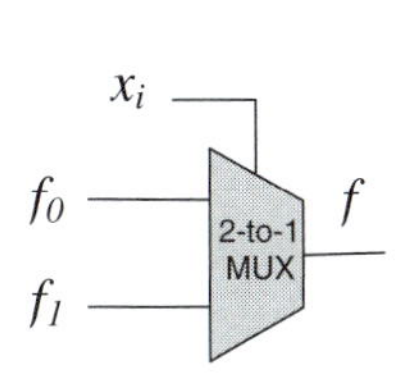

$$f = \overline{x}_i f\overbrace{(x_1, \ldots, x_{i-1} 0\, x_{i+1}, \ldots, x_n)}^{n-1\ variables}$$
$$\vee\ x_i f\underbrace{(x_1, \ldots, x_{i-1} 1\, x_{i+1}, \ldots, x_n)}_{n-1\ variables}$$
$$= \overline{x}_i f_0 \vee x_i f_1$$

where

$$f_0 = f(x_1, \ldots, x_{i-1} 0\, x_{i+1}, \ldots, x_n)$$
$$f_1 = f(x_1, \ldots, x_{i-1} 1\, x_{i+1}, \ldots, x_n)$$

Shannon expansion
with respect to a group of variables

$$f = \bigvee_{j=0}^{m} f_j\ x_1 x_2 \ldots \underbrace{x_{i_1}^{c_{i_1}} x_{i_2}^{c_{i_2}} \ldots x_{i_m}^{c_{i_m}}}_{Control\ variables} \cdots\ x_{n-1} x_n$$

where $j = c_{i_1} c_{i_2} \ldots c_{i_m}$,

$$x_j^{c_j} = \begin{cases} x_j, & c_j = 1 \\ \overline{x}_j, & c_j = 0 \end{cases}$$

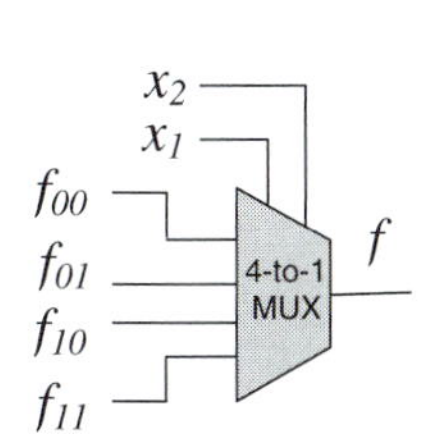

Let $m = 2^2 - 1 = 3$, and let the group of variables x_1 and x_2 be chosen for expansion. Then

$$f = f_{00} \vee f_{01} \vee f_{10} \vee f_{11},$$
$$f_{00} = \overline{x}_1 \overline{x}_2 f(0, 0, x_3, \ldots, x_n)$$
$$f_{01} = \overline{x}_1 x_2 f(0, 1, x_3, \ldots, x_n)$$
$$f_{10} = x_1 \overline{x}_2 f(1, 0, x_3, \ldots, x_n)$$
$$f_{11} = x_1 x_2\, \underbrace{f(1, 1, x_3, \ldots, x_n)}_{n-2\ inputs}$$

FIGURE 13.3
Multiplexers implement Shannon expansion with respect to a variable or a group of variables.

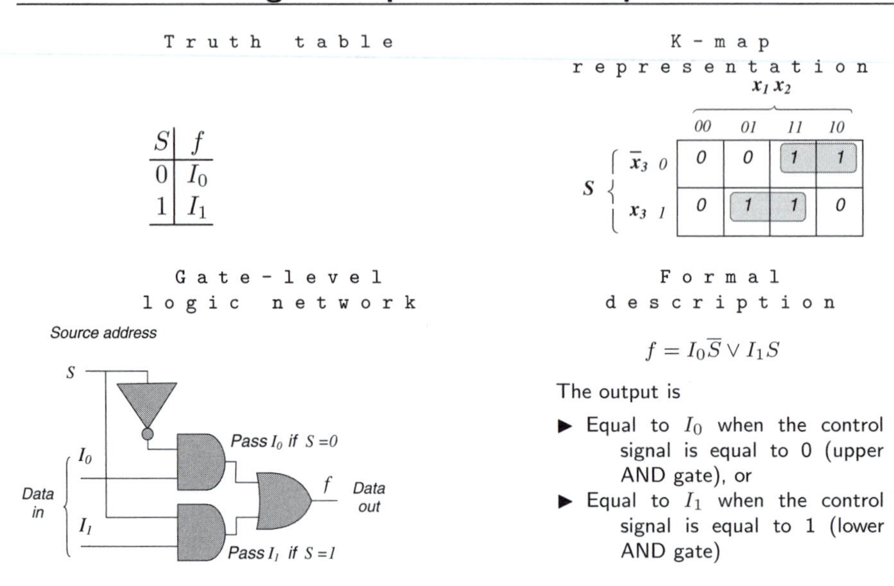

S	f
0	I_0
1	I_1

FIGURE 13.4
Single-bit (2-to-1) multiplexer: description using the truth table and K-map, and its gate-level implementation (Example 13.2).

13.2.5 Word-level multiplexer

A single word-level multiplexer generates a word **F** equal to one of the two input words, **X** and **Y**, depending on the control input bit S. There are multiplexers that allow one the selection of a word from a number of sources depending on the control signals:

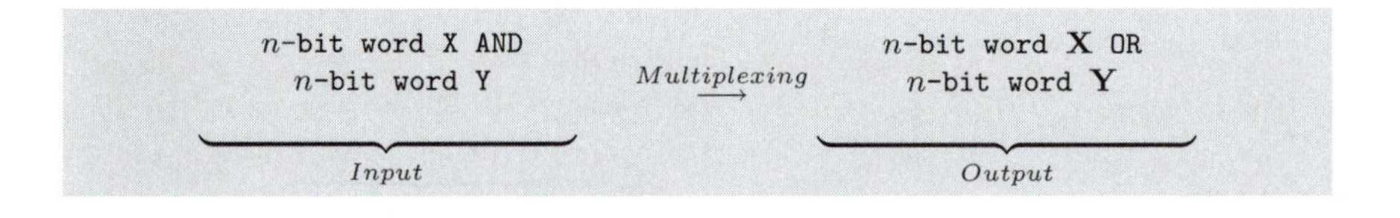

Example 13.3 (Multiplexer.) *Given the 4-bit words* $\mathbf{X} = \{x_1x_2x_3x_4\}$ *and* $\mathbf{Y} = \{y_1y_2y_3y_4\}$ *design a logic network for the selection of one of the words,* **X** *or* **Y**, *using the control signal* $S \in \{0, 1\}$ *(0 for selecting* **X** *and 1 selecting* **Y**) *and 2-to-1 multiplexers. The solution is shown in Figure 13.5a. The four logic signals that perform a common function can be grouped together to form a bus (Figure 13.5b).*

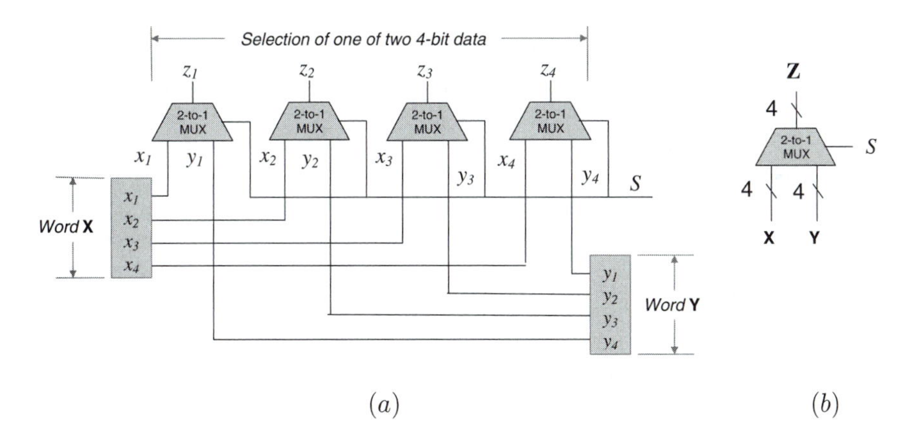

FIGURE 13.5
A logic network of 2-to-1 multiplexers for selecting one of two 4-bit words (a), and one of the multibit multiplexers (b) (Example 13.3).

In Equation 13.1, data inputs and control variables can be functions themselves. That is, they can be connected to other networks.

13.3 Implementation of Boolean functions using multiplexers

In the design of a large multiplexer-based network, the following resources can be used:

- Connections of multiplexers in a multiplexer tree,
- Extension of the number of selection (address) lines, and
- Additional logic gates.

There are two approaches to the design of networks for computing Boolean functions using multiplexers:

- Decomposition of the truth table, and
- Shannon expansion of the implemented Boolean function.

If a Boolean function is given by its truth table or a truth vector, then the first approach is appropriate. However, this approach is limited by the size of the truth table. The second approach utilizes the flexibility of the Shannon expansion, and is acceptable for algebraic representations of Boolean functions of many variables.

Approach 1: Algorithm for computing Boolean functions using multiplexers

Given: A Boolean function of n variables given by its truth vector.
Step 1: Choose k variables for selected lines.
Step 2: Divide the truth vector into 2^k groups of $2^{n-k} \times 1$ subvectors.
Step 3: These subvectors are the truth vectors of the functions that are the inputs of the multiplexer.
Step 4: Repeat steps 1-3 for each function obtained in step 3 until the new function (cofactors) contains less than $k + 1$ variables.
Output: A logic network of multiplexers

It should be noted that if the cofactors are functions of more than one variable but less than k (where $k = 2, 3, \ldots, K$), then AND, OR, and others additional logic gates can be used to form the inputs of the first level of multiplexer.

13.3.1 Multiplexer tree

The 2^n-to-n multiplexer can implement an arbitrary Boolean function of n variables. This is because any element of the input truth vector of a Boolean function can be transferred to the output using the mechanism of addressing. The disadvantage of this implementation of a Boolean functions is that the number of inputs is equal to 2^n. To resolve this problem, a 2^n-to-1 multiplexer can be represented as a multilevel network of 2^m-to-1 multiplexers, $m < n$. This network is called a *multiplexer tree.*

Example 13.4 (Connection of multiplexers.) *In Figures 13.6a,b two 2-to-1 multiplexers are combined to provide the desired functionalities. Figure 13.6c shows a multiplexer tree.*

Practice problem **13.3. (Connection of multiplexers.)** Given $f_1 = I_1\overline{S}_1 \vee I_2 S_1$ and $f_2 = I_3\overline{S}_1 \vee I_4 S_1$, construct the multiplexer-based logic network to implement the function $f = I_5\overline{f}_1\overline{f}_2 \vee I_6\overline{f}_1 f_2 \vee I_7 f_1\overline{f}_2 \vee I_8 f_1 f_2$.
Answer is given in "Solutions to practice problems."

Example 13.5 (Multiplexer tree.) *In Figure 13.7, a two-level multiplexer tree using 4-to-1 multiplexers is shown. This network realizes the function of a 16-to-1 multiplexer.*

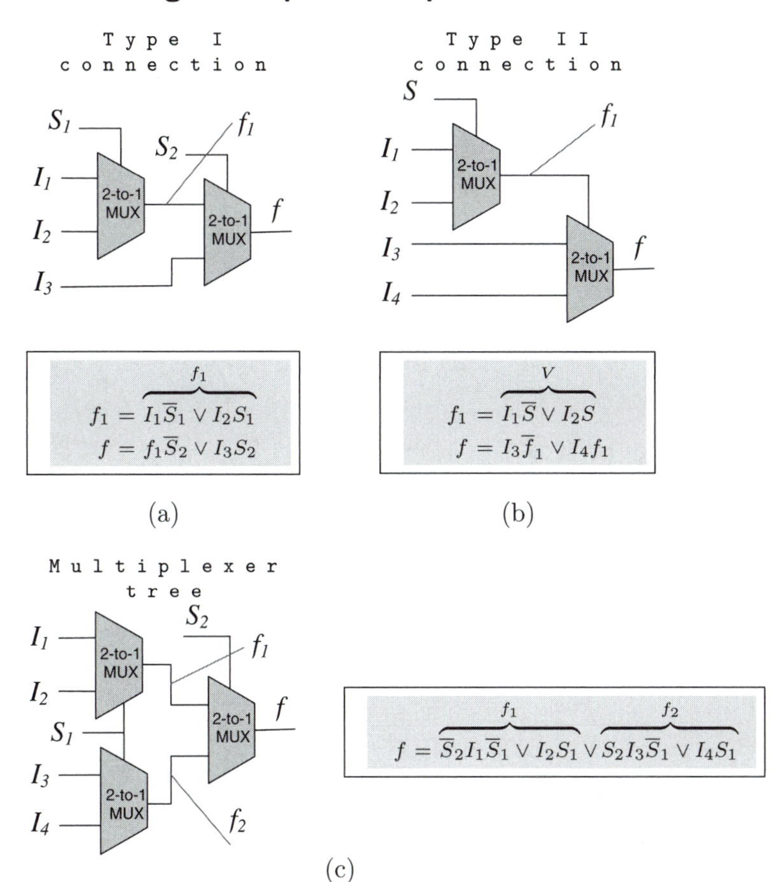

FIGURE 13.6
Connection of 2-to-1 multiplexers (Example 13.4).

13.3.2 Multiplexers and Shannon expansion theorem

Approach 2: Algorithm for computing Boolean functions using Shannon expansion

Given: A Boolean function of n variables given by its truth-vector.
Step 1: Choose k variables for selected lines.
Step 2: Divide the truth-vector into 2^k groups of $2^{n-k} \times 1$ subvectors.
Step 3: These subvectors are the truth-vectors of the functions that are the inputs of the multiplexer.
Step 4: Repeat steps 1-3 for each function obtained in step 3 until the new function (cofactors) contain less than $k + 1$ variables.
Output: A logic network of multiplexers

Design example: Multiplexer tree

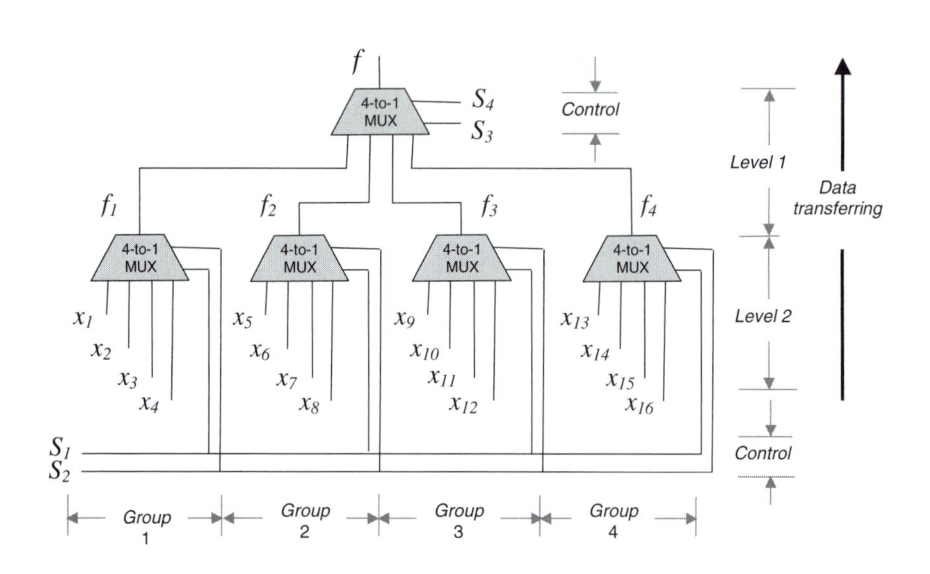

Formal description of the tree network based on 4-to-1 multiplexers

The first level of the tree network is described for each 4-to-1 multiplexer:

$$f_1 = \bigvee_{i=0}^{2^2-1} x_{i+1} S_1^{i_1} S_2^{i_2} \quad f_2 = \bigvee_{i=0}^{2^2-1} x_{i+4} S_1^{i_1} S_2^{i_2}$$

$$f_3 = \bigvee_{i=0}^{2^2-1} x_{i+8} S_1^{i_1} S_2^{i_2} \quad f_4 = \bigvee_{i=0}^{2^2-1} x_{i+11} S_1^{i_1} S_2^{i_2}$$

The second level of the tree network,

$$f = \bigvee_{i=0}^{2^2-1} f_{i+1} S_3^{i_1} S_4^{i_2}$$

FIGURE 13.7
Multiplexer tree for implementing a 16-to-1 multiplexer using 4-to-1 multiplexers (Example 13.5).

The number of data inputs to the multiplexer can be reduced if the inputs are from the set of constants and literals, $\{0, 1, x_i^{i_j}\}$. In this way, a 2^n-input multiplexer can implement a Boolean function of $n+1$ variables.

Example 13.6 (Multiplexer-based network.) *Given a Boolean function of three variables $f = \bigvee m(1, 2, 4, 6, 7)$, let us implement it using an 8-to-1 multiplexer (Figure 13.8a). The standard SOP expression for f is factored as follows:*

$$\begin{aligned} f &= \bigvee m(1,2,4,6,7) \\ &= \overline{x}_3(x_1\overline{x}_2) \vee \overline{x}_3(\overline{x}_1 x_2) \vee x_3(\overline{x}_1\overline{x}_2) \vee x_3(\overline{x}_1 x_2) \vee x_3(x_1 x_2) \\ &= \overline{x}_3(x_1\overline{x}_2) \vee \underbrace{(\overline{x}_3 \vee x_3)}_{Equal\ to\ 1}(\overline{x}_1 x_2) \vee x_3(\overline{x}_1 x_2) \vee x_3(\overline{x}_1\overline{x}_2) \vee x_3(x_1 x_2) \\ &= \overline{x}_3(x_1\overline{x}_2) \vee 1 \cdot (\overline{x}_1 x_2) \vee x_3(\overline{x}_1 x_2) \vee x_3(\overline{x}_1\overline{x}_2) \vee x_3(x_1 x_2) \end{aligned}$$

A K-map interpretation of this transform and its implementation using a 4-to-1 multiplexer are given in Figures 13.8b,c.

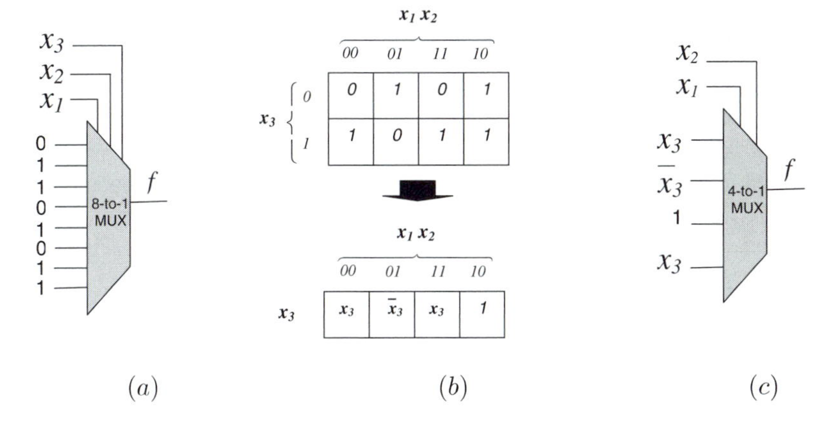

FIGURE 13.8
Interpretation of the Shannon expansion theorem for the reduction of the number of inputs in the multiplexer for implementing a Boolean function (Example 13.6).

13.3.3 Combination of design approaches using multiplexers

Example 13.7 (Design techniques.) *Design techniques for computing Boolean functions using multiplexers are given in Tables 13.2 and 13.3.*

Design example 1

Implementation of the EXOR function of two variables can be done via pairwising its truth table rows. This yields two cofactors, which are mapped into the 2-to-1 multiplexer directly. If a Boolean function is given in algebraic form, a solution is provided by the application of the Shannon expansion with respect to the variable x_1, which plays the role of the selected line of the multiplexer (Table 13.2).

Example 13.8 *(Continuation of Example 13.7.) To implement a two-input EXOR function using a 2-to-1 multiplexer, let us use the Shannon expansion* $f = x_1 \oplus x_2 = \overline{x}_1 f_0 \vee x_1 f_1$*, where* $f_0 = x_2$ *and* $f_1 = \overline{x}_1$*. The network solution is given in Table 13.2.*

Practice problem 13.4. (Design techniques.) Implement the AND function of two variables using a 2-to-1 multiplexer.
Answer is given in "Solutions to practice problems."

Design example 2

Using the first approach, a Boolean function of three variables $f = x_1x_2 \vee x_2x_3 \vee x_1x_3$ is implemented using the 4-to-1 multiplexer via decomposition of the truth table into four rows. The same result can be obtained using the Shannon expansion with respect to the variables x_1 and x_2, which become the selected lines in the 4-to-1 multiplexer. For this, the Boolean function must be represented in algebraic form, while the K-map is useful for decreasing the number of terms (Table 13.2).

Practice problem 13.5. (Design techniques.) Let a Boolean function be given by the truth vector $\mathbf{F} = [01101001]^T$. Implement this function using a 4-to-1 multiplexer.
Answer is given in "Solutions to practice problems."

Design example 3

The Boolean function of five variables $f = x_1 \oplus x_2 \oplus x_3 \oplus x_4$ is given in algebraic form; namely, in polynomial form. Shannon expansion can be applied to this

TABLE 13.2
A logic network design techniques for computing Boolean functions using multiplexers.

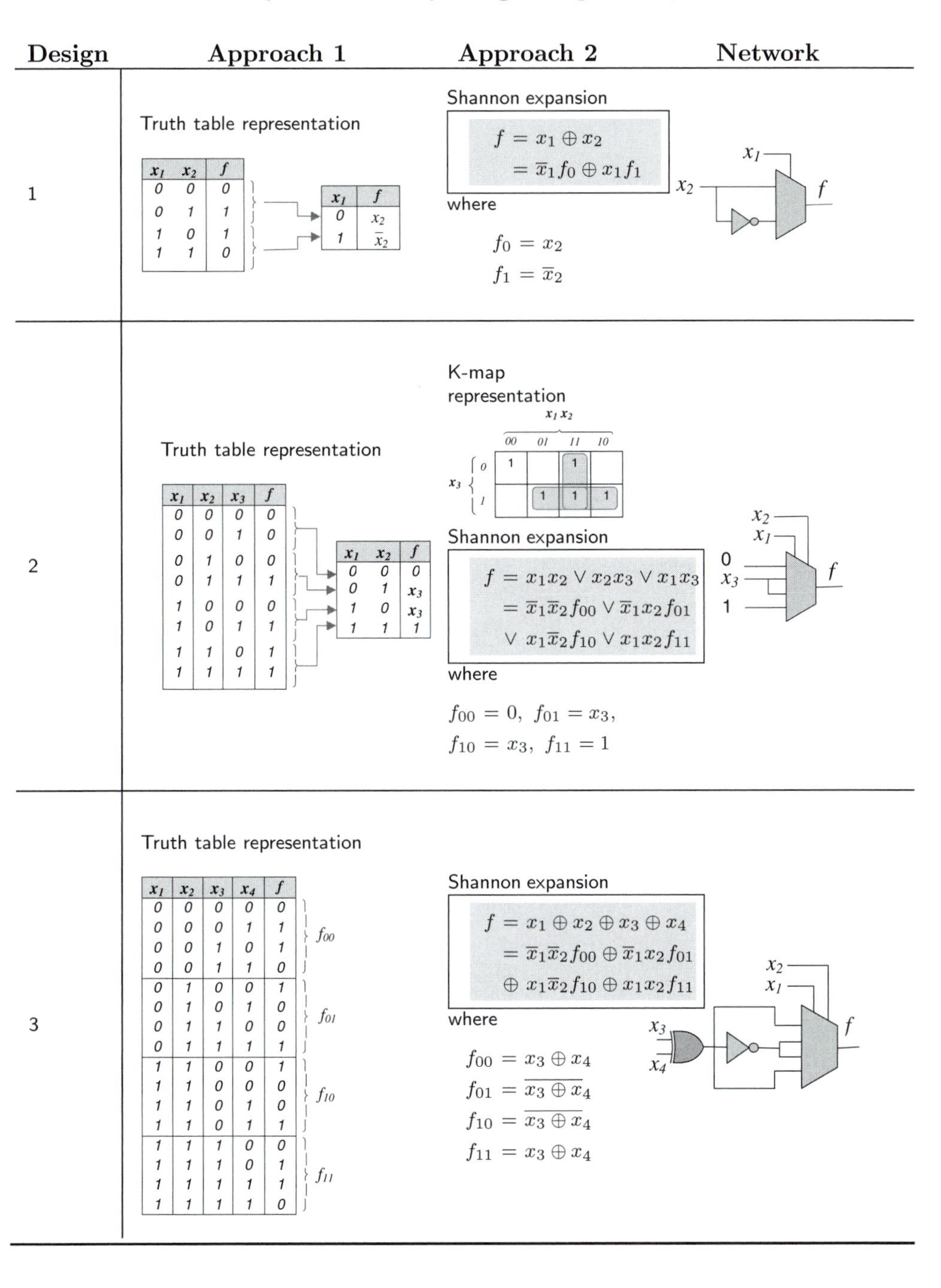

x_1	x_2	f
0	0	0
0	1	1
1	0	1
1	1	0

x_1	f
0	x_2
1	$\overline{x}_2$

x_1	x_2	x_3	f
0	0	0	0
0	0	1	0
0	1	0	0
0	1	1	1
1	0	0	0
1	0	1	1
1	1	0	1
1	1	1	1

x_1	x_2	f
0	0	0
0	1	x_3
1	0	x_3
1	1	1

x_1	x_2	x_3	x_4	f
0	0	0	0	0
0	0	0	1	1
0	0	1	0	1
0	0	1	1	0
0	1	0	0	1
0	1	0	1	0
0	1	1	0	0
0	1	1	1	1
1	1	0	0	1
1	1	0	0	0
1	1	0	1	0
1	1	0	1	1
1	1	1	0	0
1	1	1	0	1
1	1	1	1	1
1	1	1	1	0

TABLE 13.3
Logic network design techniques for computing Boolean functions using multiplexers (continuation of Table 13.2).

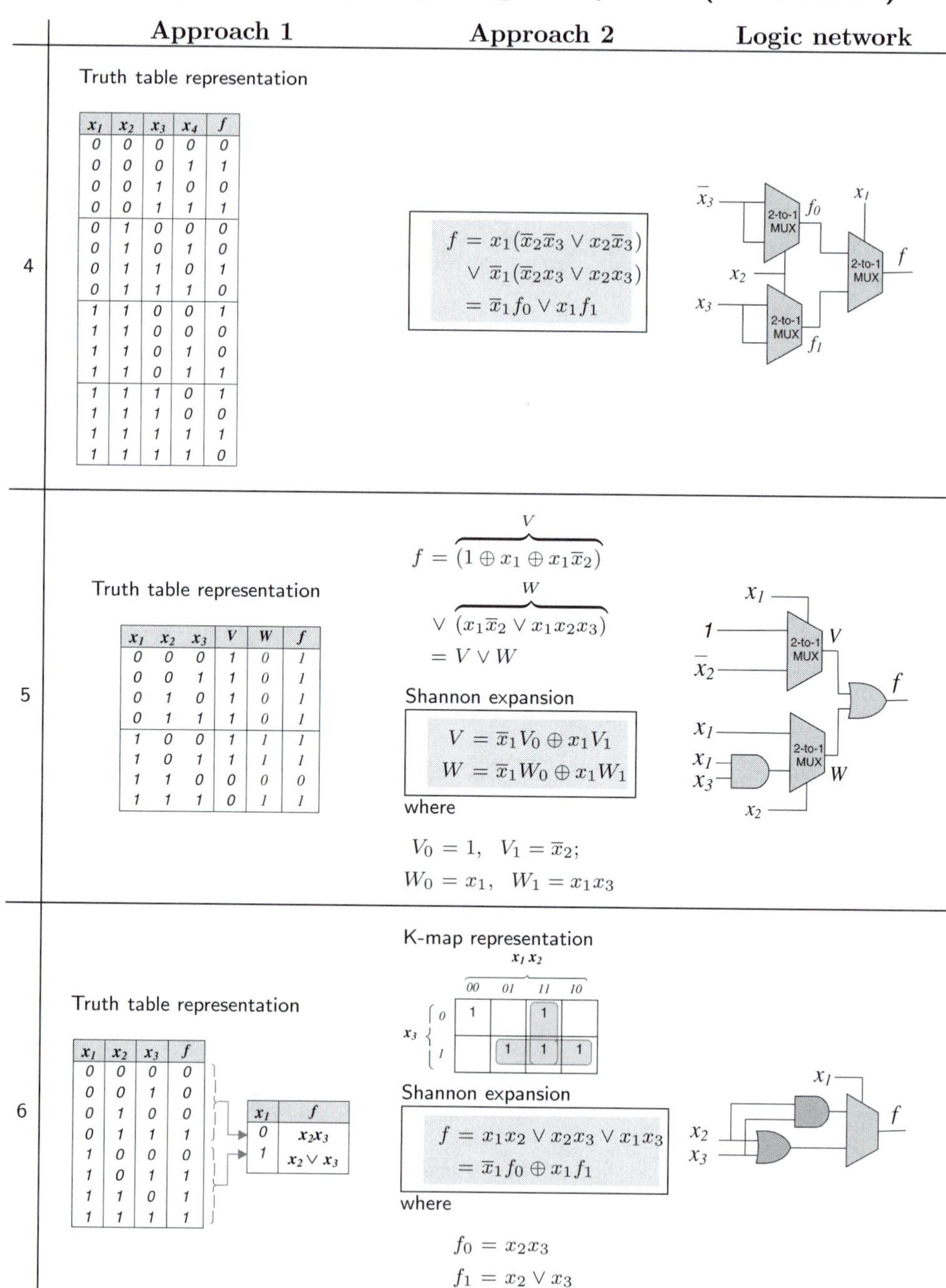

Techniques for computing using multiplexers (continuation)

Approach 1 | **Approach 2** | **Logic network**

4

Truth table representation

x_1	x_2	x_3	x_4	f
0	0	0	0	0
0	0	0	1	1
0	0	1	0	0
0	0	1	1	1
0	1	0	0	0
0	1	0	1	0
0	1	1	0	1
0	1	1	1	0
1	1	0	0	1
1	1	0	0	0
1	1	0	1	0
1	1	0	1	1
1	1	1	0	1
1	1	1	0	0
1	1	1	1	1
1	1	1	1	0

$$f = x_1(\overline{x}_2\overline{x}_3 \vee x_2\overline{x}_3) \vee \overline{x}_1(\overline{x}_2 x_3 \vee x_2 x_3) = \overline{x}_1 f_0 \vee x_1 f_1$$

5

Truth table representation

x_1	x_2	x_3	V	W	f
0	0	0	1	0	1
0	0	1	1	0	1
0	1	0	1	0	1
0	1	1	1	0	1
1	0	0	1	1	1
1	0	1	1	1	1
1	1	0	0	0	0
1	1	1	0	1	1

$$f = \overbrace{(1 \oplus x_1 \oplus x_1\overline{x}_2)}^{V} \vee \overbrace{(x_1\overline{x}_2 \vee x_1 x_2 x_3)}^{W} = V \vee W$$

Shannon expansion

$$V = \overline{x}_1 V_0 \oplus x_1 V_1$$
$$W = \overline{x}_1 W_0 \oplus x_1 W_1$$

where

$$V_0 = 1, \quad V_1 = \overline{x}_2;$$
$$W_0 = x_1, \quad W_1 = x_1 x_3$$

6

Truth table representation

x_1	x_2	x_3	f
0	0	0	0
0	0	1	0
0	1	0	0
0	1	1	1
1	0	0	0
1	0	1	1
1	1	0	1
1	1	1	1

x_1	f
0	$x_2 x_3$
1	$x_2 \vee x_3$

K-map representation

x_3 \ $x_1 x_2$	00	01	11	10
0	1		1	
1		1	1	1

Shannon expansion

$$f = x_1 x_2 \vee x_2 x_3 \vee x_1 x_3 = \overline{x}_1 f_0 \oplus x_1 f_1$$

where

$$f_0 = x_2 x_3$$
$$f_1 = x_2 \vee x_3$$

form; for example, with respect to variables x_1 and x_2. Note that in addition to the 4-to-1 multiplexer, AND and NOT gates are required (Table 13.2).

Practice problem **13.6.** **(Design techniques.)** Implement the polynomial expression $f = 1 \oplus x_1x_2 \oplus x_1x_3x_4 \oplus x_1x_2x_3x_4$ using 4-to-1 multiplexers.
Answer is given in "Solutions to practice problems."

Design example 4

Given the Boolean function $f = x_1\overline{x}_2\overline{x}_3 \vee x_1x_2\overline{x}_3 \vee \overline{x}_1\overline{x}_2x_3 \vee \overline{x}_1x_2x_3$, let us factor the function $f = x_1(\overline{x}_2\overline{x}_3 \vee x_2\overline{x}_3) \vee \overline{x}_1(\overline{x}_2x_3 \vee x_2x_3)$
$= \overline{x}_1f_0 \vee x_1f_1$ A multiplexer tree of 2-to-1 multiplexers for implementing it is shown in Table 13.3. Other solutions are also possible.

Practice problem **13.7.** **(Design techniques.)** Given a tree of multiplexers (Figure 13.9), restore the Boolean function it implements.
Answer is given in "Solutions to practice problems."

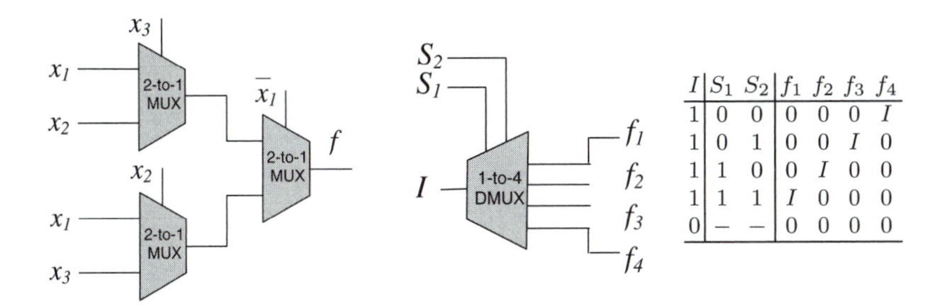

FIGURE 13.9
A tree of multiplexers to the Practice problem 13.7 (a) and demultiplexer to the Practice problem 13.11 (b).

Design example 5

Let the Boolean function f contain the polynomial expression and an SOP expression $V = 1 \oplus x_1 \oplus x_1\overline{x}_2$ and $W = x_1\overline{x}_2 \vee x_1x_2x_3$, respectively, such that $f = V \vee W$. Shannon expansion is applied to the subfunctions V and W, and then traditional design using the 2-to-1 multiplexers can be used. Finally, the function f is obtained by using the OR gate (Table 13.3).

Practice problem **13.8.** **(Design techniques.)** Implement the Boolean function $f = (x_1 \oplus x_2) \oplus (x_1\overline{x}_2x_3 \vee \overline{x}_1\overline{x}_2\overline{x}_3)$ using multiplexers and additional gates.
Answer is given in "Solutions to practice problems."

Design example 6

Given the Boolean function $f = 1 \oplus x_1x_2 \oplus x_1x_3x_4 \oplus x_1x_2x_3x_4$, implement it using 2-to-1 multiplexers and arbitrary additional gates. For this, the truth table is decomposed into two rows (Table 13.3). The same result is obtained using Shannon expansion with respect to variable x_1. The K-map is used for decreasing the number of terms (minimization is not required in this approach).

Practice problem **13.9.** **(Design techniques.)** Implement the function $f = x_1x_2 \vee x_2x_3 \vee x_1x_3$ using one 2-to-1 multiplexer and any additional gates.
Answer is given in "Solutions to practice problems."

Design example 7

An arbitrary logic gate can be modeled using a multiplexer tree. Figure 13.10 illustrates the design techniques for the AND, OR, and EXOR functions.

Practice problem **13.10.** **(Design techniques.)** Implement the NOT function using the multiplexer tree.
Answer is given in "Solutions to practice problems."

13.4 Demultiplexers

A 2^n *demultiplexer* is a combinational logic network with n selected (address) inputs, one data input, and 2^n data outputs:

$$\underbrace{\texttt{A logic signal}}_{Input} \xrightarrow{Distribution} \underbrace{2^n \texttt{ copies of logic signal}}_{Output}$$

A demultiplexer is also called a *distributor*, or selector. It performs the function inverse to that of the multiplexer. That is, a demultiplexer routes the input data to one of the outputs selected using the select lines; all other outputs are zero.

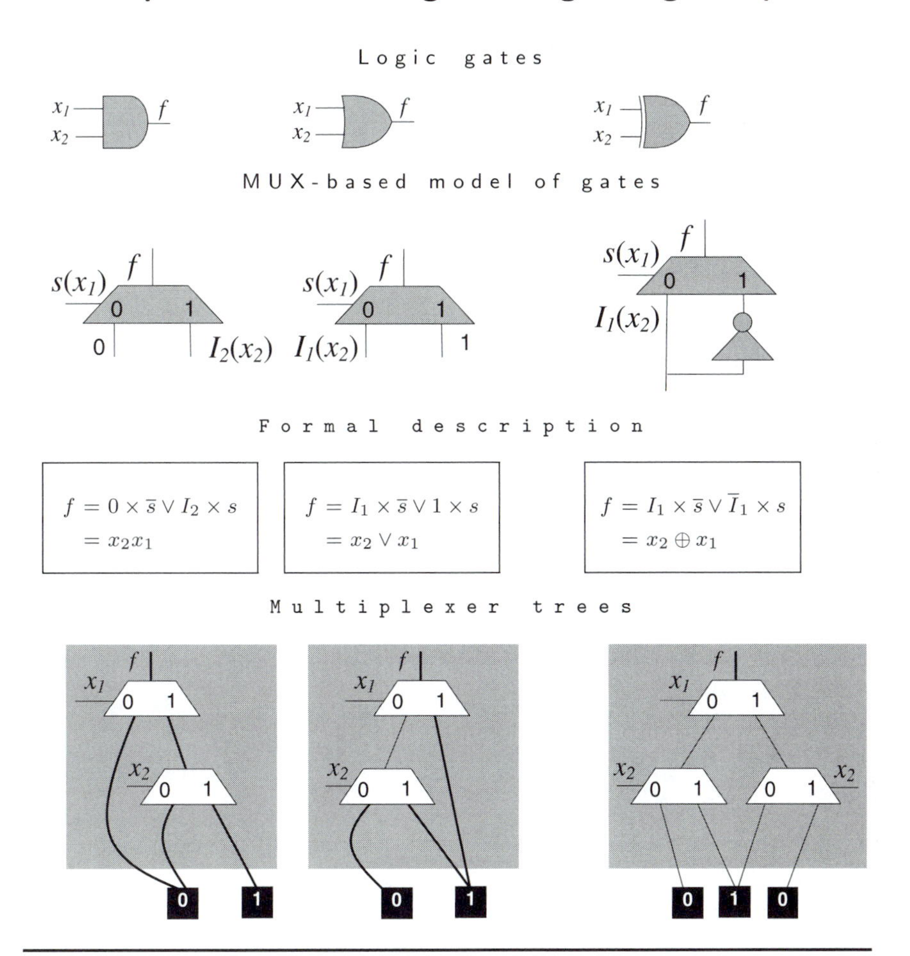

FIGURE 13.10
Multiplexer trees of the AND, OR, and EXOR functions.

Example 13.9 (Demultiplexer.) *Examples of demultiplexers denoted DMUX, are given in Table 13.4. The two-output demultiplexer requires one select input. The four-output demultiplexer has two select inputs, and 8-output demultiplexer has three select inputs.*

Formal model of a demultiplexer is defined by the following equation:

TABLE 13.4
Demultiplexers and their formal descriptions (Example 13.9).

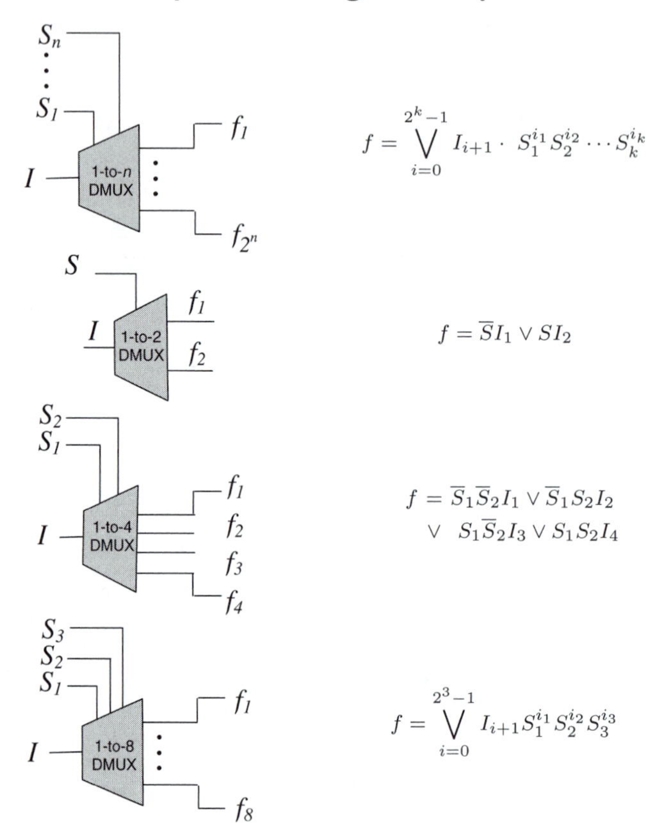

Demultiplexer

$$\texttt{Demultiplexer output } f_i = \underset{\substack{\uparrow \\ In}}{I} \cdot \overbrace{S_1^{i_1} S_2^{i_2} \cdots S_n^{i_n}}^{Address\ i}, \tag{13.3}$$

where

$$S_j^{i_j} = \begin{cases} \overline{S}_j, \text{ if } i_j = 0 \\ S_j, \text{ if } i_j = 1 \end{cases}$$

Address $S_1^{i_1} S_2^{i_2} \cdots S_n^{i_n}$ in Equation 13.3 is a minterm of k selected variables $S_j,\ j = 1, 2, \ldots, n$.

Example 13.10 (Demultiplexer.) *It follows from Equation 13.3 that 1-to-2 and 1-to-4 demultiplexers transmit their inputs to two or four outputs, respectively:*

$$n = 1: \ f = I \cdot S_1^{i_1} = I \cdot \overline{S}_1 \vee IS_2$$
$$n = 2: \ f = I \cdot S_1^{i_1} S_2^{i_2} = I \cdot \overline{S}_1 \overline{S}_2 \vee I \cdot \overline{S}_1 S_2 \vee I \cdot S_1 \overline{S}_2 \vee I \cdot S_1 S_2$$

Practice problem 13.11. (Demultiplexer.) Show the function of the multiplexer given in Figure 13.9b.
Answer is given in "Solutions to practice problems."

13.4.1 Implementation of decision diagrams on multiplexers

A binary decision diagram can be directly mapped into a multiplexer network. The multiplexer network operates in the bottom-up fashion. The bottom-up design of the network corresponds to the so-called "timed Shannon circuits with multiplexers." A decision diagram using pD and nD nodes can be mapped into a multiplexer network with additional EXOR gates (Table 13.5).

Example 13.11 (Implementation of binary decision diagrams) *Examples of multiplexer trees for implementing a binary decision diagram for a Boolean functions of three variables ($n = 3$) are given in Figure13.11.*

13.5 Decoders

A decoder is a data path device that implements functions similarly to multiplexers and demultiplexers.

Decoder

The n-to-2^n *decoder* is a logic network that decodes n inputs into 2^n outputs with the property that only one of the 2^n-output lines is asserted at a time, and each output corresponds to one valuation of the inputs (Figure 13.12a):

$$\underbrace{n \texttt{ bits}}_{Input} \overset{Decoding}{\longrightarrow} \underbrace{2^n \texttt{ bits}}_{Output}$$

An n-to-2^n decoder is a *minterm generation.*

Example 13.12 (A 3-to-8 decoder.) *The gate level implementation of a 3-to-8 decoder, its symbol, and truth table are shown in Figure 13.13.*

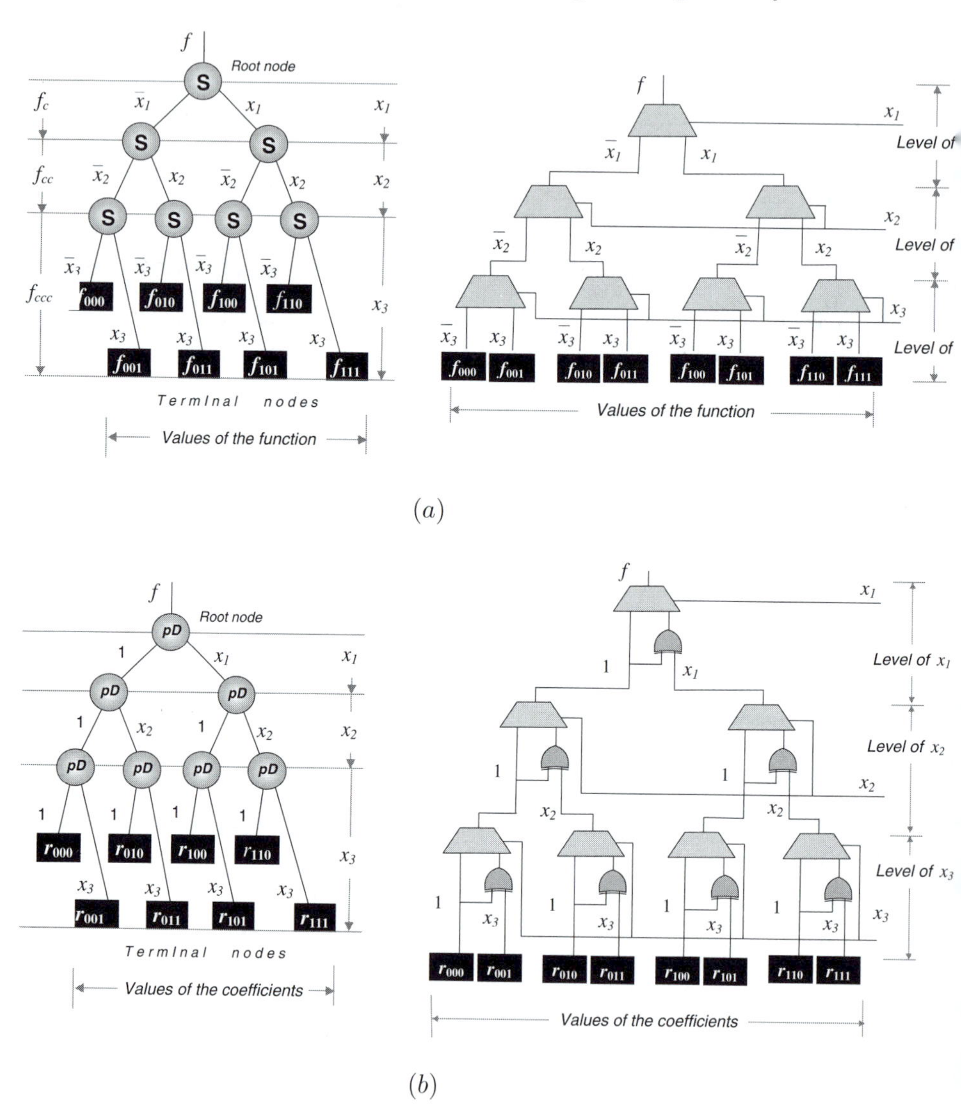

FIGURE 13.11
Implementation of binary decision trees on multiplexers (a) and multiplexers and EXOR gates (b) (Example 13.11).

TABLE 13.5
The nodes of decision trees: denotation, implementation, and formal notation.

Techniques for decision diagram design using multiplexers

Node	Realization	Algebraic form	Matrix form
		Shannon expansion	
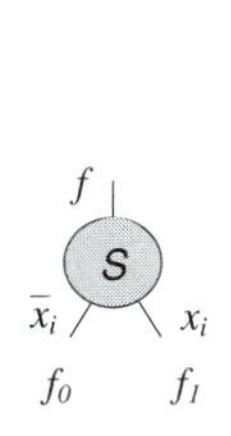	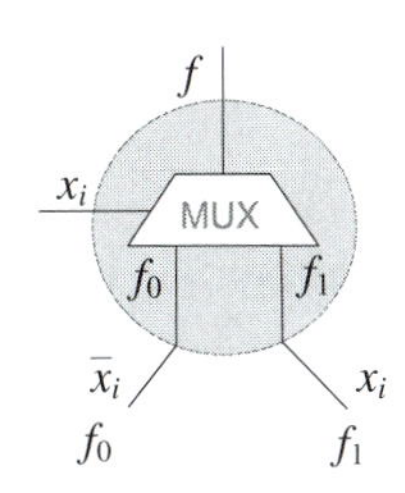	$f = \overline{x}_i f_0 \vee x_i f_1$ $f_0 = f_{x_i=0}$ $f_1 = f_{x_i=1}$	$f = [\, \overline{x}_i \; x_i \,] \begin{bmatrix} 1 & 0 \\ 0 & 1 \end{bmatrix} \begin{bmatrix} f_0 \\ f_1 \end{bmatrix}$ $= [\overline{x}_i \; x_i] \begin{bmatrix} f_0 \\ f_1 \end{bmatrix}$
		Positive Davio expansion	
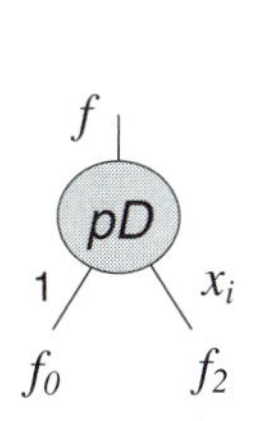	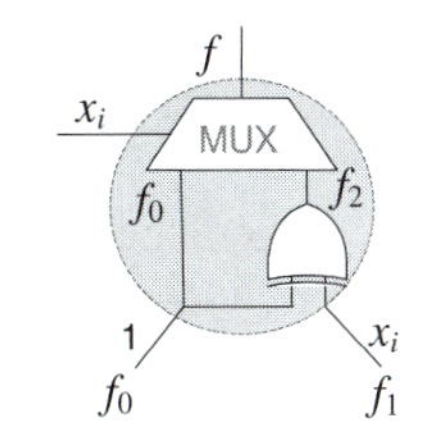	$f = f_0 \oplus x_i f_2$ $f_0 = f_{x_i=0}$ $f_2 = f_{x_i=0} \oplus f_{x_i=1}$	$f = [\, 1 \; x_i \,] \begin{bmatrix} 1 & 0 \\ 1 & 1 \end{bmatrix} \begin{bmatrix} f_0 \\ f_1 \end{bmatrix}$ $= [1 \; x_i] \begin{bmatrix} f_0 \\ f_0 \oplus f_1 \end{bmatrix}$
		Negative Davio expansion	
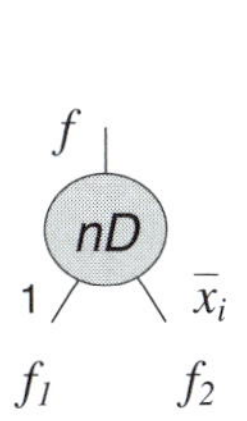	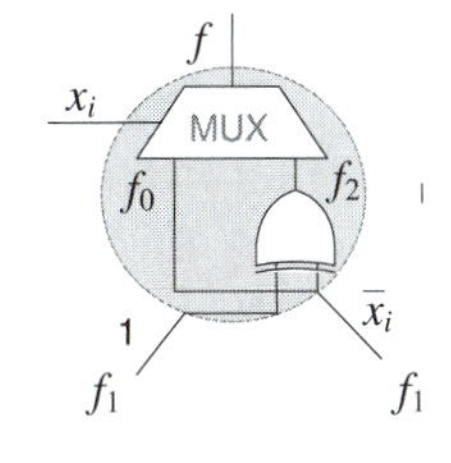	$f = f_1 \oplus \overline{x}_i f_2$ $f_1 = f_{x_i=1}$ $f_2 = f_{x_i=0} \oplus f_{x_i=1}$	$f = [\, 1 \; \overline{x}_i \,] \begin{bmatrix} 0 & 1 \\ 1 & 1 \end{bmatrix} \begin{bmatrix} f_0 \\ f_1 \end{bmatrix}$ $= [1 \; \overline{x}_i] \begin{bmatrix} f_1 \\ f_0 \oplus f_1 \end{bmatrix}$

Practice problem **13.12. (A 2-to-4-decoder.)** Derive the gate level of a 2-to-4-decoder and its truth table.
Answer is given in "Solutions to practice problems."

The cascading decoders aims at decoding larger code words.

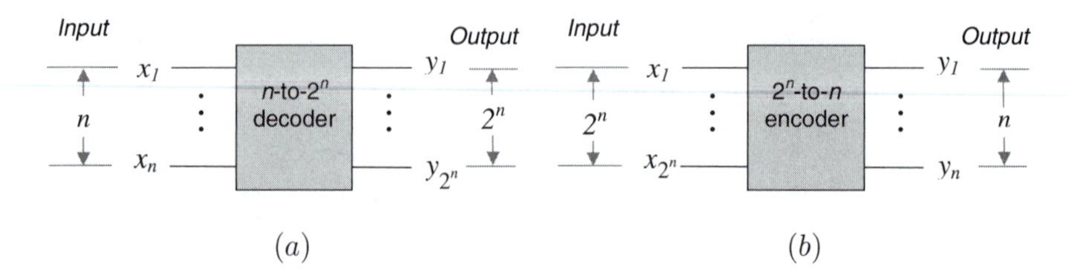

FIGURE 13.12
The n-to-2^n decoder (a) and 2^n-to-n encoder (b).

Design example: The 3-to-8 network of decoder

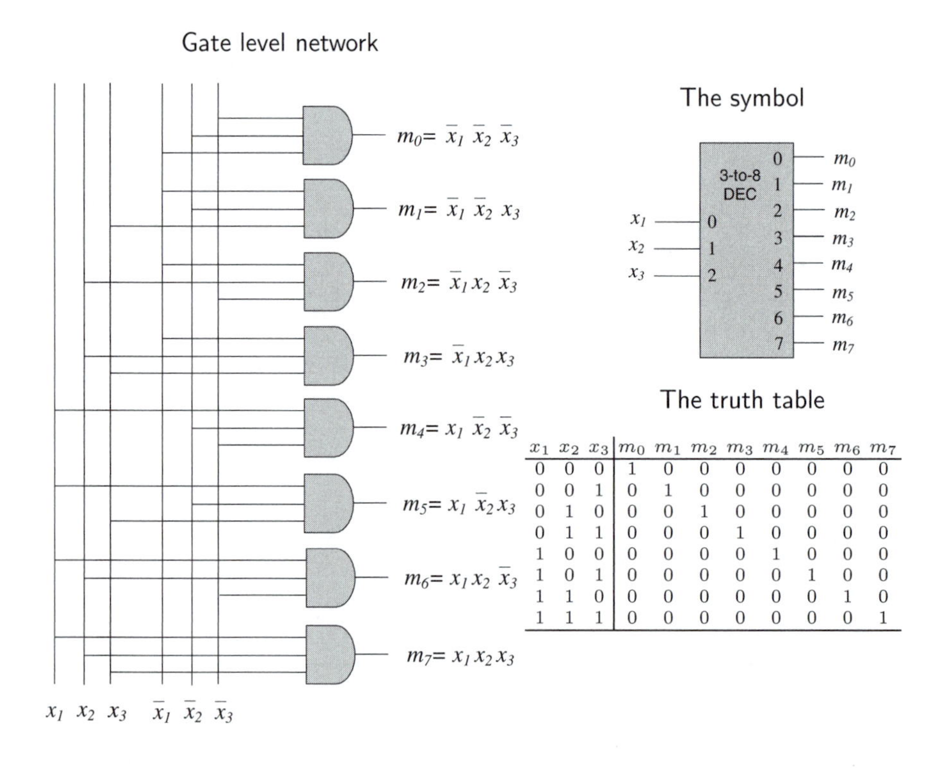

x_1	x_2	x_3	m_0	m_1	m_2	m_3	m_4	m_5	m_6	m_7
0	0	0	1	0	0	0	0	0	0	0
0	0	1	0	1	0	0	0	0	0	0
0	1	0	0	0	1	0	0	0	0	0
0	1	1	0	0	0	1	0	0	0	0
1	0	0	0	0	0	0	1	0	0	0
1	0	1	0	0	0	0	0	1	0	0
1	1	0	0	0	0	0	0	0	1	0
1	1	1	0	0	0	0	0	0	0	1

FIGURE 13.13
A 3-to-8 decoder: gate level network, symbol, and truth table (Example 13.12).

Example 13.13 (Cascading decoders) *Figure 13.14 shows how 3-to-8 decoders can be combined to make a 4-to-16 decoder. The x_4 input drives the enable (En) inputs of the two decoders. The first decoder is enabled if $x_4 = 0$, and the second decoder is enabled if $x_4 = 1$.*

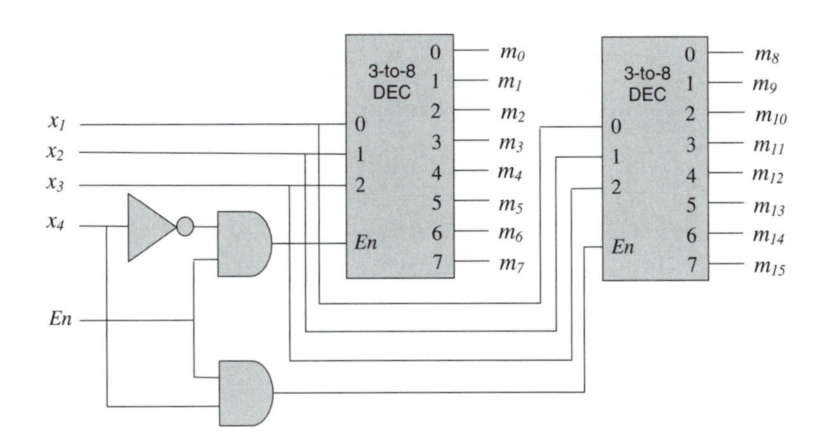

FIGURE 13.14
A 4-to-16 decoder design using 3-to-8 decoders (Example 13.13).

The concept of cascading that is introduced in Example 13.13 can be applied for the design of decoders of any size. Because of the treelike structure, the resulting networks are referred to as *decoder tree.*

Example 13.14 (Cascading decoders) *Figure 13.15 shows how 3-to-8 decoders can be combined to make a 4-to-16 decoder.*

Example 13.15 (Priority encoder.) *A block diagram and logic network of a 4-to-2 decoder is shown in Figure 13.16. The truth table shows that for $E_n = 0$, the outputs are set to 0 regardless of the input values x_1 and x_2. In this decoder,*

$$y_1 = \overline{x}_1\overline{x}_2E, \quad y_2 = \overline{x}_1x_2E,$$
$$y_3 = x_1\overline{x}_2E, \quad y_4 = x_1x_2E$$

Design example: 4-to-16 decoder tree

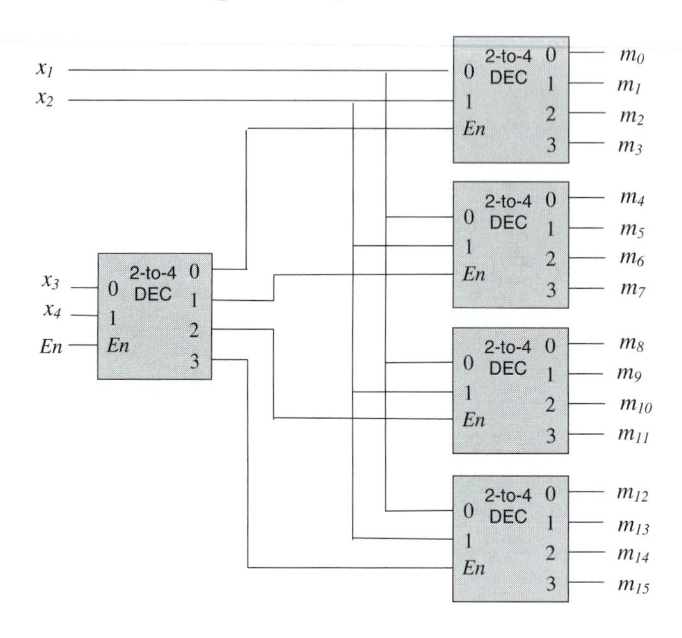

FIGURE 13.15
A 4-to-16 decoder tree (Example 13.14).

Design example: 4-to-2 decoder

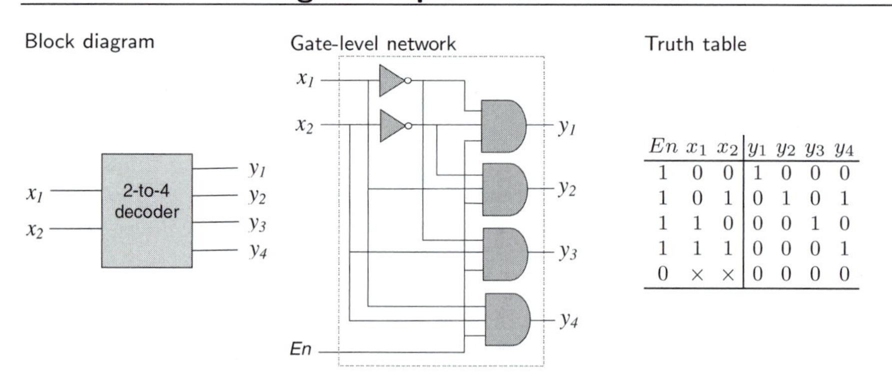

En	x_1	x_2	y_1	y_2	y_3	y_4
1	0	0	1	0	0	0
1	0	1	0	1	0	1
1	1	0	0	0	1	0
1	1	1	0	0	0	1
0	×	×	0	0	0	0

FIGURE 13.16
4-to-2 decoder (Example 13.15).

13.6 Implementation of Boolean functions using decoders

The n-to-2^n decoder can implement an arbitrary Boolean function of n variables. This is because the decoder is a minterm generator. The minterms generated by the decoder can then be combined using additional OR gates.

Given an $m - to - 2^n$ decoder, where $m < n$ and n divisible to m, the $\frac{n}{m}$ −level tree of decoders can be used.

Algorithm for computing Boolean functions using decoders

Given: A Boolean function of n variables given by its truth vector, and the size of $m - to - 2^m$ decoder.
Step 1: Divide the n variables into groups of m variables
Step 2: Design the logic network level by level.
Output: A logic network of $\frac{n}{m}$−levels of decoders (decoder tree).

In the above design technique, one decoder is in the first level, and 2^1 decoders in the next level, etc. The last level includes $2^{\frac{n}{m}}$ decoders. The output of the last-level decoder that correspond to the k minterms, are OR-ed using $k$input OR (or network of OR gates).

Note that the enable input of the first level is 1, and the enable input of the other levels are the outputs of decoders of the previous levels. It should be noted that if n is not divisible by m, then a specific design can be used.

Design example 1

Given a 2-to-4 decoder and the SOP expression for the Boolean function of two variables ($n = 2$), $f = \overline{x}_1 x_2 \vee x_1 \overline{x}_2 = \bigvee m(1,2)$, only one 2-to-4 decoder and a two-input OR gate are required. This is because $n = m = 2$ (Table 13.6).

Design example 2

Consider the Boolean function of four variables ($n = 4$), given the SOP expression (Table 13.6):

$$
\begin{aligned}
f &= \overline{x}_1\overline{x}_2\overline{x}_3 x_4 \vee \overline{x}_1\overline{x}_2 x_3\overline{x}_4 \vee \overline{x}_1 x_2 x_3\overline{x}_4 \vee \overline{x}_1 x_2 x_3 x_4 \\
&\vee x_1 x_2\overline{x}_3\overline{x}_4 \vee x_1 x_2\overline{x}_3 x_4 \vee x_1 x_2 x_3\overline{x}_4 \vee x_1 x_2\overline{x}_3 x_4 \\
&= \bigvee m(1,2,4,7,8,11,13,14)
\end{aligned}
$$

It can be implemented using a decoder tree, so that the outputs of the decoder at the first level are used as an enable inputs for the second level decoders.

Design example 3

Consider a Boolean function of three variables ($n = 3$) (Table 13.6) given the SOP expression

$$f = \overline{x}_1 x_2 x_3 \vee x_1 \overline{x}_2 x_3 \vee x_1 x_2 \overline{x}_3 \vee x_1 x_2 x_3 = \bigvee m(3, 5, 6, 7)$$

We can use two decoders and the four-input OR gate. Variable x_1 is used as enable input, so that minterm $m_3 = \overline{x}_1 x_2 x_3$ is generated by the first decoder, and the minterms $m_5 = x_1 \overline{x}_2 x_3, m_6 = x_1 x_2 \overline{x}_3$, and $m_7 = x_1 x_2 x_3$ are generated by the second decoder.

Practice problem **13.13.** **(Decoders.)** Implement the Boolean function of six variables $f = x_1 x_2 \vee x_3 x_4 \vee x_5 x_6$ using $2 - to - 4$ decoders.
Answer is given in "Solutions to practice problems."

13.7 Encoders

An encoder is a data path device that implements functions similarly to multiplexers and demultiplexers.

Encoder

A *binary encoder* encodes data from 2^n inputs into an n-bit number (Figure 13.12b):

$$\underbrace{2^n \texttt{ bits}}_{Input} \xrightarrow{Encoding} \underbrace{n \texttt{ bits}}_{Output}$$

If the input signal y_i has a logic value 1 and the other inputs are 0, then the outputs represent a binary number i.

Example 13.16 **(Priority encoder.)** *A 4-to-2 priority encoder is shown in Figure 13.17. The input patterns that have multiple inputs set to 1 are not shown in the truth table; that is, these patterns are treated as don't-care conditions.*

13.7.1 Comparators

One-bit comparator

Given two binary signals, bits x_i and x_j, the detector of their equality is defined by the equation

TABLE 13.6
Logic network design techniques for computing Boolean functions using decoders.

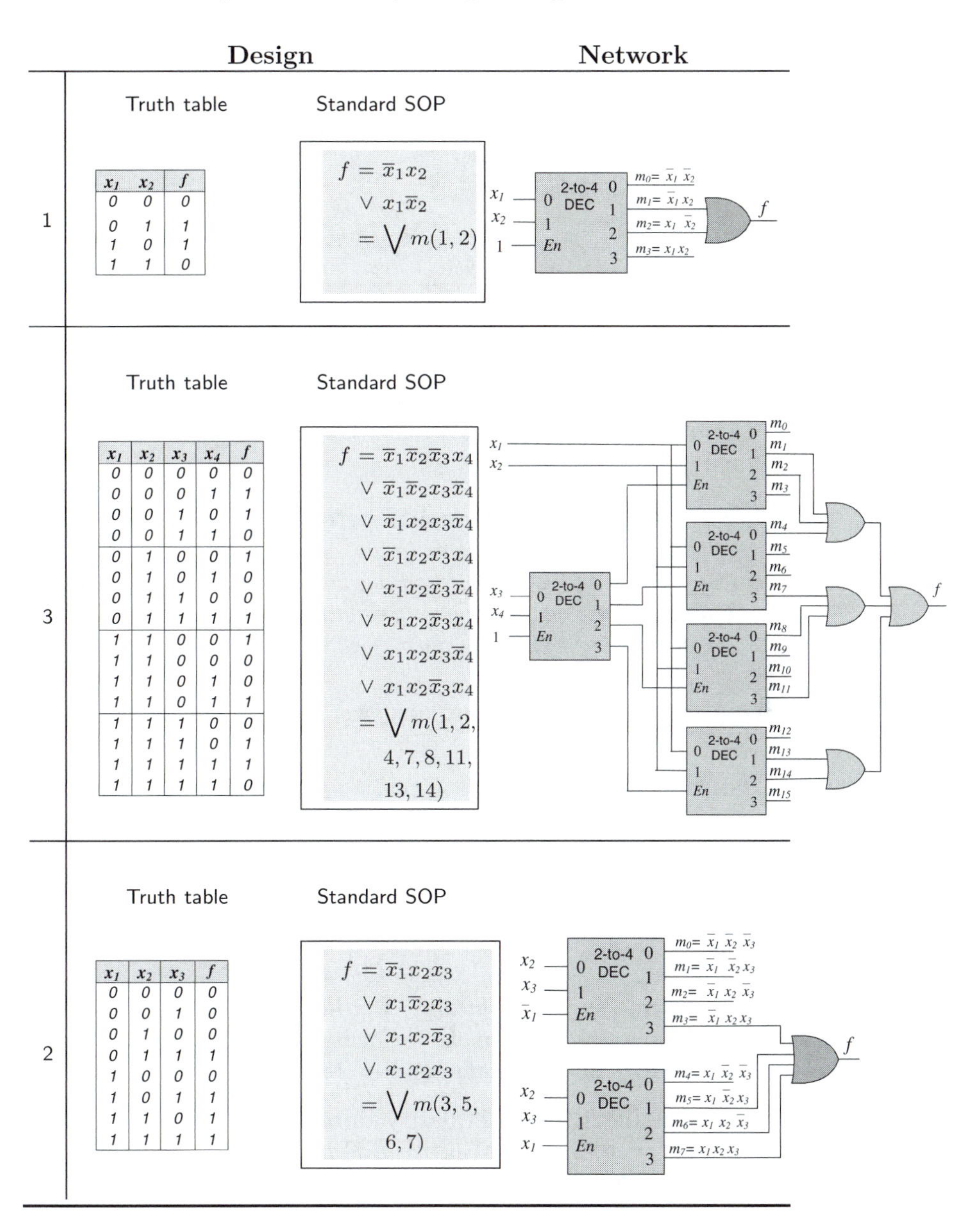

x_1	x_2	f
0	0	0
0	1	1
1	0	1
1	1	0

x_1	x_2	x_3	x_4	f
0	0	0	0	0
0	0	0	1	1
0	0	1	0	1
0	0	1	1	0
0	1	0	0	1
0	1	0	1	0
0	1	1	0	0
0	1	1	1	1
1	1	0	0	1
1	1	0	0	0
1	1	0	1	0
1	1	0	1	1
1	1	1	0	0
1	1	1	0	1
1	1	1	1	1
1	1	1	1	0

x_1	x_2	x_3	f
0	0	0	0
0	0	1	0
0	1	0	0
0	1	1	1
1	0	0	0
1	0	1	1
1	1	0	1
1	1	1	1

Design example: Priority encoder

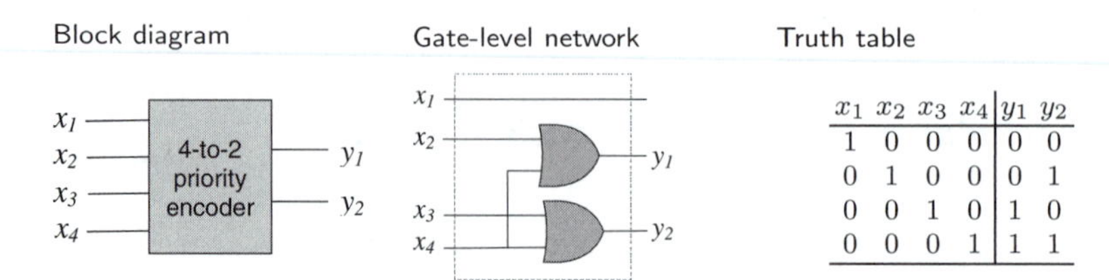

x_1	x_2	x_3	x_4	y_1	y_2
1	0	0	0	0	0
0	1	0	0	0	1
0	0	1	0	1	0
0	0	0	1	1	1

FIGURE 13.17
4-to-2 priority encoder (Example 13.16).

Comparator

$$x_i^{\sigma_i} = x_j^{\sigma_j}, \quad i \neq j, \tag{13.4}$$

where $x^1 = x$ and $x^0 = \overline{x}$

Equation 13.4 can be generalized for n signals. For example, given the three signals x_i, x_j, and x_t: $x_i^{\sigma_i} = x_j^{\sigma_j} = x_t^{\sigma_t}$, $i \neq j \neq t$.

Example 13.17 (Equality detection.) *A detector of bit equality, or a one-bit comparator, for $\sigma_i = \sigma_j$, $i = j = 1$ (Equation 13.4) is shown in Figure 13.18. It has two inputs, x_1 and x_2, and generates a single output f, such that the output is equal to 1 if either x_1 and x_2 are both 1 or are both 0.*

Practice problem **13.14. (Comparator.)** Design the bit equality detector for:
(a) Two input signals x_1 and x_2, and condition $\sigma_1 \neq \sigma_2$ in Equation 13.4
(b) Three input signals x_1, x_2, and x_3, and condition $\sigma_1 = \sigma_2$ in Equation 13.4.
Answers are given in "Solutions to practice problems."

Detector of word equality

Combinational networks that perform word-level computations are constructed using logic gates to compute the individual bits of the output word, based on the individual bits of the input word.

Example 13.18 (Equality detection.) *The detector of a 8-bit word equality is shown in Figure 13.19b. This network tests whether the two 4-bit words* **X** *and* **Y** *are equal. That is, the output is equal to 1 if and only if each bit of* **X** *is equal to the corresponding bit of* **Y**.

Design example: Equality detection

x_1	x_2	f
0	0	1 *Equality* $x_1 = x_2$
0	1	0
1	0	0
1	1	1 *Equality* $\overline{x}_1 = \overline{x}_2$

Network to test for bit equality for $\sigma_1 = \sigma_2$*:*

$$x_1^1 = x_2^1 \equiv x_1 = x_2$$
$$x_1^0 = x_2^0 \equiv \overline{x}_1 = \overline{x}_2$$

The output is equal to 1 when

- *Both inputs* x_1 *and* x_2 *are 1 (detected by the upper AND gate), or*
- *Both are equal to 0 (detected by lower AND gate)*

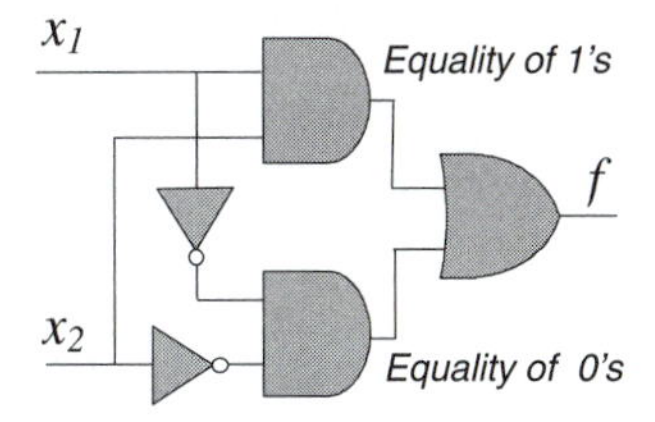

FIGURE 13.18
A combinational network for detecting bit-level equality (Example 13.17).

Practice problem 13.15. **(Word-level equality network.)** Implement the word-level equality logic network shown in Figure 13.18b using EXOR gates instead of the bit equality detector shown in Figure 13.18.
Answer is given in "Solutions to practice problems."

Practice problem 13.16. **(Comparator.)** Show that an EXOR gate can be viewed as a one-bit comparator that detects the inequality of two bits. Similarly, the XNOR gate can be viewed as a one-bit equality detector.
Answer: The output of the XNOR gate is 1 if its inputs are both 0 or both 1, and is 0 if its inputs are not equal.

13.7.2 Code detectors

Detector of word code

A bit-level equality network can be used for the detection of some specific codes, or strings, in a word.

> **Example 13.19 (Code detection.)** *The detector of the codes 0000, 0011, 1100, and 1111 in a 4-bit word is shown in Figure 13.19a.*

Practice problem 13.17. **(Code detection.)** Replace the AND gates with OR gates in Figure 13.19 and define the detected words.
Answer is given in "Solutions to practice problems."

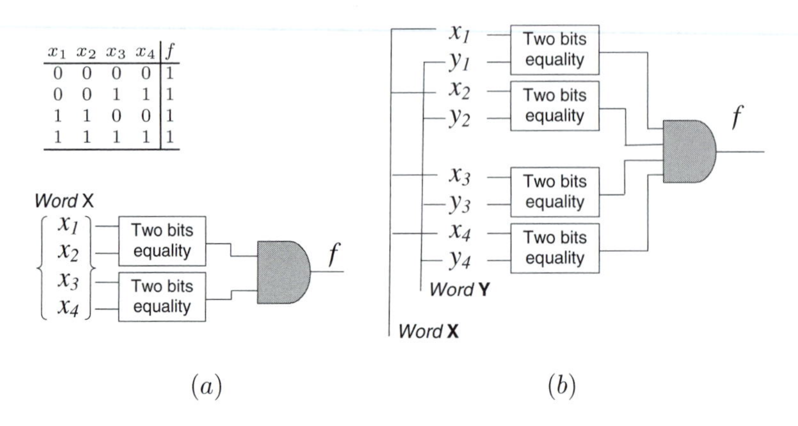

FIGURE 13.19
Combinational circuit for detecting 4-bit word equality (a), and 8-bit equality (b) of two words (Example 13.18).

13.8 Summary of standard modules of combinational logic networks

Complex digital system design is based on a modular principle; that is, systems are built using standard modules. These modules (basic blocks, components) correspond to subfunctions (subtasks) of a system's functionality. The set of subfunctions is available as ***library*** components. The key aspects of this chapter are as follows:

(*a*) Basics of the design of multiplexers, demultiplexers, encoders, decoders, code converters, comparators, shifters, and read-only memories.

(*b*) Techniques for the application of standard modules in combinational logic network design.

The main topics of this chapter are summarized as follows:

- Data transfer logic controls the basic, vital task of transferring binary data between several points in a logic network. No changes are made to the data's logic values during the transfer.

Summary (continuation)

- A ***multiplexer*** connects several sources to the same destination. A ***demultiplexer*** connects one source to one of several possible destinations.
- Multiplexers and demultiplexers are universal devices for the implementation of Boolean functions; the model for such an implementation is the Shannon expansion theorem.
- A single-bit multiplexer is defined by the equation $f = x_i\overline{S} \vee x_jS$. A multiplexer selects x_1 to connect to the output if $S = 0$, and selects x_2 if $S = 1$.
- To extend a logic network's functionality using multiplexers, the following resources can be used:

  (*a*) Connections using input-outputs,
  (*b*) Extension of selection (address) function, and
  (*c*) The size of the multiplexer.
  (*d*) The use of inputs from the set of constants and literals: $\{0, 1, x_i^{i_j}\}$.

- A ***word-level*** multiplexer generates a word $\mathbf{F}$ equal to one of the two input words, $\mathbf{X}$ and $\mathbf{Y}$, depending on the control input bit S.
- A 2^n-to-1 multiplexer can be represented as a multilevel network of 2^m-to-1 multiplexers, $m < n$. This network is called a *multiplexer tree.*
- Design techniques for computing Boolean functions using multiplexers employ two approaches:

  (*a*) An approach based on decomposing the truth table, and
  (*b*) An approach based on Shannon expansion of the implemented Boolean function.

- A 2^n demultiplexer is a combinational logic network with n control (address) inputs, one data input, and 2^n data outputs. The simplest demultiplexer is used in the implementation of decision trees and diagrams.
- A ***binary encoder*** encodes data from 2^n inputs into an n-bit code. A ***binary decoder*** is a logic network that decodes n inputs into 2^n outputs. Decoders can be used to implement logic functions.
- For advances in the design of standard modules of combinational logic networks, we refer the reader to the "Further study" section.

13.9 Further study

Historical perspective

1952: John von Neuman introduced the notion of a cellular space and discussed universal computing and constructing automata embedded in such space. The basic structure of the cellular model is infinite rectangular grid or tesselation. The work of von Neumann inspired investigations in a new field known now as the theory of cellular automata.

1956: The problem of reliable computing using unreliable logic gates was considered by John von Neuman in the paper "Probabilistic Logic and the Synthesis of Reliable Organisms from Unreliable Components," *Automata Studies, Annals of Math Studies*, 34, Princeton University Press, pp. 43–49, 1956.

Advanced topics of data structures

Topic 1: Threshold elements. A threshold function f of n variables $x_1, x_2, \ldots, x_n$ is a Boolean function which can be represented in the form: $f = 1$ iff $\sum_{i=1}^{n} w_i x_i - \theta \geq 0$, where the summation is arithmetic rather than Boolean; w_i are *weights* (which may be assumed to be positive or negative integers without loss of generality); and the integer θ is the *threshold.* While not all Boolean functions are threshold functions, any Boolean function can be composed of threshold functions. This composition corresponding to a network interconnections of two or more threshold elements, each of which realizes a particular threshold function.

Topic 2: Cellular logic is a direction of logic design which deals with the analysis and synthesis of logic networks in the form of cellular arrays. A *cellular array* consists of a 1-, 2-, or 3-dimensional of identical cells with uniform interconnection. Study of cellular arrays took on a new importance with the advent of nanotechnology. In this technology, a cell may contain a large number of logic elements. From manufacturing points of view, it is advantageous to have the "nanochip" produced in the form of an identical cells. In addition, packing density is higher, not only because of the reduction of size of the cells, but also because of elimination of much of the interconnection wiring between the nanochips. The reliability of devices based on the cellular arrays is higher compared with logic networks using another approaches. The most distinctive feature of cellular arrays is the flexibility in performance.

Further reading

"VLSI Implementation of Threshold Logic – A Comprehensive Survey" by V. Beiu, J. M. Quintana, and M. J. Avedillo, *IEEE Neural Networks*, volume 14, number 5, pages 1217–1243, 2003.

"Depth-Size Tradeoffs for Neural Computation" by K. Y. Siu, V. P. Roychowdhury, and T. Kailath, *IEEE Transactions on Computers*, volume 40, number 12, pages 1402–1411, 1991.

13.10 Solutions to practice problems

Practice problem	Solution
13.1.	16-to-1 multiplexer has 2^4 data inputs and 4 address inputs. 32-to-1 multiplexer has 2^5 data inputs and 5 address lines.
13.2.	See truth table, K-map and circuit below.
13.3, 13.4.	Problem 13.3 (circuit below). Problem 13.4: $f = x_1 x_2 = \overline{x}_1 f_0 \vee x_1 f_1$ where $f_0 = 0$ and $f_1 = x_2$.
13.5.	See truth tables and circuit below.
13.6.	$f = 1 \oplus x_1x_2 \oplus x_1x_3x_4 \oplus x_1x_2x_3x_4 = \overline{x}_1\overline{x}_2 f_{00} \vee \overline{x}_1 x_2 f_{01} \vee x_1 \overline{x}_2 f_{10} \vee x_1 x_2 f_{11}$, where $f_{00} = 1$, $f_{01} = 1$, $f_{10} = 1 \oplus x_3x_4$, $f_{11} = 0$. This expression can be implemented using a 4-to-1 multiplexer, AND gate, and EXOR gate as shown below:

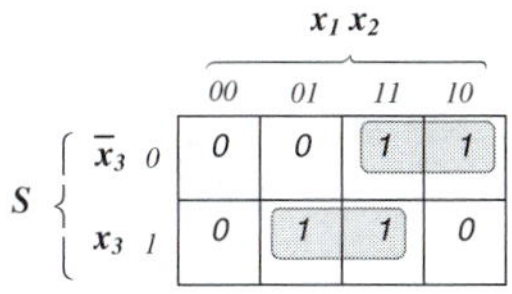

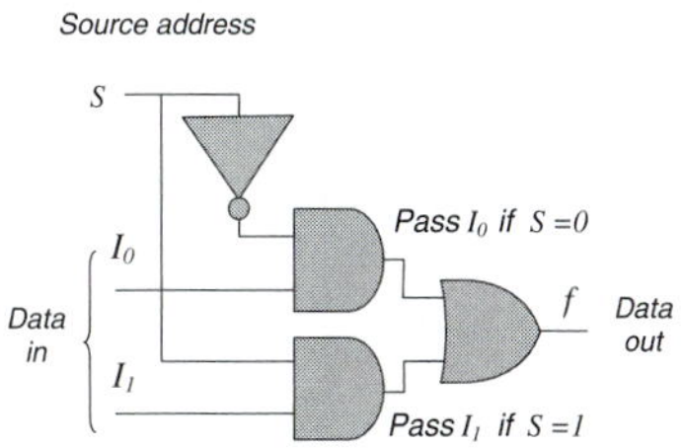

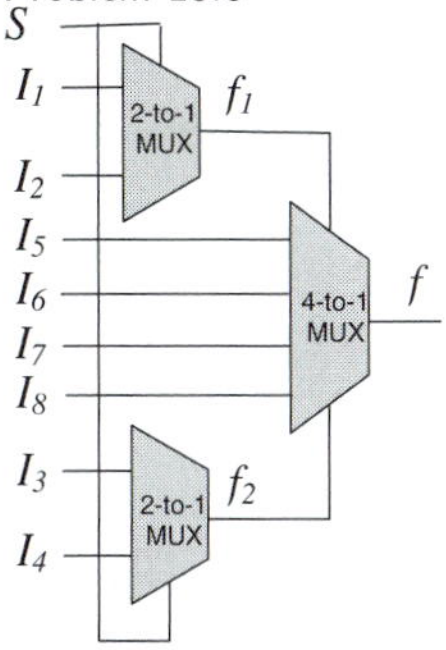

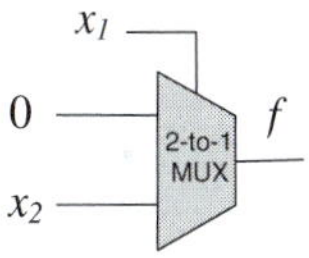

13.2.

S	f
0	I_0
1	I_1

S	x_1x_2 = 00	01	11	10
$\overline{x}_3$ 0	0	0	1	1
x_3 1	0	1	1	0

13.3, 13.4.

13.5.

x_1	x_2	x_3	f
0	0	0	0
0	0	1	1
0	1	0	1
0	1	1	0
1	0	0	1
1	0	1	0
1	1	0	0
1	1	1	1

x_1	x_2	f
0	0	x_3
0	1	$\overline{x}_3$
1	0	$\overline{x}_3$
1	1	x_3

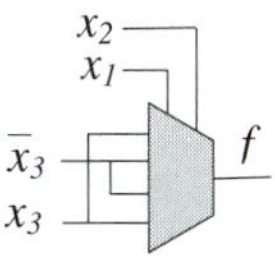

13.6.

x_1	x_2	x_3	f	
0	0	0	0	x_3
0	0	1	1	
0	1	0	1	$\overline{x}_3$
0	1	1	0	
1	0	0	1	$\overline{x}_3$
1	0	1	0	
1	1	0	0	x_3
1	1	1	1	

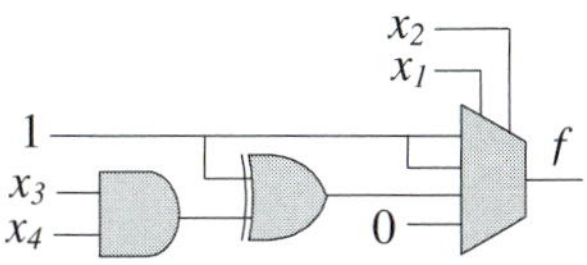

Solutions to practice problems (continuation)

Practice problem	Solution
13.7.	$f = (x_1\overline{x}_3 \vee x_2x_3)x_1 \vee (x_1\overline{x}_2 \vee x_2x_3)\overline{x}_1 = x_1\overline{x}_3 \vee x_1x_2x_3 \vee \overline{x}_1x_2x_3$ $= x_1\overline{x}_3 \vee x_2x_3$
13.8.	$f = V \oplus W$ and $V = x_1 \oplus x_2 = \overline{x}_1 f_0 \vee x_1 f_1$, where $f_0 = x_2$ and $f_1 = \overline{x}_2$. $W = \overline{x}_1\overline{x}_2x_3 \vee x_1\overline{x}_2\overline{x}_3 = \overline{x}_1(\overline{x}_2x_3) \vee x_1(\overline{x}_2\overline{x}_3)$ 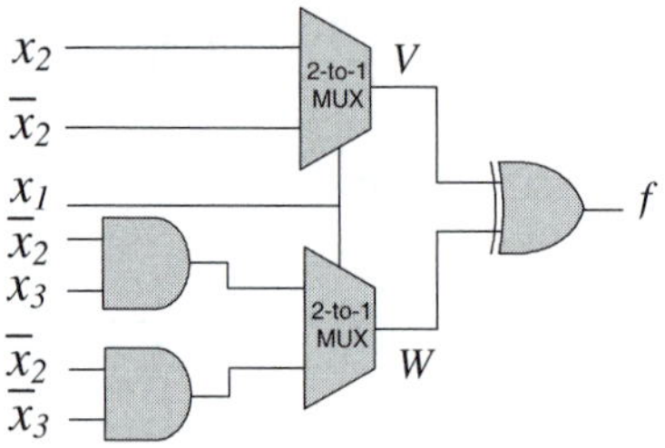
13.9.	$f = \overline{x}_1 f_0 \vee x_1 f_1$, where $f_0 = x_2x_3$ and $f_1 = x_1 \vee x_2x_3 \vee x_3 = x_1 \vee x_3$. 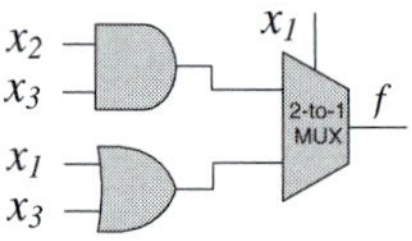
13.10.	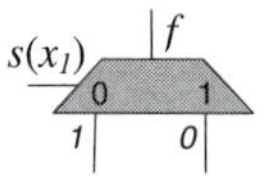 $f = 1 \times \overline{s} \vee 0 \times s$ $= \overline{x}_1$ 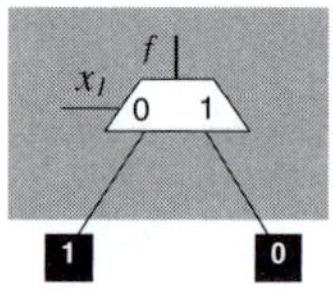
13.11.	(a) (b) (c)
13.12.	$m_0 = \overline{x}_1\,\overline{x}_2$, $m_1 = \overline{x}_1\,x_2$, $m_2 = x_1\,\overline{x}_2$, $m_3 = x_1\,x_2$

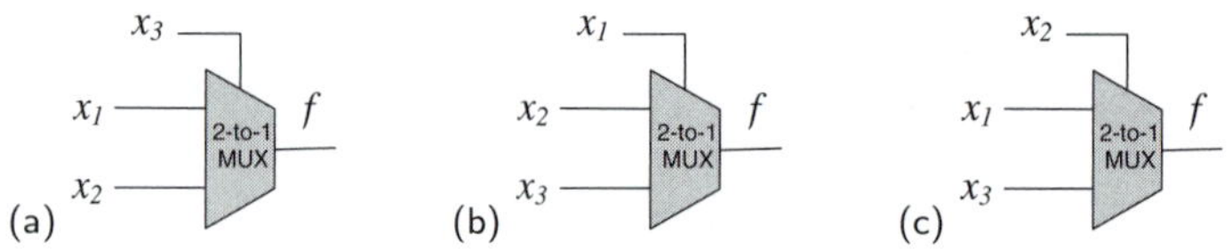

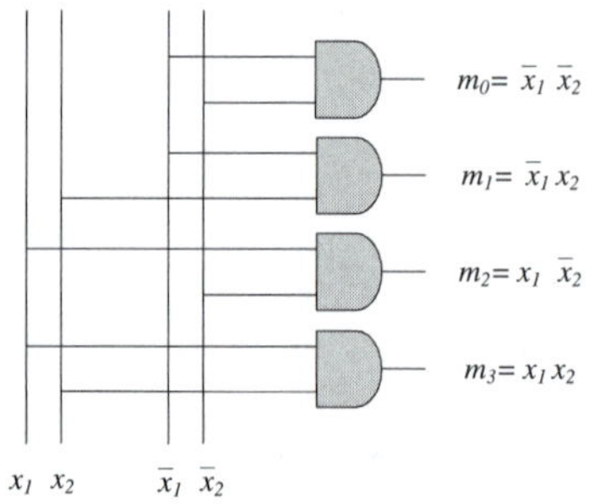

x_1	x_2	m_0	m_1	m_2	m_3
0	0	1	0	0	0
0	1	0	1	0	0
1	0	0	0	1	0
1	1	0	0	0	1

Solutions to practice problems (continuation)

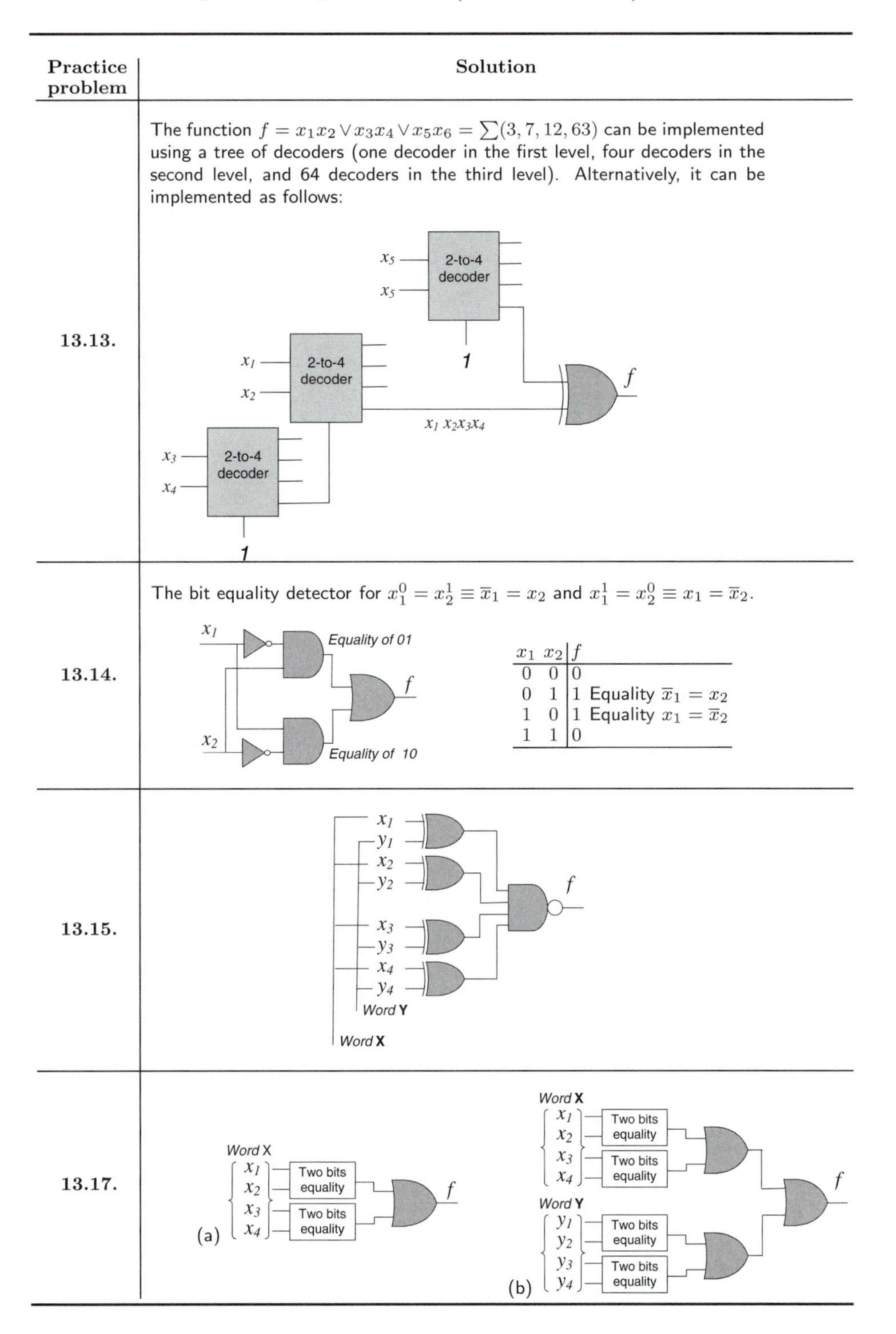

Practice problem	Solution
13.13.	The function $f = x_1x_2 \vee x_3x_4 \vee x_5x_6 = \sum(3, 7, 12, 63)$ can be implemented using a tree of decoders (one decoder in the first level, four decoders in the second level, and 64 decoders in the third level). Alternatively, it can be implemented as follows:
13.14.	The bit equality detector for $x_1^0 = x_2^1 \equiv \overline{x}_1 = x_2$ and $x_1^1 = x_2^0 \equiv x_1 = \overline{x}_2$.
13.15.	
13.17.	

x_1	x_2	f
0	0	0
0	1	1 Equality $\overline{x}_1 = x_2$
1	0	1 Equality $x_1 = \overline{x}_2$
1	1	0

13.11 Problems

Problem 13.1 Derive the 2-to-1 multiplexer tree given the following decision diagrams:

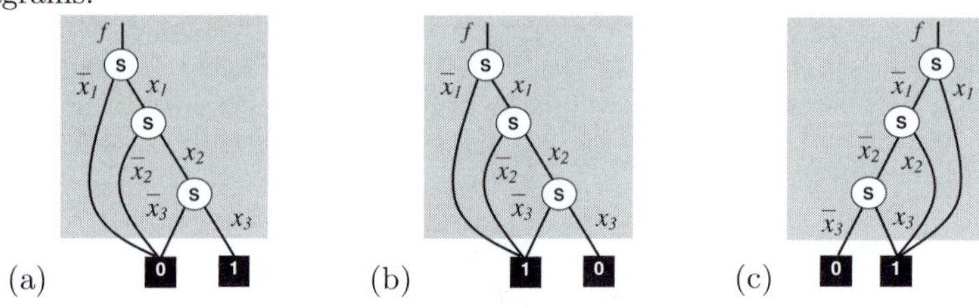

Problem 13.2 Derive the multiplexer networks to implement the functions given their following cubes:

(a) $f = [1\ 1\ 1]$ (b) $f = \begin{bmatrix} 0 & \text{x} & \text{x} \\ \text{x} & 0 & \text{x} \\ \text{x} & \text{x} & 0 \end{bmatrix}$ (c) $f = \begin{bmatrix} \text{x} & \text{x} & 1 \\ \text{x} & 1 & \text{x} \\ 1 & \text{x} & \text{x} \end{bmatrix}$

Problem 13.3 Redesign the following logic networks using 4-to-1 multiplexers:

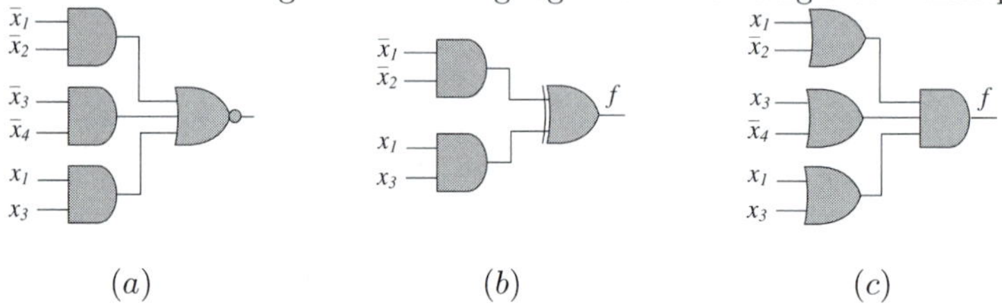

Problem 13.4 Design logic networks using 2-to-1 multiplexers and any other gates given the following Boolean functions:

(a) $f = \overline{x}_1 x_2 \vee \overline{x}_1 x_3$

(b) $f = x_1 \oplus x_2 x_3$

(c) $f = (x_1 \vee \overline{x}_1)(\overline{x}_2 \vee x_3)$

(d) $f = \overline{x}_1 \overline{x}_2 x_3 \oplus x_1 x_2 \overline{x}_3$

Problem 13.5 Redesign the following logic networks using 2-to-1 multiplexers:

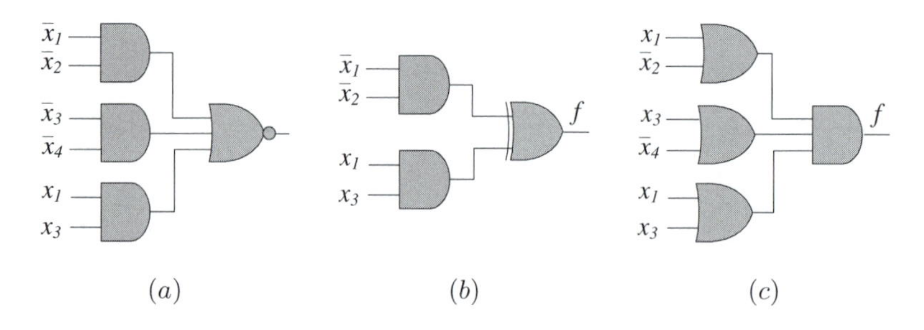

Problem 13.6 Implement the following Boolean functions given the K-maps below using 3-to-8 decoders:

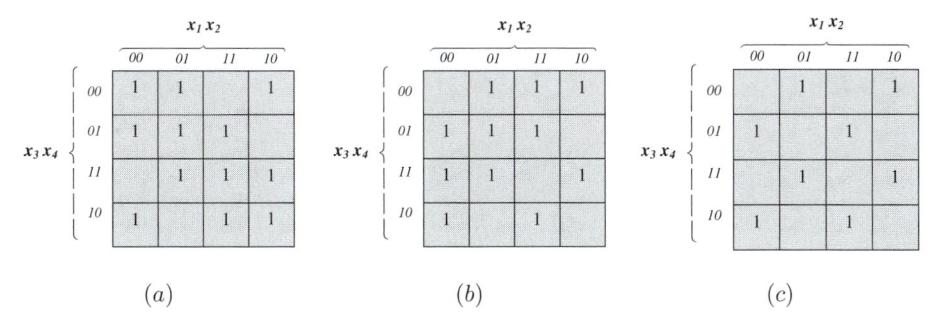

(*a*) (*b*) (*c*)

Problem 13.7 Restore the Boolean functions given the following multiplexer trees:

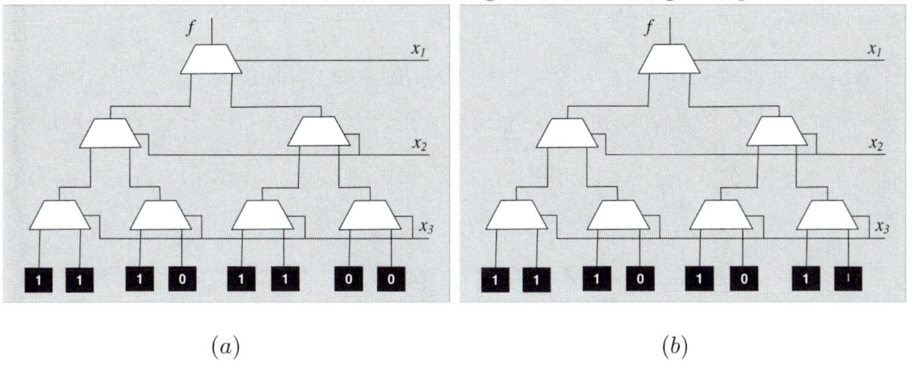

(*a*) (*b*)

Problem 13.8 Redesign the following logic networks using trees of 2-to-4 decoders:

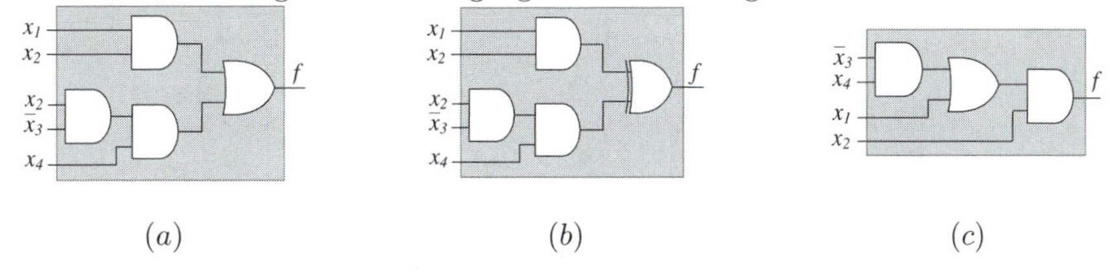

(*a*) (*b*) (*c*)

Problem 13.9 Design logic networks based on 4-to-16 multiplexers given the following truth tables of Boolean functions:

x_1	x_2	x_3	x_4	f	x_1	x_2	x_3	x_4	f
0	0	0	0	**0**	1	0	0	0	**0**
0	0	0	1	**0**	1	0	0	1	**1**
0	0	1	0	**1**	1	0	1	0	**0**
0	0	1	1	**1**	1	0	1	1	**1**
0	1	0	0	**0**	1	1	0	0	**0**
0	1	0	1	**0**	1	1	0	1	**0**
0	1	1	0	**1**	1	1	1	0	**0**
0	1	1	1	**0**	1	1	1	1	**1**

(*a*)

x_1	x_2	x_3	f
0	0	0	**1**
0	0	1	**1**
0	1	0	**1**
0	1	1	**0**
1	0	0	**1**
1	0	1	**0**
1	1	0	**0**
1	1	1	**0**

(*b*)

Problem 13.10 Design logic networks of 3-to-8 decoders given the following truth tables of Boolean functions:

(a)

x_1	x_2	x_3	f
0	0	0	**1**
0	0	1	**0**
0	1	0	**0**
0	1	1	**1**
1	0	0	**0**
1	0	1	**1**
1	1	0	**1**
1	1	1	**0**

(b)

x_1	x_2	x_3	f
0	0	0	**1**
0	0	1	**1**
0	1	0	**1**
0	1	1	**0**
1	0	0	**1**
1	0	1	**0**
1	1	0	**0**
1	1	1	**0**

(c)

x_1	x_2	x_3	f
0	0	0	**1**
0	0	1	**0**
0	1	0	**0**
0	1	1	**0**
1	0	0	**0**
1	0	1	**0**
1	1	0	**0**
1	1	1	**1**

Problem 13.11 Design a code converter using 3-to-8 decoders to convert a four-bit Gray code into a binary code.

Problem 13.12 Implement the following Boolean functions using 2-to-1 multiplexers:

(a) $f = 1 \oplus x_2 \oplus x_1 \oplus \overline{x}_1 x_2 \overline{x}_3$

(b) $f = x_1 \oplus x_2 x_3$

(c) $f = x_1 \oplus x_2 \oplus x_3$

(d) $f = 1 \oplus x_1 x_2 x_3$

14

Combinational Logic Network Design

Design examples

- Binary adder
- BCD adder
- Comparator
- Error detection and correction

Verification

- Equivalence checking
- Decision diagram-based techniques

Decomposition

- Functional
- Bi-decomposition
- Shannon decomposition

Advanced topics

- Multi-level logic network design
- Heuristic optimization

14.1 Introduction

An n-input, m-output *combinational* network is a network of logic elements that implements a set of $m \geq 1$ Boolean functions of n variables $x_1, x_2, \ldots, x_n$, $f = \{f_1, f_2, \ldots, f_n\}$. The simplest example of an n-input, 1-output combinational network is a logic gate.

> **The key property of a combinational logic network**
> The output signals of a combinational network at any time are completely determined by the combination of values assigned to the input signals $x_1, x_2, \ldots, x_n$ at that time, hence the term "combinational." Any changes in these inputs are assumed to change the network output signals.

An idealized logic network is built from elementary logic elements (gates) and standard logic blocks (multiplexers, demultiplexers, etc.), and is idealized in the sense that all signal changes are assumed to take place instantaneously. In physical networks, signals cannot be transmitted instantaneously, so there is a delay between a network input change and the output change it produces. This *delay*, or *response time*, can be measured. Typically, this time is on the order of a nanoseconds (10^{-9} s) per logic element. When addressing the functional or logical behavior of a logic network, we assume in this chapter that all lines and logic elements have zero delay. That is, combinational logic networks are equated with delay-free networks, in which all signal changes occur instantaneously.

In this chapter, the following design techniques are introduced: binary and decimal adders, comparators, and error detection and error correction networks.

14.2 Design example: Binary adder

An arithmetic network, or circuit, is a combinational logic network that performs arithmetic operations such as addition, subtraction, multiplication, and division on binary numbers or on decimal numbers coded as binary ones.

Example 14.1 (Binary adder.) *The addition suggests four possible elementary operations:* $0+0=0$, $0+1=1$, $1+0=0$, *and* $1+1=2_{10}=10_2$. *The first three operations produce a sum that requires one bit for its representation; when both the augend and addend are equal to 1, the binary sum requires two bits. Because of this, the result is generally represented by two bits, the* carry *and the* sum. *The carry obtained from the addition of two bits is added to the next higher order pair of significant bits.*

Half adder

A half-adder is an arithmetic network that generates the sum of two binary digits. The network has two inputs and two outputs. The input variables are the augend x_1 and the addend x_2 bits to be added, and the output variables produce the sum S and the carry C:

The half-adder

$$\textsc{Sum}\ S = x_1 \oplus x_2 \qquad \textsc{Carry}\ C = x_1 x_2$$

Example 14.2 **(Half-adder design.)** *In Figure 14.1, the design of a half-adder is introduced. Also, its implementation using decision diagrams is given. These diagrams implement the SOP representation; that is, the sum S is represented as* $\textsc{Sum}\ S = x_1 \oplus x_2 = x_1\overline{x}_2 \vee \overline{x}_1 x_2$.

Practice problem **14.1. (Half-adder design.)** Provide the decision tree based design for the half-adder given in Figure 14.1; that is, generate the reduced decision trees for the functions S and C.
Answer is given in "Solutions to practice problems."

Full adder

A *full adder* is a combinational logic network that forms the arithmetic sum of the following input bits (Figure 14.2): augend and addend, x_1 and x_2, and The carry from the previous (right) significant bit, C_{in}. The full adder forms the two-digit binary number $\textsc{Sum} = C_{out} \times 2^1 + S \times 2^0$, which, when expressed in bit-wise form, yields

The full adder

$$S = x_1 \oplus x_2 \oplus C_{in}$$
$$C_{out} = x_1 x_2 \vee C_{in}(x_1 \oplus x_2)$$

Example 14.3 **(Full adder design.)** *In Figure 14.2, the design of a full adder is introduced. Also, its implementation using decision diagrams is given.*

Practice problem **14.2. (Full adder design.)** Provide a step-by-step construction and reduction of the decision tree to a decision diagram for the outputs of the full adder given in Figure 14.2.
Answer is given in "Solutions to practice problems."

Design example: Half-adder

STEP 1

Formal model

Sum $S = x_1 \oplus x_2$

Carry $C = x_1 x_2$

STEP 2

Truth table

Inputs x_1	x_2	Sum S	Carry C
0	0	0	0
0	1	1	0
1	0	1	0
1	1	0	1

STEP 3

Logic network of gates

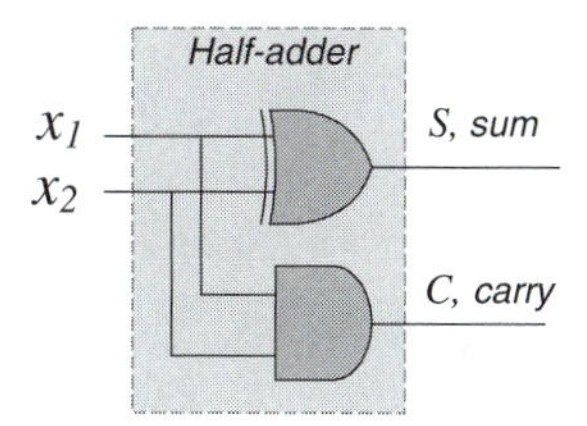

STEP 4

Decision diagram for the multiplexer-based implementation

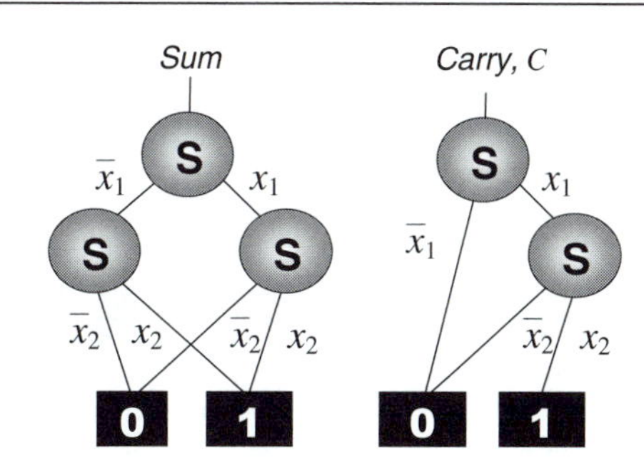

FIGURE 14.1
A half-adder design (Example 14.2).

Design example: Full adder

STEP 1

Formal model

Sum $S = x_1 \oplus x_2 \oplus C_{in}$

Carry out $C_{out} = x_1x_2 \vee C_{in}(x_1 \oplus x_2)$

STEP 2

Truth table

Inputs			Sum	Carry out
x_1	x_2	C_{in}	S	C_{out}
0	0	0	0	0
0	0	1	1	0
0	1	0	1	0
0	1	1	0	1
1	0	0	1	0
1	0	1	0	1
1	1	0	0	1
1	1	1	1	1

STEP 3

Logic network

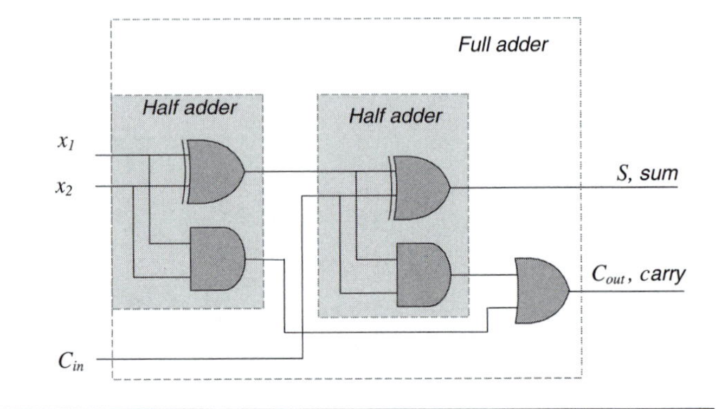

STEP 4

Decision diagram for multiplexer-based implementation

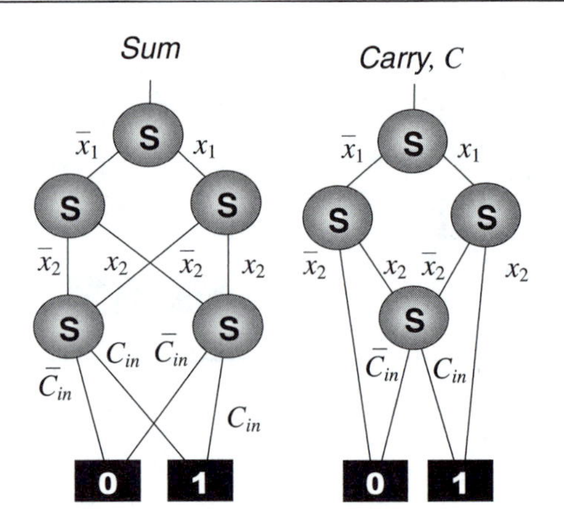

FIGURE 14.2
A full adder design (Example 14.3).

Binary cascade adder

A *binary adder* is a logic network that produces the arithmetic sum of two binary numbers using only combinational logic. To add two n-bit numbers, one-bit full adder can be connected in a chain, so that the carry-out of the previous full adder is a carry-in to the next full adder. This forms a *cascade* logic network, also called ripple-carry adder.

Example 14.4 (Cascade adder.) *In Figure 14.3, a 4-bit adder is shown, where $A = a_0a_1a_2a_3$ and $B = b_0b_1b_2b_3$. The input carry to the adder is C_0, and the output carry is C_4.*

An n-input cascade adder requires n full adders. Such a design is called a computational *array of cells*, where the full adder is referred to as a *cell.*

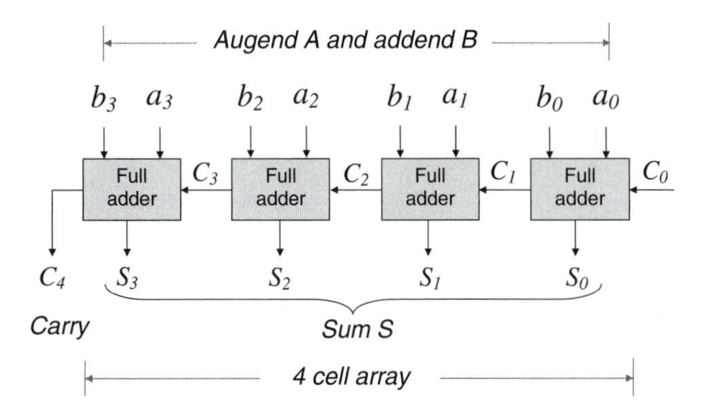

FIGURE 14.3
The 4-bit cascade binary adder (Example 14.4).

Carry-look-ahead adder

One of the drawbacks of the cascade adder is a long delay due to the many gates in the carry path from the least significant bit to the most significant bit. The delay Δ can be evaluated as $\Delta = 2n + 2$ gate delay units. The approach, called *carry-look-ahead,* allows one a reduction in the delay. Design of the carry-look-ahead adder includes: (a) the design of a *partial full adder* and (b) implementation of *carry-look-ahead logic.*

One approach to designing a full adder is formulated as follows: separate the parts of the full adder not involving the carry propagation path from those containing the path. This separation is based on the following: assuming $g_i = a_ib_i$ and $p_i = a_i \oplus b_i$, the equation for the i-th cell of the full adder can be written as follows:

$$C_{i+1} = \underbrace{a_i b_i}_{g_i} \vee C_i \underbrace{(a_i \oplus b_i)}_{p_i} = g_i \vee C_i p_i \tag{14.1}$$

It follows from Equation 14.1 that

(*a*) For the inputs $a_i = b_i = 1$, the i-th full adder *generates* a carry g_i.

(*b*) For the inputs a_i and b_i, when one of them is equal to 1, the i-th full adder *propagates* a carry p_i.

The resulting network implements the following equations:

$$g_i = a_i b_i$$
$$p_i = a_i \oplus b_i$$
$$S_i = a_i \oplus b_i \oplus C_i$$

Example 14.5 (Carry-look-ahead adder.) *The design of the 4-bit carry lookahead adder is illustrated in Figure 14.4.*

14.3 Design example: Magnitude comparator

In Chapter 13, one-bit comparators, or detectors of bit equality, were considered. Based on these, one-bit standard unit detectors of word equality were designed. However, the design of word comparators, including ones that detect words' inequality, can be implemented without using standard one-bit units.

A logic network that compares two binary numbers and indicates whether they are equal, and/or indicates an arithmetic relationship (greater or less than) between the numbers is called a *magnitude comparator*. It was shown in Chapter 13 that the EXOR and XNOR gates may be viewed as 1-bit comparators. In particular, if the input bits are equal, then the output of the XNOR gate is 1, while otherwise it is 0. Therefore, a straightforward implementation of an n-bit comparator can be accomplished using XNOR gates.

Example 14.6 (4-bit comparator.) *A logic network for a 4-bit-comparator using XNOR gates is given in Figure 14.5.*

Design example: Carry-look-ahead adder

STEP 1

Formal description

$$g_i = a_i b_i$$
$$p_i = a_i \oplus b_i$$
$$S_i = a_i \oplus b_i \oplus C_i$$

STEP 2

Carry lookahead logic description

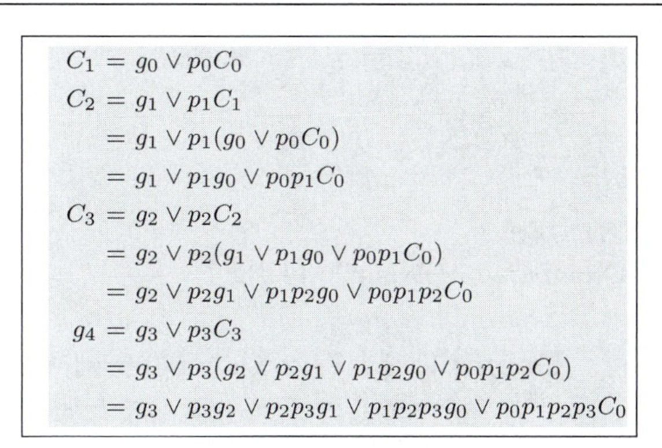

$$\begin{aligned}
C_1 &= g_0 \vee p_0 C_0 \\
C_2 &= g_1 \vee p_1 C_1 \\
&= g_1 \vee p_1(g_0 \vee p_0 C_0) \\
&= g_1 \vee p_1 g_0 \vee p_0 p_1 C_0 \\
C_3 &= g_2 \vee p_2 C_2 \\
&= g_2 \vee p_2(g_1 \vee p_1 g_0 \vee p_0 p_1 C_0) \\
&= g_2 \vee p_2 g_1 \vee p_1 p_2 g_0 \vee p_0 p_1 p_2 C_0 \\
g_4 &= g_3 \vee p_3 C_3 \\
&= g_3 \vee p_3(g_2 \vee p_2 g_1 \vee p_1 p_2 g_0 \vee p_0 p_1 p_2 C_0) \\
&= g_3 \vee p_3 g_2 \vee p_2 p_3 g_1 \vee p_1 p_2 p_3 g_0 \vee p_0 p_1 p_2 p_3 C_0
\end{aligned}$$

STEP 3

Partial full adder

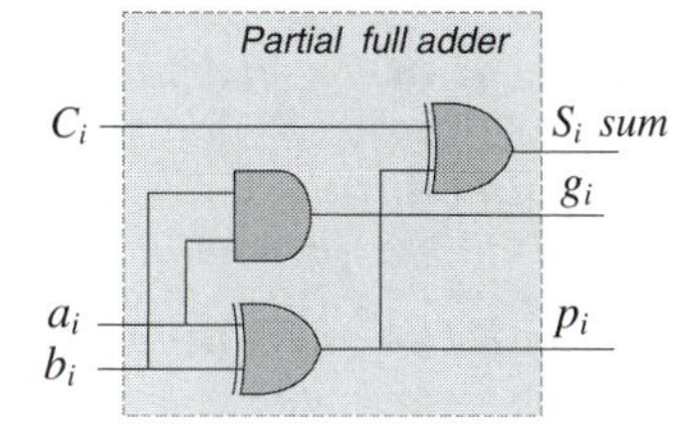

STEP 4

Lookahead logic

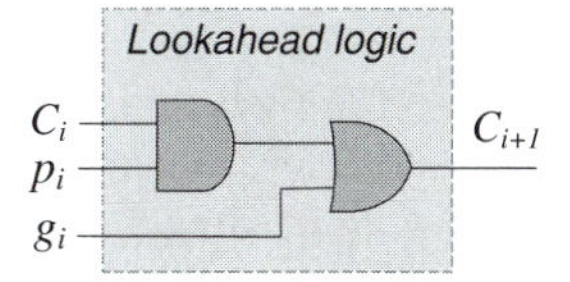

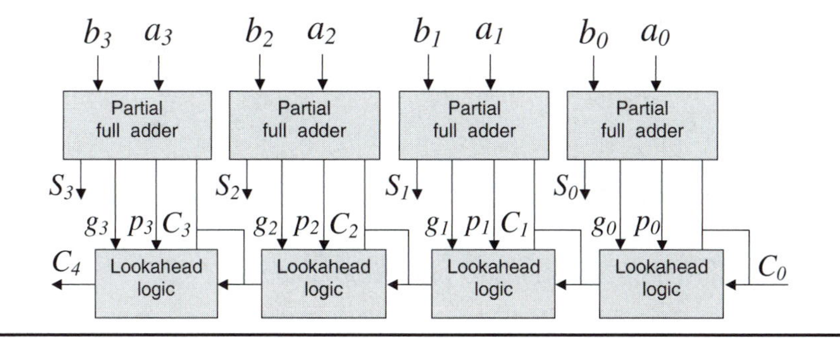

FIGURE 14.4
The 4-bit carry-look-ahead adder design (Example 14.5).

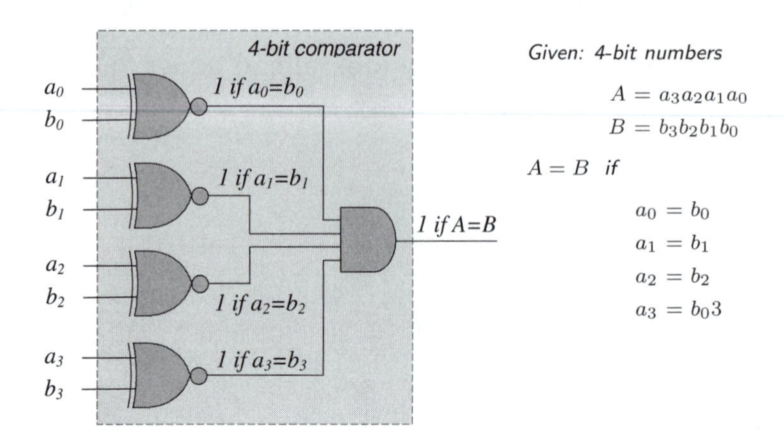

FIGURE 14.5
A 4-bit comparator (Example 14.6).

Example 14.7 (2-bit magnitude comparator.) *Given the 2-bit numbers* $A = a_1a_0$ *and* $B = b_1b_0$, *design a 2-bit magnitude comparator with three outputs:* $A = B$, $A > B$, *and* $A < B$. *The design of a 2-bit comparator includes the following steps:*

Step 1: *Formalize the problem in the form of truth tables.*
Step 2: *Formalize the problem in the form of Boolean expressions.*
Step 3: *Apply Shannon expansion (starting with the LSB of B):*

(*a*) *Determine the design of the network of 2-to-1 and 4-to-1 multiplexers for* $A = B$.
(*b*) *Find the minimal SOP expression (or POS expression) for the function* $A > B$ *and its implementation by PLA.*
(*c*) *Find the minimal SOP expression (or POS expression) for designing ROM for the implementation* $A < B$.

The function $A = B$ *is realized by two components: a 2-to-1 multiplexer and a 4-to-1 multiplexer. The function* $A > B$ *is implemented by a PLA and the function* $A < B$ *is realized by a ROM. A logic network for a 2-bit comparator is given in Figure 14.6.*

Practice problem 14.3. (Comparator.) Propose a gate network design of a 2-bit comparator that has two outputs, $A \geq B$ and $A \neq B$.
Answer is given in "Solutions to practice problems."

Design example: 2-bit magnitude comparator

STEP 1

Problem formalization in the form of truth table

Number A a_1	a_0	Number B b_1	b_0	A=B	A>B	A<B
0	0	0	0	1	0	0
0	0	0	1	0	0	1
0	0	1	0	0	0	1
0	0	1	1	0	0	1
0	1	0	0	0	1	0
0	1	0	1	1	0	0
0	1	1	0	0	0	1
0	1	1	1	0	0	1
1	0	0	0	0	1	0
1	0	0	1	0	1	0
1	0	1	0	1	0	0
1	0	1	1	0	0	1
1	1	0	0	0	1	0
1	1	0	1	0	1	0
1	1	1	0	0	1	0
1	1	1	1	1	0	0

STEP 2

Minimization using K-maps

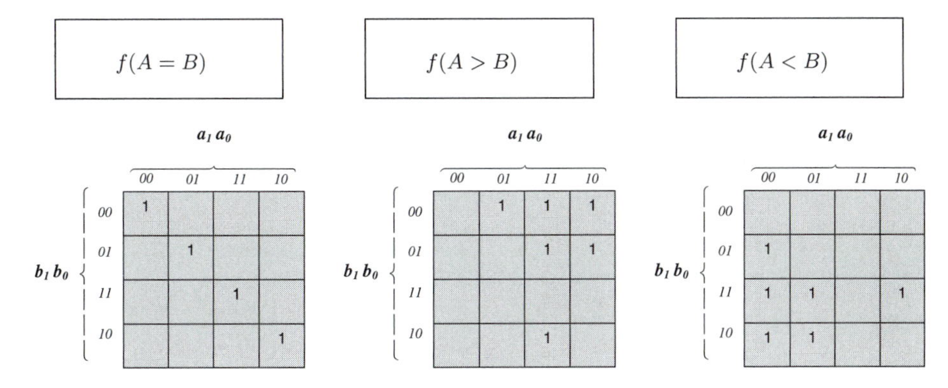

STEP 3

Formalization in the form of a minimal SOP expression

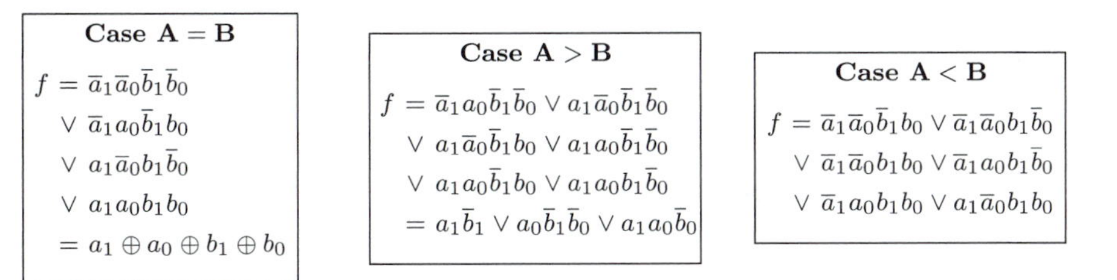

Case A = B

$$f = \overline{a}_1\overline{a}_0\overline{b}_1\overline{b}_0 \vee \overline{a}_1 a_0\overline{b}_1 b_0 \vee a_1\overline{a}_0 b_1\overline{b}_0 \vee a_1 a_0 b_1 b_0 = a_1 \oplus a_0 \oplus b_1 \oplus b_0$$

Case A > B

$$f = \overline{a}_1 a_0\overline{b}_1\overline{b}_0 \vee a_1\overline{a}_0\overline{b}_1\overline{b}_0 \vee a_1\overline{a}_0\overline{b}_1 b_0 \vee a_1 a_0\overline{b}_1\overline{b}_0 \vee a_1 a_0\overline{b}_1 b_0 \vee a_1 a_0 b_1\overline{b}_0 = a_1\overline{b}_1 \vee a_0\overline{b}_1\overline{b}_0 \vee a_1 a_0\overline{b}_0$$

Case A < B

$$f = \overline{a}_1\overline{a}_0\overline{b}_1 b_0 \vee \overline{a}_1\overline{a}_0 b_1\overline{b}_0 \vee \overline{a}_1\overline{a}_0 b_1 b_0 \vee \overline{a}_1 a_0 b_1\overline{b}_0 \vee \overline{a}_1 a_0 b_1 b_0 \vee a_1\overline{a}_0 b_1 b_0$$

FIGURE 14.6
A 2-bit magnitude comparator (Example 14.7).

Design example: 2-bit magnitude comparator (continuation)

STEP 4: IMPLEMENTATION

Multiplexer-based logic network

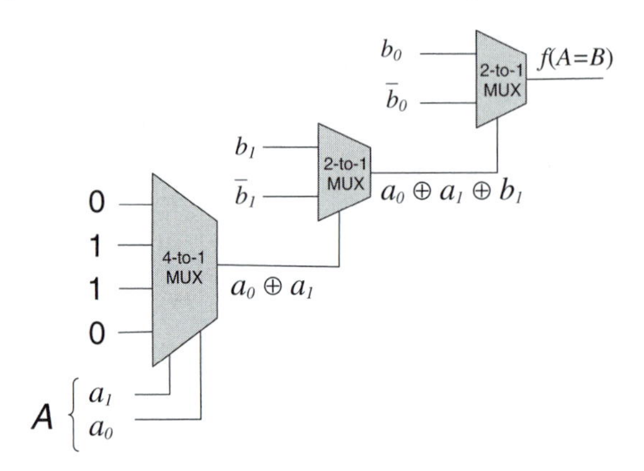

Decoder-based implementation

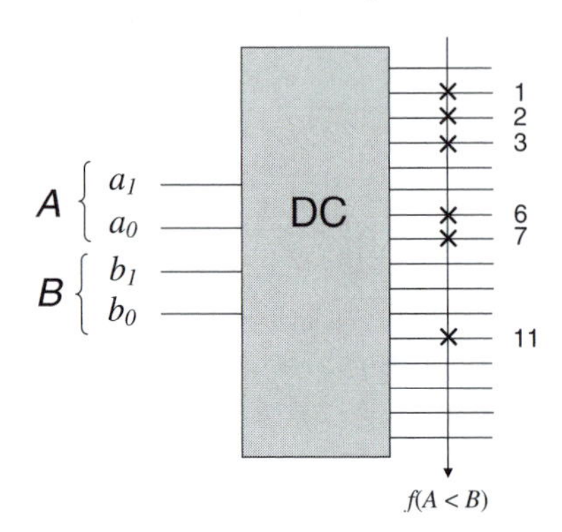

PLA-based implementation

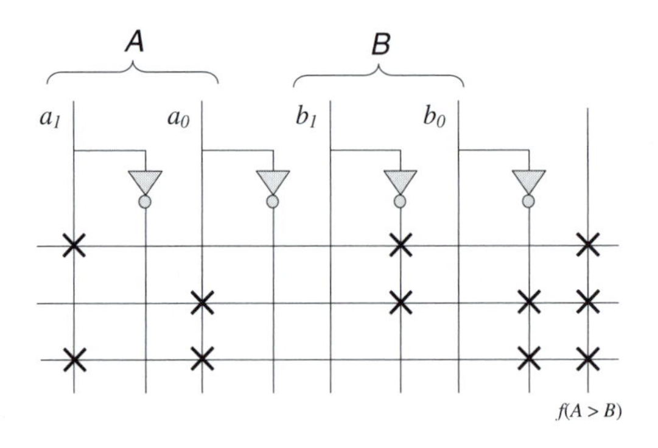

FIGURE 14.7
A 2-bit magnitude comparator, continuation (Example 14.7).

Design example: 2-bit magnitude comparator (continuation)

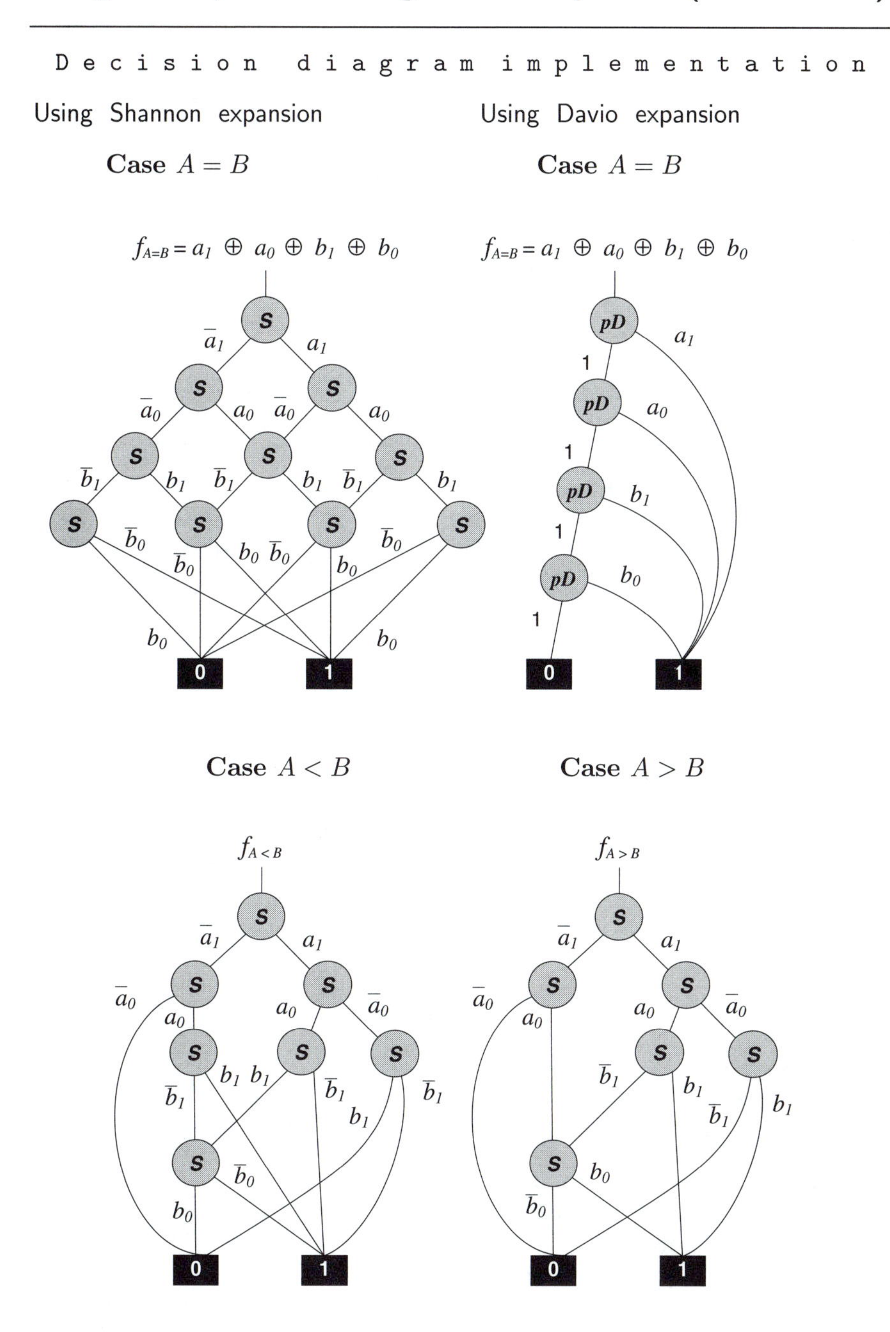

FIGURE 14.8
A 2-bit magnitude comparator, continuation (Example 14.7).

14.4 Design example: BCD adder

The code known as *binary-coded decimal* (BCD) encodes the digits 0 through 9. Because there are 10 digits to encode, it is necessary to use four bits per digit. Each digit is encoded by the binary pattern that represents its unsigned value (Table 14.1). Only 10 of the 16 available patterns are used in a BCD. This means that the remaining six patterns should not occur in logic networks that operate on BCD numbers; these patterns are usually treated as don't care conditions in the design process.

TABLE 14.1
Binary coded decimal digits.

Decimal	BCD code	Decimal	BCD code
0	0000	5	0101
1	0001	6	0110
2	0010	7	0111
3	0011	8	1000
4	0100	9	1001

Conversion between BCD and decimal representations is trivial, that is, a direct substitution of four bits for each decimal digit is all that is necessary. The addition of two BCD digits is complicated by the fact that the sum may exceed 9, in which case a correction will have to be made. Let $X = x_3x_2x_1x_0$ and $Y = y_3y_2y_1y_0$ be the BCD digits. Denote as $S = s_3s_2s_1s_0$ the desired sum of X and Y, $S = X + Y$. The algorithm for addition of two BCD digits is as follows:

BCD addition

Given: BCD X and Y
Calculate $Z = X + Y$
Step 1: If $Z \leq 9$, then the desired sum digit is equal to the result $S = Z$ and CARRY-OUT $= 0$
Step 2: (a) If $Z > 9$, then correct the desired sum digit $S = Z + 6$
(b) If $Z > 15$, and CARRY-OUT $= 1$, then correct the desired sum digit $S = Z + 6$ and keep CARRY-OUT $= 1$
Result: BCD representation of Z

The technique for adding the BCD digits is illustrated in Table 14.2.

Consider a block diagram of a BCD adder (Figure 14.9). This includes two 4-bit binary adders and a correction circuit. The first 4-bit adder produces the

TABLE 14.2
Addition of BCD digits.

Techniques for computing in BCD

Example	Comments
(4) 0100 + (5) 0101 = (9) 1001; (3) 0011 + (2) 0010 = (5) 0101	These results are correct.
(5) 0101 + (9) 1001 = (14) 1110; + 0110 Correction; (10 + 4) [1] 0100 (↑ *Carry*)	The 4-bit addition yields $Z = 5 + 9 = 14$ which is wrong in BCD format (numbers greater than 9 are not allowed). Adding 6 to the intermediate sum Z provides a necessary correction.
(8) 1000 + (9) 1001 = (17) [1] 0001; + 0110 Correction; (10 + 7) [1] 0111 (↑ *Carry*)	The 4-bit addition yields $Z = 8 + 9 = 17$ which is wrong in BCD format. Adding 6 to the intermediate sum Z provides a necessary correction.
(9) 1001 + (9) 1001 = (18) [1] 0010; + 0110 Correction; (10 + 8) [1] 1000 (↑ *Carry*)	The 4-bit addition yields $Z = 9 + 9 = 18$ which is wrong in BCD format. Adding 6 to the intermediate sum Z provides a necessary correction.

sum of two 4-bit numbers. Its result, the sum $Z = X + Y$, can be correct or incorrect. The block that detects whether $Z > 9$ generates the output signal, `ADJUST`. This signal controls the multiplexer that provides the correction of the intermediate sum when needed:

(a) If `ADJUST` $= 0$, then the output is $S = Z + 0 = Z$.

(b) If `ADJUST` $= 1$, then the output is $S = Z + 6$ and the carry-out is $C_{out} = 1$.

BCD adder design includes the following steps:

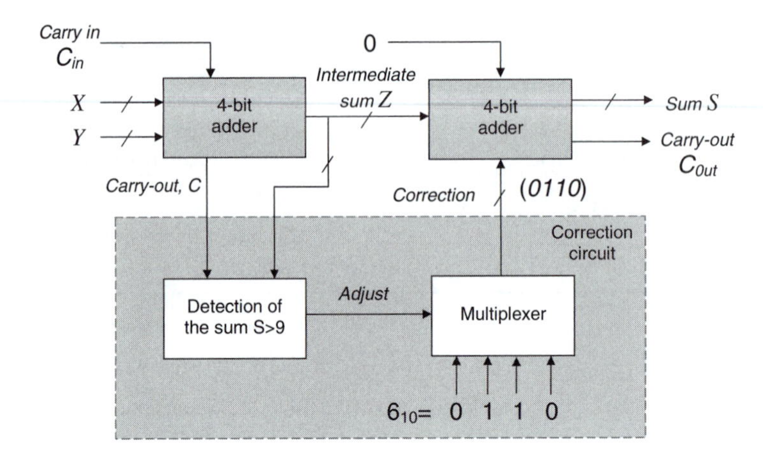

FIGURE 14.9
BCD adder.

BCD adder design

Step 1: Formalize the problem in the form of a truth table for the ADJUST function of the detection circuit using don't cares.
Step 2: Minimize the ADJUST function using a K-map.
Step 3: Design the correction circuit.

Example 14.8 (BCD adder design.) *Design the correction circuit for the 4-bit BCD adder.*
Consider the truth table of the ADJUST *output for the detector of the sum $S > 9$. The input variables are the bits $z_3z_2z_1z_0$ of sum Z, and the carry-out C of the 4-bit adder. The K-map of the* ADJUST *function is given in Figure 14.10a. The minimized function is* ADJUST $= C \vee z_1z_3 \vee z_2z_3$. *The design network is given in Figure 14.10b.*

14.5 The verification problem

Verification is the comparison of two models for consistency. There are various classes of verification techniques in logic design. In particular, they can be classified as follows:

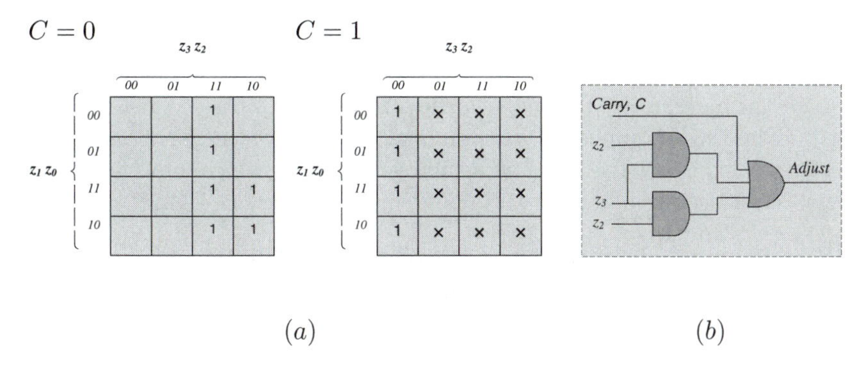

(a) (b)

FIGURE 14.10
BCD adder design: the K-maps of the ADJUST function (Example 14.8).

(a) *Implementation verification techniques*; these techniques aim to verifying the correctness of the synthesis task; and
(b) *Design verification techniques*; these techniques are related to the highest (architectural) level of design.

In logic synthesis of combinational networks, the functionality of the networks to be designed is given in terms of a network of logical gates. An important task in logic synthesis is to optimize the representation of this network on the gate level with respect to several factors. These optimization criteria include the number of gates, chip area, energy consumption, delay, clock period, etc.

14.5.1 Formal verification

Formal verification of logical networks demands that a mathematical proof of correctness implicitly covers all possible input patterns (stimuli). It has to be proven formally that the *specification* and *implementation* are functionally equivalent, i.e., that both compute exactly the same Boolean function. The verification problem is formulated as follows.

Verification problem

Given: The specifications of a network to be implemented in terms of a function f, a verified network realizing f, and a new realization which is claimed to realize f
Problem: Verify that the realization and specification are equivalent
Solution: Prove mathematically that the input-output behavior of the network and the new realization are equal

The most basic requirement of a new design is that it be *functionally correct.* That is, the design must meet the problem specification. To verify functional correctness, logic simulation software packages are used to create and test a computational model of the proposed design.

Practical design verification means validating that an implemented gate level design matches its desired behavior as specified at the register-transfer level. This is accomplished in practical design by means of *equivalence checking*. This is not the same as formal verification, which is proving that a design has a desirable property. Namely, in equivalence checking:

- Correctness is defined as the equivalence of two designs, and
- Equivalence is usually localized by finding the correspondence between latches, i.e., checking if they have the same next-state function.

14.5.2 Equivalence checking problem

Equivalence checking for combinational logic networks is formulated as follows:

Equivalence checking problem

Given: Two logic networks
Problem: Check if their corresponding outputs are equal for all possible input patterns
Solution: Since two logic networks are equivalent if and only if the canonical representations of their output functions are the same:

(*a*) Derive the canonical representation (standard SOP expressions).
(*b*) Construct the canonical graphical form, such as a full tree or reduced ordered decision diagram

Example 14.9 (Verification.) *Given the specified and optimized networks (Figure 14.12), the simplest approach to verifying that these networks are equivalent is to derive truth tables independently for each network and compare the values of the outputs for corresponding inputs. From the truth table comparison, it follows that specified and optimized networks are not equivalent. The error in optimization occurs because the input x_4 must be complemented. Correction of this error results in a network that is equivalent to the specified network (implemented network).*

Practice problem **14.4. (Verification.)** Verify using truth tables that the two logic networks given in Figure 14.11 implement the same Boolean function.
Answer is given in "Solutions to practice problems."

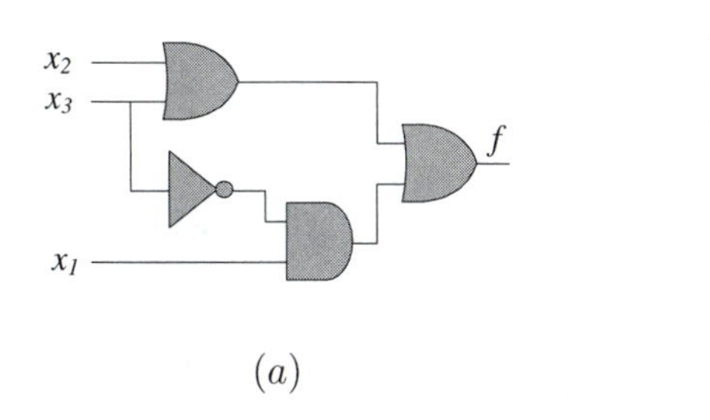

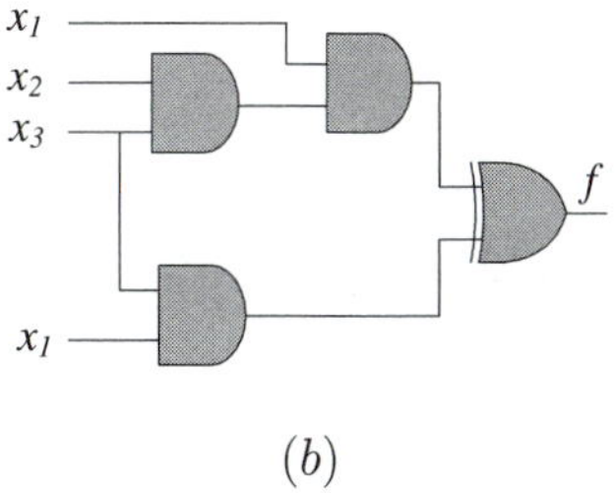

(*a*) (*b*)

FIGURE 14.11
Logic networks for Practice problem 14.4: the specified network (a) and the network after transformation (b).

A complete binary tree and a (reduced and ordered) binary decision diagram are canonical forms. Thus, two logic networks are equivalent if their decision diagrams are isomorphic.

In practice, subfunctions of two logic networks (called above specification and implementation) are transformed into a decision diagram by simulating the networks gate by gate, normally in a depth-first manner. The decision diagrams to be checked for equivalence must be reduced and ordered.

Example 14.10 (Verification.) *Figure 14.13 shows the two simplest logic networks: an AND gate and a network consisting of three NOT gates and one AND gate. Let us check their equivalence, using*

(*a*) *A comparison of the standard SOPs of both networks, and*
(*b*) *A comparison of their decision diagrams for isomorphism.*

The first network *implements the function* $f = x_1x_2$*, which is the standard SOP expression.*
The second network *implements the function* $f = \overline{\overline{x}_1 \vee \overline{x}_2}$*. Converting it into the standard SOP form yields* $g = \overline{\overline{x}_1 \vee \overline{x}_2} = xy$*.*

14.5.3 Design example 1: Functionally equivalent networks

A decision diagram is shown in Figure 14.13a. Derivation of a decision diagram requires two steps: generation of the decision diagram of the AND gate with inverted inputs, and finding the complement of this decision diagram; that is, inverting the terminal node values (Figure 14.13b).

Since SOP expressions for f and g are equal, these networks are equivalent. By inspection, it can be seen that the decision diagram of the AND gate and the last derived decision diagram are isomorphic, and, therefore, that both

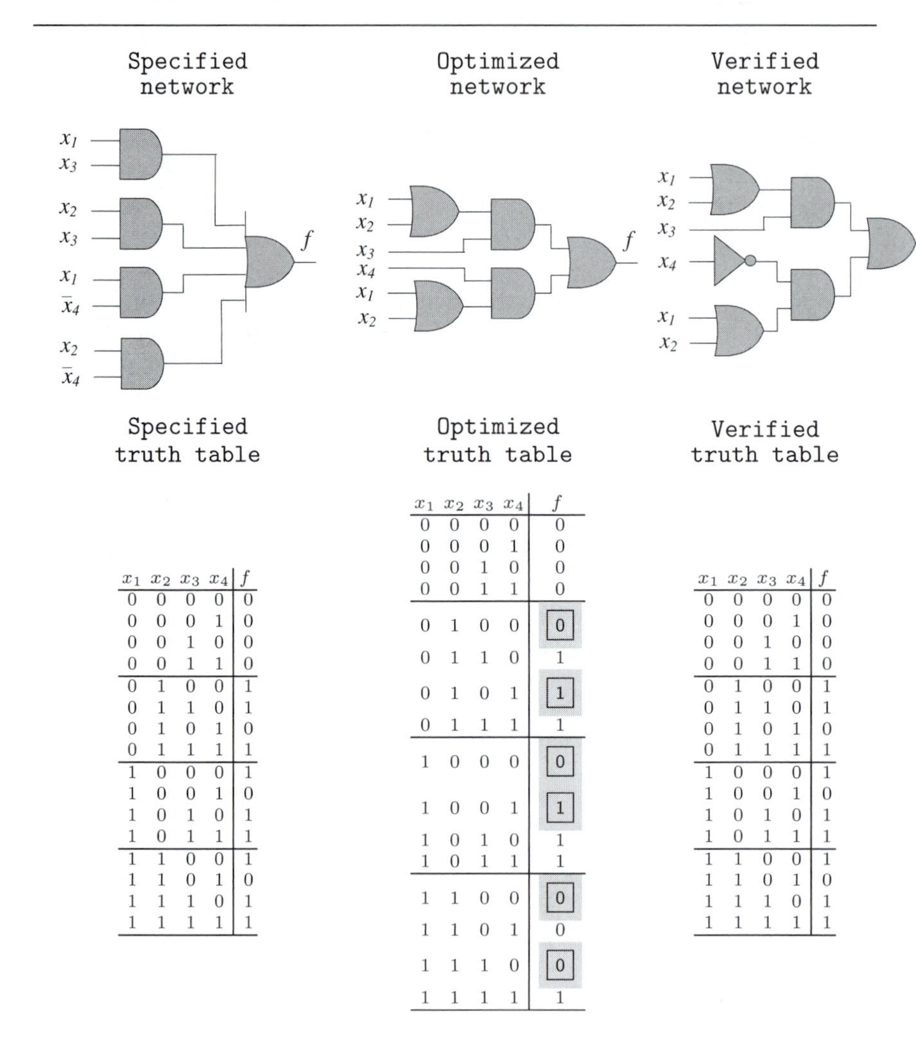

Specified truth table

x_1	x_2	x_3	x_4	f
0	0	0	0	0
0	0	0	1	0
0	0	1	0	0
0	0	1	1	0
0	1	0	0	1
0	1	1	0	1
0	1	0	1	0
0	1	1	1	1
1	0	0	0	1
1	0	0	1	0
1	0	1	0	1
1	0	1	1	1
1	1	0	0	1
1	1	0	1	0
1	1	1	0	1
1	1	1	1	1

Optimized truth table

x_1	x_2	x_3	x_4	f
0	0	0	0	0
0	0	0	1	0
0	0	1	0	0
0	0	1	1	0
0	1	0	0	0
0	1	1	0	1
0	1	0	1	1
0	1	1	1	1
1	0	0	0	0
1	0	0	1	1
1	0	1	0	1
1	0	1	1	1
1	1	0	0	0
1	1	0	1	0
1	1	1	0	0
1	1	1	1	1

Verified truth table

x_1	x_2	x_3	x_4	f
0	0	0	0	0
0	0	0	1	0
0	0	1	0	0
0	0	1	1	0
0	1	0	0	1
0	1	1	0	1
0	1	0	1	0
0	1	1	1	1
1	0	0	0	1
1	0	0	1	0
1	0	1	0	1
1	0	1	1	1
1	1	0	0	1
1	1	0	1	0
1	1	1	0	1
1	1	1	1	1

FIGURE 14.12
Verification of two networks using truth tables (Example 14.9).

networks are functionally equivalent.

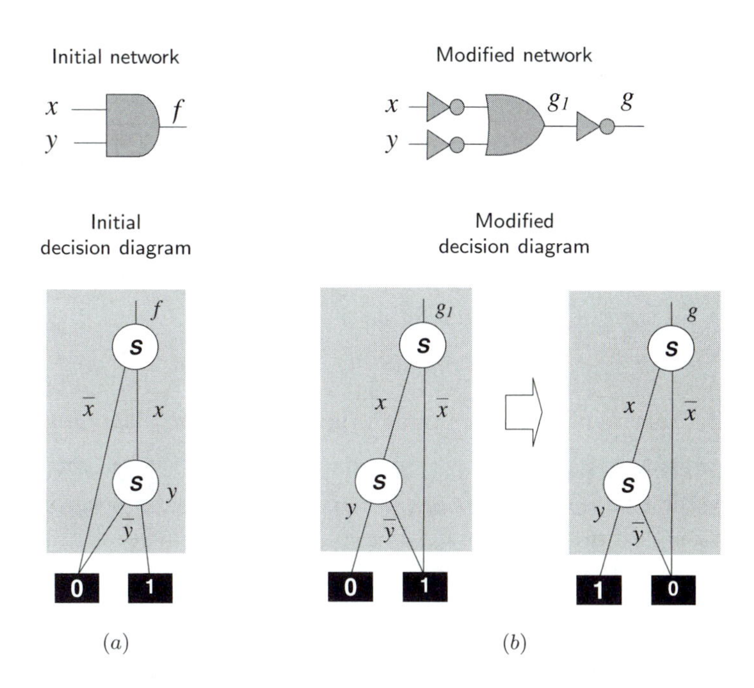

FIGURE 14.13
Two logic networks and the derivation of their decision diagrams.

Practice problem 14.5. (Verification.) Check the networks given in Figure 14.14 for equivalence using standard SOP expressions.
Answer is given in "Solutions to practice problems."

14.5.4 Design example 2: Verification of logic networks using decision diagrams

Consider the two networks shown in Figure 14.15a,b:

The first network: A decision diagram of the inverter is used to construct the decision diagram of the AND gate ($x_1\overline{x}_3$); and in the same manner the decision diagram of the other AND gate is created (x_2x_3) (Figure 14.15b). Next, their two AND functions are considered to be the inputs

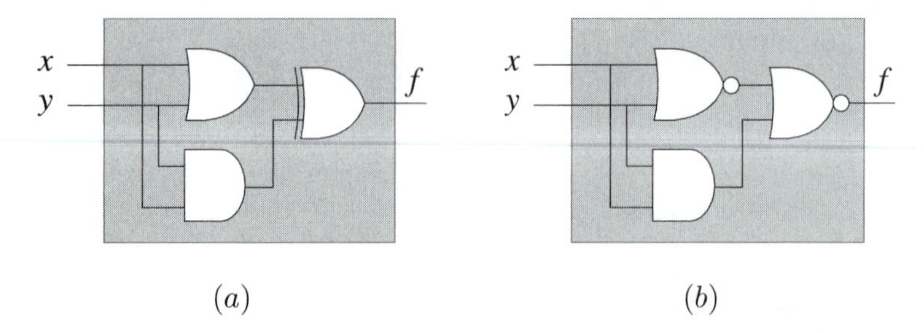

(a) (b)

FIGURE 14.14
Networks for Practice problem 14.5.

of the OR gate, which forms the function

$$f_1 = x_1\overline{x}_3 \vee x_2x_3.$$

This manipulation results in the final decision diagram for f_1.

The second network: The decision diagrams for both AND gates in the first level of the network are derived first. Next, decision diagrams for an OR gate ($x_1 \vee x_2x_3$) and for an EXOR gate are constructed

$$(x_1 \vee x_2x_3) \oplus x_1x_3.$$

The variable ordering is fixed to x_1, x_2, x_3. A check for the isomorphism of the decision diagrams presented proves their equivalence, and thus, the functional equivalence of the given networks.

Another approach is to manipulate decision diagrams without isomorphism comparison. To verify that two combinational networks with outputs F and G are equivalent, the decision diagram for $\overline{f \oplus g}$ is constructed, where f and g represent the Boolean functions for F and G, respectively. Due to the canonicity of decision diagram, the two circuits implement the same Boolean function if and only if the resulting decision diagram is identical to the terminal 1.

Example 14.11 (Verification using decision diagrams.) *Consider the functions f and g from Example 14.10. To check their equivalence, let us derive the decision diagram for the EXOR function $\overline{f \oplus g}$. Figure 14.18 illustrates the calculation using decision diagrams for f and $\overline{g}$ using the equation $\overline{f \oplus g} = f \oplus \overline{g}$. Since the resulting diagram is a constant terminal "1", both networks are considered to be equivalent.*

Practice problem 14.6 (Verification using decision diagrams.)
Prove that a complete decision tree (Figure 14.17)a and decision diagram

Design example: Verification of two logic networks

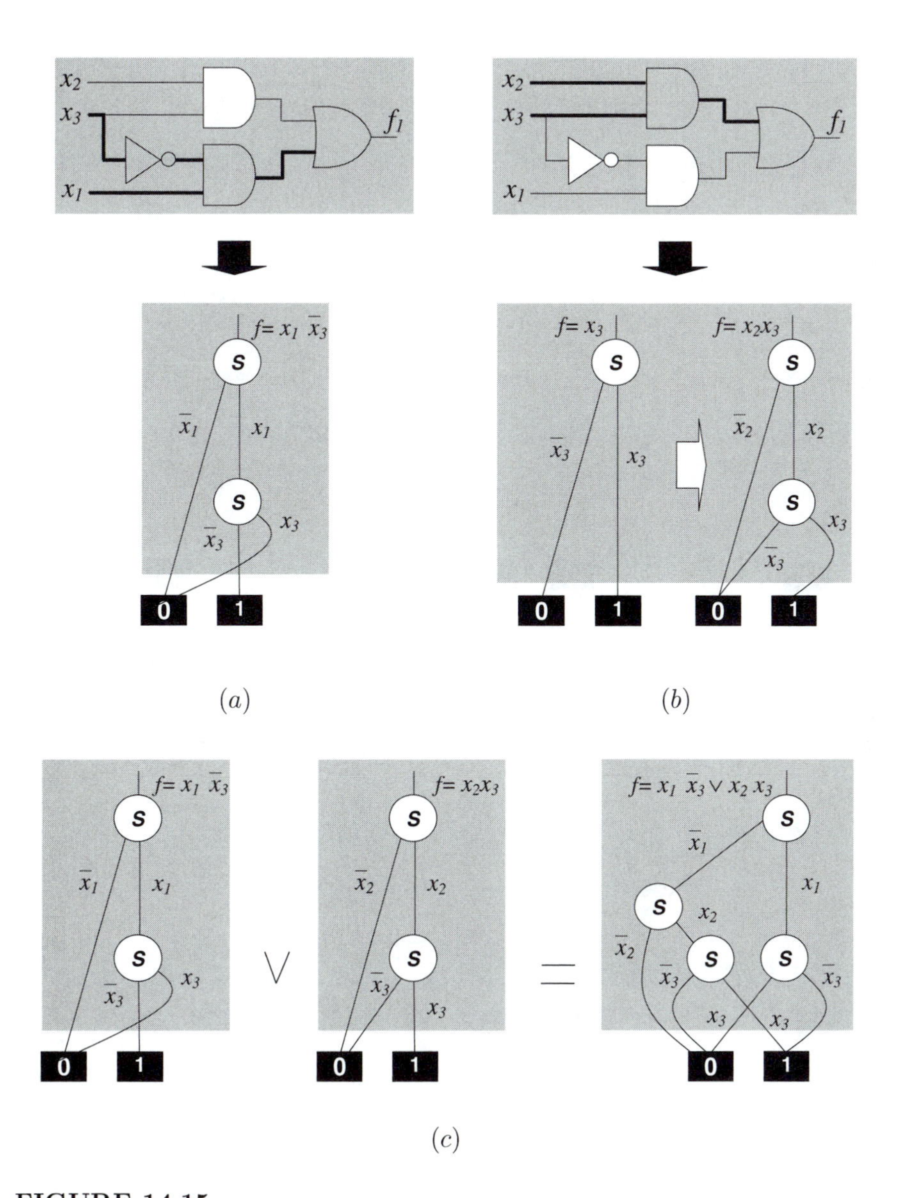

FIGURE 14.15
Construction of the decision diagram for the Boolean function $f_1 = x_1\overline{x}_3 \vee x_2x_3$.

Design example: Verification of two logic networks (continuation)

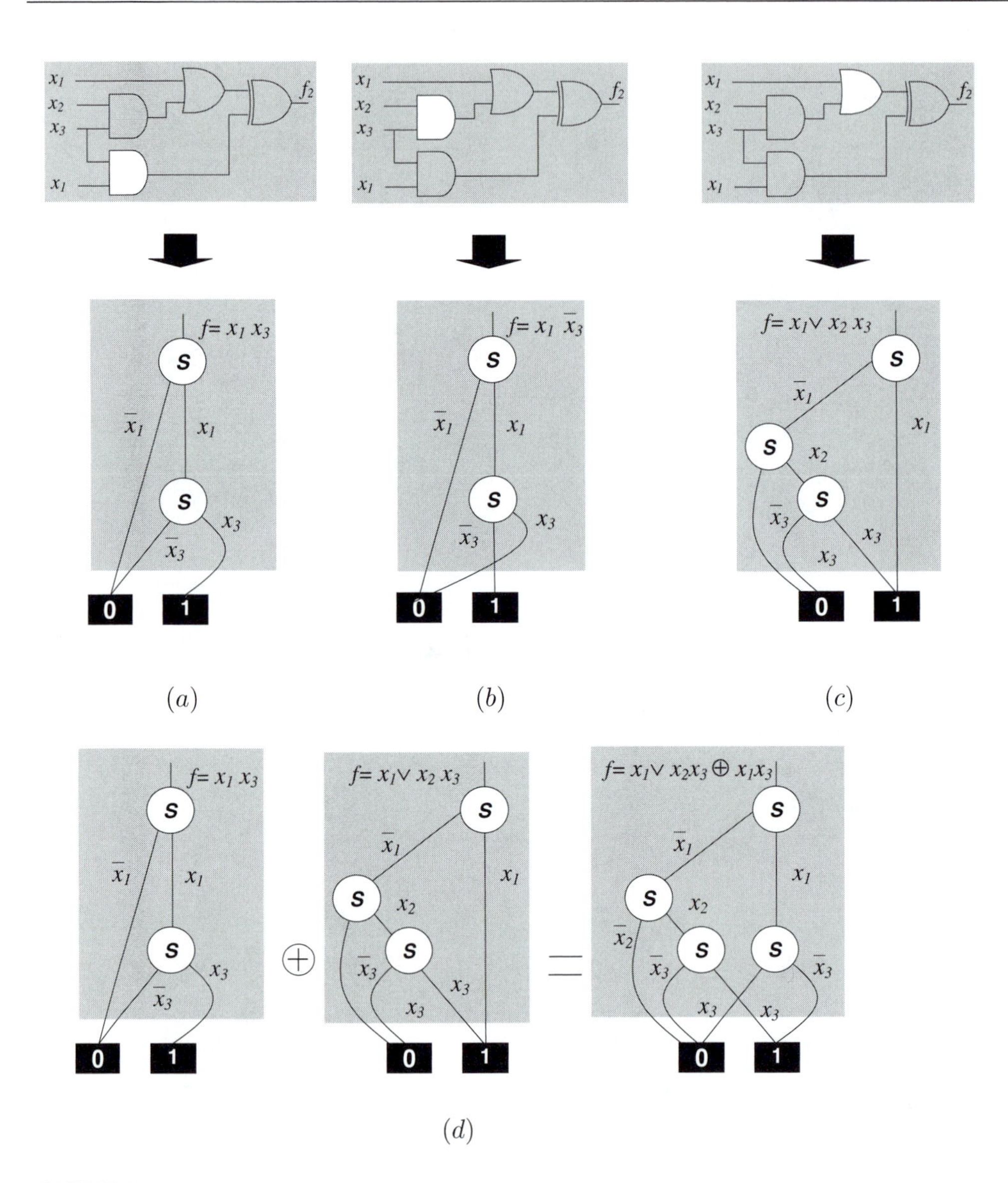

FIGURE 14.16
The logic network and the decision diagram of its two AND gates (a) and (b), and OR gate (the function $x_1 \vee x_2x_3$) (c); construction of the decision diagram for the function $(x_1 \vee x_2x_3) \oplus x_1x_3$.

(Figure 14.17)b represent the same Boolean function.
Answer is given in "Solutions to practice problems."

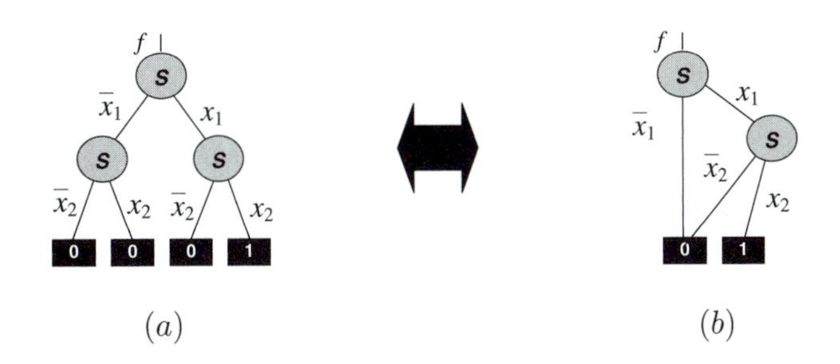

FIGURE 14.17
Decision diagrams for Practice problem 14.6.

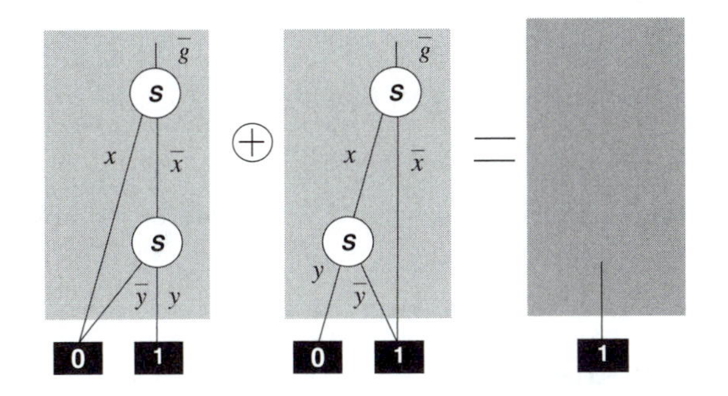

FIGURE 14.18
Calculation using decision diagrams for f and $\overline{g}$ (Example 14.11).

14.6 Decomposition

Simplification of logic networks by factoring is a satisfactory technique in many cases, but it does not have systematic methods for all cases. It is evident that a more systematic method for optimization would be very useful. This process is called *decomposition.* Decomposition is the central problem of the representation, manipulation, optimization, and implementation of Boolean functions. For example, a two-level logic network can be decomposed into a multi-level network. In some cases, the multi-level networks may reduce the cost of implementation.

Decomposability

Decomposability is a property of a Boolean function $f(\mathbf{X})$ which can be described in the form

$$f(\mathbf{X}) = h(\mathbf{X}_1, g(\mathbf{X}_2))$$

where $\mathbf{X}_1$ and $\mathbf{X}_2$ are sets of variables,

$$\mathbf{X} = \mathbf{X}_1 \cup \mathbf{X}_2,$$

and g is called a *subfunction.*

If such a decomposition exists, it is called a *functional* decomposition of f. Generally, the functions h and g are less complex than f.

14.6.1 Disjoint and non-disjoint decomposition

A decomposition is called *disjoint* if the sets of n_1 and n_2 variables, $\mathbf{X}_1$ and $\mathbf{X}_2$, respectively, are not overlapped; that is,

$$\mathbf{X}_1 \cap \mathbf{X}_2 = \emptyset.$$

A decomposition is called *non-disjoint* if the sets of variables $\mathbf{X}_1$ and $\mathbf{X}_2$ are overlapped; i.e.,

$$\mathbf{X}_1 \cap \mathbf{X}_2 \neq \emptyset.$$

Example 14.12 (Decomposition.) *Let f be a Boolean function of six variables $x_1, x_2, \ldots, x_6$.*

(a) The partition $(\mathbf{X}_1, \mathbf{X}_2)$, where $\mathbf{X}_1 = (x_1, x_2, x_6)$ and $\mathbf{X}_2 = (x_3, x_4, x_5)$ results in a disjoint decomposition because the sets $\mathbf{X}_1$ and $\mathbf{X}_2$ are not overlapped: $\mathbf{X}_1 \cap \mathbf{X}_2 = (x_1, x_2, x_6) \cap (x_3, x_4, x_5) = \emptyset$.

(b) The partition $(\mathbf{X}_1, \mathbf{X}_2)$, where $\mathbf{X}_1 = (x_1, x_2, x_3, x_6)$ and $\mathbf{X}_2 = (x_2, x_3, x_4, x_5)$ results in a non-disjoint decomposition because the sets $\mathbf{X}_1$ and $\mathbf{X}_2$ are overlapped: $\mathbf{X}_1 \cap \mathbf{X}_2 = (x_1, x_2, x_3, x_6) \cap (x_2, x_3, x_4, x_5) \neq \emptyset$.

Decomposition can be applied to various data structures: decomposition of Boolean function (in algebraic, matrix, or cube-based forms), decomposition of logic networks, and decomposition using decision diagrams.

14.6.2 Decomposition chart

The *decomposition chart* of a Boolean function f is defined as follows:

The decomposition chart

- Decomposition chart is a table with 2^{n_1} columns and 2^{n_2} rows,
- A binary number corresponding to a variable assignment is assigned to each row and column,
- One element of the table corresponds to one value of a Boolean function f.

In other words, the decomposition chart is a rearranged truth table of the function.

Example 14.13 (Decomposition chart.) *A decomposition chart for the Boolean function f of five variables is given in Figure 14.19. A set of variables* $\mathbf{X}$ *is partitioned into two sets,* $\mathbf{X}_1 = \underbrace{(x_1, x_2, x_3)}_{n_1=3}$ *and* $\mathbf{X}_2 = \underbrace{(x_4, x_5)}_{n_2=2}$.

Partition $\mathbf{X}_1 = (x_1, x_2, x_3)$

Partition $\mathbf{X}_2 = (x_4, x_5)$	000	001	010	011	100	101	110	111
00		1			1	1	1	
01								
10	1		1	1				1
11		1			1	1	1	

FIGURE 14.19
A decomposition chart for a Boolean function of five variables (Examples 14.13 and 14.14).

Practice problem 14.7. (Decomposition chart.) Given the Boolean function $f = \overline{x}_1\overline{x}_3 \vee x_2x_4$, find its decomposition charts
(a) with respect to $\mathbf{X}_1 = (x_1, x_2)$ and $\mathbf{X}_2 = (x_3, x_4)$
(b) with respect to $\mathbf{X}_1 = (x_1, x_3)$ and $\mathbf{X}_2 = (x_2, x_4)$.
Answer is given in "Solutions to practice problems."

The number of distinct columns (rows) in the decomposition chart is called the *column (row) multiplicity*.

Example 14.14 (Decomposition chart.) *The column multiplicity in the decomposition chart in Figure 14.19 is two, and the row multiplicity is three.*

14.6.3 Disjoint bi-decomposition

Consider the simplest case of decomposition, disjoint *bi-decomposition*, which means the decomposition of a Boolean function into exactly two subfunctions of variables from two partitions, $\mathbf{X}_1$ and $\mathbf{X}_2$, with respect to one of three operations:

Types of the disjoint bi-decomposition

- ▶ OR type, $f = g_1(\mathbf{X}_1) \vee g_2(\mathbf{X}_2)$,
- ▶ AND type, $f = g_1(\mathbf{X}_1) \wedge g_2(\mathbf{X}_2)$,
- ▶ EXOR type, $f = g_1(\mathbf{X}_1) \oplus g_2(\mathbf{X}_2)$

A simple check for disjoint bi-decomposition of a Boolean function can be accomplished using the decomposition chart.

Conditions of function decomposability

A Boolean function f has a disjoint bi-decomposition if and only if the row and column multiplicities of its decomposition chart are less than or equal to two.

This is a *necessary*, but *not a sufficient* condition.

Example 14.15 (Multiplicity.) *The Boolean function whose decomposition chart is given in Figure 14.19, does not have a disjoint bi-decomposition with respect to the partitions* (X_1, X_2) *since its row multiplicity is three.*

A variety of approaches to verifying this and other types of decompositions have been developed, according to various criteria of design, for single-output and multi-output functions. One of the approaches for verification of existence of OR and AND type decompositions given a single-output Boolean function is considered below.

A Boolean function has OR-type disjoint bi-decomposition if every product in the minimal SOP for this function consists of literals from $\mathbf{X}_1$ only or $\mathbf{X}_2$ only. The minimal SOP expression is defined in this case as an SOP form consisting of the prime implicants only (and some of them can be essential prime implicants), so that no product can be deleted without changing the function, represented by this expression.

Algorithm for finding OR type disjoint decomposition

Given: `a minimal SOP expression consisting of` t `products.`
Step 1 `Start with a trivial partition, so that each partition includes one variable only.`
Step 2 `Form another partition by merging two blocks of the previous partition if at least one literal from each block occurs in the first product. Repeat for all` t `products of the SOP expression.`
Step 3 `If the` t`-th partition has at least two blocks,` $\mathbf{X}_1$ `and` $\mathbf{X}_2$`, then the function has a disjoint bi-decomposition of the form`

$$f(\mathbf{X}_1, \mathbf{X}_2) = g_1(\mathbf{X}_1) \vee g_2(\mathbf{X}_2)$$

14.6.4 Design example: The OR type bi-decomposition

Consider the Boolean function $f = \overline{x}_1\overline{x}_2x_3 \vee \overline{x}_1x_2\overline{x}_3 \vee x_1x_2\overline{x}_3 \vee \overline{x}_1x_4x_5 \vee x_4x_5$. The process of finding out if the function has an OR type disjoint bi-decomposition is illustrated in Figure 14.20.

Practice problem **14.8. (OR type decomposition.)** Verify if the function $f = \overline{x}_1\overline{x}_3 \vee x_2x_4$ has an OR type disjoint decomposition with respect to the partition: (a) $(\{x_1, x_2\}, \{x_3, x_4\})$ and (b) $(\{x_1, x_3\}, \{x_2, x_4\})$
Answer is given in "Solutions to practice problems."

14.6.5 Design example: The AND type bi-decomposition

OR type decomposition is relevant to AND type decomposition as follows from the following statement:

Conditions for the existence of AND type disjoint bi-decomposition

A Boolean function f has AND type disjoint bi-decomposition with respect to $\mathbf{X}_1, \mathbf{X}_2$, if and only if $\overline{f}$ has an OR type disjoint bi-decomposition with respect to this partition.

Let $f = x_1x_2\overline{x}_4 \vee x_1x_2x_5 \vee \overline{x}_3\overline{x}_4 \vee \overline{x}_3x_5$. The decomposition chart has a column and row multiplicity of two. However, this function does not have an OR type disjoint bi-decomposition. This can be proved using the algorithm given above: the products $x_1x_2\overline{x}_4$ and $\overline{x}_3\overline{x}_4$ shows that x_1, x_2, x_3, x_4 must be included in the same set, and the products $x_1x_2x_5$ and $\overline{x}_3\overline{x}_5$ shows that x_1, x_2, x_3, x_5 must be included in the same block, so all variables must be included in one block.

Design example: OR type bi-decomposition

INPUT DATA

Boolean expression

$$f = \overline{x}_1\overline{x}_2x_3 \vee \overline{x}_1x_2\overline{x}_3 \vee x_1x_2\overline{x}_3 \vee \overline{x}_1x_4x_5 \vee x_4x_5$$

PRELIMINARY STEP

Minimization

$$f = \overline{x}_1\overline{x}_2x_3 \vee (\overline{x}_1 \vee x_1)x_2\overline{x}_3 \vee x_4x_5(\overline{x}_1 \vee 1) = \overline{x}_1\overline{x}_2x_3 \vee x_2\overline{x}_3 \vee x_4x_5$$

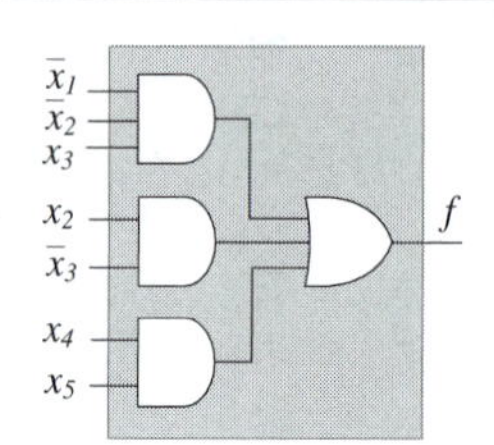

STEP 1

(a) Assign the partitions

$$f = \{(x_1), (x_2), (x_3), (x_4), (x_5)\}$$

(b) Consider the first product of the minimal SOP expression, $\overline{x}_1\overline{x}_2x_3$ Form the new partition

$$f = \{(x_1, x_2), (x_3), (x_4), (x_5)\}$$

(c) Consider the second product of the minimal SOP expression, $x_2\overline{x}_3$ and form another partition

$$f = \{(x_1, x_2, x_3), (x_4), (x_5)\}$$

(d) Consider the third product of the minimal SOP expression, x_4x_5 and form the last partition

$$f = \{(x_1, x_2, x_3), (x_4, x_5)\}$$

STEP 2

Necessity condition check

Partition $\mathbf{X}_1 = (x_1, x_2, x_3)$ (columns); Partition $\mathbf{X}_2 = (x_4, x_5)$ (rows)

	000	001	010	011	100	101	110	111
00		1	1				1	
01		1	1				1	
10		1	1				1	
11	1	1	1	1	1	1	1	1

STEP 3

OR bi-decomposition

$$f = (\overline{x}_1\overline{x}_2x_3 \vee x_2\overline{x}_3) \vee x_4x_5 = g_1\overbrace{(x_1, x_2, x_3)}^{\mathbf{X}_1} \vee g_2\overbrace{(x_4, x_5)}^{\mathbf{X}_2}$$

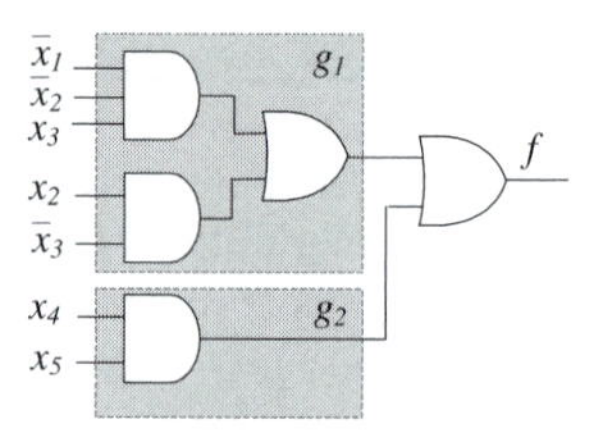

FIGURE 14.20
OR bi-decomposition for a Boolean function of five variables.

However, the complement of the function f has OR type disjoint decomposition with respect to the partition $f = (\{x_1, x_2, x_3\}, \{x_4, x_5\})$ (Figure 14.21). Therefore, the function f has AND type disjoint bi-decomposition. To find it, let us convert the SOP form of f into the POS form:

$$f = x_1x_2(\overline{x}_4 \vee x_5) \vee \overline{x}_3(\overline{x}_4 \vee x_5) = (x_1x_2 \vee \overline{x}_3)(\overline{x}_4 \vee x_5)$$

This form clearly corresponds to the AND type decomposition $f = g_1\overbrace{(x_1, x_2, x_3)}^{\mathbf{X}_1} \wedge g_2\overbrace{(x_4, x_5)}^{\mathbf{X}_2}$.

Practice problem **14.9. (AND type decomposition.)** Given the Boolean function $f = \overline{x}_1\overline{x}_3 \vee x_2x_4$, find if this function has an AND type decomposition.
Answer is given in "Solutions to practice problems."

14.6.6 Functional decomposition using decision diagrams

Decision diagrams are used in decomposition techniques as follows:

- ▶ The results of functional decomposition are converted into decision diagrams; decision diagrams are used as a data structure for implementation, they are derived from the decomposition chart; and/or
- ▶ An initial decision diagram is decomposed using the specific rules of partitioning decision diagrams; in this approach, decision diagram structure is used in all steps of the decomposition.

Table 14.3 shows two types of decomposition (out of many). Shannon decomposition is used to implement a network based on multiplexers. Davio decomposition is used in the form of the *positive Davio* (pD) expansion $f = f_0 \oplus x_i f_2$ and *negative Davio* (nD) expansion $f = f_1 \oplus \overline{x}_i f_2$, where $f_0 = f_{x_i=0}$, $f_1 = f_{x_i=1}$, and $f_2 = f_{x_i=1} \oplus f_{x_i=0}$; it is useful in the case of gate level implementation using polynomial representations of Boolean functions.

14.6.7 Design example: Shannon decomposition with respect to a subfunction

The Shannon decomposition of the Boolean function $f = x_1\overline{x}_2x_3 \vee \overline{x}_1x_2x_3 \vee \overline{x}_1\overline{x}_2x_4 \vee x_1x_2x_4$ with respect to the subfunction $g = x_1 \oplus x_2$ can be found as shown in Figure 14.22. It also shows the reduced ordered decision diagram, and decomposed decision diagram using multiplexers. The inputs of the multiplexer are the functions x_3 and x_4, and the control signal is generated by the function g, which is implemented as a decision diagram itself.

Design example: AND type bi-decomposition

INPUT DATA

Boolean function

$$f = x_1x_2\overline{x}_4 \vee x_1x_2x_5 \vee \overline{x}_3\overline{x}_4 \vee \overline{x}_3x_5$$

PRELIMINARY STEP

Find the complement of the function f:

$$\begin{aligned}\overline{f} &= \overline{x_1x_2\overline{x}_4 \vee x_1x_2x_5 \vee \overline{x}_3\overline{x}_4 \vee \overline{x}_3x_5}\\ &= \overline{x_1x_2(\overline{x}_4 \vee x_5) \vee \overline{x}_3(\overline{x}_4 \vee x_5)}\\ &= \overline{(x_1x_2 \vee \overline{x}_3)(\overline{x}_4 \vee x_5)}\\ &= \overline{x_1x_2 \vee \overline{x}_3} \vee \overline{\overline{x}_4 \vee x_5}\\ &= \overline{x_1x_2}x_3 \vee x_4\overline{x}_5\\ &= (\overline{x}_1 \vee \overline{x}_2)x_3 \vee x_4\overline{x}_5\\ &= \overline{x}_1x_3 \vee \overline{x}_2x_3 \vee x_4\overline{x}_5\end{aligned}$$

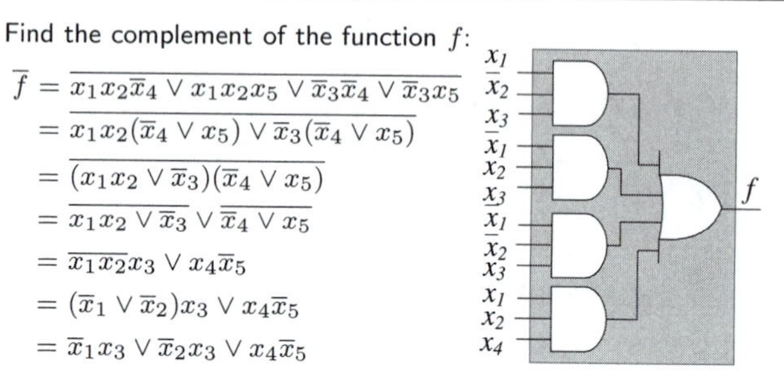

STEP 1

(*a*) Assign the partition $f = \{(x_1), (x_2), (x_3), (x_4), (x_5)\}$

(*b*) Consider the first product of the SOP expression of $\overline{f}$, $\overline{x}_1x_3$ Form the new partition $f = \{(x_1, x_3), (x_2), (x_4), (x_5)\}$

(*c*) Consider the second product of the SOP expression, $\overline{x}_2x_3$ and form another partition $f = \{(x_1, x_2, x_3), (x_4), (x_5)\}$

(*d*) Consider the third product of the SOP expression, $x_4\overline{x}_5$ and form the last partition $f = \{(x_1, x_2, x_3), (x_4, x_5)\}$

Decision: Since $\overline{f}$ has OR type decomposition, then f has AND type decomposition

STEP 2

Necessity condition check: decomposition chart

Partition $\mathbf{X}_1 = (x_1, x_2, x_3)$ (columns); Partition $\mathbf{X}_2 = (x_4, x_5)$ (rows)

	000	001	010	011	100	101	110	111
00	1		1		1		1	1
01	1		1		1		1	1
10								
11	1		1		1		1	1

Row and column multiplicity are equal to two

STEP 3

Form the POS form

$$\begin{aligned}f &= x_1x_2(\overline{x}_4 \vee x_5) \vee \overline{x}_3(\overline{x}_4 \vee x_5)\\ &= (x_1x_2 \vee \overline{x}_3)(\overline{x}_4 \vee x_5)\\ &= (x_1 \vee \overline{x}_3)(x_2 \vee \overline{x}_3)(\overline{x}_4 \vee x_5)\end{aligned}$$

STEP 4

AND type bi-decomposition

$$\overline{f} = g_1\overbrace{(x_1, x_2, x_3)}^{\mathbf{X}_1} \wedge g_2\overbrace{(x_4, x_5)}^{\mathbf{X}_2}$$

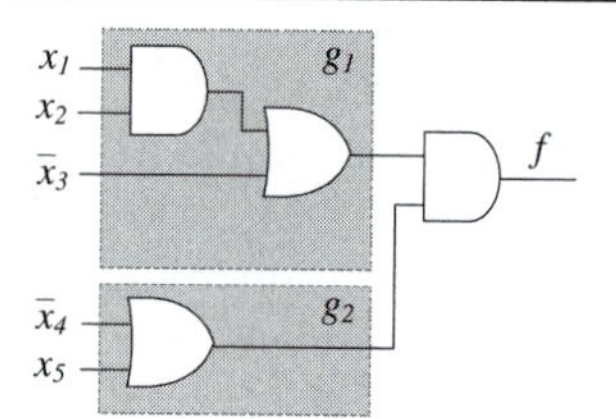

FIGURE 14.21
AND bi-decomposition for a Boolean function of five variables.

Design example: Shannon decomposition

INPUT DATA

$$f = x_1\overline{x}_2x_3 \vee \overline{x}_1x_2x_3 \vee \overline{x}_1\overline{x}_2x_4 \vee x_1x_2x_4$$

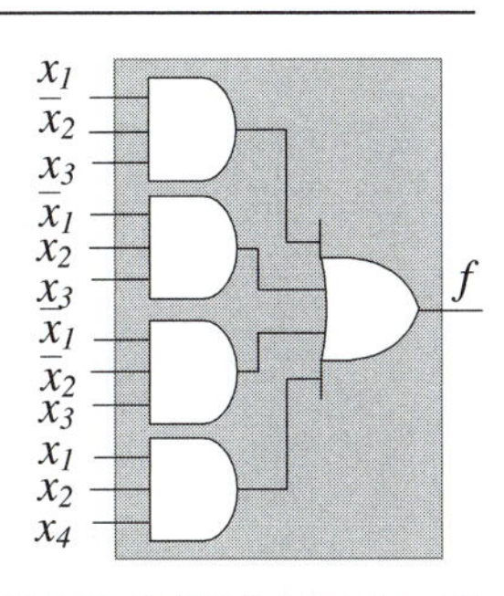

STEP 1

The Shannon decomposition of the Boolean function f with respect to the subfunction $g = x_1 \oplus x_2$ can be found via factoring:

$$f = x_3(x_1\overline{x}_2 \vee \overline{x}_1x_2) \vee x_4(\overline{x}_1\overline{x}_2 \vee x_1x_2)$$

Since

$$x_1\overline{x}_2 \vee \overline{x}_1x_2 = \overline{\overline{x}_1\overline{x}_2 \vee x_1x_2}x_1 \oplus x_2 = g,$$

then $f = x_3\overline{g} \vee x_4g$.

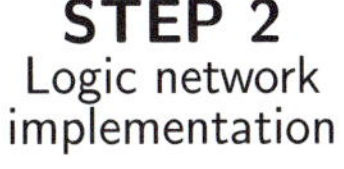

STEP 2
Logic network implementation

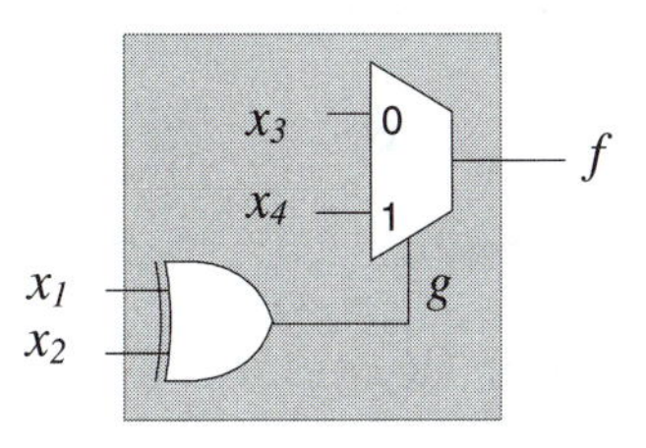

STEP 3
Decision diagram implementation

Initial decision diagram

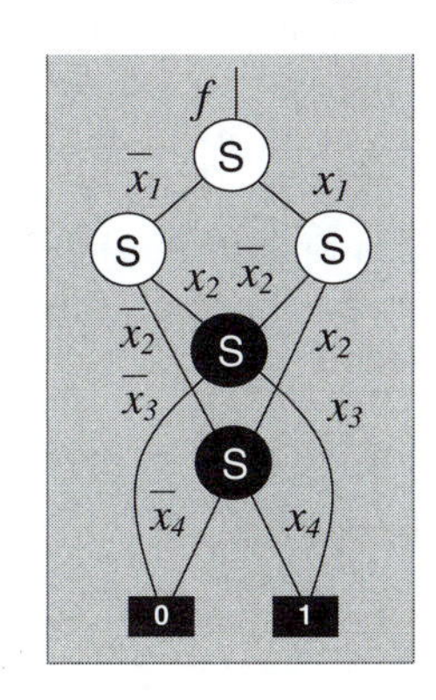

Decomposed decision diagram

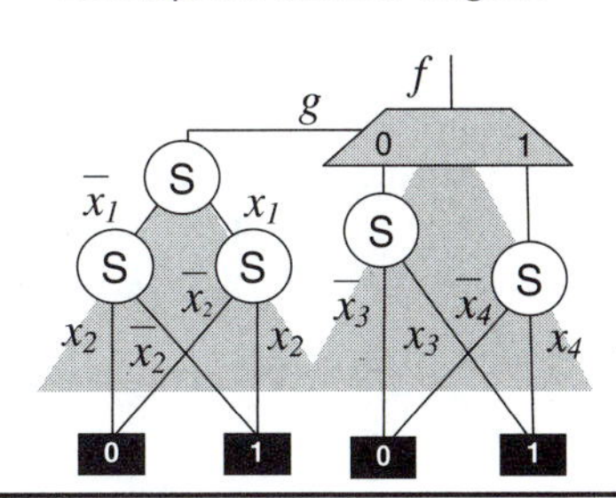

FIGURE 14.22
Shannon decomposition for a Boolean function of four variables.

TABLE 14.3
Analysis of decomposition techniques.

TYPE	DEFINITION	IMPLEMENTATION
Shannon decomposition	$f_1(X_0) = f(X_0, x_i = 0)$ $f_2(X_0) = f(X_0, x_i = 1)$ $f(f_1, f_2, x_i) = \overline{x}_i f_1 \vee x_i f_2$	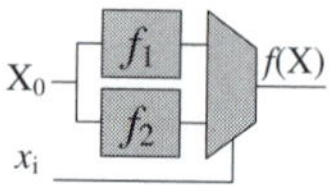
Positive Davio pD and negative Davio nD decomposition	$f_1(X_0) = f(X_0, x_i = 0)$ $f_2(X_0) = f_{x_i=0} \oplus f_{x_i=1}$ pD node: $f(f_1, f_2, x_i) = f_1 \oplus x_i f_2$ nD node: $f(f_1, f_2, x_i) = f_1 \oplus \overline{x}_i f_2$	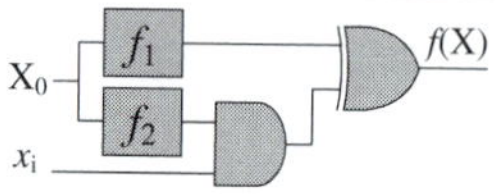 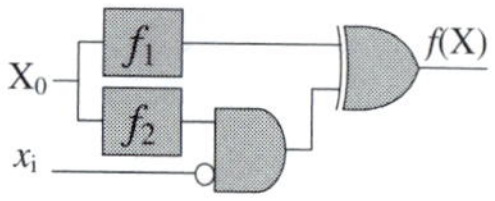

Practice problem **14.10.** **(Davio decomposition.)** Given the Boolean function $f = \overline{x}_1\overline{x}_2x_3 \vee x_1x_2$, find its positive Davio decomposition with respect to the variable x_3.
Answer is given in "Solutions to practice problems."

14.7 Error detection and error correction logic networks

Digital systems use data in the form of a group of bits for their internal operations. Since there is a possibility that during information processing or storage, data can get corrupted due to physical defects in the system, there should be some provisions in the system for detecting erroneous bits in data. It may also be necessary to correct errors in data in order to restore the system to its normal operating mode. This typically requires additional or *redundant* bits to be appended to the data or information bits for (Figure 14.23): *error detection*, and/or *error correction*. The following terminology is used in error detection and correction problems:

Error detection and error correction problem

- The number of bits in the encoded data is called a *codeword*.
- The *length* of a codeword is greater than that of the original data.
- The process of appending check bits to the information bits is called *encoding*.
- The process of extracting the original information bits from a codeword is called *decoding*.

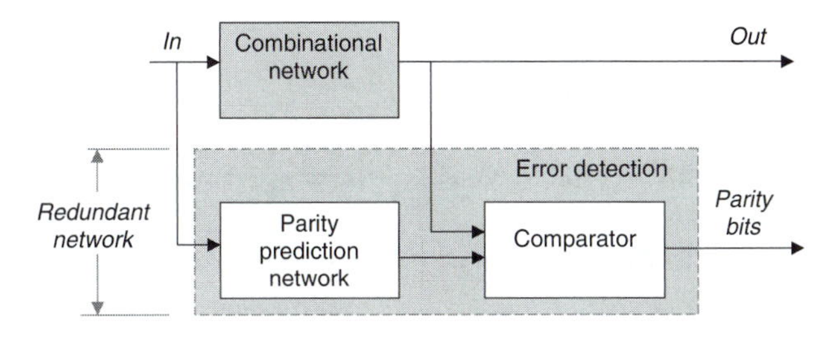

FIGURE 14.23
Principle of error detection using a parity prediction network.

A simple error detection code is capable of detecting, but not capable of correcting single errors. This code is formed by adding one check symbol to each block of k information symbols. The added symbol is called a *parity check symbol*. The parity check symbol is used as follows:

(a) If the received word contains an *even* number of errors, the decoder will not detect the errors.

(b) If the number of errors is *odd*, the decoder will detect that an odd number of errors, most likely one, has been made.

When the error is detected, the digital system can be designed to request the retransmission of the string of bits or emit a signal indicating a malfunction. Note that in the parity bit scheme, double errors are not detected since double errors do not cause the overall parity to change. However, triple errors also are detected, and, in general, it is possible to detect any odd number of errors by this method.

14.7.1 The simplest error detecting network

A linear Boolean function f of n variables is defined as a function that is represented by a linear positive polarity expression:

$$f = r_0 \oplus \bigoplus_{i=1}^{n} r_i x_i = r_0 \oplus r_1 x_1 \oplus \cdots \oplus r_n x_n$$

where $r_i \in \{0, 1\}$ is the i-th coefficient, $i = 1, 2, \ldots, n$.

Example 14.16 (Linear polynomial.) *Figure 14.24 illustrates the linear and nonlinear components of the polynomial expression of a Boolean function of two variables. Linear polynomials can be implemented using only EXOR gates.*

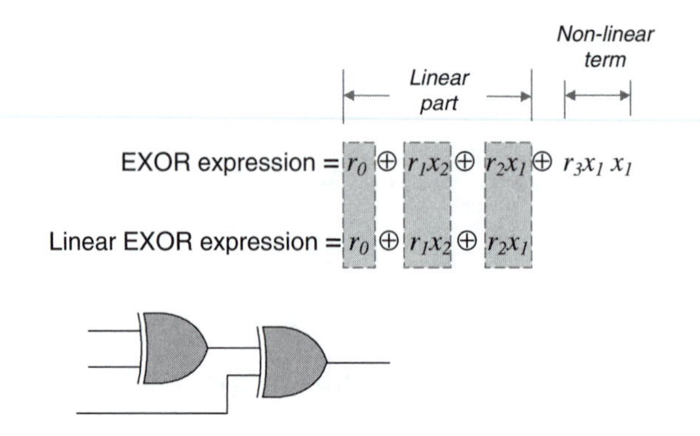

FIGURE 14.24
The linear part of an EXOR expression for a Boolean function of two variables (Example 14.16).

The equalities

$$\underbrace{x_1 \oplus x_2 \oplus \cdots \oplus x_n = 1}_{Valid\ for\ 1,3,5,...\ variables} \quad \text{and} \quad \underbrace{\overline{x_1 \oplus x_2 \oplus \cdots \oplus x_n} = 1}_{Valid\ for\ 2,4,...\ variables}$$

are valid for odd and even numbers of variables, respectively.

Details on linear polynomial expressions of Boolean functions are given in Chapters 7 and 10.

Example 14.17 (Parity function.) *The linear EXOR equality of three variables, $x_1 \oplus x_2 \oplus x_3 = 1$, is the three-input parity odd function. This function is equal to 1 when an odd number of variables (two) assume the value 1. It can be implemented by means of two-input EXOR gates (Figure 14.25a). The complement of the odd parity function is the even parity function $\overline{x_1 \oplus x_2 \oplus x_3} = 1$. The even parity function is equal to 1 when an even number of variables (one or three) is equal to 1. This function can be implemented by EXOR and XNOR gates (Figure 14.25b). Note that the output of this network is equal to 1 when none of the variables are equal to 1.*

Practice problem **14.11. (Parity function.)** Determine if the following expression is an odd or even parity function $f = x_1 \oplus x_2 \oplus \overline{x}_3 \oplus \overline{x}_4$.
Answer is given in "Solutions to practice problems."

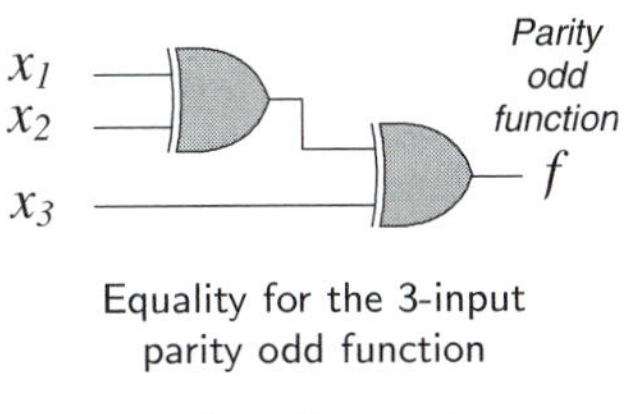

Equality for the 3-input parity odd function

$x_1 \oplus x_2 \oplus x_3 = 1$

(*a*)

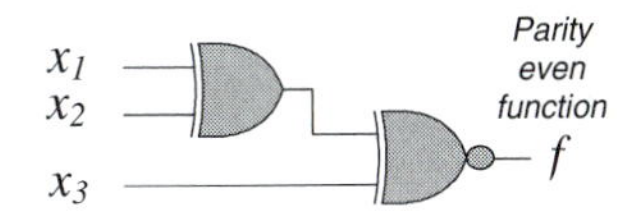

Equality for the 3-input parity even function

$\overline{x_1 \oplus x_2 \oplus x_3} = 1$

(*b*)

FIGURE 14.25
Parity odd (a) and parity even Boolean functions (b) (Example 14.17).

Components of error detection design

- The purpose of the *transmitter* is to couple the message to the *channel.*
- The network that generates the parity bit in the transmitter is called a *parity generator.*
- The *channel* is a medium through which the transmitter output is sent.
- The *receiver's* function is to extract the desired message from the received signal at the channel output. The signal is distorted in the channel. In particular, the signal is contaminated along the path by undesirable signals lumped under the broad term *noise*, which includes random and unpredictable signals.
- The network that checks the parity in the receiver is called a *parity checker.*

Example 14.18 (Error detection.) *A 3-bit message is to be transmitted together with an even parity bit (Figure 14.26). An error occurs during transmission if the four bits received have an odd number of 1s.*

A simple parity check provides error detection by ensuring that all codewords have the same parity; that is, the same number of 1s. The parity generator adds an even parity check bit to a 4-bit word. If a single error occurs in this word during transmission, the resulting received word will exhibit an odd number of 1s, and so the parity will have changed. This can be checked with a similar logic network, a parity checker, and if failure occurs, the system can be prevented from using the corrupted information.

Example 14.19 (Error detection.) *The use of a parity generator and parity checker for a 3-bit message is shown in Figure 14.27.*

Practice problem **14.12. (Error detection.)** Derive a logic network for a parity generator for a 4-bit message.
Answer is given in "Solutions to practice problems."

Design example: Parity generator and parity checker

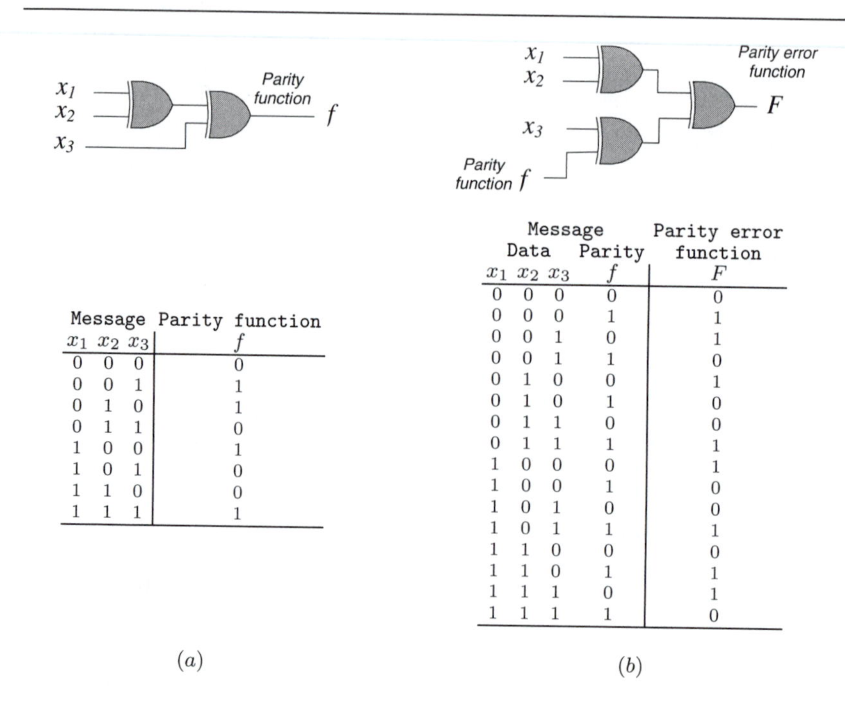

Message x_1	x_2	x_3	Parity function f
0	0	0	0
0	0	1	1
0	1	0	1
0	1	1	0
1	0	0	1
1	0	1	0
1	1	0	0
1	1	1	1

(a)

Message Data x_1	x_2	x_3	Parity f	Parity error function F
0	0	0	0	0
0	0	0	1	1
0	0	1	0	1
0	0	1	1	0
0	1	0	0	1
0	1	0	1	0
0	1	1	0	0
0	1	1	1	1
1	0	0	0	1
1	0	0	1	0
1	0	1	0	0
1	0	1	1	1
1	1	0	0	0
1	1	0	1	1
1	1	1	0	1
1	1	1	1	0

(b)

FIGURE 14.26
Parity generator (a) and parity checker for a 3-bit message (b) (Example 14.18).

14.7.2 Error correction

In the Hamming code, information bits along with several parity bits compose a code group. The values of the parity bits are determined by an even-parity scheme over selected information bits. After a code group is transmitted, the parity bits are recalculated at the receiver side to check if the correct parity still exists over their selected information bits. By comparing the recalculated parity bits against those received in the code group, it is possible to determine

- ▶ If the received code group is free of single errors, or
- ▶ If a single error has occurred, exactly which bit has erroneously changed.

If more than one bit is changed during transmission, then this coding scheme is no longer capable of determining the location of the errors. The Hamming code is constructed as described below:

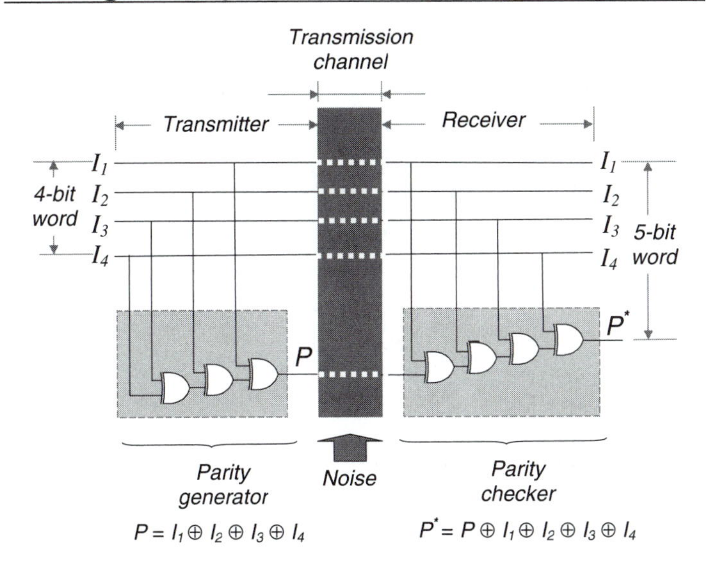

FIGURE 14.27
Using a parity generator and parity checker for a three-bit message (Example 14.19).

Principles of the Hamming code design

- A single error in a codeword will cause one or more parity checks to fail.
- The pattern of parity check failures can be used to locate the position of the error, providing the checking domains of each parity bit have been chosen correctly.
- Locating the position of an error in a binary message is sufficient for correction because only the digit found at this location must be inverted.
- The encoding and decoding logic networks for these single-error correcting Hamming codes can be performed by linear combinational networks.
- All single errors can be corrected, since the Hamming code is a parity check code having a distance of 3.

Example 14.20 (Hamming code.) *For the case of 4 information bits, 3 parity bits are included along with the 4 information bits to form a 7-bit code group. The structure of the code group in this case is given in Figure 14.28. The following rule is used for the parity bits* p_1, p_2, *and* p_3*:*

- $p_1 = 0$ *if* $b_1 \oplus b_2 \oplus b_4 =$ Even number, *and* $p_1 = 1$ *otherwise.*
- $p_2 = 0$ *if* $b_1 \oplus b_3 \oplus b_4 =$ Even number, *and* $p_2 = 1$ *otherwise.*
- $p_3 = 0$ *if* $b_2 \oplus b_3 \oplus b_4 =$ Even number, *and* $p_3 = 1$ *otherwise.*

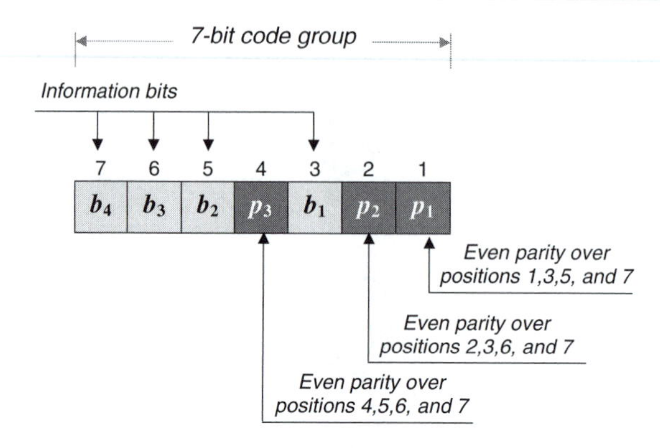

FIGURE 14.28
The structure of 7-bit Hamming code (Example 14.20).

Practice problem **14.13.** **(Hamming code.)** Given a 3-bit code, find the required number of parity bits and the total Hamming code length. ***Answer*** is given in "Solutions to practice problems."

The Hamming code for the BCD

The information bits in a Hamming code group may themselves. This can be, in particular, BCD code.

Example 14.21 **(Hamming code for BCD.)** *Figure 14.30 shows the design of the Hamming code group for ten BCD digits.*

The location of a single error in the Hamming code group

Upon the receipt of a Hamming code group, the parity bits are recalculated using the same even parity scheme:

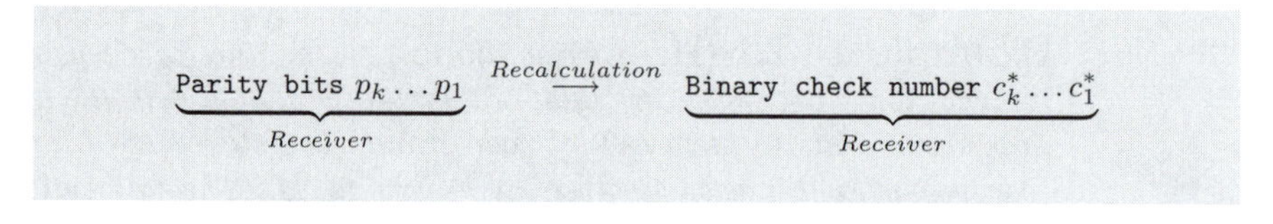

This recalculation results in a *binary check number* $c_k^* \dots c_1^*$, which is transmitted along with the code word.

Example 14.22 **(Hamming code.)** *(Continuation of Example 14.21). Design example is given in Figure 14.29.*

Design example: Hamming code for 4-bit BCD code

Let the following Hamming code group (Figure 14.30),

7	6	5	4	3	2	1
0	1	1	0	0	1	1
b_4	b_3	b_2	p_3	b_1	p_2	p_1

($p_3p_2p_1 = 011$) be transmitted (Figure 14.28). Assume that the bit b_1 is erroneously changed from 0 to 1 during transmission:

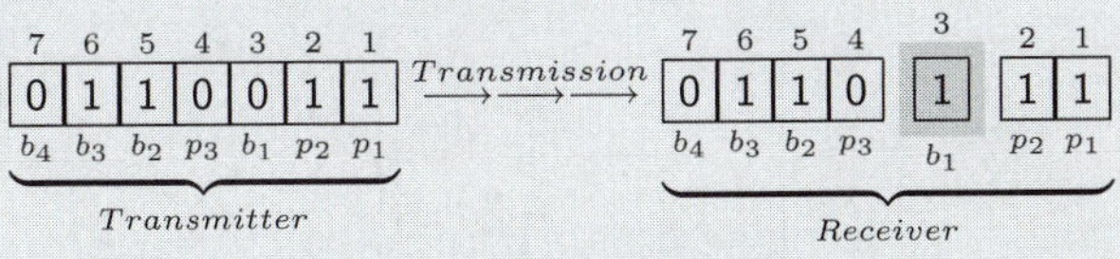

Transmitter:

7	6	5	4	3	2	1
0	1	1	0	0	1	1
b_4	b_3	b_2	p_3	b_1	p_2	p_1

$\xrightarrow{Transmission}$

Receiver:

7	6	5	4	3	2	1
0	1	1	0	1	1	1
b_4	b_3	b_2	p_3	b_1	p_2	p_1

The received Hamming code group is 0110111, which is different from the sent code group 0110011 in the third position. The binary check number is formed as follows:

- There is an odd number of 1's in the group b_4, b_2, b_1, p_1:

$$p_1 = 1, \quad \text{ODD NUMBER} = \underset{b_4}{\overset{7}{\boxed{0}}} \oplus \underset{b_2}{\overset{5}{\boxed{1}}} \oplus \underset{b_1}{\overset{3}{\boxed{1}}} \oplus \underset{p_1}{\overset{1}{\boxed{1}}} = 1, \quad c_1^* = 1$$

- There is an odd number of 1's in the group b_4, b_3, b_1, p_2:

$$p_2 = 1, \quad \text{ODD NUMBER} = \underset{b_4}{\overset{7}{\boxed{0}}} \oplus \underset{b_3}{\overset{6}{\boxed{1}}} \oplus \underset{b_1}{\overset{3}{\boxed{1}}} \oplus \underset{p_2}{\overset{2}{\boxed{1}}} = 1, \quad c_2^* = 1$$

- There is an even number of 1's in the group b_4, b_3, b_2, p_3:

$$p_3 = 0, \quad \text{EVEN NUMBER} = \underset{b_4}{\overset{7}{\boxed{0}}} \oplus \underset{b_3}{\overset{6}{\boxed{1}}} \oplus \underset{b_2}{\overset{5}{\boxed{1}}} \oplus \underset{p_3}{\overset{4}{\boxed{0}}} = 0, \quad c_3^* = 0$$

Since, $p_3p_2p_1 = 011$, the binary check number is $c_3^*c_2^*c_1^* = 011$, and it indicates that the bit in position 3 is incorrect.

FIGURE 14.29
Hamming code for 4-bit BCD code (Example 14.22).

Example 14.23 (Hamming code.) *(Continuation of Example 14.21). Design example for the case when the bit p_1 be erroneously changed from 1 to 0 during transmission is given in Figure 14.31.*

Practice problem 14.14. (Hamming code.) Consider the code 1000110 to be transmitted over a data channel. Let the code received be 0000110. Find the binary check number for checking the position of the erroneous bit.
Answer is given in "Solutions to practice problems."

Design example: Hamming code for 4-bit BCD code (continuation)

STEP 1: Hamming code

Decimal digit	b_4	b_3	b_2	p_3	b_1	p_2	p_1
0	0	0	0	0	0	0	0
1	0	0	0	0	1	1	1
2	0	0	1	1	0	0	1
3	0	0	1	1	1	1	0
4	0	1	0	1	0	1	0
5	0	1	0	1	1	0	1
6	0	1	1	0	0	1	1
7	0	1	1	0	1	0	0
8	1	0	0	1	0	1	1
9	1	0	0	1	1	0	0

STEP 2: Minimization

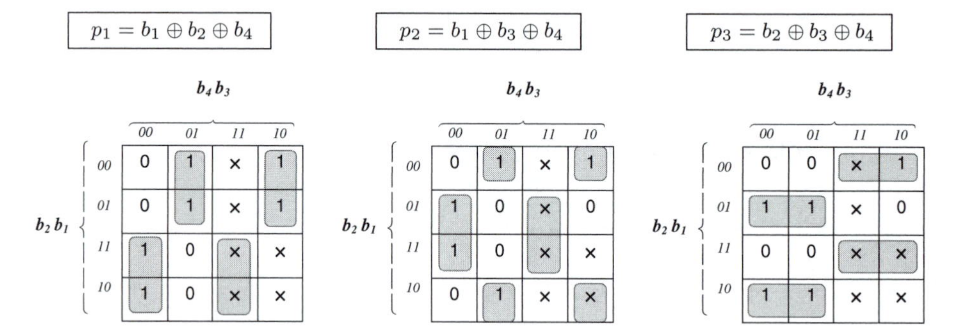

STEP 3: Gate level logic network

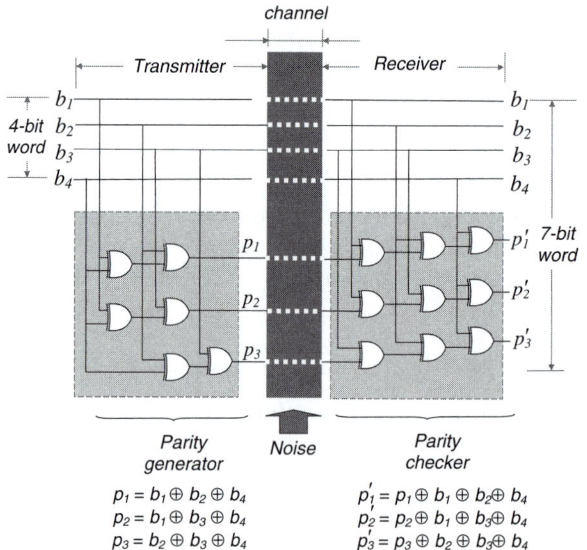

FIGURE 14.30
Hamming code for 4-bit BCD code (Example 14.22).

Design example: Hamming code for 4-bit BCD code

Let the bit p_1 be erroneously changed from 1 to 0 during transmission.

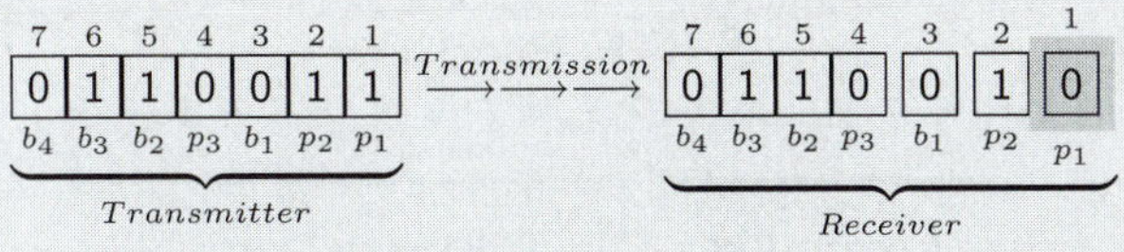

	7	6	5	4	3	2	1
Transmitter	0	1	1	0	0	1	1
	b_4	b_3	b_2	p_3	b_1	p_2	p_1
Receiver (after *Transmission*)	0	1	1	0	0	1	0
	b_4	b_3	b_2	p_3	b_1	p_2	p_1

The binary check number is formed as follows:

- There is an odd number of 1's in the group b_4, b_2, b_1, p_1:

$$p_1 = 1, \quad \text{ODD NUMBER} = \underset{b_4}{\overset{7}{\boxed{0}}} \oplus \underset{b_2}{\overset{5}{\boxed{1}}} \oplus \underset{b_1}{\overset{3}{\boxed{0}}} \oplus \underset{p_1}{\overset{1}{\boxed{0}}} = 1$$

  This requires c_1^* to be set to 1.

- There is an odd number of 1's in the group b_4, b_3, b_1, p_2:

$$p_2 = 1, \quad \text{ODD NUMBER} = \underset{b_4}{\overset{7}{\boxed{0}}} \oplus \underset{b_3}{\overset{6}{\boxed{1}}} \oplus \underset{b_1}{\overset{3}{\boxed{0}}} \oplus \underset{p_2}{\overset{2}{\boxed{1}}} = 0$$

  This implies $c_2^* = 0$.

- There is an even number of 1's in the group b_4, b_3, b_2, p_3:

$$p_3 = 0, \quad \text{EVEN NUMBER} = \underset{b_4}{\overset{7}{\boxed{0}}} \oplus \underset{b_3}{\overset{6}{\boxed{1}}} \oplus \underset{b_2}{\overset{5}{\boxed{1}}} \oplus \underset{p_3}{\overset{4}{\boxed{0}}} = 0$$

  This implies $c_3^* = 0$.

Therefore, the binary check number is $c_3^* c_2^* c_1^* = 001$ is indicating that the bit in position 1 is incorrect

FIGURE 14.31
Hamming code for 4-bit BCD code (Example 14.23).

14.7.3 Gray code

Gray code, introduced in Section 2.10, is used for encoding the indices of the nodes. There are several reasons for encoding the indices. The most important is to simplify the analysis, synthesis and embedding of topological structures. The conversion of binary code into Gray code and back (Equations 2.4 and 2.5) can be implemented using logic networks.

> **Example 14.24 (Gray code.)** *Given a 3-bit binary code, the corresponding Gray code, and its implementation are shown in Figure 14.32.*

To build a Gray code for d dimensions, one takes the Gray code for $d-1$ dimensions, reflects it top to bottom across a horizontal line just below the last element, and adds a leading one to each new element below the line of reflection.

Design example: Gray code generation

STEP 1
Binary and Gray codes

Binary code	Gray code
000	000
001	001
010	011
011	010
100	110
101	111
110	101
111	100

Binary code	Gray code
000	000
001	001
011	010
010	011
110	100
111	101
101	110
100	111

STEP 2
Equations for conversions

$$g_i = b_i \oplus b_{i+1}$$
$$i=0: g_0 = b_0 \oplus b_1$$
$$i=1: g_1 = b_1 \oplus b_2$$
$$i=2: g_2 = b_2 \oplus 0$$

$$b_i = g_0 \oplus g_1 \oplus \dots g_{3-i} = \bigoplus_{i=0}^{3-i} g_i$$
$$i=2: b_2 = g_2 \oplus 0$$
$$i=1: b_1 = g_1 \oplus b_2$$
$$= g_1 \oplus g_2$$
$$i=0: b_0 = g_0 \oplus b_1$$
$$= g_0 \oplus g_1 \oplus g_2$$

(a) (b)

STEP 3
Gate level logic network

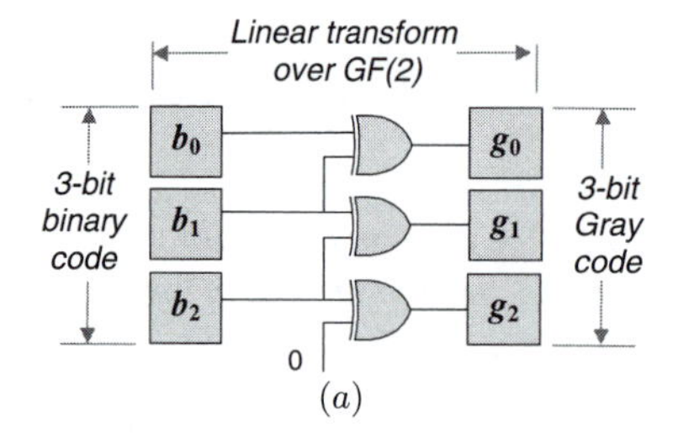

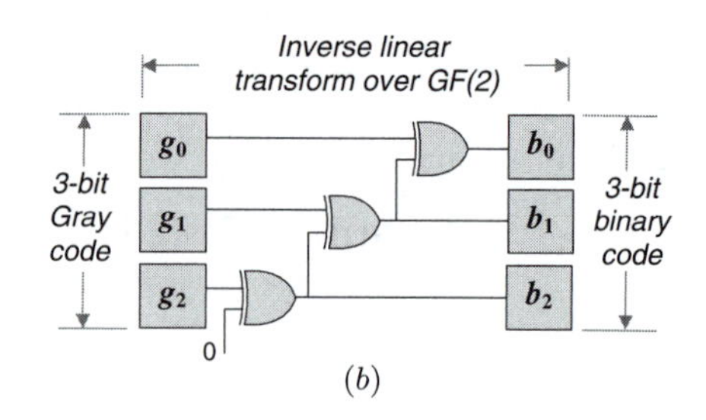

FIGURE 14.32
Gray code generation and decoding (Example 14.24).

14.8 Summary of combinational network design

This chapter introduces techniques for combinational logic network design using standard modules. The output signals of a combinational network at any time are completely determined by the combination of values assumed by the input signals $x_1, x_2, \ldots, x_n$ at that time; hence the term "combinational" network. Any changes in these inputs are assumed to change the network output signals. The key aspects of this chapter are listed below:

(*a*) Techniques for multilevel logic network design,
(*b*) Techniques for error detection and error correction, and
(*c*) Decision diagram techniques for logic network design.

The main topics of this chapter are summarized as follows:

- ▶ An n-input, m-output ***combinational*** network is a network of logic elements that realizes a set of $m \geq 1$ Boolean functions of n variables $x_1, x_2, \ldots, x_n$, $f = \{f_1, f_2, \ldots, f_n\}$. The simplest examples of an n-input, 1-output combinational network are a logic gate and a multiplexer.
- ▶ In an idealized logic network, that is built from elementary logic elements (gates) and standard logic blocks (multiplexers, demultiplexers, etc.), all signal changes are assumed to take place instantaneously. In physical networks, signals cannot be transmitted instantaneously, so there is a delay between a network input change and the output change it produces. This ***switching delay***, or ***response time***, can be measured.
- ▶ There is a possibility that during information processing or storage, data can get corrupted due to physical defects in the system. Special tools in the system are necessary for ***detecting*** erroneous bits in data and *correcting* errors in data in order to restore the system to its normal operating mode. This requires additional (redundant) bits to be appended to the data bits for error detection and/or correction. The process of appending check bits to the information bits is called ***encoding***; the opposite process – extracting the original information bits from a codeword – is known as ***decoding***.
- ▶ A simple code capable of detecting (but not capable of correcting) single errors, is formed by adding one check symbol to each block of k information symbols. If the received word contains an ***even*** number of errors, the decoder will not detect the errors. If the number of errors is ***odd***, the decoder will detect that an odd number of errors, most likely one, has been made. The parity-check control is based on the linear EXOR equalities:

Summary (continuation)

$$\underbrace{x_1 \oplus x_2 \oplus \cdots \oplus x_n = 1}_{Valid\ for\ 1,3,5,...\ variables} \quad \text{and} \quad \underbrace{\overline{x_1 \oplus x_2 \oplus \cdots \oplus x_n} = 1}_{Valid\ for\ 2,4,...\ variables}$$

which are valid for the odd and even numbers of variables, respectively.

- A ***Hamming code*** is a particular parity-check code for ***correcting*** all single errors. In this code, several parity bits are included in a code group. The values of the parity bits are determined by an even-parity scheme over certain selected bits. When a code group is received, the parity bits are recalculated. By comparing the recalculated parity bits against those received in the code group, it is possible to determine

  (a) If the received code group is free from a single error, or
  (b) If a single error has occurred, exactly which bit has erroneously changed.

  If more than one bit is changed during transmission, then this coding scheme is no longer capable of determining the location of the errors.

- *Gray code* is referred to as a ***unit-distance*** code. Let $b_n...b_1b_0$ be a binary representation of a integer positive number B and $g_n...g_1g_0$ be its Gray code. Then g_i and b_i, $i = 0, 1, 2, \ldots, n$ are generated as follows:

$$g_i = b_i \oplus b_{i+1}$$

$$b_i = g_0 \oplus g_1 \oplus \ldots g_{n-i} = \bigoplus_{i=0}^{n-i} g_i$$

  respectively, where $b_{n+1} = 0$.

- ***Decomposition*** is the process of representing a data structure as a collection of several sub-structures. The decomposition of a Boolean function in algebraic form results in a collection of several sub-functions. The decomposition of a logic network involves a partitioning of this network into several smaller ones, which can be tested, verified, and implemented easily than the initial network. In general, decomposition increases the number of levels in a logic network while decreasing the fan-in of the gates. The decomposition of a decision diagram results in a collection of smaller decision diagrams.

Summary (continuation)

- ***Verification*** is an intrinsic part of logic synthesis. Practical design verification means validating that an implemented gate-level design matches its desired behavior as specified at the register-transfer level. ***Formal verification*** of digital circuits demands that a mathematical proof of correctness implicitly covers all possible input patterns (stimuli). ***Equivalence checking*** for combinational logic networks is formulated as follows: given two networks, check if their corresponding outputs are equal for all possible input patterns.
- The alternative criterion for verifying if two logic circuits are equivalent is to check whether or not the canonical representations of their output functions are the same.
- For advances in combinational logic network design, we refer the reader to the "Further study" section.

14.9 Further study

Advanced topics of combinational logic network design

There are two main approaches to synthesis in logic design: (a) those based on formal *exact* algorithms and (b) those based on informal methods, also called *heuristic* algorithms; most of them provide quasi-optimal solutions.

Topic 1: Multi-level logic network design. Two-level logic networks based on SOP and POS expressions are sometimes impractical. Logic networks that have more than two levels often have fewer gates and meet lower fan-in and fan-out limits. The minimization of two-level networks is based on the minimization of SOP expressions. No practical and exact minimization techniques are known for general multi-level networks. The design of multi-level networks is more complex that of two-level ones. However, many *heuristic* techniques are used in practice for multi-level logic network optimization. Heuristic minimization techniques can provide minimal or at least near-minimal designs. Techniques for multi-level network design are based on the following methods: (a) decomposition, (b) factoring, and (c) local transformations.

An extension of these techniques is based on the notation functional flexibility. *Functional flexibility* is defined as the condition in which an alternative Boolean function can replace a function at a certain point in a logic network. There are many optimization methods in advanced logic design that use functional flexibility. One of them is called *set of pairs of functions to be distinguished (SPFD)* ("SPFD: A Method to Express Functional Flexibility" by S. Yamashita, H. Sawada, and A. Nagoya, *IEEE Transactions on Computer-Aided Design of Integrated Circuits and Systems*, volume 19, number 8, pages 840–849, 2000). The SPFD approach can be well understood in terms of Shannon information; that is, as the information content of the Boolean function, since it indicates what information contributes to the network performance.

Topic 2: Heuristic optimization techniques. An optimization problem is defined as a problem whose solution can be measured in terms of a cost (or objective) function. It is usually multi-iterative procedures based on heuristic approach.

Optimization methods are distinguished with respect to the type of data structure:

- Minimization algorithms for Boolean functions based on algebraic manipulations.
- Optimization algorithms for multi-level logic networks based on graphical-based network models.
- Optimization techniques for decision diagrams. The size of a decision diagram depends on the order of the variables. Optimization techniques are based on various strategies; in particular, (a) optimization of the order using the specific properties of Boolean function, and (b) dynamic reordering.

That is, the algorithms for optimization of Boolean functions in cube form and in the form of decision diagrams are different. Also different techniques are required for the optimization of logic networks of logic gates, and networks of multiplexers and threshold elements.

Exact minimization of Boolean functions is considered a classic problem of logic design; it was addressed first by *W. Quine* (1952) and *E. McCluskey* (1956). The Quine-McCluskey algorithm is the exact minimization algorithm. The major problem with this algorithm is that all prime implicants for a Boolean function have to be computed. This becomes computationally very expensive for a function with a large number of inputs. The Quine-McCluskey algorithm often fails to simplify medium size SOP expressions.

The exact algorithms are often unsuitable for the minimization of Boolean functions and optimizations of logic networks. Heuristic algorithms give acceptable solutions and sometimes even optimal ones in a short time for large size problems, but there is no guarantee that they will do so in any particular case.

Heuristic minimization of Boolean functions is motivated by the need to reduce the size of two-level forms (SOP and POS expressions). For example, the ESPRESSO package developed at IBM and the University of California at Berkeley utilizes various heuristics for avoiding the cost of generating all prime implicants. ESPRESSO can be viewed as applying a set of the following operators to minimize a Boolean function: EXPAND, REDUCE, and IRREDUNDANT. The EXPAND operator aims to maximize the size of cubes in a *cover*. A cover of a Boolean function is a set of cubes, none of which is contained by any cube in the OFF-set of the function. The ON-set of a

Boolean function is defined as a set of cubes and each cube in the set produces a logic 1 for the function; cubes that are not in the ON-set belong to the OFF-set or the DC-set (don't care set). The bigger a cube, the more minterms it covers, thereby making them redundant. The REDUCE operator decreases the size of each cube in the ON-set of a Boolean function. The IRREDUNDANT operator removes redundant implicants from the cover of a Boolean function.

ESPRESSO perform the minimization of a Boolean function specified in terms of its ON-set, OFF-set, and DC-set. The implicants in the ON-set represent the initial (non-minimum) cover of the Boolean function. By applying EXPAND, REDUCE, and IRREDUNDANT operators, ESPRESSO finds the near-minimum cover of the function.

Heuristic optimization methods can provide optimal or at least near-optimal designs for transferring two-level logic networks into multi-level networks. Two-level logic networks are sometimes impractical, usually because they require too many gates even after minimization. This is because they often fail to meet the fan-in and fan-out constraints of the implementation technology. Many-level logic networks that are obtained using heuristic algorithms often contain fewer gates than any minimal two-level design for the same function. No practical and exact optimization techniques are known for general multi-level logic networks. Heuristic techniques are usually used to solve the problem of multi-level logic network design.

Further reading

A survey of advanced logic design can be found in the special issue "Electronic Design Automation at the Turn of Century," *IEEE Transactions on Computer-Aided Design of Integrated Circuits and Systems*, volume 19, number 12, 2000.

Digital Principles and Design by D. D. Givone, McGraw-Hill, 2003.

"The Future of Logic Synthesis and Verification" by R. K. Brayton, in *Logic Synthesis and Verification* edited by S. Hassoun and T. Sasao, Consulting Editor R. K. Brayton, Kluwer, 2002.

"Multi-level Logic Optimization" by M. Fujita, Y. Matsunaga, and M. Ciesielski, pages 29–63, in *Logic Synthesis and Verification*, S. Hassoun and T. Sasao editors, R. K. Brayton, Consulting Editor, Kluwer, 2002.

"Lattice Diagrams Using Reed-Muller Logic" by M. A. Perkowski, M. Chrzanowska-Jeske, and Y. Xu, pages 85–102, in *Proceedings of the 3rd International Workshop on Applications of the Reed-Muller Expansion in Circuit Design*, Oxford University, September 1997.

"On Bi-Decomposition of Logic Functions" by T. Sasao and J. T. Butler, pages 85–102, in *Proceedings of the ACM/IEEE International Workshop on Logic Synthesis*, Tahoe City, California, May, 1997.

Switching Theory for Logic Synthesis by T. Sasao, Kluwer, 1999.

"Spatial Interconnect Analysis for Predictable Nanotechnologies" by S. N. Yanushkevich, V. P. Shmerko, and B. Steinbach, *Journal of Computational and Theoretical Nanoscience*, American Scientific Publishers, volume 4, number 8, pages 1–14, 2007.

14.10 Solutions to practice problems

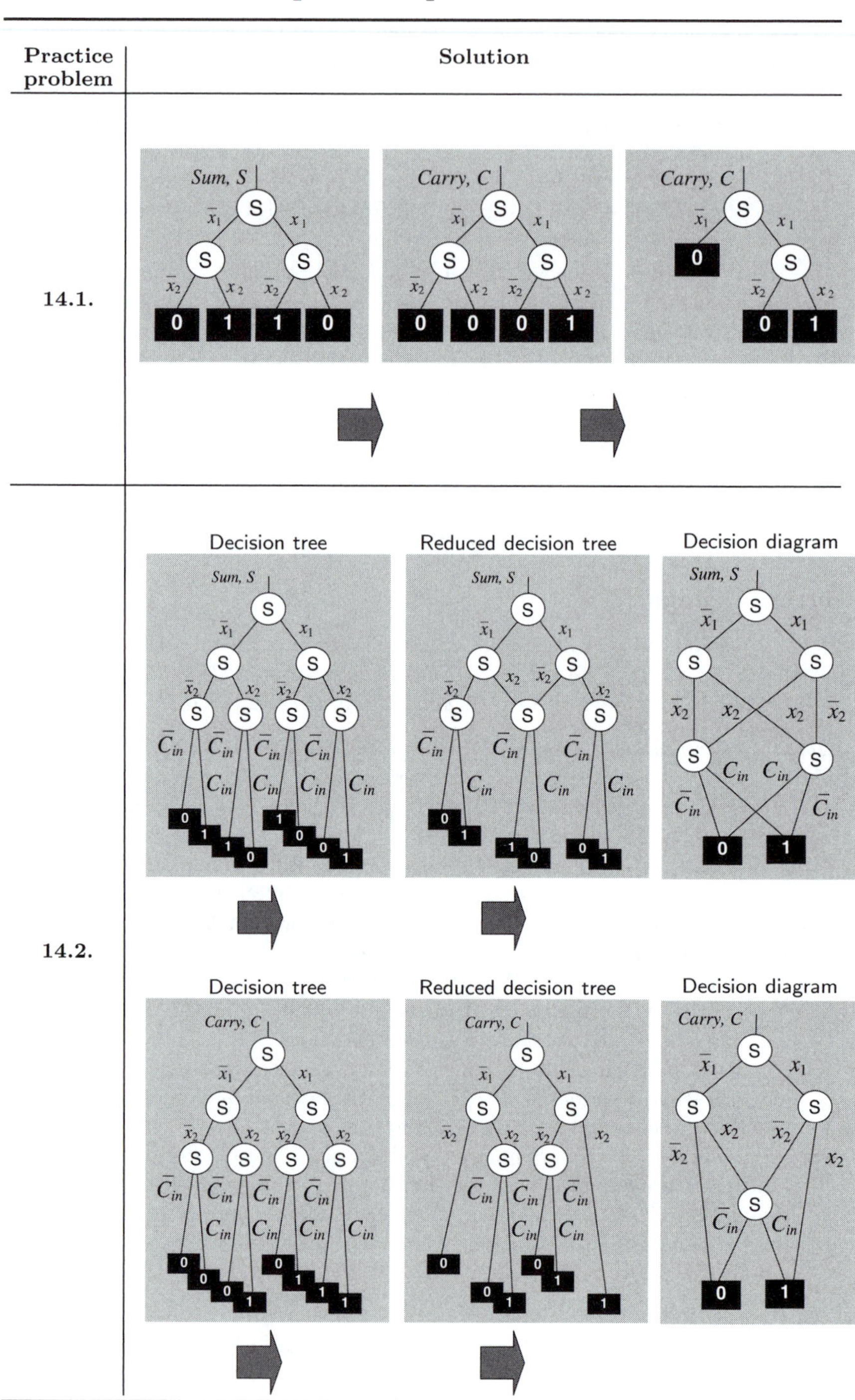

Solutions to practice problems (continuation)

Practice problem	Solution

14.3.

Number A		Number B			
a_1	a_0	b_1	b_0	$A \geq B$	$A \neq B$
0	0	0	0	1	0
0	0	0	1	0	1
0	0	1	0	0	1
0	0	1	1	0	1
0	1	0	0	1	1
0	1	0	1	1	0
0	1	1	0	0	1
0	1	1	1	0	1
1	0	0	0	1	1
1	0	0	1	1	1
1	0	1	0	1	0
1	0	1	1	0	1
1	1	0	0	1	1
1	1	0	1	1	1
1	1	1	0	1	1
1	1	1	1	1	0

$f(A \geq B)$

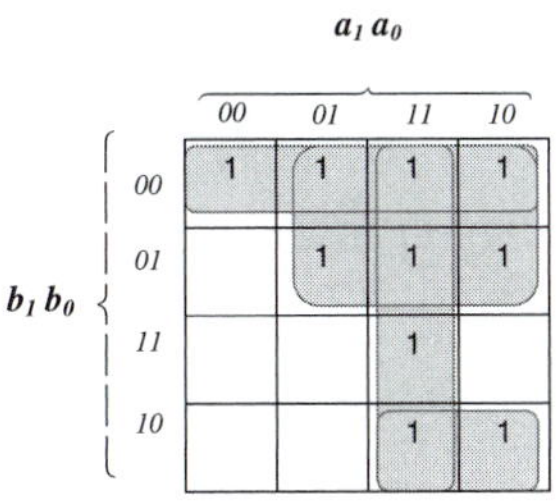

$b_1 b_0$ \ $a_1 a_0$	00	01	11	10
00	1	1	1	1
01		1	1	1
11			1	
10			1	1

$f(A \neq B)$

$b_1 b_0$ \ $a_1 a_0$	00	01	11	10
00		1	1	1
01	1		1	1
11	1	1		1
10	1	1	1	

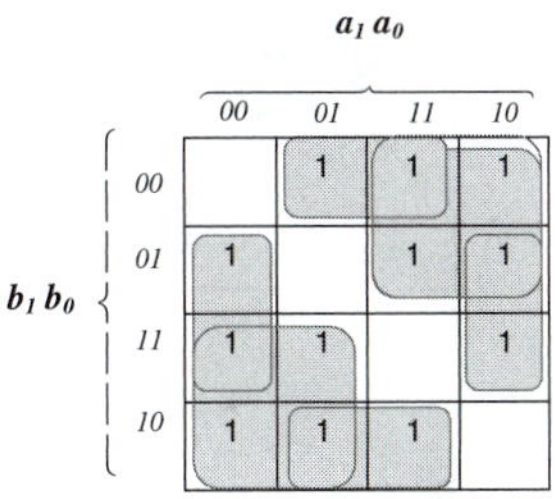

Case $A \geq B$

$$f = \overline{b}_1\overline{b}_0 \vee a_1 a_0 \vee a_0\overline{b}_1 \vee a_1\overline{b}_1 \vee a_1\overline{b}_0 = \overline{b}_1(\overline{b}_0 \vee a_0) \vee \overline{b}_1 a_1 \vee a_1(a_0 \vee \overline{b}_0)$$

Case $A \neq B$

$$f = \overline{a}_1 b_1 \vee a_1\overline{b}_1 \vee a_0\overline{b}_0 \vee \overline{a}_0 b_0 = (a_1 \oplus b_1)(a_0 \oplus b_0)$$

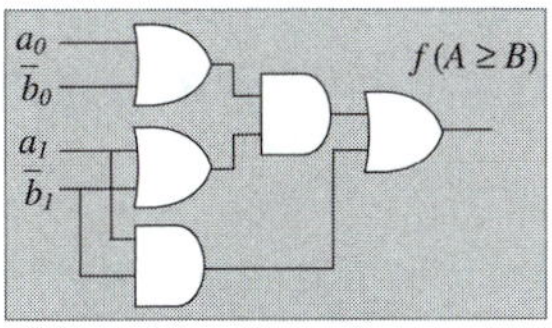

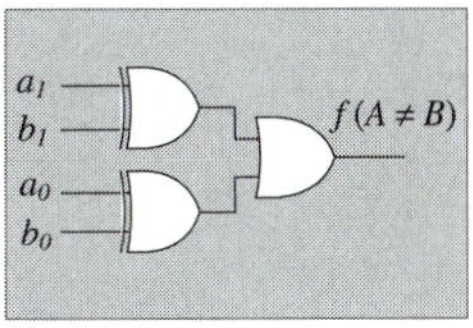

$f(A \geq B)$ $f(A \neq B)$

14.4.

Expression for the first network:

$$(x_2 \vee x_3) \vee \overline{x}_3 x_1 = x_2 \vee x_3 \vee x_1\overline{x}_3 = x_2 \vee x_3 \vee x_1$$

Expression for the second network: $x_1x_2x_3 \oplus x_1x_3 = x_1x_3(x_2 \oplus 1) = x_1\overline{x}_2x_3$. These two expressions are not equal, so the logic networks are not equivalent.

Solutions to practice problems (continuation)

Practice problem	Solution
14.5.	The SOP expression for the first network: $$(x \vee y) \oplus xy = x \oplus y \oplus xy = x \oplus y = x\overline{y} \vee \overline{x}y$$ The SOP expression for the second network: $$\overline{(\overline{x \vee y}) \vee xy} = (x \vee y)\overline{xy} = (x \vee y)(\overline{x} \vee \overline{y}) = x\overline{x} \vee x\overline{y} \vee \overline{x}y \vee y\overline{y} = x\overline{y} \vee \overline{x}y$$ These SOP expressions are equal.
14.6.	The complete decision tree can be reduced to the decision diagram as follows: 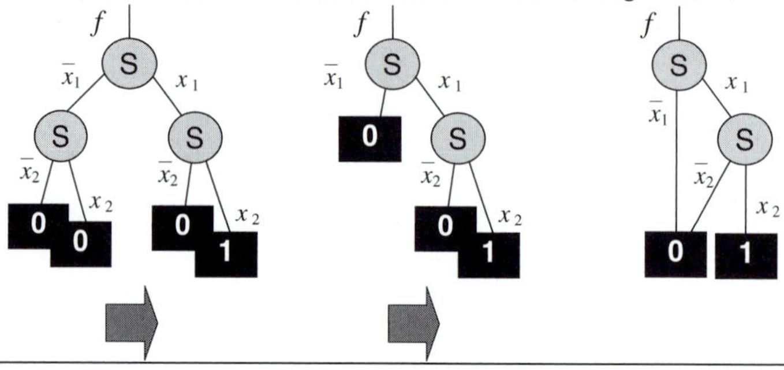
14.7.	(a) with respect to $\mathbf{X}_1 = (x_1, x_2)$ and $\mathbf{X}_2 = (x_3, x_4)$ (b) with respect to $\mathbf{X}_1 = (x_1, x_3)$ and $\mathbf{X}_2 = (x_2, x_4)$.

Solutions to practice problems (continuation)

Practice problem	Solution
14.8.	**Step 1:** $f = (\{x_1\}, \{x_2\}, \{x_3\}, \{x_4\})$ **Step 2:** $f = (\{x_1, x_3\}, \{x_2\}, \{x_4\})$ **Step 3:** $f = (\{x_1, x_3\}, \{x_2, x_4\})$ **Result:** $f = \overline{x}_1\overline{x}_3 \vee x_2x_4 = g_1(x_1, x_3) \vee g_2(x_2, x_4)$ There is no OR type decomposition with respect to the partition $\{(x_1, x_2), (x_3, x_4)\}$
14.9.	Consider the Boolean function $\overline{f} = \overline{\overline{x}_1\overline{x}_3 \vee x_2x_4} = (\overline{\overline{x}_1\overline{x}_3})(\overline{x_2x_4})$ $= (x_1 \vee x_3)(\overline{x}_2 \vee \overline{x}_4) = x_1\overline{x}_2 \vee x_1\overline{x}_4 \vee \overline{x}_2x_3 \vee x_3\overline{x}_4$ This function $\overline{f}$ does not have OR type decomposition; therefore, function f does not have AND type decomposition. Indeed, $f = \overline{x}_1\overline{x}_3 \vee x_2x_4 = (\overline{x}_1 \vee x_2x_4)(\overline{x}_3 \vee x_2x_4)$ $= (\overline{x}_1 \vee x_2)(\overline{x}_1 \vee x_4)(\overline{x}_3 \vee x_2)(\overline{x}_3 \vee x_4)$, which does not have AND type decomposition with respect to any partition (the algorithm above results in (x_1, x_2, x_3, x_4)).
14.10.	$f = \overline{x}_1\overline{x}_2x_3 \vee x_1x_2 = \overline{x}_1\overline{x}_2x_3 \oplus x_1x_2\overline{x}_1\overline{x}_2x_3 \oplus x_1x_2$ $= \overline{x}_1\overline{x}_2x_3 \oplus x_1x_2$ The expression can be represented using positive polarity expansion as follows: $f = (x_1x_2) \oplus x_3(\overline{x}_1\overline{x}_2) = f_0 \oplus x_3f_2$, where $f_0 = x_1x_2$ and $f_2 = \overline{x}_1\overline{x}_2 = \overline{x_1 \vee x_2}$. The corresponding network is shown below. 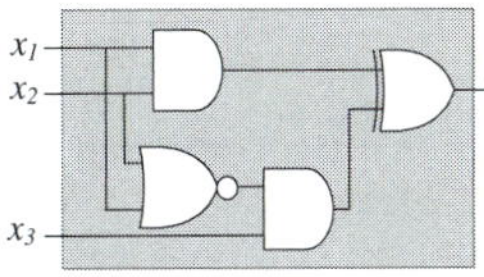
14.11.	Since $f = x_1 \oplus x_2 \oplus \overline{x}_3 \oplus \overline{x}_4 = x_1 \oplus x_2 \oplus 1 \oplus x_3 \oplus 1 \oplus x_4 = x_1 \oplus x_2 \oplus x_3 \oplus x_4$, it is an odd parity function.
14.12.	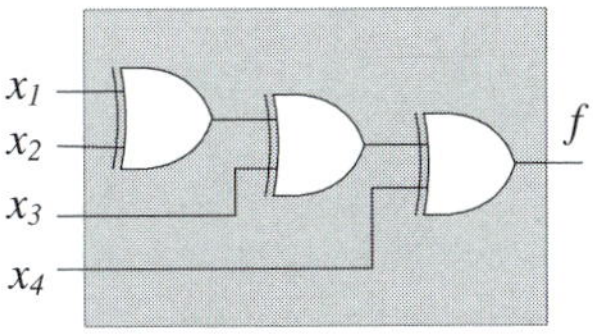 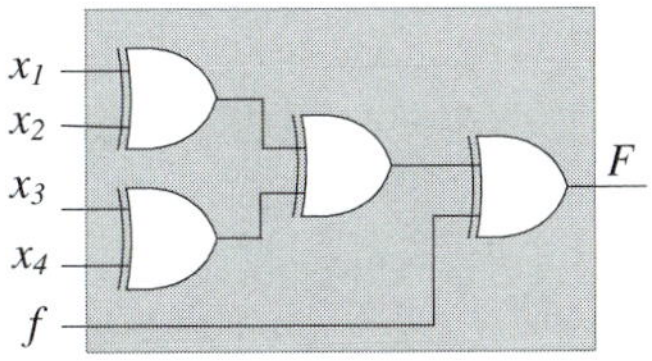
14.13.	Given a code $[b_3\ b_2\ b_1]$, the Hamming code is $[b_3\ b_2\ p_2\ b_1\ p_1]$, where $p_1 = b_1 \oplus b_2$ and $p_2 = b_1 \oplus b_3$. The number of parity bits is equal to 2. The total Hamming code length is 5.

Solutions to practice problems (continuation)

Practice problem	Solution
14.14.	Transmitter: positions 7 6 5 4 3 2 1 = 1 0 0 0 1 1 0 ($b_4\ b_3\ b_2\ p_3\ b_1\ p_2\ p_1$) $\xrightarrow{Transmission}$ Receiver: positions 7 6 5 4 3 2 1 = 0 0 0 0 1 1 0 ($b_4\ b_3\ b_2\ p_3\ b_1\ p_2\ p_1$; bit b_4, position 7, shaded) There is an even number of 1's in the group b_4, b_2, b_1, p_1: $p_1 = 0$, Even number $= \underset{b_4}{\overset{7}{0}} \oplus \underset{b_2}{\overset{5}{0}} \oplus \underset{b_1}{\overset{3}{1}} \oplus \underset{p_1}{\overset{1}{0}} = 1, \quad c_1^* = 1$ There is an odd number of 1's in the group b_4, b_3, b_1, p_2: $p_2 = 1$, Odd number $= \underset{b_4}{\overset{7}{0}} \oplus \underset{b_3}{\overset{6}{0}} \oplus \underset{b_1}{\overset{3}{1}} \oplus \underset{p_2}{\overset{2}{1}} = 1, \quad c_2^* = 1$ There is an even number of 1's in the group b_4, b_3, b_2, p_3: $p_3 = 0$, Even number $= \underset{b_4}{\overset{7}{0}} \oplus \underset{b_3}{\overset{6}{0}} \oplus \underset{b_2}{\overset{5}{0}} \oplus \underset{p_3}{\overset{4}{0}} = 0, \quad c_3^* = 1$ $c_3^* c_2^* c_1^* = 111$, which indicates that the bit in position 7 is incorrect.

14.11 Problems

Problem 14.1 Verify if the logic networks shown below perform the same Boolean function. Use appropriate transforms.

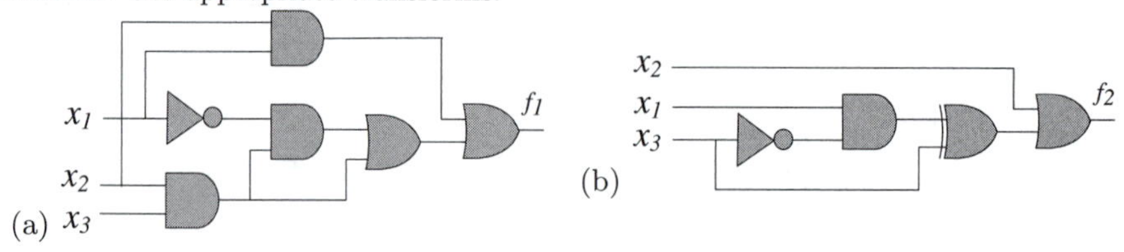

Problem 14.2 Design a logic network for performing the 9's complement of a BCD digit.

Problem 14.3 Design a two-bit BCD adder using decision diagrams.

Problem 14.4 Consider an alternative implementation of the 1-digit BCD adder given below. Verify if it does implement the same function as the BCD adder considered in the text.

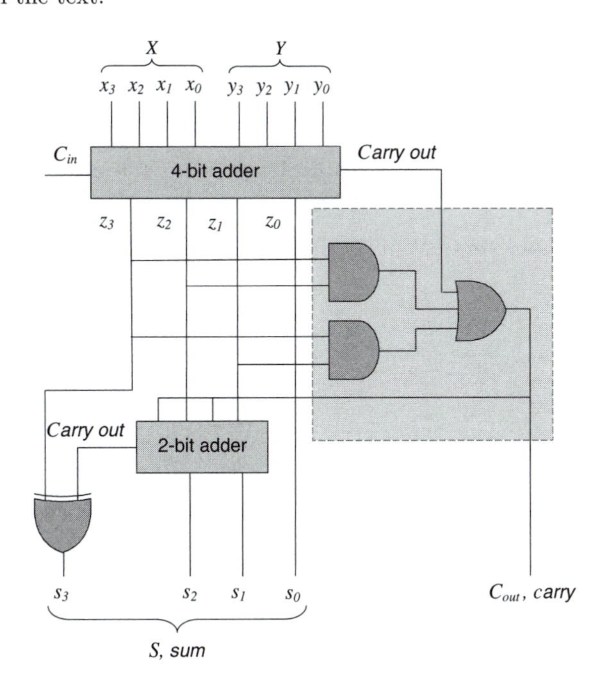

Problem 14.5 Use decision diagrams to prove or disprove the equivalence of the logic networks given in Problem 14.1.

Problem 14.6 Design logic networks of multiplexers and any other logic gates given the Boolean functions in the form of K-maps; use Shannon decomposition with respect to the variable x_1:

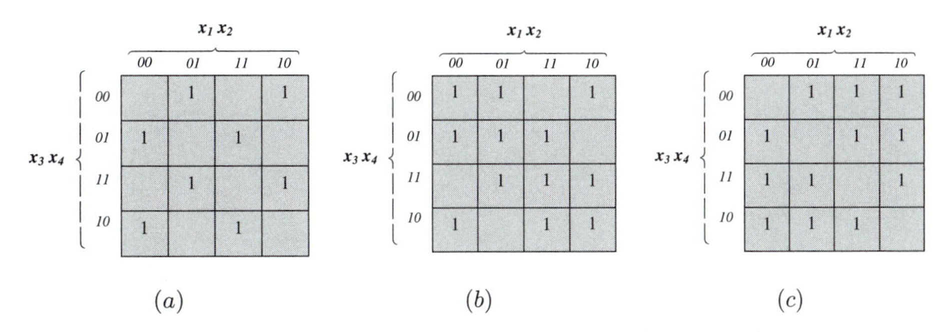

Problem 14.7 Redesign the following logic networks using Shannon decomposition with respect to the variable x_2 (use multiplexers if necessary):

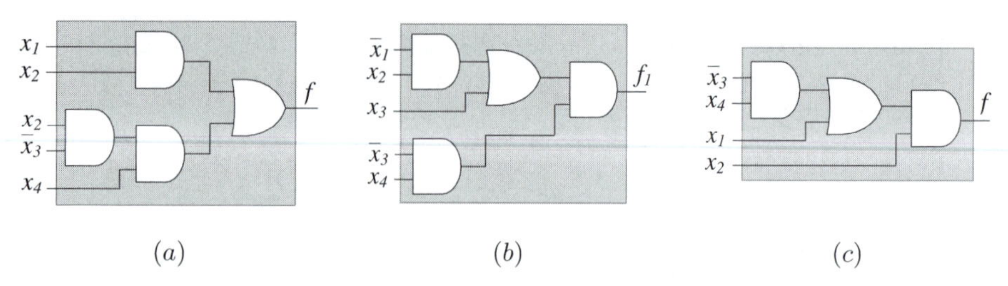

(a) (b) (c)

Problem 14.8 Redesign the logic networks given in Problem 14.7 using EXOR gates to perform positive Davio decomposition with respect to the variable x_3.

Problem 14.9 Add error correction sub-networks into the following logic networks:

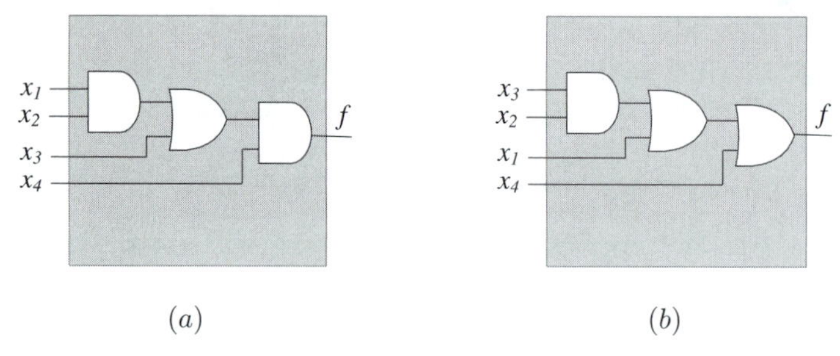

(a) (b)

Problem 14.10 Design a ternary half adder using two bits for encoding ternary digits such as $0_3 = 00_2$, $1_3 = 01_2$ and $2_3 = 10_2$. Truth table of a ternary adder is given below

x_1	x_2	Carry	Sum	x_1	x_2	Carry	Sum
0	0	0	0	2	0	0	2
0	1	0	1	2	1	1	0
0	2	0	2	2	2	1	1
1	2	1	0				

Problem 14.11 Design a ternary full adder using binary encoding of ternary digits as in Problem 14.10.

Problem 14.12 Derive the Boolean equation for the f given the following decomposition of decision diagram:

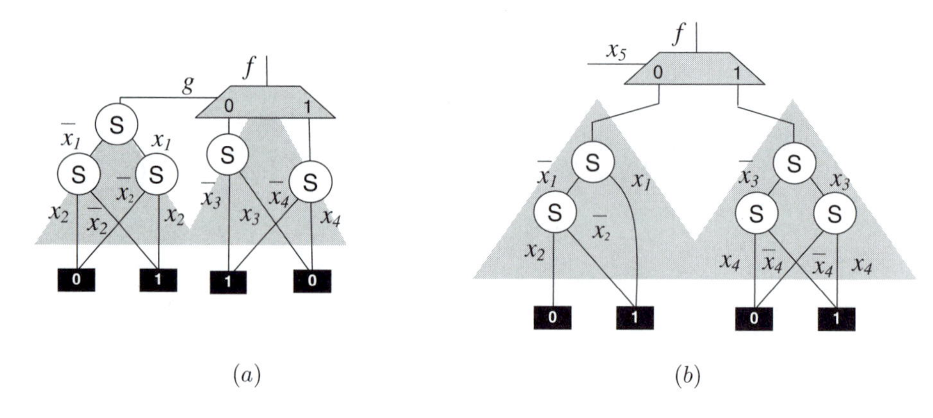

(a) (b)

Problem 14.13 Find a decomposition chart for the Boolean function $f = \overline{x}_1\overline{x}_2 \vee \overline{x}_2x_3x_4 \vee x_1x_3x_4$:

(a) $\{(x_1x_2), (x_1, x_3)\}$ (b) $\{(x_1, x_3), (x_2, x_4)\}$

Problem 14.14 Check if the following Boolean functions given decomposition charts have OR type or AND type decomposition

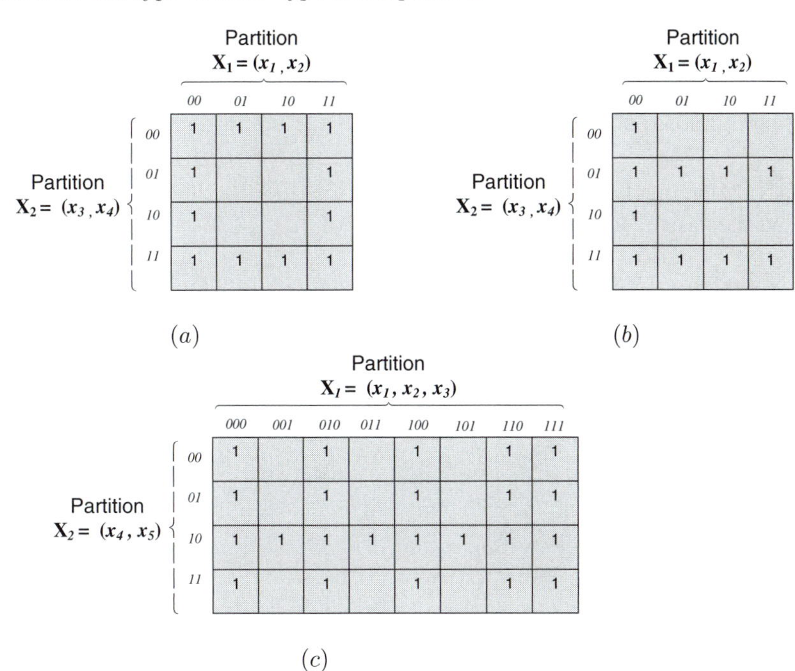

(c)

15

Standard Modules of Sequential Logic Networks

Storage as physical phenomena

- Physical effects
- Feedback

Data structures

- State equations
- Characteristic and state tables
- State diagrams

Latches and flip-flops

- SR and D latches
- D, JK, and T flip-flops

Registers

- Registers with parallel load
- Shift registers

Counters

- Ripple and synchronous counters
- Counters with unused states

Advanced topics

- Neural networks
- Multi-level memory

15.1 Introduction

The logic networks consist of two different types of logic elements: the elements that operate on data values, and the elements that contain a state. The elements that operate on data values are all combinational, which means that their outputs depend only on the current inputs. Given the same input, a combinational element always produces the same output. Other elements in the logic networks are not combinational, but instead contain a *state*. An element contains a state if it has some internal storage. These elements are called *state elements*. In a state element, the required inputs are the data values to be written into the element, and the clock, which determines when the data value is written. The output from a state element provides the value that was written in an earlier clock cycle. The ability to store data during the time between the write-enabling signal is a crucial property of a digital system.

The main property of sequential networks

Sequential networks have the property that the output depends not only on the present input but also on the past sequence of inputs. A sequential network contains combinational networks and cannot be described completely by a truth table. In sequential networks, two different types of description are used: *state* description and *time behavior*.

Sequential networks are designed using libraries of sequential and combinational standard modules. In this chapter, the standard modules of sequential networks are introduced. Standard modules that contain state are also called *sequential modules* because their outputs depend on both their inputs and the contents of the internal state. The storage properties of these modules are provided by physical phenomena. These phenomena are well studied, and widely used in modern logic design, although various new chemical and physical phenomena are being actively investigated as technology is progresses (details are given in the "Further reading" section).

15.2 Physical phenomena and data storage

Data storage can be implemented using various physical principles. The operating principle of storage elements may be based upon various physical effects and phenomena. Mechanical, electrical, magnetic, optic, acoustic, molecular, and atomic effects are utilized in storage elements.

Physical memories often store information in the form of two energy states, allowing storage and retrieval to be accomplished by a transfer of energy. In predictable technology of the future, these storage properties are studied in nanospace, that is, at the molecular and atomic levels. These phenomena can be abstracted to obtain a technology-independent model of a primitive storage element. This element can be defined as a storage device that has two configurations, or states: storing 0 and storing 1.

A binary storage element requires two stable states to represent 0 and 1, and a mechanism for writing and rewriting this information (Figure 15.1). Their storage capacity is one binary digit, or one bit. Devices with more than two states can be treated as a combination of two-state devices. In this way the "bit" may be used as a general measure for storage capacity and information content.

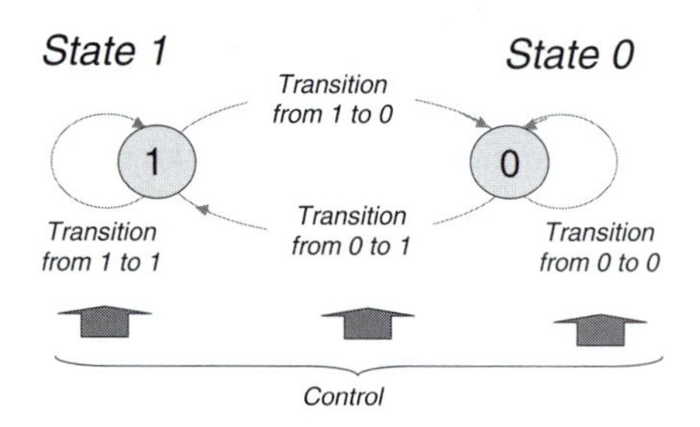

FIGURE 15.1
A binary storage element requires two stable states to represent 0 and 1, and mechanism for writing and rewriting this information.

15.3 Basic principles

In order to design sequential logic networks, memory elements are needed that have the capability of storing the value of a binary quantity Q:

- State $Q = 0$, the memory element stores the value 0, and
- State $Q = 1$, the memory element stores the value 1.

The variable Q is called the *state variable*. State $Q = 0$ and state $Q = 1$ are required to be *stable*. A change in Q requires some control in the form of additional input signals. Changing from the stable state $Q = 0$ to the state $Q = 1$ and vice versa may not be direct; that is, a memory element may pass through one or more unstable (transient) states. An element that has precisely two states is called a *bistable* element, or a bistable network.

15.3.1 Feedback

Combinational logic networks have no *feedback*; that is, no copy of an output signal goes back to the input part of the network. In simple cases, networks with feedback can be analyzed by tracing signals through the network.

> **Example 15.1 (Feedback.)** *Consider the simplest case of an inverter with feedback (Figure 15.2):*
>
> ***Step 1:*** *If, at some instance in time, the inverter input is 0, this value is propagated through the inverter and causes the output to become 1, after some delay at the inverter.*
>
> ***Step 2:*** *The logic value 1 is fed back into the input, and after the propagation delay, the inverter output will become 0.*
>
> ***Step 3:*** *When logic value 0 feeds back into the input, the output will again switch to 1.*
>
> *The inverter output will continue to oscillate back and forth between 0 and 1; that is, it never reaches a stable state.*

Cascading of two inverters in a row forms a network with two stable states called the *bistable* network. An ideal model of a bistable logic network is illustrated in Figure 15.2. It consists of two cascaded inverters with a closed loop. The states $Q = 0$ and $Q = 1$ are associated with $f(t) = 0$ and $f(t) = 1$, respectively. In this model, there is no control mechanism for changing the state from $Q = 0$ to $Q = 1$, and vice versa.

> **Example 15.2 (Feedback.)** *Consider a feedback loop with two inverters in it (Figure 15.2):*
>
> ***Stable state A:*** *If the input to the first inverter is 0, its output will be logical 1. The input to the second inverter will be 1, and its output will be 0. This 0 will feed back into the first inverter. No changes will occur, because this input is already 0. This is a stable state.*
>
> ***Stable state B:*** *The second stable state of the network occurs when the input to the first inverter is 1, and the input to the second inverter is 0.*

As follows from Example 15.2, in the cascade of two inverters there is a mechanism to get it to change from whichever initial condition it started from. No finite delay can cause it to oscillate. This feedback system is locked up at one of its two states. Extra logic is needed to set this network to a specific value; that is, the feedback path must be broken while a new value is connected to the input.

Cascaded inverters with an odd number of inverters are called *ring oscillators*. They provide useful functions, such as the generation of sequences

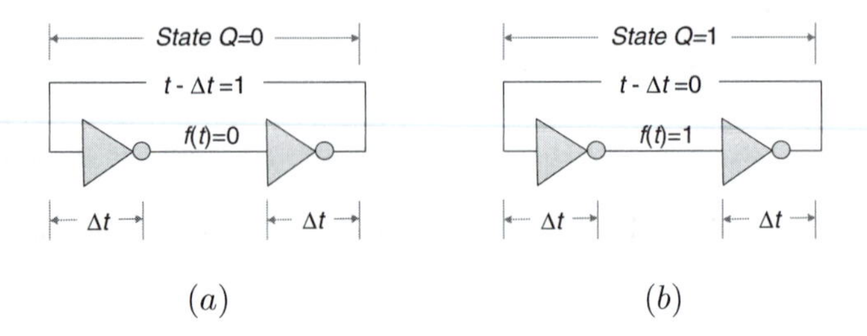

FIGURE 15.2
An ideal model of a bistable memory element with state 0 (a) and state 1 (b).

of logical 0s and 1s. This is because of oscillating behavior of this network. The signals that a ring oscillator generates are repeated every period. The odd number of inverters leads to the oscillatory behavior that repeats every $t_P = \sum_i \Delta t_i$ time units (an inverter delay is of 1 time unit). The time t_P is called the *period*. The duration of the period depends on the number of inverters in the chain.

Example 15.3 (Feedback.) *In Figure 15.3, five cascaded inverters generate sequences of logical 0s and 1s with a period of* $5\Delta t$ *time units.*

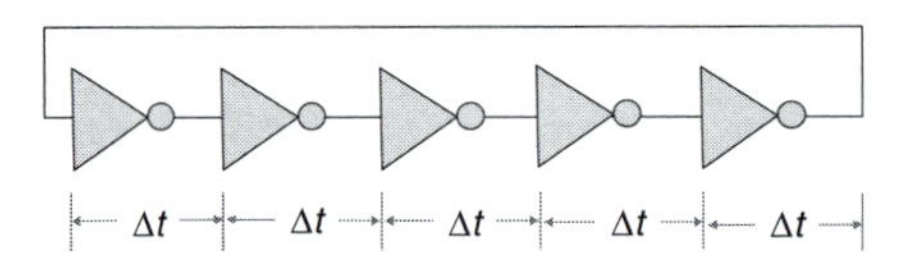

FIGURE 15.3
Cascaded inverters for generating sequences of logical 0s and 1s (Example 15.3).

15.3.2 Clocking techniques

Clocking techniques defines when signals can be read and when they can be written. It is important to specify the timing of reads and writes because, if a signal is written at the same time it is read, the value of the read could correspond to the old value, the newly written value, or even some mix of the two.

In this chapter, we assume an *edge-triggered* clocking technique. Ideally, the sides of clock pulses rise and fall in zero time (Figure 4.7 in Chapter 4).

In practice, the sides rise and fall in nonzero time and the slopes of the sides do not change from zero to nonzero (or vice versa) in zero time. One period of the clock waveform includes an interval of time when the clock pulse is 1 ("high" in positive logic) and another interval of time when it is 0 ("low" in positive logic).

An edge-triggered clocking technique means that any values stored in the network are updated only on a clock edge. Thus, the state elements all update their internal storage on the clock edge. Because only state elements can store a data value, any collection of combinational logic must have its inputs coming from a set of state elements and its outputs written into a set of state elements. The inputs are values that were written in a previous clock cycle, while the outputs are values that can be used in a following clock cycle.

15.4 Data structures for sequential logic networks

The operation of a sequential network is characterized by a sequential and continuous changing of its states, which are specified by contents of the network's registers, or memory storage. This behavior is described by the formal model called *finite state machine*. This model is represented as follows:

Finite state machine description

- A sequence of *state transitions* controlled by the network's inputs.
- *State table*, which defines the network's next-state functions.
- *State diagram*, which represents all internal states and relationships between the states.
- *Characteristic equations* that describe the behavior of the flip-flops in the network.

15.4.1 Characteristic equations

The functional behavior of a state machine can be described by *characteristic equations*, that specify their next states as functions of their current states and inputs.

Example 15.4 (Characteristic equation.) *Given a network with the input D and output Q, characteristic equation $Q^+ = D$ means that the next state of the output Q, denoted by Q^+, will be equal to the value of the input D in the present state.*

15.4.2 State tables and diagrams

The number of input/output sequences in a sequential network is infinite. However, the number of combinations of the primary input and internal state values is finite, hence, "finite state machine." This fact is utilized by the state table, which contains all input/output sequences in implicit form. The *state* of a logic network at time t is defined as the current logic values of some set of signals of interest. Two types of states are distinguished: the *internal* state, and the *total* state. The internal state represents what the network remembers from its behavior prior to time t. The total state completely determines the next action to be taken by the network.

State table

A *state table* specifies the next state and the outputs of a sequential network in terms of its present state and inputs.

State diagram

A *state diagram* is a graphical representation of the states (assigned to the nodes of the graph) and the transition between the states (denoted by the directed edges of the graph). It is a directed graph that represents all internal states and possible state transitions.

State tables and diagrams defined at the same level of abstraction contain exactly the same information about the sequential network's behavior. Given a state table, one can construct acorresponding state diagram, and vice versa.

Example 15.5 *Figure 15.4 shows the state table and the state diagram for the simplest sequential module.*

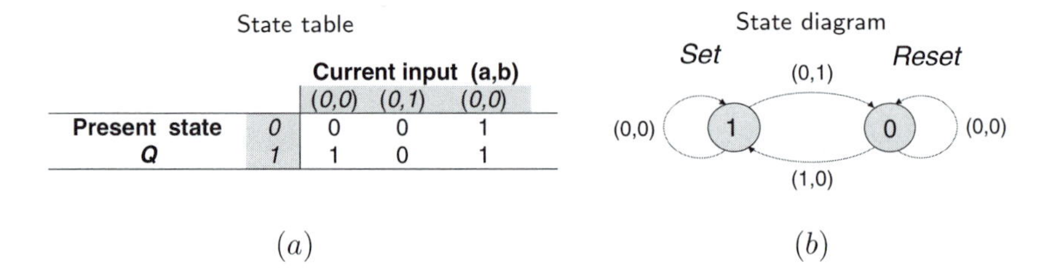

FIGURE 15.4
Excerpts from the state table (a) and the state diagram for the SR latch (b) (Example 15.5).

15.5 Latches

Latches and flip-flops are the basic building blocks of sequential logic networks. Latches provide a mechanism for the simplest control of the storage of one bit:

$$\underbrace{\texttt{1 bit memory}}_{\textit{Physical phenomenon}} \xrightarrow{\textit{Design}} \underbrace{\texttt{1 bit memory control}}_{\textit{Latch}}$$

There are several types of latches, distinguished by their method of controlling their functions. The latch is specified by a characteristic table and equation, a state table and state diagram, and a timing diagram.

15.5.1 SR latch

The *SR latch* is the basic memory element, defined as follows:

SR latch

- Is a bistable memory network,
- Has two inputs labeled *set* S and *reset* R, and
- Has two outputs labeled Q and $\overline{Q}$.

In an SR latch based on NAND gates, the NAND gates are considered to be "cross-coupled," with each NAND feeding back its output to the other NOR gate.

Example 15.6 (SR latch.) *An SR latch with two cross-coupled NAND gates is shown in Figure 15.5. The condition that is undefined for this latch is when both inputs are equal to 0 at the same time. This input combination must be avoided.*

Practice problem 15.1. (SR latch.) Use NOR gates to design SR latch. Derive the characteristic table, state diagram, and state table for the new latch. Note that the input signals do not require the complement of input values as for the NAND-based SR latch (Table 15.1).
Answer is given in "Solutions to practice problems."

The SR latch operates as follows:

(*a*) It has two stable states defined as $Q = 0$, which is called the *reset state*, and $Q = 1$, which is called the *set state*.
(*b*) Latch operation is defined by the following input combinations:

Design example: SR latch

Logic network

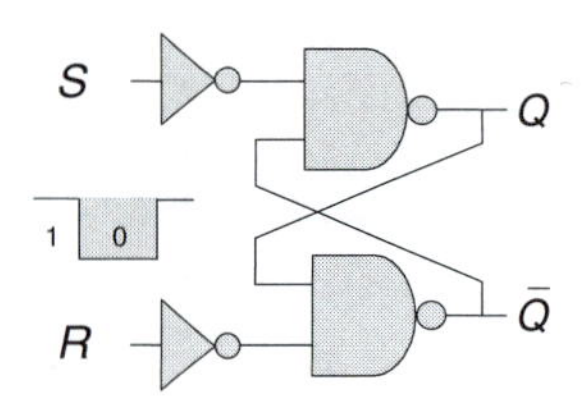

Characteristic table

S	R	Q	$\bar{Q}$	Operation
0	0	Q	$\bar{Q}$	*No change*
0	1	0	1	*Reset*
1	0	1	0	*Set*
1	1	–	–	*Undetermined*

State diagram

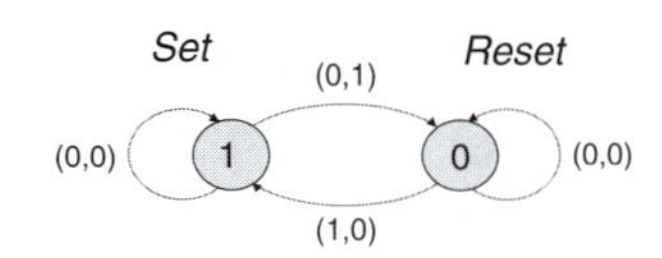

Timing diagram

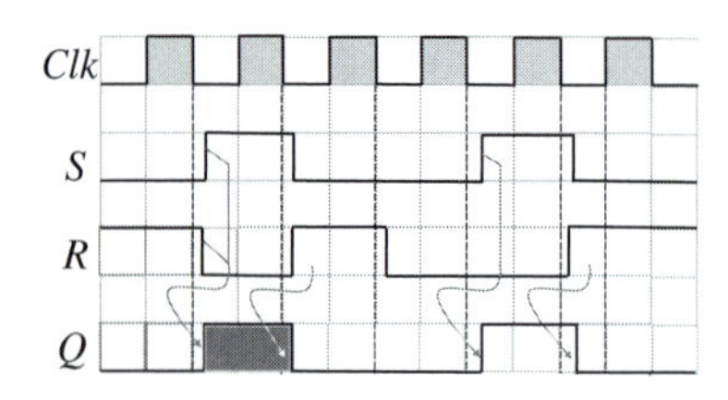

State table

		Current input (S, R)		
		(0,0)	(0,1)	(1,0)
Present state Q	0	0	0	1
	1	1	0	1

FIGURE 15.5
The SR latch using NAND gates, timing diagram (failing-edge clock-controlled), characteristic table, state diagram, and state table (Example 15.6).

- $(S, R) = (0, 0)$ leaves the latch in either of its stable states indefinitely. This state may be either $(Q, \overline{Q}) = (0, 1)$ or $(Q, \overline{Q}) = (1, 0)$ which is indicated in the truth table by stating that the $Q, \overline{Q}$ outputs have values 0/1 and 1/0, respectively.
- $(S, R) = (1, 0)$ sets the latch by changing its state from $Q = 0$ to $Q = 1$. If the initial state is already $Q = 1$, setting the latch has no effect.

$(S, R) = (0, 1)$ resets the latch by changing its state from $Q = 1$ to $Q = 0$. If the initial state is already $Q = 0$, setting the latch has no effect.

- The input combination $(S, R) = (1, 1)$ must be avoided by making sure that 1's are not applied to both inputs simultaneously. This is because an SR latch has unpredictable behavior under this operating condition.

The outputs Q and $\overline{Q}$ of a latch are complementary only after the state has stabilized, because unstable states can appear only temporarily during state transition.

15.5.2 Gated SR latch

The operations of the SR latch can be modified by providing an additional control input. This input determines when the state of the latch can be changed. This latch is called a *gated SR latch.* A gated SR latch is defined as follows:

Gated SR latch

- ▶ Is a bistable memory network.
- ▶ Includes two NAND gates and, in addition, has two other gates controlled by S, R, and clock Clk signals.
- ▶ The control input Clk acts as an enable signal for the S and R inputs.
- ▶ The output of the NAND gates stays at logical 1 as long as the control input remains at 0.
- ▶ When the control input goes to 1, data from the S or R input is allowed to affect the SR latch.

An indeterminate condition occurs when all three inputs are equal to logical 1.

Example 15.7 (Gated SR latch.) *The gated SR latch with NAND gates and its data structures is given in Figure 15.6.*

15.5.3 D latch

Assume that the inputs S and R of the SR latch are fed into one data input, D (Figure 15.7). A D latch is defined as follows:

D latch

- ▶ Is a bistable memory network with a single input,
- ▶ Is based on the gated SR latch with connected inputs,
- ▶ Has the data input, D, and control signal, Clk,
- ▶ Has two outputs, Q and $\overline{Q}$.

Practice problem **15.2. (D latch.)** Design a clocked D latch using SR latch and additional AND gates and inverters.
Answer is given in "Solutions to practice problems."

In the D latch, it is impossible to have a troublesome situation such as in the SR latch when $S = R = 1$. This is because the output Q merely tracks

Design example: Gated SR latch

Characteristic table

Clk	*S*	*R*	*Q*	$\bar{Q}$	*Operation*
0	×	×	Q	$\bar{Q}$	*No change*
1	0	0	Q	$\bar{Q}$	*No change*
1	0	1	0	1	*Reset*
1	1	0	1	0	*Set*
1	1	1	–	–	*Undetermined*

State diagram

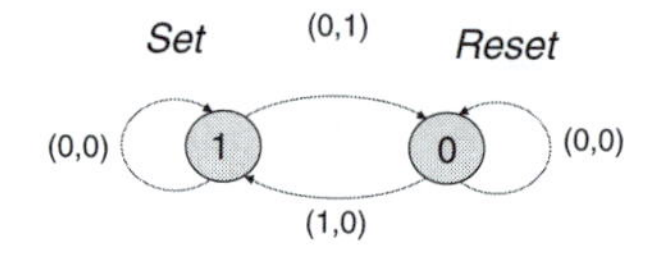

Logic network

State table

Present state Q		Current input (S, R) (0,0)	(0,1)	(1,0)
	0	0	0	1
	1	1	0	1

Timing diagram

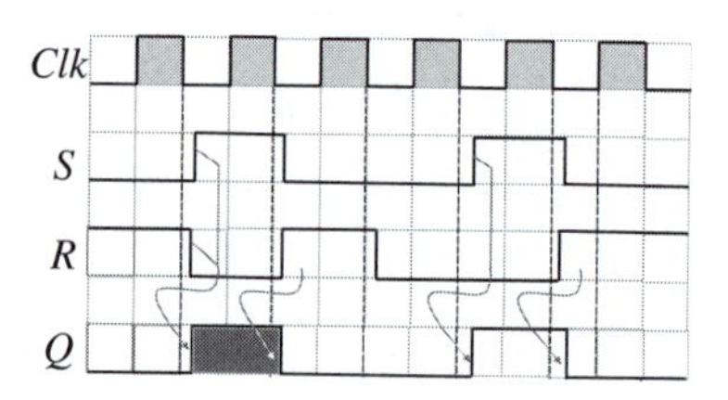

FIGURE 15.6
The SR gated latch is based on latch with additional control logic (Example 15.7).

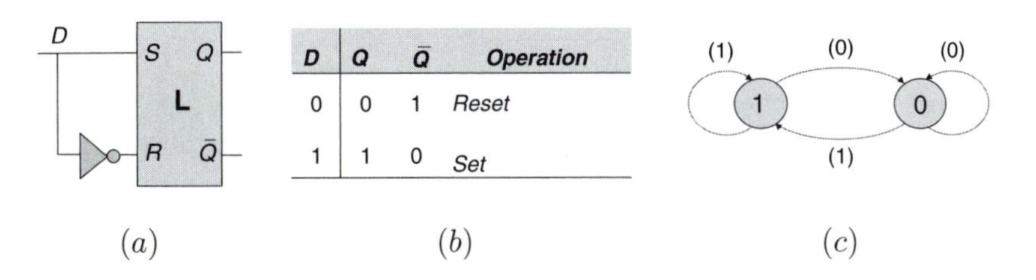

FIGURE 15.7
The D latch based on SR latch.

the value of the input D while $Clk = 1$. As soon as Clk goes to 0, the state of the latch is "frozen" until the next time the clock signal goes to 1. Therefore, the D latch scores the value of the D input seen at the time the clock changes from 1 to 0.

Example 15.8 (Multiplexer-based D latch.) *Figure 15.8 shows an implementation of D latches using multiplexers. For a positive-edge triggered latch, the D input is selected when the clock signal Clk is high, and the output is held using feedback when the clock signal is low. For a negative-edge triggered latch, input 0 of the multiplexer is selected when the clock is low, and the D input is passed to the output. When the clock signal is high, input 1 of the multiplexer is selected.*

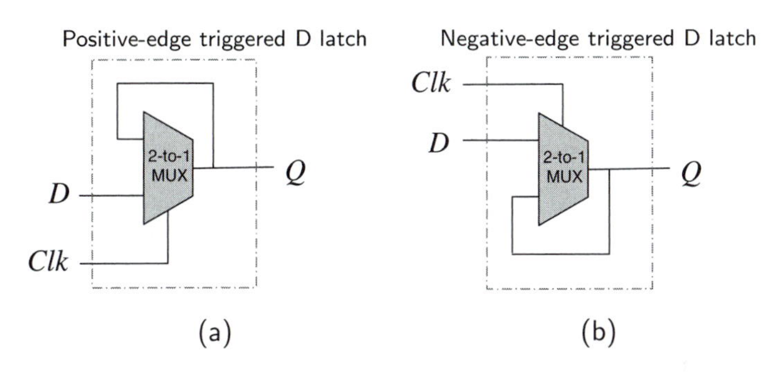

FIGURE 15.8
Multiplexer-based D latches: negative-edge triggered (a) and positive-edge triggered D latch (Example 15.8).

15.6 Flip-flops

The most important property of sequential logic networks is that they have global feedback. The implementation of the global feedback requires a flexible control functions; that is, the memory elements must efficiently elaborate timing logic. Latches do not satisfy this requirement. They have a common property of immediate output responses (to within the propagation delay times). This property is an undesirable property in certain applications: it is necessary that output changes occur only coincident with changes on a control input line. However, a combination of two latches called *flip-flops*, can produce the desired behavior. Flip-flops employ the clock signal to precisely control the times at which data are transferred.

A flip-flop is defined as follows:

Flip-flop

- Is a bistable memory device, with inputs, that remains in a given state as long as power is applied and until input signals are applied to cause its output to change.
- Consists of a latch, a basic bistable element, to which appropriate logic is added in order to control its state.
- Employs special control signals in order to specify:
  - (a) The times at which the network responds to changes to its input data signals,
  - (b) The times at which the network changes its output data signal.

The process of storing a logic 1 into a flip-flop is called *setting* the flip-flop. The process of storing a logic 0 into flip-flop is called *resetting* or *clearing* the flip-flop.

15.6.1 The master-slave principle in flip-flop design

The property of having the timing of a flip-flop response being related to a control input signal is achieved with *master-slave* and *edge-triggered* flip-flops.

A master-slave flip-flop consists of two cascaded sections, each capable of storing either a 0 or a 1. The first section is referred to as the *master* and the second section as the *slave*. Information is entered into the master on one edge of a control signal and is transferred to the slave on the next edge of the control signal.

$$\underbrace{\texttt{1 bit memory}}_{Physical\ phenomenon} \xrightarrow{Design} \underbrace{\texttt{Master}}_{Latch} \xrightarrow{Control} \underbrace{\texttt{Slave}}_{Latch}$$

The momentary change is called a *trigger*. The transition a trigger causes is refereed to *trigger* the flip-flop. The key to the proper operation of a flip-flop is to trigger it only during a signal transition. Two types of edge triggered flip-flops are distinguished:

- ▶ *Positive-edge triggered* (change takes place when the clock goes from 0 to 1, that is, rising edge) flip flops, and
- ▶ *Negative-edge triggered* (change takes place when the clock goes from 1 to 0, that is, falling edge) flip flops.

The positive transition is defined as the positive-edge and the negative transition as the negative-edge. In positive-edge triggered flip-flop, the positive or rising edge of the clock to initiate the entire state transition process is used. A negative-edge triggered flip-flop behaves like the positive analog, except that the negative or falling edge of the clock initiates the state transition process. The stored value after the transition depends on the inputs and what was stored prior to the transition.

Example 15.9 (Multiplexer-based master-slave flip-flop.) *Figure 15.9 shows a positive edge-triggered flip-flop based on a master-slave configuration. The flip-flop is built of the cascaded negative latch (master) with a positive latch (slave). A negative edge-triggered flip-flop can be constructed by using the positive latch first.*

Practice problem 15.3. (Multiplexer-based master-slave flip-flop.) Design a negative edge-triggered flip-flop using multiplexers.
Answer: By analogy with positive edge-triggered flip-flop (Example 15.9) but use the positive edge-triggered latch first.

15.6.2 D flip-flop

The D flip-flop is the counterpart of the D latch. The D flip-flop can be constructed from a pair of D latches connected as shown in Figure 15.10. This particular configuration of a network is known as a *master-slave* network. The D flip-flop is defined as follows:

D flip-flop

- ▶ Is a bistable memory network with data input and control clock input signals.
- ▶ Is based on D latches, one of which is called a *master* and is controlled by the clock, *Clk*, while the other is called *slave*.
- ▶ Is negatively or positively clocked; the first case corresponds to the inverted clock *Clk* for the slave latch, and the second case corresponds to the inverted clock $\overline{Clk}$ for the master latch.

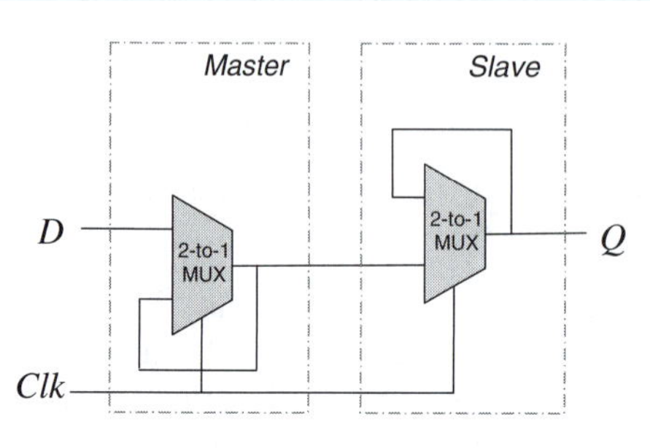

- The master is transparant on the low phase of the clock, and the D input is passed to the master output
- The slave is in the hold mode, keeping its previous value by using feedback
- The master stops sampling the input, and the slave starts sampling
- The slave samples the output of the master during the high phase of the clock while the master remains in a hold mode
- The value Q is the value of D right before the rising edge of the clock, achieving the positive edge-triggeed effect.

FIGURE 15.9
Multiplexer-based master-slave positive edge-triggered flip-flop (Example 15.9).

The D flip-flop operates as follows:

(a) When the clock signal *Clk* is 1, the master latch is disabled and its output remains stable; the data output of the slave latch is stable too.

(b) When the clock signal *Clk* changes to 0, the slave latch is disabled and its data outputs remain unchanged. The master latch is enabled and begins to respond to the input data D.

Example 15.10 (D flip-flops.) *The behavior of both negative- and positive-edge triggered D flip-flops is shown in Figure 15.10. The initial value Q is unknown. In particular, for the positive-edge triggered D flip-flop, when the first raising edge of the clock occurs, the state of the D flip-flop is established. Since* $D = 0$ *at this time, Q goes to 0. There is a slight delay in the output. Usually the input D changes shortly after transition. The output* $\overline{Q}$ *is the opposite of the output Q. At the second raising edge,* $D = 1$*, and* $Q = 1$ *for the next clock period. At the third raising edge,* $D = 1$*, and* $Q = 1$ *for another clock period. If the signal D were to go back and forth between clock transitions, the output Q would not be affected. The characteristic equations, state tables, and the state diagram are given in Figure 15.10 as well.*

Design example: D flip-flops

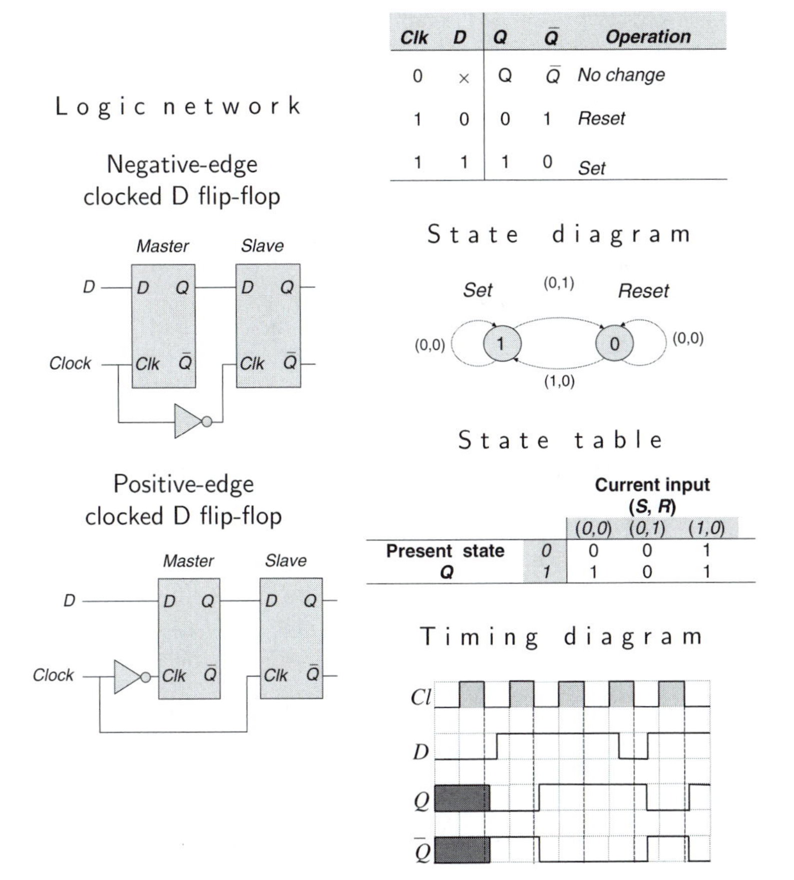

FIGURE 15.10
A D flip-flop is based on the master-slave network configuration (Example 15.10).

15.6.3 JK flip-flop

The functioning of the JK flip-flop is identical to that of the SR flip-flop in the SET, RESET, and no change conditions of operation. The difference is that the JK flip-flop has no invalid state as does the SR flip-flop. The JK flip-flop is defined as follows:

JK flip-flop

- Is a bistable memory network with data input and control clock input signals.
- Can be constructed from both of SR or D latches in a master-slave configuration, with additional logic.
- Is negative- or positive-edge triggered.

The next state of each JK flip-flop is evaluated from the corresponding J and K inputs and characteristic equation $Q = J\overline{Q} \vee \overline{K}Q$. There are four cases:

$JK = 00$: no change, the next state is same as the present state, `Next state=Present state`;

$JK = 11$: the next state is the complement of the present state, `Next state` $=$ $\overline{\texttt{Present state}}$.

$JK = 01$: the next state is 0;

$JK = 10$: the next state is 1;

Example 15.11 (Negative edge-triggered JK flip-flop.) *Figure 15.11 shows the waveforms applied to the J, K, and clock Clk inputs of the negative edge-triggered JK flip-flop. The Q output is determined, assuming that the flip-flop is initially in RESET. The Q output is determined, assuming that the flip-flop starts out RESET and the clock is active LOW.*

Example 15.12 (JK flip-flop.) *A negative-edge triggered JK flip-flop based on SR latches is shown in Figure 15.12.*

Practice problem **15.4. (JK flip-flop.)** Design a negative-edge triggered JK flip-flop based on the D latches and any additional logic gates. ***Answer*** is given in "Solutions to practice problems."

15.6.4 T flip-flop

The T flip-flop is a complementing flip-flop which is defined as follows:

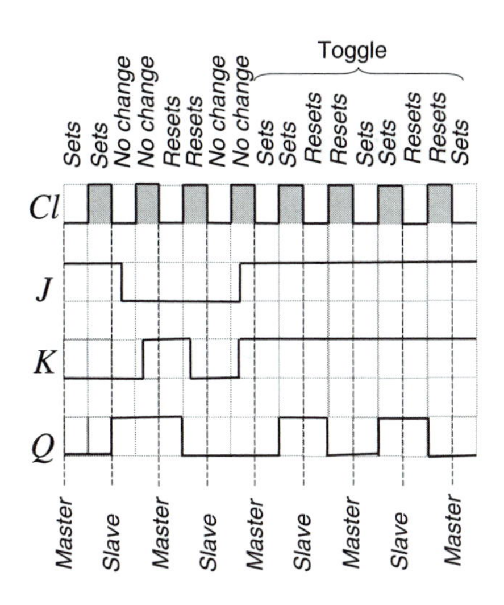

- *The master latch is assumed to be in the state determined by the J and K inputs beginning at the leading (positive going) edge of the clock pulse.*
- *The state of the master latch is transferred to the slave latch on the trailing edge of the clock pulse*
- *The state of the slave latch appears on the Q and $\overline{Q}$outputs at the trailing edge.*
- *The feedback specifies the characteristic toggle operation when $J = K = 1$.*
- *The J and K inputs cannot change while the clock pulse is active because the state of the master latch can change during this time.*

FIGURE 15.11
Timing diagram of the master-slave JK flip-flop (Example 15.12).

T flip-flop

- Is a bistable memory network with data input and control clock input signals.
- Toggles the output signal values on each active clock edge (negative for negative-edge triggered flip-flop, and positive for positive-edge triggered flip-flop) when the input $T = 1$.
- Can be obtained from a JK flip-flop when the inputs J and K are tied together, or using a D flip-flop and an EXOR gate.

The T flip-flop is operated as follows:

- Its characteristic equation is $Q^+ = Q \oplus T$.
- When $T = 0$ ($J = K = 0$), a clock edge does not change its output.
- When $T = 1$ ($J = K = 1$), a clock edge complements its output.

Example 15.13 **(T flip-flop.)** *A negative-edge triggered T flip-flop is shown in Figure 15.13.*

Practice problem **15.5.** **(T flip-flop.)** Design a positive-edge triggered T flip-flop based on the D flip-flop.
Answer is given in "Solutions to practice problems."

Design example: JK flip-flop

Characteristic table

Clk	*J*	*K*	*Q*	$\bar{Q}$	*Operation*
0	×	×	Q	$\bar{Q}$	*No change*
1	0	0	Q	$\bar{Q}$	*No change*
1	0	1	0	1	*Reset*
1	1	0	1	0	*Set*
1	1	1	Q	$\bar{Q}$	*Toggle*

State diagram

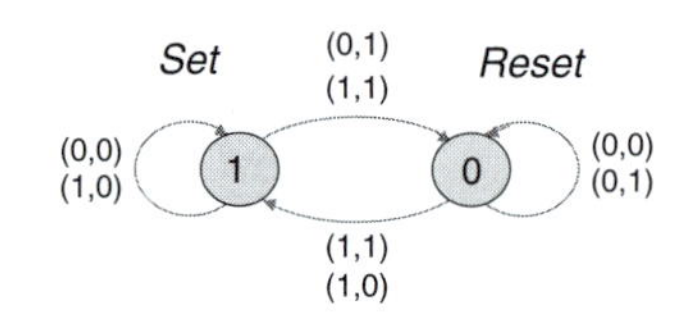

Logic network

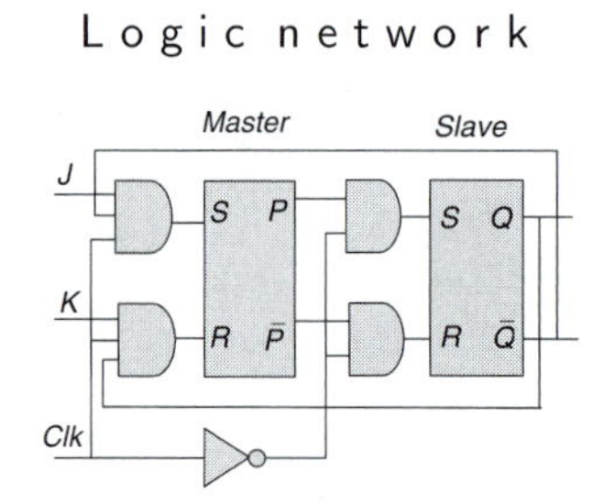

State table

		Current input (*J, K*)			
		(0,0)	*(0,1)*	*(1,0)*	*(1,1)*
Present state	*0*	0	0	1	1
Q	*1*	1	0	1	0

Timing diagram

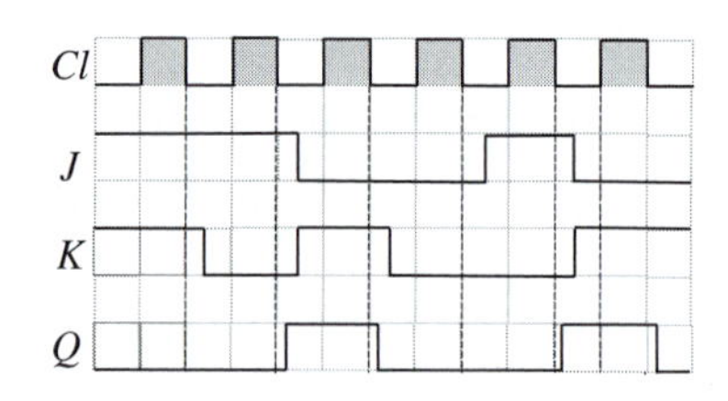

FIGURE 15.12
A negative-edge triggered JK flip-flop based on SR latches in master-slave configuration.

Design example: T flip-flop

Characteristic table

Clk	D	Q	$\bar{Q}$	Operation
0	×	Q	$\bar{Q}$	*No change*
1	0	Q	$\bar{Q}$	*No change*
1	1	1	0	*Toggle*

State diagram

Set (1) Reset

(0) 1 0 (0)

(1)

Logic network

T flip-flop based on JK flip-flop

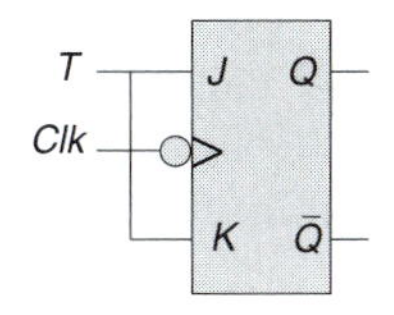

State table

Present state Q		Current input (0)	Current input (1)
	0	0	0
	1	1	0

Timing diagram

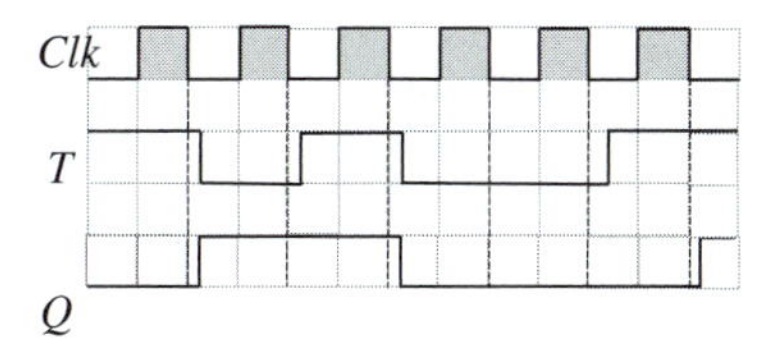

FIGURE 15.13
A negative-edge clocked T flip-flop based on the JK flip-flop (Example 15.13).

15.7 Registers

A flip-flop stores one bit of information. A set of n flip-flops stores n bits of information; for instance, an n-bit number. Registers are classified as

- Storing and
- Shift registers.

15.7.1 Storing register

An n-bit storing *register* with parallel load is defined as follows (Figure 15.14):

Storing n-bit register

- Is a network of flip-flops for storing an n-bit number or n-bit vector.
- Has:
  - n inputs $I = I_1, I_2, \ldots, I_n$
  - n outputs $I = Z_1, Z_2, \ldots, Z_n$
  - Control inputs *Load*, *Clear*, and *Clock* (*Clk*)

The control input *Clear* is asynchronous; that is, it affects the output immediately rather than when the clock signal is received. The control signal *Clear* forces the value $00\ldots0$ into the register. This operation is useful for initialization purposes to guarantee that the register contains a predefined value.

The cell design principle provides the possibility for design and implementation using copies.

Example 15.14 (Registers.) *In Figure 15.15, a 4-bit register using D flip-flops and 2-to-1 multiplexers is given. The register can be constructed using four copies of 1-bit registers.*

Practice problem **15.6. (Register.)** Derive the state table and the state diagram given a 1-bit register, with input D load, clock, and CLEAR signals.
Answer is given in "Solutions to practice problems."

15.7.2 Shift register

A register that provides the ability to shift its contents is called a *shift register*. A n-bit shift register is capable of transferring data among adjacent flip-flops. These transfers can be

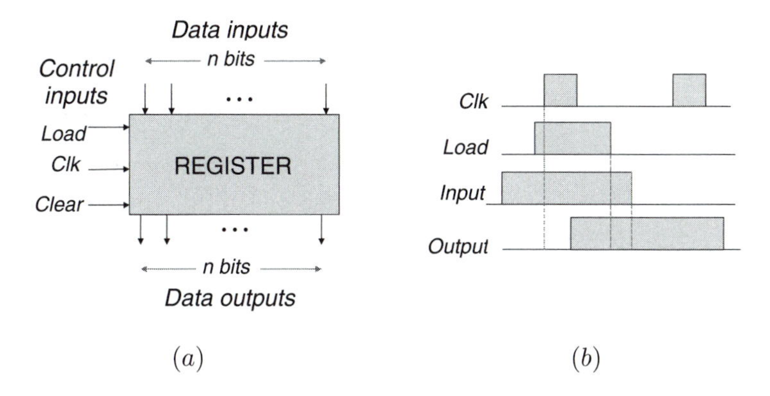

FIGURE 15.14
An n-bit register module (a) and time-behavior diagram (b).

- Bidirectional or
- Unidirectional (either to the left or to the right).

An n-bit shift register is defined as follows (Figure 15.16):

Shift register

- n inputs $I = I_1, I_2, \ldots, I_n$
- n outputs $I = Z_1, Z_2, \ldots, Z_n$
- Control inputs *Load*, *Left*, *Right*, and *Clock* (*Clk*)

Example 15.15 (Shift register.) *In Figure 15.15, a four-bit parallel-in, parallel-out bidirectional shift register using D flip-flops and 4-to-1 multiplexers is given.*

The functionality of shift registers can vary. For example, an n-bit sequential-in, parallel-out shift register has one input to shift data in, and n outputs.

15.7.3 Other shift registers: FIFO and LIFO

Two special but very often used types of register with shift function are first-in-first-out (FIFO) and last-in-first-out (LIFO) registers. The FIFO register acts as a pipeline, or a queue, as the data enter the register sequentially, bit-by-bit, and exits sequentially, starting from the first bit entered. The LIFO register acts as a stack, since the bit pushed first into the LIFO is fetched last. FIFO and LIFO registers can form word-wide arrays. The control of such arrays is not trivial, and requires additional control circuitry.

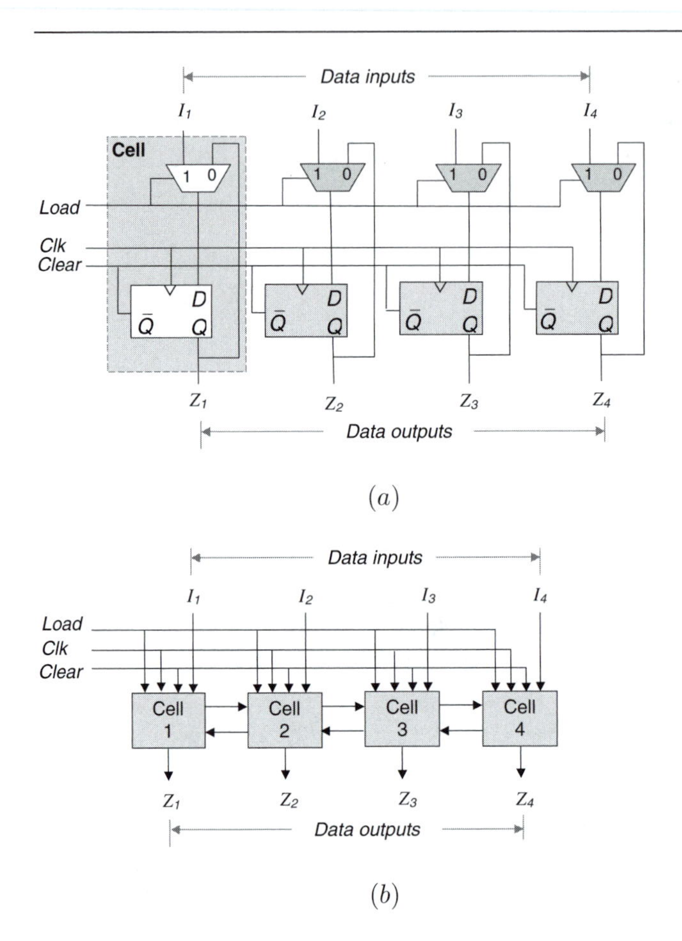

FIGURE 15.15
A 4-bit register (a) and its use for storing data in the terminal nodes of decision tree (b) (Example 15.14).

15.8 Counters

A *counter* is defined as a sequential network with n binary outputs and $p \leq 2^n$ states. Counters are classified using *functional* (direction of count, coding of binary sequences, etc.) and *implementation* (ripple, or asynchronous counter,

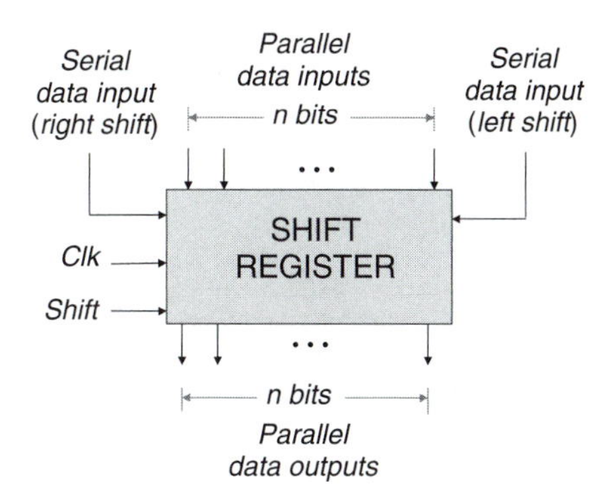

FIGURE 15.16
An n-bit parallel-in, parallel-out bidirectional shift register module.

synchronous counter, etc.) criteria.

The following types of counters are distinguished with respect to the direction of the count:

- *Upward* counter; state i is followed by state $(i+1)$ *mod* p,
- *Downward* counter; state i is followed by state $(i-1)$ *mod* p,
- *Up/down* counters count both ways; control input determines the direction.

Counters are also classified with respect to (a) number of states and (b) type of output encoding. Using these criteria, the following counters are distinguished:

- *Binary* counter (2^n states, binary representation of integers $0, 1, \ldots, 2^n-1$);
- *Decimal* counter (10 states, decimal code),
- *Gray-code* counter (2^n states, Gray code), and
- *Ring* counter.

15.8.1 Binary counters

A counter that follows the binary number sequence is called a *binary* counter. An n-bit binary counter consists of n flip-flops and can count in binary from 0 through $2^n - 1$. Binary counters can be implemented using *asynchronous* or *synchronous* design principles.

Asynchronous binary counters

An asynchronous, or *ripple* counter, is a cascade of T flip-flops such that the input of the i-th flip-flop is connected to the output of the $(i-1)$ the flip-flop, $i = 1, 2, ..., n$, and the input for the first flip-flop is connected to the clock signal.

Design example: Cell design principle

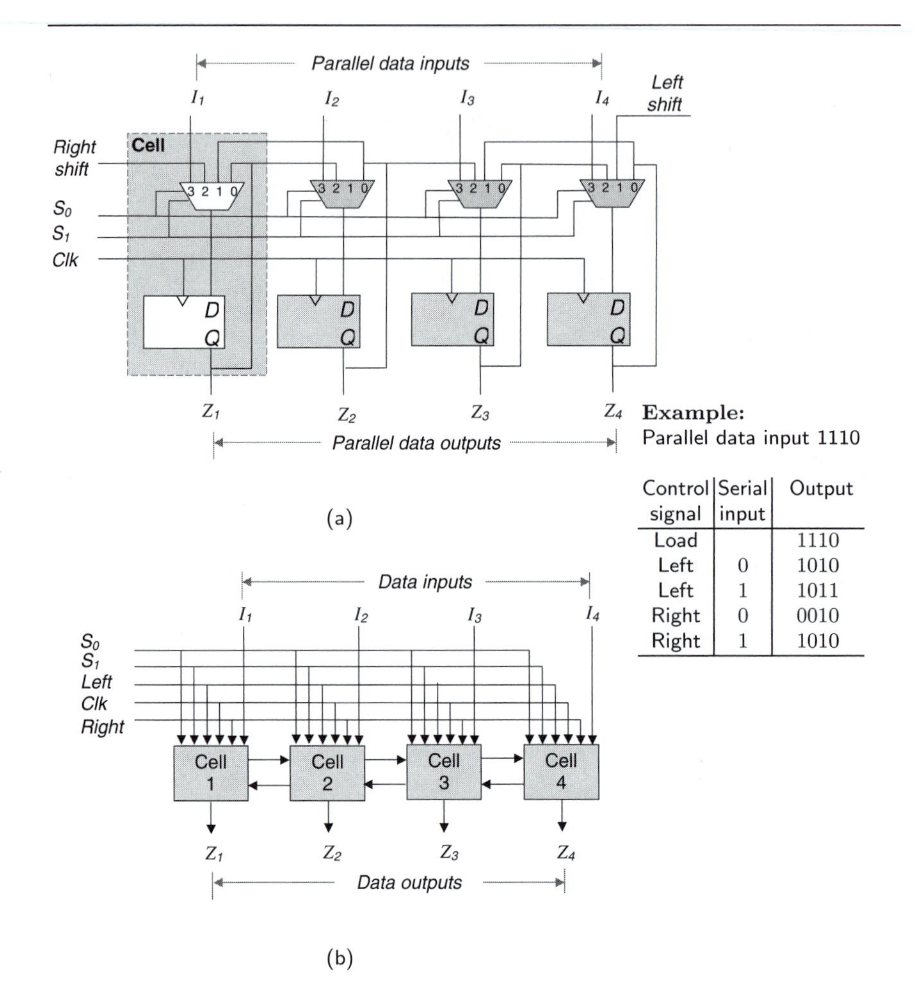

Control signal	Serial input	Output
Load		1110
Left	0	1010
Left	1	1011
Right	0	0010
Right	1	1010

FIGURE 15.17
A 4-bit bidirectional shift register (Example 15.15).

Example 15.16 **(Ripple counter.)** *In Figure 15.18, a 4-bit binary ripple counter is shown. In this counter, T flip-flops are used because a T flip-flop changes state (toggles) on every rising edge of its clock input. Thus, each bit of the counter toggles if and only if the immediately preceding bit changes from 1 to 0.*

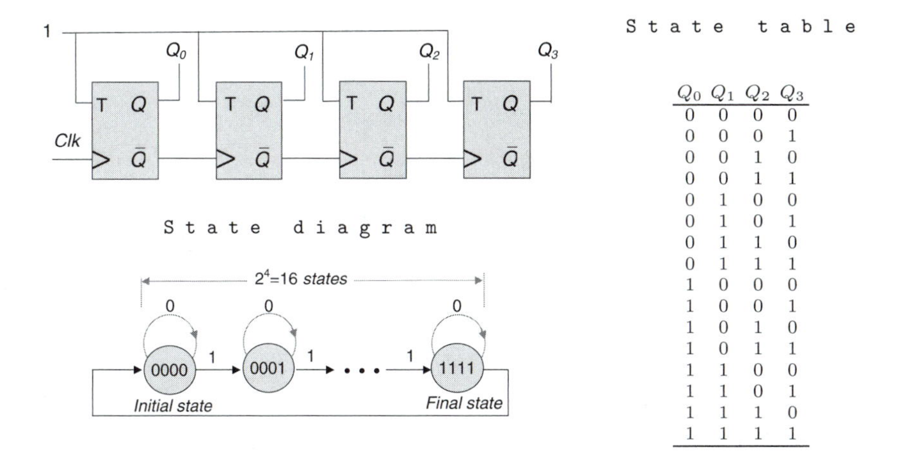

Q_0	Q_1	Q_2	Q_3
0	0	0	0
0	0	0	1
0	0	1	0
0	0	1	1
0	1	0	0
0	1	0	1
0	1	1	0
0	1	1	1
1	0	0	0
1	0	0	1
1	0	1	0
1	0	1	1
1	1	0	0
1	1	0	1
1	1	1	0
1	1	1	1

FIGURE 15.18
A 4-bit binary ripple counter (Example 15.16).

Practice problem **15.7. (Ripple counter.)** Design a 4-bit ripple counter on D flip-flops.
Answer is given in "Solutions to practice problems."

The disadvantage of ripple counters is that the propagation delay between the input and the output of a flip-flop is summed up to maximum n times, since toggle signals must propagate through the entire counter. The worst case occurs when the counter outputs change from 11...1 to 00...0.

Synchronous binary counter

In a synchronous binary counter, all of the flip-flops' clock inputs are connected to the same common clock signal and all of the flip-flops' outputs change at the same time. This allows the avoidance of the delays present in asynchronous counters. All the flip-flops in a synchronous counter change simultaneously after a single flip-flop propagation delay.

Example 15.17 **(Synchronous counter.)** *Figure 15.19, shows a 4-bit synchronous binary counter. The design of the counter starts with the derivation of the state diagram, and then the excitation equations are specified for each T flip-flop.*

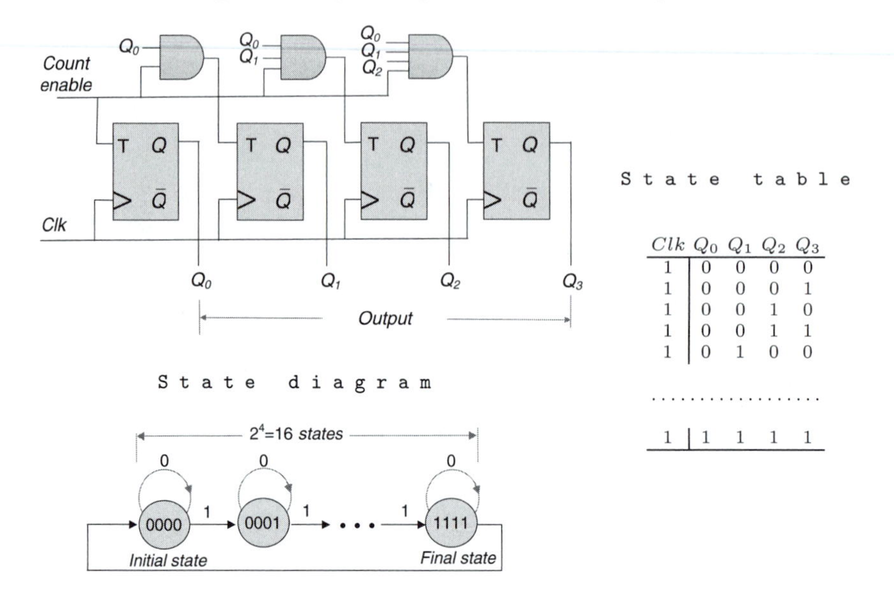

FIGURE 15.19
A 4-bit synchronous binary counter (Example 15.17).

Practice problem **15.8. (Synchronous counter.)** Design a 4-bit synchronous counter based on JK flip-flops.
Answer is given in "Solutions to practice problems."

The binary counters are also called *modulo-*2^n *counters*. For example, a 4-bit binary counter is a modulo-16 counter.

Modulo-m counters

Both asynchronous and synchronous counters can be modified to count up to a certain number, $m \leq 2^n$. Such a counter is built on the n-bit binary counter, which can be re-designed to accommodate counting from 0 to $m-1$ and then wrap around.

Example 15.18 (Synchronous modulo-10 counter.) *Figure 15.20, shows a 4-bit synchronous modulo-10 counter. The counter is built on the 4-bit binary counter, with a special connection of its data, load and count inputs. This connection ensures that when the counter output reaches the value* $1001_2 = 9_{10}$*, the counter should be reset to 0000.*

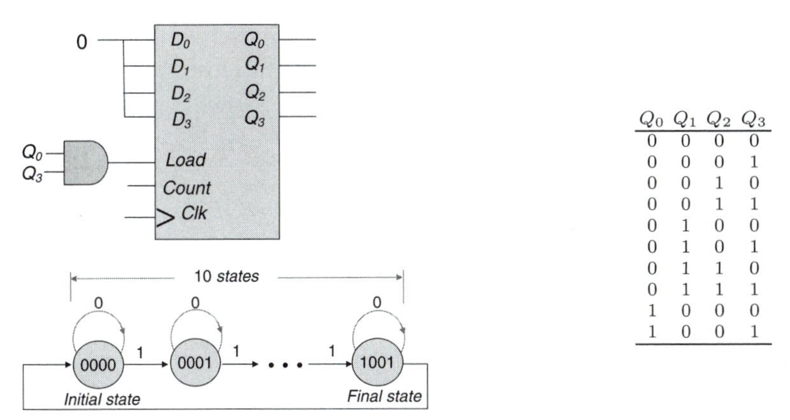

Q_0	Q_1	Q_2	Q_3
0	0	0	0
0	0	0	1
0	0	1	0
0	0	1	1
0	1	0	0
0	1	0	1
0	1	1	0
0	1	1	1
1	0	0	0
1	0	0	1

FIGURE 15.20
A modulo-10 counter (Example 15.18).

Practice problem **15.9.** **(Modulo counter.)** Design a modulo-3 counter using T flip-flops.
Answer is given in "Solutions to practice problems."

Multi-bit counters can be assembled from smaller counters, connected as a cascade. For example, an 8-bit binary synchronous counter can be built based on two 4-bit synchronous counters.

Counters based on shift registers

Shift registers can be used for the design of special types of counters, called *ring counters* and *Johnson counter*. The ring counter includes a shift register that is initialized so that only one of its flip-flops is set to 1, while the others are set to 0. Upon each count pulse, the single 1 is shifted right. An n-bit ring counter has n states in its counting sequence; this counter circulates a single bit among the flip-flops to provide n distinguishable states.

> **Example 15.19** **(Ring counter.)** *Figure 15.21 shows a 4-bit synchronous ring counter.*

Practice problem **15.10.** **(Ring counter.)** Derive the state table for the 4-bit ring counter.
Answer is given in "Solutions to practice problems."

Design example: Modulo 4 ring counter

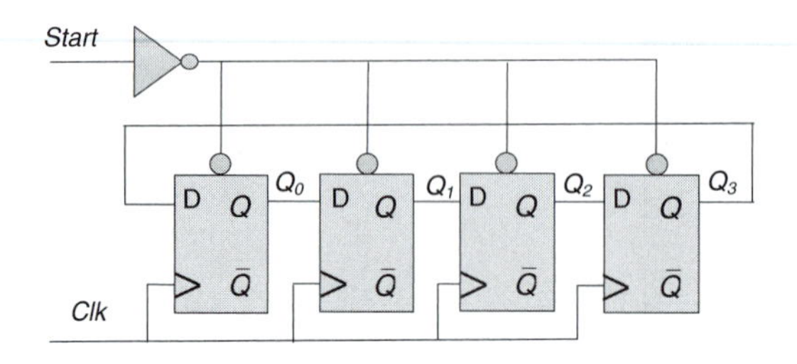

FIGURE 15.21
A 4-bit ring counter (Example 15.19).

Design example: Johnson counter

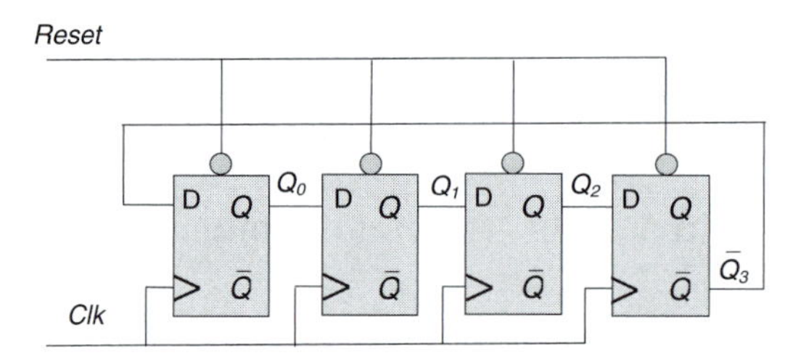

FIGURE 15.22
A 4-bit Johnson counter (Example 15.20).

Example 15.20 (Johnson counter). *A particular case of the ring counter can be obtained if, instead of the output Q output, the $\overline{Q}$ output of the last stage is used (Figure 15.22). An n-bit synchronous Johnson counter generates a counting sequence of length 2n. To initialize the Johnson counter, it is necessary to reset all flip-flops.*

Practice problem 15.11. (Johnson counter.) Derive the state table a for the 4-bit Johnson counter.
Answer is given in "Solutions to practice problems."

15.8.2 Countdown chains

Cascaded counters are often used to divide a high-frequency clock signal to obtain highly accurate pulse frequencies. Cascaded counters used for such purposes are called *countdown chains*.

Example 15.21 (Countdown chains.) *Let a basic clock frequency be of 1MHz. To generate signals of 100kHz, 10kHz, and 1kHz, three cascaded decade counters can be used (forming a divide-by-1000 with intermediate divide-by-10 and divide-by-100 outputs).*

Practice problem **15.12. (Countdown chain.)** Determine the overall modulus of the following cascaded counters: (a) divide-by-8, divide-by-12, and divide-by-16; (b) divide-by-2, divide-by-4, and divide-by-8;
Answer: (a) $8 \times 12 \times 16 = 1536$; (b) $2 \times 4 \times 8 = 64$.

15.9 Summary of standard modules of sequential logic networks

This chapter introduces two basic types of memory elements, latches and flip-flops. The storage capacity of one latch or one flip-flop is one bit. ***Sequential standard modules*** have the property that their output depends not only on the present input but also on the past sequence of inputs. In sequential standard modules, two different types of description are used: ***state*** description and ***time behavior*** (Tables 15.1, 15.2, and 15.3). The key aspects covered in this chapter are:

(*a*) Standard modules for storing one bit of information, such as ***latches*** and ***flip-flops*** (construction and description).

(*b*) Standard modules for storing more than one bit of information, such as registers and counters.

(*c*) Data structures for describing standard modules of sequential network, such as characteristic tables and equations, state tables and diagrams, and timing diagrams.

The main topics of this chapter are summarized as follows:

- ***Sequential networks*** have the quality of memory. The ability to store information distinguishes sequential logic networks from combinational networks. A sequential network contains ***feedback***, which can create outputs that depend not only on the network present input, but on the sequence of inputs that led up to the present.

Summary (continuation)

- The ***SR latch*** is a basic memory element, defined as follows: (a) it is a bistable memory network, (b) it has two inputs labeled set S and reset R, and (c) it has two outputs labeled Q and $\overline{Q}$.
- A ***D latch*** is defined as follows: (a) it is a bistable memory network with a single input, and (b) it is based on the gated SR latch,
- A ***flip-flop*** is defined as follows: (a) It is a bistable memory network, and (b) it employs special control signals to specify the times at which the network responds to changes to its input data signals, and the times at which the network changes its output data signal.
- The ***D flip-flop*** is the counterpart of the ***D latch***, and is defined as follows: (a) it is a bistable memory network with data input and control clock input signals, and (b) Iit is based on D latches, forming a ***master-slave*** pair.
- A ***JK flip-flop*** is defined as follows: (a) it is a bistable memory network with data input and control clock input signals, and (b) it is based on D latches, forming a ***master-slave*** pair.
- A ***T flip-flop*** can be obtained from the JK flip-flop, when the inputs J and K are tied together, or can be constructed from a D flip flop and an EXOR gate.
- Standard modules of sequential networks are specified by a ***characteristic table***, a ***characteristic equation***, ***state table***, ***state diagram***, and ***timing diagram***.
- The ***characteristic equation*** specifies the next state of a sequential network as a function of the current states and inputs. The characteristic equations for basic sequential modules are as follows:

SR latch:	$Q^+ = S \vee \overline{R}Q \quad (\text{SR}=0)$
D latch:	$Q^+ = \overline{Clk} \cdot Q \vee Clk \cdot D$
D flip-flop:	$Q^+ = D$
SR flip-flop:	$Q^+ = \overline{Clk} \cdot Q \vee Clk \cdot D$
JK flip-flop:	$Q^+ = J\overline{Q} \vee \overline{K}Q$
T flip-flop:	$Q^+ = T \oplus Q = T\overline{Q} \vee \overline{T}Q$

- A ***state table*** specifies the next state and output of a sequential network in terms of its present state and input.
- An n-***bit storing register*** is defined as a set of flip-flops for storing an n-bit number, or n-bit vector. It has n sequential inputs and outputs, and control inputs *Load*, *Clear*, and *Clock* (Clk).

Summary (continuation)

- A register that is capable of transferring data among adjacent flip-flops is called a ***shift register***. This transfer can be bidirectional or unidirectional (either to the left or to the right). It has n inputs and outputs, and control inputs *Load*, *Left*, *Right*, and *Clock* (Clk).
- An ***n-bit counter*** is a sequential network with one binary input and n outputs. The following types of counters are distinguished:

  (*a*) ***Upward*** counter: state i is followed by state $(i+1)$ *mod* p,
  (*b*) ***Downward*** counter: state i is followed by state $(i-1)$ *mod* p,
  (*c*) ***Up/down*** counters that counts both ways, and the control input determines the direction.

- Counters are also classified with respect to the code for representing the output:

  (*a*) ***Binary*** counter (2^n states, binary representation of integers $0, 1, \ldots, 2^n - 1$);
  (*b*) ***Decimal*** counter (10 states, decimal code);
  (*c*) ***Gray-code*** counter (2^n states, Gray code); and
  (*d*) ***Ring*** counter and other special counters.

- For advances in standard modules of sequential logic networks, we refer the reader to the "Further study" section.

15.10 Further study

Historical perspective

1955: Sequential logic network models, finite state machines, are named Mealy and Moore machines, after *G. H. Mealy* ("A Method for Synthesizing Sequential Circuits", Bell Syst. Tech. Journal, vol 34, pp. 1045–1079, 1955) and *E. F. Moore* ("Gedanken Experiments on

TABLE 15.1
Standard sequential elements.

Techniques for using standard sequential elements

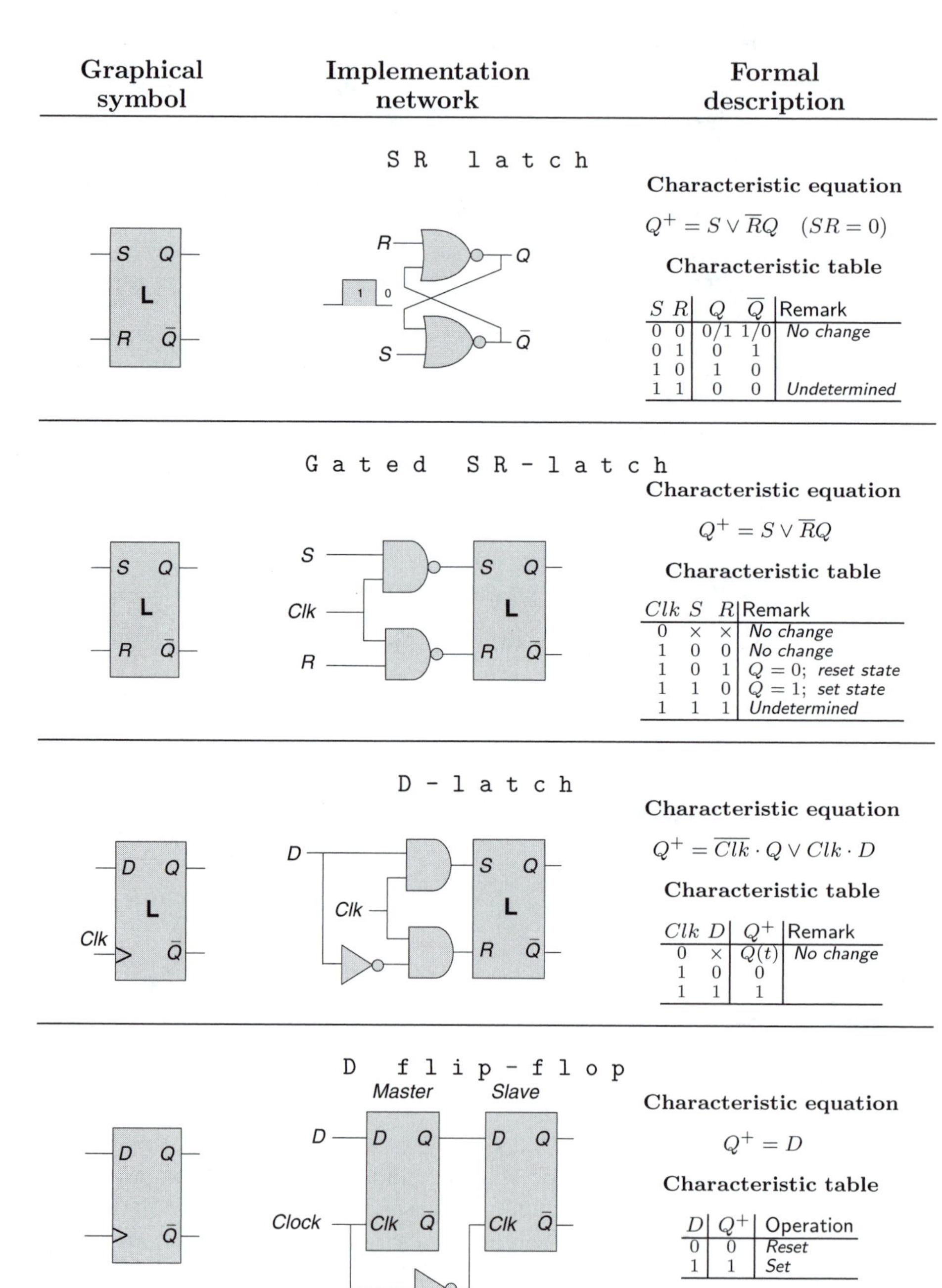

Graphical symbol	Implementation network	Formal description

SR latch

Characteristic equation

$$Q^+ = S \vee \overline{R}Q \quad (SR = 0)$$

Characteristic table

S	R	Q	$\overline{Q}$	Remark
0	0	0/1	1/0	*No change*
0	1	0	1	
1	0	1	0	
1	1	0	0	*Undetermined*

Gated SR-latch

Characteristic equation

$$Q^+ = S \vee \overline{R}Q$$

Characteristic table

Clk	S	R	Remark
0	×	×	*No change*
1	0	0	*No change*
1	0	1	$Q = 0$; *reset state*
1	1	0	$Q = 1$; *set state*
1	1	1	*Undetermined*

D-latch

Characteristic equation

$$Q^+ = \overline{Clk} \cdot Q \vee Clk \cdot D$$

Characteristic table

Clk	D	Q^+	Remark
0	×	$Q(t)$	*No change*
1	0	0	
1	1	1	

D flip-flop

Characteristic equation

$$Q^+ = D$$

Characteristic table

D	Q^+	Operation
0	0	*Reset*
1	1	*Set*

TABLE 15.2
Standard sequential elements (continuation).

Techniques for using standard sequential elements

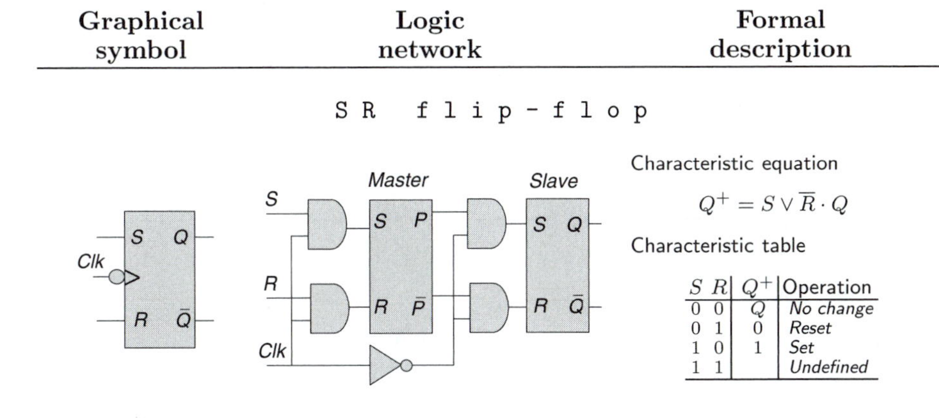

Characteristic equation

$$Q^{+} = S \vee \overline{R} \cdot Q$$

Characteristic table

S	R	Q^{+}	Operation
0	0	Q	*No change*
0	1	0	*Reset*
1	0	1	*Set*
1	1		*Undefined*

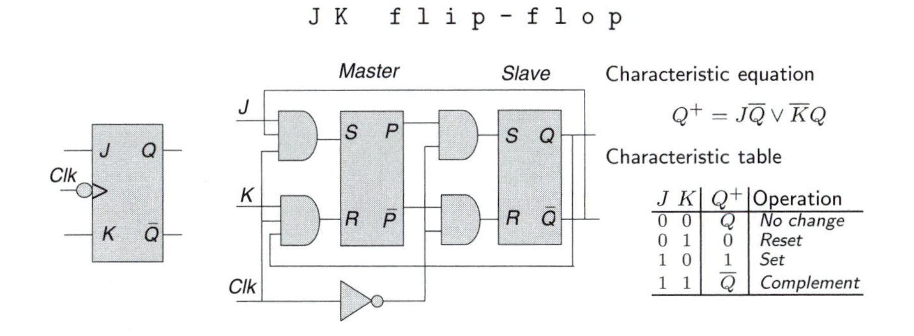

Characteristic equation

$$Q^{+} = J\overline{Q} \vee \overline{K}Q$$

Characteristic table

J	K	Q^{+}	Operation
0	0	Q	*No change*
0	1	0	*Reset*
1	0	1	*Set*
1	1	$\overline{Q}$	*Complement*

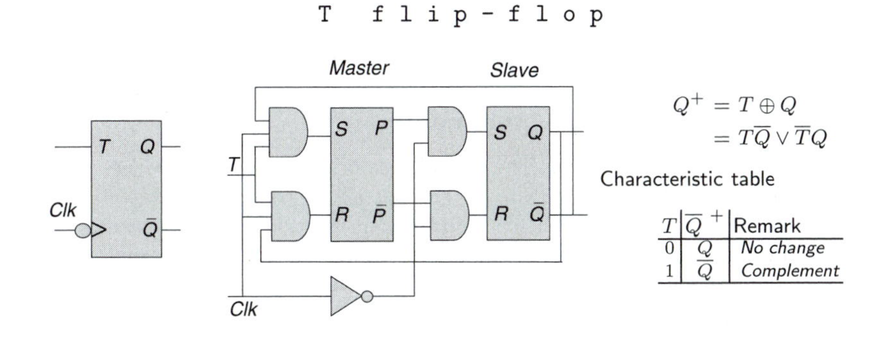

$$Q^{+} = T \oplus Q = T\overline{Q} \vee \overline{T}Q$$

Characteristic table

T	$\overline{Q}^{+}$	Remark
0	Q	*No change*
1	$\overline{Q}$	*Complement*

TABLE 15.3
Description of flip-flops in terms of sequential networks

Techniques for using standard sequential elements

Flip-flop	Characteristic equation	State table	State diagram

D flip-flop

$Q^+ = D$

D	Q	Q^+
0	0	0
0	1	0
1	0	1
1	1	1

T flip-flop

$Q^+ = T \oplus Q$

T	Q	Q^+
0	0	0
0	1	1
1	0	1
1	1	0

JK flip-flop

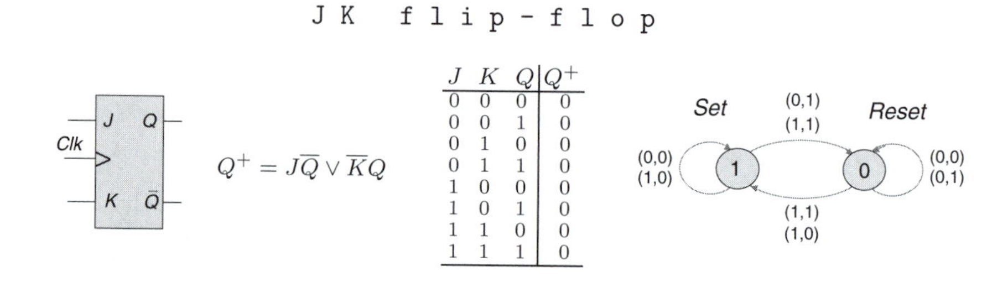

$Q^+ = J\overline{Q} \vee \overline{K}Q$

J	K	Q	Q^+
0	0	0	0
0	0	1	0
0	1	0	0
0	1	1	0
1	0	0	0
1	0	1	0
1	1	0	0
1	1	1	0

Sequential Machines", Automata Studies, Annals of Math Studies 34, Princeton University Press, pp. 120–153, 1956), who first studied their behavior at AT&T Bell Laboratories in the 1950s.

Advanced topics of sequential modules

Topic 1: Neural networks. The neural network such as Hopfield networks are capable of storing the state of the system. The advantages of Hopfield networks for the computing and implementation of an elementary logic function include the small number of neuron cells and the iterations to achieve minimum energy, simple interconnect topology.

Topic 2: Multi-level memory. In 1992, Intel began a research effort to reduce the amount of silicon required to store a bit of data to a fraction of transistor by storing more than 1 bits of information per cell. The research resulted in manufacturing in 1997 a four-valued 64 Mbit memory device StrataFlash storing two bits of information per cell. Intels innovation was followed by a series of announcement of development of multi-level flash memories, including Samsung Electrons 128Mbit 3.3V four-level NAND flash memory, and Hitachi and Mitsubishis 256Mbit four-level NAND flash memory. In 1997 NEC has announced developing of a 4Gbit four-level DRAM with four-level sensing and restoring operations. An overview of CMOS-related multiple-valued memory technologies has been done in "A review of multiple-valued memory technology" by Glenn Gulak, *28th International Symposium on Multiple-Valued Logic*, pages 222–233, 1998. Mon-volatile multiple-valued memory technologies have been considered in "Non-Volatile Multilevel Memories for Digital Applications" by B. Ricco et al., *Proceedings of the IEEE*, volume 86, issue 12, pages 2399-2421, 1998. Recent advances in nanotechnology led to a series of prospective proposals on multiple-valued memory, in particular, single-electron multiple-valued memory, as well as analogue information storage using self-assembled nanoparticle films.

Further reading

Contemporary Logic Design by R. H. Katz and G. Borriello, Prentice Hall, 2005.

Digital Design by M. Morris Mano, Prentice Hall, 2001.

Digital Principles and Design by D. D. Givone, McGraw-Hill, New York, 2003.

Introduction to Logic Design by Alan B. Marcovitz, McGraw Hill, 2007.

"Neural Networks and Physical Systems with Emergent Collective Computational Abilities" by J. J. Hopfield, *Proceedings of the National Academy of Sciences, USA*, volume 79, pages 2554–2558, 1982.

"Reliable Circuits Using Less Reliable Relays" by E. F. Moore and C. E. Shannon, *J. Franklin Institute*, volume 262, pages 191–208, Sept, 1956, and pages 281–297, Oct. 1956.

Synthesis of Finite State Machines: Logic Optimization by T. Villa, T. Kam, R. K. Brayton, A. Sangiovanni-Vincentelli, Kluwer, 1997.

15.11 Solutions to practice problems

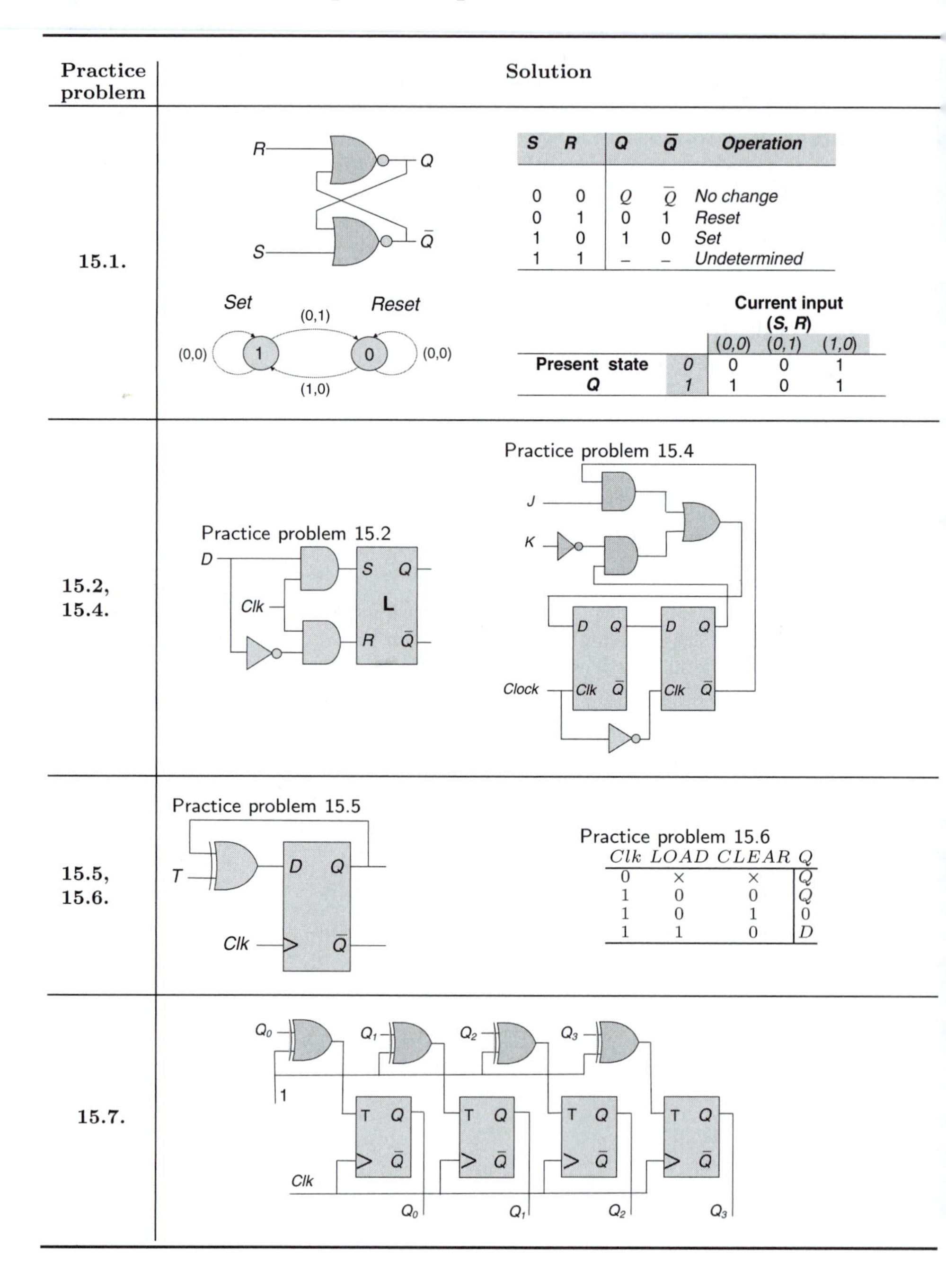

S	R	Q	$\bar{Q}$	Operation
0	0	Q	$\bar{Q}$	No change
0	1	0	1	Reset
1	0	1	0	Set
1	1	–	–	Undetermined

		Current input (S, R)		
		(0,0)	(0,1)	(1,0)
Present state Q	0	0	0	1
	1	1	0	1

Clk	LOAD	CLEAR	Q
0	×	×	Q
1	0	0	Q
1	0	1	0
1	1	0	D

Solutions to practice problems (continuation)

Practice problem	Solution
15.8.	
15.9.	
15.10, 15.11.	

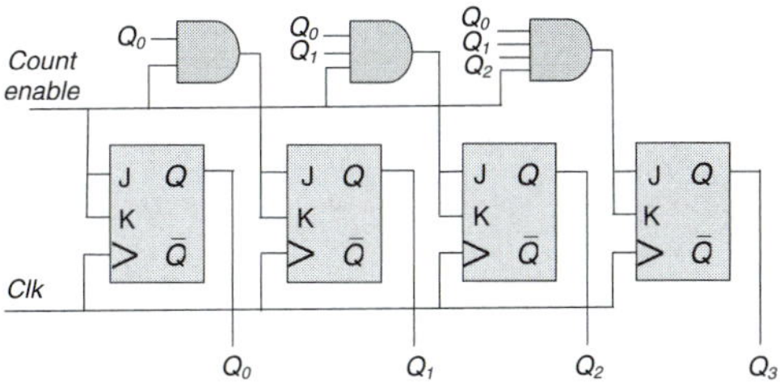

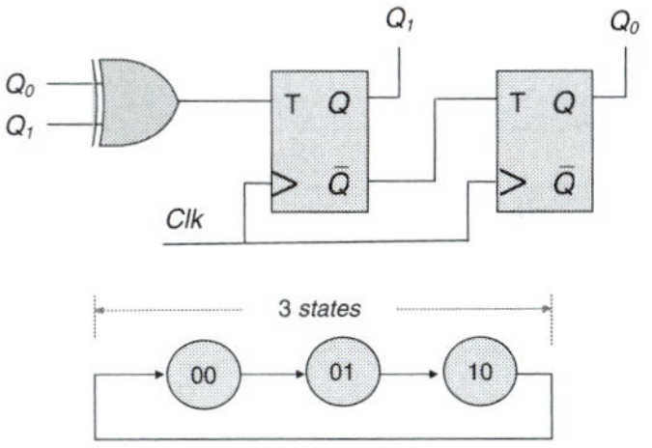

15.9.

Q_1	Q_0	Q_1^+	Q_0^+	T_1	T_0
0	0	0	1	0	1
0	1	1	0	1	1
1	0	1	1	1	0
1	1	×	×	×	×

15.10, 15.11.

Problem 15.10

Q_0	Q_1	Q_2	Q_3	Q_0^+	Q_1^+	Q_2^+	Q_3^+
1	0	0	0	0	1	0	0
0	1	0	0	0	0	1	0
0	0	1	0	0	0	0	1
0	0	0	1	1	0	0	0

Problem 15.11

Q_0	Q_1	Q_2	Q_3	Q_0^+	Q_1^+	Q_2^+	Q_3^+
0	0	0	0	1	0	0	0
1	0	0	0	1	1	0	0
1	1	0	0	1	1	1	0
1	1	1	0	1	1	1	1
1	1	1	1	0	1	1	1
0	1	1	1	0	0	1	1
0	0	1	1	0	0	0	1
0	0	0	1	0	0	0	0

15.12 Problems

Problem 15.1 Sketch the Q and $\overline{Q}$ output signals for the following SR lathes:

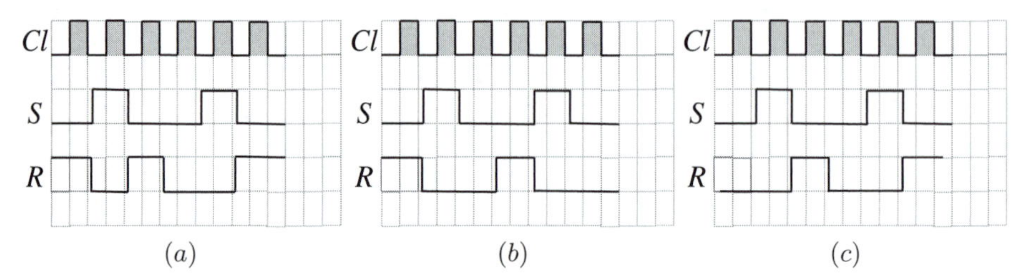

Problem 15.2 Sketch the Q and $\overline{Q}$ output signals for the following flip-flops:

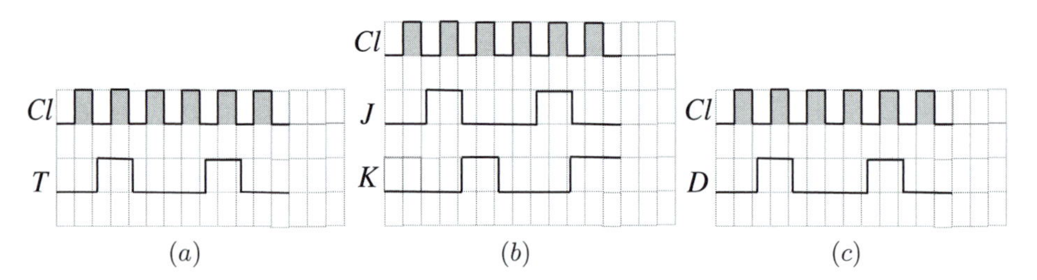

Problem 15.3 Assume the shift register initially contains 1101. What is the content of the register after the following positive edge clock signal:
(a) 110101 (b) 110101 (c) 110101

Problem 15.4 Design a 4-bit binary ripple up-counter using positive edge-triggered JK flip-flop.

Problem 15.5 Design a 4-bit binary ripple down-counter using positive edge-triggered T flip-flop.

Problem 15.6 Design a mod-5 counter whose counting sequences are as follows:
(a) 0000,0001,. . . ,0100,0000,. . .
(b) 0010,0011,. . . ,0110,0010,. . .
(c) 1011,1100,. . . ,1111,1011,. . .
(d) 1000,1001,. . . ,1100,1000,. . .

Problem 15.7 Design a mod-10 counter using the following positive edge-triggered flip-flops:
(a) D flip-flop (b) JK flip-flop (c) T flip-flop

16

Memory and Programmable Devices

16.1 Introduction

A memory unit is a collection of cells capable of storing a large quantity of binary data. Binary data is transferred to memory for storage. When data is needed for processing, they are transferred from memory to selected registers in the processing unit. Similarly, binary data, received from an input device, are stored in memory, and data, transferred to an output device, are taken from memory.

There are two common operations on data once memory is accessed:

- The process of storing new information into memory is referred to as a memory *write* operation, and
- The process of transferring the stored information out of memory is referred to as a memory *read* operation.

Medium and large scale integration electronic devices are implemented using the concept of arrays of logic elements, or memory cells. In term of organization, memory arrays are classified as follows:

- Read-only memory (ROM), and
- Random access memory (RAM).

Details of memory organization will be considered in this chapter.

Random-access memory (ROM)

Random-access memory can perform both the write and read operations. Read-only memory can perform only the read operation. This means that suitable binary information is already stored inside the memory, which can be retrieved or read at any time. However, the existing information cannot be altered by writing because read-only memory can only read; it cannot write.

Read-only memory (ROM)

The read-only memory is implemented as an array of homogeneous cells connected in a regular manner and configured to implement some logic. It is, therefore, a *programmable* logic device. The binary data that is stored within a programmable logic device is embedded within the hardware, and this process is referred to as *programming* the device. The word "programming" here refers to a hardware procedure that specifies the bits that are inserted into the hardware configuration of the device.

16.2 Programmable devices

Combinational logic arrays, extended sometimes using sequential elements, are organized based on the following concepts:

Combinational logic array concepts

- Fine-granularity devices, or the sea of gates concept, which enables the implementation of any Boolean function using logic cells such as NAND gates only;
- Medium-granularity devices, which are based on logic blocks, look-up tables and programmable input-output blocks with flip-flops (medium-grain field-programmable gate arrays (FPGAs)); and
- Large-granularity devices such as complex programmable logic devices (CPLDs) and sequential programmable logic devices (SPLDs) which include AND/OR arrays and universal input/output logic blocks.

Nowadays, mostly medium- and large-granularity programmable devices are employed in electronic designs, and they will be considered in detail in this chapter. It should be noted that ROM is sometimes classified as programmable array logic; it is used to implement the look-up-table parts of FPGAs.

A device based on the principle of *programmable logic* is defined as a general-purpose chip containing logic gates and programmable switches for implementing logic network. It contains a collection of logic network elements. Programmable switches allow the logic gates inside the chip to be connected together to implement desired functions.

Simple programmable logic devices include:

- *Programmable logic array (PLA)*,
- *Programmable array logic (PAL)*,
- *Gate array logic (GAL)*,
- *Programmable logic devices (PLD)*, and
- *Sequential programmable logic devices (SPLD).*

As the system design evolves in response to the system requirements, some functions may be identified for implementation in complex electronic devices such as:

- *Field Programmable Gate Arrays (FPGA)*,
- *Complex Programmable Logic Devices (CPLD)*,
- *Application Specific Integrated Circuits (ASIC)*,
- *System-on-Chip (SoC), and*
- *Field Programmable System Chip (FPSC).*

Example 16.1 (Arrays.) *Figure 16.1 shows how conventional symbols of logic AND gates can be replaced by array logic symbols. This simplification of a graphical representation can be used for other logic gates.*

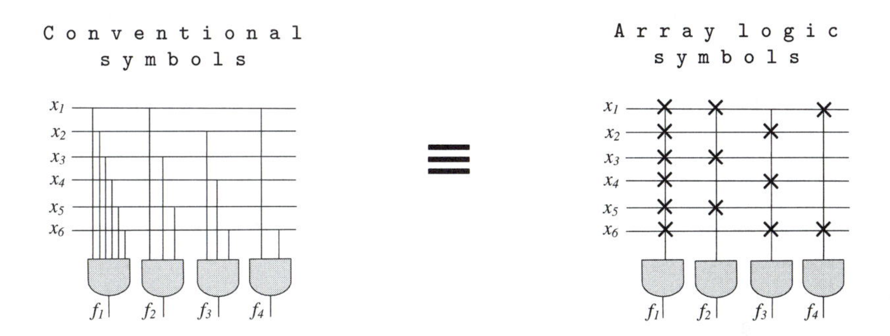

FIGURE 16.1
Excerpt of a network, where conventional symbols of logic gates are replaced by array logic symbols (Example 16.1).

Field-programmable gate array (FPGA)

A field-programmable gate array is a semiconductor device containing programmable logic components called "logic blocks," and programmable interconnects. Logic blocks can be programmed to perform the function of basic logic gates such as AND and EXOR, or more complex combinational functions such as decoders or simple mathematical functions. In most FPGAs, the logic blocks also include memory elements, which may be simple flip-flops or more complete blocks of memories.

A hierarchy of programmable interconnects allows logic blocks to be interconnected as needed by the designer. Logic blocks and interconnects can be programmed by the customer/designer, after the FPGA is manufactured, to implement any logical function (hence the name "field-programmable").

FPGAs are usually slower than their ASIC counterparts, as they are not optimized for particular applications, and draw more power. But their advantages include a shorter time to market, ability to be re-programmed, and lower costs.

The ASIC methodology is the *gate array*. Gate arrays are characterized by their topologies, some of which are shown in Figure 16.2.

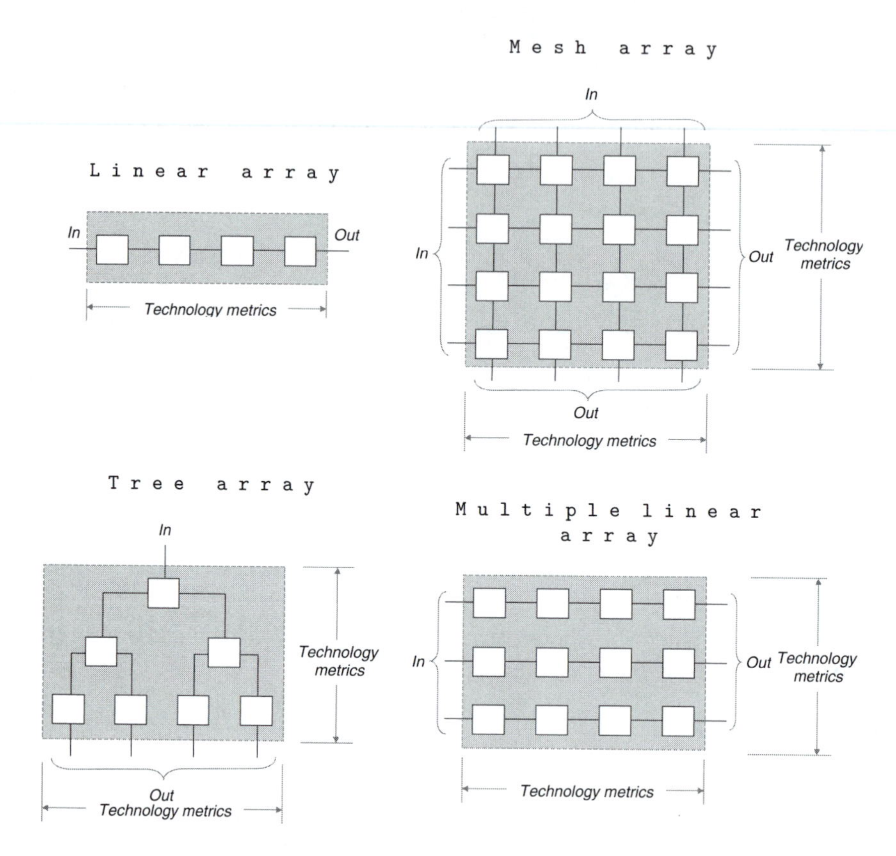

FIGURE 16.2
Array topologies: linear, mesh, tree, and multiple linear.

16.3 Random-access memory

RAM can be thought of as a collection of storage cells, or words of cells (registers) with associated logic networks needed to transfer information in and out of the device. The term *random-access memory* remains to distinguish the sequential type of access in early memories (tape and drum drives) from the immediate access to any address in contemporary transistor-based memories.

16.3.1 Memory array

Each storage element in a memory retains either a 1 or a 0 and is called a *binary storage cell.*

Example 16.2 (RAM.) *An example of a binary storage cell is shown in Figure 16.3. The storage part of the cell is modeled by an SR latch with associated gates. The cell stores one bit in its internal latch.*

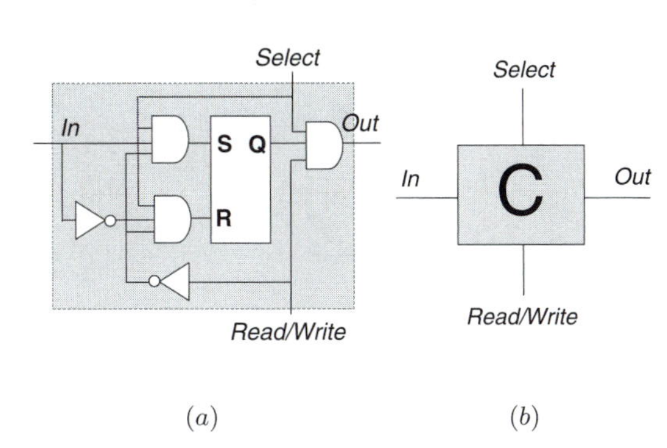

The selected input enables the cell for reading or writing and the read/write input determines the cell operation when it is selected:
►A 1 in the read/write input provides the read operation by forming a path from the latch to the output terminal
►A 0 in the read/write input provides the write operation by forming a path from the input terminal to the latch.

FIGURE 16.3
A memory cell (a) and its symbol in the memory array (b) (Example 16.2).

A memory consists of $k \times m$ binary storage cells, where k is the number of words and m is the number of bits per word.

Example 16.3 (RAM.) *A 4×4 RAM is shown in Figure 16.4a. A memory with four words needs two address lines. The two address inputs are applied to a 2×4 decoder to select one of the four words. During the read operation, the four bits of the selected word are transmitted through OR gates to the output terminals. During the write operation, the data available in the input lines are transferred into the four binary cells of the selected word. The cells, that are not selected, are disabled, and their previous binary values remain unchanged. When the memory select input, which goes into the decoder, is equal to 0, none of the words are selected, and the contents of all cells remain unchanged regardless of the value of the read/write input.*

16.3.2 Words

Memory stores binary data in groups of bits called *words*. A word in memory is an entity of bits that move in and out of storage as a unit. A memory word is a group of 1s and 0s and may represent a number, an instruction, one

Design example: Random-access memory (RAM)

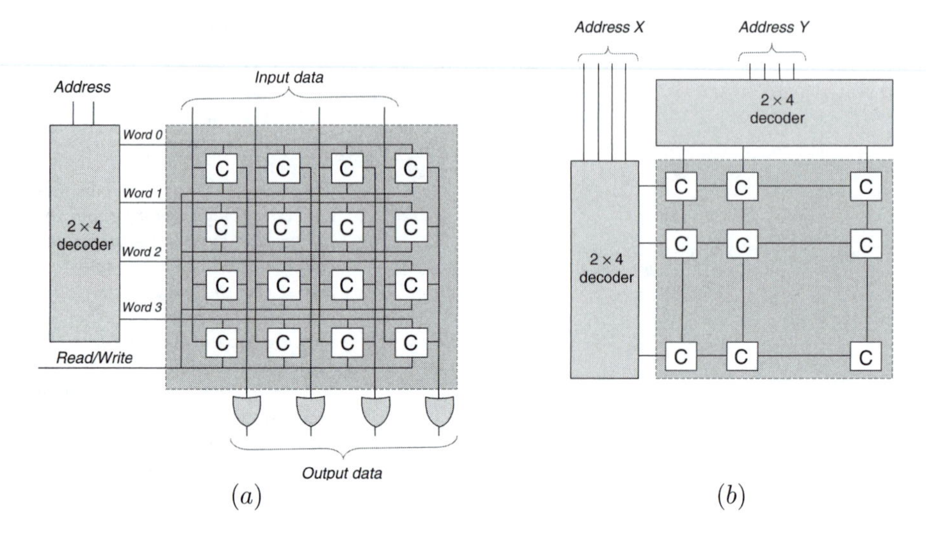

FIGURE 16.4
A 4 × 4 RAM (a) (Example 16.3) and coincident decoding (b) (a) (Example 16.7).

or more alphanumeric characters, or any other binary-coded information. A group of eight bits is called a *byte*. Most computer memories use words that are multiples of eight bits in length.

> **Example 16.4 (Memory words.)** *A 16-bit word contains two bytes, a 32-bit word is made up of four bytes, and a 64-bit word contains eight bites.*

Practice problem **16.1.(Memory words.)** How many bits and bytes are contained in 1Mb memory?
Answer: A memory of 1 Mb includes $2^{10} = 1024$ bits, or $2^7 = 128$ bytes.

16.3.3 Address

Communication between a memory and the connected units is achieved through (Figure 16.5):

- Data *input* and *output* lines;
- *Addresses*, or selection lines; and
- *Control* lines that specify the direction of transfer, and other functions.

The memory unit is specified by the number of words it contains and the number of bits in each word. The address lines select one particular word. The location of a unit of data in a memory array is called its *address*. Each word in memory is assigned an identification number; that is, its address, starting from 0 up to $2^n - 1$, where n is the number of address lines. The selection of a specific word inside memory is done by applying the n-bit address to the address lines. A decoder accepts this address and opens the paths needed to select the specified word.

It is customary to refer to the number of words (or bytes) in a memory with one of the letters

- K (kilo), meaning 2^{10};
- M (mega), meaning 2^{20};
- G (giga), meaning 2^{30}.

The k data input lines provide the information to be stored in memory, and the m data output lines supply the information coming out of memory. The n address lines specify the particular word chosen among the many available. The two control inputs specify the direction of transfer:

- The *write* input causes binary data to be transferred into memory, and
- The *read* input causes binary data to be transferred out of memory.

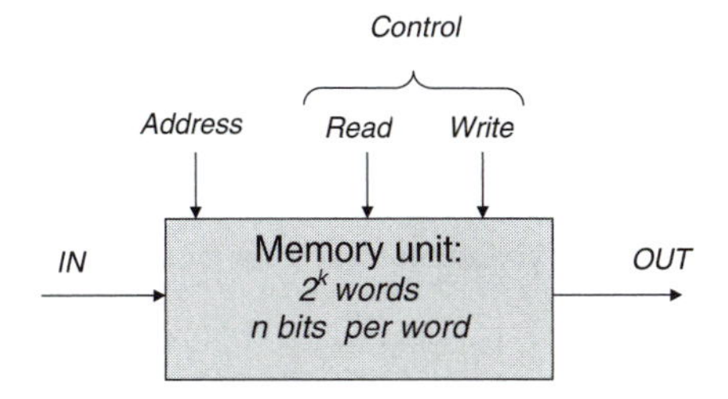

FIGURE 16.5
A generic memory unit consists of n data input and n data output lines, k address lines, and control lines to specify the direction of transfer.

16.3.4 Memory capacity

A memory which contains M m-bit words is referred to as memory of size $M \times m$. The number of address lines required to address M words is $log_2 M$.

Example 16.5 (Memory capacity.) *A memory of size* $1M \times 16$ *includes* $2^{10} = 1,024$ *words and requires an address of 10 bits.*

Practice problem **16.2.(Memory capacity.)** How many words and what size are included in 6G×64 memory, and what is the capacity of the address line?
Answer: A $4G \times 64$ memory includes 2^{32} 64-bit words and requires 32 address bits.

The *capacity* of a memory is the total number of data units that can be stored. It is usually stated as the total number of bits, and sometimes bytes that it can store.

Example 16.6 (Memory capacity.) *A* $1M \times 16$ *memory accommodates* $2^{10} \times 16 = 2^{14}$ *bits, or* 2^{11} *bytes.*

Practice problem **16.3.(Memory capacity.)** Determine the capacity of the following memories: (a) $2M \times 32$, and (b) $1G \times 64$.
Answer: (a) A $2M \times 32$ memory contains $2^{21} \times 32 = 2^{26}$ bits, or 2^{23} bytes, and (b) a $1G \times 64$ memory has the capacity $2^{3} \times 64 = 2^{36}$ bits, or 2^{33} bytes.

16.3.5 Write and read operations

The two operations that memory performs are the write and read operations. The write signal specifies a *transfer-in* operation. The read signal specifies a *transfer-out* operation. Writing, or transferring a new word to be stored in memory, is performed as follows:

Write to memory

Step 1: Apply the binary address of the desired word to the address lines.
Step 2: Apply the data bits that must be stored in memory to the data input lines.
Step 3: Activate the ''write'' input.
Step 4: The bits are taken from the input data lines and are stored in the word, specified by the address lines.

Reading, or transferring a stored word out of memory, is implemented as follows:

Read from memory

Step 1: Apply the binary address of the desired word to the address lines.
Step 2: Activate the ''read'' input.
Step 3: The bits are taken from the word that has been selected by the address, and are applied to the output data lines.

Normally, the content of the selected word does not change after reading.

16.3.6 Address management

A decoder with n inputs requires 2^n AND gates with k inputs per gate. The total number of gates and the number of inputs per gate can be reduced by employing two decoders in a 2-dimensional selection scheme. The basic idea in 2-dimensional decoding is to arrange the memory cells in an array that is as close as possible to square. In this configuration, two $n/2$-input decoders are used instead of one n-input decoder. One decoder performs the row selection and the column selection in a 2-dimensional matrix configuration.

Example 16.7 (Memory addressing.) *Figure 16.4a shows regular word address decoding (the whole word is read from such a memory), and Figure 16.4b illustrates 2-dimensional, or coincident decoding of the address (a single bit can be accessed in such a memory).*

Practice problem **16.4.(Memory addressing.)** A 64-cell array can be organized in several ways: an 8×8 array, which is viewed as a 64-bit memory or an 8-byte memory; a 16×4 array, and an 64×1 array.
Answer: (a) Two 3-to-8 decoders, 6-bit address (3 and 3) and 8-bit data, (b) one 4-to-18 decoder and one 2-to-4 decoder, 6-bit address (4 and 2), and 4-bit data, (c) one 6-to-64 decoder, 6-bit address, 1-bit data.

16.4 Read-only memory

Read-only memory (ROM) is a combinational logic array in which permanent or semipermanent binary data are stored. The data are then embedded in the array using an interconnection pattern. This pattern is established by programming the array, and remains unchanged ("stored"), even when power is turned off and on again. A ROM has:

(*a*) n address lines and
(*b*) m outputs, from which the data bits of the stored word are read when selected by the address.

The number of words in a ROM that can be addressed by n address lines is 2^n. Therefore, the ROM size is $2^n \times m$. A $2^n \times m$ ROM has:

(*a*) An $n \times 2^n$ decoder, and
(*b*) A $2^n \times m$ OR array.

Note that ROM does not have data inputs because it does not have a write operation. However, some types of ROM allow the reconfiguration of the array pattern and, thus, the "re-writing" the memory. These include:

- Programmable ROM (PROM),
- Erasable programmable ROM (EPROM),
- Electrically erasable programmable ROM (EEPROM).

The internal binary storage of a ROM is specified by a truth table, or a *connection matrix*. They determine the word content at each address.

> **Example 16.8 (ROM.)** *A* 32×8 *ROM consists of 32 words of 8 bits each (Figure 16.6a). There are five address lines that are decoded into 32 outputs by means of a* 5×32 *decoder. The 32 lines from the decoder's output form memory addresses, called* word lines, *which are binary numbers from 0 through 31. The horizontal word lines together with the vertical lines form the OR array. To reflect this fact, the OR gates are shown at the bottom end of each vertical line, so that each OR gate must be considered as having 32 inputs (the gate itself does not physically exist).*

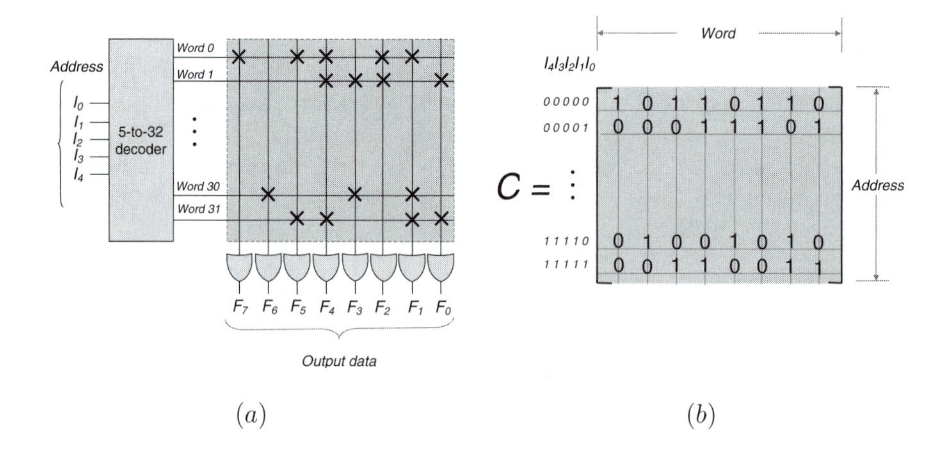

FIGURE 16.6
Internal logic and programming of the 32×8 ROM (a) (Example 16.8) and connection matrix C (b) (Example 16.9).

16.4.1 Programming ROM

ROM interconnections are programmable. A programmable connection between two lines is logically equivalent to a switch that can be altered to either *closed* (meaning that the two lines are connected), or *open* (meaning that the two lines are disconnected).

The programmable intersection between two lines is also called a *crosspoint.* To implement crosspoint switches, several technologies can be used. In particular, a fuse can be employed, which normally connects the two points, but is open or "blown" by applying a high-voltage pulse to the fuse.

Example 16.9 (ROM.) *Figure 16.6b shows the connection matrix C corresponding to the ROM in Figure 16.6a. This matrix shows that at each address, an 8-bit word is stored. In the connection matrix C, every 0 specifies no connection, and every 1 specifies a path that is obtained by connection. The complete matrix includes 8 columns and 32 rows, thus storing 32 words (only the first and last two words in the ROM are shown in Figure 16.6b).*

Practice problem **16.5.(ROM.)** Sketch the ROM for storing 8 words representing the 4-bit binary numbers 1 through 8, and the corresponding connection matrix.
Answer is given in "Solutions to practice problems."

16.4.2 Programming the decoder

The decoder in a ROM can be programmed as well. The programming of the decoder corresponds to the implementation of an AND type array, that forms all possible products (minterms) of the decoder's inputs.

Example 16.10 *Figure 16.7a shows the programmed 3-to-8 decoder. This array is comprised of three input lines and three complemented input lines, to form an input variable and its complement for each of the inputs, as well as $2^3 = 8$ output lines that must be seen as 8 AND gates (which do not physically exist) with six inputs each. The AND gates form 8 three-variable products, which are all possible minterms, $\overline{I}_2, \overline{I}_1, \overline{I}_0$ through I_2, I_1, I_0.*

16.4.3 Combinational logic network implementation

The $n \times 2^n$ decoder of the ROM generates all possible 2^n products of its n binary inputs. These products can be interpreted as the minterms of n input variables. By programming a line in the OR array of the ROM, some of the

Design example: Read-only memory (ROM)

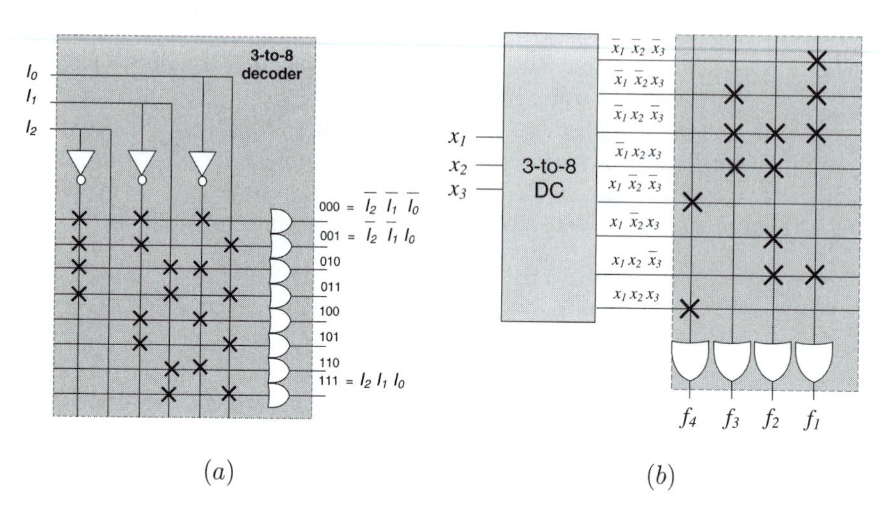

FIGURE 16.7
Programming the 3-to-8 decoder for a ROM (a) (Example 16.10) and implementation of four Boolean functions on a ROM with three inputs and four outputs (b) (Example 16.11).

minterms can be logically added to form a canonical SOP. In this way m arbitrary logic functions can be implemented using the decoder and $2^n \times m$ OR array. The algorithm for mapping a system of m Boolean functions of n variables into a ROM is given below:

ROM implementation of m Boolean functions

Step 1: `Represent each of the` m `Boolean functions in a canonical SOP form.`

Step 2: `Specify the` $2^n \times m$ `decoder.`

Step 3: `Determine the OR connection matrix in accordance with SOP expressions of the given` m `functions.`

Example 16.11 (Functions on ROM.) *A ROM with 3 inputs and 4 outputs is given in Figure 16.7b. The decoder with three inputs has* $2^3 = 8$ *output lines, which together with the OR gates form an* 8×4 *OR matrix that implements the four canonical SOPs:* $f_1 = \bigvee m(0,1,2,6)$, $f_2 = \bigvee m(2,3,5,6)$, $f_3 = \bigvee m(1,2,3)$, $f_4 = \bigvee m(4,7)$.

Practice problem **16.6.(Function on ROM.)** Implement the following Boolean functions on ROM: $f_1 = x_1x_2 \vee x_1\overline{x}_3 \vee \overline{x}_1\overline{x}_2x_3$, $f_2 =$

$x_1x_2 \vee \overline{x}_1\overline{x}_2x_3 \vee x_1x_3$, and $f_3 = x_1 \vee x_2 \vee x_3$.
Answer is given in "Solutions to practice problems."

16.5 Memory expansion

Available ROM or RAM memory can be expanded to increase

- ▶ Word length (the number of bits in a word at each address), or
- ▶ Word capacity (number of different addresses) or
- ▶ Both word length and capacity.

Memory expansion is accomplished by adding an appropriate number of memory arrays while arranging additional address and control lines.

Example 16.12 **(Word length expansion.)** *Consider an $8K \times 8$ RAM. To design $8K \times 16$ RAM, two $8K \times 8$ chips can be connected as shown in Figure 16.8.*

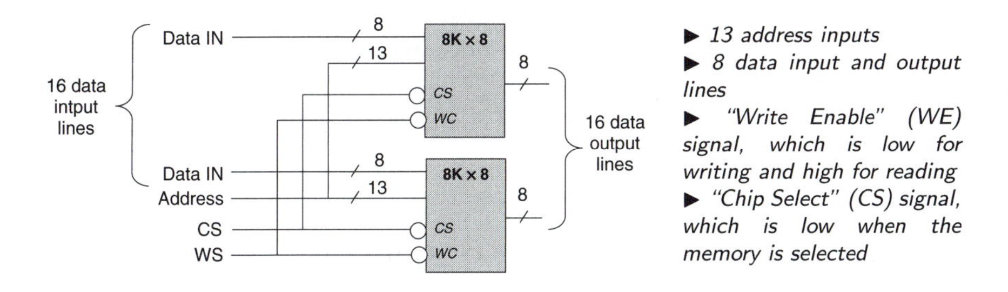

FIGURE 16.8
Implementation of $8K \times 16$ RAM using $8K \times 8$ RAM (Example 16.12).

Example 16.13 **(Word capacity expansion.)** *Design a $32K \times 8$ RAM given an $8K \times 8$ RAM with Write Enable (WE) and Chip Select (CS) control inputs. The desired word capacity can be achieved by using four $8K \times 8$ chips (Figure 16.9).*

Expansion of ROM capacity can be achieved by using 2-dimensional decoding of the address and multiplexing the data output.

Design example: Word capacity expansion

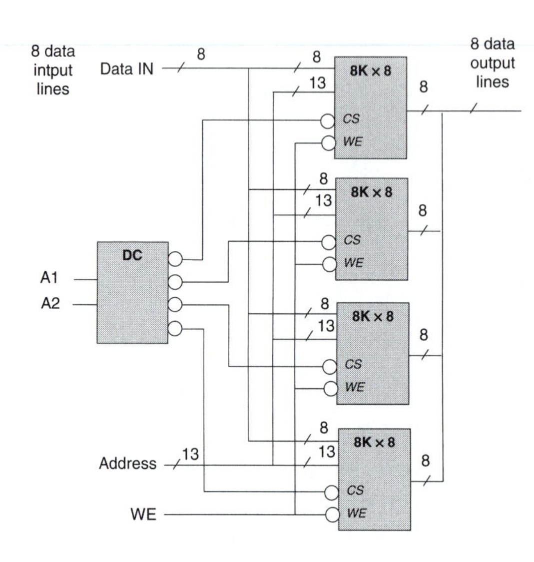

- *To address 32K words, an address length of 2^{15} is required*
- *Using 13 bits to address the words in each chip, the remaining 3 bits of the address can be used to select between four chips*
- *To read from the chip that is selected, a low WE signal activates the data output for one of four chips only, and those data are transferred to the 8-bit bus using three-state buffers.*

FIGURE 16.9
Implementation of $32K \times 8$ RAM with $8K \times 8$ RAM (Example 16.13).

Example 16.14 (Capacity expansion.) *Design a $16K \times 8$ ROM. The maximum allowed size of the decoder in a ROM is $8 - to - 256$, and the maximum allowed multiplexer is $64 - to - 1$. Choose the appropriate size of the OR array. A solution with 2-dimensional addressing is shown in Figure 16.10.*

Practice problem **16.7.(Capacity expansion.)** Given 128×32 ROM arrays, design a $4K \times 32$ ROM.
Answer is given in "Solutions to practice problems."

16.6 Programmable logic

Programmable logic devices, PLAs and PALs, are multi-input, multi-output devices, typically organized into an AND gate array, and an OR gate array. Both devices are suited to the SOP representation of Boolean functions. The AND array is used to compute particular product terms of the inputs,

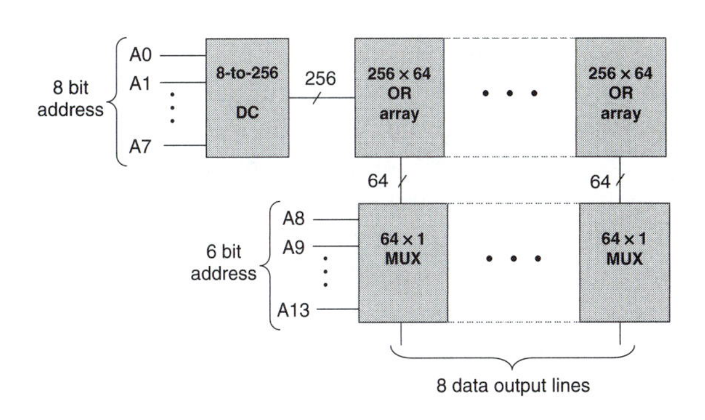
Design example: Word capacity expansion
8 bit address
A0
A1
A7
8-to-256 DC
256
256 × 64 OR array
256 × 64 OR array
64
64
6 bit address
A8
A9
A13
64 × 1 MUX
64 × 1 MUX
8 data output lines
▶ 14 address lines are required, since 16K = 2^14
▶ The upper part of the address is decoded using an 8-to-256 decoder
▶ The lower part is used to select one of the 64 outputs of each of 8 256 × 64 OR arrays
▶ The desired memory capacity of 16K × 8 = 256 × 64 × 8 is achieved

FIGURE 16.10
Implementation of $16K \times 8$ ROM with 2-dimensional addressing (Example 16.14).

depending on the programmed connections. The OR array logically adds these terms together to produce the final SOP expression. The difference between PLA and PAL is the following:

- ▶ In PLA, both the AND and OR arrays are programmed, and
- ▶ In PAL, the AND array is programmed and OR array is fixed.

It should be noted that, according to the above classification, ROM has a fixed AND array and a programmed OR array.

16.6.1 Programmable logic array (PLA)

A PLA implements Boolean functions in SOP form. A PLA comprises a collection of (Figure 16.11):

- ▶ AND gates, called an *AND array*, and
- ▶ OR gates, called an *OR array*.

The AND array consists of literal lines and product lines. It generates a set of k products $P_1, P_2, \ldots P_k$ on its product lines using n inputs $x_1, x_2, \ldots x_n$ and their complements $\overline{x}_1, \overline{x}_2, \ldots \overline{x}_n$ available on the literal lines. The product terms serve as the inputs to an OR array, which produces m outputs $f_1, f_2, \ldots f_m$ on its output lines. Each output can be configured to realize any sum of products, and, hence, any SOP representation of a Boolean function. This SOP, in general, is not canonical, but rather optimized, to get a PLA of minimal size.

The size of the PLA is defined as follows:

$$\texttt{PLA size} = (2 \times n + m) \times k$$

where n is the number of data inputs (input variables), m is the number of data outputs, and k is the number of product lines. The upper bound for k is 2^n, but since implementation is intended to minimize the size of the PLA, k is equal to the sum of the number of products (minus the number of shared products) in the SOPs of the m implemented functions.

The PLA is characterized by the following features: (a) Regular structure (this feature is intended to simplify the manufacturing process); (b) Minimal area of the array needed for implementation, compared to ROM implementation; and (c) Efficiency in implementation of systems of functions (not single functions) with shared products.

Design example: Programmable logic array (PLA)

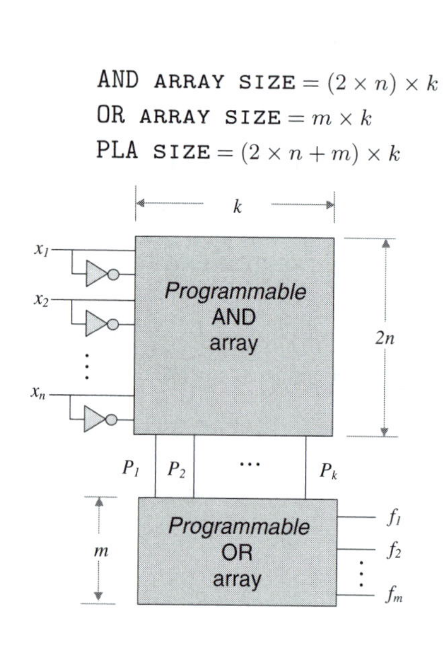

Programmable AND array:

- n *inputs* $x_1, x_2, \ldots x_n$
- n *complemented inputs* $\overline{x}_1, \overline{x}_2, \ldots \overline{x}_n$
- k *output products* $P_1, P_2, \ldots P_k$

Programmable OR array:

- k *inputs* $P_1, P_2, \ldots P_k$
- m *outputs* $f_1, f_2, \ldots f_m$

Binary connection matrix for AND array:

$$C_{AND} = \begin{matrix} & x_1 & \overline{x}_1 & \cdots & x_n & \overline{x}_n \\ P_1 & & & & & \\ P_2 & & & & & \\ \vdots & & & & & \\ P_k & & & & & \end{matrix}$$

Binary connection matrix for OR array:

$$C_{OR} = \begin{matrix} & f_1 & f_2 & \cdots & f_m \\ P_1 & & & & \\ P_2 & & & & \\ \vdots & & & & \\ P_k & & & & \end{matrix}$$

FIGURE 16.11

Structure of the n-input m-output PLA and connection matrices for AND gate and OR gate arrays.

16.6.2 The PLA's connection matrices

A convenient way to describe Boolean functions in the form acceptable for PLA implementation is by a *connection matrix*. There are two types of connection matrices:

- ▶ The binary $(2 \times n) \times k$ matrix C_{AND} for specifying AND gate array connections; the matrix C_{AND} describes which literal should be included in each product term (the rows determine product terms and the columns represent literals (inputs)).
- ▶ The binary matrix C_{OR} $m \times k$ for specifying OR gate array connections; matrix C_{OR} describes which product should be included in each output to form the desired functions (the rows determine product terms and the columns represent outputs).

If there are multiple 1s in a row in the connection matrix, it means that the corresponding product term participates in more than one function. Sharing products is a resource for minimizing the PLA's area.

Example 16.15 (Connection matrix.) *Let us derive the connection matrix for the 4-output Boolean function of three variables:* $f_1 = x_1 \vee \overline{x}_2\overline{x}_3$, $f_2 = x_1\overline{x}_3 \vee x_1x_2$, $f_3 = \overline{x}_2\overline{x}_3 \vee x_1x_2$, *and* $f_4 = \overline{x}_2x_3 \vee x_1$. *These functions contain the following product terms:* $P_1 = x_1$, $P_2 = \overline{x}_2\overline{x}_3$, $P_3 = x_1\overline{x}_3$, $P_4 = x_1x_2, P_5 = \overline{x}_2\overline{x}_3$, $P_6 = x_1x_2$, $P_7 = \overline{x}_2x_3$, $P_8 = x_1$. *The set of product terms after deleting the shared products is:* $P_1 = x_1$, $P_2 = \overline{x}_2\overline{x}_3$, $P_3 = x_1\overline{x}_3$, $P_4 = x_1x_2$, $P_5 = \overline{x}_2x_3$. *The connection matrices for AND,* C_{AND}, *and OR,* C_{OR}, *gate arrays:*

$C_{AND} =$

	x_1	$\overline{x}_1$	x_2	$\overline{x}_2$	x_3	$\overline{x}_3$	x_4	$\overline{x}_4$
P_1	1						1	
P_2				1		1		
P_3	1					1		
P_4	1		1					
P_5				1	1			

$C_{OR} =$

	f_1	f_2	f_3	f_4
P_1	1			1
P_2	1		1	
P_3		1		
P_4		1	1	
P_5				1

Practice problem 16.8. (Connection matrix.) Derive the connection matrix for the 3-output Boolean function of three variables, $f_1 = \bigvee m(0,1,2,4)$, $f_2 = \bigvee m(2,4,5,7)$, and $f_3 = \bigvee m(0,2,3,7)$.
Answer is given in "Solutions to practice problems."

16.6.3 Implementation of Boolean functions using PLAs

Given an m-output Boolean function of n variables, or a system of m functions, it can be implemented on a PLA, according to the following algorithm.

PLA implementation of Boolean functions

Step 1: `Specify a Boolean function in SOP form (minimize the SOP representation if necessary).`
Step 2: `Determine the PLA connection matrix.`
Step 3: `Map the matrix into the topology of AND and OR arrays.`

Example 16.16 (Functions on PLA.) *The PLA with three inputs, four product terms, and two outputs is given in Figure 16.12. Each AND gate (which physically does not exist) in the AND plane has six inputs, corresponding to the true and complemented versions of the three input signals. Programming the connections is shown as follows: A signal that is connected to an AND gate is indicated with a wavy line, and a signal that is not connected to the gate is shown with a broken line.*

It follows from Example 16.16 that the circuitry is designed such that any unconnected AND-gate inputs do not affect the output of the AND gate.

Practice problem **16.9. (Functions on PLA.)** Implement the Boolean functions (a) $f_1 = x_1x_2 \vee x_1\overline{x}_3 \vee \overline{x}_1\overline{x}_2x_3$ and (b) $f_2 = x_1x_2 \vee \overline{x}_1\overline{x}_2x_3 \vee x_1x_3$. using a PLA.
Answer is given in "Solutions to practice problems."

16.6.4 Programmable array logic (PAL)

The programmable switches in large PLA arrays may reduce the speed-performance of the implemented circuits. This drawback led to the development of a similar device known as a *programmable array logic (PAL)*, in which (Figure 16.13a):

- The AND arrays are programmable, and
- The OR arrays are fixed.

Example 16.17 (PAL.) *Figure 16.13b shows the PAL with three inputs, four product terms, and two outputs that implements a system of two functions:*

$$f_1 = \underbrace{x_1x_2}_{P_1} \vee \underbrace{x_1\overline{x}_3}_{P_2} \quad \textit{and} \quad f_2 = \underbrace{\overline{x}_1\overline{x}_2x_3}_{P_3} \vee \underbrace{x_1x_3}_{P_4}$$

The product terms P_1 and P_2 are hardwired to one OR gate, and P_3 and P_4 are hardwired to the other OR gate.

PALs, in general, offer less flexibility in the implementation of Boolean functions than PLAs.

Design example: Programmable logic array (PLA)

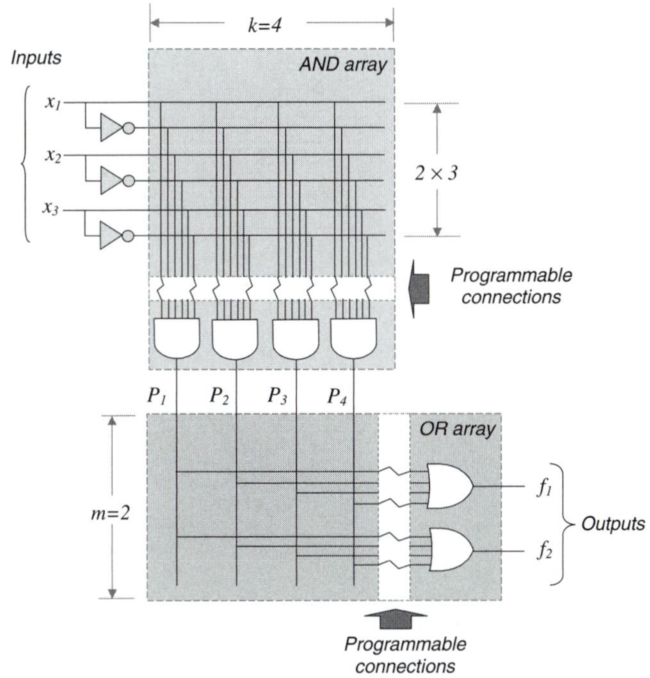

FIGURE 16.12
Gate level PLA with three inputs, four product terms, and two outputs (Example 16.16).

16.6.5 Using PLAs and PALs for EXOR polynomial computing

Using PLAs or PALs with internal EXOR gates, it is possible to compute EXOR expressions of Boolean functions using PLAs and PALs.

Example 16.18 (PAL for EXOR functions.) *Figure 16.14 shows how to compute the linear Boolean function $x_1 \oplus x_2 \oplus x_3 \oplus x_4$ using a PAL containing an internal EXOR gate.*

Example 16.19 (Boolean difference.) *Figure 16.15 shows how to compute the Boolean difference of the function $f = x_1x_2 \vee \overline{x}_3x_4$ using a PAL containing an internal EXOR gate.*

Design example: Programmable logic array (PLA)

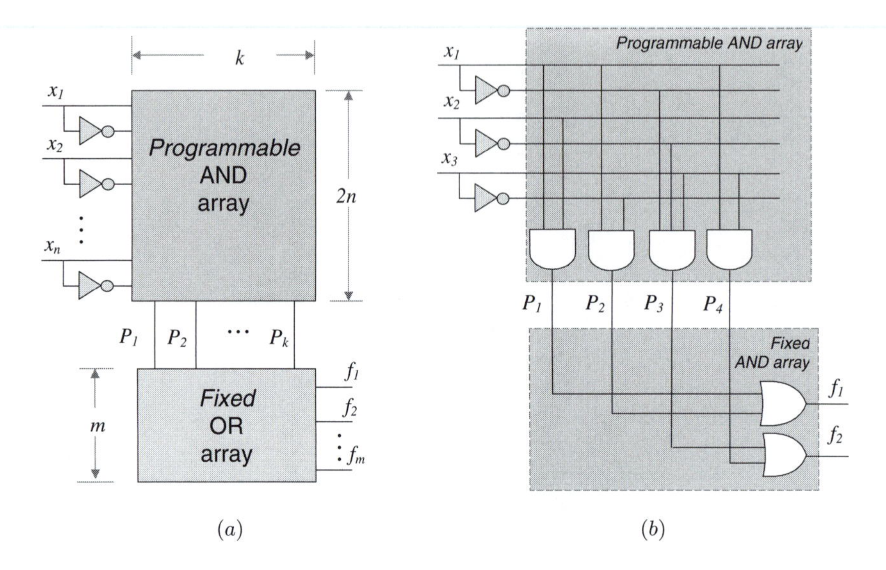

FIGURE 16.13
Structure of an n-input m-output PAL (a), and example of the implementation of a Boolean function (b) (Example 16.17).

Practice problem **16.10. (PLA for EXOR functions.)** Prove that a PLA in Figure 16.16 realizes the linear Boolean function $x_1 \oplus x_2 \oplus x_3 \oplus x_4$. Use approaches based on (a) manipulation in algebraic form and (b) a truth table.
Answer is given in "Solutions to practice problems."

16.7 Field programmable gate arrays

A *field programmable gate array (FPGA)* is a programmable logic device that supports the implementation of large logic networks. The primary advantage of FPGAs is their ability to implement any circuit by appropriate programming. The use of a standard FPGA, rather than custom technologies such as application-specific integrated circuits (ASICs), has two key benefits: Lower cost and faster time-to-market.

The topological structure of an FPGA is defined as a mesh array (Figure

Design example: Programmable logic array (PLA)

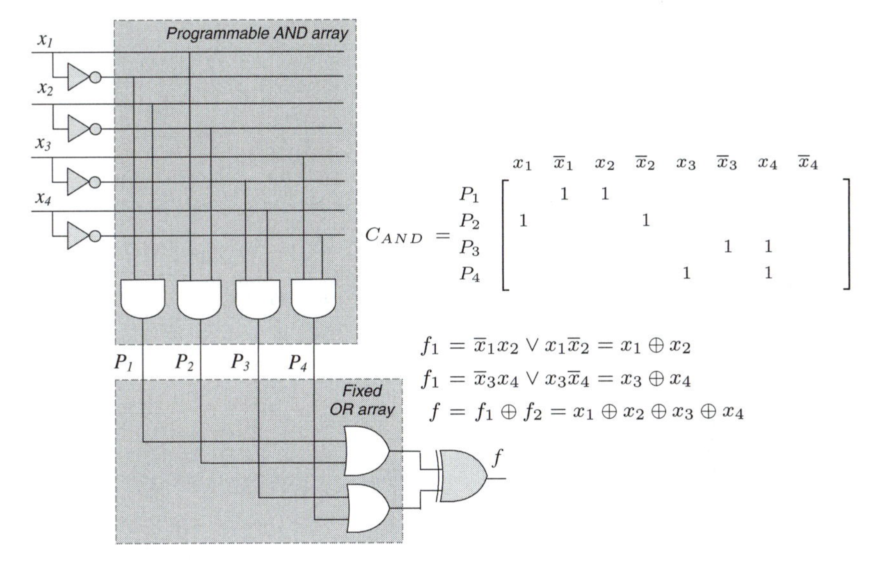

FIGURE 16.14
Computing the linear Boolean function $x_1 \oplus x_2 \oplus x_3 \oplus x_4$ (Example 16.18).

16.17a). An FPGA includes:

- *Logic blocks*, denoted as □,
- *Interconnection switches*, denoted as ■, and
- Input/Output (I/O) blocks.

The logic blocks are arranged in a 2-dimensional array. Input/Output blocks are used for connecting the pins of the package. The interconnection wires are organized as horizontal and vertical *routing channels* between rows and columns of logic blocks. In Figure 16.17a, two locations for programmable switches are shown:

- Switches (black boxes) adjacent to logic blocks (white boxes) hold switches that connect the logic block input and output terminals to the interconnection wires.
- Switches (black boxes) that are diagonally between logic blocks connect one interconnection wire to another (such as a vertical wire to a horizontal wire).

Programmable connections also exist between the input/output blocks and the interconnection wires.

Design example: Computing Boolean differences using PLAs

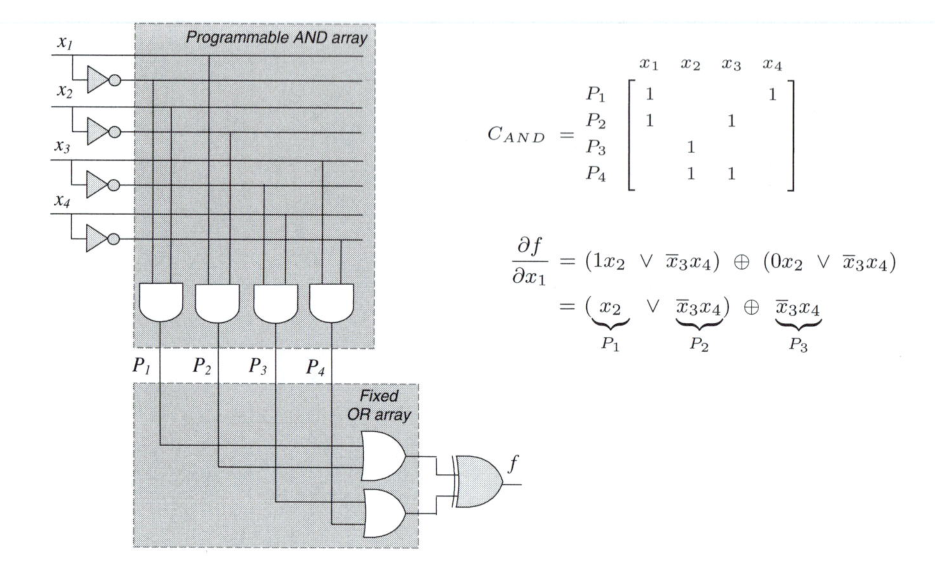

FIGURE 16.15
Computing Boolean difference (Example 16.19) using a PAL with an EXOR gate extension.

16.7.1 Logic blocks

Each logic block has a small number of inputs and one output. The most commonly used logic block is a *lookup table (LUT)*. FPGAs connect circuitry via programmable switches. These switches add significant capacitance to connections, reducing circuit speed. A circuit implemented in an FPGA is typically 10 times larger and roughly 3 times slower than the same circuit implemented on programmed arrays such as PLAs or PALs. The larger size of FPGA circuitry makes FPGA implementations more expensive than other programmable designs.

16.7.2 FPGA architecture

All FPGAs consist of a large number of programmable logic blocks, each of which implements a small amount of digital logic, and programmable routing that allows the logic block inputs and outputs to be connected to form larger circuits. The global routing architecture of an FPGA specifies the relative width of the various wiring channels within the chip.

Logic blocks are groups of LUTs and flip-flops along with local routing to

Design example: Computing linear Boolean functions using PLAs

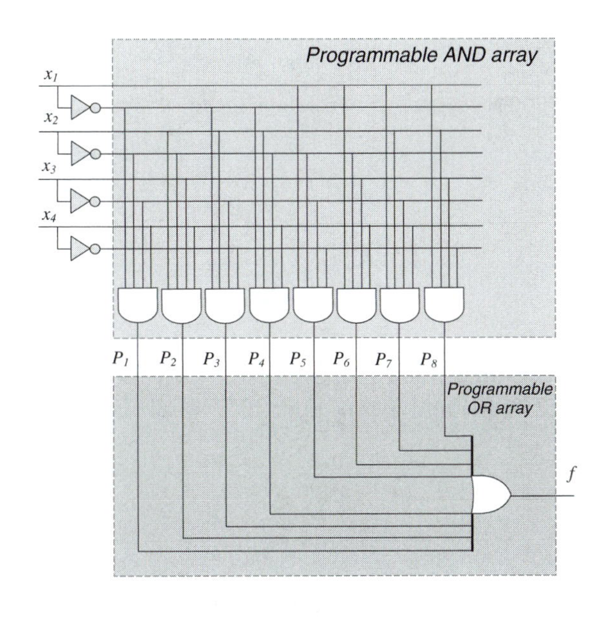

Connection matrix:

$$C = \begin{array}{c} \\ P_1 \\ P_2 \\ P_3 \\ P_4 \\ P_5 \\ P_6 \\ P_7 \\ P_8 \end{array} \begin{array}{c} \begin{array}{cccc} x_1 & x_2 & x_3 & x_4 \end{array} \\ \left[\begin{array}{cccc} 1 & & & 1 \\ 1 & & 1 & \\ & 1 & & \\ & 1 & 1 & \\ & & & 1 \\ & & & 1 \\ & & & 1 \\ & & & 1 \end{array}\right] \end{array}$$

Product terms:

$$P_1 = \overline{x}_1\overline{x}_2\overline{x}_3x_4$$
$$P_2 = \overline{x}_1\overline{x}_2x_3\overline{x}_4$$
$$P_3 = \overline{x}_1x_2\overline{x}_3\overline{x}_4$$
$$P_4 = \overline{x}_1x_2x_3x_4$$
$$P_5 = x_1\overline{x}_2\overline{x}_3\overline{x}_4$$
$$P_6 = x_1\overline{x}_2x_3x_4$$
$$P_7 = x_1x_2\overline{x}_3x_4$$
$$P_8 = x_1x_2x_3\overline{x}_4$$

FIGURE 16.16
Computing the linear Boolean function $x_1 \oplus x_2 \oplus x_3 \oplus x_4$ using PLA (Practice problem 16.10).

interconnect the LUTs within a group. The logic block used in an FPGA strongly influences the FPGA speed and area efficiency. Most FPGAs use logic blocks based on LUTs.

> **Example 16.20 (FPGA architecture.)** *Figure 16.17b shows how a 2-input LUT can be implemented in a RAM-based FPGA.*

A k-input LUT requires 2^k RAM cells and a 2^k-input multiplexer. A k-input LUT can implement any function of k-inputs; one simply programs the 2^k RAM cells to be the truth table of the desired function.

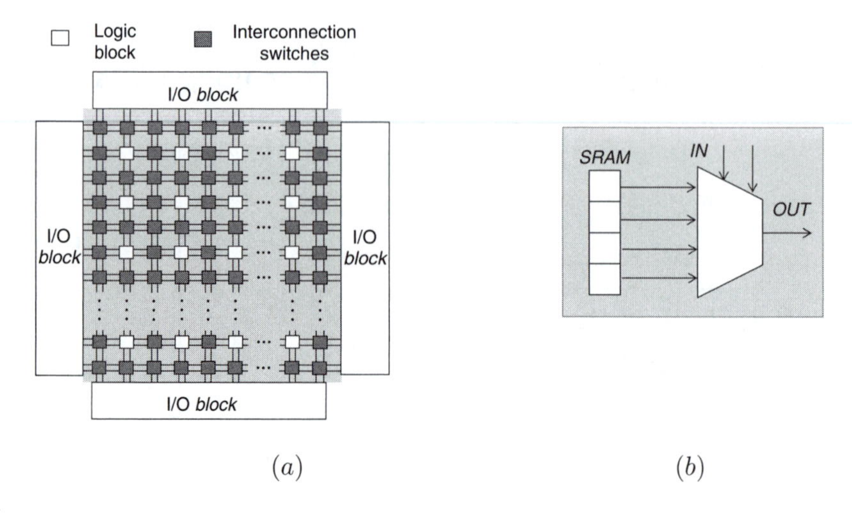

FIGURE 16.17
Topological structure of an FPGA (a) and a two-input LUT implemented in a RAM-based FPGA (b).

16.8 Summary of design techniques using memory and programmable devices

This chapter introduces memory units and programmable devices. A memory unit is a device to which binary information is transferred for storage and from which information is available when needed for processing. A programmable device is a combinational logic device consisting of arrays of logic gates and programmable switches. The switches allow the logic gates inside the array to be connected together to implement any Boolean function. The key aspects considered in this chapter are:

(a) The design and application of memory devices, and
(b) The design and application of programmable devices.

The main topics of the chapter are listed below:

- ▶ A memory unit is a collection of cells capable of storing a large quantity of binary information. The main types of memories in digital systems are (a)***Random-access memory (RAM)***, and (b)***Read-only memory (ROM)***.

Summary (continuation)

- RAM is a type of memory in which all addresses are accessible in an equal amount of time and can be selected in any order for *Read* or *Write* operations. RAM comes in two varieties: ***Static*** RAM (SRAM), and ***Dynamic*** RAM (DRAM). SRAM is faster and more expensive than DRAM. DRAM stores each bit as charge on a capacitor. DRAM and SRAM lose their stored data when power is turned off; that is, they are ***volatile***.
- ROM is a type of memory in which data are stored permanently or semipermanently. Data can be read from ROM, but there is no write operation as in RAM. ROM retains stored data even if the power is turned off; that is, it is ***nonvolatile*** memory.
- Chips using programmable logic are commercially available as ***Read-only memory (ROM)***, ***Programmable logic array (PLA)***, and ***Programmable array logic (PAL)***. ROMs, PLAs and PALs are multi-input multi-output devices, typically organized into an AND gate array, and an OR gate array. This structure is suited to the SOP representation of Boolean functions. The AND array is used to compute particular product terms of the inputs, while the OR array takes these terms and ORs them together to produce the final SOP expression.
- A PLA can implement a modest collection of Boolean functions of considerable complexity. This complexity is determined by the number of inputs; the number of product terms (that is, the number of AND gates) and the number of outputs (that is, the number of OR gates that the PLA can support).
- ROM, PLA, and PAL are distinguished as follows:

  (*a*) In a ROM, the AND array is fixed, and the OR array is programmable.
  (*b*) In a PLA both the AND and OR arrays are programmable.
  (*c*) In a PAL, the AND array is programmable and the OR array is fixed.

- A ***field-programmable gate array (FPGA)*** is a programmable logic device that supports the implementation of large logic networks. An FPGA consists of ***logic blocks***, ***interconnection switches***, and ***Input/Output (I/O)*** blocks.
- Each logic block of an FPGA has a small number of inputs and one output. The most commonly used logic block is a ***lookup table (LUT)*** which is based on a ROM or other programmable array.
- New horizons in design techniques using memory and programmable devices are introduced in the "Further study" section.

16.9 Further study

1950: The earliest work on non-volatile memory, also called core memory, was carried out by An Wang and Way-Dong Woo, who created the pulse transfer controlling device in 1949, in which the magnetic field of the cores was used to control the switching of current in electro-mechanical systems. Core memory was unaffected by electro-magnetic pulses and radiation, it was applied in many military installations and vehicles like fighter aircraft, as well as spacecraft, and led to core being used for a number of years after availability of semiconductor MOS memory (see also MOSFET). For example, the Space Shuttle flight computers initially used core memory, which preserved the contents of memory even through the Challenger's explosion and subsequent plunge into the sea in 1986.

1970: First semiconductor ROM called mask ROM have been developed. Programmable ROM (PROM), invented in 1956, allowed users to program its contents exactly once by physically altering its structure with the application of high-voltage pulses.

A non-volatile computer memory that uses a thin film of a magnetic material to hold small magnetized areas, known as bubbles, which each store one bit of data. A promising technology developed by Andrew Bobeck and called *bubble* memory started out in the 1970s. Bubble memory is a type of This technology is associated with the development of the first magnetic core memory system driven by a transistor-based controller. Bubble memory failed commercially as hard disk prices fell rapidly in the 1980s.

The 1971 invention of Erasable PROM (EPROM) brought the possibility to repeatedly reset it to its unprogrammed state by exposure to strong ultraviolet light.

1980: EEPROM, invented in 1983, were able to be programmed in-place if the containing device provides a means to receive the program contents from an external source, for example, a personal computer via a serial cable. Flash memory, invented at Toshiba in the mid-1980s, and commercialized in the early 1990s, is a form of EEPROM that makes very efficient use of chip area and can be erased and reprogrammed thousands of times without damage. Flash memory allows its memory to be written only in blocks, which greatly simplifies the internal wiring and allows for higher densities. Flash memory has evolved into one of the lowest cost solid-state memory devices available. New multi-bit techniques appear to be able to double or quadruple the density.

The first programmable integrated circuits were developed. These devices contain circuits whose logical function and connectivity can be programmed by the user, rather than being fixed by the integrated circuit manufacturer. This allows a single chip to be programmed to implement different logic functions. Current devices, called FPGAs (Field Programmable Gate Arrays) can now implement tens of thousands of LSI circuits in parallel and operate at up to 550 MHz.

The most recent development is NAND flash, also invented by Toshiba. It offers throughput comparable to hard disks, higher tolerance of physical shock, extreme miniaturization (in the form of USB flash drives and tiny microSD memory cards, for example), and much lower power consumption.

Advanced topics of design techniques using memory and programmable devices

Topic 1: Hierarchical programmable devices.

Advanced topics involve the design of specific topologies, including three-dimensional structures. A hierarchical, or binary-tree-like FPGA is based on single-input two-output switches. A 2×2 cluster of processing elements can be connected using switches, and four copies of the cluster organized into a "macro" cluster. The structure of this FPGA is represented by a binary decision tree of depth 4, in which the root and levels correspond to switching blocks, and 16 terminal nodes correspond to processing elements.

Another architecture of the hierarchical FPGA is based on a cluster of logic blocks connected by single-input 4-output switch blocks. Possible configurations can be described by a complete binary tree or a multi-rooted k-ary tree. This tree can be embedded into hypercube-like structures.

In Figure 16.18, two topologies of FPGA are illustrated, where ■ denotes a processing element and ∘ denotes a switch block. Four copies of the cluster are organized into a "macro cluster" of different configurations. The first topology (Figure 16.18a) is based on H-tree construction and described by a complete binary tree.

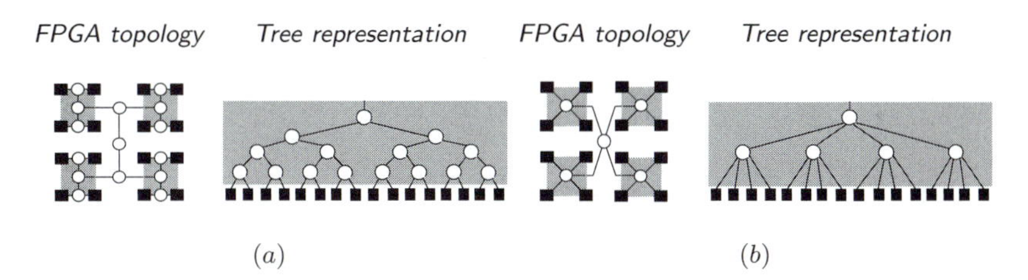

FIGURE 16.18
Topological representations of hierarchical FPGAs by 2D structure, tree, and hypercube-like structure.

Topic 2: Alternative memory devices. Currently, the new technologies for non-volatile memories are being developed to achieve a real RAM-like role, or to erase the border between the both types, ROM and RAM.

One example is Ferroelectric RAM (FeRAM). FeRAM uses a ferroelectric layer in a cell that is otherwise similar to conventional DRAM, this layer holding the charge in a 1 or 0 even with the power removed. To date, FeRAM are still larger than Flash devices.

Another example is Magnetoresistive Random Access Memory, or MRAM, which uses magnetic elements and generally operates in a fashion similar to core. Another technique, known as STT-MRAM, appears to allow for much higher densities.

One more new development is Phase-change RAM, or PRAM. PRAM is based on the same storage mechanism as writable CDs and DVDs, but reads them based on their changes in electrical resistance rather than changes in their optical properties.

A number of more esoteric devices have been proposed, including Nano-RAM based on quantum dots, carbon nanotubes and nanowires, as well as molecular-scale memory devices have been developed recently, but these are currently far from commercialization.

Further reading

Digital Principles and Design by D. D. Givone, McGraw-Hill, New York, 2003.

Fundamentals of Digital Logic with VHDL Design by S. Brown and Z. Vranesic, McGraw-Hill, 2000.

"Hierarchical Interconnection Structures for Field Programmable Gate Arrays" by Y. T. Lai and P. T. Wang, *IEEE Transactions on VLSI, Systems*, volume 5, number 2, pages 186–196, 1997.

Switching Theory for Logic Synthesis by T. Sasao, Kluwer, 1999.

16.10 Solutions to practice problems

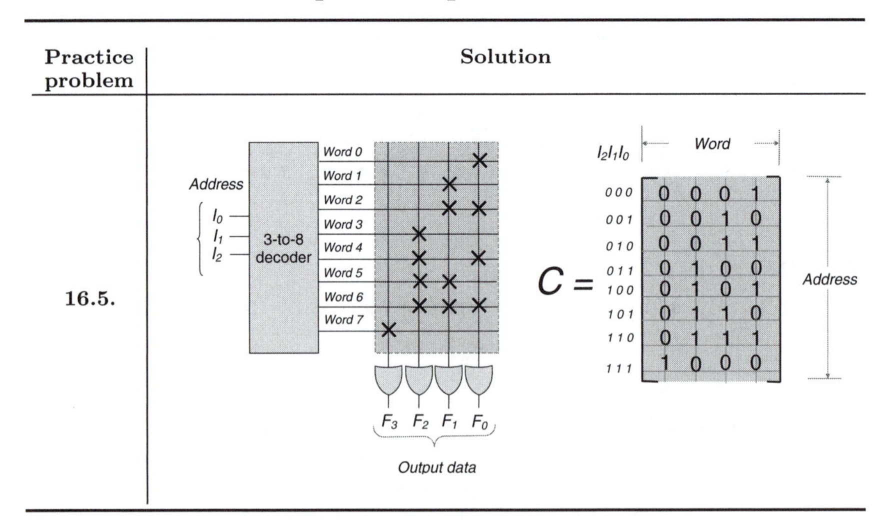

Solutions to practice problems (continuation)

Practice problem	Solution
16.6.	First, the canonical SOPs must be formed: $f_1 = x_1x_2(\overline{x}_3 \vee x_3) \vee x_1(\overline{x}_2 \vee x_2)\overline{x}_3 \vee \overline{x}_1\overline{x}_2x_3$ $= x_1x_2\overline{x}_3 \vee x_1x_2x_3 \vee x_1\overline{x}_2\overline{x}_3 \vee x_1x_2\overline{x}_3 \vee \overline{x}_1\overline{x}_2x_3$ $= \bigvee m(0,3,6,7)$ $f_2 = x_1x_2(\overline{x}_3 \vee x_3) \vee \overline{x}_1\overline{x}_2x_3 \vee x_1(\overline{x}_2 \vee x_2)x_3$ $= x_1x_2\overline{x}_3 \vee x_1x_2x_3 \vee \overline{x}_1\overline{x}_2\overline{x}_3 \vee x_1\overline{x}_2x_3 \vee x_1x_2x_3$ $= \bigvee m(0,5,6,7)$ $f_3 = x_1 \vee x_2 \vee x_3 = \overline{\overline{x}_1\overline{x}_2\overline{x}_3}$ $= \prod M(0) = \bigvee m(1,2,3,4,5,6,7)$ Address; x_1; x_2; x_3; 3-to-8 decoder; Word 0; Word 1; Word 2; Word 3; Word 4; Word 5; Word 6; Word 7; f_3 f_2 f_1; Output data
16.7.	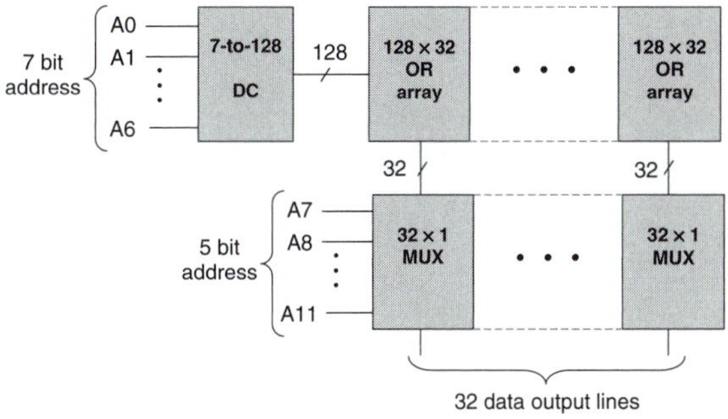 7 bit address; A0; A1; A6; 7-to-128 DC; 128; 128 × 32 OR array; . . .; 128 × 32 OR array; 32; 32; 5 bit address; A7; A8; A11; 32 × 1 MUX; . . .; 32 × 1 MUX; 32 data output lines

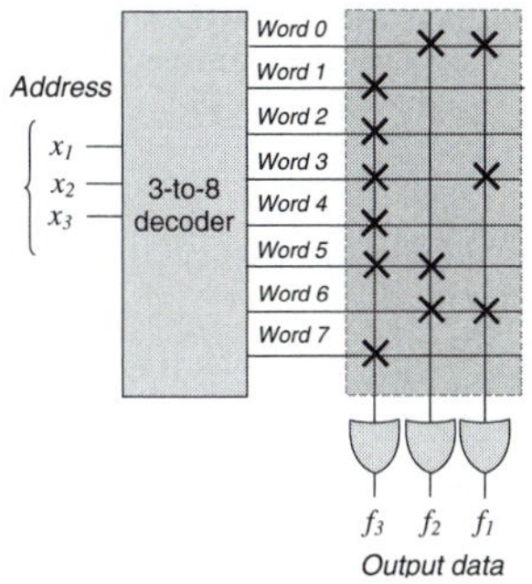

Solutions to practice problems (continuation)

Practice problem	Solution

16.8.

First, the functions can be minimized:

K-map for the function f_1:

x_3 \ x_1x_2	00	01	11	10
0	1	1		1
1	1			

K-map for the function f_2:

x_3 \ x_1x_2	00	01	11	10
0		1		1
1			1	1

K-map for the function f_3:

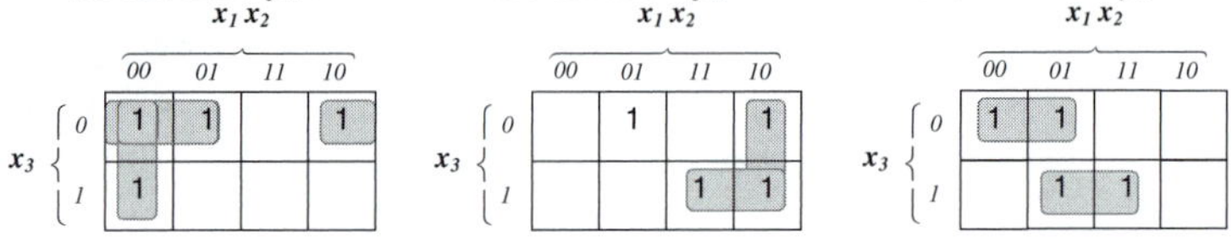

x_3 \ x_1x_2	00	01	11	10
0	1	1		
1		1	1	

$$f_1 = \overline{x}_1\overline{x}_3 \vee \overline{x}_2\overline{x}_3 \vee \overline{x}_1\overline{x}_2$$
$$f_2 = \overline{x}_1x_2\overline{x}_3 \vee x_1\overline{x}_2 \vee x_1x_3$$
$$f_3 = \overline{x}_1\overline{x}_3 \vee x_2x_3$$

These SOP expressions contain the following products:

$$P_1 = x_1x_2, \quad P_2 = x_1\overline{x}_3, \quad P_3 = \overline{x}_1\overline{x}_2x_3, \quad P_4 = x_1x_3$$

Connection matrices for AND and OR array are as follows:

Connection matrix for AND array

$$C_{AND} = \begin{array}{c} \\ P_1 \\ P_2 \\ P_3 \\ P_4 \\ P_5 \\ P_6 \\ P_7 \\ P_8 \end{array} \begin{array}{c} \begin{array}{cccccc} x_1 & \overline{x}_1 & x_2 & \overline{x}_2 & x_3 & \overline{x}_3 \end{array} \\ \left[\begin{array}{cccccc} & 1 & & & & 1 \\ & & 1 & & & 1 \\ & 1 & & 1 & & \\ & 1 & 1 & & & 1 \\ 1 & & & 1 & & \\ 1 & & & & 1 & \\ 1 & & & & 1 & \\ & & 1 & & 1 & \end{array}\right] \end{array}$$

Connection matrix for OR array

$$C_{OR} = \begin{array}{c} \\ P_1 \\ P_2 \\ P_3 \\ P_4 \\ P_5 \\ P_6 \\ P_7 \\ P_8 \end{array} \begin{array}{c} \begin{array}{ccc} f_1 & f_2 & f_3 \end{array} \\ \left[\begin{array}{ccc} 1 & & 1 \\ 1 & & \\ 1 & & \\ & 1 & \\ & 1 & \\ & 1 & \\ & 1 & \\ & & 1 \end{array}\right] \end{array}$$

16.9.

$$P_1 = x_1x_2, \quad P_2 = x_1\overline{x}_3$$
$$P_3 = \overline{x}_1\overline{x}_2x_3, \quad P_4 = x_1x_3$$

x_1 x_2 x_3 P_1 P_2 P_3 P_4 f_1 f_2

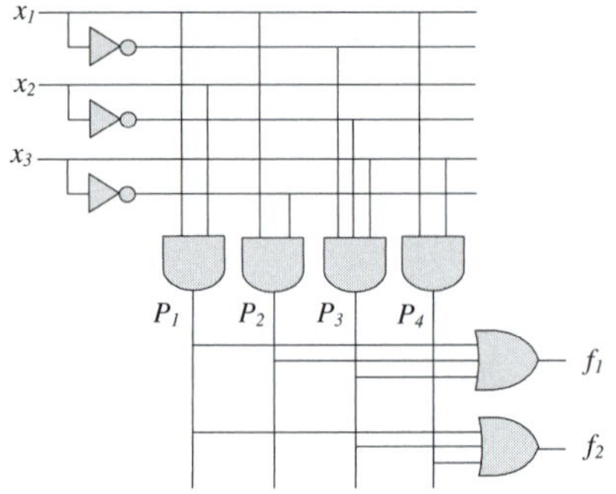

Solutions to practice problems (continuation)

Practice problem	Solution
16.10.	

An approach based on manipulation in algebraic form:

$$f = x_1 \oplus x_2 \oplus x_3 \oplus x_4 = \underbrace{(x_1 \oplus x_2 \oplus x_3)}_{A} \oplus x_4 = \overline{A}x_4 \vee A\overline{x}_4$$

$$\overline{A} = \overline{x_1 \oplus x_2 \oplus x_3} = 1 \oplus x_1 \oplus x_2 \oplus x_3 = \underbrace{(1 \oplus x_1 \oplus x_2)}_{B} \oplus x_3 = \overline{B}x_3 \vee B\overline{x}_3$$

$$\overline{B} = \overline{1 \oplus x_1 \oplus x_2} = x_1 \oplus x_2 = \overline{x}_1x_2 \vee x_1\overline{x}_2$$

$$B = \overline{\overline{B}} = \overline{\overline{x}_1x_2 \vee x_1\overline{x}_2} = (x_1 \vee \overline{x}_2)(\overline{x}_1 \vee x_2) = x_1\overline{x}_2 \vee x_1x_2 \vee \overline{x}_1\overline{x}_2 \vee \overline{x}_2x_2$$
$$= x_1x_2 \vee \overline{x}_1\overline{x}_2$$

$$\overline{A} = \overline{B}x_3 \vee B\overline{x}_3 = (\overline{x}_1x_2 \vee x_1\overline{x}_2)x_3 \vee (x_1x_2 \vee \overline{x}_1\overline{x}_2)\overline{x}_3$$
$$= \overline{x}_1x_2x_3 \vee x_1\overline{x}_2x_3 \vee x_1x_2\overline{x}_3 \vee \overline{x}_1\overline{x}_2\overline{x}_3$$

$$A = \overline{\overline{A}} = \overline{\overline{x}_1x_2x_3 \vee x_1\overline{x}_2x_3 \vee x_1x_2\overline{x}_3 \vee \overline{x}_1\overline{x}_2\overline{x}_3}$$
$$= (x_1 \vee \overline{x}_2 \vee \overline{x}_3)(\overline{x}_1 \vee x_2 \vee \overline{x}_3)(\overline{x}_1 \vee \overline{x}_2 \vee x_3)(x_1 \vee x_2 \vee x_3)$$
$$= \overline{x}_1\overline{x}_2x_3 \vee \overline{x}_1x_2\overline{x}_3 \vee x_1\overline{x}_2\overline{x}_3 \vee x_1x_2x_3$$

Finally

$$\begin{aligned} f &= \overline{A}x_4 \vee A\overline{x}_4 \\ &= \underbrace{\overline{x}_1x_2x_3x_4}_{P_4} \vee \underbrace{x_1\overline{x}_2x_3x_4}_{P_6} \vee \underbrace{x_1x_2\overline{x}_3x_4}_{P_7} \vee \underbrace{\overline{x}_1\overline{x}_2\overline{x}_3x_4}_{P_1} \\ &\vee \underbrace{\overline{x}_1\overline{x}_2x_3\overline{x}_4}_{P_2} \vee \underbrace{\overline{x}_1x_2\overline{x}_3\overline{x}_4}_{P_3} \vee \underbrace{x_1\overline{x}_2\overline{x}_3\overline{x}_4}_{P_5} \vee \underbrace{x_1x_2x_3\overline{x}_4}_{P_8} \end{aligned}$$

An approach based on the truth table

x_1	x_2	x_3	x_4	f	Minterm
0	0	0	0	0	
0	0	0	1	1	$\overline{x}_1\overline{x}_2\overline{x}_3x_4 = P_1$
0	0	1	0	1	$\overline{x}_1\overline{x}_2x_3\overline{x}_4 = P_2$
0	0	1	1	0	
0	1	0	0	1	$\overline{x}_1x_2\overline{x}_3\overline{x}_4 = P_3$
0	1	0	1	0	
0	1	1	0	0	
0	1	1	1	1	$\overline{x}_1x_2x_3x_4 = P_4$
1	0	0	0	1	$x_1\overline{x}_2\overline{x}_3\overline{x}_4 = P_5$
1	0	0	1	0	
1	0	1	0	0	
1	0	1	1	1	$x_1\overline{x}_2x_3x_4 = P_6$
1	1	0	0	0	
1	1	0	1	1	$x_1x_2\overline{x}_3x_4 = P_7$
1	1	1	0	1	$x_1x_2x_3\overline{x}_4 = P_8$
1	1	1	1	0	

$$\begin{aligned} f &= x_1 \oplus x_2 \oplus x_3 \oplus x_4 \\ &= P_1 \vee P_2 \vee P_3 \vee P_4 \vee P_5 \vee P_6 \vee P_7 \vee P_8 \end{aligned}$$

16.11 Problems

Problem 16.1 Design $16K \times 16$ RAM using (a) $8K \times 8$ chips, (b) $16K \times 8$ chips, and $8K \times 16$ chips.

Problem 16.2 Design $2K \times 32$ ROM using (a) 9-to-512 decoder and 32-to-1 multiplexers, (b) 8-to-256 decoder and 64-to-1 multiplexers.

Problem 16.3 Implement the Boolean functions represented by the following logic networks using ROM, PLA, and PAL:

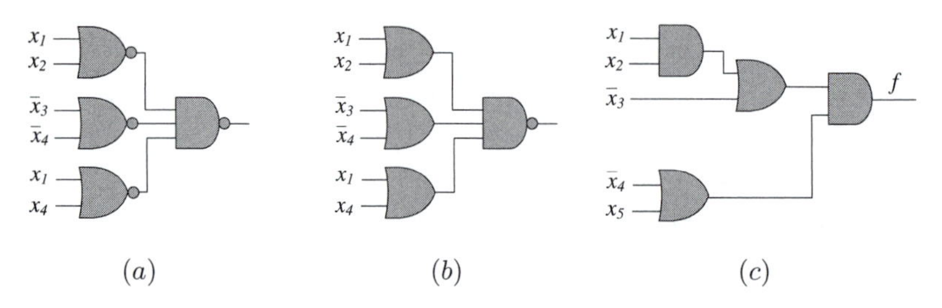

(a) (b) (c)

Problem 16.4 Implement the following Boolean functions of four variables using PLA:
(a) $f_1 = \bigvee m(0,1,2,5,11)$ (b) $f_1 = \prod M(1,3,4,5)$ (c) $f_1 = \prod M(0,4)$

Problem 16.5 Implement the following three Boolean functions, given by their K-maps, on a PAL:

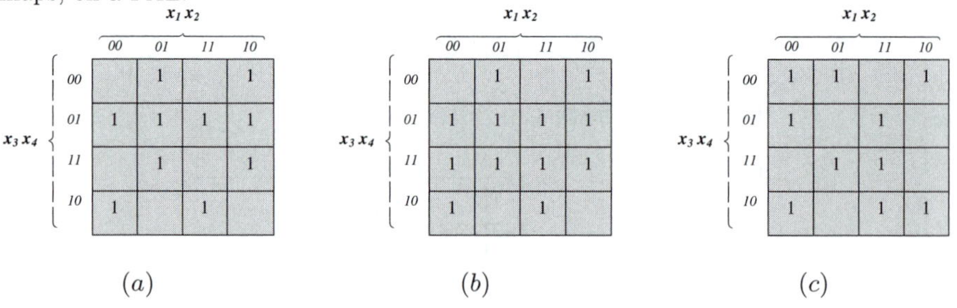

(a) (b) (c)

Problem 16.6 Given the LUT, implement this function using ROM:

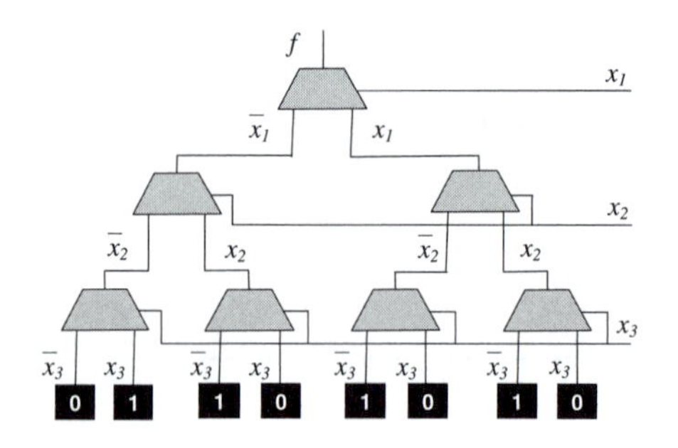

Problem 16.7 Restore the Boolean functions implemented by the following PLAs:

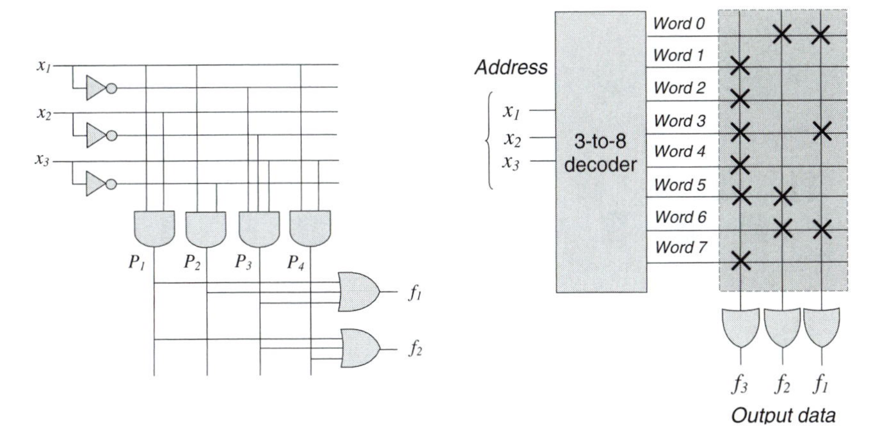

Problem 16.8 Restore the Boolean functions implemented by the following PLAs:

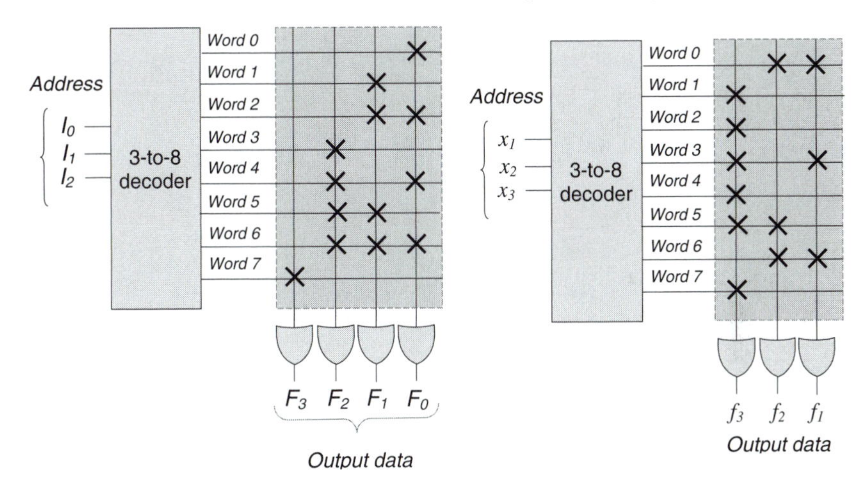

Problem 16.9 Restore the Boolean functions f_1, f_2, and f_3 implemented by PLA, using the following connection matrixes:

C_{AND} =

	x_1	$\overline{x}_1$	x_2	$\overline{x}_2$	x_3	$\overline{x}_3$
P_1		1				1
P_2			1			1
P_3		1		1		
P_4		1	1			1
P_5	1			1		
P_6	1				1	
P_7	1				1	
P_8	1		1		1	

C_{OR} =

	f_1	f_2	f_3
P_1	1		1
P_2		1	
P_3	1	1	1
P_4			1
P_5	1	1	1
P_6	1	1	1
P_7	1		1
P_8	1		1

Problem 16.10 Implement the Boolean functions represented by the following decision diagrams, using ROM, PLA, and PAL:

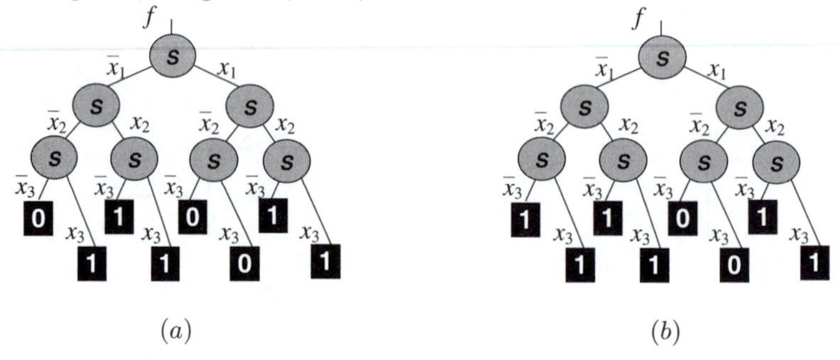

17

Sequential Logic Network Design

Analysis

- Mealy and Moore models
- Analysis of networks with D flip-flops
- Analysis of networks with JK flip-flops
- Analysis of networks with T flip-flops

Synthesis

- Synthesis using D flip-flops
- Synthesis using JK flip-flops
- Synthesis using T flip-flops

Design techniques

- State reduction
- State assignment
- Redesign

Advanced topics

- Lower-power sequential logic network design
- Verification of sequential circuits

17.1 Introduction

In combinational network designs, the design process typically starts with the problem statement, which is often a verbal description of the intended behavior of the system. The goal is to develop a logic network utilizing the available components and meeting the design objectives and constraints.

Sequential network design is based on the concept of assembling a complete sequential network from basic simple elements; both elements with storage properties and combinational elements. Therefore, circuits are at the top level in the design hierarchy, and their design supersedes the design of combinational components.

Sequential networks

The operation of a sequential network is described as a sequence of *state transitions* controlled by the inputs. *State tables* define the network's next-state functions. The *state diagram* represents all internal states and relationships between the states. The state behavior of the sequential circuits is described by the state table, state diagram, as well as excitation and output equations.

Comparison of data structures that are used in analysis and synthesis of combinational and sequential logic networks is given in Table 17.1.

TABLE 17.1
Data structures in logic network design.

Combinational networks based on SOP forms	Combinational network based on polynomial forms	Sequential networks
Truth table	functional table	State table
SOP expressions	Polynomial expressions	State equation, excitation equation, and output equation
Decision trees and diagrams	Functional decision trees and diagrams	State diagrams
Gates	Gates	Gates and flip flops
Cubes	Functional cubes	

Analysis and synthesis of sequential networks

The two facets of sequential logic network design are analysis and synthesis. The analysis problem is formulated as follows:

Analysis problem

Given a sequential network, provide a tabular description of this network. The sequential networks can be synchronous (globally clocked) or asynchronous (locally clocked). A sequential network at any given time is described by:

- Its inputs,
- Its outputs, and
- The state of its flip-flops at this time.

The flip-flop's outputs changed at every clock pulse, and the outputs and the next state of the flip-flops are both a function of the inputs and the present state. The analysis of a sequential network consists of obtaining a table or a diagram for the time sequence of inputs, outputs, and internal states. It is also possible to write Boolean expressions that describe the behavior of the sequential network. These expressions must include the necessary time sequence.

The steps involved in the synthesis of sequential networks are basically the reverse of those involved in analysis. Synthesis involves the establishment of a sequential network realization that satisfies a set of input/output specifications:

Synthesis problem

Given a word specification of a network, design a sequential network.

17.2 Mealy and Moore models of sequential networks

Finite state machines are the formal models of sequential logic networks, and they are divided into two classes: the *Mealy* model, and the *Moore* model. The Mealy model is characterized as follows (Figure 17.1):

Mealy model of sequential networks

- The output is a function of both the present state and the inputs.
- The outputs of a sequential network may change if the inputs change during the clock cycle; the new outputs are available by the next clock edge.

The specific properties of the Moore model are listed below (Figure 17.1):

Moore model of sequential networks

- The output is a function of the present state only.
- The outputs are synchronized with the clock; i.e., the new outputs are not available by the next clock edge.

Both models are used for the analysis and synthesis of sequential networks, and are represented using various data structures.

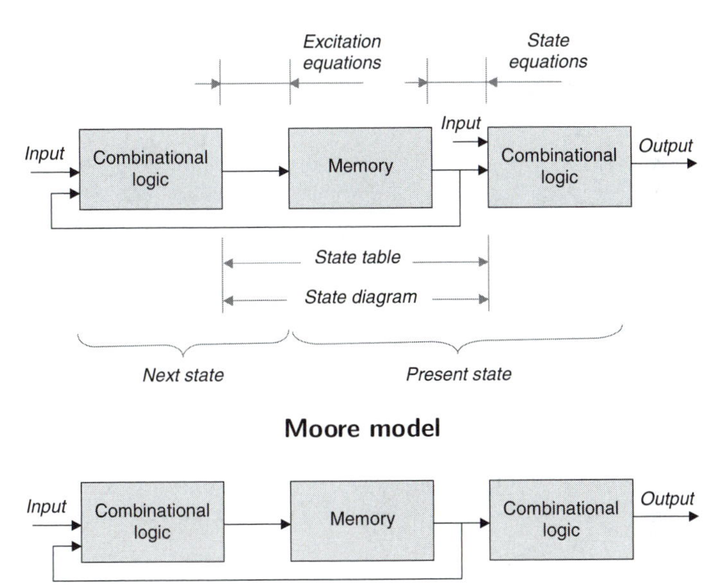

FIGURE 17.1
Mealy and Moore models of a synchronous sequential network.

17.3 Data structures for analysis of sequential networks

A sequential network consists of

- A network with memory properties using flip-flops, and
- A combinational network using available logic gates.

Data structures for analyzing combinational networks were considered in the previous chapters. For analyzing a network consisting of a combinational network and a memory-based network, specific data structures are used, such as

- *State equations*,
- *State tables*, or *transition tables*,
- *State diagrams*,
- *Excitation equations*, and
- *Output equations*.

17.3.1 State equations

The behavior of a clocked sequential network can be described in algebraic form by means of state equations. The state equations, also called *transition equations*, specify the next state as a function of the present state and the inputs:

Past state $\overset{Transition}{\longrightarrow}$ Present state $\overset{Transition}{\longrightarrow}$ Next state

17.3.2 Excitation and output equations

Knowledge of the type of flip-flop present provides necessary information for deriving a description of the memory part of a sequential network. Combinational networks in sequential networks are distinguished with respect to their functions. They are described as follows:

(*a*) The part of the combinational network that generates external outputs; this part is described by a set of Boolean equations called *output equations*.

(*b*) The part of the combinational network that generates the input signals to flip-flops; this part is described by a set of Boolean equations called *excitation equations*.

Example 17.1 (Excitation equations.) *The sequential network shown in Figure 17.2 consists of two D flip-flops, input x, output y, and the clock signal Clk. This network is a Mealy model because the output y is a function of both input x and the present state of D flip-flops. Let us derive the state and excitation equations for the sequential network:*

Step 1: *Derive the excitation which are the input functions for both flip-flops, $D_1 = Q_1x \vee Q_2x$ and $D_2 = \overline{Q}_1x$.*

Step 2: *Derive the output equation: $y = (Q_1 \vee Q_2)\overline{x}$*

Step 3: *Since the characteristic equations for D flip-flops are: $D_1 = Q_1^+$ and $D_2 = Q_2^+$, then the state equations are as follows: $Q_1^+ = Q_1x \vee Q_2x$ and $Q_2^+ = \overline{Q}_1x$.*

Design example: Analysis of D flip-flop based logic network

STEP 1

Initial sequential network

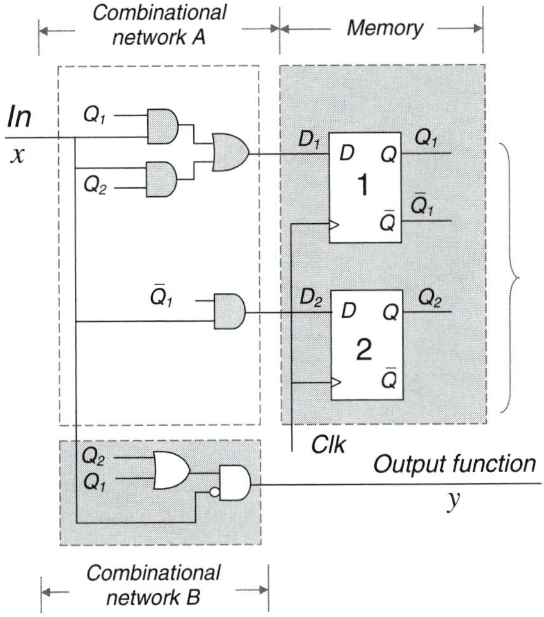

STEP 2

State equations

$$Q_1^+ = Q_1 x \vee Q_2 x$$
$$Q_2^+ = \overline{Q_1} x$$

Output function

$$y = (Q_1 \vee Q_2)\overline{x}$$

Excitation equations

$$D_1 = Q_1 x \vee Q_2 x$$
$$D_2 = \overline{Q_1} x$$

STEP 3

State table

Present state Q_1	Q_2	Input x	Next state Q_1^+	Q_2^+	Output y
0	0	0	0	0	0
0	0	1	0	1	0
0	1	0	0	0	1
0	1	1	1	1	0
1	0	0	0	0	1
1	0	1	1	0	0
1	1	0	0	0	1
1	1	1	1	0	0

STEP 4

State diagram

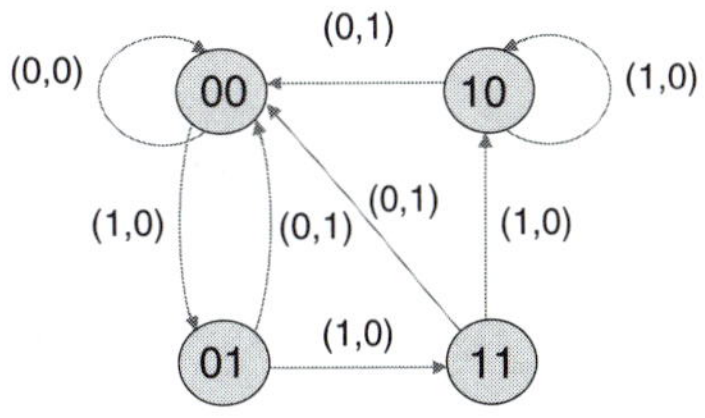

FIGURE 17.2
Sequential network analysis (Examples 17.1, 17.2, and 17.3).

Practice problem **17.1.** **(Sequential network.)** Given the sequential network (Figure 17.3a), derive the state, output, and excitation equations.
Answers are given in "Solutions to practice problems."

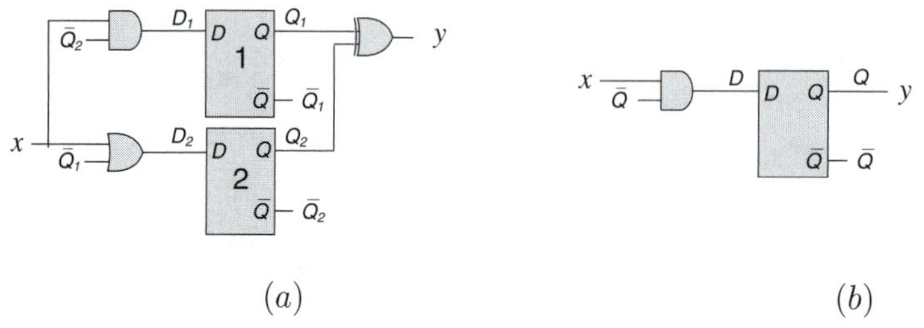

FIGURE 17.3
Sequential networks for Practice problems 17.1 (a) and 17.4 (b).

17.3.3 State table

Derivation of the state table requires listing all possible binary combinations of the present states and inputs. This creates the truth table for the functions of the next-states and outputs, whose values are calculated for the table from the state and output equations.

> **Example 17.2** *(Continuation of Example 17.1). Since there are three variables on the right side of the state equations (two present states, Q_1 and Q_2 and one input; Figure 17.2), there are 8 binary combinations of them. The values of the next-states of the first and second flip-flops and the output can be calculated from the state and output equations: $Q_1^+ = Q_1x \vee Q_2x$, $Q_2^+ = \overline{Q_1}x$ and $y = (Q_1 \vee Q_2)\overline{x}$, respectively, by substitution of all 8 assignments into them.*

Practice problem **17.2.** **(Sequential network.)** Derive the state table for the network given in Practice problem 17.1.
Answer is given in "Solutions to practice problems."

17.3.4 State diagram

The state diagram is a graphical data structure (directed graphs) for the representation of sequential logic networks. It provides the same information as a state table. The binary number inside each node identifies the state of the flip-flops. The direct links are labeled with two binary numbers:

- The input value during the present state, and
- The output during the present state with the given input.

Example 17.3 *(Continuation of Example 17.1). Given: the state table and the assignments of the flip-flop outputs,* Q_1 *and* Q_2*. The number of positive codes specifies the number of states. This example deals with the codes 00, 01, 10, and 11, and therefore four states are distinguished. This implies a state diagram, consisting of four nodes, and links between them labeled by pairs (Input, Output). For example, the directed link from state 00 to 01 is labeled (1,0), meaning that when the network is in the present state 00 and the input is 1, the output is 0. The network makes transition to the next state 01 after the next clock cycle:*

- *If the input changes to 0, then the output becomes 1.*
- *If the input remains at 1, the output stays at 0.*

Practice problem **17.3. (Sequential network.)** Derive the state diagram for the network given in Practice problem 17.1.
Answer is given in "Solutions to practice problems."

The above examples and practice problems have dealt, so far, with networks with D flip-flops. In general, the derivation of state equations depends on the type of flip-flops in the network. This analysis is considered in the next section.

17.4 Analysis of sequential networks with various types of flip-flops

A logic network with n flip-flops represent a finite state machine with 2^n different states. Analysis of this state machine can be performed based on knowledge of the types of flip-flops and a list of the Boolean expressions describing the combinational parts of the logic network. It provides the necessary information for deriving the state, output, and excitation equations, as well as the state diagrams. Summarizing, the algorithm for the analysis of a sequential network includes four major steps, as shown below.

Analysis of sequential networks with flip-flops

Given: A logic network with flip-flops.
Step 1: Determine the flip-flops' input equations by inspection of the network, in terms of the present state and input variables. These form the excitation equations.
Step 2: Determine the next-state equation using the corresponding flip-flop characteristic equations.
Step 3: Create the truth table of the values of the input equations for each combination of the present state and inputs (state table).
Step 4: Build the state diagram using the state table.
Output: The state table and the state diagram of the network.

17.4.1 Analysis of a sequential network with D flip-flops

A logic network with D flip-flops can be analyzed using the general steps described above. The specific feature of the network with D flip-flops is that the characteristic equation for D flip-flop is $Q^+ = D$.

Example 17.4 (Analysis of a sequential network.) *Consider a sequential network with D flip-flop (Figure 17.4). Its analysis involves the following steps:* ***Step 1:*** *Inspection of the network and derivation of the input equation for the D-flip-flops.* ***Step 2:*** *Determination of the next-state values using the D flip-flop characteristic equation $Q^+ = D$.* ***Step 3:*** *Creation of the state table using the next-state equation determined in Step 2.* ***Step 4:*** *Drawing the state diagram.*

Practice problem 17.4. **(Analysis of a sequential network.)**
Provide the analysis of the sequential network given in Figure 17.3b.
Answer is given in "Solutions to practice problems."

17.4.2 Analysis of a sequential network with JK flip-flops

The analysis of a network with JK flip-flops is performed in a slightly different way than the previous techniques for networks with D flip-flops. For a D flip-flop, the state equation is the same as the input equation. Deriving the state equations for JK flip-flops involves the following steps:

Deriving the state equations for JK flip-flops

Step 1: Determine the excitation equations.
Step 2: Substitute the excitation equation into the characteristic equation for JK flip-flop, or, instead, list all combinations for the values of each excitation equation and determine the next-state values using the JK flip-flop characteristics table.

Design example: Analysis of a D flip-flop based logic network

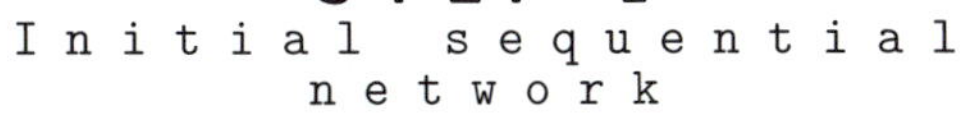

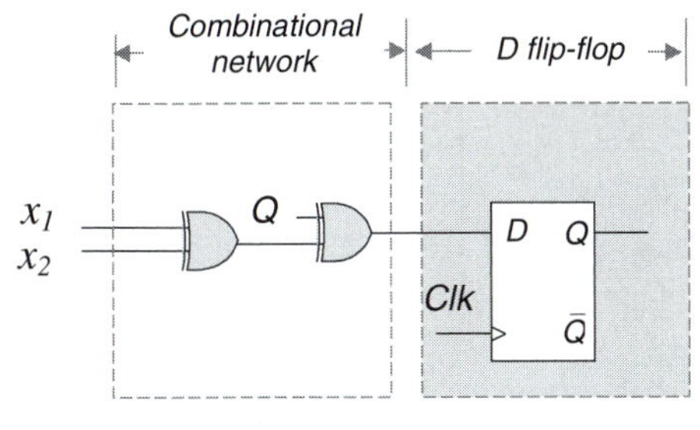

STEP 2

State equation

$$Q^+ = Q \oplus x_1 \oplus x_2$$

Excitation equation

$$D = Q \oplus x_1 \oplus x_2$$

STEP 3

State table

Present state Q	Input x_1	Input x_2	Next state Q
0	0	0	0
0	0	1	1
0	1	0	1
0	1	1	0
1	0	0	1
1	0	1	0
1	1	0	0
1	1	1	1

STEP 4

State diagram

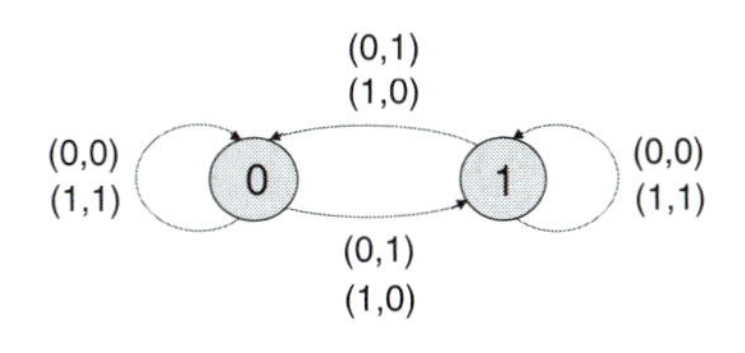

FIGURE 17.4
Analysis of a sequential network with D flip-flops (Example 17.4).

Example 17.5 (Analysis of a sequential network.) *Consider the sequential network with JK flip-flops (Figure 17.5). This is a Moore model, since the output is a function of the present state only. By inspection of the network, the inputs of the JK flip-flop are derived as follows:* $J_1 = Q_2$, $K_1 = Q_2\overline{x}$, *and* $J_2 = \overline{x}$, $K_2 = Q_1 \oplus x$. *The state equations are derived by substituting the above expressions into the characteristic equation for JK flip-flop* ($Q^+ = J\overline{Q} \vee \overline{K}Q$):

$$Q_1^+ = J_1\overline{Q_1} \vee \overline{K_1}Q_1 = \overline{Q_1}Q_2 \vee Q_1\overline{Q_2} \vee Q_1 x$$
$$Q_2^+ = J_2\overline{Q_2} \vee \overline{K_2}Q_2 = \overline{Q_2}\overline{x} \vee Q_1 Q_2 x \vee \overline{Q_1}Q_2\overline{x}$$

The state diagram is derived from the state equations. Note that at the links of the state diagram, only input values are indicated as (`Input`).

Design example: Analysis of JK flip-flop based logic network

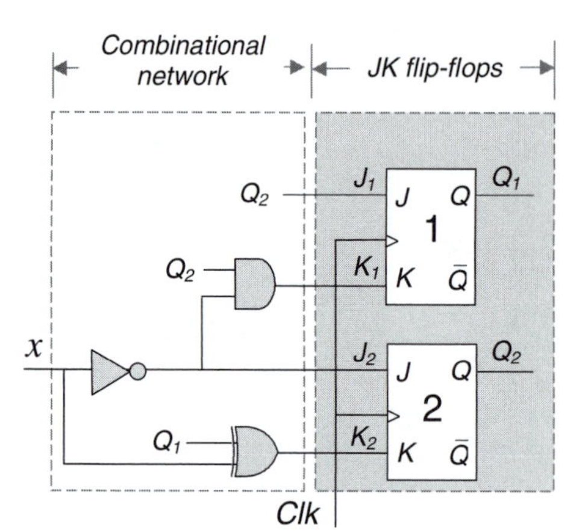

STEP 2
State equations

$$Q_1^+ = J_1\overline{Q_1} \vee \overline{K_1}Q_1$$
$$= \overline{Q_1}Q_2 \vee Q_1\overline{Q_2} \vee Q_1x$$
$$Q_2^+ = J_2\overline{Q_2} \vee \overline{K_2}Q_2$$
$$= \overline{Q_2}\overline{x} \vee Q_1Q_2x \vee \overline{Q_1}Q_2\overline{x}$$

Excitation equations

$$J_1 = Q_2$$
$$K_1 = Q_2\overline{x}$$
$$J_2 = \overline{x}$$
$$K_2 = Q_1 \oplus x$$

STEP 3
State table

Present state Q_1	Q_2	Input x	Next state Q_1^+	Q_2^+	JK flip-flop Inputs J_1	K_1	J_2	K_2
0	0	0	0	1	0	0	1	0
0	0	1	0	0	0	0	0	1
0	1	0	1	1	1	1	1	0
0	1	1	1	0	1	0	0	1
1	0	0	1	1	0	0	1	1
1	0	1	1	0	0	0	0	0
1	1	0	0	0	1	1	1	1
1	1	1	1	1	1	0	0	0

STEP 4
State diagram

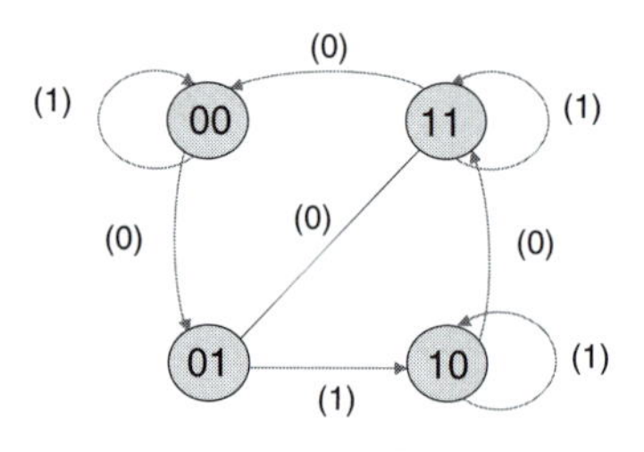

FIGURE 17.5
Analysis of a sequential network with JK flip-flops (Example 17.5).

Practice problem **17.5. (Analysis of a sequential network.)**
Perform analysis of the network given in Figure 17.6a.
Answer is given in "Solutions to practice problems."

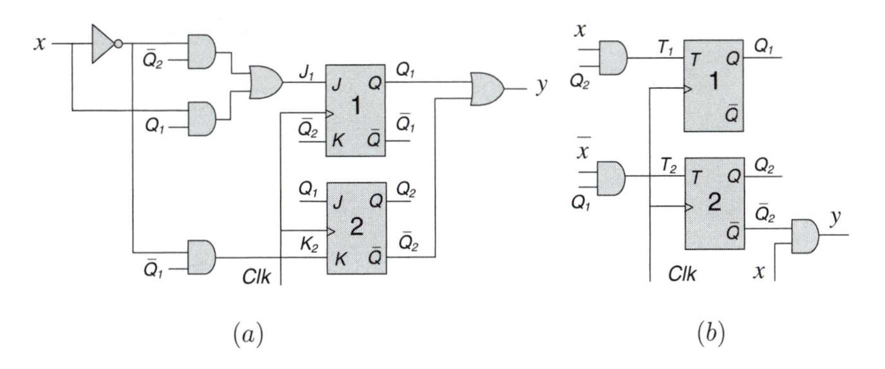

FIGURE 17.6
Sequential networks for Practice problems 17.5 and 17.6.

17.4.3 Analysis of a sequential network with T flip-flops

Analysis of a network with T flip-flops is similar to the analysis of the networks with JK flip-flops.

Deriving the state equations for T flip-flops

Step 1: `Determine the excitation equations.`
Step 2: `Substitute the excitation equation into the characteristic equation of T flip-flops, or, instead, list all combinations for the values of each excitation equation and determine the next-state values using the T flip-flop characteristic table.`

Example 17.6 (Analysis of a sequential network.)
Analysis of a sequential network with T flip-flops requires knowledge of the characteristic equations of the T flip-flops and includes the following main steps (Figure 17.7).
This is a Moore model. The output depends only on the T flip-flop values, that is, the output is a function of the present state only. The state equations and the state diagram, whose links are labeled by input values (`Input`), are shown in Figure 17.7. Note that if the input x is equal to 1, the network behaves as a binary counter with states 00, 01, 10, 11, and back to 00. When $x = 0$, the network remains in the same state.

Practice problem **17.6. (Analysis of a sequential network.)**
Analyze the sequential network shown in Figure 17.6b.
Answer is given in "Solutions to practice problems."

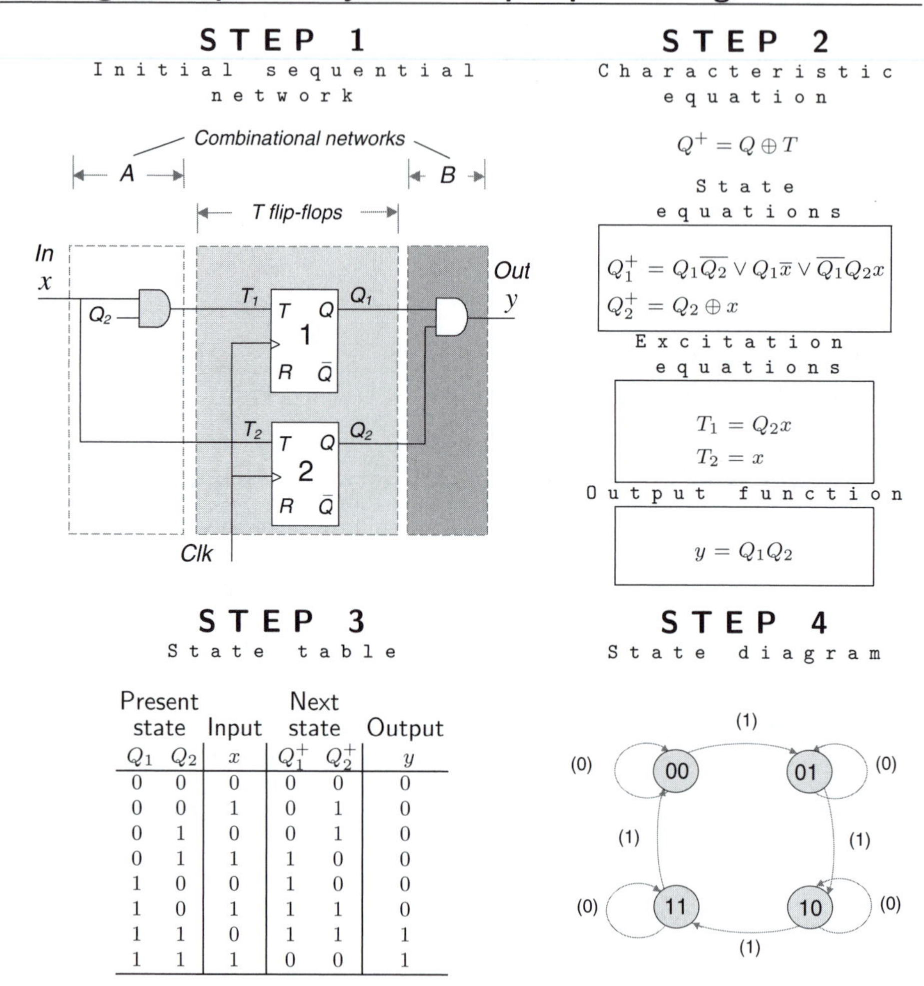

Present state		Input	Next state		Output
Q_1	Q_2	x	Q_1^+	Q_2^+	y
0	0	0	0	0	0
0	0	1	0	1	0
0	1	0	0	1	0
0	1	1	1	0	0
1	0	0	1	0	0
1	0	1	1	1	0
1	1	0	1	1	1
1	1	1	0	0	1

FIGURE 17.7
Analysis of a sequential network with T flip-flops (Example 17.6).

17.5 Techniques for the synthesis of sequential networks

The main steps to design a sequential network are as follows: (a) state behavior specification, (b) state assignment, and (c) combinational function specification and combinational network design:

State behavior	$\xrightarrow{Specification}$	State assignment	$\xrightarrow{Design}$	Logic network

The design procedure for synchronous sequential networks consists of the following main steps:

Design procedure for synchronous sequential networks

Step 1: Derive a state diagram using the word description of the problem.
Step 2: Reduce the number of states, if necessary.
Step 3: Assign binary values to the states and obtain the binary-coded state table.
Step 4: Choose the type of flip-flops and derive excitation equations and the output function.
Step 5: Draw the sequential logic network.

17.5.1 Synthesis of a sequential network using D flip-flops

Given a state diagram or the state table, a network can be synthesized using the above algorithm, while taking into account the specified type of the flip-flop.

Example 17.7 **(Synthesis of a sequential network.)**
Design a sequential network with D flip-flops given the state diagram (Figure 17.9).
The given diagram represents the Moore model. Using this diagram, we derive a state table with four states S_0, S_1, S_2, and S_3. We choose to encode the states as follows: $S_0 = 00$, $S_1 = 01$, $S_2 = 10$, $S_3 = 11$. Note that the equations are derived for the functions Q_1^+, Q_2^+, and y from the table using K-maps for minimization. Note that $Q_1^+ = D_1$ and $Q_2^+ = D_2$.

Practice problem **17.7. (Synthesis of a sequential network.)**
Design a logic network with D flip-flops given the state diagram shown in Figure 17.8a.
Answer is given in "Solutions to practice problems."

17.5.2 Synthesis of sequential networks using JK flip-flops

In design of the networks with JK flip-flops, the state table and equations are derived using the JK flip-flop characteristic equations.

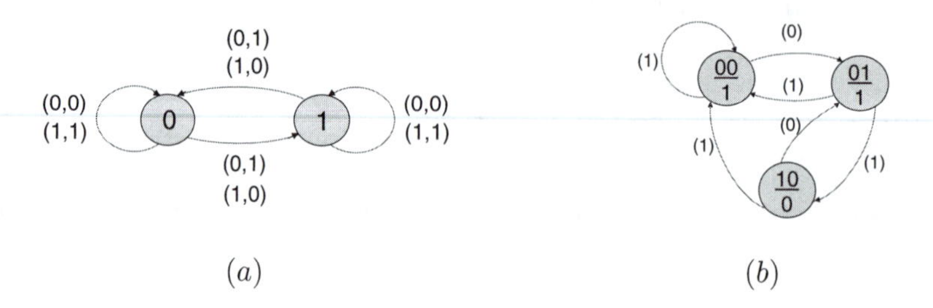

FIGURE 17.8
State diagrams for Practice problems 17.7 and 17.8.

Example 17.8 (Synthesis of a sequential network.)
Design a sequential network using JK flip-flops given the state diagram shown in Figure 17.10. The states $S_0, S_1, S_2,$ *and* S_3 *are encoded as 00,01,10, and 11, respectively. The state table is derived using the characteristic table for JK flip-flops. The table for the next states,* Q_1^+ *and* Q_2^+*, is derived first Using the pairs* $Q_1Q_2^+$ *and* $Q_1^+Q_2^+$*, the corresponding* J_1, K_1, J_2, K_2 *are placed in the last four columns of the state table* J_1, K_1, J_2, K_2 *are determined using the characteristic table.*

Practice problem 17.8. (Synthesis of a sequential network.)
Design a logic network using JK flip-flops given the state diagram in Figure 17.8b.
Answer is given in "Solutions to practice problems."

17.5.3 Synthesis of sequential networks using T flip-flops

Designing a network with T flip-flops is similar to the design of a network with JK flip-flops.

Example 17.9 (Synthesis of a sequential network.)
Design a sequential network using T flip-flops given the state diagram (Figure 17.11).
This is the state diagram of a 3-bit binary counter. State transitions occur during a clock edge, so the counter remains in its present state if no clock is applied. The number of states is 8, so $\log_2 8 = 3$ *T flip-flops are needed.*
The excitation equations are derived from the excitation table of the T flip-flop and from inspection of the state transition of the present state to the next state. Next, the excitation equations are simplified using K-maps, and, finally, the sequential network is drawn.

Design example: Synthesis using D flip-flops

STEP 1 State diagram

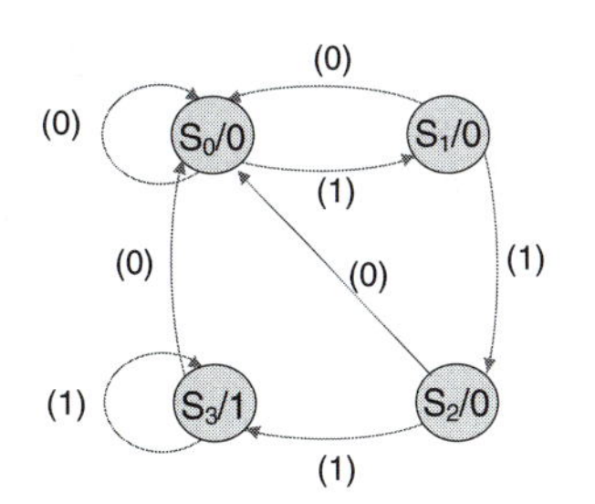

STEP 2 State table

Present state Q_1	Present state Q_2	Input x	Next state Q_1^+	Next state Q_2^+	Output y
0	0	0	0	0	0
0	0	1	0	1	0
0	1	0	0	0	0
0	1	1	1	0	0
1	0	0	0	0	0
1	0	1	1	1	0
1	1	0	0	0	1
1	1	1	1	1	1

STEP 3 Excitation equations

The first D flip-flop

$$D_1 = Q_1^+ = \overline{Q}_1 Q_2 x \vee Q_1 \overline{Q}_2 x \vee Q_1 Q_2 x$$

The second D flip-flop

$$D_2 = Q_2^+ = \overline{Q}_1 \overline{Q}_2 x \vee Q_1 \overline{Q}_2 x \vee Q_1 Q_2 x$$

Output function

$$y = Q_1 Q_2 \overline{x} \vee Q_1 Q_2 x$$

STEP 4 Minimization of excitation equations

$$D_1 = Q_1 x \vee Q_2 x$$

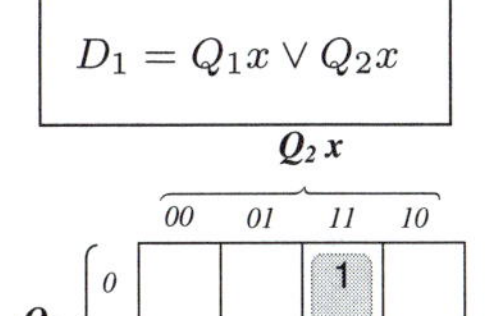

$$D_2 = Q_1 x \vee \overline{Q}_2 x$$

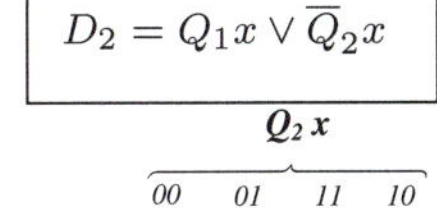

$$y = Q_1 Q_2$$

STEP 5 Synthesized sequential network

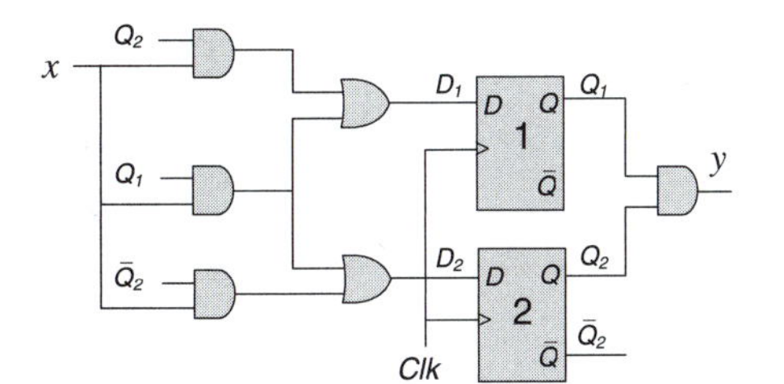

State equation

$$Q_1^+ = \overline{Q}_1 Q_2 x \vee Q_1 \overline{Q}_2 x \vee Q_1 Q_2 x$$

$$Q_2^+ = \overline{Q}_1 \overline{Q}_2 x \vee Q_1 \overline{Q}_2 x \vee Q_1 Q_2 x$$

FIGURE 17.9
Sequential network design using D flip-flops (Example 17.7).

Design example: Synthesis using JK flip-flops

STEP 1
State diagram

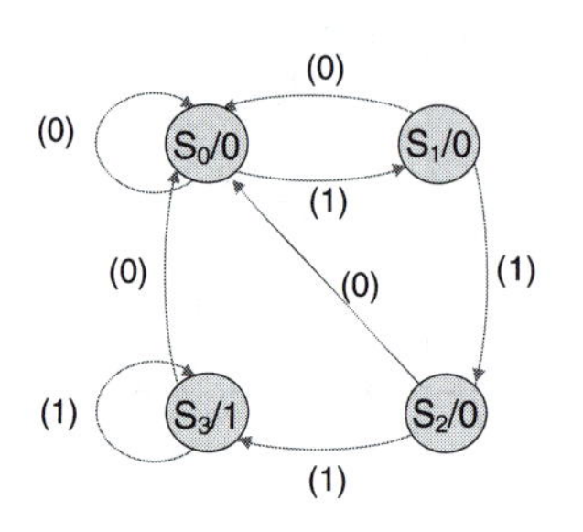

STEP 2
State table

Present state Q_1	Q_2	Input x	Next state Q_1^+	Q_2^+	J_1	K_1	J_2	K_2
0	0	0	0	0	0	×	0	×
0	0	1	0	1	0	×	1	×
0	1	0	1	0	1	×	×	1
0	1	1	0	1	0	×	×	0
1	0	0	1	0	×	0	0	×
1	0	1	1	1	×	0	1	×
1	1	0	1	1	×	0	×	0
1	1	1	0	0	×	1	×	1

(JK flip-flop Inputs: J_1, K_1, J_2, K_2)

STEP 3
Excitation equations

The first JK flip-flop

$$J_1 = Q_1 x \vee Q_2 x$$
$$K_1 = Q_2 x$$

The second JK fl ip-flop

$$J_2 = Q_1 x \vee \overline{Q}_2 x$$
$$K_2 = Q_2 x \vee \overline{Q}_2 \overline{x}$$

Output function

$$y = Q_1 Q_2$$

STEP 4
Minimization of excitation equations

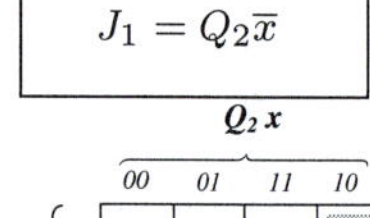

$$J_1 = Q_2 \overline{x}$$

Q_1 \ $Q_2 x$	00	01	11	10
0				1
1	×	×	×	×

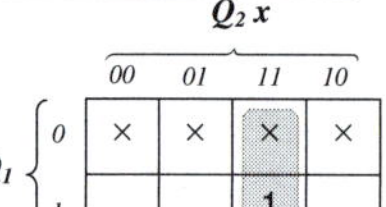

$$K_1 = Q_2 x$$

Q_1 \ $Q_2 x$	00	01	11	10
0	×	×	×	×
1			1	

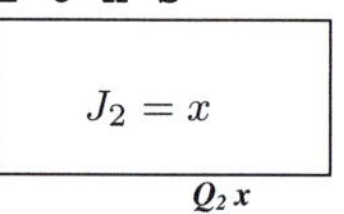

$$J_2 = x$$

Q_1 \ $Q_2 x$	00	01	11	10
0		1	×	×
1		1	×	×

$$K_2 = \overline{Q_1 \oplus x}$$

Q_1 \ $Q_2 x$	00	01	11	10
0	×	×		1
1	×	×	1	

STEP 5
Synthesized sequential network

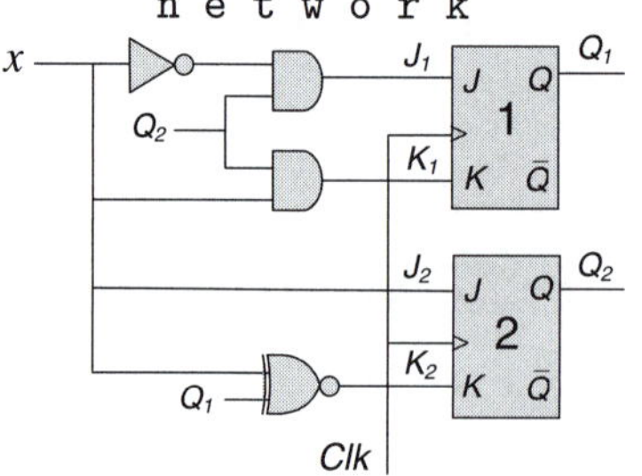

State equation

$$Q_1^+ = Q_1 Q_2 \overline{x} \vee \overline{Q}_1 Q_2 x$$
$$Q_2^+ = \overline{Q}_1 \overline{Q}_2 \overline{x} \vee Q_1 \overline{Q}_2 x \vee Q_2 x$$

FIGURE 17.10
Sequential network using JK flip-flops (Example 17.8).

Practice problem **17.9**. Consider the state diagram of the counter modulo 5 of the number of 1s in the sequence. Design a sequential network based on T flip-flops.
Answer is given in "Solutions to practice problems."

17.6 Redesign

Sometimes a design need a particular type of flip-flop for a specific application, but all you have available is another type. This often happens with an application needing T flip-flops, since these are not generally available in commercial packages.

This can be accomplished through redesigning the flip-flops. This means to re-wire an available type to perform as a targeted flip-flop. We have already seen that a JK flip-flop with its J and K inputs connected to a logic 1 will operate as a T flip-flop. Converting a D flip-flop to T operation is quite similar; the $\overline{Q}$ output is connected back to the D input. To convert a D flip-flop into JK operation, some gates must be added to implements the logical truth that

$$D = J\overline{Q} + \overline{K}Q$$

CMOS flip-flops are typically constructed as D types because of the nature of their internal operation. Commercial CMOS JK flip-flops then add this circuit to the input in order to get JK operation. This approach eliminates the internal latching effect, that occurs with the general JK master-slave flip-flop: The J and K input signals must be present at the time the clock signal falls to logic 0, in order to affect the new output state.

The other approach is the total circuit redesign, that may reduce the general number of logic gates in the network.

The redesign problem is formulated as follows:

The redesign problem

Given a sequential logic network based on a particular type of flip-flops, redesign it using another type of flip-flops.

Example 17.10 (Redesign.) *Consider a network with two D flip-flops (Figure 17.12, Step 1). Redesigning of this network using JK flip-flops is shown at the fifth step in Figure 17.12.*

Design example: Synthesis using T flip-flops

STEP 1
State diagram

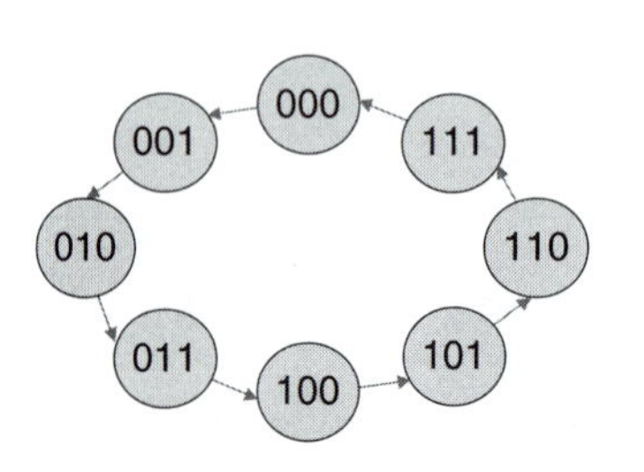

STEP 2
State table

Present state			Next state			T flip-flop Inputs		
Q_3	Q_2	Q_1	Q_3	Q_2	Q_1	T_3	T_2	T_1
0	0	0	0	0	1	0	0	1
0	0	1	0	1	0	0	1	1
0	1	0	0	1	1	0	0	1
0	1	1	1	0	0	1	1	1
1	0	0	1	0	1	0	0	1
1	0	1	1	1	0	0	1	1
1	1	0	1	1	1	0	1	1
1	1	1	0	0	0	1	1	1

STEP 3
Excitation equations

The third T flip-flop

$$T_3 = \overline{Q}_1 Q_2 Q_3 \vee Q_1 Q_1 Q_1$$

The second T flip-flop

$$T_2 = \overline{Q}_1 \overline{Q}_2 Q_3 \vee Q_1 \overline{Q}_1 \overline{Q}_1 \vee Q_1 Q_2 \overline{Q}_3 \vee Q_1 Q_2 Q_3$$

The first T flip-flop

$$T_1 = 1$$

STEP 4
Minimization of excitation equations

$$T_3 = Q_2 Q_1$$

$$T_2 = Q_1$$

$$T_1 = 1$$

Q_1 \ Q_2Q_3	00	01	11	10
0			1	
1			1	

Q_1 \ Q_2Q_3	00	01	11	10
0		1	1	
1		1	1	

Q_1 \ Q_2Q_3	00	01	11	10
0	1	1	1	1
1	1	1	1	1

STEP 5
Synthesized sequential network

State equation

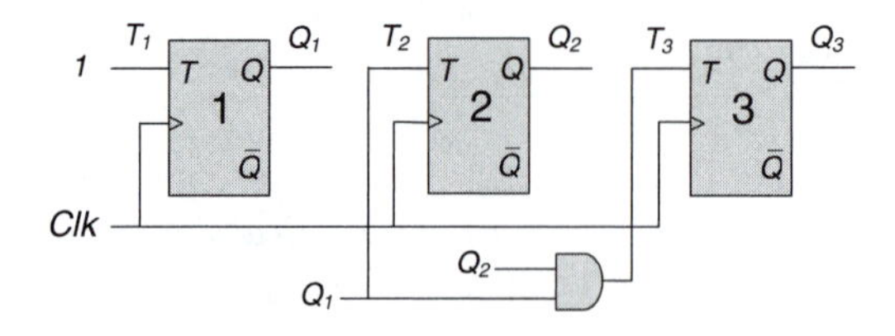

FIGURE 17.11
Sequential network design using T flip-flops (Example 17.9).

Design example: Redesign

STEP 1
Initial sequential network

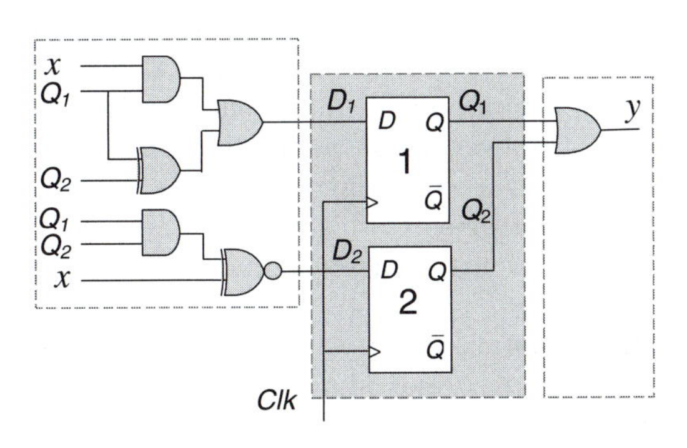

STEP 2
State equations

$$Q_1^+ = D_1 = (Q_1 \oplus Q_2) \vee Q_1 X$$
$$Q_2^+ = D_2 = \overline{x \oplus Q_1 Q_2}$$

Excitation equations

$$D_1 = Q_1$$
$$D_2 = Q_2$$

Output equations

$$y = Q_1 \vee Q_2$$

STEP 3
State table

Present state		Input	Next state		JK flip-flop Inputs			
Q_1	Q_2	x	Q_1^+	Q_2^+	J_1	K_1	J_2	K_2
0	0	0	0	1	0	0	1	0
0	0	1	0	0	0	0	0	1
0	1	0	1	1	1	1	1	0
0	1	1	1	0	1	0	0	1
1	0	0	1	1	0	0	1	1
1	0	1	1	0	0	0	0	0
1	1	0	0	0	1	1	1	1
1	1	1	1	1	1	0	0	0

STEP 4
Excitation equations

$$J_1 = Q_2$$
$$K_1 = Q_2\overline{x}$$
$$J_2 = \overline{x}$$
$$K_2 = Q_1 \oplus x$$

Output equations

$$y = Q_1 \vee Q_2$$

STEP 5
Redesigned network

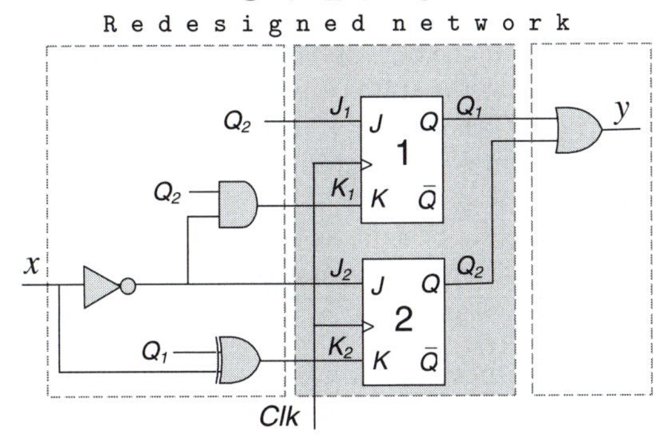

FIGURE 17.12
Redesign of a sequential network with D flip-flops to a network with JK flip-flops (Example 17.10).

17.7 State table reduction and assignment

The behavior of a sequential network is defined by the relationship between its input and output signals. In the case of combinational logic networks, this behavior can be expressed, for example, in the form of a truth table; that is, the list of all input signal possibilities along with output signals. In sequential networks, it is necessary to list all possible input/output sequences that can be applied to the network. However, there is an infinite number of such sequences. Moreover, the sequences can be of arbitrary length. Hence, an explicit listing of states is impractical. To resolve this problem, approaches to the reduction of states have been developed in sequential logic network analysis and synthesis.

The equivalent states in the state diagram can be eliminated, so that the total number of states is reduced. Two states p and q have *equivalent* behavior $p \equiv q$ if for all input combinations their outputs are the same and they change to the same or equivalent next states. If the states p and q do not have *equivalent* behavior ($p \neq q$), then the states are not equivalent.

To check each pair of states for possible equivalence, a tabulated data structure called an *implication table*, or *chart*, is used. The chart has a square for every possible pair of states. A square in column i and row j corresponds to the state pair (i, j). The square in the first column corresponds to the state pairs $(a, b), (a, c)$, etc. The following procedure is used for implication chart construction:

Implication chart construction

Step 1: To fill in the first column of the chart, the first row a of the state table must be compared with each of the other rows. If the output for row a is different than the output for row b, $a \neq b$, place the symbol $\times$ in the (a, b) square of the chart. If the outputs are the same, place the pair a, b in the square, indicating that the states are quasi-equivalent.

Step 2: Repeat step 1 for other rows. If square $a - b$ contains the implied pair, $C - d$, and the square $C - d$ contains an $\times$, then $a \neq b$, and $\times$ should be placed in square $a - b$.

Step 3: If any $\times$ are added in Step 2, repeat Step 2 until no more $\times$ are added. For each square $a - b$, which does not contain an $\times$, assign $a = b$.

Note that the squares above the diagonal are not included in the chart because if $i = j$ and $j = i$, and only one of the state pairs (i, j) and (j, i) is needed. The squares corresponding to pairs (a, a), (b, b), etc., are omitted.

The implication chart is used to determine the equivalent states, and reduce the number of states of the equivalent states are found.

Determining state equivalence

Given: A state table of a sequential network

Step 1: Construct a chart that contains a square for each pair of states.

Step 2: Compare each pair of rows in the state table.

- If the outputs associated with states i and j are different, place an $\times$ in square (i,j) to indicate that $i \neq j$. If the outputs are the same, place the implied pairs in square (i,j).
- If the outputs and next states are the same (or (i,j) only implies itself), place a check (✓) in square (i,j) to indicate that $i \equiv j$.

Remark: If the next states of i and j are m and n for some input x, then (m,n) is an implied pair.

Step 3: Go through the table square-by-square. If square (i,j) contains the implied pair (m,n), and square (m,n) contains an $\times$, then $i \neq j$, and an $\times$ should be placed in square (m,n).

Step 4: If any $\times$'s were added in step 3, repeat step 3 until no more $\times$'s are added.

Step 5: For each square (i,j) that does not contain an $\times$, $i \equiv j$.

Output: Reduced state table

Example 17.11 (Implication chart.) *Given a state table, its implication chart is given in Figure 17.13.*

Present state	Next state $x=0$	Next state $x=1$	Present outputs
a	d	c	0
b	f	h	0
c	e	d	1
d	a	e	0

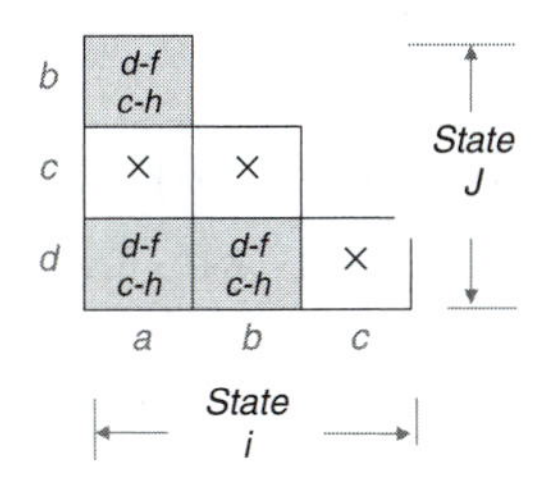

FIGURE 17.13
The simplest implication chart construction (Example 17.11).

Practice problem 17.10. (Implication chart.) Construct the implication chart for a sequential network given its state table (Figure 17.14a). ***Answer*** is given in "Solutions to practice problems."

Present state	Next state $x=0$	$x=1$	Present outputs
a	d	c	0
b	f	h	0
c	e	d	1
d	a	e	0

(a)

Present state	Next state $x=0$	$x=1$	Present outputs
a	d	c	0
b	f	h	0
c	e	d	1
d	a	e	0
e	c	a	1
f	f	b	1
g	b	h	0
h	c	g	1

(b)

FIGURE 17.14
State tables for Practice problems 17.10 (a) and 17.11 (b).

Example 17.12 (State equivalence.) *Consider the state table given in Figure 17.15. Determine equivalent states if any:*
Step 1: *Analyze the structure of a state table. Construct* 7×7 *chart and fill in* $\times$*'s for non-implied pairs.*
Step 2: *Create the other charts using the above algorithm.*
Step 3: *Repeat step 2 until no more* $\times$*'s are added*
Step 4: *Reconstruct the initial state table by replacing state "d" with state "a", state "e" with state "c", and eliminate rows "d" and "e". The state table is reduced to six rows. This means that the state machine with 8 states is reduced to 6 states.*

Practice problem 17.11. (State equivalence.) Use the state table shown in Figure 17.14b, and its implication chart to determine the equivalent states.
Answer is given in "Solutions to practice problems."

Design example: State equivalence

STEP 1

Present state	Next state $x=0$	Next state $x=1$	Present outputs
a	*d*	*c*	0
b	*f*	*h*	0
c	*e*	*d*	1
d	*a*	*e*	0
e	*c*	*a*	1
f	*f*	*b*	1
g	*b*	*h*	0
h	*c*	*g*	1

STEP 2

	a	*b*	*c*	*d*	*e*	*f*	*g*
b	*d-f* *c-h*						
c	×	×					
d	*a-d* *c-e*	*a-f* *e-h*	×				
e	×	×	*a-e* *a-d*	×			
f	×	×	*a-f* *b-d*	×	*c-f* *a-b*		
g	*b-d* *c-h*	*b-f*	×	*a-b* *e-h*	×	×	
h	×	×	*c-e* *a-g*	×	*a-g*	*c-f* *b-g*	×

STEP 3

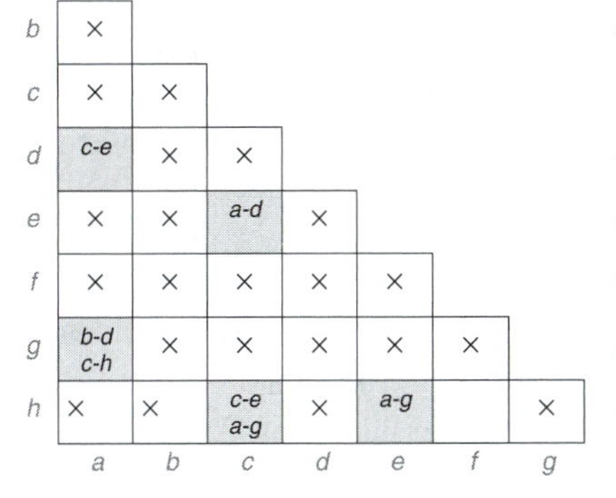

STEP 4

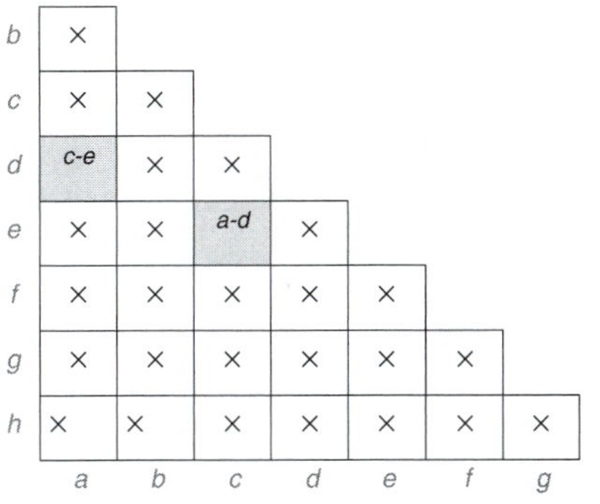

STEP 5

Present state	Next state $x=0$	Next state $x=1$	Present outputs
a	*d*	*c*	0
b	*f*	*h*	0
c	*e*	*d*	1
f	*f*	*b*	1
g	*b*	*h*	0
h	*c*	*g*	1

FIGURE 17.15
Determination of state equivalence using an implication table (Example 17.12).

17.8 Summary of sequential logic network design

This chapter introduces design techniques for sequential logic networks. Sequential logic networks differ from combinational networks in that their output signals depend upon past as well as present input signals. This effect is implemented by using memory that stores data derived from past input signals. The key aspects of this chapter include design techniques, and data structures for design. The main topics of this chapter are summarized as follows:

- ***Synchronous*** sequential logic networks contain clocked flip-flops as memories. These networks use periodic clock signals to synchronize the times at which flip-flops change their states. ***Asynchronous*** logic networks do not use global clocks.
- The operation of a sequential network is described as a sequence of ***state transitions*** controlled by input signals. ***State tables*** define the network next-state functions. ***State diagrams*** represent all internal states and the relationships between these states. The state behavior of flip-flops is described by a state table, a state diagram, or a ***characteristic equation***.
- A model for describing a sequential network is a finite state machine. The types of finite state machines are ***Moore*** and ***Mealy*** machines. In the Moore model, the outputs are functions of the internal states only. In the Mealy model, the outputs are functions of both the internal states and the primary input signals.
- The main goal of sequential logic *analysis* is to analyze the behavior of an existing logic network. The main goal of *synthesis* is to design a sequential logic network.
- Two states have ***equivalent*** behavior if for all input combinations their outputs are the same and they change to the same, or equivalent next states.
- Decision diagrams are used for sequential logic networks for representing their next state and output functions.
- New horizons in sequential logic network design are introduced in the "Further study" section.

17.9 Further study

Historical perspective

1950: In the late 50's, Edward Moore and George Mealy made fundamental contributions to automata theory. G. H. Mealy wrote "A Method for Synthesizing Sequential Circuits" in 1955, in which he introduced a finite state automaton where the outputs are determined by the current state and the input. E. F. Moore wrote "Gedanken-experiments on Sequential Machines", describing a machine as a finite state automaton, where the outputs are determined by the current state alone (and do not depend directly on the input).

1960: Behavioral models for level-sensitive and edge-sensitive logic have been developed, and synchronous and asynchronous circuits models have been studied.

1990: Binary decision diagrams have been used to represent state graphs, as well as to find equivalence between deterministic finite automata.

Advanced topics of sequential logic network design

Topic 1: Performance and lower-power sequential logic network design remains an open problem. Much of the previous work in sequential logic synthesis have targeted minimization of the implementation area. The state encoding was mostly fixed, and the logic optimization attempted to improve the logic implementation under the given state assignment. Further exploration of the design space have proved that a proper state encoding can improve design quality and performance.

Topic 2: Verification of sequential circuits is much more complex that that of combinational networks. It is aggravated by the increasing design complexity, which means that the number of states is too large to be represented by the finite state machine. Decomposition of the design to the smaller state machines is the approach used in practice.

Further reading

"Automatic verification of finite-state concurrent systems using temporal logic specifications" by E. M. Clarke, E. A. Emerson, and A. P. Sistla, *ACM Trans. Prog. Lang. Syst.*, volume 8, number 2, pages 244–263, 1986.

"Energy Aware Computing Through Probabilistic Switching: A Study of Limits" by K. V. Palem, *IEEE Transactions on Computers*, volume 54, number 9, pages 1123–1137, 2005.

"Binary Decision Diagrams" by S. B. Akers, *IEEE Transactions on Computers*, volume C-27, number 6, pages 509–516, 1978.

"Gedanken-experiments on Sequential Machines" by Edward F. Moore, *Automata Studies, Annals of Mathematical Studies* volume 34, pages 129-153, Princeton, New York, Princeton University Press, 1956.

An Introduction to Mathematical Machine Theory by S. Ginsburg, Addison-Wesley, 1962.

Introduction to the Theory of Finite-state Machines by A. Gill, McGraw-Hill, 1962.

Switching and Finite Automata Theory by Z. Kohavi, McGraw-Hill, 1978.

Theory of Finite Automata with an Introduction to Formal Languages by J. Carroll and D. T. Long, Prentice Hall, Englewood Cliffs, 1989.

Symbolic Model Checking by Ken L. McMillan, Kluwer, 1993.

Synthesis of Finite State Machines: Logic Optimization by Tiziano Villa, Kluwer, 1997.

17.10 Solutions to practice problems

<table>
<tr><th>Practice problem</th><th>Solution</th></tr>
<tr><td>17.1.</td><td>Excitation equations: $D_1 = x\overline{Q}_2$, $D_2 = \overline{Q}_2 \vee x$, and $y = xQ_1 \oplus Q_2$. The output function is $y = xQ_1 \oplus Q_2$. This is a Moore type model.</td></tr>
<tr><td>17.2
17.3.</td><td>Practice problem 17.2
<table>
<tr><th colspan="2">Present state</th><th>Input</th><th colspan="2">Next state</th><th>Output</th></tr>
<tr><th>Q_1</th><th>Q_2</th><th>x</th><th>Q_1</th><th>Q_2</th><th>y</th></tr>
<tr><td>0</td><td>0</td><td>0</td><td>0</td><td>1</td><td>0</td></tr>
<tr><td>0</td><td>0</td><td>1</td><td>1</td><td>1</td><td>0</td></tr>
<tr><td>0</td><td>1</td><td>0</td><td>0</td><td>1</td><td>1</td></tr>
<tr><td>0</td><td>1</td><td>1</td><td>0</td><td>1</td><td>1</td></tr>
<tr><td>1</td><td>0</td><td>0</td><td>0</td><td>0</td><td>1</td></tr>
<tr><td>1</td><td>0</td><td>1</td><td>1</td><td>1</td><td>1</td></tr>
<tr><td>1</td><td>1</td><td>0</td><td>0</td><td>0</td><td>0</td></tr>
<tr><td>1</td><td>1</td><td>1</td><td>0</td><td>1</td><td>0</td></tr>
</table>
Practice problem 17.3
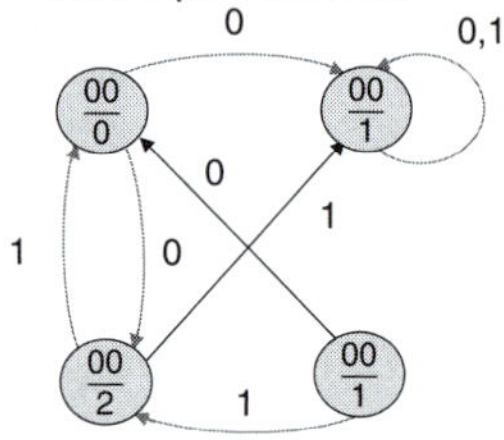
</td></tr>
<tr><td>17.4.</td><td>Excitation equations: $D = x\overline{Q}$. The output function: $y = Q$. State equation: $D = x\overline{Q}$.
<table>
<tr><th>Present state</th><th>Input</th><th>Next state</th><th>Output</th></tr>
<tr><th>Q</th><th>x</th><th>Q^+</th><th>y</th></tr>
<tr><td>0</td><td>0</td><td>0</td><td>0</td></tr>
<tr><td>0</td><td>1</td><td>1</td><td>0</td></tr>
<tr><td>1</td><td>0</td><td>0</td><td>1</td></tr>
<tr><td>1</td><td>1</td><td>0</td><td>1</td></tr>
</table>
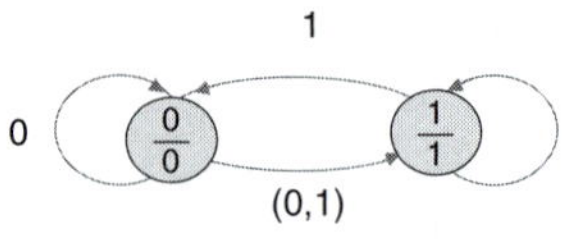
</td></tr>
</table>

Solutions to practice problems (continuation)

Practice problem	Solution

17.5.

Excitation equations: $J_1 = \overline{x}\overline{Q_2} \vee xQ_1$, $K_1 = Q_2$
$J_2 = Q_1$, $K_2 = \overline{x}\overline{Q_1}$. The output function: $y = Q_1 \vee \overline{Q}_2$. This is a Moore model. State equations:

$$Q_1(t \vee 1) = Q_1\overline{K}_1 \vee \overline{Q}_1 J_1 = Q_1\overline{Q}_2 \vee \overline{Q}_1(\overline{x}\overline{Q}_2 \vee xQ_1) = \overline{Q}_2(Q_1 \vee \overline{x})$$
$$Q_2(^+ = Q_2\overline{K}_2 \vee \overline{Q}_2 J_2 = Q_2^+ \vee \overline{Q}_2 Q_1 = Q_2 x \vee Q_1$$

State table and state diagram:

Present state Q_1	Q_2	Input x	Next state Q_1^+	Q_2^+	Output y
0	0	0	1	0	1
0	0	1	0	0	1
0	1	0	0	0	0
0	1	1	0	1	0
1	0	0	1	1	1
1	0	1	0	1	1
1	1	0	0	1	1
1	1	1	0	1	1

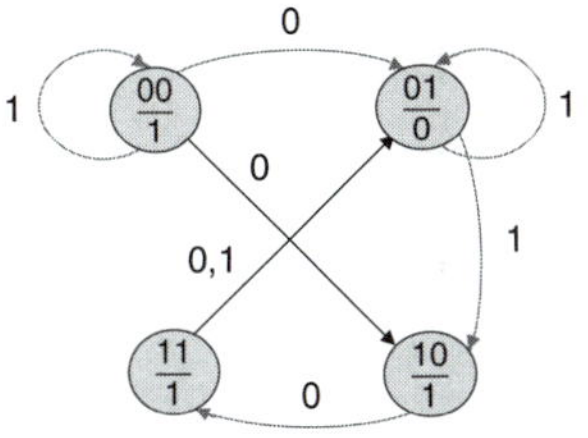

17.6.

Excitation equations: $T_1 = x_1 Q_2$ and $T_2 = x_0 Q_1$. State equations (since $Q^+ = Q \oplus T$): $Q_1^+ = Q_1 \oplus x_1 Q_2$ and $Q_2^+ = Q_2 \oplus x_0 Q_1$. The output function: $y = x_1\overline{Q}_2$. State table and state diagram:

Present state Q_1	Q_2	Input x_0	x_1	Next state Q_1^+	Q_2^+	Output y
0	0	0	0	0	0	0
0	0	0	1	0	0	1
0	0	1	0	0	0	0
0	0	1	1	0	0	1
0	1	0	0	0	1	0
0	1	0	1	1	1	0
0	1	1	0	0	1	0
0	1	1	1	1	1	0
1	0	0	0	1	0	0
1	0	0	1	1	0	1
1	0	1	0	1	1	0
1	0	1	1	1	1	1
1	1	0	0	1	1	0
1	1	0	1	0	1	0
1	1	1	0	1	0	0
1	1	1	1	0	0	0

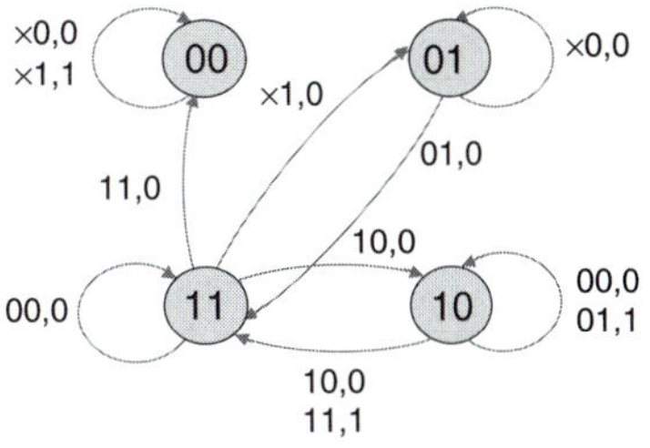

17.7.

Excitation equations: $D = Q^+ = \overline{x}$. The output function: $y = \overline{Q}$.

Present state Q	Input x	Next state Q^+	Output y
0	0	1	1
0	1	0	1
1	0	1	0
1	1	0	0

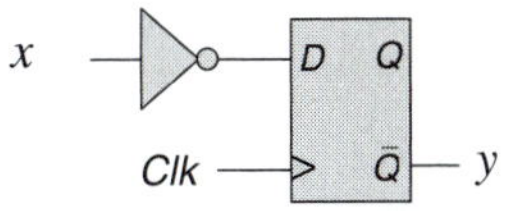

Solutions to practice problems (continuation)

Practice problem	Solution
17.8.	(see below)

Present state Q_1	Q_2	Input x	Next state Q_1^+	Q_2^+	y	JK flip-flop Inputs J_1	K_1	J_2	K_2
0	0	0	0	1	1	0	×	1	×
0	0	1	0	0	1	0	×	0	×
0	1	0	1	0	1	1	×	×	0
0	1	1	0	0	1	0	×	×	0
1	0	0	0	0	0	×	0	0	×
1	0	1	0	1	0	×	0	1	×
1	1	0	×	×	×	×	×	×	×
1	1	1	×	×	×	×	×	×	×

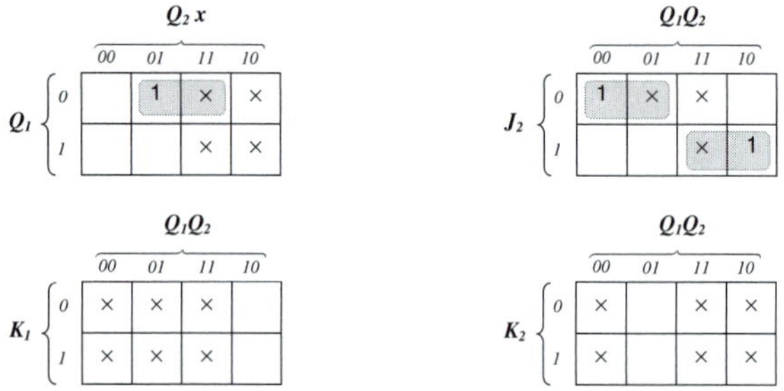

Excitation equations $J_1 = \overline{x}Q_2$, $K_1 = 0$, $J_2 = \overline{x}\overline{Q}_1 \vee xQ_1$, $K_2 = 0$. Output function $y = \overline{Q}_1$.

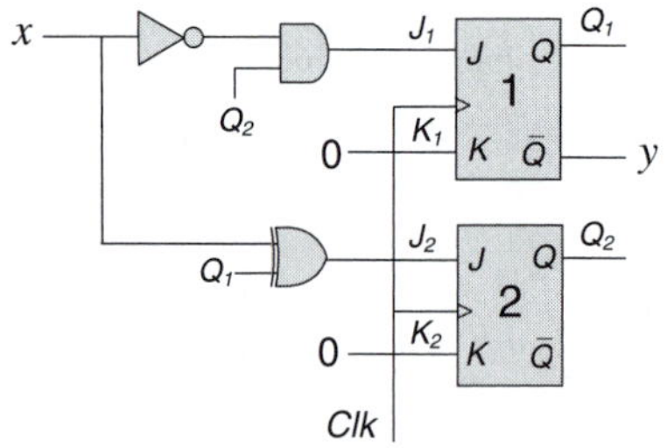

17.9.

It follows from the state diagram that the counter counts 0,1,2,3,4,0,1,2,3,4,0,... Hence, the output set requires three binary variables. Let the values of the output y be coded as a binary (radix-2) number system:

y	y_2	y_1	y_0
0	0	0	0
1	0	0	1
2	0	1	0
3	0	1	1
4	1	0	0

Solutions to practice problems (continuation)

Practice problem	Solution

17.9.

(Continuation) The state table and the T flip-flop input functions are as follows:

Present state Q_3	Q_2	Q_1	Next state Q_3	Q_2	Q_1	T_3	T_2	T_1
0	0	0	0	0	1	0	0	1
0	0	1	0	1	0	0	1	1
0	1	0	0	1	1	0	0	1
0	1	1	1	0	0	1	1	1
1	0	0	1	0	1	0	0	1
1	0	1	1	1	0	0	1	1
1	1	0	1	1	1	0	1	1
1	1	1	0	0	0	1	1	1

$T_2 = xQ_2 \vee xQ_1Q_0$

Q_2Q_1 \ Q_0x	00	01	11	10
00	0	0	0	0
01	0	0	1	0
11	×	×	×	×
10	0	1	×	×

$T_1 = xQ_0$

Q_2Q_1 \ Q_0x	00	01	11	10
00	0	0	1	0
01	0	0	1	0
11	×	×	×	×
10	0	0	×	×

$T_1 = x\overline{Q}_2$

Q_2Q_1 \ Q_0x	00	01	11	10
00	0	1	1	0
01	0	1	1	0
11	×	×	×	×
10	0	0	×	×

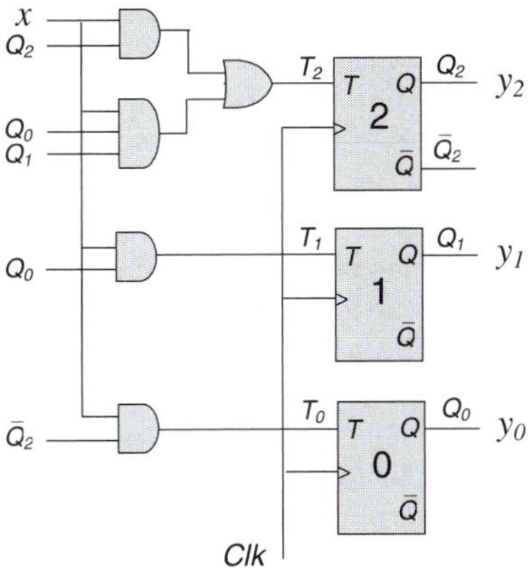

17.10.

	a	b	c	d
b	e-a c-b			
c	×	×		
d	×	×	d-e a-b	
e	e-a	c-b	×	×

Solutions to practice problems (continuation)

Practice problem: 17.11.

Solution

STEP 1

b	e-a c-b			
c	×	×		
d	×	×	d-e a-b	
e	e-a	c-b	×	×
	a	b	c	d

STEP 2

b	×			
c	×	×		
d	×	×	×	
e	e-a	×	×	×
	a	b	c	d

STEP 3

Present state	Next state		Output
	x=0	x=1	
a	a	c	1
b	a	b	1
c	d	a	0
d	a	b	0

17.11 Problems

Problem 17.1 A sequential logic network consists of two D flip-flops, D_1 and D_2, inputs x_1 and x_2, and output y, and specified by the following input equations: $D_1 = \overline{x}_1 Q_1 \vee x_1 x_2$, $D_2 = \overline{x}_1 Q_1 \vee x_1 Q_2$, and $y = x_1 Q_2$. Draw the logic network and derive the state table and state diagram.

Problem 17.2 A sequential logic network consists two JK flip-flops, inputs x_1 and x_2, and output y, and specified by the following equations: $J_1 = xQ_1$, $K_1 = xQ_1$, $J_2 = Q_1 \oplus Q_2$, $K_2 = Q_2$, and $y = \overline{Q}_1$. Draw the logic network and derive the state table and state diagram.

Problem 17.3 A sequential logic network contains two flip-flops, inputs x_1 and x_2, and output y, and specified by the following equations: $T_1 = Q_1 \oplus x$, $D_2 = Q_2$, and $y = xQ_2$. Draw the logic network, and derive the state table and the state diagram.

Problem 17.4 A sequential logic network has two D flip-flops D_1 and D_2, and input x, and is specified by the following equations: $D_1 = (Q_1\overline{Q}_2 \vee \overline{Q}_1 Q_2)x \vee (Q_1 Q_2 \vee \overline{Q}_2\overline{Q}_3)\overline{x}$, $D_2 = Q_1$, and $y = Q_2$. Redesign the circuit on T flip-flops.

Problem 17.5 A sequential logic network contains two D flip-flops D_1 and D_2, and input x, and is specified by the following equations: $D_1 = Q_1 Q_2$, $D_2 = Q_2 \oplus x$, and $y = Q_1 \vee Q_2$. Redesign the network using JK flip-flops.

Problem 17.6 Draw the state diagrams for the sequential logic networks specified by the following state tables:

Present state Q_1	Q_2	Input x	Next state Q_1^+	Q_2^+	Output y
0	0	0	0	0	0
0	0	1	0	1	0
0	1	0	0	0	1
0	1	1	1	1	0
1	0	0	0	0	1
1	0	1	1	0	0
1	1	0	0	0	1
1	1	1	1	0	0

(*a*)

Present state Q_1	Q_2	Input x	Next state Q_1^+	Q_2^+	Output y_1	y_2
0	0	0	0	0	0	1
0	0	1	0	1	0	0
0	1	0	0	0	1	1
0	1	1	1	1	0	0
1	0	0	0	0	1	1
1	0	1	1	0	0	0
1	1	0	0	0	1	1
1	1	1	1	0	0	0

(*b*)

Present state Q_1	Q_2	Input x_1	x_2	Next state Q_1^+	Q_2^+	Output y
0	0	0	0	0	0	0
0	0	0	1	0	1	0
0	0	1	0	0	0	1
0	0	1	1	1	1	0
0	1	0	1	0	0	1
0	1	1	1	1	0	0
0	1	0	0	0	0	1
0	1	1	0	0	0	0
1	0	0	0	1	0	1
1	0	0	1	1	1	1
1	0	1	0	1	0	1
1	0	1	1	1	0	1
1	1	0	1	1	1	1
1	1	0	1	1	1	1
1	1	0	0	0	0	1
1	1	1	0	0	0	1

(*c*)

Present state Q_1	Q_2	Input x	Next state Q_1^+	Q_2^+	Output y_1	y_2	y_3
0	0	0	0	0	0	0	0
0	0	1	0	1	0	0	1
0	1	0	0	0	1	0	0
0	1	1	1	1	0	0	1
1	0	0	0	0	1	1	0
1	0	1	1	0	0	1	1
1	1	0	0	0	1	1	0
1	1	1	1	0	0	1	1

(*d*)

Problem 17.7 Design the sequential logic networks using D, T, and JK flip-flops for the following state diagrams:

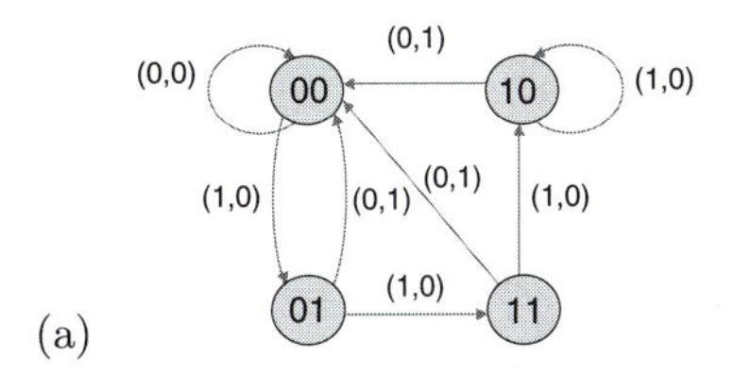

(a)

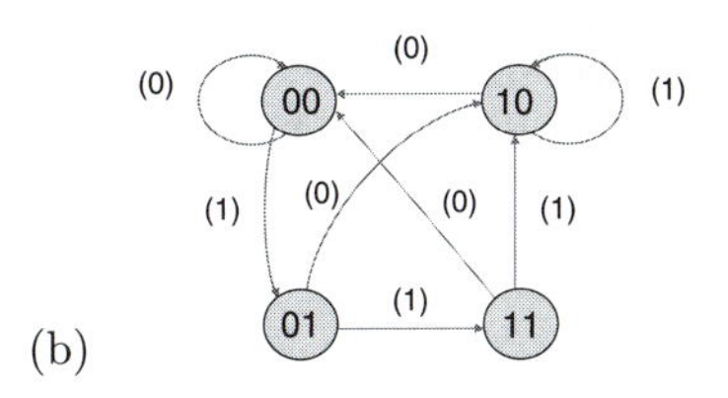

(b)

Problem 17.8 Derive the state equations, excitation equations, state tables, and state diagrams for the following sequential networks:

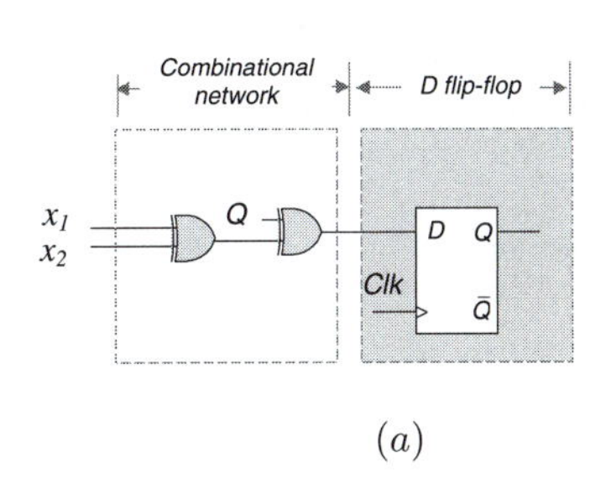

(*a*)

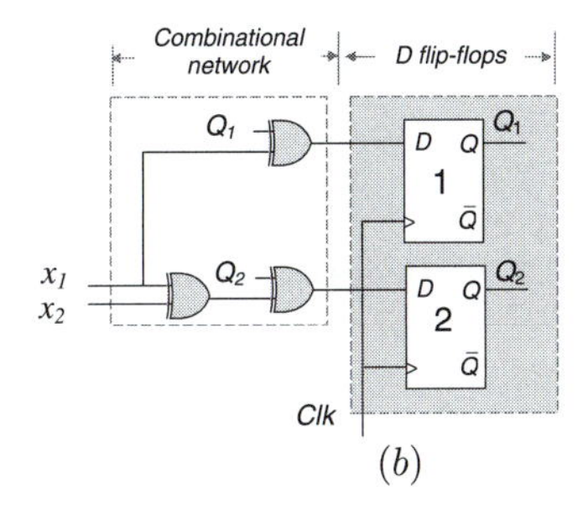

(*b*)

Problem 17.9 Derive the state equations, excitation equations, state tables, and state diagrams for the following sequential networks:

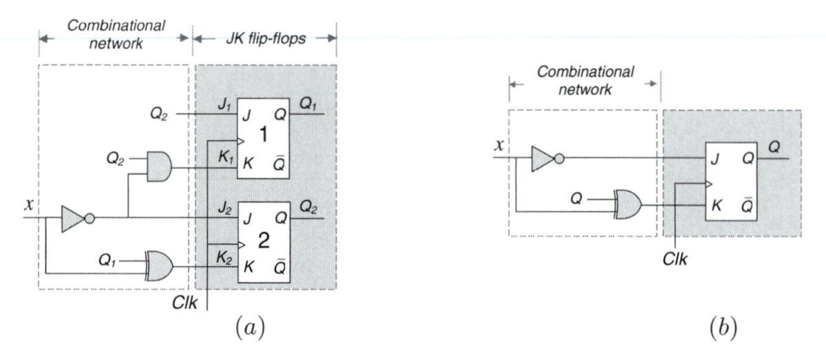

Problem 17.10 Derive the state equations, excitation equations, state tables, and state diagrams for the following sequential networks:

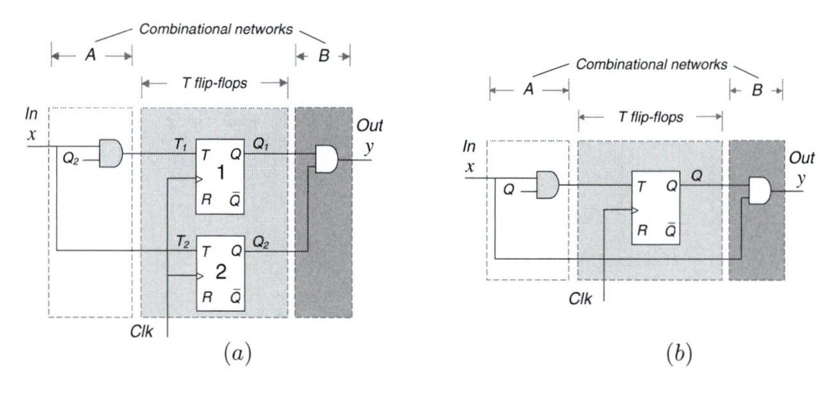

Problem 17.11 Derive the state equations, excitation equations, state tables, and state diagrams for the following sequential networks:

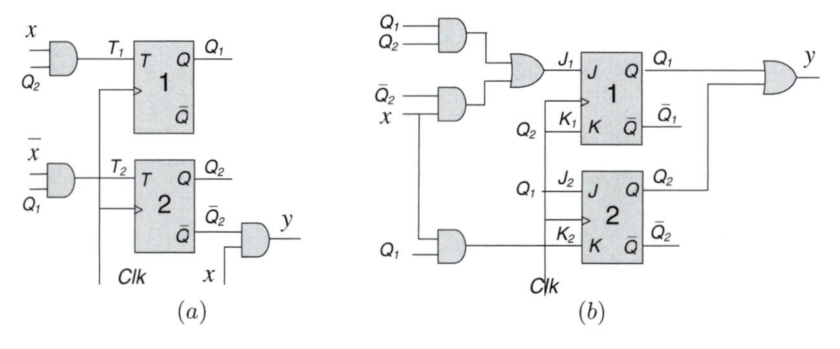

Problem 17.12 Redesign the networks given in Problem 17.11 using D flip-flops.

18

Design for Testability

Fault models

- Fault sources
- Stuck-at model
- Single and multiple faults
- Models based on Boolean differences

Fault detection

- Observability
- Controllability
- Detection using Boolean differences

Improving testability

- Using local transformation
- Using decomposition
- Built-in self-test concept

Advanced topics

- Automatic test-pattern generation (ATPG)
- Testability of decision diagrams
- Fault tolerant systems

18.1 Introduction

Logic networks can fail to perform their assigned functions, if:

- They have been designed incorrectly, or
- Physical defects occur during manufacture or use.

Design errors are addressed by verification of correctness of the network function, and redesign if necessary. The incidence of physical faults can be reduced by quality control in the manufacturing process, by ensuring proper conditions of operation, and by the use of reliable components. Unavoidable failures are addressed by the use of test procedures that determine the presence and location of faults. Testing of the network is performed at the logical and physical levels.

Physical faults have numerous sources: (a) faults that occur during manufacturing, such as missing or defective parts and improper assembly; and (b) faults occurring after manufacturing, such as network failure due to various wear-and-tear phenomena. Logic testing is concerned with *logic models of physical faults*. In this chapter, only *logic testing* is introduced. At the logical level, we are primarily concerned with testing the logic network's behavior, and for this purpose the test patterns and responses are viewed as sequences of 0s and 1s.

The logic network being tested is called the *network under test*. Sequences of the network input assignments are called *test patterns*, or *test vectors*. During testing, a test pattern is used as the stimulus for detecting the presence of a particular fault. A test pattern is used for *controlling* the logic network so that the presence of a fault in a network can be *observed*, for example, on at least one of the network's external pins. Therefore, the testability of a logic network is characterized by:

(*a*) The capability to propagate a test pattern from the primary inputs to a test point of the network, called *controllability*, and

(*b*) The possibility of observing the responses of a correct and faulty logic network at the primary outputs, called *observability*.

The network output is called the *test response*, $R_{network}$ (Figure 18.1). This response is compared to the expected *fault-free response*, $R_{fault\ free}$. If

$$R_{network} = R_{fault\ free}$$

the network passes the test and is assumed to be fault-free. If

$$R_{network} \neq R_{fault\ free}$$

the test fails, and the network is known to be faulty. If the test is appropriately designed, the test response $R_{network}$ contains enough information to locate or isolate faults.

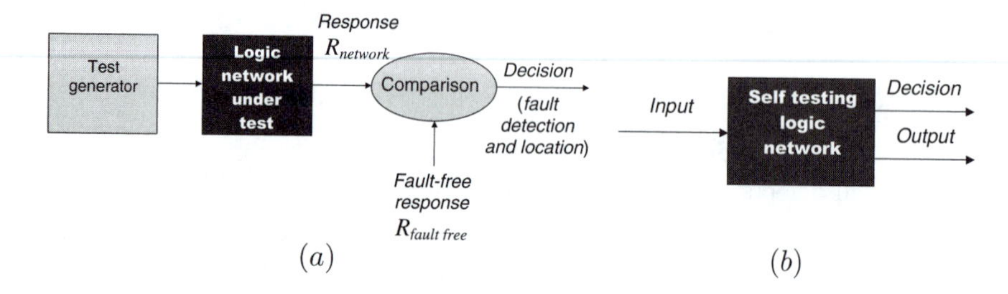

FIGURE 18.1
Testing techniques based on external (a) and internal (b) test resources.

Testing techniques can be classified with respect to the way they exploit test resources:

- Techniques in which *external* test resources are used (Figure 18.1a), and
- Techniques in which *internal* test resources are used; these techniques are called *self-testing* techniques (Figure 18.1b).

The cost of testing grows with increasing density of components and the limited number of input and output pins. For many digital devices the cost of testing is higher, than the cost of design and manufacture.

Consider a combinational logic network. The correctness of the network can be validated by exhaustive application of all possible input patterns and observation of the responses. For an n-input network, this requires the application of 2^n patterns. The use of all 2^n input patterns is called *exhaustive testing.*

> **Example 18.1 (Exhaustive testing.)** *For a combinational logic network with $n = 20$ inputs, more than 1 million patterns are needed. If the application and observation of a single pattern takes 1 μs, the complete testing of the network requires 1 s.*

Because the number of test patterns of a combinational network with n inputs, 2^n, grows exponentially, exhaustive testing is limited in practice. The testability of a logic network can be improved using various techniques, such as local transformations and decomposition. Efficient testing is accomplished by using automatic test pattern generation (ATPG). ATPG techniques can be divided into *deterministic* and *random* techniques.

18.2 Fault models

Logic testing recognizes gates, latches, and flip-flops as the primitive components, and 0s and 1s as the primitive signals. At this level of complexity, *logic models* of faults are required. These models allow faulty behavior to be defined in terms of the logic components and signals:

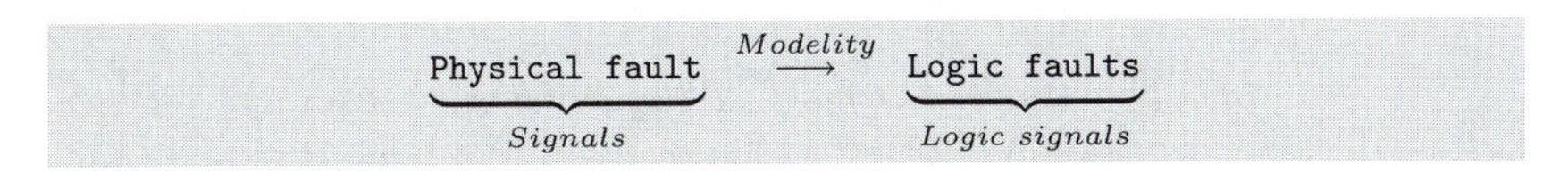

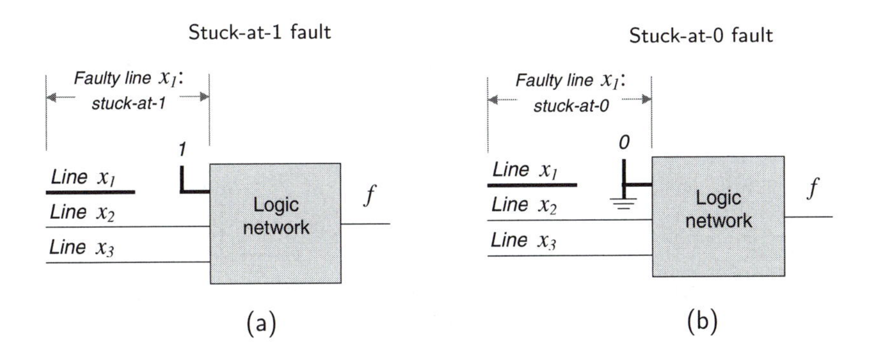

FIGURE 18.2
Single stuck-line models of faults: stuck-at-1 (a) and stuck-at-0 (b) faults.

18.2.1 The single stuck-at model

The *single stuck-at* model is used for gate level testing. The single stuck-at model defines two types of faults for each line x in a logic network (Figure 18.2):

Two types of faults

- The *stuck-at-0* fault, which causes the line x to be permanently stuck at the logic value 0, and
- The *stuck-at-1* fault, which causes the line x to be permanently stuck at the logic value 1.

In this model, only one line is assumed to be faulty at a time (the single-fault assumption), and all gates and other components operate correctly. Presumably, the *stuck-at* model does not cover the complete range of faults that can occur in a complex integrated circuit. However, it has been established that a large number of specific faults, such as *stuck-at-open* and *stuck-at-short*, are covered by the s-a-0 and s-a-1 model. Even though the s-a-0 and s-a-1 models are not perfect, the ease of use and relatively large coverage of the fault space have made it the standard model.

Example 18.2 (Single stuck-at model.) *Figure 18.3 shows the effect of stuck-at faults on the line x_1, which is an input of the two-input AND gate:*

(a) The stuck-at-0 fault at the input x_1 causes the output of the AND gate to change from $f = x_1x_2$ to $f = 0$, and

(b) The stuck-at-1 fault at the input x_1 causes the output of the AND gate to change from $f = x_1x_2$ to $f = x_2$.

Note that the other lines are assumed to be unaffected by faults on the line x_1.

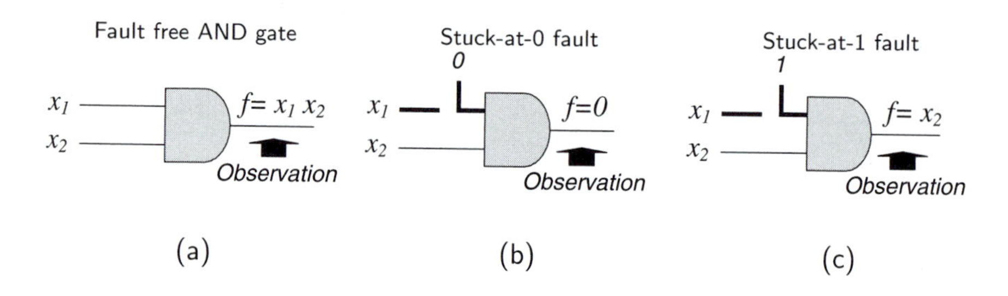

FIGURE 18.3
A fault-free AND gate (a), stuck-at-0 fault (b), and stuck-at-1 fault (c) (Example 18.2).

18.2.2 Fault coverage

The quality of a test is measured by the *fault coverage*:

$$\texttt{FAULT COVERAGE} = \frac{\texttt{TOTAL NUMBER OF FAULTS DETECTED}}{2 \times \texttt{NUMBER OF NODES IN THE NETWORK}}$$

In this equation, the total number of faults detected by the test sequence is divided by two times the number of nodes in the network because each node can give rise to s-a-0 and s-a-1 faults. The coverage number obtained is defined by the fault model employed. In the s-a-0 and s-a-1 model, some of the bridge and short faults are not covered and may not appear in the coverage statistics.

Example 18.3 (Fault-coverage.) *Figure 18.4 shows the potential stuck-at faults in the two-input NAND gate. Each row of the table corresponds to a test pattern and each column corresponds to a potential fault. An* $\times$ *is placed in the table if and only if the test pattern detects a fault; that is, the test pattern covers the (specific) fault. For example, the test pattern 00 detects only a s-a-0 fault in the line* f*; the test pattern 11 covers s-a-0 faults in the lines* x_1 *and* x_2*, as well as the fault s-a-1 in the line* f*. For test patterns 00 and 10, fault coverage is* $\frac{1}{2\times 1} = \frac{1}{2}$.

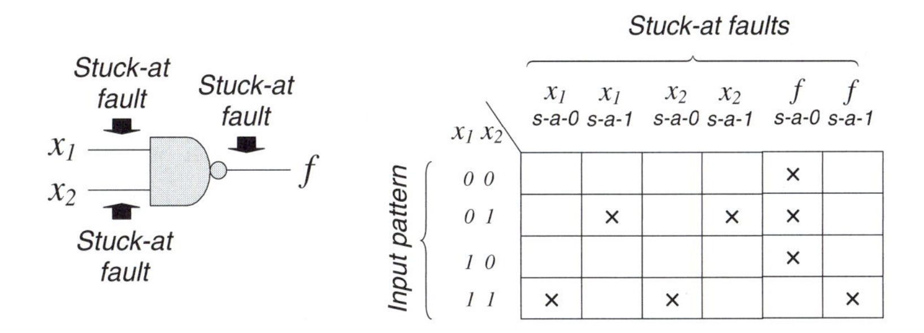

$x_1 x_2$	x_1 s-a-0	x_1 s-a-1	x_2 s-a-0	x_2 s-a-1	f s-a-0	f s-a-1
0 0					×	
0 1		×		×	×	
1 0					×	
1 1	×		×			×

FIGURE 18.4
The two-input NAND gate and its fault coverage table (Example 18.3).

Practice problem **18.1. (Fault-coverage.)** Determine the fault coverage for the NAND gate for test patterns 01 and 11.
Answer is given in "Solutions to practice problems."

Stuck-at faults can be detected under appropriate controllability and observability conditions.

18.3 Controllability and observability

To generate a test to detect a fault in a combinational network, the following steps are required:

- Path sensitization: the inputs must be specified so as to sensitize, or generate the appropriate value (0 for stuck-at-1 and 1 for stuck-at-0 faults) at the site of the fault.

▶ Error propagation: a path must be selected from the faulty line to an output along with remaining signal values, to propagate the fault signal to the output, which must be observable.

The path sensitization is provided by network controllability, and error propagation is ensured by network observability.

Controllability and observability

Controllability is the ability of a network to have a test pattern such that a faulty logic value is produced at the fault line. A node is *easily controllable* if it can be brought to any condition with only a single input vector. A node (or a network) with *low controllability* needs a long sequence of vectors to be brought to a desired state.

Observability is an ability of a logic network to provide the conditions (usually specified by input patterns) for observation of the specified network output. A node with a *high observability* can be monitored directly on the output pins. A node with a *low observability* needs a number of cycles before its state appears on the outputs.

Controllability is provided by setting the values at the node or line under testing to the value opposite to the assumed fault values. For example, if we are testing a line for the stuck-at-0 fault, the line must be set to 1.

Example 18.4 (Conditions for controllability.) *Consider the fault s-a-1 in line x_1 of the AND gate (Table 18.1). To generate the opposite value (0) at the output of the AND gate, conditions $x_1 = x_2 = 1$ must be held. These conditions mean that the fault s-a-1 in line x_1 is propagated to the output of the AND gate. The corresponding conditions for OR and EXOR gates can be specified by analogy.*

Practice problem 18.2. (Conditions for controllability.) Specify conditions for propagating s-a-0 and s-a-1 faults from the line x_1 to the outputs of NAND, NOR, and NOT gates.

Answer: NAND gate: $x_1 = x_2 = 1$; NOR gate: $x_1 = x_2 = 0$; NOT gate: none.

Observability in logic networks refers to the possibility of observing the outputs of the logic network via external outputs (observation points):

Observability problem

Given: `The logic network output`

Find: `Conditions for observability`

Step 1: `Derive a logic equation for the input variables using the observable values of the output`

Step 2: `Generate inputs from the logic equation`

TABLE 18.1
Conditions for error propagation through logic gates (Example 18.4).

Design example: Conditions for controllability

Gate	Fault s-a-0	Fault s-a-1	Condition for propagation
Fault, Propagated fault; x_1, x_2, f; $f = x_1 \wedge x_2$	✓		$x_1 = x_2 = 1$
		✓	$x_1 = x_2 = 1$
x_1, x_2, f; $f = x_1 \vee x_2$	✓		$x_1 = x_2 = 0$
		✓	$x_1 = x_2 = 10$
x_1, x_2, f; $f = x_1 \oplus x_2$	✓		$x_1 = x_2 = 1$
		✓	$x_1 = x_2 = 1$

Example 18.5 (Gate observability.) *Conditions for observing the output $f = 1$ for the two-input AND gate can be found as shown in Figure 18.5. The solution to the equation $x_1x_2 = 1$ is $x_1 = x_2 = 1$. That is, in order to observe a logic "1" at the output f, both input signals must be equal to 1. The observed value of the fault-free output must be 1. The alternate value (0) is an indication of a fault at the output line.*

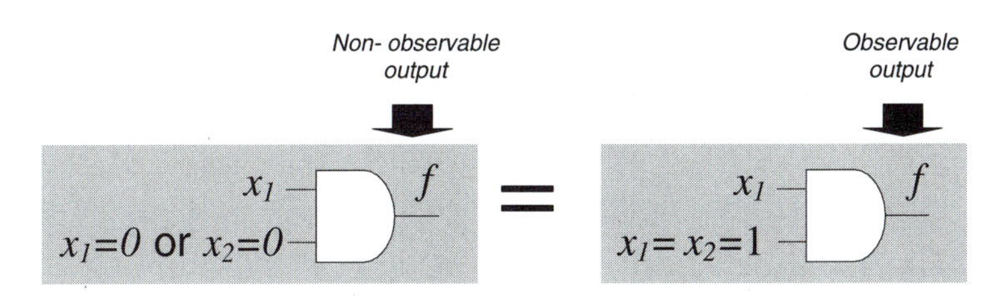

FIGURE 18.5
Observability for the two-input AND gate is defined by the logic equation $x_1x_2 = 1$ (Example 18.5).

Example 18.6 (Observability.) *Conditions for observing the output $f = 1$ for the two-input EXOR gate can be found using the following manipulations (Figure 18.6):*

$$f = x_1 \oplus x_2 = 1 \ \Rightarrow x_1 = x_2 \oplus 1 \text{ and } x_1 = \overline{x}_2$$

That is, to observe a logic "1" at the output, the input signals must satisfy the logic equation $\boxed{x_1 = \overline{x}_2.}$ *The solutions to this equation, $x_1 = 0$, $x_2 = 1$, and $x_1 = 1$, $x_2 = 0$, are the desired input signals. The observed value of the output must be 1. The alternate value (0) is an indication of a fault at one of the inputs.*

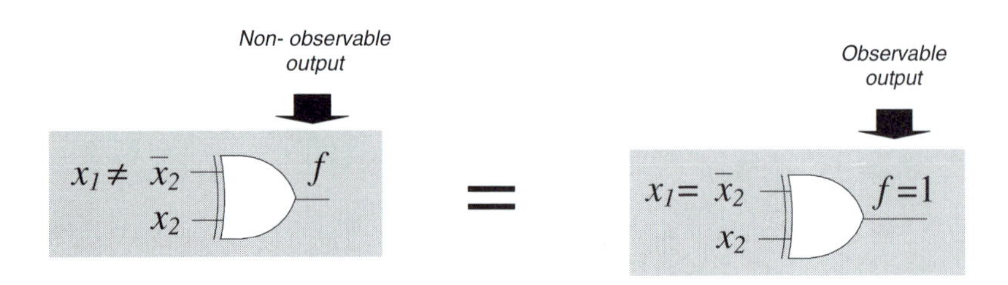

FIGURE 18.6
Observability for the two-input EXOR gate is defined by the logic equation $x_1 = \overline{x}_2$ (Example 18.6).

Practice problem 18.3. (Gate observability.) Find the conditions for observability of the output $f = 1$ for the three-input NOR gate.
Answer is given in "Solutions to practice problems."

Example 18.7 (Network observability.) *In Table 18.2, the observability conditions for some logic networks are given. For example, for the first network, a logic value of 1 can be observed at the output only if $x_1 = x_2 = 0$. For the second two-output network, the logic equation $x_1 x_2 = 1$ implies that the values 1 and 0 are generated at the outputs f_1 and f_2, respectively, only if $x_1 = x_2 = 1$.*

Practice problem 18.4. (Network observability.) Find the observability conditions for the two-output network shown in the second row of Table 18.2 given $f_1 = f_2 = 1$.
Answer is given in "Solutions to practice problems."

TABLE 18.2
Observability conditions in logic networks (Example 18.7).

Design example: Observability in logic networks

Network	Description	Observability
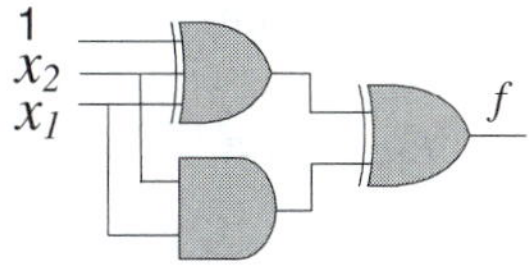	$f = 1 \oplus x_1 \oplus x_2 \oplus x_1x_2$	$1 \oplus x_1 \oplus x_2 \oplus x_1x_2 = 1$ $x_1 \oplus x_2 \oplus x_1x_2 = 1 \oplus 1 = 0$ $x_1 \oplus x_2 = x_1x_2$ $x_1 = x_2 = 0$
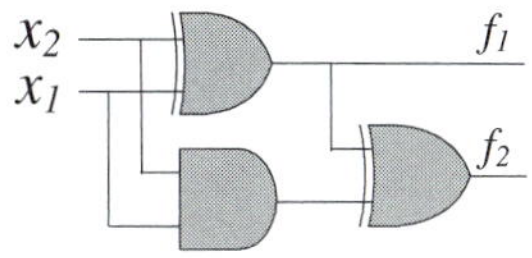	$\begin{cases} f_1 = x_1 \oplus x_2 \\ f_2 = x_1 \oplus x_2 \oplus x_1x_2 \end{cases}$	$\begin{cases} f_1 = x_1 \oplus x_2 = 1 \\ f_2 = x_1 \oplus x_2 \oplus x_1x_2 = 0 \end{cases}$ $f_1 \oplus x_1x_2 = 0$ $1 \oplus x_1x_2 = 0$ $x_1x_2 = 1$ $x_1 = x_2 = 1$
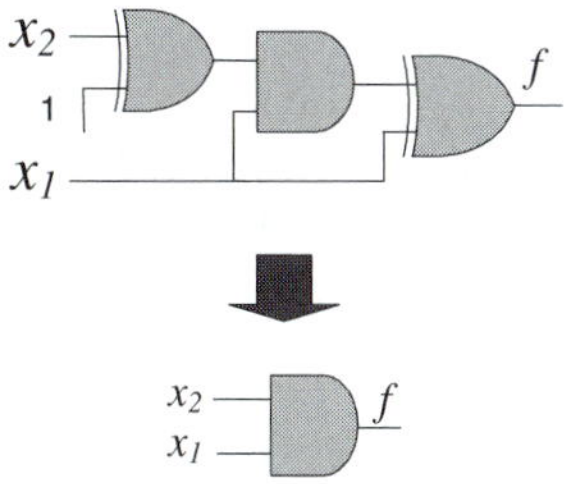	$f = x_1(x_2 \oplus 1) \oplus x_1$ After local transformations: $f = x_1x_2$	$x_1x_2 = 1$ $x_1 = x_2 = 1$

The controllability and observability of a logic network can be improved using local transformation, decomposition, and additional gates. Combinational logic networks fall into the class of easy observable and controllable networks, since any node can be controlled and observed in a single cycle.

18.3.1 Observability and Boolean differences

A signal in a binary system is represented by two logical levels, 0 and 1. Let us formulate the task as the detection of a change in this signal. Let f be the output of a fault-free logic network, and let f_{fault} be the output of the faulty logic network. The *difference* function, D, which distinguishes the two functions, is defined using the EXOR operation:

$$D = f \oplus f_{fault} \tag{18.1}$$

Given a type of fault, the assignment of variables for the function f_{fault} for which $f \oplus f_{fault} = 1$ is called a *test pattern* input for this fault. If not change itself, but direction of change is the matter as issue, then two logical values 0 and 1 can characterize the behavior of the logic signal in terms of change, where 0 means no change in a signal, and 1 indicates that one of two possible changes has occurred $0 \rightarrow 1$ or $1 \rightarrow 0$.

Let a fault at the i-th input of a Boolean function cause a change from the value x_i to the opposite value, $\overline{x}_i$. This causes the circuit output to be changed from the initial value $f(x_i)$ to $f(\overline{x}_i)$. Note that the values $f(x)$ and $f(\overline{x}_i)$ are not necessarily different. To recognize whether or not they are different is to find $f(x_i) \oplus f(\overline{x}_i)$.

The latter is the Boolean difference of a Boolean function f of n variables with respect to a variable x_i defined by the equation

$$\frac{\partial f}{\partial x_i} = \underbrace{f(x_1, \ldots, x_i, \ldots, x_n)}_{Initial\ function} \oplus \underbrace{f(x_1, \ldots, \overline{x}_i, \ldots, x_n)}_{Function\ with\ complemented\ x_i}$$

If $\frac{\partial f}{\partial x_i} = 0$, f does not depend on x_i. Details on Boolean differences are given in Chapter 5.

Example 18.8 (Boolean difference.) *The Boolean difference specifies the observability of the fault at the input x_1. The Boolean differences with respect to both input variables x_1 and x_2 specify the conditions for observing changes in the output function $f = x_1 \vee x_2$ with respect to changes in the inputs x_1 and x_2. (Figure 18.7).*

Practice problem 18.5. (Boolean difference and observabi-lity.)

Given a logic network with a faulty line x_1 (Figure 18.8), find the observability conditions using Boolean differences.

Answer is given in "Solutions to practice problems."

Example 18.9 (Fault detection using Boolean differences.) *Consider the logic network implementing the Boolean function $x_1x_2 \vee x_3$. Figure 18.9 shows how to specify controllability and observability conditions for stuck-at faults using Boolean difference with respect to the variable x_3.*

18.3.2 Enhancing observability and controllability

A *test point* is an input or output signal for controlling or observing intermediate signals in a logic network. Suitable locations for the test point in

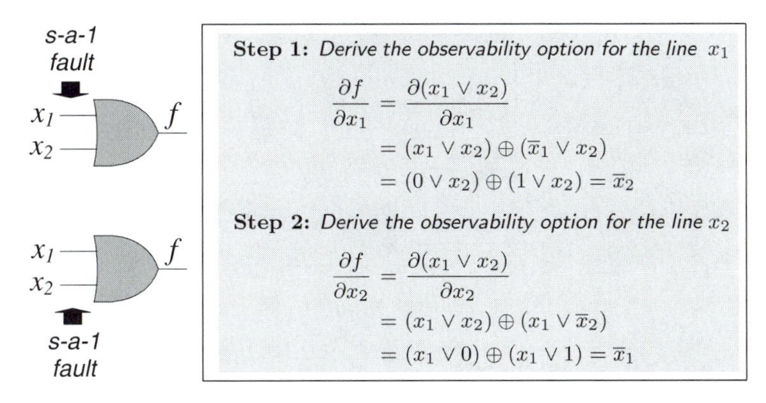

Design example:
Specification of observability conditions
s-a-1
fault
x_1
x_2
f
s-a-1
fault
Step 1: Derive the observability option for the line x_1
$\frac{\partial f}{\partial x_1} = \frac{\partial(x_1 \vee x_2)}{\partial x_1}$
$= (x_1 \vee x_2) \oplus (\overline{x}_1 \vee x_2)$
$= (0 \vee x_2) \oplus (1 \vee x_2) = \overline{x}_2$
Step 2: Derive the observability option for the line x_2
$\frac{\partial f}{\partial x_2} = \frac{\partial(x_1 \vee x_2)}{\partial x_2}$
$= (x_1 \vee x_2) \oplus (x_1 \vee \overline{x}_2)$
$= (x_1 \vee 0) \oplus (x_1 \vee 1) = \overline{x}_1$

FIGURE 18.7
Observability options for a two-input OR gate in terms of Boolean differences (Example 18.8).

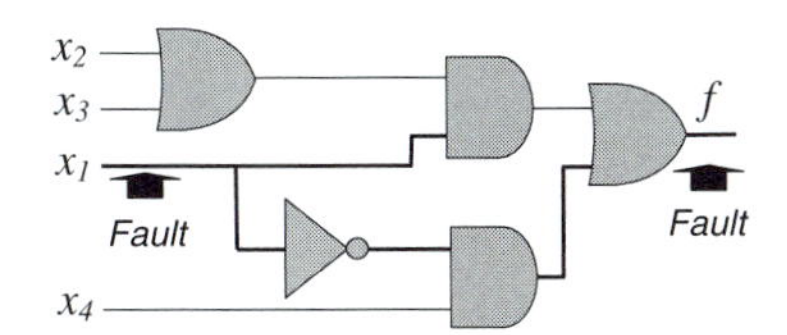

FIGURE 18.8
Faulty logic network for Practice problem 18.5.

a logic network are defined as places where test points involve: (a) signals that are hard to control or observe, or (b) signals with values that are particularly useful during test application. In practice, the number of test points is limited. An arbitrary line x in a logic network can be made controllable with respect to either 0 or 1.

Example 18.10 (Observability.) *Given the non-observable line x (Figure 18.10), this line can be made observable as follows. Suppose two external control signals C_1 and C_2 force a line to 0 and 1, respectively. The task is to replace the line x by a logic network with inputs x, C_0, and C_1 that produces an observable x denoted as* OBSERVABLE x. *The Boolean equation for the observable line x is as follows (Figure 18.10):*

$$\text{OBSERVABLE } x = \overline{C}_0(x \vee C_1) \tag{18.2}$$

Design example:
Specification of observability conditions

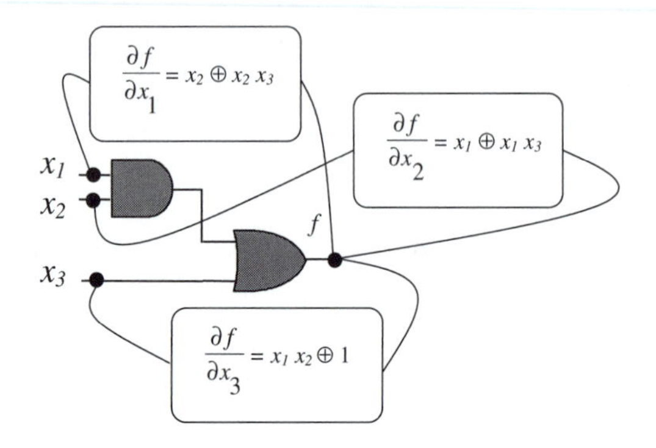

Step 1: *Observability condition for the line* x_1: $\frac{\partial f}{\partial x_1} = x_2\overline{x}_3$.
Step 2: *Specify the observability conditions for detecting the changes at* x_1 *by solving the logic equation* $x_2\overline{x}_3 = 1$: $x_2x_3 = \{10\}$. *When* $x_2x_3 = \{10\}$, *a change at* x_1 *cause a change of either* $0 \rightarrow 1$ *or* $1 \rightarrow 0$ *at* f.
Step 3: *Observability condition for the line* x_2: $\frac{\partial f}{\partial x_1} = x_1\overline{x}_3$.
Step 4: *Specifies the observability conditions for detecting the changes at* x_2 *by solving the logic equation* $x_2\overline{x}_3 = 1$. *When* $x_2x_3 = \{10\}$, *a change of* x_1 *is observable at the output*
Step 5: *Observability condition for line* x_3: $\frac{\partial f}{\partial x_1} = \overline{x_1x_2}$.
Step 6: *Specify the observability conditions for detecting changes at* x_3 *by solving the logic equation* $\overline{x_1x_2} = 1$. *When* $x_1x_2 = \{00, 01, 10\}$, *any change in* x_3 *causes changes in the output* f.

FIGURE 18.9
Specification of observability conditions using Boolean differences and logic equations (Example 18.9).

The implementation of Equation 18.2 is called *testability enhancing logic* for the test point.

Practice problem **18.6.** **(Observability.)** Consider the logic network consisting of a combinational module and memory (Figure 18.11). The memory is accessible only through the combinational module, since writing and reading operations are the functions of a combinational module. In this network, the memory is non-observable directly from the Input/Output bus. Make this memory observable from the Input/Output bus using additional combinational modules.
Answer is given in "Solutions to practice problems."

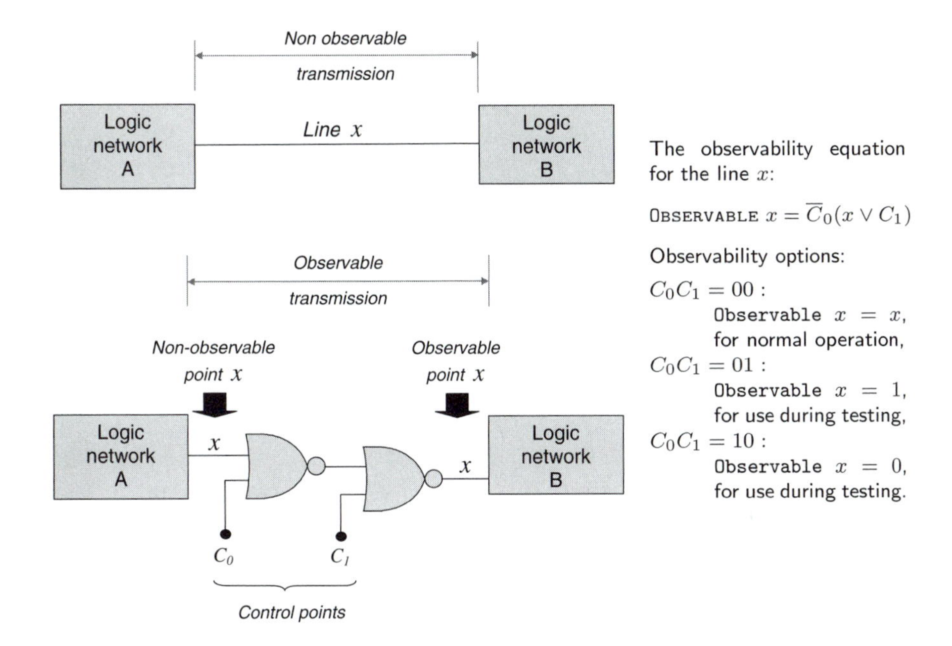

FIGURE 18.10
Non-observable and observable transmission line x between two logic networks A and B (Example 18.10).

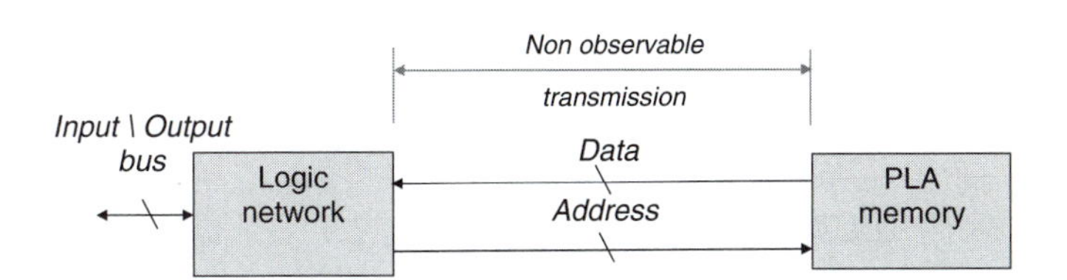

FIGURE 18.11
Non-observable transmission between logic network and memory for Practice problem 18.6.

18.3.3 Detection of stuck-at faults

Consider the output f of a logic network with the stuck-at-0 fault at the input line x_i. This network's output is then described by the Boolean function

$$f_{fault} = f(x_1, \ldots, x_{i-1}, 0, x_{i+1} \ldots, x_n) = f_i(0)$$

The difference (Equation 18.1) that detects a fault in the network caused by the stuck-at-0 fault at the input line x_i is calculated as follows:

$$\begin{aligned} f \oplus f_{fault} &= f \oplus f_i(0) \\ &= \overline{x}_i f_i(0) \oplus x_i f_i(1) \oplus f_i(0) = \overline{x}_i f_i(0) \oplus x_i f_i(1) \oplus (\overline{x}_i \oplus 1) f_i(0) \\ &= x_i f_i(1) \oplus x_i f_i(0) = x_i (f_i(0) \oplus f_i(1)) = x_i \frac{\partial f}{\partial x_i} \end{aligned}$$

Therefore, all test patterns for the stuck-at-0 fault at the input line x_i can be found as solutions to the Boolean equation:

Conditions for detecting the stuck-at-0 fault at the input line x_i

$$\underset{\substack{\uparrow \\ Controllability}}{x_i} \times \underset{\substack{\uparrow \\ Observability}}{\frac{\partial f}{\partial x_i}} = 1 \qquad (18.3)$$

Equation 18.3 combines the controllability and observability conditions.

Example 18.11 (The stuck-at-0 fault.) *Assume that we suspect the logic network shown in Figure 18.12 has the stuck-at-0 fault on the line x_2. The test for detecting this fault is derived as shown in Figure 18.12.*

Practice problem 18.7. (The stuck-at-0 fault.) Consider the logic network in Figure 18.13. Derive test patterns for detecting the stuck-at-0 fault on line x_2.
Answer is given in "Solutions to practice problems."

Design example: Detection of the input stuck-at-0 fault

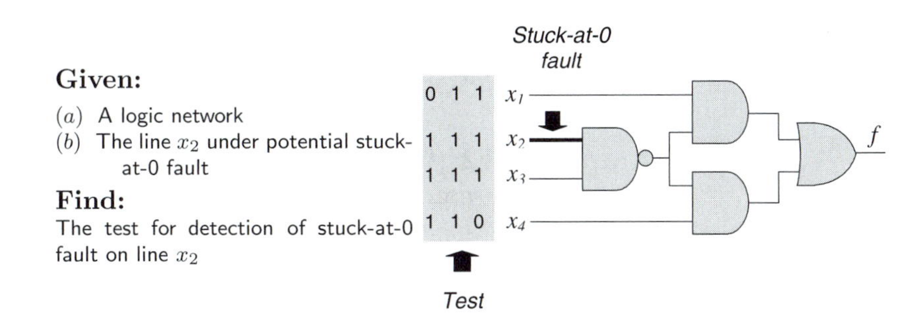
Given:
(a) A logic network
(b) The line x_2 under potential stuck-at-0 fault
Find:
The test for detection of stuck-at-0 fault on line x_2
Stuck-at-0 fault
0 1 1
1 1 1
1 1 1
1 1 0
x_1
x_2
x_3
x_4
f
Test

Procedure for the test generation for the stuck-at-0 fault

Step 1: Derive the Boolean expression for the output of the network

$$f = \overline{\overline{x}_2 \overline{x}_3}(x_1 \vee x_4)$$

Step 2: Compute the Boolean difference with respect to variable x_2

$$\begin{aligned}
\frac{\partial f}{\partial x_2} &= f_{x_2=0} \oplus f_{x_2=1} \\
&= (x_1 \vee x_4) \oplus \overline{x}_3(x_1 \vee x_4) \\
&= \overline{(x_1 \vee x_4)}\overline{x}_3(x_1 \vee x_4) \vee (x_1 \vee x_4)\overline{\overline{x}_3(x_1 \vee x_4)} \\
&= \overline{x}_1\overline{x}_3\overline{x}_4(x_1 \vee x_2) \vee (x_1 \vee x_4)(x_3 \vee \underbrace{\overline{(x_1 \vee x_4)}}_{\overline{x}_1\overline{x}_4}) \\
&= \underbrace{\overline{x}_1\overline{x}_3\overline{x}_4 x_1}_{0} \vee \underbrace{\overline{x}_1\overline{x}_3\overline{x}_4 x_4}_{0} \vee x_1 x_3 \vee \underbrace{x_1\overline{x}_1\overline{x_4}}_{0} \vee x_3 x_4 \vee \underbrace{x_4\overline{x}_1 x_4}_{0} \\
&= x_3(x_1 \vee x_4)
\end{aligned}$$

Step 3: Derive the logic equation

$$x_2 \frac{\partial f}{\partial x_2} = x_3(x_1 \vee x_4) = 1$$

Step 4: Solve this equation and find the test patterns. The test inputs for the stuck-at-0 fault on line x_2 are:

$$\texttt{Test patterns} = x_2 x_3(x_1 \vee x_2) = 1$$
$$(x_1, x_2, x_3, x_4) = (0,1,1,1), (1,1,1,0), (1,1,1,1)$$

FIGURE 18.12
Detection of the input stuck-at-0 fault on line x_2 (Example 18.11).

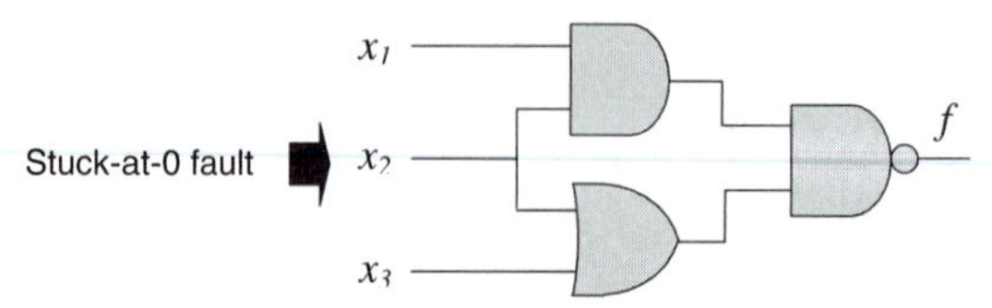

FIGURE 18.13
Logic network under testing for Practice problem 18.7.

Example 18.12 (Controllability.) *Determine the input excitation that exposes an s-a-0 fault, occurring at line A, at the output f of the logic network shown in Figure 18.14. The first requirement of such an excitation is that it should force the fault to occur (Step 1). In this case, we look for a pattern that would set A to 1 under normal circumstances. The only option here is $x_1 = 1$ and $x_2 = 1$. The faulty signal has to propagate to the output f (Step 2). For any change in line A to propagate, it is necessary for line B to be set to 1 and line x_5 to 0. Finally, the test for A_{s-a-0} can now be assembled: $x_1 = x_2 = x_3 = x_4 = 1, x_5 = 0$ (Step 3).*

Practice problem **18.8. (Controllability.)** Specify the controllability and observability conditions for the detection of a s-a-0 fault in line B (Figure 18.14). Determine the test for this node.
Answer is given in "Solutions to practice problems."

Consider the output f of a logic network with the stuck-at-1 fault at the input line x_i. This network is described by the Boolean function

$$f_{fault} = f(x_1, \ldots, x_{i-1}, 1, x_{i+1} \ldots, x_n) = f_i(1)$$

The fault difference function (Equation 18.1) that detects a fault in the network caused by the stuck-at-1 fault at the input line x_i is calculated as follows:

$$\begin{aligned} f \oplus f_{fault} &= f \oplus f_i(1) = \overline{x}_i f_i(0) \oplus x_i f_i(1) \oplus f_i(1) \\ &= \overline{x}_i f_i(0) \oplus (x_i \oplus 1) f_i(0) = \overline{x}_i f(0) \oplus \overline{x}_i f(1) \\ &= \overline{x}_i (f_i(0) \oplus f_i(1)) = \overline{x}_i \frac{\partial f}{\partial x_i} \end{aligned}$$

Therefore, all tests for the stuck-at-1 fault at the input line x_i are solutions of the Boolean equation below

Design example: Controllability and observability

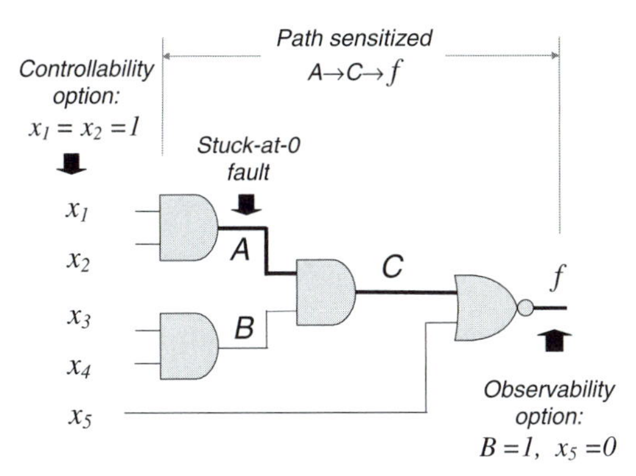

Step 1: Controllability for A. Set A to 1 under normal conditions ($x_1 = 1$ and $x_2 = 1$)

Step 2: Observability for A. The faulty signal has to propagate to output node f, so that it can be observed (B to be set to 1 and node x_5 to be set to 0):

$$\begin{aligned}\frac{\partial f}{\partial A} &= f_{A=0} \oplus f_{A=1} \\ &= \overline{0 \cdot x_3x_4 \vee x_5} \oplus \overline{1 \cdot x_3x_4 \vee x_5} \\ &= \overline{x}_5 \oplus \overline{x_3x_4 \vee x_5} = \overline{x}_5 \oplus \overline{x_3\overline{x}_4\overline{x}_5} \\ &= \overline{x}_5(1 \oplus \overline{x_3\overline{x}_4}) = x_3x_4\overline{x}_5\end{aligned}$$

Step 3: Combining Steps 1 and 2: $x_1x_2\frac{\partial f}{\partial A} = x_1x_2x_3x_4\overline{x}_5$. Test vector for A: $x_1 = x_2 = x_3 = x_4 = 1, x_5 = 0$

FIGURE 18.14
Controllability and observability (Example 18.12).

Conditions for detecting the stuck-at-1 fault at the input line x_i

$$\underset{\substack{\downarrow\\ \textit{Controllability}}}{\overline{x}_i} \times \underset{\substack{\uparrow\\ \textit{Observability}}}{\frac{\partial f}{\partial x_i}} = 1 \qquad (18.4)$$

Example 18.13 (Stuck-at-1 fault.) *The logic network in Figure 18.15 implements the Boolean function $f = x_1x_2x_3 \vee \overline{x}_1\overline{x}_2\overline{x}_3$. Assume a stuck-at-1 fault at the input terminal x_1 of the AND gate occurs. The fault difference is equal to 1 (Equation 18.1) once the test pattern is (011). That is, for the assignments $x_1 = 0, x_2 = 1$, and $x_3 = 1$ the value of f is 0, and the value of f_{fault} is 1.*

Practice problem 18.9. (Test pattern.) Find the test pattern for the network given in Figure 18.15 if the stuck-at-1 fault is observed at the input terminal x_1.

Answer is given in "Solutions to practice problems."

Design example: Detection of the stuck-at-1 fault

Given:

(a) A logic network

(b) Boolean function for the fault-free network

$$f = x_1x_2x_3 \vee \overline{x}_1\overline{x}_2\overline{x}_3$$

(c) The line x_1 under potential stuck-at-1 fault

Find:

The test for detection of the stuck-at-1 fault on line x_1

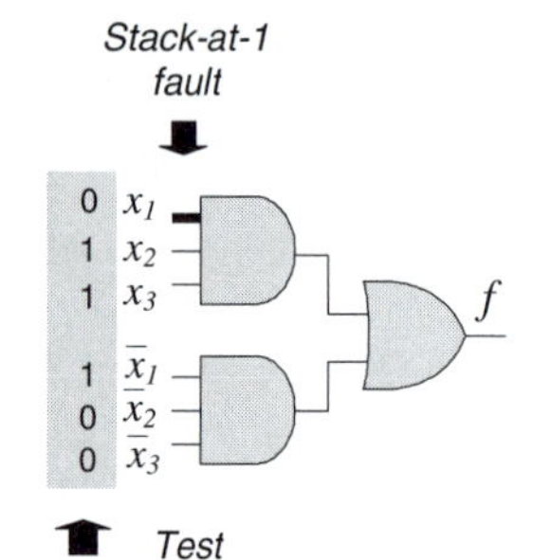

Step 1: Controllability conditions: $x_1 = 0$ (that is, $\overline{x}_1 = 1$). Since a stuck-at-1 fault at the input terminal x_1 causes $x_1 = 1$:

$$f_{fault} = \boxed{1} \cdot x_2x_3 \vee \boxed{0} \cdot \overline{x}_2\overline{x}_3 = x_2x_3$$

Step 2: Observability conditions:

$$\frac{\partial f}{\partial x_1} = \overline{x}_2\overline{x}_3 \oplus x_2x_3 = \overline{x_2 \oplus x_3}$$

Step 3: The equation to test for the stuck-at-1 fault at input terminal x_1 is:

$$\overline{x}_1\overline{x_2 \oplus x_3} = 1$$

The solutions are the assignments

(a) $x_1 = 0,\ x_2 = 1,\ x_3 = 1$

(b) $x_1 = 0,\ x_2 = 0,\ x_3 = 0$

That is, TEST PATTERNS = (0,1,1), (0,0,0)

FIGURE 18.15
Logic network with a stuck-at-1 fault at the input terminal x_1 (Example 18.13).

Practice problem **18.10. (Stuck-at-1 fault.)** Consider the logic network shown in Figure 18.16a. Assume a stuck-at-1 fault on the line x_2 is suspected. Derive the test for this fault.
Answer is given in "Solutions to practice problems."

Example 18.14 (Boolean difference.) *Figure 18.17 shows the Boolean function in the form of its truth vector* **F**, *and conditions for detecting stuck-at-0 and stuck-at-1 faults. The tests are shown as well.*

18.3.4 Testing decision tree-based logic networks

Decision trees and diagrams are multiplexer networks. Nodes of these trees and diagrams are implemented using multiplexers. The stuck-at model can be applied to the input and output of the node.

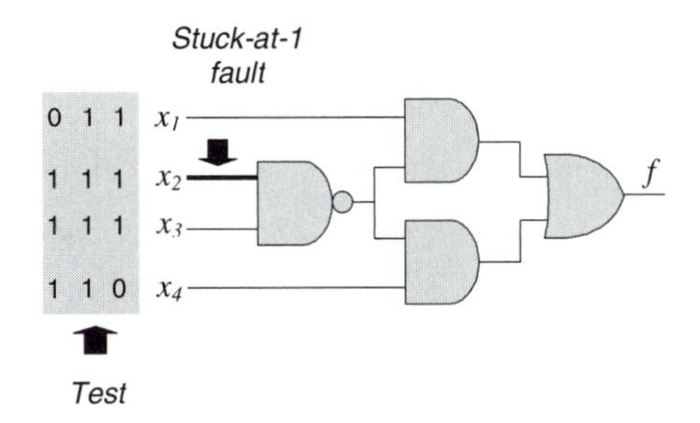

FIGURE 18.16
The logic network under test for Practice problem 18.10

Example 18.15 (Fault node of a decision diagram.) *The fault nodes are shown in Figure 18.18. Under normal conditions, the node performs the Shannon expansion* $f = \overline{x}_i f_{x_i=0} \vee x_i f_{x_i=0}$. *If the s-a-0 fault causes the* f *line, the output is* $f = 0$. *If the s-a-0 fault causes the* x_i *line, the node produces the output* $f_{x_i=0}$.

Practice problem 18.11. (Faulty node.) Specify the node function under s-a-1 faults (Figure 18.18).
Answer is given in "Solutions to practice problems."

Example 18.16 (Single stuck-at model.) *Figure 18.19 shows how stuck-at-0 and stuck-at-1 faults can be analyzed using decision diagrams. For example, if a stuck-at-0 fault occurs at the input line* x_1 *of the AND gate, it causes the function (and its decision diagram) to be reduced to the constant* $f = 0$.

Practice problem 18.12. (Single stuck-at model.) Analyze the effect of stuck-at-0 and stuck-at-1 faults at the first input line for the EXOR gate using decision diagrams.
Answer is given in "Solutions to practice problems."

18.4 Functional decision diagrams for computing Boolean differences

Boolean differences are useful for deriving controllability and observability options, as well for detecting faults. However, Boolean differences must

Design example: Deriving tests to detect stuck-at-0 and stuck-at-1 faults

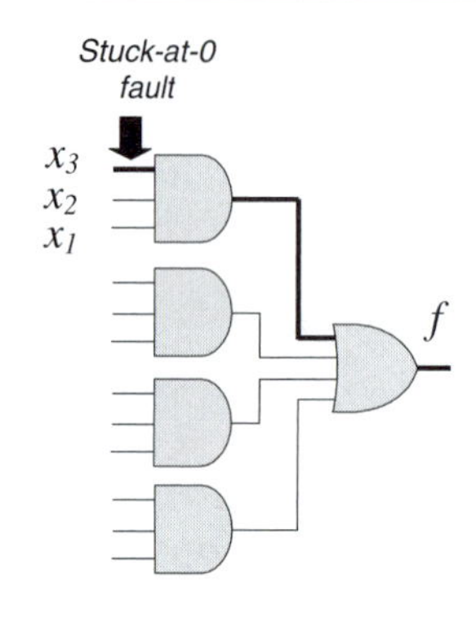

$x_1x_2x_3$	F	$x_3\frac{\partial F}{\partial x_3}$	$\overline{x}_3\frac{\partial F}{\partial x_3}$
0 0 0	0	0	1
0 0 1	1	1	0
0 1 0	1	0	1
0 1 1	0	1	0
1 0 0	1	0	1
1 0 1	0	1	0
1 1 0	0	0	1
1 1 1	1	1	0

Given:

(*a*) A logic network and the location of potential faults;

(*b*) The truth-vector **F** of this network; and

(*c*) Conditions for detecting stuck-at-0 and stuck-at-1 faults, $x_3\frac{\partial F}{\partial x_3}$ and $\overline{x}_3\frac{\partial F}{\partial x_3}$, respectively

Step 1: Derive a logic equation for the detection of stuck-at-0 faults. Stuck-at-0 at x_3 causes each value $f_{x_3=1}$ to be changed to $f_{x_3=0}$ Equation to find the test for detecting stuck-at-0 faults:

$$x_3\frac{\partial F}{\partial x_3} = 1$$

Step 2: Find the test for the detection of stuck-at-0 faults as the solution of the logic equation:

$$\texttt{Test}= \{001, 011, 101, 111\}$$

Step 3: Derive a logic equation for the detection of stuck-at-1 faults. Stuck-at-1 at x_3 causes each value $f_{x_3=0}$ to be changed to $f_{x_3=1}$ Equation to find the test for detecting stuck-at-1 faults:

$$\overline{x}_3\frac{\partial F}{\partial x_3} = 1$$

Step 4: Find the test for the detection of stuck-at-0 faults as the solution of the logic equation:

$$\texttt{Test}= \{000, 010, 100, 110\}$$

FIGURE 18.17
Deriving tests to detect stuck-at-0 and stuck-at-1 faults (Example 18.14).

be computed. Decision diagrams can be used for: (a) computing Boolean differences for modeling logic networks under the fault attack, and (b) computing Boolean functions as a multiplexer-based networks. In this section, we show that functional decision diagrams can be used not only for computing polynomial forms of Boolean functions, but also for computing Boolean differences. It is remarkable that the same decision diagram can perform these two functions: they are are distinguished by different interpretations of the results obtained.

Let us rewrite the positive Davio expansion in the form

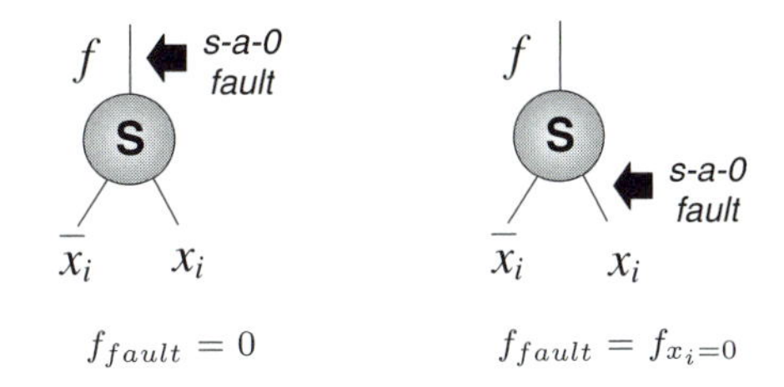

FIGURE 18.18
Fault models of decision diagrams (Example 18.15).

$$f = f|_{x_i=0} \oplus x_i \, (f|_{x_i=0} \oplus f|_{x_i=1}) = \underbrace{f|_{x_i=0}}_{Left\ branch} \oplus \underbrace{x_i \frac{\partial f}{\partial x_i}}_{Right\ branch}$$

It follows from this form that:

Functional decision tree for computing of Boolean differences

- Branches of the functional decision tree carry information about Boolean differences;
- Terminal nodes are the values of Boolean differences for corresponding variable assignments.
- The functional decision tree includes the values of all single and multiple Boolean differences given a variable assignment $x_1x_2 \ldots x_n = 00 \ldots 0$. This assignment corresponds to the calculation of a polynomial expansion of polarity 0.

Example 18.17 (Functional decision tree.) *Figure 18.20 shows a functional decision tree for an arbitrary Boolean function of two and three variables. The values of the terminal nodes can be interpreted both as the values of Boolean differences and of the polynomial expression.*

Practice problem 18.13. (Functional decision tree.) Compute Boolean differences for the function $f = x_1 \vee x_2$ using a functional decision tree.
Answer is given in "Solutions to practice problems."

Practice problem 18.14. (Boolean differences.) Compute Boolean differences with respect to the variables x_1, x_2 and x_3 for the function $f = x_1x_2 \vee x_3$ using a functional decision tree.
Answer is given in "Solutions to practice problems."

Design example: Analysis of faults using decision diagrams

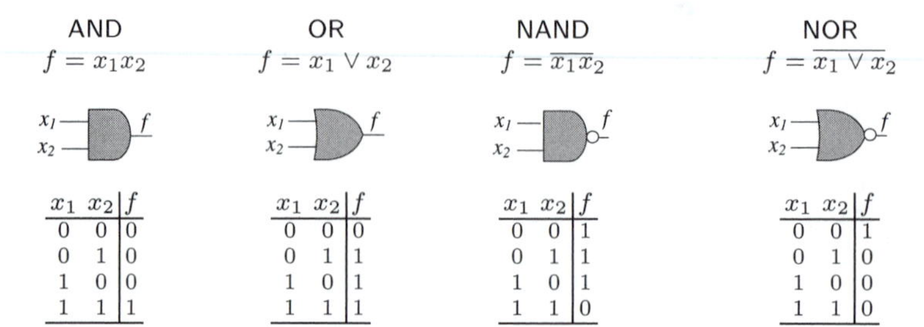

AND:

x_1	x_2	f
0	0	0
0	1	0
1	0	0
1	1	1

OR:

x_1	x_2	f
0	0	0
0	1	1
1	0	1
1	1	1

NAND:

x_1	x_2	f
0	0	1
0	1	1
1	0	1
1	1	0

NOR:

x_1	x_2	f
0	0	1
0	1	0
1	0	0
1	1	0

Free-fault decision diagram

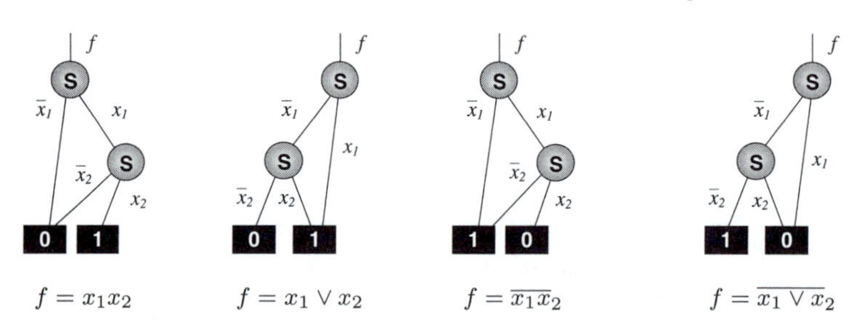

Stuck-at-0 fault at line x_1

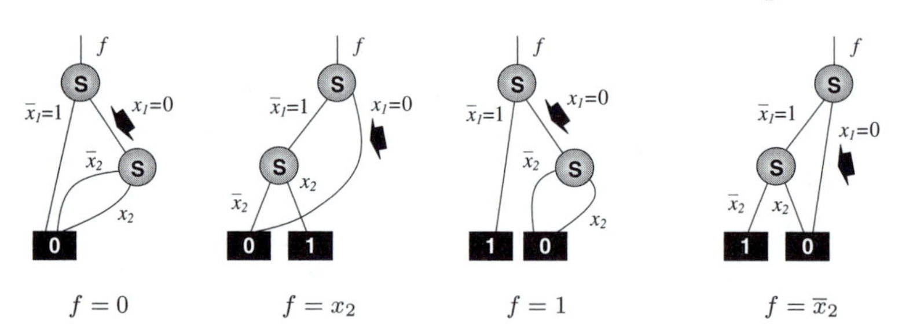

Stuck-at-1 fault at line x_1

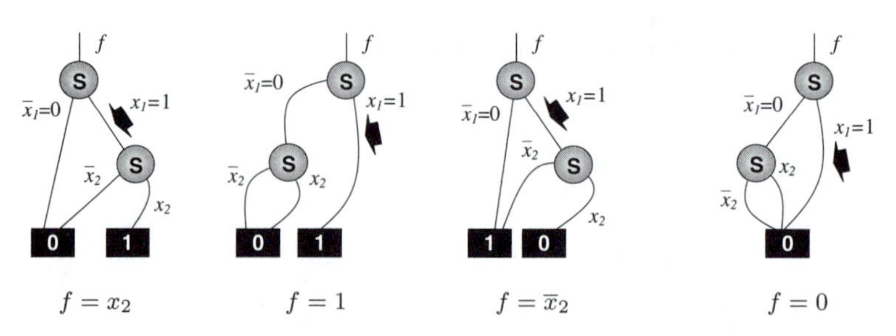

FIGURE 18.19
Interpretation of stuck-at-0 and stuck-at-1 faults for the input of an elementary logic gate using decision diagrams (Example 18.16).

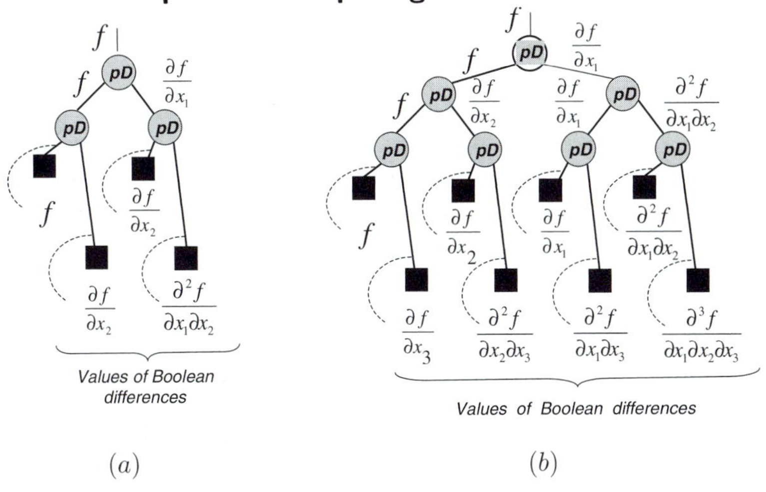

FIGURE 18.20
Computing Boolean differences by functional decision tree for a Boolean function of two (a) and three (b) variables (Example 18.17).

18.5 Random testing

Given a logic network with a large number n of inputs, the exhaustive application of all possible 2^n input patterns and observation the responses, becomes infeasible in terms of time complexity. An alternative approach is required. The test patterns can be generated using *random* test techniques which are based on the random selection of tests from the set patterns. Such an approach is based on the following premises:

Premises of random test techniques

Premise 1. An exhaustive enumeration of all possible input patterns is *redundant* since a single fault in the logic network is usually covered by a number of inputs patterns.

Premise 2. A substantial reduction in the number of patterns can be obtained by relaxing the condition that all faults must be detected. The cost of detecting the last single percentage of possible faults might be larger than the eventual replacement cost. For that reason, typical test procedures only attempt a 95–99% fault coverage.

By eliminating redundancy and providing a reduced fault coverage, it is possible to test most combinational logic networks with a limited set of input vectors. Given a logic network under test,

Example 18.18 (Random testing.) *The desired function $f = x_1 \oplus x_2$ is implemented by the logic network shown in Figure 18.21. This network is tested under possible stuck-at fault attack on wires a, b, c, d and e. Each of these single faults causes a new function that is different from the desired EXOR function of two variables. For example, a stuck-at-1 fault on the line d causes the OR function instead of the desired EXOR function.*
Choose an arbitrary assignment of the variables x_1 and x_2 as the first test; say, $x_1x_2 = (0, 1)$. This test can detect faulty networks which are caused by the faults $a = 0$ and $b = 1$.
The fragment of a fault detection using random testing is given in the table. The effect of other stuck-at faults, that is, the values of faulty functions, is shown in the left column of the table.

Design example: Random test

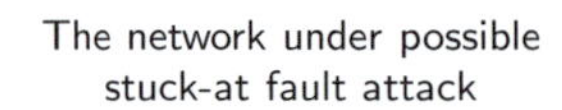

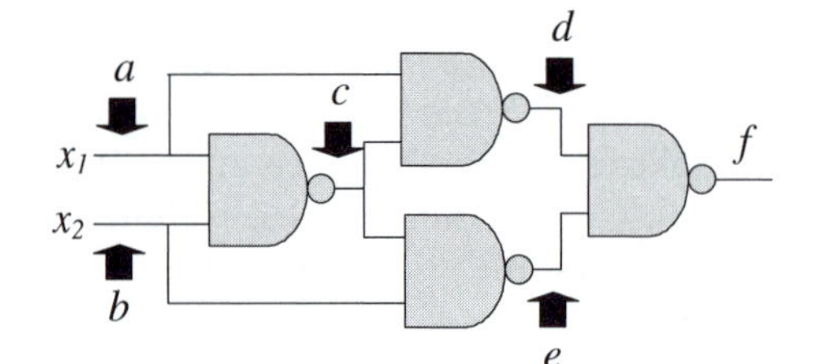

The effect of stuck-at fault attack

Stuck-at a	b	c	d	e	Faulty function
0					x_2
1					$\overline{x}_2$
	0				x_1
	1				$\overline{x}_1$
		0			0
		1			$x_1 \vee x_2$
			0		1
			1		$\overline{x}_1 x_2$
				0	1
				1	$x_1 \overline{x}_2$

R a n d o m t e s t i n g (f r a g m e n t)

Fault free x_1	x_2	f	Fault	fault detection f_{fault}	Detection
0	0	0	$a = 0$	$f = x_2 = 0$	$0 \oplus 0 = 0$
0	1	1	$a = 0$	$f = x_2 = 1$	$1 \oplus 1 = 0$
1	0	1	$a = 0$	$f = x_2 = 0$	$1 \oplus 0 = 1$
1	1	0	$a = 0$	$f = x_2 = 1$	$0 \oplus 1 = 1$

Fault free x_1	x_2	f	Fault	Fault detection f_{fault}	Detection
0	0	0	$c = 1$	$f = x_2 = 0$	$0 \oplus 0 = 0$
0	1	1	$c = 1$	$f = x_2 = 1$	$1 \oplus 1 = 0$
1	0	1	$c = 1$	$f = x_2 = 0$	$1 \oplus 0 = 1$
1	1	0	$c = 1$	$f = x_2 = 1$	$0 \oplus 1 = 1$

FIGURE 18.21
Random test against stuck-at attack (Example 18.18).

Practice problem **18.15. (Random testing.)** Evaluate the effect of random testing of the network shown in Figure 18.22. Assume a single-line stuck-at faults $a = 0, a = 1, b = 0$, and $b = 1$
Answer is given in "Solutions to practice problems."

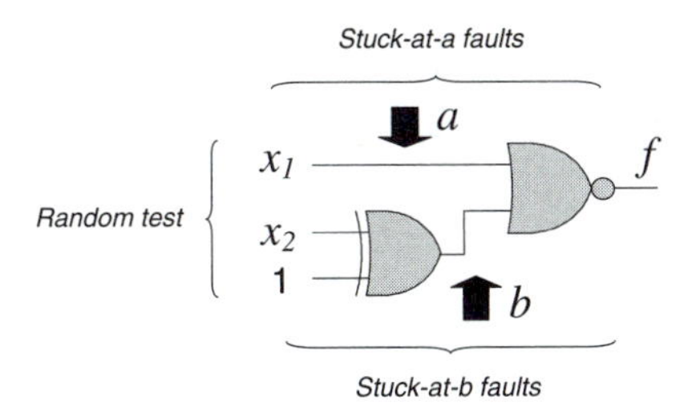

FIGURE 18.22
The logic network for random testing for Practice problem 18.15.

18.6 Design for testability techniques

Design for testability means that the logic network is designed so that it can be easily tested, or has the ability to test itself. These techniques include the following approaches:

- Built-in self-test technique (BIST),
- Scan-path techniques,
- Signature analysis, and
- Boundary-scan techniques.

18.6.1 Self-checking logic networks

An attractive approach to testability is to have the network itself generate the test patterns instead of requiring the application of external patterns. Even more appealing is a technique where the logic network itself decides if the obtained results are correct. Depending upon the nature of the network, this might require the addition of extra circuitry for generation and analysis of the patterns. Some excising hardware can be used, so that the size overhead of the self-test is not big.

Self-checking logic networks

A logic network can be designed to make it possible to determine, from the outputs of the network, whether a certain fault exists within the network. In this case it is unnecessary to explicitly test for this fault. Such a logic network is said to be *self-checking* for this fault.

An additional network, called a *checker*, can be designed to generate a warning signal whenever the outputs of network indicate the presence of a fault within this network. Ideally, a logic networks must be self-checking for as many faults as possible. The checker circuit must be designed to be self-checking too.

In previous chapters, various aspects of the application of error detection and error correcting codes were introduced. For instance, the parity check circuit uses Hamming code for error detection. In the parity check code, the Hamming distance is equal to two (the minimum number of bits by which any two code words differ), and the number of check bits is one (independent of the number of outputs in the original network). There are two types of parity checks, even and odd. Any error, affecting a single bit, causes the output to have an even (odd) number of bits, and, hence is automatically detected. Additional hardware, added to the design, can automatically detect and correct errors in the network, though not in the checker itself.

18.6.2 Built-in self-test (BIST)

Built-in self-test technique

Built-in self-test technique (BIST) is a testing technique, in which external test resources are not required to apply test patterns and check a logic network's response to those patterns.

In BIST techniques, the test patterns are preliminary loaded into the network or generated by the network itself. A BIST design approach is given in Figure 18.23. There are three main components: logic network under test, stimulus generator, and response analyzer.

A stimulus generator generates stimuli using *exhaustive* and the *random* approaches. In the exhaustive approach, the test length is 2^n, where n is the number of inputs to the network, and is intended to detect all detectible faults, given the space of the available input signals. An example of an exhaustive pattern generator is an n-bit counter. For networks with large n, the time to cycle through the complete input space is unacceptable. An alternative approach is to use *random* testing, which implies the application of a randomly chosen sub-set of 2^n possible input patterns. This subset should be selected so that a reasonable fault coverage is obtained. An example of a pseudorandom pattern generator is the *linear-feedback shift register*. Some of the outputs are connected to the inputs of EXOR gates, whose output is fed

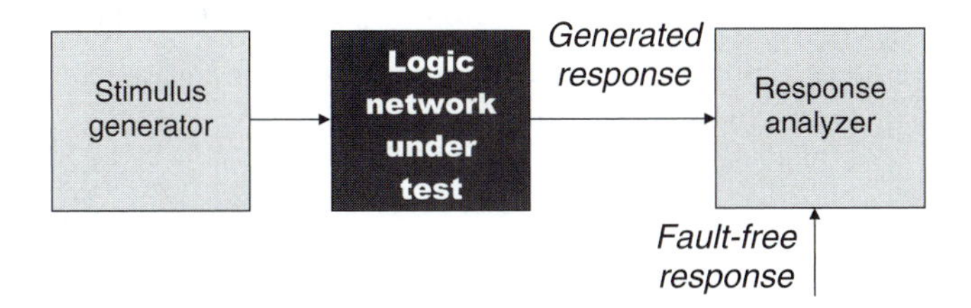

FIGURE 18.23
A built-in self-test (BIST) design approach.

back to the input of the shift register. An n-bit linear-feedback shift register cycles through $2^n - 1$ states before repeating the sequence, which produces a seemingly random pattern.

A response analyzer compares the generated response against the expected response stored in an on-chip memory, though this approach requires impractical area overhead. A cheaper technique called *signature analysis* is to *compress* the responses before comparing them. Storing the compressed response (the signature) of the network requires a minimal amount of memory. Thus, the response analyzer consists of circuitry that compresses the output of a network under test, and a comparator.

18.6.3 Easily testable EXOR logic networks

An easily testable logic network

An *easily testable* logic network is defined as a network that has a small test set.

Easily testable logic networks can be designed using the appropriate representations (SOP or polynomial forms), decomposition, and local transformations. An example of an easily testable logic network is the network of AND and EXOR gates that implements a polynomial form of a Boolean function.

Example 18.19 (Easily testable network.) *Figure 18.24 shows the logic network that implements the polynomial expression $f = 1 \oplus x_2 \oplus x_1x_3 \oplus x_2x_3 \oplus x_1x_2x_3$. Any change at a single line x_1, x_2 or x_3 implies a change at the output f if a single stuck-at-0 or stuck-at-1 fault occurs at the inputs and outputs of any gate. For example, if x_2 has changed from the correct value 0 to 1, then applying the test pattern (0000) or (1111) yields a change of $f = 0$ to $f = 1$. Applying the test patterns (1000) or (1111) causes a change in $f = 1$ to $f = 0$.*

Practice problem **18.16.** **(Easily testable network.)** Let $f = x_1x_2x_3 \oplus x_1x_2 \oplus x_3$. Show that this is an easy testable network.
Answer is given in "Solutions to practice problems."

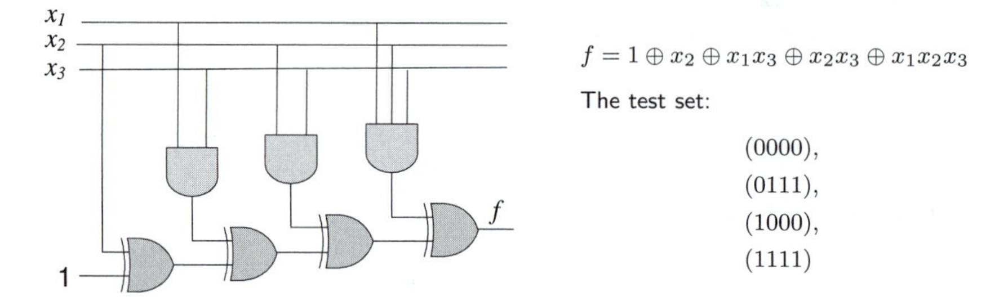

FIGURE 18.24
An easily testable logic network that implements the polynomial form of a Boolean function of three variables (Example 18.19).

18.6.4 Improving testability using local transformations

Local transformations can improve the conditions for testability.

> **Example 18.20 (Improving testability.)** *Figure 18.25 illustrates how the testability of a logic network can be improved using local transformations:*
> *The area A:* *The inverter is replaced by an EXOR gate using the identity rule for variables and constants* $\overline{x}_2 = x_2 \oplus 1$.
> *The area B:* *The inverter is replaced by the EXOR gate using the rule* $\overline{x}_1 = x_1 \oplus 1$
> *The areas A and B:* $x_1(x_2 \oplus 1) \oplus 1 = x_1x_2 \oplus x_1 \oplus 1$.

Practice problem **18.17.** **(Improving testability.)** Apply local transforms to the areas A, B, and C of a logic network given in Figure 18.26
Answer is given in "Solutions to practice problems."

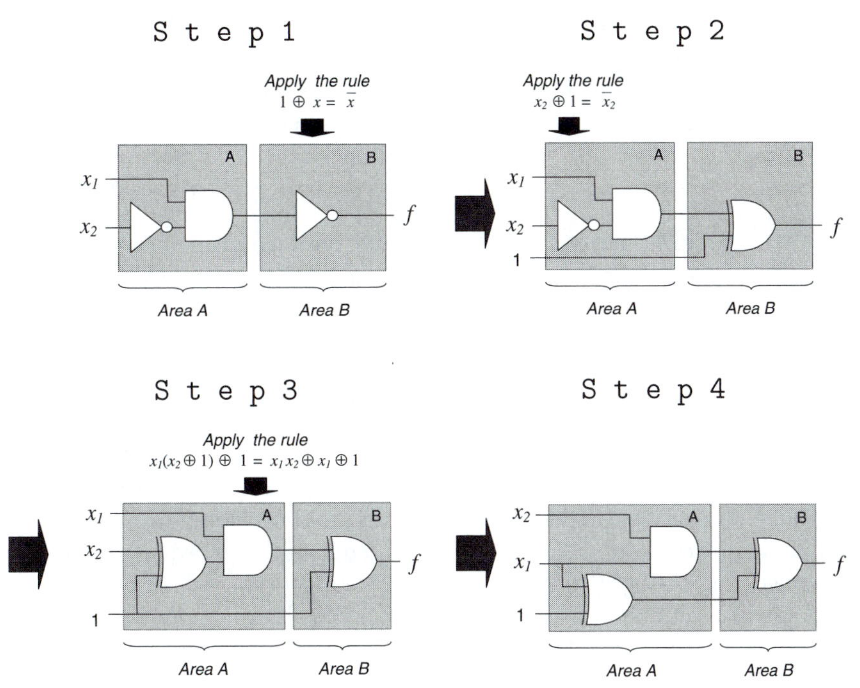

FIGURE 18.25
Improving testability using local transformations in logic network (Example 18.20).

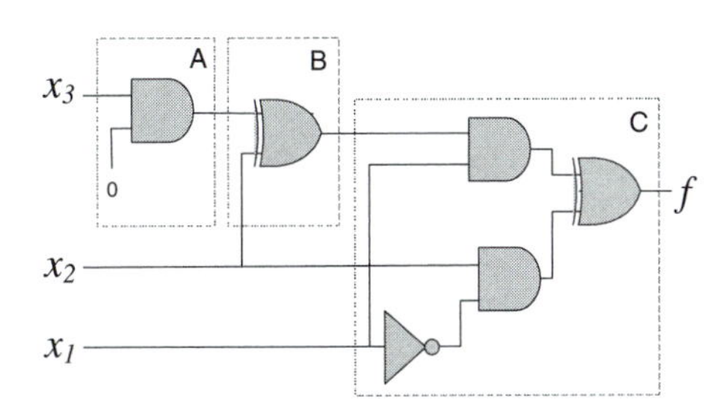

FIGURE 18.26
Logic network for the Practice problem 18.17.

18.6.5 Testing sequential logic networks

While testing a combinational network by applying all possible 2^n input patterns and observing the responses is not acceptable for large n, the situation becomes more dramatic for a sequential network. A combinational logic network can be tested using techniques based on either deterministic or random principles. In the testing of combinational networks, the response of a network under test is compared with the desired functionality. This approach is not applicable for sequential networks. In a sequential network, the output depends not only upon the inputs applied, but also upon the value of the state. To exhaustively test this finite state machine, the application of 2^{n+m} input patterns is required, where m is the number of state registers.

> **Example 18.21 (Exhaustive testing)** *For a state machine of moderate size (say $m = 10$) this means that 1 billion patterns must be evaluated, which may take 16 minutes. Modeling a modern microprocessor as a state machine translates into an equivalent model with over 50 state registers. Exhaustive testing of such an engine would require over a billion years.*

Obviously, that a more feasible approach is required. To test a given fault in a state machine, the engine must be brought to the desired state first before applying the correct input excitation. This requires that a sequence of inputs be applied, and the network response be propagated to one of the output pins, which might require another sequence of patterns. In other words, testing for a single fault in a finite state machine requires a sequence of vectors. This process may be time-expensive.

One approach to addressing the problem is to turn the sequential network into a combinational one by breaking the feedback loop in the course of the test. This is the key concepts of the so-called *scan-test* methodology. Another approach is to let the network test itself. This approach is called the *self-test* concept.

18.7 Summary of design for testability

Design for testability techniques refers to the design of logic networks so that they are testable rather than designed without consideration of testing. The key aspects of these techniques include the following:

(*a*) The combinational part can be tested using combinational tests.
(*b*) In the sequential part, registers must be able to be reconfigured into shift registers.
(*c*) In test mode, a test vector can be shifted into the logic network, and a test can be performed by checking the output patterns.

The main principles of combinational network testing are summarized as follows:

- ▶ Testing is done by applying input patterns (***test patterns***) to the network under test and observing the resulting output signals (***test responses***).
- ▶ The ***observability*** problem consists in identifying an input pattern such that a signal change at a line can propagate to the output.
- ▶ Identifying input patterns such that a faulty logic value is produced at the fault line is called the ***controllability*** problem.
- ▶ The standard fault model for gate-level testing is the ***single stuck-line model***. It defines two types of faults for each line in a logic network: the ***stuck-at-0*** fault and the ***stuck-at-1*** fault, which cause the line to be permanently stuck at the logic values 0 or 1, respectively.
- ▶ To generate tests for stuck-at faults, simultaneous solution of the observability and controllability is required. Observability in logic networks refers to the ability to observe the changes at a line of the logic network via external outputs (observation points).
- ▶ A fault can be modeled by detecting changes in the value of a switching function f in response to a change in input. The function used to detect this is Boolean difference.
- ▶ All tests for the stuck-at-0 and stuck-at-1 fault at the input line x_i can be obtained as solutions to the Boolean equation $x_i \frac{\partial f}{\partial x_i} = 1$ and $\overline{x}_i \frac{\partial f}{\partial x_i} = 1$, respectively.
- ▶ New horizons in design for testability are introduced in the "Further study" section.

18.8 Further study

Historical perspective

1950: R. W. Hamming introduced a class of error detection/correction codes now called *Hamming Codes* ("Error-Detecting and Error-Correcting Codes" by R. W. Hamming, *Bell Systems and Technology Journal*, volume 29, pages 147–160, April 1950). This approach allows any code to be modified by adding parity or checking bits. With enough parity or check bits added to the code, a logical network can detect as well as correct errors in the data being received.

1966: The path sensitization approach known as D algorithm was developed by J. P. Roth ("Diagnosis of Automata Failures: A Calculus and a Method" J. P. Roth, *IBM Journal of Research and Development*, volume 10, pages 278–291, July 1966).

1967: W. H. Kautz studied the fault detection testing problems of one- and two-dimensional arrays of identical cells arranged in rectilinear patterns ("Testing for Faults in Cellular Logic Arrays" by W. H. Kautz, *Proceedings of the Switching and Automata Theory Symposium*, pages 161–174, 1967).

Advanced topics of testability

Topic 1: Logic network test generation. A wide range of techniques has been developed for combinational logic network test generation.

At one end of the spectrum are *exhaustive* and *random* techniques.

Exhaustive techniques. If the number of primary inputs of the network is small, generating an exhaustive test that applies to all possible input vectors, may be possible utilized.

Random techniques. Random test generation means probabilistically generating input vectors and determining whether they detect any faults in the network. If a random vector detects any fault, then it is retained as a test. To maintain high fault coverage, the number of random vectors can be very large in some cases.

Signature analysis techniques are based on the compression of a sequence of test results into a single signature. In *signature analysis*, a set of known inputs is dynamically applied to the logic network under test.

Scan-path techniques are based on the isolation of the flip-flops and using them independently during sequential network operation. For this purpose multiplexers can be used on flip-flop inputs.

Boundary-scan techniques assume that the primary inputs and outputs of a logic network under test are observable.

At the other end of the spectrum are *deterministic* techniques, which can be partitioned into two groups.

Algebraic algorithms use the Boolean differences formulation to symbolically solve the problem. These algorithms have not been very practical, yet serve the purpose of illuminating the fundamental nature of the testing problem.

Path sensitization algorithms exploit the gate-level representation of the logic network. In the path sensitization approach, rather than testing the whole logic network, one of alternatives is to deal with paths. Each path can be activated so that the changes in signal that propagates along the path has a direct impact on the output signal. These algorithms systematically enumerate the search space using the *branch-and-bound* method and employ heuristics to guide the search process. For example, in the D-algorithm developed by Roth at IBM, every signal in the network is considered as a *decision point* in the search process; if the search leads to some signal being assigned both values 0 and 1, the decision is retracted and an alternative choice is made. This process is called *backtracking*. The D-algorithm may require an excessive amount of backtracking for some networks. Its modification, the *Path Oriented Decision Making (PODEM)* algorithm, offers a significant improvement. Here, only primary input signals are considered as decision points in the search process. Heuristics are used to dynamically determine the next decision point. *SOCKRATES* incorporate heuristics to accelerate the basic PODEM algorithm.

Fault simulation is used to determine the fault coverage in order to measure the quality of a test program. The most common approach to fault simulation is the parallel fault-simulation technique, in which the correct network is simulated concurrently with a faulty one, each of which has a single fault injected. The results are compared, and a fault is labeled as detected for a given test vector set if the outputs diverge. Most simulators employ a number of techniques, such as selecting the faults with a higher chance of detection first, to expedite the simulation process. Fault-simulation can also be performed on hardware devices called *accelerators*, which provide a substantial increase in speed over software-based simulations.

Ad hoc testing combines various techniques, such as the partitioning of logic networks and the addition of extra test points in order to increase the observability and controllability of design. The applicability of most of these techniques depends upon the application and architecture.

Topic 2: Fault tolerant systems. Assuming reliable computing in the presence of faults is called *fault-tolerance*. In *fault-tolerant* computing paradigms, the effects of faults are mitigated, and correct computations are guaranteed with a certain level of reliability. *Redundancy* is one fault-tolerant techniques that can be used if additional resources are available.

Fault tolerance is introduced to a system through utilizing many design and testing techniques; in particular, error detecting and correcting codes, self checking, and massive redundancy. Fault detection techniques supply warnings of faulty results. However, the use of fault detection techniques does not provide fault tolerance.

Several techniques have been proposed for gate-level fault masking. *Fault masking* employs redundancy, which provides fault tolerance by isolating or correcting fault effects before they are used in computing.

The problem is formulated as: Given an unreliable logic gate, achieve the correct output values. remains unsolved. For example, *Von Neumann's approach* states that it is possible theoretically to achieve an acceptable level of fault-tolerance for the simplest imperfect logic gates using massive redundancy under certain constraints (J. von Neumann, "Probabilistic Logics and the Synthesis of Reliable Organisms from Unreliable Components" edited by C. E. Shannon and J. McCarthy, , *Automata Studies*, Princeton University Press, Princeton, New York, pages 43–98, 1956).

Topic 3: Automatic test-pattern generation (ATPG). The task of the automatic test-pattern generation (ATPG) is to determine a minimum set of excitation vectors that covers a sufficient portion of the faults set defined by the adopted fault model. It mainly relies on a random set of test patterns, which is used in fault simulation to determine how many of the potential faults are detected. Using the obtained results as a guide, extra vectors can be added or removed. More advanced approaches rely knowledge of the functionality of a logic network to derive a suitable test vector for a given fault.

Topic 4: Testability of decision diagrams. The *cellular fault model* is often used in testing decision diagrams. In this model, it is assumed that a fault modifies the behavior of exactly one node.

Further reading

"Analyzing Errors with the Boolean Difference" by E. F. Sellers, M. Y. Hsiao, and L. W. Bearnson, *IEEE Transactions on Computers*, volume C-17, number 7, pages 676–683, 1981.

Coding and Information Theory by R. W. Hamming, Prentice-Hall, New York, 1980.

"Design for Testability – A Survey" by T. W. Williams and K. P. Parker, *IEEE Transactions on Computers*, volume C-31, pages 2–15, January 1982.

Digital Systems Testing and Testable Design by M. Abramovici, M. A. Breuer, and A. D. Friedman, Rockville, Computer Science Press, 1990.

"Easily Testable Realization for Logic Functions" by S. M. Reddy, *IEEE Transactions on Computers*, volume 21, pages 1183–1188, December 1972.

"Easily Testable Realization for Generalized Reed-Muller Expansions" by T. Sasao, *IEEE Transactions on Computers*, volume 46, number 6, pages 709–716, 1997.

"An Implicit Enumeration Algorithm to Generate Tests for Combinational Logic Circuits" by P. Goel, *IEEE Transactions on Computers*, volume C-30, number 3, pp. 215–222, 1981.

Logic Testing and Design for Testability by H. Fujiwara, MIT Press, Cambridge, MA, 1985.

"SOCRATES: A Highly Efficient Automatic Test Pattern Generation System" by M. H. Schulz, E. Trischler, and T. M. Sarfert, *IEEE Transactions on Computer-Aided Design*, volume 7, number 1, pages 126–136, 1988.

The Theory of Error-Correcting Codes by F. J. MacWilliams and N. J. A. Sloan, Noth-Holland Publishing Company, 1978.

"Fault-Tolerance in Nanocomputers: A Cellular Array Approach" by F. Peper, et al., *IEEE Transactions on Nanotechnology*, volume 3, number 1, pages 187–202, 2004.

18.9 Solutions to practice problems

Practice problem	Solution
18.1.	$\text{FAULT COVERAGE} = \frac{3}{2\times 1} = \frac{3}{2} = 1\frac{1}{2}$
18.3.	x_1, x_2, x_3, f $f = \overline{x_1 \vee x_2 \vee x_3} = 1$ $x_1 \vee x_2 \vee x_3 = 0$ The condition for observability of $f = 0$: $x_1 = x_2 = x_3 = 0$
18.4.	$f_1 = x_1 \oplus x_2 = 1$ and $f_2 = x_1 \oplus x_2 \oplus x_1x_2 = 1$. $f_1 \oplus x_1x_2 = 1 \Rightarrow 1 \oplus x_1x_2 = 1 \Rightarrow x_1x_2 = 0$ $\Rightarrow x_1 = x_2 = 0$ or $x_1 = 0, x_1 = 1$ or $x_1 = 1, x_2 = 0$
18.5.	The output of the given logic network is described by the equation $f = x_1(x_2 \vee x_3) \vee \overline{x}_1x_4$. The controllability at the line x_1 is defined by the Boolean equation $\frac{\partial f}{\partial x_1} = f(0, x_2, x_3, x_4) \oplus f(1, x_2, x_3, x_4)$ $= x_4 \oplus (x_2 \vee x_3) = \overline{x}_2\overline{x}_3x_4 \vee x_2\overline{x}_4 \vee x_3\overline{x}_4 = 1$ There are five solutions to this equation: $x_2 = x_3 = 0, x_4 = 1$; $x_2 = 1, x_3 = x_4 = 0$; $x_2 = x_3 = 1, x_4 = 0$; $x_2 = 0, x_3 = 1, x_4 = 0$; and $x_2 = x_3 = 1, x_4 = 0$. These solutions determine the controllability conditions.
18.6.	Multiplexers are added on the data and address busses. During normal operation mode, these multiplexers direct the memory ports to the combinational module. During test, the data and address ports are connected directly to the Input/Output bus, and testing the memory can proceed more efficiently. 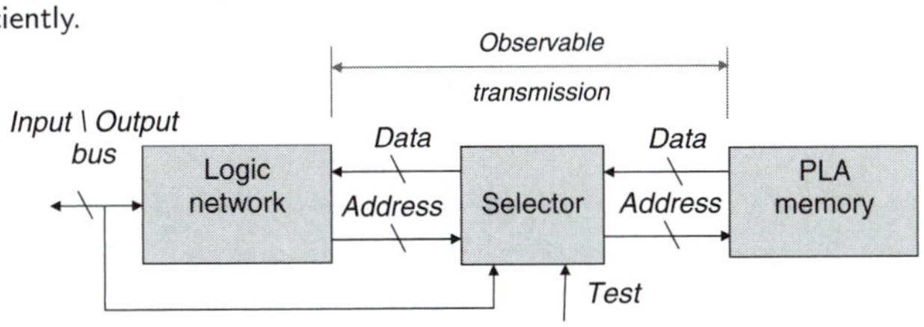 Observable transmission Input \ Output bus Logic network Data Address Selector Data Address PLA memory Test
18.7.	Stuck-at-0 fault at the input terminal x_2: $f_{fault} = \overline{x_1 \cdot \boxed{0}\,(x_3 \vee \boxed{1})} = \overline{0} = 1$ The fault difference function: $D = f \oplus f_{fault} = \overline{x_1x_2(x_2 \vee x_3)} \oplus 1$ $= \overline{x_1x_2 \vee x_1x_2x_3} \oplus 1 = \overline{x_1x_2} \oplus 1 = x_1x_2$ Testing conditions for the stuck-at-0 fault at the input terminal x_2: $x_1\overline{x}_2 = 1$, which yields $x_1 = 1$, $x_2 = 1$, and x_3 can be either 0 or 1. Test pattern: $(1, 1, 0)$ and $(1, 1, 1)$.

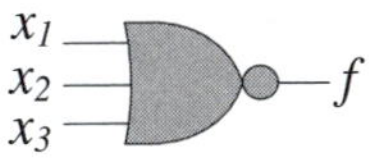

Solutions to practice problems (continuation)

Practice problem	Solution
18.8.	**Step 1: Controllability for B:** $B = x_3x_4 = 0 \Rightarrow x_3 = x_4 = 0$ or $x_3 = 0, x_4 = 1$ or $x_3 = 1, x_4 = 0$ **Step 2: Observability for B:** $\frac{\partial f}{\partial B} = f_{B=0} \oplus f_{B=1} = x_1x_2\overline{x}_5$ **Step 3:** Combination of Steps 1 and 2: $\overline{B}\frac{\partial f}{\partial B} = 1 \Rightarrow \overline{x_3x_4} \cdot x_1x_2\overline{x}_5 = 1 \Rightarrow (\overline{x}_3 \vee \overline{x}_4)x_1x_2\overline{x}_5 = 1$ $\Rightarrow x_3 = x_4 = x_5 = 0, x_1 = x_2 = 1$ or $x_3 = x_5 = 0, x_1 = x_2 = x_4 = 1$ or $x_4 = x_5 = 0, x_1 = x_2 = x_3 = 1$
18.9.	Stuck-at-1 fault at the input terminal x_2: $f_{fault} = x_1 \cdot 1 \cdot x_3 \vee \overline{x}_1\overline{x}_2\overline{x}_3$. The fault difference function $$D = f \oplus f_{fault} = (x_1x_2x_3 \vee \overline{x}_1\overline{x}_2\overline{x}_3) \oplus (x_1x_3 \vee \overline{x}_1\overline{x}_2\overline{x}_3)$$ Simplification yields: $$D = (x_1x_2x_3 \oplus \overline{x}_1\overline{x}_2\overline{x}_3) \oplus ((x_2 \vee \overline{x}_2)x_1x_3 \vee \overline{x}_1\overline{x}_2\overline{x}_3)$$ $$= (x_1x_2x_3 \oplus \overline{x}_1\overline{x}_2\overline{x}_3) \oplus (x_1x_2x_3 \vee x_1x\overline{x}_2x_3 \vee \overline{x}_1\overline{x}_2\overline{x}_3)$$ $$= (x_1x_2x_3 \oplus \overline{x}_1\overline{x}_2\overline{x}_3) \oplus (x_1x_2x_3 \oplus \overline{x}_1x_2x_3 \oplus \overline{x}_1\overline{x}_2\overline{x}_3)$$ $$= x_1\overline{x}_2x_3$$ Test condition for the stuck-at-1 fault at the input terminal x_2: $D = x_1\overline{x}_2x_3 = 1$ for the assignment $x_1 = 1, x_2 = 0, x_3 = 1$, that is, TEST PATTERN $= (1, 0, 1)$.
18.10.	The test for detecting this fault is derived by analogy with Example 18.11 as follows. The test patterns for the stuck-at-0 fault on the line x_2 are generated using Equation 18.4, $\overline{x}_2\frac{\partial f}{\partial x_2} = 1$. Boolean difference with respect to the variable x_2 is $\frac{\partial f}{\partial x_2 = x_3(x_1 \vee x_4)}$. The equation $$x_2\frac{\partial f}{\partial x_2} = \overline{x}_2x_3(x_1 \vee x_4) = 1$$ This implies the following test patterns: $(x_1, x_2, x_3, x_4) = (1010), (1011), (0011)$.
18.11.	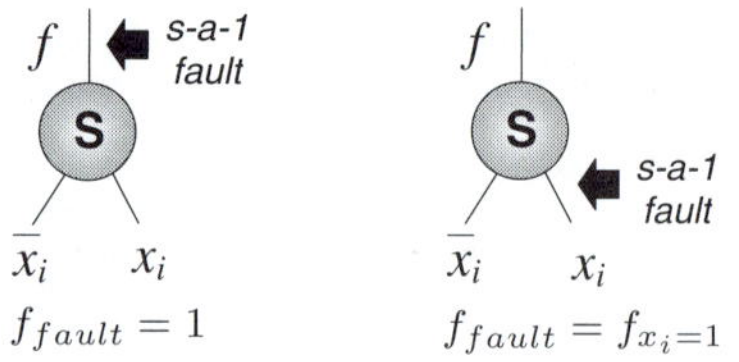
18.12.	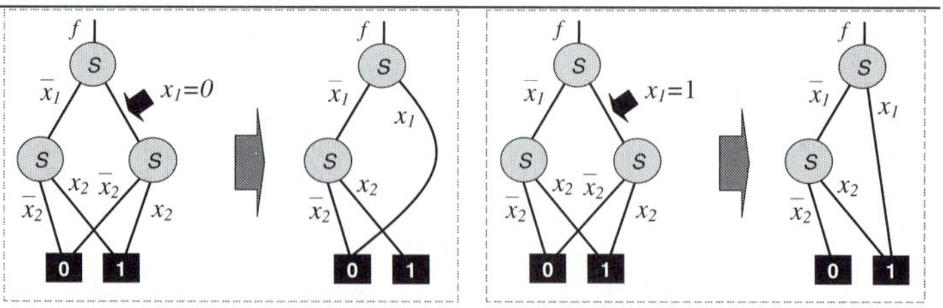

Solutions to practice problems (continuation)

Practice problem	Solution
18.13.	The values of Boolean differences for the function $f = x_1 \vee x_2$ given the assignment $x_1x_2 = \{00\}$ are: $f(00) = 0, \quad \frac{\partial f(00)}{\partial x_1} = x_2 = 0,$ $\frac{\partial f(00)}{\partial x_2} = x_1 = 0, \quad \frac{\partial^2 f(00)}{\partial x_1 \partial x_2} = 1$ These correspond to the terminal nodes of the functional tree.
18.14.	Complete decision tree / Reduced decision tree — Values of Boolean differences. Decision trees for Boolean differences: $\frac{\partial f}{\partial x_1} = x_2 \oplus x_2x_3$, $\frac{\partial f}{\partial x_1} = x_1 \oplus x_1x_3$, $\frac{\partial f}{\partial x_1} = x_1x_2 \oplus 1$

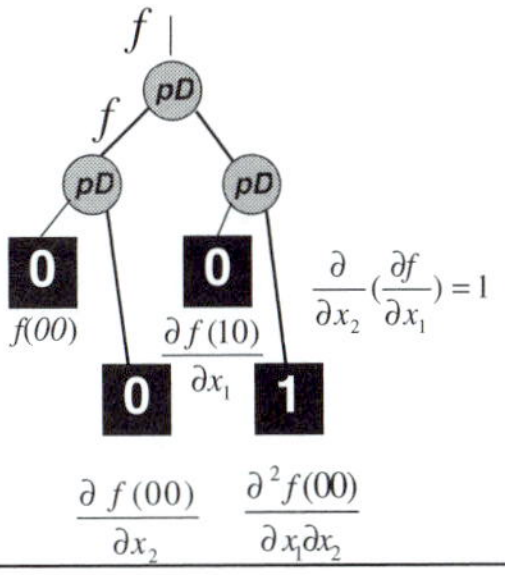

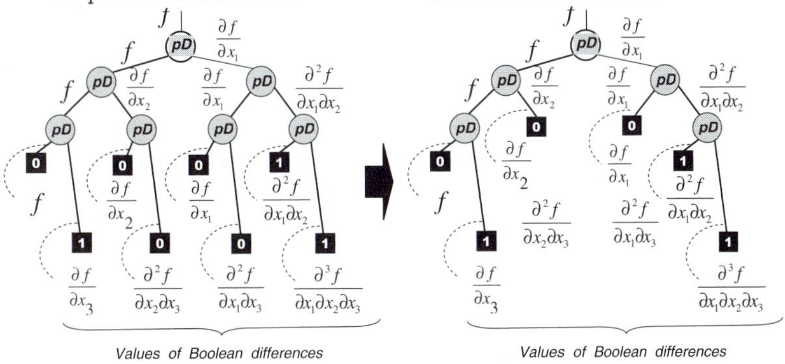

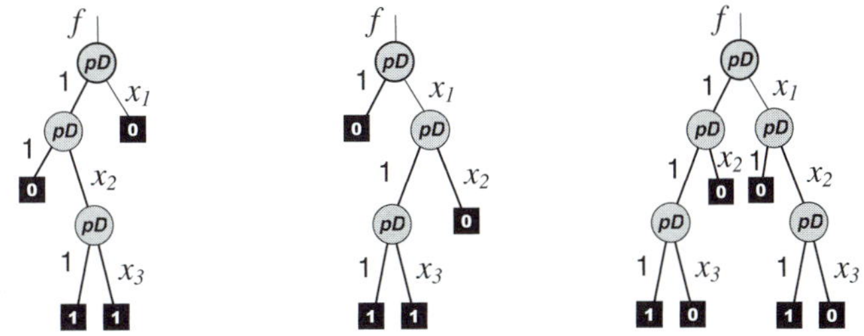

Boolean difference with respect to x_1

$$\frac{\partial f}{\partial x_1} = f(0, x_2, x_3) \oplus f(1, x_2, x_3)$$
$$= (0x_2 \vee x_3) \oplus (1x_2 \vee x_3) = x_3 \oplus \overline{\overline{x_2 \vee x_3}}$$
$$= x_3 \oplus (x_2 \oplus 1)(x_3 \oplus 1) \oplus 1 = x_2 \oplus x_2x_3 = x_2\overline{x}_3$$

By analogy, Boolean differences with respect to x_2 and x_3

$$\frac{\partial f}{\partial x_2} = x_1 \oplus x_1x_3 = x_1\overline{x}_3 \quad \text{and} \quad \frac{\partial f}{\partial x_3} = x_1x_2 \oplus 1 = \overline{x_1x_2}$$

Solutions to practice problems (continuation)

Practice problem	Solution

18.15.

Stuck-at a	Stuck-at b	Faulty function
0		x_2
1		0
	0	x_1
	1	0

Fault free x_1	x_2	f	Fault	Fault detection f_{fault}	Detection
0	0	0	$a=0$	$f=x_2=0$	$1\oplus 0=1$
0	1	1	$a=0$	$f=x_2=1$	$1\oplus 1=0$
1	0	0	$a=0$	$f=x_2=0$	$0\oplus 0=0$
1	1	1	$a=0$	$f=x_2=1$	$1\oplus 1=0$
0	0	1	$a=1$	$f=0$	$1\oplus 0=1$
0	1	1	$a=1$	$f=0$	$1\oplus 0=1$
1	0	0	$a=1$	$f=0$	$0\oplus 0=0$
1	1	1	$a=1$	$f=0$	$1\oplus 0=1$
0	0	0	$b=0$	$f=x_1=0$	$1\oplus 0=1$
0	1	1	$b=0$	$f=x_1=0$	$1\oplus 0=1$
1	0	1	$b=0$	$f=x_1=1$	$0\oplus 1=1$
1	1	0	$b=0$	$f=x_1=1$	$1\oplus 1=0$
0	0	1	$b=1$	$f=0$	$1\oplus 0=1$
0	1	1	$b=1$	$f=0$	$1\oplus 0=1$
1	0	0	$b=1$	$f=0$	$0\oplus 0=0$
1	1	1	$b=1$	$f=0$	$1\oplus 0=1$

18.16.

If $x_1=0$, then $f=x_3$. If $x_1=1$, then $f=x_2\vee x_3$.
If $x_2=0$, then $f=x_3$. If $x_2=1$, then $f=x_1\vee x_3$.
If $x_3=0$, then $f=x_1x_2$. If $x_3=1$, then $f=1$.

18.17.

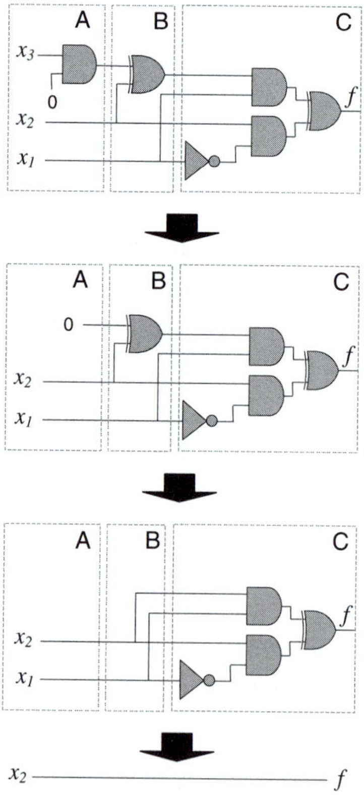

A B C
x_3 0 x_2 x_1 f

Improving the testability using local transformations:

Step 1: Area A

$$x_3\cdot 0=0$$

Result: the AND gate is deleted

Step 2: Area B:

$$0\oplus x_2=x_2$$

Result: the EXOR gate is deleted

Step 3: Area C:

$$f=x_1x_2\oplus \overline{x}_1x_2=x_2$$

Result: two AND gates, EXOR gate, and the inverter are deleted

A B C
0 x_2 x_1 f

A B C
x_2 x_1 f

x_2 —— f

18.10 Problems

Problem 18.1 Detect all test patterns for stuck-at-0 faults at the input line x_2 for the logic network (a), and at the input line x_4 for the logic network (b):

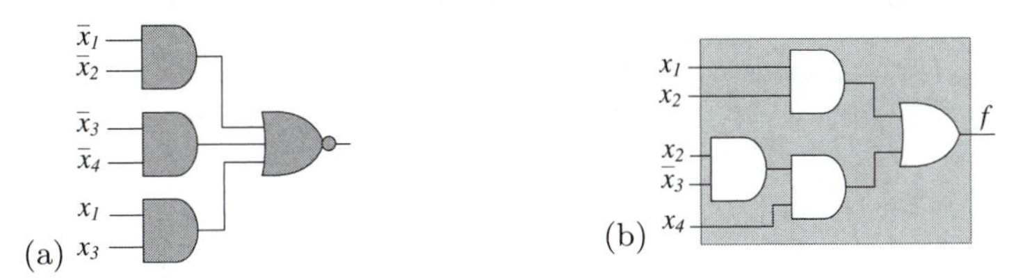

Problem 18.2 Detect all test patterns for stuck-at-1 faults at the input line x_2 for the logic network given in Problem 18.1a, and at the input line x_4 for the logic network given in Problem 18.1b.

Problem 18.3 Calculate the Boolean differences with respect to variable x_2 for the logic networks given in Problem 18.1, (a) and (b).

Problem 18.4 Derive the Boolean differences of the Boolean functions given the following truth tables:

(a)

x_1	x_2	x_3	f
0	0	0	**0**
0	0	1	**0**
0	1	0	**0**
0	1	1	**1**
1	0	0	**0**
1	0	1	**1**
1	1	0	**1**
1	1	1	**0**

(b)

x_1	x_2	x_3	f
0	0	0	**1**
0	0	1	**1**
0	1	0	**0**
0	1	1	**1**
1	0	0	**0**
1	0	1	**1**
1	1	0	**1**
1	1	1	**0**

(c)

x_1	x_2	x_3	f
0	0	0	**0**
0	0	1	**0**
0	1	0	**0**
0	1	1	**1**
1	0	0	**1**
1	0	1	**1**
1	1	0	**1**
1	1	1	**1**

Problem 18.5 Use Boolean differences to find all the tests in the logic network below for:

(a) Line a_1, stuck-at-0
(b) Line a_6, stuck-at-0
(c) Line a_4, stuck-at-1
(d) Line a_7, stuck-at-1

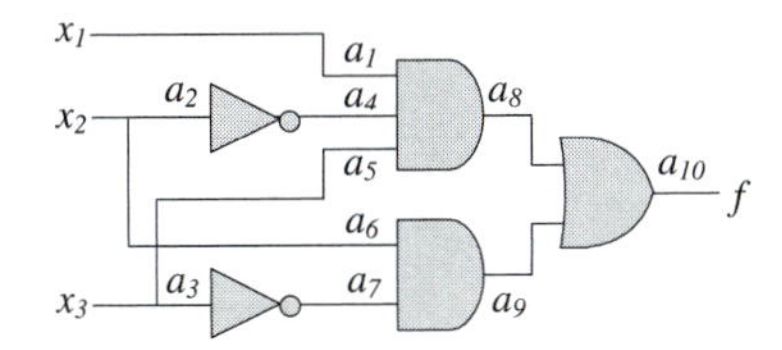

Problem 18.6 Given the following decision diagrams, compute the Boolean difference with respect to variable x_1:

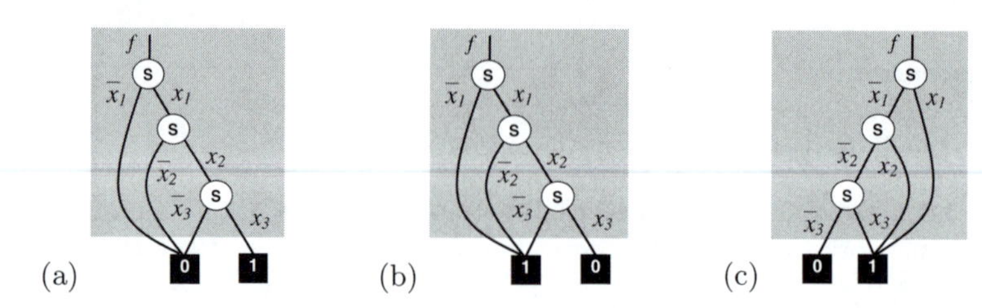

Problem 18.7 Evaluate the effect (the values of the faulty output function) of single-line stuck-at faults on wires a, b, c, d, e, f, and g for the logic network given below:

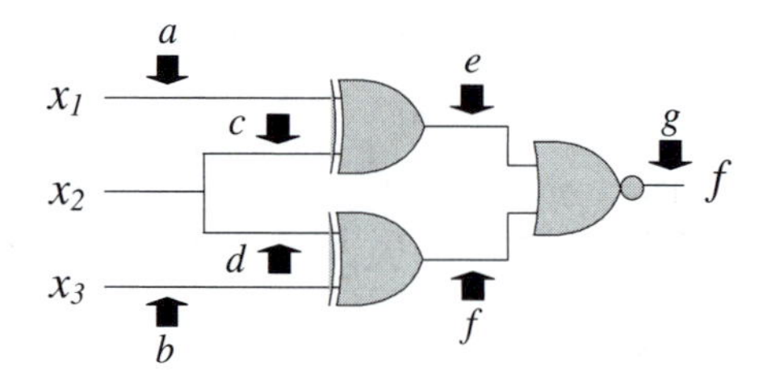

Index